Multispecies and Watershed Approaches to Freshwater Fish Conservation

Multispecies and Watershed Approaches to Freshwater Fish Conservation

Edited by

Daniel C. Dauwalter
Trout Unlimited
910 Main Street, Suite 342, Boise, Idaho 83702, USA

Timothy W. Birdsong
Texas Parks and Wildlife Department, Inland Fisheries Division
4200 Smith School Road, Austin, Texas 78744, USA

Gary P. Garrett
Texas Natural History Collections, Department of Integrative Biology
University of Texas at Austin
10100 Burnet Road, Austin, Texas 78758, USA

American Fisheries Society Symposium 91

Proceedings of the Symposia

"Multispecies and Watershed Approaches to Freshwater Fish Conservation:
Science, Planning, and Implementation"
Held in Tampa, Florida, USA
August 23–24, 2017

"Landscape Conservation Cooperatives and Fish Habitat Partnerships
Facilitate Networks and Tools for Ecologically Connected Landscapes and Seascapes"
Held in Tampa, Florida, USA
August 21, 2017

American Fisheries Society
Bethesda, Maryland, USA
2019

Suggested citation formats follow.

Entire book

Dauwalter, D. C., T. W. Birdsong, and G. P. Garrett, editors. 2019. Multispecies and watershed approaches to freshwater fish conservation. American Fisheries Society, Symposium 91, Bethesda, Maryland.

Chapter in book

Sievert, N., J. B. Whittier, and C. Paukert. 2019. Influence of conservation planning methodology on species representation and spatial distribution of priority areas. Pages 85–108 *in* D. C. Dauwalter, T. W. Birdsong, and G. P. Garrett, editors. Multispecies and watershed approaches to freshwater fish conservation. American Fisheries Society, Symposium 91, Bethesda, Maryland.

Front cover photo: Westslope Cutthroat Trout *Oncorhynchus clarkii lewisi*, Glacier National Park, Montana. Credit: Pat Clayton. Back cover photo: Kayakers paddle the Devils River near Del Rio, Texas. Credit: Earl Nottingham, Texas Parks and Wildlife Department.

Printed in the United States of America on acid-free paper.

Library of Congress Control Number 2019944988

ISBN 978-1-934874-57-8
ISSN 0892-2284

American Fisheries Society Web site: www.fisheries.org

American Fisheries Society
425 Barlow Place, Suite 110
Bethesda, Maryland 20814
USA

Contents

Preface: Multispecies and Watershed Approaches to Conservation—Thinking and Acting Holistically ix
Daniel C. Dauwalter, Timothy W. Birdsong, and Gary P. Garrett

Symbols and Abbreviations xiii

Inspiring Photography and Artwork of Watersheds and Freshwater Fishes xv

Multispecies and Watershed Conservation: Increasing Efficiency and Effectiveness for Native Fish Conservation 1
Jack E. Williams

Partnering on Multispecies Aquatic Assessments to Inform Efficient Conservation Delivery 11
Daniel C. Dauwalter, Stephanie L. Vail-Muse, Therese Race Thompson, Joanna B. Whittier, Kevin M. Johnson, and Megan G. Bean

The Really Big Picture for Fish Habitat: The Conceptual Underpinnings and Vision for the National Fish Habitat Partnership's National Fish Habitat Assessment 33
Gary E. Whelan

Who's Asking? Interjurisdictional Conservation Assessment and Planning for Great Plains Fishes 57
Ben J. Labay, Joshuah S. Perkin, Dean A. Hendrickson, Arthur R. Cooper, Gary P. Garrett, and Timothy W. Birdsong

Influence of Conservation Planning Methodology on Species Representation and Spatial Distribution of Priority Areas 85
Nicholas Sievert, Joanna B. Whittier, and Craig Paukert

Prioritizing Hawaii's Stream Habitats to Conserve Native Species with Changing Climate 109
Ralph W. Tingley III, Dana M. Infante, Wesley M. Daniel, Arthur R. Cooper, Gordon C. Smith, and Kyle Herreman

Identification and Utility of Native Fish Conservation Areas in the Upper Snake River Basin, USA 135
Richard N. Williams, Daniel C. Dauwalter, Russell F. Thurow, David P. Philipp, Jack E. Williams, and Chris A. Walser

Spatial Conservation Assessment for Balancing Avoidance of Impacts of Tilapia Introduction on Imperiled Fish Biodiversity with Economic Impacts to the Aquaculture Industry 161
Monica E. McGarrity

Texas Native Fish Conservation Areas Network: Strategic Investments in Restoration and Preservation of Freshwater Fish Diversity 183
Timothy W. Birdsong, Gary P. Garrett, Ben J. Labay, Megan G. Bean, Preston T. Bean, John Botros, Melissa J. Casarez, Adam E. Cohen, Thomas G. Heger,

Arlene Kalmbach, Dean A. Hendrickson, Stephan J. Magnelia, Kevin B. Mayes, Monica E. McGarrity, Ryan McGillicuddy, M. Melissa Parker, and Sarah Robertson

Chihuahuan Desert Native Fish Conservation Areas: A Multispecies and Watershed Approach to Preservation of Freshwater Fish Diversity 231
Gary P. Garrett, Timothy W. Birdsong, Megan G. Bean, and Ben J. Labay

Developing a Geospatial Decision Support Tool for Protecting and Restoring Environmental Flows in Texas Rivers and Streams 253
Johanna Valente, David Bradsby, Kevin B. Mayes, Cindy Loeffler, Lynne Hamlin, Dakus Geeslin, Kimberly Horndeski, David Young, Joe Trungale, Ryan Smith, Kyle Garmany, and Tom Hayes

Four Decades of Conserving Native Fish in the Colorado River Watershed, Texas 269
Stephan J. Magnelia, Kevin B. Mayes, Megan G. Bean, Cindy L. Loeffler, and David D. Bradsby

Developing the Science and Public Support Needed to Preserve the Devils River: A Case Study in Collaborative Conservation 293
Sarah Robertson, Brad D. Wolaver, Todd G. Caldwell, Timothy W. Birdsong, Ryan Smith, Thomas Hardy, Julie Lewey, and Joe Joplin

Watershed-Scale Conservation of Native Fishes in the Brazos River Basin, Texas 315
Kevin B. Mayes, Gene R. Wilde, Monica E. McGarrity, Brad D. Wolaver, and Todd G. Caldwell

Instream Flow Restoration and Watershed Conservation in the Cypress Basin, Texas 345
Ryan Smith, Joe Trungale, Rick Lowerre, Tom Hayes, Mike Montagne, Tim Bister, Laura-Ashley Overdyke, and Marcia Hackett

Lessons from the Driftless Area: The Importance of Human Dimensions Research for Stream Restoration and Fisheries Management 367
Caroline Gottschalk Druschke and Eric G. Booth

Incorporating Predictive Models into a Resilient Fisheries Management Framework: A Brook Trout Perspective 389
Eric R. Merriam, J. Todd Petty, and Cory T. Trego

Building a Network of Partners Working to Reconnect Fragmented Streams in the Southeastern United States through Actionable Science 401
Jessica Graham and Kathleen Hoenke

Little Tennessee Native Fish Conservation Partnership: Aquatic Conservation on a Landscape Scale 415
Andrea Leslie, Erin McCombs, and Fred Harris

An Approach for Implementing Large-Scale Watershed Restoration Efforts: Abandoned Mine Drainage Restoration in the West Branch Susquehanna River Watershed, Pennsylvania 431
Shawn M. Rummel and Amy G. Wolfe

Restoration of Diadromous Fish Populations in Tributaries of Waquoit Bay, Cape Cod, Massachusetts 455
Stephen T. Hurley

Challenges and Opportunities to Apply Process-Based Restoration at Scales Appropriate to Anadromous Fishes 479
Kris M. Homel, Chris M. Lorion, and Benjamin J. Clemens

New Approaches to Providing Streamflow for Fisheries in the American West: Embracing Prior Appropriation and the Marketplace 515
Rob Van Kirk, Brandon Hoffner, Amy Verbeten, and Scott Yates

The Weber River Partnership: How Fish Gained Relevance through a Recently Formed Watershed Group 565
Paul D. Thompson and Paul C. Burnett

Collaboration, Commitment, and Adaptive Learning Enable Eradication of Nonnative Trout and Establishment of Native Westslope Cutthroat Trout into One-Hundred Kilometers of Cherry Creek, a Tributary to the Madison River, Montana 589
Patrick T. Clancey, Bradley B. Shepard, Carter G. Kruse, Scott A. Barndt, Lee Nelson, Bruce C. Roberts, and R. Beauregard Turner

Blackfoot River Restoration: A Thirty-Year Review of a Wild Trout Conservation Endeavor 649
Ron Pierce, W. Ladd Knotek, Craig Podner, and Don Peters

It's Not Easy to Save a River 683
Sky Jones-Lewey

Preface: Multispecies and Watershed Approaches to Conservation—Thinking and Acting Holistically

Freshwater systems in the United States have been altered dramatically, resulting in degradation of fish habitats and declines in native freshwater fishes. Groundwater wells used for agricultural irrigation and other purposes have depleted aquifer resources, leading to reductions in spring flows and stream base flows. Dams have severely fragmented the connectivity of freshwater systems and have altered natural flow regimes. Urbanization, climate change, water quality degradation, and nonindigenous species add to the list and threaten watershed function, fish habitat quality, and freshwater fish diversity.

Innovative conservation approaches are needed to restore watershed processes for freshwater fish conservation while simultaneously supporting human needs (e.g., flood control, water supply). This need has driven development of innovative multispecies and watershed-based concepts, assessments, prioritizations, planning, and delivery—such as native fish conservation areas—that focus conservation efforts on entire aquatic communities at watershed scales while incorporating species life history needs and acknowledging compatible human uses. These approaches have yielded multiagency partnerships and large-scale funding programs focused on operationalizing conservation plans and supporting meaningful and transformative conservation delivery for freshwater fishes and their habitats. This book, which was borne out of a symposium titled "Multispecies and Watershed Approaches to Native Fish Conservation: Science, Planning, and Implementation" and inspired by a second titled "Landscape Conservation Cooperatives and Fish Habitat Partnerships Facilitate Networks and Tools for Ecologically Connected Landscapes and Seascapes," both held at the 2017 American Fisheries Society Annual Meeting in Tampa, Florida, highlights these innovative approaches to freshwater fish conservation, profiling case studies from freshwater systems throughout the United States that include diverse partnerships encompassing state, federal and local agencies, watershed councils, nongovernmental organizations, and fish habitat partnerships.

The science and software tools underpinning the identification of watersheds across broad landscapes for conservation of multiple species in freshwater systems are now fully developed and freely available. Multiple chapters in this book demonstrate the application of conservation planning principles and software to identify important watersheds for conservation of freshwater fish diversity. There are examples of identifying priority watersheds based on sensitive fish species in conservation need, illustrations of how various analytical decisions and assessment goals and objectives that inform spatial analyses can impact the perceived importance of watersheds for fish conservation, and discussions of how landscape-based assessments of freshwater fishes help to inform funding decisions for habitat protection and restoration projects.

The book is also replete with chapters exemplifying the importance of thoughtful planning and prioritization to guide and inform efficient conservation delivery. They encompass an array of methodologies from systematic conservation planning for ranking and prioritization to basic coordination among partners for goal development and distribution of duties. Examples can be found throughout the book that discuss the development of species distribution models and how they can synthesize complex parameters into observable patterns, as well as how conservation planning and prioritization is sensitive to selection of data inputs. There are also details on coordinating broad-based conservation planning with a diverse group of conservation partners to yield restoration and preservation of native fishes and the habitats they need to thrive. These include the value of multiagency conservation planning and delivery for increasing the scope and scale of conservation investments; utilization of multidiscipline expertise to help ensure success of large-scale restoration initiatives; the importance of coordination among government agencies, private landowners, nongovernmental partners and university researchers for logistically and politically complex projects that require long-term commitments; and the benefits of connecting expertise on native fishes, habitat restoration, land protection, and threat abatement with local knowledge for successful conservation planning.

Inspired by seminal papers (e.g., Dauwalter et al. 2011; Williams et al. 2011) that prescribe the need for watershed-scale approaches to freshwater fish conservation and facilitated through cooperative, multiagency partnerships (e.g., fish habitat partnerships and Landscape Conservation Cooperatives), the concept of collaborative stewardship has now been embraced and operationalized throughout the United States. A litany of transformative case studies in delivery of multispecies, watershed-scale conservation are now available that identify strategies and approaches transferrable to other areas and systems. Many of those case studies are included as chapters in this book and demonstrate solutions to conservation challenges for native freshwater fishes, such as alteration of natural stream flows, physical habitat degradation, stream fragmentation, and introduction of nonindigenous species. These issues are commonly implicated as causes in the decline or loss of native freshwater fish populations. They are ubiquitous to rivers and watersheds throughout the United States, and although innovative solutions have been demonstrated, those have not been widely disseminated throughout the fisheries conservation community. Herein, numerous chapters demonstrate holistic, multifaceted, and integrated frameworks to confront pressing conservation challenges that can only be effective when addressed at watershed scales. These include case studies that share strategies and approaches to restore natural stream flow patterns that meet the life history requirements of freshwater fishes. Others profile pathways for building strategic focus, partner engagement, and public support to restore connectivity in fragmented streams. Successful efforts to implement large-scale management of nonindigenous species are also profiled, along with innovative strategies and solutions to proactively and preemptively implement conservation measures to preserve pristine systems threatened by human development. Additional chapters communicate big picture perspectives that help reflect, recalibrate, and provide a vision for the future as we continue to refine and enhance the tools and strategies employed in the conservation of native freshwater fishes and their habitats. Collectively, we believe the chapters of this book demonstrate effective solutions and offer hopeful, optimistic, and inspirational case studies that identify foundational strategies for which future conservation programs and initiatives can be established to effectively conserve native freshwater fishes throughout the United States and beyond.

This book also profiles highly effective and impactful conservation programs and initiatives, several of which have spanned entire careers and represent decades of unwavering commitment and passion by agencies, organizations, and individuals to restore and preserve freshwater systems. We are truly humbled and honored to have the opportunity to collaborate on this book with so many distinguished conservation professionals who we consider to be true conservation heroes, luminaries, and mentors. These individuals have left a lasting conservation legacy through their incredibly productive and impactful careers, and have offered figurative road maps to guide and inform the efforts of current and future conservation professionals. In particular, we would like to celebrate the conservation legacy of Sky Jones-Lewey (2019, this volume), author of "It's Not Easy to Save a River." Sky's innate ability to recognize problems, educate stakeholders, build partnerships, and plan, coordinate, and deliver meaningful solutions to conservation challenges facing the Nueces River, Texas has simply been remarkable. Sky has spent more than 50 years as a tireless conservation advocate for her home waters. Over that time frame, she pioneered approaches that have since been transferred and replicated to conserve rivers throughout Texas and the nation. The reach of Sky's impact and influence has transcended the Nueces River watershed, and as with so many of the contributing authors of this book, her deep-seated sense of purpose, passion, vision, and stalwart leadership has been motivating and inspirational.

We offer our sincere appreciation to the extensive list of individuals and organizations who contributed to the publication of this book, including the authors, their respective organizations, and the editorial staff at the American Fisheries Society. We especially thank the following organizations who made substantial financial contributions to book publication costs: the Southeast Aquatic Resources Partnership; U.S. Fish and Wildlife Service; Wildlife Management Institute; Education, Fisheries Management, and Habitat sections of the American Fisheries Society; Oregon Chapter and Western Division of the American Fisheries Society; Desert Fish Habitat Partnership; Trout Unlimited; Richard King Mellon Foundation; Inland Fisheries Division of the Texas Parks and Wildlife Department; Texas Parks and Wildlife Foundation; Fisheries Division of the Michigan Department of Natural Resources; Oregon Department of Fish and Wildlife; and Western Native Trout Initiative.

If one thing is clear from this book, it is that those involved in conservation of freshwater fishes have been thinking bigger and more holistically when it comes to conservation assessment, planning, and delivery. People are thinking about how to look at entire landscapes with an eye towards investing in watersheds that will benefit the highest number of freshwater species most efficiently, they are delivering conservation at watershed and larger scales, and they are using networks of strategically identified watersheds to drive agency programs that span both aquatic and terrestrial realms to best ensure conservation of regional fish diversity. The chapters in this book simply highlight case studies that show it is possible to successfully effect change at watersheds scales for multiple species and set a high bar for the next generation of aquatic conservationists and fisheries managers focused on the conservation of freshwater fishes.

References

Dauwalter, D. C., J. S. Sanderson, J. E. Williams, and J. R. Sedell. 2011. Identification and implementation of native fish conservation areas in the upper Colorado River basin. Fisheries 36:278–288.

Jones-Lewey, S. 2019. It's not easy to save a river. Pages 683–693 *in* D. C. Dauwalter, T. W. Birdsong, and G. P. Garrett, editors. Multispecies and watershed approaches to freshwater fish conservation. American Fisheries Society, Symposium 91, Bethesda, Maryland.

Williams, J. E., R. N. Williams, R. F. Thurow, L. Elwell, D. P. Philipp, F. A. Harris, J. L. Kershner, P. J. Martinez, D. Miller, G. H. Reeves, C. A. Frissell, and J. R. Sedell. 2011. Native fish conservation areas: a vision for large-scale conservation of native fish communities. Fisheries 36:267–277.

Daniel C. Dauwalter
Timothy W. Birdsong
Gary P. Garrett
2019

Symbols and Abbreviations

The following symbols and abbreviations may be found in this book without definition. Also undefined are standard mathematical and statistical symbols given in most dictionaries.

A	ampere
AC	alternating current
Bq	becquerel
C	coulomb
°C	degrees Celsius
cal	calorie
cd	candela
cm	centimeter
Co.	Company
Corp.	Corporation
cov	covariance
DC	direct current; District of Columbia
D	dextro (as a prefix)
d	day
d	dextrorotatory
df	degrees of freedom
dL	deciliter
E	east
E	expected value
e	base of natural logarithm (2.71828...)
e.g.	(exempli gratia) for example
eq	equivalent
et al.	(et alii) and others
etc.	et cetera
eV	electron volt
F	filial generation; Farad
°F	degrees Fahrenheit
fc	footcandle (0.0929 lx)
ft	foot (30.5 cm)
ft^3/s	cubic feet per second (0.0283 m^3/s)
g	gram
G	giga (10^9, as a prefix)
gal	gallon (3.79 L)
Gy	gray
h	hour
ha	hectare (2.47 acres)
hp	horsepower (746 W)
Hz	hertz
in	inch (2.54 cm)
Inc.	Incorporated
i.e.	(id est) that is
IU	international unit
J	joule
K	Kelvin (degrees above absolute zero)
k	kilo (10^3, as a prefix)
kg	kilogram
km	kilometer
l	levorotatory
L	levo (as a prefix)
L	liter (0.264 gal, 1.06 qt)
lb	pound (0.454 kg, 454g)
lm	lumen
log	logarithm
Ltd.	Limited
M	mega (10^6, as a prefix); molar (as a suffix or by itself)
m	meter (as a suffix or by itself); milli (10^{-3}, as a prefix)
mi	mile (1.61 km)
min	minute
mol	mole
N	normal (for chemistry); north (for geography); newton
N	sample size
NS	not significant
n	ploidy; nanno (10^{-9}, as a prefix)
o	ortho (as a chemical prefix)
oz	ounce (28.4 g)
P	probability
p	para (as a chemical prefix)
p	pico (10^{-12}, as a prefix)
Pa	pascal
pH	negative log of hydrogen ion activity
ppm	parts per million
qt	quart (0.946 L)
R	multiple correlation or regression coefficient

r	simple correlation or regression coefficient
rad	radian
S	siemens (for electrical conductance); south (for geography)
SD	standard deviation
SE	standard error
s	second
T	tesla
tris	tris(hydroxymethyl)-aminomethane (a buffer)
UK	United Kingdom
U.S.	United States (adjective)
USA	United States of America (noun)
V	volt
V, Var	variance (population)
var	variance (sample)
W	watt (for power); west (for geography)
Wb	weber
yd	yard (0.914 m, 91.4 cm)
α	probability of type I error (false rejection of null hypothesis)
β	probability of type II error (false acceptance of null hypothesis)
Ω	ohm
μ	micro (10^{-6}, as a prefix)
′	minute (angular)
″	second (angular)
°	degree (temperature as a prefix, angular as a suffix)
%	per cent (per hundred)
‰	per mille (per thousand)

Photography and Artwork

The following inspirational photographs represent the watershed and multispecies themes and diverse ecosystems depicted throughout the book. The artwork illustrates the various aesthetic, recreational, and conservation components of pristine, restored, and degraded freshwater tributaries within a watershed.

Sam's Throne overlooking Big Creek, Buffalo River watershed, Ozark National Forest, Arkansas. Credit: Isaac Szabo.

Striped Shiner *Luxilus chrysocephalus* with Hornyhead Chub *Nocomis biguttatus* and Duskystripe Shiner *L. pilsbryi* in Big Springs, White River watershed, Arkansas. Credit: Isaac Szabo.

Smith River, upper Missouri River watershed, western Montana. Credit: Pat Clayton.

Comanche Springs Pupfish *Cyprinodon elegans* (male and female courtship) in San Soloman Springs, Pecos River watershed, Balmorhea State Park, Texas. Credit: Isaac Szabo.

Looking Glass Falls, Pisgah National Forest, North Carolina. Credit: Isaac Szabo.

River Chub *Nocomis micropogon* with Saffron Shiners *Notropis rubricroceus* and Mountain Redbelly Dace *Chrosomus oreas in Whitetop Laurel Creek, Tennessee River watershed, George Washington and Jefferson National Forest, Virginia*. Credit: Isaac Szabo.

Commemorative artwork by artist Alice Best commissioned for the book, illustrating a conceptual watershed with pristine (left), restored (middle), and degraded (right) tributary streams.

American Fisheries Society Symposium 91:1–9, 2019

Multispecies and Watershed Conservation: Increasing Efficiency and Effectiveness for Native Fish Conservation

JACK E. WILLIAMS*
Trout Unlimited
4393 Pioneer Road, Medford, Oregon 97501, USA

Abstract.—Traditional approaches to fish conservation have focused on the protection of small habitat patches or on individual species at risk of extinction. These strategies have been important yet largely have been too little and too late for widespread protection of aquatic faunas. Such small-scale and reactive approaches also are costly in terms of recovery programs and aggressive in terms of regulatory controls. Further, the linear nature of streams and the networked configuration of drainage systems suggest that a fundamentally different approach to reserve design and protected areas is necessary for effective conservation of freshwater communities when compared to terrestrial systems. Larger-scale, multispecies approaches to native fish conservation offer a more efficient and effective conservation strategy because entire fish communities and the ecological processes that support maintenance of habitat diversity can be sustained before the status of individual species deteriorates to critical levels. Protecting entire communities and watersheds also offers some resistance to climate change impacts, which rapidly are altering flow regimes and disturbance dynamics in aquatic systems. Identification and protection of high-value aquatic communities will provide an important supplement to current conservation strategies during times of increasing threats and future uncertainty.

Introduction

Despite large investments in aquatic habitat restoration and endangered and threatened fish recovery, the status of native fishes and other aquatic species in the United States continues to decline. The American Fisheries Society lists 39% of native freshwater and diadromous fishes at risk of extinction (Jelks et al. 2008), with 89% of these taxa deteriorating in condition and only 6% improving in condition since the previous assessment (Williams et al. 1989; Jelks et al. 2008). Freshwater mussels are at even greater risk, with more than two-thirds of North American species considered to be rare or endangered (Williams et al. 1993).

The decline of North American freshwater diversity has been attributed to a myriad of threats, including large-scale physical alteration of habitats, continued pollution of streams and lakes, modification of river flows, and the introduction of nonnative species (Arthington et al. 2010). The U.S. Environmental Protection Agency's (U.S. EPA 2016) national assessment of stream condition found that 28% of streams were in good biological condition, 25% in fair condition, and 46% in poor condition. Chemical pollution, caused primarily by high levels of phosphorus and nitrogen from livestock and other ag-

* Corresponding author: jwilliams@tu.org

ricultural operations, impacts slightly more than 40% of stream length in the nation, and poor riparian condition impacts 20% (U.S. EPA 2016). Rapid climate change is making stream habitat conditions less suitable for native freshwater species as temperatures rise; runoff patterns change; erosion increases; storms become larger; and drought, flooding, and wildfires are more widespread and severe (Kovach et al. 2015; Myers et al. 2017).

Spring habitats also are threatened. Springs typically are small and easily degraded by overgrazing or water diversion, with subsequent loss of biodiversity (Perla and Stevens 2008). Despite their small size, springs often harbor endemic fishes and invertebrates but often are poorly monitored compared to streams and lakes. Springsnails, which are a diverse group occupying western spring systems, are known for their high proportion of recently described and imperiled species (Hershler et al. 2014).

In addition to chemical and physical alterations, waters of the United States have become a dumping ground for nonnative species. The U.S. Geological Survey's nonindigenous aquatic species database lists 798 nonnative fish taxa recorded from U.S. waters (USGS 2018). More insidious has been the growing homogenization of our fish fauna as rivers have been simplified, polluted, and dammed, creating conditions suitable for a relatively small number of widely introduced fish species, such as Common Carp *Cyprinus carpio*, Golden Shiner *Notemigonus crysoleucas*, Green Sunfish *Lepomis cyanellus*, and others, that have a broad tolerance for poor environmental conditions (Rahel 2000). Once invasive fishes are established, the interconnected nature of stream and lake habitats renders them very difficult to remove.

Because of the problems described above, conservation of native North American aquatic species has been an ongoing challenge, despite large investments in stream restoration and endangered fish recovery efforts (Williams et al. 2011). Aquatic species in North America are at much greater risk of extinction than are their terrestrial counterparts. The proportion of freshwater fishes, amphibians, crayfishes, and freshwater mussel taxa at risk of extinction is two to four times greater than comparable terrestrial animal groups (Figure 1). The actual extinction rate for aquatic species also appears to be greater than for terrestrial species. Burkhead (2012) estimated a rate of extinction for North American freshwater fishes during the past century that was twice as high as that recorded for birds or mammals.

In this chapter, I focus on three areas of aquatic species conservation: (1) how traditional conservation efforts that focus on single species protection have had limited effectiveness for aquatic species, (2) how aquatic and terrestrial ecosystems differ in their structure and how this should inform conservation efforts, and (3) why broader watershed-scale conservation efforts are needed during these times of increasing threats to aquatic ecosystems.

Traditional Conservation Efforts: Too Little and Too Late

The Endangered Species Act of 1973 (ESA) has been an effective tool in preventing the extinction of species, but it has proven to be less effective in achieving their recovery (Doremus and Pagel 2001; Williams et al. 2005). Over the years, the number of species listed pursuant to the ESA has grown as more species have been added, while few have been removed because of their recovery. As of 2018, 166 U.S. fish taxa and evolutionarily significant units have been listed, but only six taxa have been delisted (Figure 2). Of the delisted taxa, two were delisted because of recovery (Oregon Chub *Oregonichthys crameri* and Modoc Sucker *Catostomus microps*), one because of taxonomic revision and three because of extinction (Figure 3). Although it is worth noting, that one of the three extinct taxa, the Tecopa Pupfish *Cyprinodon nevadensis calidae*, may have been extinct prior

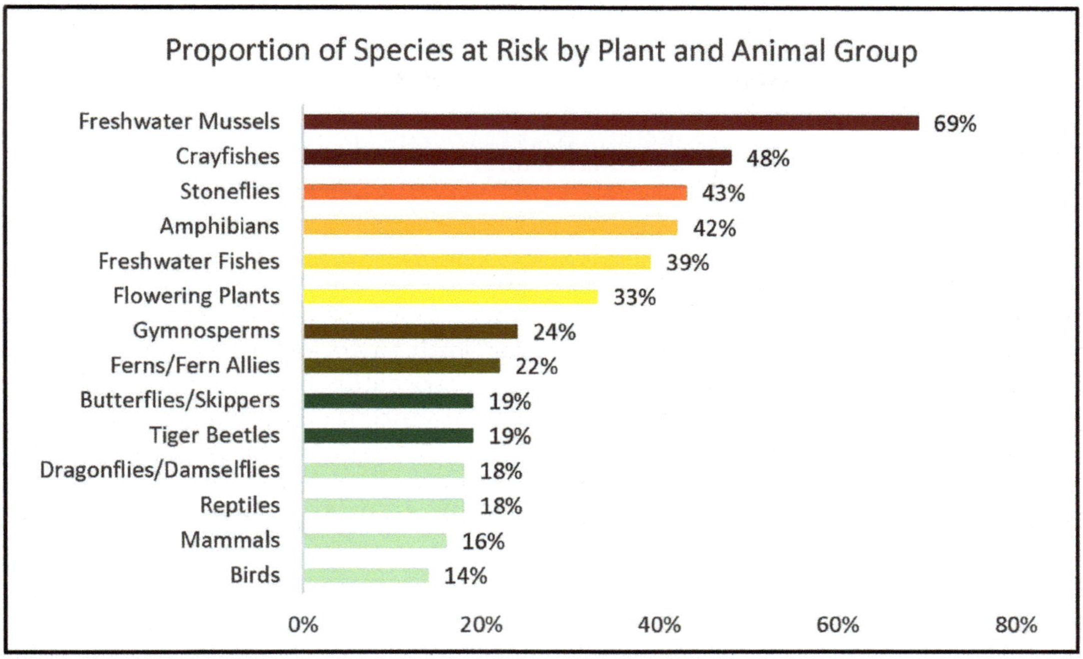

Figure 1. Percentage of at-risk species within various animal and plant groups within the United States. Figure modified from Master et al. (2000) and updated with data from Bradford (2005), Jelks et al. (2008), and Taylor et al. (2007).

to listing under the ESA, and some of the narrowly endemic fish taxa may be inherently vulnerable and unlikely to be recovered at any time (Williams et al. 2005). The ESA has been effective, but primarily for preventing extinctions rather than achieving recovery.

Funding for endangered and threatened fish recovery has been substantial. The U.S. Fish and Wildlife's fiscal year 2016 report on federal and state spending for threatened and endangered species (U.S. FWS 2016) reported that the top ten taxa ranked in order of expenditures were all fishes (mostly anadromous fishes), with the total spent for these 10 fishes during fiscal year 2016 amounting to about US$445 million. Bull Trout *Salvelinus confluentus*, the sixth-ranked species on the funding list, accounted for slightly more than $40 million in fiscal year 2016. Clearly, few fish are being recovered despite large investments in recovery efforts.

The ESA is primarily a reactive tool for conservation in that species are not added to the list until their condition has degraded to the point that they are in danger of extinction or likely to become endangered in the foreseeable future, a point at which considerable regulatory effort and financial commitments are needed to reverse damaging trends. It is a time-consuming and expensive prospect to recover species once they are poised on the edge of extinction.

The U.S. Fish and Wildlife Service occasionally develops candidate conservation agreements with public or private parties to protect species that are likely candidates for future listing. These candidate conservation agreements are intended to be more proactive approaches to conservation than adding threatened or endangered species to the ESA, but this approach has been underutilized. Recovery plans may include multiple species from a single area, which can be a more proactive approach but may also simply signal widespread declines among multiple species within a limited area.

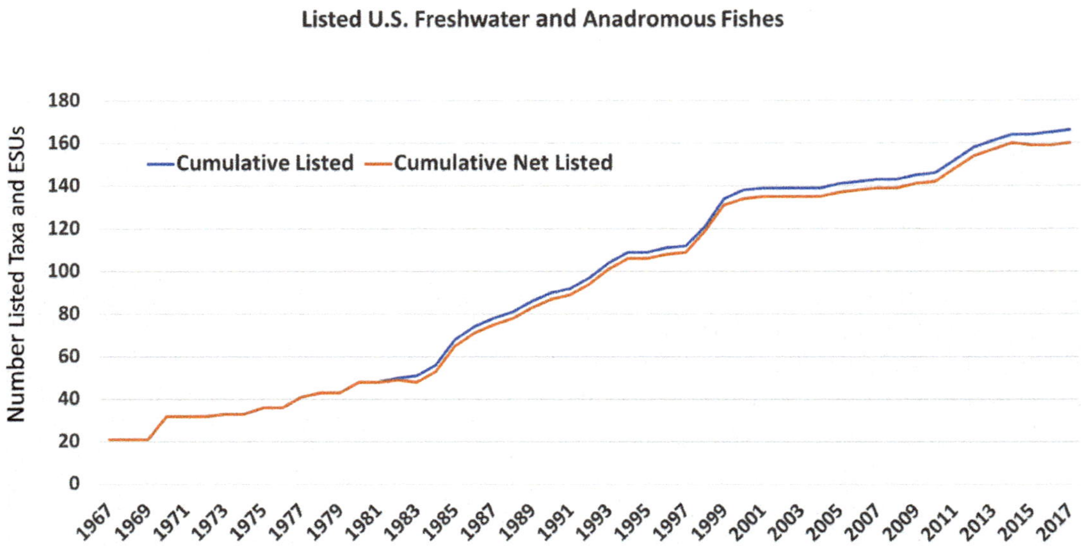

Figure 2. Number of U.S. freshwater and anadromous fish taxa and evolutionarily significant units (ESUs) listed pursuant to the Endangered Species Preservation Act of 1966 and Endangered Species Act of 1973. Cumulative listed includes total listings. Cumulative net listings reflect both taxa and ESUs listed and delisted from year to year. The small difference between the two lines reflects the small number of taxa that have been delisted over time. Data from the U.S. Fish and Wildlife Service Environmental Conservation Online System (https://ecos.fws.gov/ecp/).

Another primary tool for species conservation has been establishment of national wildlife refuges. National wildlife refuges were initially established to protect wildlife populations and especially migratory waterfowl following passage of the Migratory Bird Treaty Act and Migratory Bird Conservation Act, laws that have been broadly used to acquire and protect habitat for migratory birds (Bean 1983). With a few notable exceptions, national wildlife refuges largely have been ineffective for protection of riverine-dependent aquatic species whose habitat flows in and out of refuges and whose boundaries typically are based on terrestrial features rather than watershed boundaries. Grantham et al. (2016) described similar issues for protected areas of all sorts in California, where many rare fish taxa are represented in protected areas but most of their ranges extend beyond preserve boundaries making existing preserves mostly ineffective for fish conservation. Exceptions include places like Ash Meadows National Wildlife Refuge in Nevada, which was established to protect spring-dwelling fishes and aquatic invertebrates and largely has been effective as springs provide discrete conservation targets and are more readily included within refuge boundaries. Unless protected area boundaries correspond with watershed boundaries, the protection of stream- and river-dwelling aquatic species is problematic.

Aquatic Reserve Considerations

Establishing reserves for aquatic species poses a fundamentally different set of challenges than do reserves for terrestrial species. River systems are linear and are fed by a network of interconnected tributaries. Watersheds are the primary boundaries for aquatic systems, yet many nature reserves are defined by geologic, political, or landownership boundaries that typically do not

Figure 3. On February 17, 2015, the Oregon Chub became the first fish species removed from the list of endangered and threatened species because of recovery. When the Oregon Chub was listed as endangered in 1993, there were eight known populations in the Willamette River drainage. Because of reintroduction efforts and discovery of previously unknown populations, there currently are about 50 populations. The Oregon Chub is pictured here with a juvenile Chinook Salmon *Oncorhynchus tshawytscha*, another listed species in the Willamette River. Image courtesy of Jeremy Monroe and Freshwaters Illustrated.

correspond with borders of river basins, watersheds, or smaller catchments.

If habitats are not fragmented by human development, fishes, both native and introduced, often move freely within lakes, tributaries, and stream networks. The ability of fishes to move freely within aquatic systems is critical to species survival under changing environment conditions where droughts, floods, or wildfires may render certain streams uninhabitable until they recover from disturbances. As climate change results in more variable streamflows and more severe disturbances, it increasingly is important that stream systems are intact and interconnected if fishes are to persist for the long term (Haak and Williams 2012; Isaak et al. 2012). On the other hand, in many aquatic habitats, the presence of nonnative fishes complicates the desire for an interconnected stream system and can result in managers constructing instream barriers to isolate native species from downstream invasives (Fausch et al. 2009).

Flow and temperature regimes produced in stream environments are master variables that drive the diversity and vitality of riverine, riparian, and floodplain ecosystems (Poff et al. 1997; Arthington et al. 2010). Maintenance of these ecosystem variables are important considerations in aquatic reserve design and may require reserves that include habitats of sufficient quality and watershed-scales so that these ecosystem processes continue to produce desired habitat conditions over time.

The location of potential aquatic reserves is driven primarily by existing or potential habitat condition and native fish persistence. Many stream, lake, and spring habitats are so degraded that suitable conditions for native species exist only in small habitat patches that are unsustainable without significant investments in habitat restoration and maintenance. Restoration projects may expand suitable habitats by restoring normative flow and temperature regimes and by reconnecting fragmented stream networks, connecting headwaters with valley bottoms, and reconnecting rivers with their floodplains.

The impacts of climate change are unlikely to progress evenly across the landscape and may pose both opportunities and constraints on aquatic reserve design. Isaak et al. (2014) identified a subset of streams in the Northern Rocky Mountains that are likely to remain cold enough to support native trout populations and resist nonnative fish invasions despite climate warming. While periodic floods, droughts, and wildfires can be beneficial to aquatic habitat formation and diversity, stream systems subject to chronic disturbances of high intensity may exceed the capacity for resilience within the system and cause fish, amphibian, and aquatic invertebrate populations to decline.

Because of the factors described above, aquatic habitats located within terrestrial-based reserve systems may not be adequate to sustain native aquatic communities. Clearly, many factors should be considered in the design of aquatic reserves, but a complete discussion of these is beyond the scope of this brief introduction. Readers are referred to Moyle and Sato (1991), Dauwalter et al. (2011), and Haak and Williams (2013) for further insights and recommendations.

Watershed and Community Conservation Approaches

The concept of watershed-scale efforts to protect native fishes has been around for many decades yet has been implemented relatively infrequently compared to species-level conservation. Proactive conservation measures that focus on protection and management of existing strongholds and intact native aquatic communities are needed to complement the important, but more reactive approaches common to endangered and threatened species protection. Simply stated, protecting intact aquatic communities within a well-functioning habitat template will be a much more efficient and cost-effective approach to conservation than are efforts aimed at recovery of individual imperiled species.

In order to address the continuing decline of California's freshwater fish fauna, Moyle and Yoshiyama (1994) proposed a multi-tier conservation approach that coupled protection of threatened and endangered species with creation of a network of aquatic diversity management areas and programs to protect the watersheds where high aquatic diversity occurs. More recent analyses have identified priority watersheds across the state that are important to protect aquatic biodiversity and could guide future conservation planning (Howard et al. 2018).

A similar approach to protect the best remaining native fish communities was proposed by Williams et al. (2011), and potential native fish conservation areas were identified in the upper Colorado River basin by Dauwalter et al. (2011). Distributional data are lacking for many fish species, but Dauwalter et al. (2011) found adequate data for four fishes, including the native Colorado River Cutthroat Trout *Oncorhynchus clarkii pleuriticus* and three candidate fishes (Roundtail Chub *Gila robusta*, Bluehead Sucker *Catostomus discolobus*, and Flannelmouth Sucker *C. latipinnis*) to identify priority areas for native fish conservation. Further examples of watershed-scale proactive efforts to protect native aquatic communities are described in this volume.

Ideally, freshwater reserves contain the following principal elements: intact watersheds and aquatic communities, natural hydrological regimes, an absence of nonnative species, and an aquatic management emphasis (Saunders et al. 2002). Identifying intact remaining aquatic strongholds is still possible in some river basins, but it is more likely that one or more elements will be degraded and require some form of stream restoration (either directly through stream habitat restoration or indirectly through adoption of best management practices within the broader watershed) and/or nonnative control programs to achieve desired conditions. As human populations continue to grow and land and water resources are consumed at an ever-increasing rate, resource managers should become more proactive in their approaches to biodiversity conservation by protecting entire communities before their component species become threatened or endangered. Creativity and persistence may be among the most valued assets of aquatic reserve designers as they struggle to find, restore, and protect native freshwater communities and their watersheds.

Acknowledgments

Over the years, my thoughts about large-scale conservation planning benefitted by conversations with a number of colleagues, especially Dan Dauwalter, Amy Haak, Peter Moyle, Chris Wood, and the late Jim Sedell. I would also like to thank Cindy Deacon Williams and two anonymous reviewers for their comments and improvements to this article.

References

Arthington, A. H., R. J. Naiman, M. E. McClain, and C. Nilsson. 2010. Preserving the biodiversity and ecological services of rivers: new challenges and research opportunities. Freshwater Biology 55:1–16.

Bean, M. J. 1983. The evolution of national wildlife law (revised and expanded edition). Praeger Publishers, New York.

Bradford, D. F. 2005. Factors implicated in amphibian population declines in the United States. Pages 915–925 *in* M. J. Lannoo, editor. Amphibian declines: the conservation status of United States species. University of California Press, Berkeley.

Burkhead, N. M. 2012. Extinction rates in North American freshwater fishes, 1900–2010. BioScience 62:798–808.

Dauwalter, D. C., J. S. Sanderson, J. E. Williams, and J. R. Sedell. 2011. Identification and implementation of native fish conservation areas in the upper Colorado River basin. Fisheries 36:278–288.

Doremus, H., and J. E. Pagel. 2001. Why listing may be forever: perspectives on delisting under the U.S. Endangered Species Act. Conservation Biology 15:1258–1268.

Fausch, K. D., B. E. Rieman, J. B. Dunham, M. K. Young, and D. P. Peterson. 2009. Invasion versus isolation: trade-offs in managing native salmonids with barriers to upstream movement. Conservation Biology 23:859–870.

Grantham, T. E., K. A. Fresenmyer, R. Peek, E. Holmes, R. M. Quiñones, A. Bell, N. Santos, J. K. Howard, J. H. Viers, and P. B. Moyle. 2016. Missing the boat on freshwater fish conservation in California. Conservation Letters 10:77–85.

Haak, A. L., and J. E. Williams. 2012. Spreading the risk: native trout management in a warmer and less-certain future. North American Journal of Fisheries Management 32:387–401.

Haak, A. L., and J. E. Williams. 2013. Using native trout restoration to jumpstart freshwater conservation planning in the Interior West. Journal of Conservation Planning 9:38–52.

Hershler, R., H-P Liu, and J. Howard. 2014. Springsnails: a new conservation focus in western North America. BioScience 64:693–700.

Howard, J. K., K. A. Fesenmyer, T. E. Grantham, J. H. Viers, P. R. Ode, P. B. Moyle, S. J. Kupferburg, J. L. Furnish, A. Rehn, J. Slusark, R. D. Mazor, N. R. Santos, R. A. Peek, and A. N. Wright. 2018. A freshwater conservation blueprint for California: prioritizing watersheds for freshwater biodiversity. Freshwater Science 37:417–431.

Isaak, D. J., C. C. Muhlfeld, A. S. Todd, R. Al-Chokhachy, J. Roberts, J. L. Kershner, K. D. Fausch, and S. W. Hostetler. 2012. The past as prelude to the future for understanding 21st-century climate effects on Rocky Mountain trout. Fisheries 37:542–556.

Isaak, D. J., M. K. Young, D. Nagel, and D. Horan. 2014. Coldwater as a climate shield to preserve native trout through the 21st century. Pages 110–116 *in* R. F. Carline and C. LoSapio, editors. Proceedings of the Wild Trout XI symposium: looking back and moving forward. Wild Trout Symposium, Bozeman, Montana.

Jelks, H. L., S. J. Walsh, N. M. Burkhead, S. Contreras-Balderas, E. Díaz-Pardo, D. A. Hendrickson, J. Lyons, N. E. Mandrak, F. McCormick, J. S. Nelson, S. P. Platania, B. A. Porter, C. B. Renaud, J. J. Schmitter-Soto, E. B. Taylor, and M. L. Warren, Jr. 2008. Conservation status of imperiled North American freshwater and diadromous fishes. Fisheries 33:372–407.

Kovach, R. P., C. C. Muhlfeld, R. Al-Chokhachy, J. B. Dunham, B. H. Letcher, and J. L. Kershner. 2015. Impacts of climate variation on trout: a global synthesis and path forward. Reviews in Fish Biology and Fisheries 26:135–151.

Master, L. L., B. A. Stein, L. S. Kutner, and G. A. Hammerson. 2000. Vanishing assets: conservation status of U.S. species. Pages 93–118 *in* B. Stein, L. Kutner, and J. Adams, editors. Precious heritage: the status of biodiversity in the United States. Oxford University Press, New York.

Moyle, P. B., and G. M. Sato. 1991. On the design of preserves to protect native fishes. Pages 155–169 *in* W. L. Minckley and J. E. Deacon, editors. Battle against extinction: native fish management in the American West. University of Arizona Press, Tucson.

Moyle, P. B., and R. M. Yoshiyama. 1994. Protection of aquatic biodiversity in California: a five-tiered approach. Fisheries 19:6–18.

Myers, B. J. E., A. J. Lynch, D. B. Bunnell, C. Chu, J. A. Falke, R. P. Kovach, T. J. Krabbenhoft, T. J. Kwak, and C. P. Paukert. 2017. Global synthesis of the documented and projected effects of climate change on inland fishes. Reviews in Fish Biology and Fisheries 27:339–361.

Perla, B. S., and L. E. Stevens. 2008. Biodiversity and productivity at an undisturbed spring in comparison with adjacent grazed riparian and upland habitats. Pages 230–243 *in* L. E. Stevens and V. J. Meretsky, editors. Aridland springs in North America: ecology and conservation. University of Arizona Press, Tucson.

Poff, N. L., J. D. Allan, M. B. Bain, J. R. Karr, K. L. Prestegaard, B. D. Richter, R. E. Sparks, and J. C. Stromberg. 1997. The natural flow regime: a paradigm for river conservation and restoration. BioScience 47:769–784.

Rahel, F. J. 2000. Homogenization of fish faunas across the United States. Science 288:854–856.

Saunders, D. L., J. J. Meeuwig, and A. C. J. Vincent. 2002. Freshwater protected areas: strategies for conservation. Conservation Biology 16:30–41.

Taylor, C. A., G. A. Schuster, J. E. Cooper, R. J. DiStefano, A. G. Eversole, P. Hamr, H. H. Hobbs, III, H. W. Robison, C. E. Skelton, and R. F. Thoma. 2007. A reassessment of the conservation status of crayfishes of the United States and Canada after 10+ years of increased awareness. Fisheries 32:372–389.

U.S. EPA (U.S. Environmental Protection Agency). 2016. National rivers and stream assessment 2008–2009: a collaborative survey. U.S. EPA, EPA/841/R-16/007, Washington, D.C.

U.S. FWS (U.S. Fish and Wildlife Service). 2016. Federal and state endangered and threatened species expenditures: fiscal year 2016. US FWS, Washington, D.C.

USGS (U.S. Geological Survey). 2018. Nonindigenous aquatic species database. Available: https://nas.er.usgs.gov/taxgroup/fish/default.aspx. (April 2018).

Williams, J. D., M. L. Warren, Jr., K. S. Cummings, J. S. Harris, and R. J. Neves. 1993. Conservation status of freshwater mussels of the United States and Canada. Fisheries 18(9):6–22.

Williams, J. E., J. E. Johnson, D. A. Hendrickson, S. Contreras-Balderas, J. D. Williams, M. Navarro-Mendoza, D. E. McAllister, and J. E. Deacon. 1989. Fishes of North America endangered, threatened, or of special concern: 1989. Fisheries 14(6):2–20.

Williams, J. E., C. A. Macdonald, C. Deacon Williams, H. Weeks, G. Lampman, and D. W. Sada. 2005. Prospects for recovering endemic fishes pursuant to the U.S. Endangered Species Act. Fisheries 30:24–29.

Williams, J. E., R. N. Williams, R. F. Thurow, L. Elwell, D. P. Philipp, F. A. Harris, J. L. Kershner, P. J. Martinez, D. Miller, G. H. Reeves, C. A. Frissell, and J. R Sedell. 2011. Native fish conservation areas: a vision for large-scale conservation of native fish communities. Fisheries 36:267–277.

American Fisheries Society Symposium 91:11–32, 2019

Partnering on Multispecies Aquatic Assessments to Inform Efficient Conservation Delivery

DANIEL C. DAUWALTER*[1]
Trout Unlimited
910 Main Street, Suite 342, Boise, Idaho 83702, USA

STEPHANIE L. VAIL-MUSE
Desert Fish Habitat Partnership and U.S. Fish and Wildlife Service
Post Office Box 39, Pinetop, Arizona 85935, USA

THERESE RACE THOMPSON
Western Native Trout Initiative
134 Union Boulevard Suite 665, Lakewood, Colorado 80228, USA

JOANNA B. WHITTIER
School of Natural Resources, University of Missouri
203P Anheuser-Busch Natural Resources Building
1105 East Rollins Street, Columbia, Missouri 65211, USA

KEVIN M. JOHNSON
Southern Rockies Landscape Conservation Cooperative and U.S. Fish and Wildlife Service
134 Union Boulevard, Lakewood, Colorado 80228, USA

MEGAN G. BEAN[2]
Texas Parks and Wildlife Department
5103 Junction Highway, Mountain Home, Texas 78058, USA

Abstract.—Efficient conservation can require making strategic decisions across large landscapes. For example, two fish habitat partnerships—the Desert Fish Habitat Partnership and the Western Native Trout Initiative (WNTI)—fund conservation and restoration projects across the western United States. The Desert Fish Habitat Partnership alone serves 11 states and nearly 180 native fish species, and the Western Native Trout Initiative covers 21 salmonid species across 12 western states. Because of the large landscapes they represent, the partnerships are using multispecies aquatic assessments developed for specific river basins to aid in conservation delivery. These assessments yield a conservation value for every catchment in a basin based on known and modeled native fish distributions (including salmonids), riverine connectivity, and threats to aquatic habitats. The conservation values are scaled between 0 (low) and 1 (high) and have been used to evaluate the landscape context of conservation projects submitted for funding through the National Fish Habitat Partnership. While assessments are complete for some basins (e.g., upper and lower Colorado basins, upper Rio Grande basin),

* Corresponding author: ddauwalter@tu.org

[1] Chair, Science and Data Committee, Desert Fish Habitat Partnership

[2] Past co-chair of Desert Fish Habitat Partnership

the partnerships are currently working with additional partners to fund aquatic assessments in new geographies (e.g., Bonneville and Lahontan basins). Multispecies assessments are used in conjunction with the knowledge of field biologists to inform on-the-ground conservation across large landscapes and make conservation delivery more efficient for the many imperiled native fishes in the western United States.

The Conservation Need

Freshwater fishes are threatened globally. Only 46% of 7,301 freshwater fish species have had their distribution mapped and threats identified, and 31% of those are threatened with extinction (Darwall and Freyhof 2016). North America alone has more than 700 species of freshwater fish (Page and Burr 1991), which, along with mollusks, snails, and crayfishes, are highly imperiled, and many species are listed as endangered or are known to have gone extinct (Taylor et al. 2007; Jelks et al. 2008; Haag and Williams 2014). The imperilment status of freshwater organisms exceeds that of their terrestrial counterparts (Master et al. 2000). The rich aquatic biodiversity in North America is threatened by, and imperiled due to, habitat degradation and loss, invasive species, overharvest, chemical and organic pollution, and global climate change, which are a result of myriad human activities (Allan and Flecker 1993; Closs et al. 2016; Darwall and Freyhof 2016).

Restoration of aquatic resources is now a multibillion dollar per year enterprise (Bernhardt et al. 2005). Restoration programs exist in every corner of North America, as well as in many parts of the world (Cowx and Welcomme 1998; Palmer et al. 2007). Much early restoration was focused on restoring water quality degraded from point source pollutants, but efforts today are more often focused on integrated restoration of ecosystem processes and ecological integrity (Karr and Dudley 1981; Beechie et al. 2013; Jones et al. 2018). Restoration can vary in scope from singular projects implemented at one location during one year to watershed- or landscape-scale programs that implement hundreds of projects over decades (Bernhardt et al. 2007; Pierce et al. 2019, this volume).

Implementation of a landscape-scale program requires efficient allocation of scarce resources to maximize conservation benefits, as there are many different sites or watersheds in which to potentially implement conservation actions (Williams et al. 2007; Roni et al. 2013). Maximizing conservation benefit can be guided by effective goal development, resource assessment, planning, and prioritization (Knight et al. 2006; Ferrier and Wintle 2009). Advances in spatial data and assessment methods over the past 40 years have helped to facilitate a better understanding of resources across broad landscapes (Ferrier and Wintle 2009), and this understanding has allowed spatial conservation assessment and prioritization to become more available as a formal part of conservation planning and decision making (Knight et al. 2006).

In this chapter, we (1) introduce the Desert Fish Habitat Partnership (DFHP) and the Western Native Trout Initiative (WNTI)—two regional partnerships in the western United States under the National Fish Habitat Partnership (NFHP), (2) present an overview of multispecies assessment and prioritization and its recent convergence on a common analytical framework developed for aquatic systems, (3) demonstrate how the DFHP and WNTI have partnered to acquire funding to develop multispecies aquatic assessments specific to large river basins, and (4) give examples of how these assessments are currently used to inform efficient conservation decision making and partnership collaboration.

Fish Habitat Partnerships

The National Fish Habitat Action Plan was developed in 2006 to address degradation and loss of fish habitat across the United States (www.fishhabitat.org). The plan brought together anglers; conservation groups; scientists; tribal governments; state, territorial, and federal agencies; and industry leaders that saw the need to foster voluntary, nonregulatory, science-based action to protect, restore, and enhance aquatic systems across the United States. This group is now recognized as the NFHP. Since 2006, the NFHP Board has approved 20 self-organized regional and resource-based fish habitat partnerships (FHPs) covering a multitude of aquatic habitat types across all 50 states and the District of Columbia (Figure 1). The partnerships share a responsibility to pass along healthy fish habitats and intact aquatic systems to future generations. Federal, state, tribal, and privately raised funds are leveraged under the NFHP through regional FHPs to address the nation's biggest fish habitat challenges.

The DFHP was formed in 2005 to conserve native desert fishes by protecting, restoring, and enhancing their habitats in cooperation with state and tribal fish and wildlife agencies, federal resource agencies, research and private organizations, and engaged individuals (www.desertfhp.org; DFHP 2015). The DFHP supports on-the-ground projects that protect the most underserved, imperiled desert fish species by addressing critical fish and aquatic habitat conservation needs in the Great Basin and Mohave, Sonoran, and

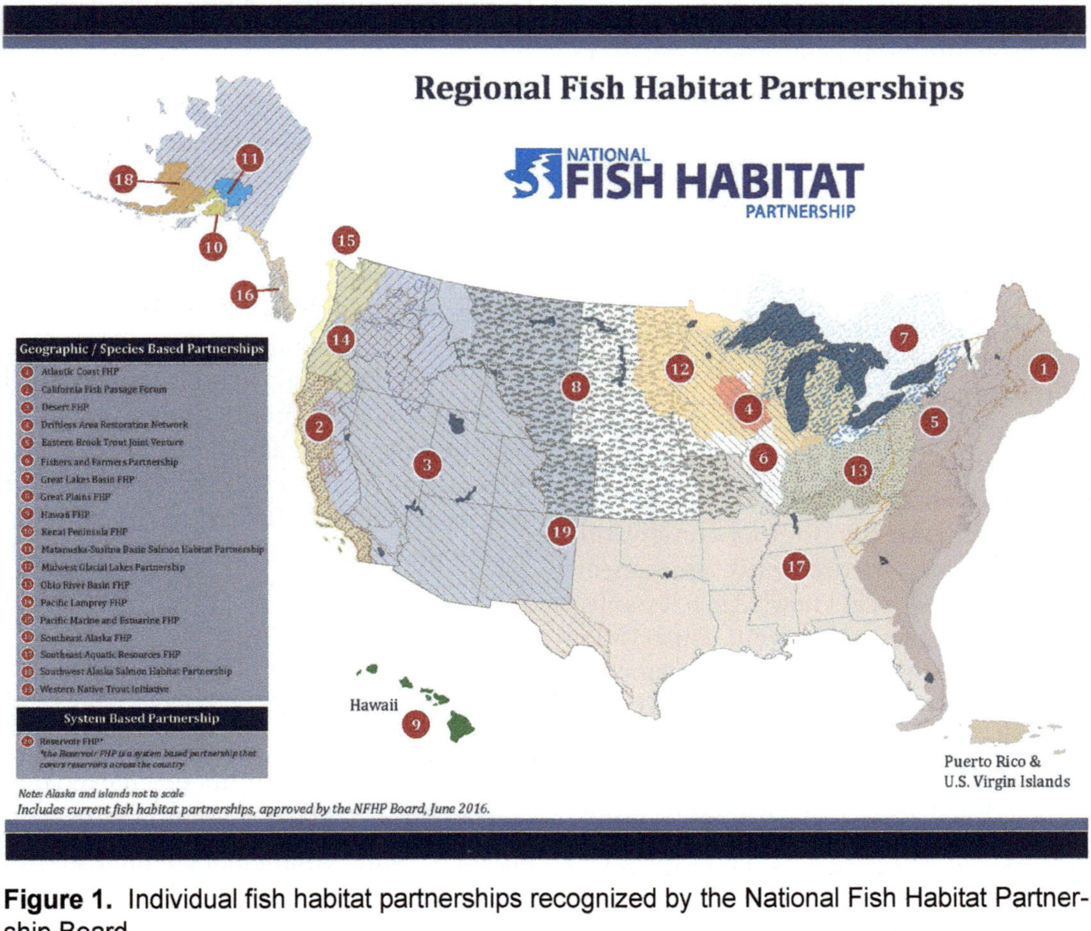

Figure 1. Individual fish habitat partnerships recognized by the National Fish Habitat Partnership Board.

Chihuahuan deserts in the western United States. These lands support 179 nonsalmonid native fish taxa prioritized for conservation by the DFHP under the guidance of state wildlife action plans and the NFHP (DFHP 2015). By identifying priority species and habitats, working across geopolitical boundaries, integrating and applying the best available science, and promoting community involvement, the DFHP identifies and prioritizes necessary conservation actions to protect and restore desert fish habitats. Through 2018, the DFHP has directed more than US$2.6 million in federal NFHP funds, leveraged with matching contributions, towards 28 habitat protection and restoration projects to benefit desert fishes.

The WNTI works collaboratively across 12 western U.S. states to conserve, protect, restore, and recover 21 native trout and char species across their historical ranges. Operating under the guidance of the Western Association of Fish and Wildlife Agencies since 2006, and a recognized FHP since 2007, the WNTI is governed by a steering committee that represents 12 western state fish and wildlife agencies, five federal natural resource management agencies, tribes, and nonprofit conservation organizations. The WNTI and its partners invest private and public resources toward completing the highest-impact on-the-ground projects led by local communities and resource agencies across the western United States. To achieve its mission, the WNTI works together with its partners to establish joint priorities for conservation by combining science-based assessments with expert and local knowledge to establish joint priorities for native trout conservation at a landscape scale. Through its Campaign for Western Native Trout, the WNTI catalyzes education, outreach, and stewardship by raising awareness for the importance of healthy watersheds and facilitating greater public support for native trout conservation within local communities. Between 2006 and 2017, the WNTI directed almost $5.5 million in federal NFHP funds leveraged with just over $25 million public and private matching dollars for 141 priority conservation projects. The WNTI and its partners have removed 87 barriers to fish passage, reconnected or improved 1,817 km (1,129 mi) of native trout habitat, assessed 671 watersheds or populations, and placed 30 protective fish barriers to conserve important native trout conservation populations (WNTI 2008, 2016). Details for these projects can be found at www.westernnativetrout.org.

Multispecies Aquatic Conservation Assessments

Assessment of aquatic systems and species can take many forms. It can range from assessment of water quality or single species in individual water bodies (Zale et al. 2012) to broadscale surveys designed to draw regional inferences on the status and trend of ecological integrity (Karr 1993). More recently, assessment of aquatic systems has included spatially explicit assessments of habitat condition, ecological threats, and species diversity using existing data sets (Kuehne et al. 2017). This trend has been facilitated by recognition of the need to go beyond single-species approaches to focus on ecosystems and entire communities for many applications (Franklin 1993), as well as continued advancement of spatially explicit data sets representing various aquatic ecosystem components and improved approaches to predict species distributions (Olden 2003; Leathwick et al. 2010; Mainali et al. 2015; Dauwalter et al. 2017). Single-species assessment approaches are sometimes necessary, especially for species listed under the Endangered Species Act. However, biodiversity applications necessarily focus on multiple (sometimes hundreds) species, and some conservation planning decisions are most efficient if based on multispecies assessment information (Kuehne et al. 2017). Some contemporary multispecies aquatic conservation assessment frameworks now integrate conservation biology principles underpinning protected area or reserve selection (representation, complementarity, etc.; Table 1) with ecological integrity and threat assessment information in a spatially explicit framework. The assessment output is a conservation rank or value for each spatial planning unit (catchment, watershed, hydrologic unit) across entire river basins that are used to guide conservation planning (Moilanen et al. 2008; Hermoso et al. 2012).

Methods for assigning conservation value to spatial planning units can generally be categorized as scoring-based or complementarity-based approaches (Ferrier and Wintle 2009). Scoring-based approaches are relatively straightforward as they assign an independent score for each planning unit based on specified factors of interest (e.g., habitat quality, presence–absence of focal species, species richness, and threats). Multiple factors are often scored and summed (or multiplied) into a final composite score intended to reflect conservation value. For example, if each planning unit in Figure 2A receives one point for each species present, planning unit 1 would have the highest score (and thus highest conservation value). Planning units 2 and 3 would have the same score (tied conservation value) despite different species composition and not accounting for a species not yet represented in planning unit 5. While scoring approaches have utility for some applications, they fail to account for which species are present (community composition) and complementarity, that is, how different planning units complement each other when considered as a set (McKinney 1997), which is one of the conceptual underpinnings of contemporary systematic conservation planning (Margules and Pressey 2000; Sarkar and Illoldi-Rangel 2010; Linke et al. 2011).

Complementarity-based approaches provide an efficient pathway to maximize the number of species represented in a minimum number (or a set number) of planning units. For example, in Figure 2A, planning unit 1 has the highest conservation value because it represents five species. The next highest conservation value would be assigned to planning unit 5 because it is the only planning unit that contains two new species not represented in planning unit 1; that is, planning unit 5 complements planning unit 1, and together, they represent the greatest number of species across a set of two planning units. Thus, the value of an individual planning unit depends

Table 1. Definitions of terms and concepts used in spatial conservation planning and prioritization (Kukkala and Moilanen 2013).

Definitions
Systematic conservation planning: a structured approach to identify priority areas based on their complementarity
Complementarity: the contribution of a spatial planning unit toward a measure of biodiversity (i.e., functional, community, species) that complements other units
Comprehensiveness: the representation of many biodiversity features across all planning units in a set
Efficiency: representation of the highest amount of biodiversity features in the fewest number of planning units
Irreplaceability: a measure of uniqueness associated with a spatial planning unit based on the biodiversity features represented
Representation: the occurrence of a biodiversity feature in a selected set of spatial planning units
Representativeness: the total number biodiversity features represented in a selected set of spatial planning units
Redundancy: the replication of the measure of biodiversity across spatial planning units

A)

	Planning Unit					
Species	1	2	3	4	5	**Range**
A	1	1	1	1	0	**4**
B	1	0	1	1	0	**3**
C	1	1	0	0	1	**3**
D	1	0	1	1	0	**3**
E	1	1	1	0	0	**3**
F	0	1	0	0	1	**2**
G	0	0	0	0	1	**1**
Richness	**5**	**4**	**4**	**3**	**3**	

B)

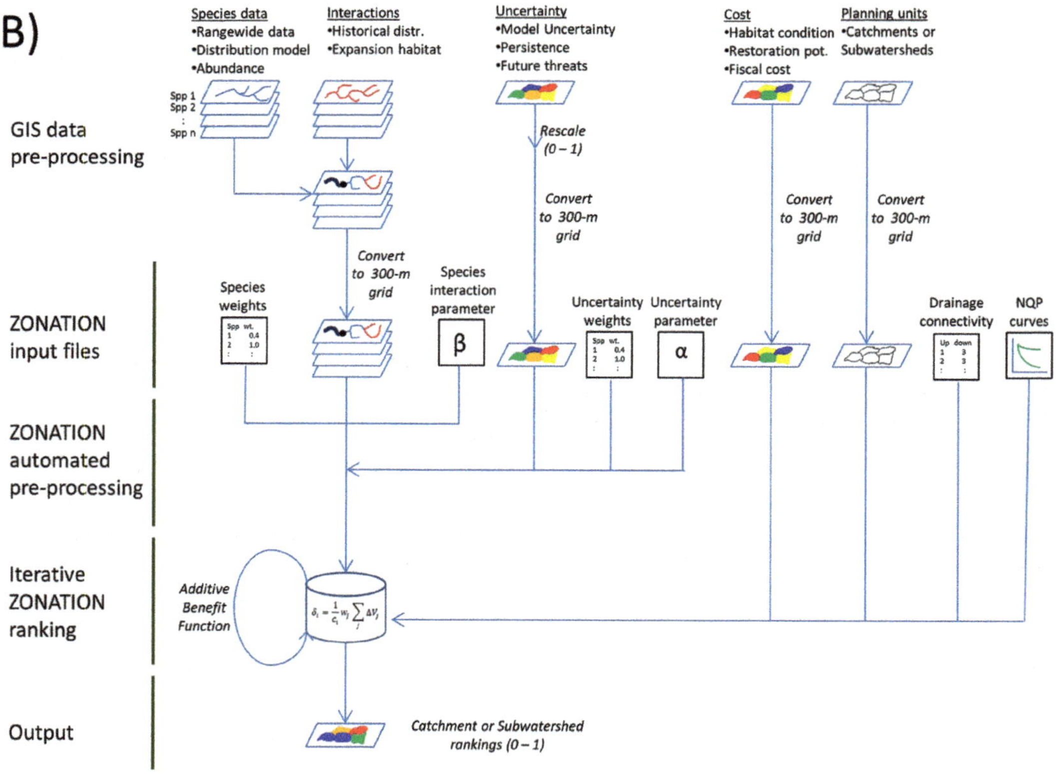

Figure 2. Simple set of planning units and species presence data showing **(A)** complementarity and species representation versus **(B)** one example of a more complex analysis workflow that incorporates different types of species data, connectivity among planning units, habitat condition, species interactions, and other ecological factors in a complementary-based algorithm.

on what is represented in the other planning units and whether they complement one another. A set of planning units with high complementarity will have the highest benefit : cost (number of species : number of units) ratio for conservation purposes (Nel et al. 2009). Complementarity is often the fundamental basis for the complex algorithms implemented in popular spatial conservation prioritization software (e.g., Marxan, Zonation, ConsNet, and C-Plan; Moilanen et al. 2009).

Quantitative approaches to conservation assessment can also incorporate more than just species representation by including factors that contribute to species persistence, such as habitat types and condition, human threats, stream connectivity, dispersal capabilities, or minimum home range size (Margules and Pressey 2000; Nel et al. 2011). Figure 2B depicts one example of the complexity and ecological reality that can now be incorporated into aquatic conservation assessments, which includes species representation, abundance, and persistence; historical distributions; interactions between species; minimum habitat requirements based on home range size; and the effect of current habitat conditions or future threats on the landscape. Each of these factors can be integrated into a comprehensive and quantitative assessment of entire river basins where each planning unit can be assigned a conservation value between 0 (low value) to 1 (high value). The exact nature of their integration depends on how the analysis is constructed, which can vary considerably. High conservation values reflect important planning units for the representation of native fish species balanced by the habitat condition or threat level, watershed connectivity, and, in some cases, proximity to protected areas (e.g., national parks). For example, two watersheds with different threat levels can be valued differently despite equal species representation (composition) because watersheds with high threat value (e.g., urbanization) are risky places to invest in conservation and therefore should be down weighted in their value. The availability of these approaches has led to their increased use in freshwater conservation planning applications (Wenger et al. 2009; Dauwalter et al. 2011; Howard et al. 2018; Birdsong et al. 2019, this volume).

Partnering on multispecies aquatic conservation assessments

Because of the large landscapes they cover, the DFHP and WNTI have together pursued development and use of multispecies aquatic assessments that identify hydrologic units important for native fish diversity to inform partnership decision making within river basins (the WNTI has also supported development of species-specific rangewide status assessments; e.g., Gresswell 2011; Muhlfeld et al. 2015). As described above, these assessments produce a rank for all catchments (land area draining a ~1-km National Hydrography Dataset stream segment; USEPA and USGS 2005) or subwatersheds (Hydrologic Unit Code 12, ~30,000 ha; NRCS et al. 2008) that effectively represents the conservation value of that unit based on fish species richness and representation of rare species (representation, complementarity, and comprehensiveness), aquatic connectivity relative to species biology, and threats to aquatic systems. Each assessment was also developed individually using data sets and analysis decisions unique to the focal basin based on input from stakeholder workgroups (details can be found in respective assessment reports or publications). Despite some differences, each assessment produced a conservation value for each spatial planning unit that was scaled to range from 0 (low conservation value) to 1 (high conservation value). The initial assessment effort focused on the lower Colorado River basin, whereby the U.S. Geological Survey initiated an aquatic gap analysis to identify riverine fishes that were inadequately represented (gaps) within the existing network of protected lands (e.g., national parks; Whittier et al. 2006, 2011). The analytical framework used known and modeled species distributions, riverine connectivity, species-specific home range sizes, and an ecological threat index (Paukert et al. 2011). The utility of the lower Colorado River assessment resulted in interest in an equivalent assessment in the upper Colorado River basin, which was funded through the U.S. Fish and Wildlife Service Multistate Conservation Grant program to the National Fish Habitat Partnership Board, administered by the Western Association of Fish and Wildlife Agencies and WNTI and completed by the University of Missouri (Whittier and Sievert 2014). This led the partnerships to pursue development of additional basin-specific assessments. The upper Rio Grande assessment was funded by the Southern Rockies Landscape Conservation Cooperative, administered by the Western Association of Fish and Wildlife Agencies and WNTI and completed by Siglo Group (Labay et al. 2018). The Bonneville, and Lahontan and Central Nevada basin assessments have been pursued but not initiated due to lack of funding (Table 2; Figure 3). The diverse partners represented in each assessment effort represents a synergy towards combining resources to accomplish a common goal, that is, developing the assessment and shepherding it towards meaningful and targeted delivery of aquatic conservation on the landscape.

The convergence of multispecies aquatic assessments around a general analytical framework resulted in additional assessments being developed within DFHP and WNTI geographies using methods similar to those used for the lower and upper Colorado River basins. Williams et al. (2019, this volume) explored the utility of the native fish conservation area concept in the upper Snake River basin with an analytical framework that incorporated known and modeled species distributions, aquatic connectivity interrupted by large dams, land protection status, and information on integrity and future security of habitats at the scale of subwatersheds (Hydrologic Unit Code 12). Likewise, similar analysis methods were used to develop a similar aquatic assessment (fish and other aquatic organisms) for the state of California (Howard et al. 2018). The California analysis was also at the subwatershed scale and used species distribution data, aquatic connectivity, and proximity to protected areas as used

Table 2. Funding sources and lead entities in developing the multispecies aquatic assessment for use in fish habitat partnership decision making.

Region/basin	Funding	Assessment lead
Upper Colorado River	U.S. Fish and Wildlife Service Multistate Conservation Grant/ Western Native Trout Initiative	University of Missouri
Lower Colorado River	U.S. Geological Survey (USGS) National Gap Program	USGS/Kansas State University
Upper Snake River	National Fish and Wildlife Foundation	Federation of Fly Fishers/ Trout Unlimited
Rio Grande	Southern Rockies Landscape Conservation Cooperative	Siglo Group
California	The Nature Conservancy	The Nature Conservancy/ Trout Unlimited
Bonneville basin	Not completed	
Lahontan and central Nevada basin	Not completed	
Mid-Columbia	Not completed	

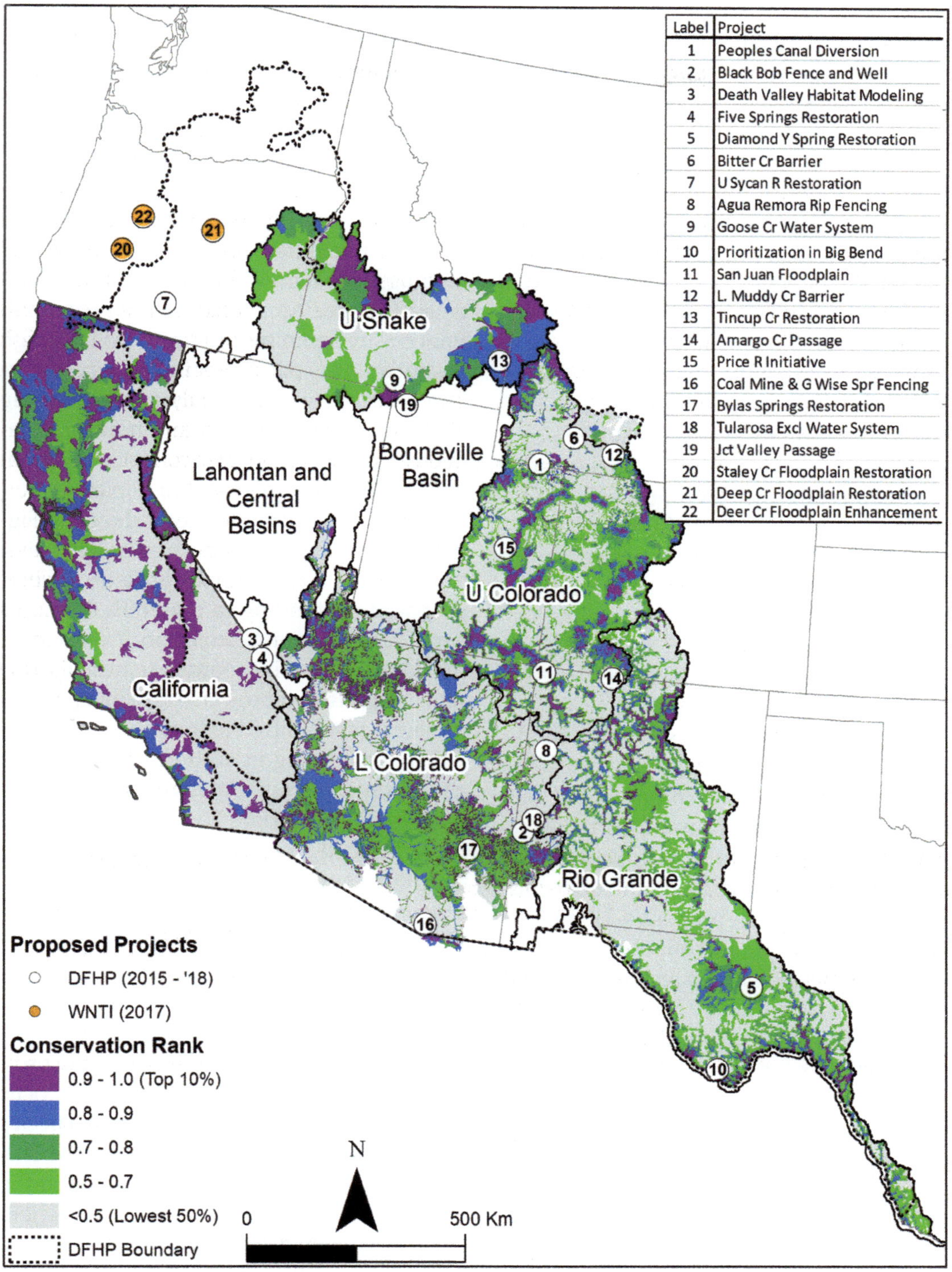

Figure 3. Multispecies aquatic assessments completed within the Desert Fish Habitat Partnership domain and projects submitted to the Desert Fish Habitat Partnership for funding from 2015 to 2018. Projects submitted to the Western Native Trout Initiative in 2017 with a strong nontrout element that were evaluated for cosponsorship are also shown.

previously in other basins. While these two assessments used the common analytical framework as previous assessments, the analyses were also slightly tailored to their goals, focal species, and focal landscapes.

Assessment use in decision making

The primary objective of the DFHP is to focus funding on imperiled habitats and fishes unique to the deserts of North America that are underrepresented by current conservation efforts. By prioritizing funding for these nearly 180 species, the DFHP provides immediate benefits to these species and their habitats, and the partnership directly assists in achieving priority conservation needs that have been established by state, federal, private, and tribal partners. Scientific assessments as described above are used by the DFHP to help inform decisions on funding habitat conservation, restoration, and enhancement projects.

From 2015 to 2018, the DFHP received 22 project proposals for funding (Figure 3). The project funding process begins with development of a request for proposals that is distributed to partners. Proposals are then submitted to the DFHP, usually through coordination with one of the four basin coordinators. These coordinators then review and rank the basin-specific proposals. Upon completion of the reviews and initial project ranking, the proposals are then reviewed, ranked, and voted on by the DFHP Steering Committee. The DFHP coordinator then submits the ranked projects for funding to the U.S. Fish and Wildlife Service. As part of the project review process, the DFHP Science and Data Committee overlays project locations on the basin-specific aquatic assessments and extracts the catchment or subwatershed rank (0 = low value, 1 = high value) to provide an important landscape context on which to evaluate each proposal (i.e., to help answer the question "how important is the watershed to native fish conservation relative to the others in the basin given ecological threats and connectivity?") (Figure 3; Table 3). Because the assessments are focused on rivers and streams and only partially incorporate data on spring and ciénega habitats, and some assessments have not yet been completed (Bonneville, Lahontan and central Nevada, and interior Columbia basins), the conservation values from the assessments are not yet formally integrated into the DFHP project ranking criteria on which proposals are evaluated (but see Appendix IV in DFHP 2015). Projects are submitted to the WNTI through a similar and parallel annual request-for-proposals process. Those that have a strong nonsalmonid component have been simultaneously submitted to the DFHP for potential DFHP–WNTI co-sponsorship (Figure 3).

Three examples of projects shown in Figure 3 illustrate use of assessment information to contextualize the importance of projects proposed for funding to the FHPs. The Tincup Creek Stream Restoration Project near the Idaho–Wyoming border represents one such cosponsored project. The Tincup Creek project was submitted to both the WNTI and DFHP for funding in 2016 (for 2017 funding allocation) by Trout Unlimited and the Caribou-Targhee National Forest. The proposed project aimed to arrest streambank erosion and reconstruct the stream channel (Box 1). Tincup Creek supports a diverse fish assemblage that includes both native Yellowstone Cutthroat Trout *Oncorhynchus clarkii bouvieri* and a rich assemblage of nongame species, including the imperiled Northern Leatherside Chub *Lepidomeda copei*. Thus, the project benefited a native trout that is the focus of the WNTI and native nongame species ranked highly by the DFHP. The diverse Tincup Creek fish assemblage is reflected in the upper Snake River basin assessment, where the subwatershed in which the project was located was highly ranked (Table 3; Figure 4; Williams et al. 2019).

The Black Bob Allotment project was proposed in 2015 (for 2016 funding) by the U.S. Forest Service to help protect the ripar-

Table 3. The conservation value of a catchment/subwatershed in which a proposed project is located as determined from the regional multispecies aquatic assessments for streams and rivers. Conservation values range from 1 (high value) to 0 (low value).

Year	Proposed project	Assessment region	Conservation value
2015	Peoples Canal Diversion	Upper Colorado	0.77
	Black Bob Fence and Well	Lower Colorado	0.77
	Death Valley Habitat Modeling	–	[ab]
	Five Springs Restoration	Lower Colorado	[ab]
	Diamond Y Spring Restoration	Rio Grande	0.71[b]
2016	Bitter Creek Drop Structure	Upper Colorado	0.51
	Upper Sycan River Restoration	Interior Columbia	[a]
	Agua Remora Riparian Fencing	Lower Colorado	0.15
	Goose Creek Group Allotment Pipeline Project	Upper Snake	0.98
	Big Bend Restoration Prioritization	Rio Grande	[a]
	San Juan River Floodplain Wetland Restoration	Upper Colorado	0.91
2017	Lower Muddy Creek Barrier	Upper Colorado	0.85
	Tincup Creek Restoration	Upper Snake	0.90
	Amargo Creek Connectivity and Habitat Enhancement	Upper Colorado	0.77
2018	Price River Restoration Initiative	Upper Colorado	0.63
	Coal Mine & George Wise Spring Fencing	Lower Colorado	0.26[b]
	Bylas Springs Restoration	Lower Colorado	0.75[b]
	Tularosa Exclosure and Water System	Lower Colorado	0.79
	Junction Valley Fish Passage	Upper Snake	0.93

[a] Scientific assessment not available or was not available the year project was proposed.
[b] Spring system. Stream and river assessment may be irrelevant.

ian vegetation and streambanks of the San Francisco River in New Mexico. The proposed project was to restrict livestock from the river channel by constructing 5.5 km of fence along the river channel to complement 5.5 km of natural features restricting livestock access to the river. The project also installed a well to supply an off-stream water source. The San Francisco River is inhabited by the Loach Minnow *Rhinichthys cobitis*, an endangered nongame species, and other species of greatest conservation need such as Sonora Sucker *Catostomus insignis* and Desert Sucker *C. clarkii*. The value of San Francisco River habitat to native fishes is reflected by the project catchment being in the top 23% of all catchments included in the lower Colorado River basin assessment (Table 3; Figure 4; Whittier et al. 2011).

The Goose Creek Group Allotment project was proposed by Trout Unlimited and the U.S. Bureau of Land Management in 2015 (for 2016 funding) to update off-stream watering infrastructure and improve riparian health and instream habitat in the Goose Creek watershed in Idaho. The proposed project, when completed, will install more than 21 km of new pipeline and 15 water troughs to reduce livestock reliance on streams for water. The project will improve riparian vegetation, instream habitat, and water quality in Goose Creek and its tributaries. The main stem provides habitat for Northern Leatherside Chub, Bluehead Sucker *Catostomus discobolus*, and

Box 1. Restoration of Tincup Creek in the Upper Snake River Basin

Lee Mabey, Caribou-Targhee National Forest, 1405 Hollipark Drive, Idaho Falls, Idaho 83401, USA

Leslie Bahn Steen, Trout Unlimited, 185 Center Street, Suite B, Post Office Box 5002, Jackson, Wyoming 83001, USA

The Tincup Creek Stream Restoration Project began in 2017 and is a large-scale, multiphased project led by Trout Unlimited and the U.S. Forest Service, Caribou-Targhee National Forest. The project provides a unique opportunity to improve habitat conditions for Yellowstone Cutthroat Trout *Oncorhynchus clarkii bouvieri*, Northern Leatherside Chub *Lepidomeda copei*, boreal toad *Bufo boreas boreas*, and western pearlshell mussel *Margaritifera falcata*, all native species that have been designated by federal and state management agencies as species of concern. There are also other native aquatic species present within the project area that together form a diverse assemblage of native species. Because of the native fish assemblage, the Tincup Creek watershed ranks high for native fish conservation value in the upper Snake River basin as a potential native fish conservation area (Williams et al. 2019), and the project has been named as one of the 2017 "Waters to Watch" by the National Fish Habitat Partnership.

The project goals are to restore channel processes and floodplain function on 6.5 km of stream that were degraded by accidental spraying of floodplain willows in the late 1950s. The design reconnects the stream to its floodplain and improves habitat by narrowing the stream, elevating riffles, stabilizing eroding banks on outside meanders, and reconnecting historical meanders. Project work is being accomplished with on-site native materials such as gravel, whole willow clumps, and sod mats. In addition, 500 large trees with rootwads have been brought in to provide stability and habitat complexity. The project's focus on healthy stream function and processes is expected to increase habitat complexity and quality through time and benefit all life stages of a diverse assemblage of native fishes. For example, rootwads with wood cover are expected to benefit Northern Leatherside Chub as well as juvenile trout.

To successfully complete a project of this magnitude, it has been necessary to enlist the help of a wide array of partners, including financial and volunteer support from local Trout Unlimited chapters; in-kind support from Agrium, the Bear Lake Grazing Association, the Idaho Department of Fish and Game, local grazing permittees, Caribou County, and the Idaho Transportation Department; and funding support to date from the Desert Fish Habitat Partnership, Caribou-Targhee National Forest, Jackson Hole One Fly, the Idaho office of the U.S. Fish and Wildlife Service, and the Western Native Trout Initiative. Making a measurable impact has required cooperation and engagement from many sources as well as a project design that focuses on stream function to meet a suite of conservation objectives for multiple native species.

Box continues

Box 1. Continued

Box Figure 1. Tincup Creek before channel restoration (June 6, 2017).

Box Figure 2. Tincup Creek after channel restoration (September 14, 2017).

Box continues

Box 1. Continued

Box Table 1. Funding sources as of 2017 for Tincup Creek restoration project.

Contributor	In-kind	Cash
U.S. Forest Service	✓	✓
National Trout Unlimited	✓	
Jackson Hole Trout Unlimited		✓
Snake River Cutthroats Trout Unlimited		✓
Jackson Hole One Fly		✓
Western Native Trout Initiative		✓
Agrium/Bear Lake Grazing Association	✓	
Idaho Fish and Game Department	✓	
Desert Fish Habitat Partnership		✓
U.S. Fish and Wildlife Service		✓
Caribou County	✓	
Idaho Transportation Department	✓	
National Forest Foundation		✓
Idaho Department of Environmental Quality		✓
Trout Unlimited–Orvis Embrace-A-Stream		✓
Subtotal	$69,000.00	$185,500.00
Grand total	$254,500.00	

Yellowstone Cutthroat Trout—all species of greatest conservation need in Idaho—as well as a diverse set of other native fishes linked to healthy riparian and diverse instream habitats (Dauwalter et al. 2014; Walrath et al. 2016; Dauwalter et al. 2018). The diverse fish assemblage and the occurrence of two rare species in Goose Creek and its tributaries is why this subwatershed ranks within the top 2% of all subwatersheds in the upper Snake River basin (Table 3; Williams et al. 2019).

Informing Efficient Conservation Delivery

Organizations and partnerships representing large geographies often have the difficult task of deciding where to focus conservation efforts with limited resources. Numerous approaches to aquatic conservation assessment have been developed to help with this task (Kuehne et al. 2017). Assessment frameworks range from simple overlays of spatial data on environment and human stressors to very quantitative optimization algorithms focused on the conservation principles of comprehensiveness, representation, and others while accounting for species biology (home ranges), landscape connectivity, threats, climate change, and other factors, as described herein and elsewhere (Pressey and Cowling 2001; Moilanen et al. 2009). Algorithms, and the assessment outputs they produce (rankings and priority areas), are not a panacea, but instead serve to guide and inform rather than prescribe planning decisions. They are part of the planning process and not the process itself. The assessments simply highlight important areas and become part of a decision support system, in which humans are integral, for conservation decision mak-

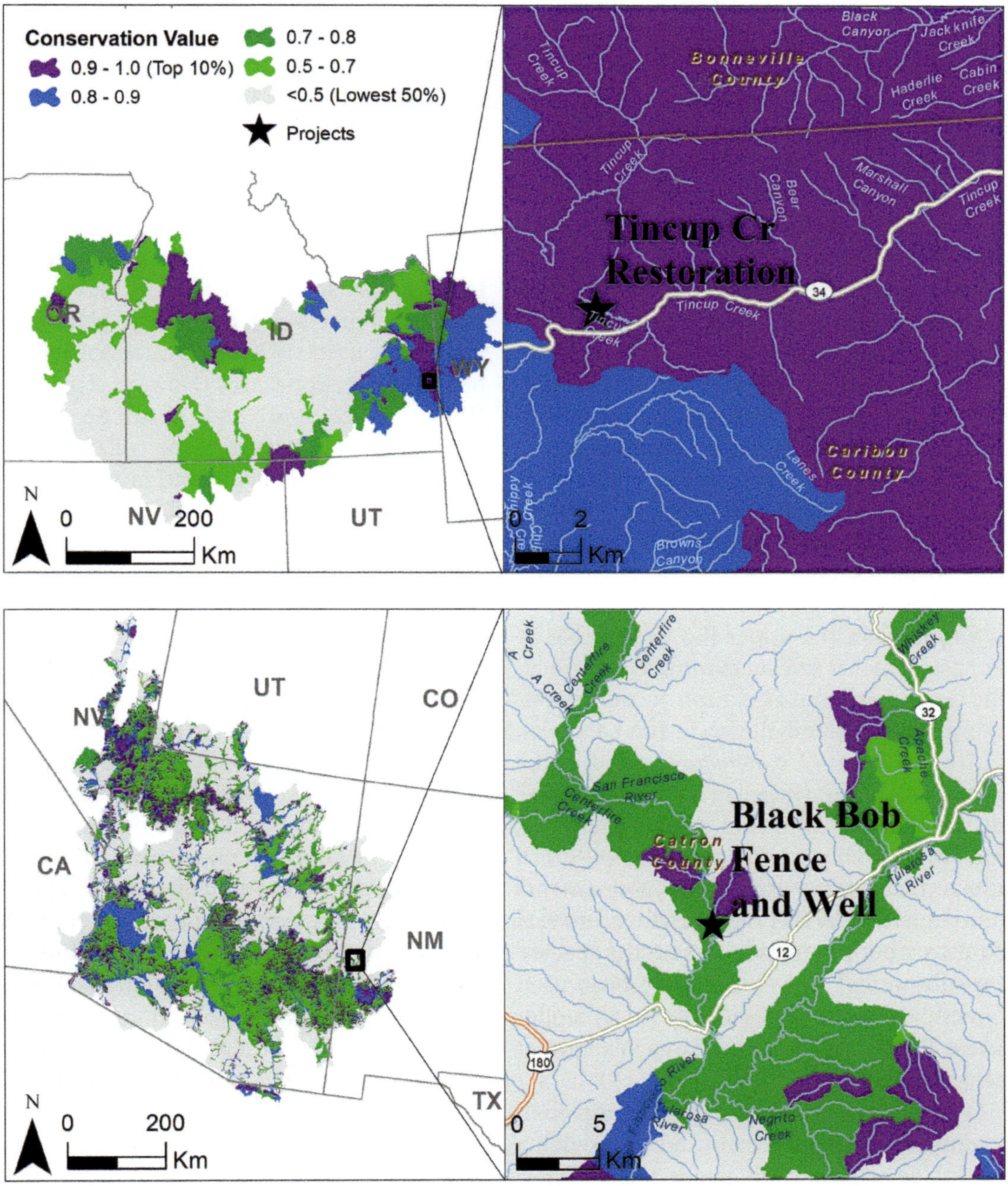

Figure 4. Catchment conservation values (ranks) for the upper Snake River basin (top left) and lower Colorado River basin (bottom left) and specifically for the Tincup Creek watershed proposed for restoration (top right; conservation value = 0.90) and San Francisco River where the Black Bob Allotment project was proposed (bottom right; conservation value = 0.77) (Table 3).

ing (Pressey and Cowling 2001). When good data are available, quantitative assessment and prioritization can enhance the explicitness, repeatability, and scientific credibility of conservation decision making (Ferrier and Wintle 2009).

As partnerships representing large geographies, the DFHP and WNTI have had to be creative in completing aquatic assessments in focal river basins (and WNTI focal species and subspecies) that represent both the scale at which the partnerships operate and a scale appropriate to the conservation priorities and historical ranges of native fishes. As demonstrated above, the partnerships have had to work together to identify diverse funding sources and diverse partners to pursue development of basin-specific assessments. Likewise, basin-specific aquatic assessments naturally align with the regional biogeography of fishes. For example, the historical distribution of native trouts largely align with the individual basins for which assessments were completed: Colorado River Cutthroat Trout *O. c.pleuriticus* in the upper Colorado River basin, Apache Trout *O. apache* and Gila Trout *O. gilae* in the lower Colorado River basin; Rio Grande Cutthroat Trout *O. c. virginalis* in the upper Rio Grande basin. For assessments not yet completed, Bonneville Cutthroat Trout *O. c. utah* occur naturally only in the Bonneville basin and Lahontan Cutthroat Trout *O. c. henshawi* occur in the Lahontan basin (Behnke 2002). Nongame fish distributions are also, mostly, unique to individual basins at the same scale (Smith 1981). Trout-based rangewide assessments funded by the WNTI have produced spatial data on trout populations across the species' (or subspecies') ranges that have been used directly in some assessments (e.g., upper Snake River; Muhlfeld et al. 2015; Williams et al. 2019).

The aquatic assessments completed, to date, have largely been used to provide a landscape or basinwide perspective for projects proposed to the FHPs for funding. Both FHPs use a scoring rubric to rank proposed projects. As mentioned earlier, projects proposed to the DFHP for funding include projects on streams and rivers—habitats for which the assessments discussed herein focus—but they also include projects on springs and ciénegas where assessments have not been completed. The lack of consistent information across habitat types and geographies prohibits the assessments from being formally integrated into project-scoring rubrics at this time. However, completed assessments (1) do give important basinwide context to proposed projects, (2) highlight hydrologic units with high value for conserving native fish diversity at a landscape scale because of high species richness or representation of rare species, (3) identify habitat and protection needs by overlaying high-value hydrologic units with human stressor indices or land protection status (Howard et al. 2018; Williams et al. 2019), and (4) can be used to identify focal watersheds for collaboration, efficient use of resources, and targeted long-term conservation efforts (Dauwalter et al. 2011; Birdsong et al. 2015; Labay et al. 2018; Birdsong et al. 2019).

Social, economic, and political considerations also influence where conservation is implemented on the ground. Landscape-scale aquatic assessments, such as those presented here, can be used in conjunction with socioeconomic and political information to ensure that opportunities for conservation action are balanced with biological priorities as a form of informed opportunism (Noss et al. 2002; Pressey and Bottrill 2008). There is a role, then, for assessments to inform conservation and be part of the collaborative decision-making process among diverse stakeholder groups (Barmuta et al. 2011; Souder 2013). The Tincup Creek project is a good example where project managers submitted project funding proposals to both the WNTI and DFHP, and the FHPs used the upper Snake River basin assessment to confirm that the project occurred in a watershed with high conservation value for both native Yellowstone Cutthroat Trout and native nongame fishes (Figure 4). In other cases, strategic opportunities for cosponsorship of projects are clear without assessment information. The DFHP and WNTI cofunded a fish passage project in the Weber River, Utah that is habitat for fluvial Bonneville Cutthroat Trout and the only population of Bluehead Sucker in the Bonneville basin and has a formal watershed advisory group, thus confirming the importance of that watershed despite the lack of a formal multispecies assessment. The Weber River Partnership is highlighted in Thompson and Burnett (2019, this volume).

Others also have used multispecies aquatic assessments to identify focal areas. The Rio Grande basin assessment included a specific step of using native fish-based conservation values to identify focal planning units referred to as native fish conservation areas (NFCAs; Williams et al. 2011; Labay et al. 2018). The authors suggested that identified NFCAs represent units to focus planning, communication, cooperation, and coordination among multiple stakeholders. A similar approach to identify NFCAs based on multispecies aquatic assessment results has been used in Texas to organize facilitated workshops and catalyze cooperative conservation of aquatic resources through a data-driven process (Birdsong et al. 2019). Successful application of this approach in Texas led to a similar assessment and planning approach for the Great Plains under the Great Plains Fish Habitat Partnership and Great Plains Landscape Conservation Cooperative (Labay et al. 2019, this volume). A similar process could be undertaken with the assessments presented herein (Figure 5).

Efficient conservation delivery is a balance between pragmatism, socioeconomic and political forces, and maximizing the representation and persistence of focal biodiversity. There are simply not enough conservation resources available to be wasteful, and many freshwater fishes are imperiled (Darwall and Freyhof 2016). Luckily, spatial conservation assessment methods are more accessible than ever and are being applied to

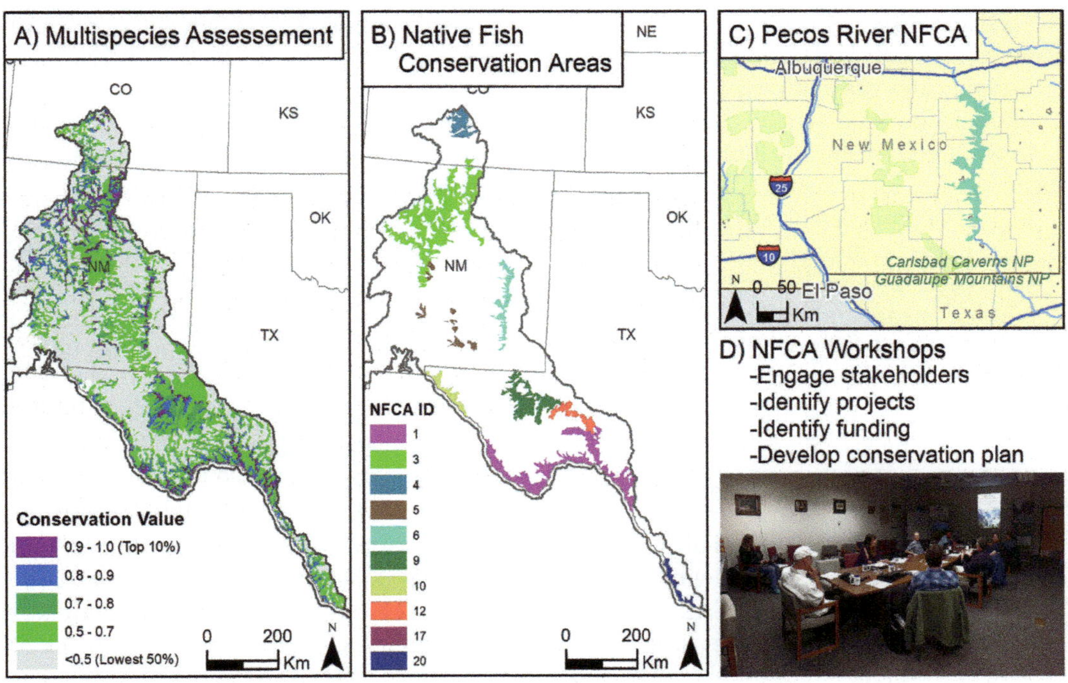

Figure 5. Conceptual process for using **(A)** multispecies aquatic assessments to identify **(B, C)** native fish conservation areas (or focal watersheds) that can be used to organize **(D)** workshops with stakeholders to develop watershed-based conservation plans emphasizing protection and restoration of fish habitats and native fish species.

aquatic systems with watersheds as the focal units. Completed assessments facilitate conservation planning at landscape scales through integration with the local knowledge and pragmatism of field biologists that have expertise on native fishes and in habitat restoration, land protection, and threat abatement. This naturally lends itself to a dual-pronged approach that is both top-down (assessment driven) and bottom-up (local knowledge) to deliver informed and efficient conservation across broad landscapes and large river basins.

Acknowledgments

The authors would like to thank K. Fesenmyer and two reviewers for improving earlier versions of the manuscript.

References

Allan, J. D., and A. S. Flecker. 1993. Biodiversity conservation in running waters. BioScience 43:32–43.

Barmuta, L. A., S. Linke, and E. Turak. 2011. Bridging the gap between "planning" and "doing" for biodiversity conservation in freshwaters. Freshwater Biology 56:180–195.

Beechie, T., J. S. Richardson, A. M. Gurnell, and J. Negishi. 2013. Watershed processes, human impacts, and process-based restoration. Pages 11–49 *in* P. Roni and T. Beechie, editors. Stream and watershed restoration: a guide to restoring riverine processes and habitats. Wiley, West Sussex, UK.

Behnke, R. J. 2002. Trout and salmon of North America. The Free Press, New York.

Bernhardt, E. S., M. A. Palmer, J. D. Allan, G. Alexander, K. Barnas, S. Brooks, J. Carr, S. Clayton, C. Dahm, J. Follstad-Shah, D. Galat, S. Gloss, P. Goodwin, D. Hart, B. Hassett, R. Jenkinson, S. Katz, G. M. Kondolf, P. S. Lake, R. Lave, J. L. Meyer, T. K. O'Donnell, L. Pagano, B. Powell, and E. Sudduth. 2005. Synthesizing U.S. river restoration efforts. Science 308:636–637.

Bernhardt, E. S., E. B. Sudduth, M. A. Palmer, J. D. Allan, J. L. Meyer, G. Alexander, J. Follastad-Shah, B. Hassett, R. Jenkinson, R. Lave, J. Rumps, and L. Pagano. 2007. Restoring rivers one reach at a time: results from a survey of U.S. river restoration practitioners. Restoration Ecology 15:482–493.

Birdsong, T. W., M. S. Allen, J. E. Claussen, G. P. Garrett, T. B. Grabowski, J. Graham, F. Harris, A. Hartzog, D. A. Hendrickson, R. A. Krause, J. K. Leitner, J. M. Long, C. K. Metcalf, D. P. Philipp, W. F. Porak, S. Robinson, S. M. Sammons, S. Shaw, J. E. Slaughter, IV, and M. D. Tringali. 2015. Native Black Bass Initiative: implementing watershed-scale approaches to conservation of endemic black bass and other native fishes in the southern United States. Pages 363–378 *in* M. D. Tringali, J. M. Long, T. W. Birdsong, and M. S. Allen, editors. Black bass diversity: multidisciplinary science for conservation. American Fisheries Society, Symposium 82, Bethesda, Maryland.

Birdsong, T. W., G. P. Garrett, B. J. Labay, M. G. Bean, P. T. Bean, J. Botros, M. J. Casarez, A. E. Cohen, T. G. Heger, A. Kalmbach, D. A. Hendrickson, S. J. Magnelia, K. B. Mayes, M. E. McGarrity, R. McGillicuddy, M. M. Parker, and S. Robertson. 2019. Texas Native Fish Conservation Areas Network: strategic investments in restoration and preservation of freshwater fish diversity. Pages 183–229 *in* D. C. Dauwalter, T. W. Birdsong, and G. P. Garrett, editors. Multispecies and watershed approaches to freshwater fish conservation. American Fisheries Society, Symposium 91, Bethesda, Maryland.

Closs, G. P., P. L. Angermeier, W. R. T. Darwall, and S. R. Balcombe. 2016. Why are freshwater fishes so imperiled? Pages 37–75 *in* G. P. Closs, M. Krkosek, and J. D. Olden, editors. Conservation of freshwater fishes. Cambridge University Press, Cambridge, UK.

Cowx, I. G., and R. L. Welcomme. 1998. Rehabilitation of rivers for fish. Blackwell Scientific Publications, Carlton, Victoria, Australia.

Darwall, W. R. T., and J. Freyhof. 2016. Lost fishes, who is counting? The extent of the threat to freshwater fish biodiversity. Pages 1–36 *in* G. P. Closs, M. Krkosek, and J.

D. Olden, editors. Conservation of freshwater fishes. Cambridge University Press, Cambridge, UK.

Dauwalter, D. C., K. A. Fesenmyer, R. Bjork, D. R. Leasure, and S. J. Wenger. 2017. Satellite and airborne remote sensing applications for freshwater fisheries. Fisheries 42:526–537.

Dauwalter, D. C., K. A. Fesenmyer, S. W. Miller, and T. Porter. 2018. Response of riparian vegetation, instream habitat, and aquatic biota to riparian grazing exclosures. North American Journal of Fisheries Management 38:1187–2000.

Dauwalter, D. C., J. S. Sanderson, J. E. Williams, and J. R. Sedell. 2011. Identification and implementation of native fish conservation areas in the upper Colorado River basin. Fisheries 36:278–288.

Dauwalter, D. C., S. J. Wenger, and P. Gardner. 2014. The role of habitat complexity in habitat use and selection by stream fishes in a Snake River basin tributary. Transactions of the American Fisheries Society 143:1177–1187.

DFHP (Desert Fish Habitat Partnership). 2015. Framework for strategic conservation of desert fishes. DFHP, Salt Lake City, Utah.

Ferrier, S., and B. Wintle. 2009. Quantitative approaches to spatial conservation prioritization: matching the solution to the need. Pages 1–15 *in* A. Moilanen, K. A. Wilson, and H. P. Possingham, editors. Spatial conservation prioritization. Oxford University Press, Oxford, UK.

Franklin, J. F. 1993. Preserving biodiversity: species, ecosystems, or landscapes? Ecological Applications 3:202–205.

Gresswell, R. E. 2011. Biology, status, and management of the Yellowstone Cutthroat Trout. North American Journal of Fisheries Management 31:782–812.

Haag, W. R., and J. D. Williams. 2014. Biodiversity on the brink: an assessment of conservation strategies for North American freshwater mussels. Hydrobiologia 735:45–60.

Hermoso, V., M. J. Kennard, and S. Linke. 2012. Integrating multidirectional connectivity requirements in systematic conservation planning for freshwater systems. Diversity and Distributions 18:448–458.

Howard, J. K., K. A. Fesenmyer, T. E. Grantham, J. H. Viers, P. R. Ode, P. B. Moyle, S. J. Kupferburg, J. L. Furnish, A. Rehn, J. Slusark, R. D. Mazor, N. R. Santos, R. A. Peek, and A. N. Wright. 2018. A freshwater conservation blueprint for California: prioritizing watersheds for freshwater biodiversity. Freshwater Science 37:417–431.

Jelks, H. L., S. J. Walsh, S. Contreras-Balderas, E. Díaz-Pardo, N. M. Burkhead, D. A. Hendrickson, J. Lyons, N. E. Mandrak, F. McCormick, J. S. Nelson, S. P. Platania, B. A. Porter, C. B. Renaud, J. J. Schmitter-Soto, E. B. Taylor, and M. L. Warren. 2008. Conservation status of imperiled North American freshwater and diadromous fishes. Fisheries 33:372–407.

Jones, H. P., P. C. Jones, E. B. Barbier, R. C. Blackburn, J. M. Rey Benayas, K. D. Holl, M. McCrackin, P. Meli, D. Montoya, and D. M. Mateos. 2018. Restoration and repair of Earth's damaged ecosystems. Proceedings of the Royal Society B [online serial] 285:20172577.

Karr, J. R. 1993. Defining and assessing ecological integrity: beyond water quality. Environmental Toxicology and Chemistry 12:1521–1531.

Karr, J. R., and D. R. Dudley. 1981. Ecological perspective on water quality goals. Environmental Management 5:55–68.

Knight, A. T., A. Driver, R. M. Cowling, K. Maze, P. G. Desmet, A. T. Lombard, M. Rouget, M. A. Botha, A. F. Boshoff, J. G. Castley, P. S. Goodman, K. MacKinnon, S. M. Pierce, R. Sims-Castley, W. I. Stewart, and A. von Hase. 2006. Designing systematic conservation assessments that promote effective implementation: best practice from South Africa. Conservation Biology 20:739–750.

Kuehne, L. M., J. D. Olden, A. L. Strecker, J. J. Lawler, and D. M. Theobald. 2017. Past, present, and future of ecological integrity assessment for fresh waters. Frontiers in Ecology and the Environment 15:197–205.

Kukkala, A. S., and A. Moilanen. 2013. Core concepts of spatial prioritisation in systematic conservation planning. Biological Reviews 88:443–464.

Labay, B., B. Prince, J. Ogren, and K. Tasker. 2018. Rio Grande fishes conservation assessment and mapping. Report to Southern Rockies Landscape Conservation Cooperative and Western Native Trout Initiative. Siglo Group, Austin, Texas, Available: http://nativefishconservation.org/network/rio-grande/. (April 2019).

Labay, B. J., J. S. Perkin, D. A. Hendrickson, A. R. Cooper, G. P. Garrett, and T. W. Birdsong. 2019. Who's asking? Interjurisdictional conservation assessment and planning for Great Plains fishes. Pages 57–83 *in* D. C. Dauwalter, T. W. Birdsong, and G. P. Garrett, editors. Multispecies and watershed approaches to freshwater fish conservation. American Fisheries Society, Symposium 91, Bethesda, Maryland.

Leathwick, J. R., A. Moilanen, S. Ferrier, and K. Julian. 2010. Complementarity-based conservation prioritization using a community classification, and its application to riverine ecosystems. Biological Conservation 143:984–991.

Linke, S., E. Turak, and J. Nel. 2011. Freshwater conservation planning: the case for systematic approaches. Freshwater Biology 56:6–20.

Mainali, K. P., D. L. Warren, K. Dhileepan, A. McConnachie, L. Strathie, G. Hassan, D. Karki, B. B. Shrestha, and C. Parmesan. 2015. Projecting future expansion of invasive species: comparing and improving methodologies for species distribution modeling. Global Change Biology 21:4464–4480.

Margules, C. R., and R. L. Pressey. 2000. Systematic conservation planning. Nature (London) 405:243–253.

Master, L. L., B. A. Stein, L. S. Kutner, and G. A. Hammerson. 2000. Vanishing assets: conservation status of U.S. species. Pages 93–118 *in* B. A. Stein, editor. Precious heritage: the status of biodiversity in the United States. Oxford University Press, New York.

McKinney, M. L. 1997. Extinction vulnerability and selectivity: combining ecological and paleontological views. Annual Review of Ecology and Systematics 28:495–516.

Moilanen, A., J. Leathwick, and J. Elith. 2008. A method for spatial freshwater conservation prioritization. Freshwater Biology 53:577–592.

Moilanen, A., K. A. Wilson, and H. P. Possingham. 2009. Spatial conservation prioritization: quantitative methods and computational tools. Oxford University Press, New York.

Muhlfeld, C. C., S. E. Albeke, S. L. Gunckel, B. J. Writer, B. B. Shepard, and B. E. May. 2015. Status and conservation of Interior Redband Trout in the western United States. North American Journal of Fisheries Management 35:31–53.

Nel, J. L., B. Reyers, D. J. Roux, D. Impson, and R. M. Cowling. 2011. Designing a conservation area network that supports the representation and persistence of freshwater diversity. Freshwater Biology 56:106–124.

Nel, J. L., D. J. Roux, R. Abell, P. J. Ashton, R. M. Cowling, J. V. Higgins, M. Thieme, and J. H. Viers. 2009. Progress and challenges in freshwater conservation planning. Aquatic Conservation: Marine and Freshwater Ecosystems 19:474–485.

Noss, R. F., C. Carroll, K. Vance-Borland, and G. Wuerthner. 2002. A multicriteria assessment of the irreplaceability and vulnerability of sites in the Greater Yellowstone Ecosystem. Conservation Biology 16:895–908.

NRCS (Natural Resources Conservation Service), USGS (U.S. Geological Survey), and USEPA (U.S. Environmental Protection Agency). 2008. Watershed boundary dataset. NRCS, Washington, D.C.

Olden, J. D. 2003. A species-specific approach to modeling biological communities and its potential for conservation. Conservation Biology 17:854–863.

Page, L. M., and B. M. Burr. 1991. Freshwater fishes. Houghton Mifflin Company, New York.

Palmer, M., J. D. Allan, J. Meyer, and E. Bernhardt. 2007. River restoration in the twenty-first century: data and experiential knowledge to inform future efforts. Restoration Ecology 15:472–481.

Paukert, C. P., K. L. Pitts, J. B. Whittier, and J. D. Olden. 2011. Development and assessment of a landscape-scale ecological threat index for the lower Colorado River basin. Ecological Indicators 11:304–310.

Pierce, R., W. L. Knotek, C. Podner, and D. Peters. 2019. Blackfoot River restoration: a thirty-year review of a wild trout conservation endeavor. Pages 649–682 *in* D. C. Dauwalter, T. W. Birdsong, and G. P. Garrett, editors. Multispecies and watershed approaches to freshwater fish conservation. American Fisheries Society, Symposium 91, Bethesda, Maryland.

Pressey, R. L., and M. C. Bottrill. 2008. Opportunism, threats, and the evolution of systematic conservation planning. Conservation Biology 22:1340–1345.

Pressey, R. L., and R. M. Cowling. 2001. Reserve selection algorithms and the real world. Conservation Biology 15:275–277.

Roni, P., T. Beechie, S. Schmutz, and S. Muhar. 2013. Prioritization of watersheds and restoration projects. Pages 189–214 *in* P. Roni and T. Beechie, editors. Stream and river restoration: a guide to restoring riverine processes and habitats. Wiley, West Sussex, UK.

Sarkar, S., and P. Illoldi-Rangel. 2010. Systematic conservation planning: an updated protocol. Natureza & Conservação 8:19–26.

Smith, G. R. 1981. Late Cenozoic freshwater fishes of North America. Annual Review of Ecology and Systematics 12:163–193.

Souder, J. 2013. The human dimensions of stream restoration: working with diverse partners to develop and implement restoration. Pages 114–143 *in* P. Roni and T. Beechie, editors. Stream and watershed restoration: a guide to restoring riverine processes and habitats. Wiley, West Sussex, UK.

Taylor, C. A., G. A. Schuster, J. E. Cooper, R. J. DiStefano, A. G. Eversole, P. Hamr, H. H. I. Hobbs, H. W. Robison, C. E. Skelton, and R. F. Thoma. 2007. A reassessment of the conservation status of crayfishes of the United States and Canada after 10+ years of increased awareness. Fisheries 32:372–389.

Thompson, P. D., and P. C. Burnett. 2019. The Weber River Partnership: how fish gained relevance through a recently formed watershed group. Pages 565–588 *in* D. C. Dauwalter, T. W. Birdsong, and G. P. Garrett, editors. Multispecies and watershed approaches to freshwater fish conservation. American Fisheries Society, Symposium 91, Bethesda, Maryland.

USEPA (U.S. Environmental Protection Agency) and USGS (U.S. Geological Survey). 2005. National Hydrography Dataset Plus - NHDPlus (1:100,000 scale). USEPA, Washington, D.C. and USGS, Reston, Virginia.

Walrath, J. D., D. C. Dauwalter, and D. Reinke. 2016. Influence of stream condition on habitat diversity and fish assemblages in an impaired upper Snake River Basin watershed. Transactions of the American Fisheries Society 145:821–834.

Wenger, S. J., M. M. Hagler, and B. J. Freeman. 2009. Prioritizing areas of the Conasauga River sub-basin in Georgia and Tennessee for preservation and restoration. Proceedings of the Southeastern Fishes Council 51:31–38.

Whittier, J., C. Paukert, and K. Gido. 2006. Development of an aquatic GAP for the lower Colorado River basin. GAP Analysis Bulletin 14:65–68.

Whittier, J., C. Paukert, J. Olden, K. Pitts, and A. Strecker. 2011. Lower Colorado River Basin Aquatic Gap Analysis Project. U.S. Geological Survey, Reston, Virginia.

Whittier, J., and N. Sievert. 2014. Conservation assessment for native fish in the upper Colorado River basin, Report to Western Native Trout Initiative and Western Association of Fish and Wildlife Agencies. University of Missouri, Columbia.

Williams, J. E., A. L. Haak, N. G. Gillespie, and W. T. Colyer. 2007. The conservation success index: synthesizing and communicating salmonid condition and management needs. Fisheries 32:477–492.

Williams, J. E., R. N. Williams, R. F. Thurow, L. Elwell, D. P. Philipp, F. A. Harris, J. L. Kershner, P. J. Martinez, D. Miller, G. H.

Reeves, C. A. Frissell, and J. R. Sedell. 2011. Native fish conservation areas: a vision for large-scale conservation of native fish communities. Fisheries 36:267–277.

Williams, R. N., D. C. Dauwalter, R. F. Thurow, D. P. Philipp, J. E. Williams, and C. A. Walser. 2019. Identification and utility of native fish conservation areas in the upper Snake River basin, USA. Pages 135–159 *in* D. C. Dauwalter, T. W. Birdsong, and G. P. Garrett, editors. Multispecies and watershed approaches to freshwater fish conservation. American Fisheries Society, Symposium 91, Bethesda, Maryland.

WNTI (Western Native Trout Initiative). 2008. A plan for strategic action. WNTI, Denver, Colorado. Available: www.westernnativetrout.org/media/a-plan-for-strategic-action.pdf. (April 2019).

WNTI (Western Native Trout Initiative). 2016. A plan for strategic action: update. WNTI, Denver, Colorado. Available: www.westernnativetrout.org/media/final-update-to-wnti-strategic-plan-approved-by-sc-november-10-2016.pdf. (April 2019).

Zale, A. V., D. L. Parrish, and T. M. Sutton. 2012. Fisheries techniques, 3rd edition. American Fisheries Society, Bethesda, Maryland.

American Fisheries Society Symposium 91:33–55, 2019

The Really Big Picture for Fish Habitat: The Conceptual Underpinnings and Vision for the National Fish Habitat Partnership's National Fish Habitat Assessment

Gary E. Whelan*
Michigan Department of Natural Resources-Fisheries Division
Post Office Box 30446, Lansing, Michigan 48909, USA

Abstract.—Over much of the history of fisheries management, fisheries biologists challenged with the conservation of degraded fisheries habitats have primarily focused on addressing the symptoms of habitat degradation as opposed to confronting the overarching processes and factors that control fish habitat condition. This is often attributable to the substantial amount of inaccessible or unorganized data that confound resource management decisions. The National Fish Habitat Partnership (NFHP) was formed in 2006 to provide a science-based, holistic, and voluntary-based approach to address the trillions of U.S. dollars in damages that have been inflicted on fish habitats in the United States. The NFHP uses a periodically measured, landscape-level national fish habitat assessment to identify intact systems that need conservation or protection and to assess the root causes of aquatic habitat degradation in altered systems. Categories of data and information contained within the NFHP national fish habitat assessment consist of hydrology, connectivity, water quality, material transport and recruitment, geomorphology, and aquatic organisms' effect on habitat and energy flow. These processes are critically important in controlling fish habitat condition in all types of aquatic systems, with the key differences being the relative importance and the rates in which the processes and factors operate. Data and information on fish and aquatic organisms and social data are the other components needed to build a comprehensive assessment and decision support framework for fish habitats in the United States. A framework for a model national fish habitat assessment (model assessment) is outlined herein, with each category described in measurable subcomponents that are actionable by fisheries biologists or other aquatic resource managers. Key variables for each process and factor, along with needed data and information for development of dose–response relationships and social data for societal importance indication, are also provided. Although much of the data to fully populate a model assessment are not available currently, it is important to establish a vision for the future. Many of the envisioned data necessary for a model assessment are available on a localized or regional basis to enable the detailed analyses to occur on those spatial scales, allowing the testing of the robustness of the framework. Once the model assessment is fully developed, aquatic resource managers will

* Corresponding author: whelang@michigan.gov

have a powerful tool to prioritize the trillions of dollars needed to conserve intact and rehabilitate degraded aquatic habitats to build self-sustaining and resilient fish communities. The tool will also help facilitate the NFHP's goals to maintain intact systems and to move degraded system processes and factors back to within 25% of the expected norms for those watersheds.

Introduction

A pressing challenge for aquatic resource managers is that we are swimming in billions of dollars (U.S.) of fish and aquatic community and fish habitat data in the United States that are unorganized and generally unavailable, except for some state and individual watershed efforts. To address this issue, the development of a complete and comparable picture of the condition of U.S. fisheries habitat that can assist in large-scale prioritization of habitat restoration efforts has been a long-time dream of fisheries biologists. This dream was not technically feasible until the 2000s with the advent of geographic information systems and powerful desktop computers. The need for such decision support tools is clear as many U.S. fisheries and their supporting populations are in decline, with about 20% of the biota that rely on aquatic systems considered imperiled or critically imperiled (H. John Heinz III Center for Science, Economics and the Environment 2002). In addition, approximately 37% of fish fauna are considered at risk or vulnerable, with nearly 4 of every 10 native freshwater fishes at risk of extinction (Stein et al. 2000). Declines in at-risk species are found throughout the United States, with the highest proportion in Hawaii and the southeastern United States (H. John Heinz III Center for Science, Economics and the Environment 2002). Miller et al. (1989) stated of the multiple causes involved with the extinction of America's freshwater fishes that physical habitat changes are the most common cause implicated (73% of extinctions), followed by introduced species effects (68%), chemical alteration of habitat (38%), hybridization (38%), and overharvest (15%). Similar trends have been noted in coastal systems in the Pacific Northwest, where 214 salmonid stocks representing genetically distinct populations are rare or threatened and another 106 populations have disappeared due to habitat alteration, dam construction, deforestation, and inappropriate uses of hatchery fish (Nehlson et al. 1991). Similar declines have been documented on the eastern seaboard of the United States (Waldham 2013). However, these initial large-scale fish habitat analyses cannot provide aquatic resource managers with sufficient detail to take actions on specific aquatic systems as the spatial scale was inappropriately large and available data and information were not specific to individual systems or watersheds. Furthermore, these analyses did not provide direct information on the impaired processes or functions to focus conservation actions.

In response to this clear need, the National Fish Habitat Partnership Board (Board) developed a National Fish Habitat Assessment Program (Assessment Program) designed to provide more concise aquatic habitat information to guide and inform conservation actions. The National Fish Habitat Partnership (NFHP), a voluntary and nonregulatory approach to conserving and improving fish habitat, was established in 2006 to bring new resources and attention to the decline in fish stocks linked to long-term aquatic habitat degradation. The Board's latest approved October 2017 vision for the Assessment Program is a comprehensive, comparable, and connected assessment of all U.S. fish habitat from the mountaintop to the shelf. Ultimately, the Assessment Program should provide a connected habitat portrait of the entire nation and, in the future, include waters that influence U.S. aquatic habitats from Canada and Mexico. The Assessment Program's purpose is to support the conservation, rehabilitation, and improvement of fish habitats by providing (1) fish habitat partnerships with national and regional data sets and accompanying analytical procedures that assess the condition of U.S. fish habitats using the best available data, and (2) the Board and stakeholders with communication products to highlight the status and importance of the nation's fish habitats.

The Assessment Program has produced two national assessments that provided the initial national fine-scale fish habitat portraits (National Fish Habitat Board 2010; Crawford et al. 2016). These assessments established the geospatial framework based on the National Hydrography Dataset Version 2 (NHD+; 1:100,000 scale) for inland waters and initiated the geospatial framework development for coastal areas. Both assessments used available standardized national data sets, consisting of landscape-scale land use and land cover data layers attributed to individual stream segments and estuaries. By the 2016 assessment, fish community data for the conterminous United States were developed from single pass electrofishing information from more than 39,000 stream reaches and from trawl information from selected coastal estuaries with a focus on the Gulf Coast (Crawford et al. 2016). Significant analytical improvements were made in the 2016 assessment, including inclusion of statistically defensible, dose–response relationships. Significant assessment gaps were present in the 2010 and 2016 assessments, including fish habitats in Alaska, lakes and most reservoirs, most coastal areas, and deepwater habitat from all coasts and Great Lakes. These habitats were not assessed due to insufficient staff resources along with the unavailability of NHD+ in Alaska; peer-reviewed spatial frameworks, particularly for lakes, reservoirs, and the inshore marine systems; consistently sampled fish community data; and analytical frameworks for the unassessed habitats. All of the Assessment Program data are housed at http://assessment.fishhabitat.org.

Although all landscape data layers (e.g., percent urban area, percent pasture and grazing) were attributed at the appropriate scale for assessed inland waters and estuaries in both assessments, most of the information was very difficult for aquatic resource managers to take direct action on. Additionally, most of the data layers have effects on a range of habitat-forming processes and functions, again making direct action difficult. For example, how can a fisheries manager effectively change the hydrologic alteration from the high percentage of impervious surfaces within urban areas, a key habitat condition driver in the 2010 and 2016 assessments, just one of many effects of urban land use on fish habitats. Given the lessons learned by aquatic resource managers from both assessments and the difficulty of influencing landscape-scale variables, the objective of this chapter is to provide a list of recommended variables to use and a template of what a model national fish habitat assessment (model assessment) would look like to allow it to be used as a decision support tool for aquatic resource managers. The fully implemented model assessment will provide fine-scale information to allow managers to take direct action on the causative habitat processes and factors, assuming sufficient resources are available, and to allow improved prioritization and allocation of habitat funding across large regions of or the entire United States.

Key Tenets and Assumptions for a Model National Fish Habitat Assessment

Aquatic habitat assessments must have a set of clear foundational concepts that provide analytical boundaries and set an overall direction for habitat assessment efforts. The following is a set of foundational tenets proposed for the model assessment:

- Information from the model assessment focuses on the controlling physical processes, watershed functions, and factors not the symptoms of degradation.
- Model assessment data provide clear direction on habitat conservation, rehabilitation, and improvement priorities to get the best return on incremental habitat investments, along with displaying locations where systems are intact and require conservation or protection from degradation.
- Model assessment uses spatially based information from all three fishery components as follows:
    - o fish and aquatic communities—to allow for dose-response relationships that detail the effects of variables on fish and aquatic organisms to be built,
    - o habitat—to build fish-community relationships and to focus conservation efforts, and
    - o people—to allow social information to be incorporated in the decision support system.
- Model assessment incorporates spatially based data on the key system constraints, such as physical geography controls (i.e., surficial geology, climate, and slope) on each system being evaluated that cannot be changed by management actions.
- The assessed variables are consistently measured across the landscape or at least have transfer or conversion functions to allow seamless data use between databases and sampling methods, can be influenced through management actions, and have a measurable effect on fish and other aquatic resources.
- Model assessment uses the expected value and range for all processes, factors, or variables as action criteria.
- The socioeconomics and societal values for waters should be incorporated spatially and used as another set of prioritization variables.
- Model assessment employs and mines existing data sets.
- Information should be available at any needed spatial scale or template (i.e., individual lakes, river segment(s), watershed(s), estuary complexes, congressional districts, and state boundaries).

Key Processes, Factors, and Functions

A model assessment focuses on key processes, factors, and functions, and these need to be robust enough to be useable for all aquatic habitats (i.e., rivers and streams, lakes and reservoirs, estuarine, and coastal systems). The Board's Science and Data Committee developed six processes, factors, and function categories that incorporate the concepts documented by the Instream Flow Council in Annear et al. (2004) and Locke et al. (2008). The following categories are recommended in the model assessment: connectivity; hydrology; geomorphology and bottom form; material recruitment and transport to include woody debris and sediment; water quality; and aquatic community effects on habitat and energy flow, including living habitat, invasive species, and size spectra of organisms that control energy flow.

Expected values

All model assessment processes, factors, and functions should either have (1) an expected condition estimate and associated variance based on a designated baseline date, preferably from an unaltered or least impaired period as possible; or (2) a range of estimated expected values leading to variance calculation in the future as new data and information become available. This will allow thresholds to be determined when intervention is required. These thresholds will focus habitat efforts on protection or conservation in systems that are within expected values and on rehabilitation or improvement for systems that are not within the expected values. The Board's Science and Data Committee recommends action be taken on any process or function variable that is 25% beyond the expected range or variation in a system. This threshold aligns with the estimates from Booth and Jackson (1997), Carlisle et al. (2010), Poff and Zimmerman (2010), Grantham et al. (2014), and Wagner and Midway (2014), who showed system changes with deviations of 10–50% for a range of parameters. In addition to measuring threshold points, it is equally important to measure rates of change in processes, factors, and functions temporally with the intent to target habitat efforts towards the most rapidly changing or threatened landscapes.

Model assessment category—hydrology

Hydrology in the model assessment is broadly defined in this chapter to include water discharge characteristics, water surface elevations, and velocity and current vectors to ensure applicability to all aquatic systems. Annear et al. (2004) states that the central character of rivers is driven by hydrology with clear contributions from the other processes and factors. This is equally true for all other types of aquatic habitats. This is a master variable that drives other key processes and factors that directly influence many others in the model assessment (Poff et al. 1997). The patterns of discharge or currents often dictate structure and function of systems more than the absolute amount of flow or currents (Arthington et al. 2006).

There are a broad range of potential ecologically useful variables to examine for hydrology, ranging from 67 in the Indicators of Hydrologic Alteration tool kit (Richter et al. 1996) to 171 in the Hydrologic Assessment Tool (Cade 2006). It is critical to set a baseline date for all systems to allow for a time-series analysis from that date, and the baseline date should be as unaltered or least impaired as possible, except for reservoirs, which are by definition man-made systems. While the many hydrologic variables in these tools provide insights into ecological questions and system changes over time, the focus should be on two sets of variables: the mean and the variance in the annual and daily hydrology and water surface elevation, and the timing of key hydrologic events. This focus generally follows the suggested approach or the analysis of hydrologic variables on aquatic communities in Poff and Allan (1995), Keough et al. (1999), Bunn and Arthington (2002), Sommer et al. (2004), Magilligan and Nislow (2005) , Helms et al. (2009), Konrad et al. (2008), Poff et al. (2006), Poff and Zimmerman (2010), and McManamay and Frimpong (2015).

Hydrology is one of the few processes with consistently measured data available as a result of the U.S. Geological Survey (USGS) national gaging system and its standardization. Assuming societal and legal approval allowing a watershed hydrology change, it is also one of the simplest processes to correct and rehabilitate. An example of how rapid changes can occur as a result of improvements in hydrologic conditions was demonstrated by Auer (1996) for Lake Sturgeon *Acipenser fulvescens* with a change in a hydropower project from hydropeaking operations to strict run-of-river operation. The Prickett Dam Hydroelectric Project on the Sturgeon River (Baraga County, Michigan, USA) operated daily with high flows, at or above bank-full flows, during generating periods and essentially no flow during off-generating periods, similar to extreme drought conditions. Lake Sturgeon, a state-listed threatened species, used this river for spawning and rearing and peaking flows resulted in poor recruitment. Once the flow regime from the powerhouse was changed to run of the river (i.e., inflows equaling outflows), successful spawning and recruitment of Lake Sturgeon was immediately documented by Auer (1996).

U.S. hydrology data continue to be limited at this time with key gaps regionally and for the state of Alaska, along with high altitude and small watersheds in other areas throughout the United States (Deweber et al. 2014). Miller et al. (2018) addressed many gaps in hydrology data for the conterminous United States using random forest modeling to estimate natural discharges from 1950 to 2015.

For each habitat type, the following category variables are recommended for the model assessment:

- *River and stream flows* using mean daily discharge data on an annual time step
  - o average monthly hydrograph and variance
  - o average and variance or range of the timing for and volume of peak and base flows
  - o duration and variance for or range of peak and base flows
- *River and stream flows* using instantaneous discharge data on a monthly time-step
  - o mean daily flow and variance or range
  - o mean daily flow fluctuation and its variance or range as defined by the difference in minimum and maximum flow
- *Lake and reservoir* levels using mean daily elevation data on an annual time step
  - o average monthly water surface elevation and variance
  - o average and variance of the timing for and volume of peak and lowest water surface elevation
  - o duration and variance for peak and lowest water surface elevation
- *Reservoir and impoundment* storage using mean daily data on a monthly time step
  - o monthly average storage change and variance
- *Lake and reservoir* levels using instantaneous elevation data on a monthly time step
  - o mean daily water surface elevation and variance or range
  - o mean daily water surface elevation fluctuation and its variance or range as defined by the difference in minimum and maximum water surface elevations
- *Coastal systems including the Great Lakes* current patterns using mean daily velocity and flow pattern daily data on a monthly time step
  - o average flow patterns and velocities, direction, and their variance
  - o average and variance of the timing for and volume of peak and lowest flow patterns and velocities
  - o annual duration and variance for peak and lowest flow patterns and velocities
- *Coastal systems including the Great Lakes* water surface elevations using mean daily data on a monthly time step
  - o average monthly water surface elevation and its variance
  - o average and variance of the timing for and volume of peak and lowest water surface elevation
  - o duration and variance for the peak and lowest water surface elevation
- *Wetlands*
  - o number and acreage of wetlands by type within watersheds to complete system water storage estimates along with providing data on a key habitat class for many fish species

Model assessment category—connectivity

Highly connected systems are needed to move sediment and woody debris from the landscape. Connected systems are essential to maximize the productivity of aquatic systems by providing the best available habitat for all life stages and life history strategies of aquatic life (Schlosser 1991; Able 2005). Fish communities lose species and productivity when fragmented (Winemiller and Rose 1992).

Fish movement is critical to the movement of nutrients and energy from large to small water bodies, and this transport mechanism is well documented (Pacific: Cederholm et al. 1999; Gende et al. 2002; Naiman et al. 2002; Schlinder et al. 2003; Wipfli et al. 2003; Twining et al. 2017; Atlantic: Durbin et al. 1979; Twining et al. 2013; Waldham 2013; Great Lakes: Childress and McIntyre 2016; Jones and Mackereth 2016; Europe: Jonssen and Jonssen 2003). Large water body subsidies to smaller waters from migrating fish come in the form of excretion products, carcasses, eggs, fry and young-of-year fish, which are absolutely required to overcome the low productivity from nutrient and energy poor systems frequently found in interior systems.

Another benefit of system connectivity is as a recruitment refugia, allowing for genetically related fish populations to exist in separate parts of watersheds while ensuring connections to common reproductive habitat. Some part of the overall fish population will reproduce in any given year, despite any system catastrophes. Pacific coast examples of this benefit include Rainbow Trout *Oncorhynchus mykiss* and steelhead (anadromous Rainbow Trout) and Sockeye Salmon *O. nerka* and kokanee (lacustrine Sockeye Salmon). In the Great Lakes, nearly all the Great Lakes fish species historically used tributary streams for spawning with a subpopulation remaining resident in the tributary stream with examples, including Round Whitefish *Prosopium cylindraceum* and Smallmouth Bass *Micropterus dolomieu*.

Connectivity is one of the most feasible processes to correct, if impaired, and provides very quick results in most watersheds. This process also has a substantial amount of data available. There are major efforts at this time nationally to survey watersheds for their current connectivity status, including work done on northeastern dams (Martin and Apse 2011), northeastern road crossings (North Atlantic Aquatic Connectivity Collaborative, http://northatlanticlcc.org/products/north-atlantic-aquatic-connectivity-collaborative), southeastern barriers (http://southeastaquatics.net/sarps-programs/southeast-aquatic-connectivity-assessment-program-seacap/prioritization-connectivity-tools-and-other-resources/connectivity-resources), Great Lakes barriers (https://greatlakesconnectivity.org), California barriers (Statewide Barrier Inventory, www.cafishpassageforum.org/statewide-barrier-inventory), northwest barriers(www.streamnet.org/data/interactive-maps-and-gis-data/), and the Kenai River watershed, Alaska (https://kenaiwatershed.org/science-in-action/fish-barriers/culvert-assessment/). There are also a number of decision support models available, including the Northeast Aquatic Barriers Prioritization Tool (http://maps.freshwaternetwork.org/northeast/), the Chesapeake Bay Fish Passage Prioritization Tool (http://maps.freshwaternetwork.org/chesapeake/), the North Carolina Barrier Prioritization Tool (http://portal.ncdenr.org/web/cpt/), Fishwerks—Great Lakes Basin (https://greatlakesconnectivity.org), and FishPass—Anadromous Fish Optimization Tool (www.cafishpassageforum.org/fishpass). Given the amount of currently available and generally consistently measured information, the development of this layer of the model assessment is well along for the United States and will likely be available soon.

To assess connectivity, the following variables are recommended for the model assessment:

- *Natural barrier* determination using spatially referenced locations of waterfalls, high-gradient rapids, and shoots. The intent of the model assessment is to ensure natural connectivity patterns are maintained or re-established. Natural barrier data should include the slope, length, and height of each barrier to allow for additional hydrologic analysis of passability in extreme flow conditions along with the analysis of natural passage requirements of each target fish species.

- *Anthropogenic barrier* determination using spatially referenced locations of
    - fisheries management barriers such as weirs
    - dams
    - culverts
    - fishways, which are still barriers with fish passage delays
    - concrete channels and conveyance channels that are often shallow, wide, and without any roughness, creating effective velocity and depth movement barriers
    - water-quality impairments that will create fish avoidance conditions
    - water-quantity impairments, such as large diversions that can create dewatered habitat downstream blocking fish movement

  Each barrier should have data collected on the slope, length, and height of each barrier, along with current velocity distribution information at and within the barrier. For fishways, information should include design parameters, attraction flow, and location on the structure, fallback estimates, and fish passage delay times to allow for a passability estimate to be generated.
- *Coastal velocity barriers* determined using spatially referenced locations of developments that have altered natural flow patterns, particularly in inshore and estuarine systems, including constructed navigation channels and structures, and tidal pond structures and gates. Each velocity barrier should have data collected on the length, height (only for tidal pond structures), and depth of each barrier along with any current velocity distribution information.

Model assessment category—material transport and recruitment

Watersheds are conveyance systems for the movement of soil and vegetative materials from the landscape, with rivers being the conveyors and lakes along with coastal areas being storage locations for these materials. Rivers, lakes, and coastal areas have temporary and long-term recruitment, storage, and internal processing rates for materials. These processes can be rapidly altered by hydrology and land use changes, either from natural or anthropogenic causes. Starting with Leopold et al. (1964), many geomorphologists have detailed the importance of the balance between input and movement rates of materials in transport. Destabilizing this balance leads to degraded river habitat (Collier et al. 1996; Van Steeter and Pitlick 1998; Kondolf et al. 2006). Rivers along with their associated fish habitat respond quickly to changes in this process. Fish habitat relies on this process working within geologic and climatic bounds.

Channel and shoreline processes, energy processing, and physical fish habitat rely on coarse organic material and large woody debris inputs being maintained within expected amounts for materials from the forest to the sea (Maser and Sedell 1996; Martin 2001; Gregory et al. 2003). In systems with large woody debris inputs, these are channel geomorphology controlling processes (Montgomery et al. 2003; Mutz 2003). Understanding the recruitment of materials to waters from riparian zones and ensuring the rate is appropriate for the forest type are key variables for fish and aquatic habitat for which measurable methods are available (Benda et al. 2003; Gurnell 2003; Piegay 2003).

Material transport and recruitment inputs and rates are possible to manage (Gregory et al. 2003; Kondolf et al. 2014; Rhienheimer and Yarnell 2017), but efforts will have a time delay, particularly if being rehabilitated on a watershed basis. This is an expensive process to correct, thus keeping systems within expected background values is critical. Currently, data on this process are only available locally or for specific point locations, with generalized data available through such sources as the USGS SPARROW (Spatially Referenced Regression on Watershed Attributes) models, which also serve as decision support models.

While sediment measurement is mostly standardized, woody debris and course organic material have not been nationally standardized, and efforts will be needed to make these data comparable nationally using the guidance in Gregory et al. (2003) to fully incorporate these data into the model assessment. It is expected that this category will take at least a decade to be incorporated into the model assessment, except in small-scale analyses.

To assess material transport and recruitment, the following variables are recommended for the model assessment:

- *All materials*
    - natural undisturbed annual rates and variances of sediment, woody debris, and coarse organic material recruitment and transport based on the forest type, geology, and climate for each watershed and coast
- *Sediment*
    - annual erosion rates and variance for each soil type in each watershed and coast
    - annual anthropogenic erosion rates and variance by land use
    - annual sediment transport and storage rates and variance for both in-channel and within-lake/reservoir/coastal storage areas. Data on how these rates are affected by land use changes are also desirable.
- *Woody debris*
    - annual riparian forest growth dynamics and expected variance, with an emphasis on the potential area of recruitment based on valley slope
    - annual woody debris recruitment rates and variance based on forest type and mortality rates
    - annual woody debris and course organic material transport and storage rates and variance for both in-channel and within-lake/reservoir/coastal storage areas. Estimated rate effects from land use changes are another desirable variable.

Model assessment category—water quality

While having appropriate quantities of water is clearly important to fish and aquatic communities, it must have the appropriate chemical characteristics to support all life stages and to maximize productivity of the expected aquatic community (Magnuson et al. 1979; Mainstone and Gulson 1990; Whitfield and Elliott 2005; Bain and Jia 2012; Hamilton et al. 2015; Payne et al. 2015; Speers-Roesch and Norin 2016). There is a substantial number of potential variables, and all of these are measured using standardized techniques such as the U.S. Environmental Protection Agency Clean Water Act standard methods (USEPA 2017). Guidance is available on the appropriate collection of river and stream temperature (Heck et al. 2018) and dissolved oxygen (Rounds et al. 2013). The more challenging issue is to develop the expected value for each variable with and without human land use changes. An additional quandary is which variables will provide the largest return on investment.

Similar to material transport and recruitment, water quality is possible to manage, but efforts usually will have a significant time delay, particularly if being rehabilitated on a watershed basis. This category is often an expensive process to correct; thus, preventing degradation and keeping systems within expected values is critical. Currently, data are available broadly across the United States, but generally only for specific point locations within a watershed. More generalized data are available through such sources as SPARROW models. Given the fragmented nature of these data, it is not likely that this category can be fully incorporated into a model assessment for another decade, except in small-scale analyses.

The following water quality variables are recommended for the model assessment, with other variables, such as contaminant or chloride concentrations, being important in specific applications:

- *River, stream, and coastal temperature* is the key driving factor that structures fish and aquatic communities and should be measured similar to hydrologic factors, using both daily and annual time steps.
  - annual summary temperature variables to measure using daily mean data on a monthly time step are average monthly temperature and variance, average and variance of the timing for the monthly maximum and minimum temperatures, and duration and variance for maximum and minimum temperatures for each month
  - monthly summary temperature variables to measure on a daily time step are average daily temperature and variance, and average daily temperature flux (daily maximum minus the minimum temperature) and variance
- *Lake and reservoir temperature* using daily data on a monthly or seasonal time step.
  - average monthly or seasonal temperature profile and variance
  - average and variance of the timing for spring and fall turnover events
- *River, stream, and coastal dissolved oxygen*—Given the relationship of temperature and altitude to dissolved oxygen concentrations, the key variable to understand is percent saturation.
  - annual summary dissolved oxygen variables to measure using mean daily data on a monthly time step are average monthly percent saturation and variance, average and variance of the timing for maximum and minimum percent saturation, and dura tion and variance for maximum and minimum percent saturation for each month
  - monthly summary dissolved oxygen variables to measure on a monthly time step using daily data are average percent saturation temperature and variance, and average daily percent saturation fluctuation and its variance
- *Lake and reservoir dissolved oxygen*
  - annual summary dissolved oxygen variables to measure on a monthly or seasonal time step using daily data are average monthly dissolved oxygen profile and variance, and annual average and variance of the timing for spring and fall turnover events
- *Nutrients*—While there are a host of potential nutrients to examine, the two key nutrients are total phosphorus and nitrogen, which control many of the primary production rates across fish habitats (Mueller and Helsel 1996 and Cloern 2001).
  - annual summary nutrient variables to measure for phosphorus and nitrogen on a monthly or seasonal time step using daily mean data are average monthly loadings of total phosphorus and nitrogen and their variance, and average and variance of the amount and timing for maximum phosphorus and nitrogen loading events
- *pH*—The acidity of water controls a broad range of chemical processes both in freshwater and saltwater and is known to affect aquatic organisms from plankton to shellfish to salmonids (Gunn 1986; Ikuta et al. 2003; Logan 2010; Talmage and Gobler 2010; Mackenzie et al. 2014).

  - annual summary pH variables to measure using daily mean data on a monthly time step are average monthly pH and variance, average and variance of the timing for the monthly maximum and minimum pH, and duration and variance for maximum and minimum pH for each month
  - monthly summary pH variables to measure on a daily time step are average daily pH and variance, and average daily pH flux (daily maximum minus the minimum temperature) and variance
- *Salinity*—Nearly every estuarine and coastal species has very specific tolerances and ranges for salinity, which is a controlling factor for fish and aquatic species distributions in these environments (Peterson and Meador 1994; Boeuf and Payan 2001; Martino and Able 2003; Elliott et al. 2007). In nearshore and estuarine systems, it is highly influenced by inflow hydrology patterns.
  - annual summary salinity variables to measure for salinity on a monthly time step using daily mean data are average and variance of the monthly salinity, average and variance of the timing for maximum and minimum salinities, and duration and variance for maximum and minimum salinities
  - monthly summary salinity variables to measure on a daily time step are average daily salinity and variance for each month, and average daily salinity flux and its variance for each month

Model assessment category—geomorphology

Bottom shape or rugosity and system geomorphology or channel pattern, regardless of whether in rivers, lakes, reservoirs, or coastal systems, has a key influence shaping fish and aquatic communities, as it often controls current velocity distribution and sediment types (Imhof et al. 1996; Lamouroux et al. 2002; Walters et al. 2003; Kuffner et al. 2007; Dauwalter et al. 2008; Wright and Heyman 2008; D'Ambrosio et al. 2014; Trebilco et al. 2015). Geomorphology and bottom shape is directly related to, connected to, and controlled by hydrology and material recruitment and transport with indirect relationships to connectivity and water quality. It is also a habitat factor that has been greatly altered by humans, whose actions generally simplify habitat.

Similar to material transport and recruitment and water quality, it is certainly possible to manage geomorphology, but efforts usually will have a significant time delay, particularly if being rehabilitated on a watershed basis. This category is often an expensive process to correct; thus, keeping systems from being degraded and within undisturbed expected values is critical. A current barrier is that data are only available for localized areas and broadly acceptable standardized approaches for collecting and analyzing this information are generally not available, except for rivers and streams. Additional research is needed to fully develop the methodology for this category for inclusion into the model assessment and is unlikely to be fully incorporated into the model assessment for another decade, except in small-scale analyses.

To evaluate this factor, it is important to understand what the unaltered bottom and channel shape was to comprehend what changes have been made. This is a significant information gap across the continent, which is not likely to be filled until far into the future. Understanding the current deviation of the geomorphic variables from expected background values is a focus of the model assessment in this category, as pointed out in Airoldi and Beck (2007). The geomorphic variables recommended for the model assessment are

- *Rivers and streams*
  - channel sinuosity, pool-riffle-run ratios, bottom substrate, and bottom shape and depth diversity by river segment, as defined by NHD+ or, where available, National Hydrography Dataset High Definition (NHD+HD; 1:24,000 scale) for both undeveloped and current time periods
  - amount of channel modification, including channelization
- *Lakes, reservoirs, and coastal systems*
  - areal measures of rugosity, bottom shape, bottom substrate, and depth diversity for both undeveloped and current time periods
  - anthropogenic affect and analysis require data on channelization and navigation channel construction (lineal distance and area); levee development (height and distance from thalweg); hardened shoreline amount in lineal distance; jetty numbers, mean size and variance, and density per kilometer; and dredged area and volume

Model assessment category—aquatic organisms effect on habitat and energy flow

This category measures the size distribution of all biota in the ecosystem, which controls energy flow in the system (Sheldon et al. 1972; Kerr 1974; Kerr 1977; Borgmann 1987), along with how organisms' control and form physical habitat. Organisms are known to control water chemistry and fish habitat through top-down and bottom-up control of system processes and factors.

The Great Lakes provide recent examples of both bottom-up and top-down control of ecosystems, as summarized below and detailed in Mills et al. (1994), Strayer (2009), and Lin and Guo (2016). Historically, the Great Lakes were dominated by Lake Trout *Salvelinus namaycush* (the apex predator), a diverse coregonid community that were key pelagic prey items for Lake Trout, inshore percids, and a diverse group of benthic fish. This community structure was altered by rapid declines in predators, particularly Lake Trout, in the 1940s, caused by a combination of intensive harvest, invasions, mortality by invasive and parasitic Sea Lamprey *Petromyzon marinus*, and the broad expansion of Alewife *Alosa pseudoharengus*, an invasive planktivorous fish that has high levels of thiaminase, which is harmful to Lake Trout recruitment. By the mid-1960s, Alewives comprised up to 95% of the fish biomass in the Great Lakes and were able to selectively alter the plankton size structure, which in turn changed water quality and primary production. The introduction of Pacific salmon into the Great Lakes in the mid-1960s reduced Alewife populations and moved plankton size distributions and primary producers toward pre-Alewife conditions. Bottom-up effects can be seen a few decades later by the changes incurred by zebra mussel *Dreissena polymorpha* and quagga mussel *D. bugensis* introductions into the Great Lakes in the late 1980s. These mussels filtered much of the suspended materials and plankton in Great Lakes waters, which in turn caused large Alewife reductions in Lake Michigan and a complete loss of this key pelagic prey in Lake Huron. This forage change resulted in pelagic predators, in particular Chinook Salmon *O. tshawytscha*, responding in the 2000s with lake-wide population declines in Lake Michigan and a collapse of Chinook Salmon populations in Lake Huron.

The biota and flora of a system provide direct physical habitat, as evidenced by oyster and coral reefs in inshore marine areas and submerged aquatic vegetation in all habitat types. One of key mechanisms biota and flora affect habitat is by reducing water velocities through increased bottom roughness, which in turn changes the deposition rates of suspended and bed load materials and the distribution of fish communities.

The particle-size spectrum hypothesis indicates that each system has an overall community size distribution that is the most effective at moving energy from one productivity level to the next (Sheldon et al. 1972; Kerr 1974; Borgmann 1982). When size spectra are disrupted by catastrophic events, anthropogenic alteration, or fish harvest, the system shows cascading ecosystem effects that change fish habitat through water quality and potentially physical habitat changes.

Another component of this category is the direct and indirect effects of invasive species. These species can directly alter fish habitat (e.g., Common Carp *Cyprinus carpio* increasing water turbidity [Weber and Brown 2009]) and particle size spectrum, as discussed above, with the Alewife and zebra and quagga mussel invasions of the Great Lakes and invasions of other inland lakes (Strayer 2009; Lin and Guo 2016). Invasive species can indirectly alter fish habitat through control measures such as barrier construction to stop Sea Lampreys in the Great Lakes from reaching spawning habitat, which in turn fragments watersheds.

Given the low amount of information available on total particle size distributions in ecosystems and the lack of measurement consistency, it will be difficult to make progress on this subvariable for decades. The movement of energy is a key structuring factor for fish habitat in each ecosystem and is likely one with the least amount of information on it, as it is very rare to have biomass data from bacteria to the largest fish or mammals in a system. Partial information on size structure of invertebrates and fish communities is more readily available, but unless excessive extraction or perturbation is an issue preventing the expression of expected size frequencies, it is a similarly difficult variable to affect. Extraction is a fairly straightforward issue to deal with, particularly by using size limits or total allowable harvest regulations.

Standardized invasive species distribution and population information is available for some invasive species in parts of their range. Information on the edges of range expansions is not likely to be fully available. Data on control measures are available as these actions permitted by regulatory entities although their direct effect on fish habitat may not be fully known in every instance, but overall invasive species effects are theoretically possible to manage in some instances at this time.

Information on living habitat as defined above has not yet been nationally standardized but is available at some level in some locations. There are clear methods available to rehabilitate degraded living habitats, so this is an area where progress can be made in the near time. Given the many data gaps in this category, it is not likely that these data can be included in the model assessment for at least a decade.

Again, the starting point is generating data for both expected and current conditions on each variable, as this difference is the key to rehabilitation or protection efforts. This is also one of the more difficult categories to obtain standardized data on, particularly for particle size distribution in a system.

The category variables recommended for the model assessment are

- *Particle size distribution*
  - Mean monthly particle size distribution and variance; since this is likely to be lacking for some decades, the available invertebrate and fish community size structure and its variance would be a good starting place; even a consistently measured fish size distribution data set that is available for all or part of the fish community is a good initial measurement for this factor with an adjustment for gear bias
- *Living habitat distribution* using data on the seasonal location and density of living habitat.

- Freshwater systems—annual areal estimates and variances for the distribution and density of submerged aquatic vegetation, mussel beds, submerged and emergent wetland complexes, and water resident trees such as bald cypress *Taxodium distichum*
- Coastal systems—annual areal estimates and variances for the distribution and density of submerged aquatic vegetation, oyster, clam and mussel beds, coral reefs, submerged and emergent wetland complexes, and water resident trees such as mangroves

- *Invasive species*—By definition, this group of variables has no invasive species as the null baseline condition, but information should be developed to ensure that organisms were properly classified as introduced or invasive and not native to the system; additionally, first dates of occurrence should be noted.
  - seasonal location and density of aquatic invasive species along with the variation in the patch occurrence
  - seasonal effects of aquatic invasive species along with the variation in these effects
  - location and frequency of control measures, including physical, biological, and chemical measures

Fisheries and Aquatic Organism Data

To develop actionable strategies to protect, conserve, and improve fish habitat, a dose–response analysis between habitat variables and fish or aquatic communities, as detailed by Greene et al. (2015), Daniel et al. (2015), and Crawford et al. (2016), are required. The dose–response analysis includes a change-point analysis to determine greatest differences in community responses to stressors and a piecewise linear regression to allow threshold response determination (Qian and Cuffney 2011). The dose–response relationships use the stressor variables as predictor variables and measures of fish or aquatic community occurrence, abundance, or biomass as the response variable. Significant inflection points in the relationship provide information on where the level of the stressor has no effect and the value that reduces the fish or aquatic organism variables to their lowest level.

Consistently measured fisheries and aquatic organism data are needed to allow appropriate development of these relationships. Crawford et al. (2016) used survey data that targeted entire fish communities in rivers and streams using single-pass electrofishing data and trawl survey data in estuarine systems for the development of dose–response relationships for landscape-scale variables, as these were the only standardized data set available on a national scale. Ecological and habitat assemblages (e.g., lithophilic spawning guilds) for aggregated ecoregions are used to summarize data and improve samples sizes. Standardized fisheries and other aquatic organism data are needed for lakes, reservoirs, and coastal areas to enable dose–response relationship development. One possible approach for fish in lentic systems is collecting new data and using and accessing existing data from Bonar et al. (2009). This publication, currently being updated, details standardized methods for lakes and reservoirs and provides a compendium of data following their methods.

While a large amount of fisheries and aquatic organism data have been collected nationally, little is broadly available in centralized databases at this time, and even less of these data were collected in a standardized manner. The lack of standardization in collection methodologies is currently an acknowledged issue in many fisheries and natural resource agencies. Investment in developing transfer or conversion functions between gear types and methods, along with collecting new standardized information, is needed and will likely not be fully available for another decade. Investigators who worked on Crawford et al. (2016) faced communication and coordination issues in obtaining available data, resulting in data gaps, which also need to be improved for the model assessment and illustrate the need for centralized data repositories.

To fully build the needed dose–response relationships for all habitats, the fish and aquatic organism variables and data needed for the model assessment are

- the seasonal means and variances for fish and aquatic organism species composition variables, including species abundance, size distribution of the community or key species, and species movement dynamics (e.g., home range and life history stage-specific movements). These variables should be developed at the smallest spatial scale, which would be individual waters for lentic systems, each stream reach for lotic systems, and, for each classified unit, typically a specified grid size such as 30 m^2 or 1 km^2 for large lakes (e.g., Great Lakes) and estuarine and marine systems.
- mean annual timing of fish and aquatic organism movements or key activities along with the variance of these events
- the ecological and habitat fish and aquatic organism guilds

Socioeconomics Information

What society views and values as critical aquatic habitat are key pieces of information for aquatic resource managers and directly determine where habitat investments in conservation, rehabilitation, and improvement will occur. These beliefs are expressed by what society indicates the habitat priorities and issues are in each water, with the number of entities and plans that hold those views providing insights on the societal importance of that water and its habitat. Spatially recording each of the documented habitat issues and concerns from individual entities and planning documents for a particular location or water can make this information actionable by and informative to resources managers. One can reasonably assume the larger the number of entities and cumulative number of priorities that a water has is an indication of a higher societal profile and need.

Unlike many of the other processes, factors, and variables, there is a lot of available information in this category but little spatially attributed. Priorities are available from a wide range of entities, including state, federal, and tribal natural resource agencies to nongovernmental organizations (e.g., Trout Unlimited, The Nature Conservancy, and the Bass Anglers Sportsmen's Society) to local habitat rehabilitation groups (e.g., resource conservation authorities and watershed groups) to local and regional governments. These existing data need to be organized by aggregating the data and attributing it to the model assessment spatial framework. These data are available and can be immediately incorporated into a model assessment.

Another source of socioeconomic information is the estimated historic, current, and future economic value of fisheries and aquatic communities in each watershed. This information provides a clear indication of the importance of systems to society. Where available, economic values are usually restricted to extraction values or replacement values, with very few waters systems having existence or ecosystem service values. In inland areas where fishing or fisheries extraction are important, some valuation information is available from spot creel census data, but these data usually have a limited temporal and spatial range. Larger systems often have broad-ranging creel census information, such as the Great Lakes and coastal systems, to support harvest allocation work, and these data provide the ability to examine the fisheries value of these waters or locations or regions within these waters. Commercial fisheries on the coasts are additional sources of economic value information that can be used to support habitat prioritization efforts. The fragmented nature of these data makes them unlikely to be fully incorporated into the model assessment for at least another decade, except in small-scale analyses.

Protected lands are a piece of socioeconomic data that are available and should be included as a variable in the model assessment. This information provides key insights into public values for given systems and at least some information on ecosystem service values. These lands, as listed in order of protective class, include

1. Preserved lands and natural areas
   a. national and some state and tribal parks
   b. designated wilderness and natural areas
   c. some federal and state wild and scenic rivers
   d. some national and state wildlife refuges
2. Multiple use/purpose designations
   a. national and state forests
   b. Bureau of Land Management lands
   c. Native American and Alaska lands
   d. some national and state wildlife refuges
   e. some state parks
   f. some federal and state wild and scenic rivers
3. Private lands with conservation easements
4. Focused recreational use areas—state, county, and city parks
5. Private lands

These land designations provide an indication of the amount of management needed to protect and rehabilitate fish habitat, as lower numbers indicate protected areas with low rehabilitation costs. Protected lands are likely to have higher social values. This information is spatially available and can be immediately incorporated into the model assessment.

The final socioeconomic variables in the model assessment are the estimated costs of protecting/conserving intact systems and repairing damage to degraded waters. This information, which is generally not yet available nationally but is available on a local level in some areas, provides key information on the highest return on investments. Efforts should be made to develop national data sets on the costs to protect intact systems, through either direct purchase or easement acquisition, and on projects that are rehabilitating the processes and factors detailed above. Given the fragmented nature of these data, they are unlikely to be fully incorporated into a model assessment for a decade, except in small-scale analyses.

The socioeconomic variables, spatially attributed for each water or segment, for the model assessment are

- number of priorities and actions called for by individual entities and planning documents, with higher counts having higher scores or weighting
- economic value of the aquatic community, with the greater values having the higher scores or weighting
- area of each protected habitat, with the highest amount of Category 1 lands having the highest scores or weighting
- costs of habitat protection or rehabilitation measures with systems, with the lowest costs having the higher score or weighting

The best methods of incorporating and aggregating the above socioeconomic variables into water and system scores need further development.

Assessing Aquatic Habitat Condition

Model assessment habitat condition scoring for waters follows the methodology detailed in Esselman et al. (2013), Greene et al. (2015), Daniel et al. (2015), and Crawford et al. (2016) and is summarized in this section. After developing dose–response relationships between all stressor variables for each fish and aquatic organism aggregation, boosted regression tree analysis (Elith et al. 2008) is conducted to control the variance from natural variables and to allow the selection of the stressors that explain the largest amount of variance seen in the selected fish and aquatic organism assemblage values. The statistically significant stressor variables have their responses divided into effect scores using the threshold analysis of inflection points discussed previously, ranging from no effect to complete loss of fish and aquatic organism aggregation variable value. Crawford et al. (2016) used five effect scores from 5 for stressor levels of no effect to 1 for lowest fish community values documented at the stressor level. Values between 5 and 1 were evenly spaced along the response relationship.

To score waters, only statistically significant stressors are used for each fish and aquatic organism species assemblage for each ecological region. Scores are computed for each significant stressor by assemblage for each water or part of a water within an aggregated ecosystem using the appropriate dose–response relationship. Using the assumption that the stressor causing the most disturbance to the assemblage is the limiting factor for that assemblage group, the lowest score is used for that assemblage. Assuming that all assemblages are of equal value, the scores across all assemblages in the ecological region being evaluated are averaged to provide scores for each water or part of a water. As more information becomes available, scores for both stressors and assemblages should be weighted to improve the model realism.

As demonstrated in Crawford et al. (2016), there are different responses from assemblages to stressor spatial scales. Analyses should be done at both local (NHD+ segment or individual lake or estuary-shed) and network (cumulative effect of all waters upstream, including the local area) scales. The lowest stressor score, representing the most disturbance, between the spatial scales for a water or part of a water is used for final scoring.

The model assessment provides habitat scoring for all waters or parts of waters (e.g., river segment). Scores can then be aggregated by averaging individual water or segment scores for any spatial extent desired. When more data are available on how stressors effect network systems and aquatic species aggregations, scores should be weighted for each part of a system being analyzed. The socioeconomic data and information are one source of weighting system scores, with the highest socioeconomic score providing overall higher scores.

Conclusion

The attribution of the above listed data into the model assessment with the subsequent development of dose–response relationships, regression tree analysis of these relationships for overall habitat condition model building, scoring of waters and parts of waters, and incorporation of socioeconomic information as a key information weighting factor will provide aquatic resource managers with an information synthesis and decision support tool. When fully developed, the model assessment will determine the condition of any water from the mountains to the continent shelf, determine the expected outcome of changing a process or factor along with its effects on fish communities, and provide estimates of conservation or rehabilitation costs, including the societal value of a particular water. Waters can then be scaled and scored to allow for a much-improved mechanism to prioritize fish habitat conservation across a continental landscape. While the visualized data and the actual model assessment may be decades away on a continental framework, many of these data exist today or will be accessible soon for smaller geographic areas or for individual categories that can inform resource management decisions, if the data are gathered and displayed in a compelling fashion. For the first time, we can visualize a reliable mechanism to compile and organize the mass of available aquatic data and information spatially and have the analytical tools to develop all required supporting relationships to build a true decision support system to help direct the trillions of dollars of investment needed to protect, enhance, and conserve our fisheries habitats.

Acknowledgments

For the ideas and concepts behind this chapter, I owe a huge debt of gratitude to all of the past and current members of the NFHP Board Science and Data Committee along with the Inland and Coastal National Fish Habitat Assessment Teams, many of whose papers were cited in this chapter. In particular, I would like to thank Dana Infante (Michigan State University), Wes Daniel (U.S. Geological Survey), and Kristan Blackhart (National Oceanic and Atmospheric Administration–National Marine Fisheries Service), along with the previous NFHP Inland and Coastal Assessment Leaders, for their many contributions. Without the long discussions on how to build a system that was not possible a generation ago but is attainable now, this chapter would not be possible or even conceived. I also need to acknowledge the Board and Board staff along with the primary funders of the assessments, the U.S. Fish and Wildlife Service, the National Oceanic and Atmospheric Administration–National Marine Fisheries Service, and the U.S. Geological Survey, which have provided this unique opportunity to try to build this none like it in the world assessment. I owe all of them my deepest thanks.

References

Able, K. W. 2005. A re-examination of fish estuarine dependence: evidence for connectivity between estuarine and ocean habitats. Estuarine, Coastal and Shelf Science 64:5–17.

Airoldi, L., and M. W. Beck. 2007. Loss, status and trends for coastal marine habitats of Europe. Oceanography and Marine Biology: An Annual Review 45:345–405.

Annear, T., I. Chisholm, H. Beecher, A. Locke, P. Aarrestad, C. Coomer, C. Estes. J. Hunt, R. Jacobson, G. Jobsis, J. Kauffman, J. Marshall, K. Mayes, G. Smith, R. Wentworth, and C. Stalnaker. 2004. Instream flows for riverine resource stewardship, revised edition. Instream Flow Council, Cheyenne, Wyoming.

Arthington, A. H., S. E. Bunn, N.L. Poff, and R. J. Naiman. 2006. The challenge of providing environmental flow rules to sustain river ecosystems. Ecological Applications 16:1311–1318.

Auer, N. A. 1996. Response of Lake Sturgeons to change in hydroelectric facility operation. Transactions of the American Fisheries Society 125:66–77.

Bain, M. B., and H. Jia. 2012. A habitat model for fish communities in large streams and small rivers. International Journal of Ecology [online serial] 2012:962071.

Benda, L., D. Miller, J. Sias, D. Martin, R. Bilby, C. Veldhyuisen, and T. Dunne. 2003. Wood recruitment processes and wood budgeting. Pages 49–74 *in* S. V. Gregory, K. L. Boyer, and A. M. Gurnell, editors. The ecology and management of wood in world rivers. American Fisheries Society, Symposium 37, Bethesda, Maryland.

Boeuf, G., and P. Payan. 2001. How should salinity influence fish growth? Comparative Biochemistry and Physiology Part C 130:411–423.

Bonar, S. A., W. A. Hubert, and D. W. Willis, editors. 2009. Standard methods for sampling North American Freshwater Fishes. American Fisheries Society, Bethesda, Maryland.

Booth, D. B., and C. R. Jackson. 1997. Urbanization of aquatic systems: degradation thresholds, stormwater detention, and the limits of mitigation. Journal of the American Water Resources Association 22:1077–1090.

Borgmann, U. 1987. Models on the slope of, and biomass flow up, the biomass size spectrum. Canadian Journal of Fisheries and Aquatic Sciences 44(Supplement 2):136–140.

Bunn, S. E., and A. H. Arthington. 2002. Basic principles and ecological consequences of altered flow regimes for aquatic biodiversity. Environmental Management 30:492–507.

Cade, B. S. 2006. National hydrologic assessment tool (NATHAT). U.S. Geological Survey, Reston, Virginia.

Carlisle, D. M., J. Falcone, D. M. Wolock, M. R. Meador, and R. H. Norris. 2010. Predicting the natural flow regime: models for assessing hydrological alteration in streams. River Research and Applications 26:118–136.

Cederholm, C. J., M. D. Kunze, T. Murota, and A. Sibatani. 1999. Pacific salmon carcasses: essential contributions of nutrients and energy for aquatic and terrestrial ecosystems. Fisheries 24(10):6–15.

Childress, E. S., and P. B. McIntyre. 2016. Life history traits and spawning behavior modulate ecosystem level effects of nutrient subsidies from fish migrations. Ecosphere [online serial] 7(6):e01301.

Cloern, J. E. 2001. Our evolving conceptual model of the coastal eutrophication problem. Marine Ecology Progress Series 210:223–253.

Collier, M., R. H. Webb, and J. C. Schmidt. 1996. Dams and rivers: primer on the downstream effects of dams. U.S. Geological Survey Circular 1126.

Crawford, S., G. Whelan, D. M. Infante, K. Blackhart, W. M. Daniel, P. L. Fuller, T. Birdsong, D. J. Wieferich, R. McClees-Funinan, S. M. Stedman, K. Herreman, and P. Ruhl. 2016. Through a fish's eye: the status of fish habitats in the United States 2015. National Fish Habitat Partnership Board. Available: http://assessment.fishhabitat.org. (April 2019).

Daniel, W. M., D. M. Infante, R. M. Hughes, Y. Tsang, P. C. Esselman, D. Weiferich, K. Herreman, A. R. Cooper, L. Wang, and W. W. Taylor. 2015. Characterizing coal and mineral mines as a regional source of stress to stream fish assemblages. Ecological Indicators 50:50–61.

Dauwalter, D. C., D. K. Splinter, W. L. Fisher, and R. A. Marston. 2008. Biogeography, ecoregions and geomorphology affect fish species composition in streams of eastern Oklahoma, USA. Environmental Biology of Fishes 82:237–249.

D'Ambrosio, J. L., L. R. Williams, M. G. Williams, J. D. Witter, and A. D. Ward. 2014. Geomorphology, habitat, spatial location influences on fish and macroinvertebrate communities in modified channels of an agriculturally-dominated watershed in Ohio, USA. Ecological Engineering 68:32–46.

Deweber, J. T., T. Tsang, D. M. Krueger, J. B. Whittier, T. Wagner, D. M. Infante, and G. Whelan. 2014. Importance of understanding landscape biases in USGS gage locations: implications and solutions for managers. Fisheries 39(4):155–163.

Durbin, A. G., S. W. Nixon, and C. A. Oviatt. 1979. Effects of the spawning migration of the Alewife, *Alosa pseudoharengus*, on freshwater ecosystems. Ecology 60:8–17.

Elith, J., J. R. Leathwick, and T. Hastie. 2008. A working guide to boosted regression trees. Journal of Animal Ecology 77:802–813.

Elliott, M., A. K. Whitfield, I. C. Potter, S. J. M. Blaber, D. P. Cyrus, F. G. Nordlie, and T. D. Harrison. 2007. The guild approach to categorizing estuarine fish assemblages: a global review. Fish and Fisheries 8:241–268.

Esselman, P. C., D. M. Infante, L. Wang, D. Wu, A. R. Cooper, and W. W. Taylor. 2013. An index of cumulative disturbance to river fish habitats of the conterminous United States from landscape anthropogenic activities. Ecological Restoration 29:133–151.

Gende, S. M., R. T. Edwards, M. F. Willson, and M. S. Wipfli. 2002. Pacific salmon in aquatic and terrestrial ecosystems. Bioscience 52:917–928.

Grantham, T. E., J. H. Viers, and P. B. Moyle. 2014. Systematic screening of dams for environmental flow assessment and implementation. Bioscience 64:1006–1018.

Greene, C. M., K. Blackhart, J. Nohner, A. Candelmo, and D. M. Nelson. 2015. A national assessment of stressors to estuarine habitats in the contiguous USA. Estuaries and Coasts 38:782–799.

Gregory, S. V., K. L. Boyer, and A. M. Gurnell, editors. 2003. The ecology and management of wood in world rivers. American Fisheries Society, Symposium 37, Bethesda, Maryland.

Gunn, J. M. 1986. Behavior and ecology of salmonid fishes exposed to episodic pH depressions. Environmental Biology of Fishes 17:241–252.

Gurnell, A.M. 2003. Wood storage and mobility. Pages 75–92 *in* S. V. Gregory, K. L. Boyer, and A. M. Gurnell, editors. The ecology and management of wood in world rivers. American Fisheries Society, Symposium 37, Bethesda, Maryland.

Hamilton, P. B., I. G. Cowx, M. F. Oleksiak, A. M. Griffiths, M. Grahn, J. R. Stevens, G. R. Cavalho, E. Nicol, and C. R. Tyler. 2015. Population-level consequences for wild fish exposed to sublethal concentrations of chemicals—a critical review. Fish and Fisheries 17:545–566.

Heck, M. P., L. D. Schultz, D. Hockman-Wert, E. C. Dinger, and J. B. Dunham. 2018. Monitoring stream temperatures—a guide for non-specialists. U.S. Geological Survey Techniques and Methods, book 3, chapter A25. Available: https://pubs.er.usgs.gov/publication/tm3A25. (April 2019).

H. John Heinz III Center for Science, Economics and the Environment. 2002. The state of the nation's ecosystems: measuring the lands, waters, and living resources of the United States. Cambridge University Press, Cambridge, UK.

Helms, B. S., J. E. Schoonover, and J. W. Feminella. 2009. Assessing influences of hydrology, physicochemistry, and habitat on stream fish assemblages across a changing landscape. Journal of the American Water Resources Association 45:157–169.

Ikuta, K., Y. Suzuki, and S. Kitamura. 2003. Effects of low pH on the reproductive behavior of salmonids fishes. Fish Physiology and Biochemistry 28:407–410.

Imhof, J. G., J. Fitzgibbon, and W. K. Annable. 1996. A hierarchical evaluation system for characterizing watershed ecosystems for fish habitat. Canadian Journal of Fisheries and Aquatic Sciences 53(Supplement 1):312–326.

Keough, J. R., T. A. Thompson, G. R. Guntenspergen, and D. A. Wilcox. 1999. Hydrogeomorphic factors and ecosystem responses in coast wetlands of the Great Lakes. Wetlands 19:821–834.

Kerr, S. R. 1974. Theory of size distribution in ecological communities. Canadian Journal of Fisheries and Aquatic Sciences 31:1859–1862.

Kerr, S. R. 1977. Structure and transformation of fish production systems. Canadian Journal of Fisheries and Aquatic Sciences 34:1989–1993.

Kondolf, G. M., A. J. Boulton, S. O'Daniel, G. C. Poole, F. J. Rahel, E. H. Stanley, E. Wohl, A. Bang, J. Carlstrom, C. Cristoni, H. Huber, S. Koljonen, P. Louhi, and K. Nakamura. 2006. Process-based ecological river restoration: visualizing three-dimensional connectivity and dynamic vectors to recover lost linkages. Ecology and Society [online serial] 11(2):5.

Kondolf, G. M., Y. Gao, G. W. Annandale, G. L. Morris, E. Jiang, J. Zhang, Y. Cao, P. Carling, K. Fu, Q. Guo, R. Hotchkiss, C. Peteuil, T. Sumi, H. Wang, Z. Wei, B. Wu, C. Wu, and C. Yang. 2014. Sustainable sediment management in reservoirs and regulated rivers: experiences from five continents. Earth's Future 2:256–280.

Konrad, C. P., A. M. D. Brasher, and J. T. May. 2008. Assessing streamflow characteristics as limiting factors on benthic invertebrate assemblages in streams across the western United States. Freshwater Biology 53:1983–1998.

Kuffner, I. B., J. C. Brock, R. Grober-Dunsmore, V. E. Bonito, T. D. Hickey, and C. W. Wright. 2007. Relationships between reef fish communities and remotely sensed rugosity measurements in Biscayne National Park, Florida, USA. Environmental Biology of Fishes 78:71–82.

Lamouroux, N., N. L. Poff, and P. L. Angermeier. 2002. Intercontinental convergence of stream fish community traits along geomorphic and hydraulic gradients. Ecology 83:1792–1807.

Leopold, L. B, M. G. Wolman, and J. P. Miller. 1964. Fluvial processes in geomorphology. W.H. Freeman, San Francisco, California.

Lin, P., and L. Guo. 2016. Do invasive quagga mussels alter CO_2 dynamics in the Laurentian Great Lakes? Science Reports [online serial] 6:39078.

Locke, A., C. Stalnaker, S. Zellmer, K. Williams, H. Beecher, T. Richards, C. Robertson, A. Wald, A. Paul, and T. Annear. 2008. Integrated approaches to riverine resource management: case studies, science, law, people, and policy. Instream Flow Council, Cheyenne, Wyoming.

Logan, C. A. 2010. A review of ocean acidification and America's response. Bioscience 60:819–828.

Jones, N. E., and R. W. Mackereth. 2016. Resource subsidies from adfluvial fishes increase stream productivity. Freshwater Ecology 61:991–1005.

Jonssen, B., and N. Jonssen. 2003. Migratory Atlantic Salmon as vectors for the transfer of energy and nutrients between freshwater and marine environments. Freshwater Biology 48:21–17.

Magilligan, F. J., and K. H. Nislow. 2005. Changes in hydrologic regime by dams. Geomorphology 71:61–78.

Magnuson, J. J., L. B. Crowder, and P. A. Medvick. 1979. Temperature as an ecological resource. American Zoologist 19:331–343.

Mainstone, C. P., and J. Gulson. 1990. The effects of water quality on freshwater fish populations. Final Report to National Rivers Authority, London.

Martin, D. J. 2001. The influence of geomorphic factors and geographic region on large woody debris loading and fish habitat in Alaska coastal streams. North American Journal of Fisheries Management 21:429–440.

Martin, E. H., and C. D. Apse. 2011. Northeast aquatic connectivity: an assessment of dams on northeastern rivers. The Nature Conservancy, Eastern Freshwater Program, Brunswick, Maine.

Martino, E. J., and K. W. Able. 2003. Fish assemblages across the marine to low salinity transition zone of a temperate estuary. Estuarine, Coastal and Shelf Science 56:969–987.

Maser, C., and J. R. Sedell. 1996. From the forest to the sea: the ecology of wood in streams, rivers, estuaries and oceans. St. Lucie Press, Boca Raton, Florida.

Mackenzie, C. L., G. A. Orondroyd, S. F. Curling, R. J. Ball, N. M. Whiteley, and S. K. Malham. 2014. Ocean warming, more than acidification, reduces shell strength in a commercial shellfish species during food limitation. PLOS (Public Library of Science) ONE [online serial] 9(1):e86764.

McManamay, R. A., and E. A. Frimpong. 2015. Hydrologic filtering of fish life history strategies across the United States: implications for stream flow alteration. Ecological Applications 25:243–263.

Mills, E. L., J. H. Leach, J. T. Carlton, and C. L. Secor. 1994. Exotic species and the integrity of the Great Lakes. Bioscience 44:666–676.

Miller, P. M., D. M. Carlisle, D. M. Wolock, and M. Wieczorek. 2018. A database of natural monthly streamflow estimates from 1950 to 2015 for the conterminous United States. Journal of the American Water Resources Association 54:1258–1269.

Miller, R. R., J. D. Williams, and J. E. Williams. 1989. Extinctions of North American fishes during the past century. Fisheries 14(6):22–38.

Montgomery, D. R., B. D. Collins, J. M. Buffington, and T. B. Abbe. 2003. Geomorphic effects of wood in rivers. Pages 21–48 *in* S. V. Gregory, K. L. Boyer, and A. M. Gurnell, editors. The ecology and management of wood in world rivers. American Fisheries Society, Symposium 37, Bethesda, Maryland.

Mueller, D. K., and D. R. Helsel. 1996. Nutrients in the nation's waters—too much of a good thing? U.S. Geological Survey Circular 1136.

Mutz, M. 2003. Hydraulic effects of wood in

streams and rivers. Pages 93–108 *in* S. V. Gregory, K. L. Boyer, and A. M. Gurnell, editors. The ecology and management of wood in world rivers. American Fisheries Society, Symposium 37, Bethesda, Maryland.

Naiman, R. J., R. E. Bilby, D. E. Schlinder, and J. M. Helfield. 2002. Pacific salmon, nutrients, and the dynamics of freshwater and riparian ecosystems. Ecosystems 5:399–417.

National Fish Habitat Board. 2010. Through a fish's eye: the status of fish habitats in the United States 2010. Association of Fish and Wildlife Agencies, Washington, D.C.

Nehlson, W., J. E. Williams, and J. A. Lichatowich. 1991. Pacific salmon at the crossroads: stocks at risk from California, Oregon, Idaho and Washington. Fisheries 16(2):4–21.

Payne, N. L., J. A. Smith, D. E. van der Meulen, M. D. Taylor, Y. Y. Watanabe, A. Takahashi, T. A. Marzullo, C. A. Gray, G. Cadiou, and I. M. Suthers. 2015. Temperature dependence of fish performance in the wild: link with species biogeography and physiological thermal tolerance. Functional Ecology 30:903–912.

Peterson, M. S., and M. R. Meador. 1994. Effects of salinity on freshwater fishes in the coastal plain drainages in the southeastern U.S. Reviews in Fisheries Science 2(2):95–121.

Piegay, H. 2003. Dynamics of wood in large rivers. Pages 109–134 *in* S. V. Gregory, K. L. Boyer, and A. M. Gurnell, editors. The ecology and management of wood in world rivers. American Fisheries Society, Symposium 37, Bethesda, Maryland.

Poff, N. L., and J. D. Allan. 1995. Functional organization of stream fish assemblages in relation to hydrological variability. Ecology 76:606–627.

Poff, N. L., J. D. Allan, M. B. Bain, J. R. Karr, K. L. Pretegaard, B. D. Richter, R. E. Sparks, and J. C. Stromberg. 1997. The natural flow regime: a paradigm for river conservation and restoration. BioScience 47:769–784.

Poff, N. L., B. P. Bledsoe, and C. O. Cuhaciyan. 2006. Hydrologic variation with land use across the contiguous United States: geomorphic and ecological consequences for stream ecosystems. Geomorphology 79:264–285.

Poff, N. L., and J. K. Zimmerman. 2010. Ecological responses to altered flow regimes: a literature review to inform the science and management of environmental flows. Freshwater Biology 55:194–205.

Qian, S. S., and T. F. Cuffney. 2011. To threshold or not to threshold? That's the question. Ecological Indicators 15:1–9.

Rhienheimer, D. E., and S. M. Yarnell. 2017. Tools for sediment management in rivers. Pages 237–263 *in* A. C. Horne, J. A. Webb, M. J. Stewardson, B. Richter, and M. Acreman, editors. Water for the environment from policy and science to implementation and management, Academic Press, Cambridge, Massachusetts.

Richter, B., J. V. Baumgartner, J. Powell, and D. P. Braun. 1996. A method for assessing hydrologic alteration within ecosystems. Conservation Biology 10:1163–1174.

Rounds, S. A., F. D. Wilde, and G. F. Ritz. 2013. Dissolved oxygen (version 3.0). U.S. Geological Survey Techniques of Water Resources Investigations, book 9, chapter A6, section 6.2. Available: http://water.usgs.gov/owq/FieldManual/Chapter6/6.2_v3.0.pdf. (April 2019)

Schlinder, D. E., M. D. Scheuerell, J. W. Moore, S. M. Gende, T. B. Francis, and W. J. Palen. 2003. Pacific salmon and the ecology of coastal ecosystems. Frontiers in Ecology and Environment 1(1):31–37.

Schlosser, I. J. 1991. Stream fish ecology: a landscape perspective. Bioscience 41:704–712.

Sheldon, R. W., A. Prakash, and W. H. Sutcliffe, Jr. 1972. The size distribution of particles in the ocean. Limnology and Oceanography 17:327–340.

Sommer, T. R., W. C. Harrell, A. Mueller Solger, B. Tom, and W. Kimmerer. 2004. Effects of flow variation on channel and floodplain biota and habitats of the Sacramento River, California, USA. Aquatic Conservation: Marine and Freshwater Ecosystems 14:247–261.

Speers-Roesch, B., and T. Norin. 2016. Ecological significance of thermal tolerance and performance in fishes: new insights from integrating field and laboratory approaches. Functional Ecology 30:842–844.

Strayer, D. L. 2009. Twenty years of zebra mussels: lessons from the mollusk that made headlines. Frontiers of Ecology and Environment 7(3):135–141.

Stein, B. L., L. S. Kutner, and J. S. Adams, editors. 2000. Precious heritage: the status of biodiversity in the United States, Oxford University Press, New York.

Talmage, S. C., and C. J. Gobler. 2010. Effects of past, present, and future ocean carbon dioxide concentrations on the growth and survival of larval shellfish. Proceedings of the National Academy of Science of the United States of America 107:17246–17251.

Trebilco, R., N. K. Dulvy, H. Stewart, and A. K. Salomon. 2015. The role of habitat complexity in shaping the size structure of a temperate reef fish community. Inter-Research Marine Ecology Progress Series 532:197–211.

Twining, C. W., D. C. West, and D. M. Post. 2013. Historical changes in nutrient inputs from humans and anadromous fish in New England's coastal wetlands. Limnology and Oceanography 58:1286–1300.

Twining, C. W., E. P. Palkovacs, M. A. Friedman, D. J. Hasselman, and D. W. Post. 2017. Nutrient loading by anadromous fishes: species-specific contributions and effects of diversity. Canadian Journal of Fisheries and Aquatic Sciences 74:609–619.

USEPA (U.S. Environmental Protection Agency). 2017. Approved Clean Water Act chemical test methods. Available: www.epa.gov/cwa-methods/approved-cwa-chemical-test-methods. (May 2018).

Van Steeter, M. M., and J. Pitlick. 1998. Geomorphology and endangered fish habitats of the upper Colorado River: 1. Historic changes in streamflow, sediment load and channel morphology. Water Resources Research 34:287–302.

Wagner, T., and S. R. Midway. 2014. Modeling spatially varying landscape change points in species occurrence thresholds. Ecosphere [online serial] 5(11):145.

Waldham, J. 2013. Running silver: restoring Atlantic rivers and their great fish migrations. Lyons Press, Guilford, Connecticut.

Walters, D. M., D. S. Leigh, M. C. Freeman, B. J. Freeman, and C. M. Pringle. 2003. Geomorphology and fish assemblages in a Piedmont river basin, USA. Freshwater Biology 48:1950–1970.

Weber, M. J., and M. L. Brown. 2009. Effects of common carp on aquatic ecosystems 80 years after "carp as a dominant": ecological insights for fisheries management. Reviews in Fisheries Science 17:524–537.

Whitfield, A. K., and M. Elliott. 2005. Fishes as indicators or environmental and ecological changes within estuaries: a review of progress and some suggestions for the future. Journal of Fish Biology 61(sA):229–250.

Winemiller, K. O., and K. A. Rose. 1992. Patterns of life history diversification in North American fishes: implications for population regulation. Canadian Journal of Fisheries and Aquatic Resources 49:2196–2218.

Wipfli, M. S., J. P. Hudson, J.P. Caoette, and D. T. Charloner. 2003. Marine subsidies in freshwater ecosystems: salmon carcasses increase growth rates of stream-resident salmonids. Transactions of the American Fisheries Society 132:371–381.

Wright, D. J., and W. D. Heyman. 2008. Introduction to the special issue: marine and coastal GIS for geomorphology, habitat mapping and marine reserves. Marine Geodesy 31:1–8.

American Fisheries Society Symposium 91:57–83, 2019

Who's Asking? Interjurisdictional Conservation Assessment and Planning for Great Plains Fishes

BEN J. LABAY*
Siglo Group
1609 West 6th Street, Austin, Texas 78703, USA

JOSHUAH S. PERKIN
Department of Wildlife and Fisheries Sciences, Texas A&M University
2258 TAMU, College Station, Texas 77843, USA

DEAN A. HENDRICKSON
Texas Natural History Collections, Department of Integrative Biology
University of Texas at Austin
10100 Burnet Road, Austin, Texas 78758, USA

ARTHUR R. COOPER
Department of Fisheries and Wildlife, Michigan State University
1405 South Harrison Road, East Lansing, Michigan 48823, USA

GARY P. GARRETT
Texas Natural History Collections, Department of Integrative Biology
University of Texas at Austin
10100 Burnet Road, Austin, Texas 78758, USA

TIMOTHY W. BIRDSONG
Texas Parks and Wildlife Department, Inland Fisheries Division
4200 Smith School Road, Austin, Texas 78744, USA

Abstract—Aquatic biodiversity is threatened by anthropogenic activities operating across jurisdictional and conservation area boundaries. Strategic conservation planning for broad, multispecies and multijurisdictional landscapes benefits from data-driven approaches emphasizing persistence of priority species while accounting for human uses and stakeholder priorities. This study presents such an assessment for conservation of priority fishes of the U.S. Great Plains. Distribution models for 28 priority fishes were incorporated into a prioritization framework using the open-source software Zonation. A series of assessments were produced, including (1) identification of distinct conservation areas based on connectivity and compositional similarity of priority streams, (2) perspectives for fish habitat condition prioritized towards undisturbed habitat (indicating protection potential) and disturbed habitat (indicating restoration potential), (3) ranking species conservation values at local (state) and global scales, and (4) development of "bang-for-buck" perspectives emphasizing richness of species at state, basin, and study region scales. Assessment highlights include prioritizations primarily among unfragmented main-stem reaches,

* Corresponding author: blabay@siglogroup.com

considerable state-boundary-based edge effects for rankings when using state-based conservation values, and identification of eight distinct regions containing natural communities of priority taxa. Further, we integrate an assessment product into a tiered framework for conservation implementation that facilitates coordination among stakeholders across jurisdictions and increases efficiency of conservation efforts. This set of analyses thus provides varying perspectives to

Introduction

Aquatic biodiversity is threatened globally by human appropriation of freshwater resources at scales and rates that ecosystems cannot sustain (Dudgeon et al. 2006). Consequently, freshwater resource conservation is required to maintain both human and freshwater organism populations (Vorosmarty et al. 2010). These conservation actions must occur at scales relevant for species persistence and survival, an operationally challenging caveat given the current and projected limited availability of resources allocated to conservation activities (Strayer and Dudgeon 2010). In fact, resources are generally too limited to protect all freshwater ecosystem services in the face of increasing pressures from population growth and changing climates (Dodds et al. 2013), and there is growing recognition that no project-, species-, or geographic-based conservation focus fully accounts for the innate complexities associated with tradeoffs between ecosystem health and social and economic development (Martinuzzi et al. 2014).

Shifting demands on global landscapes related to both human population growth and resource allocation are driving new approaches to aquatic conservation prioritization frameworks. An earlier "fortress" approach to species protection (e.g., nature preserves, marine reserves) has transitioned into broader, more integrated planning approaches that seek efficiencies in managing both species protection and human uses (Williams et al. 2011; Palomo et al. 2014). This shift is driven primarily by three major themes. First, the need for monetary efficiencies is magnified because of the stark contrast between the amount of conservation work needed and the limited funding allotted to solving environmental problems (Groves et al. 2002; Balmford and Whitten 2003). Second, research indicates that proactive and systematic approaches to conservation are more effective than reactionary or ad hoc approaches, and a general forward-looking approach to conservation is necessary (Margules and Pressey 2000). Finally, conservation frameworks must be built with an understanding of the biological processes that underpin biodiversity at broad scales (Pressey et al. 2007). Thus, conservation constituents across the globe seek to develop economically strategic conservation plans anchored by proactive management and cast in the light of human ecosystem needs and interactions. Internationally, more than 190 countries are committed to implementing conservation planning to reduce biological loss (Balmford et al. 2005), and the Intergovernmental Platform on Biodiversity and Ecosystem Services was recently established to facilitate conservation of biodiversity and sustainable development (Díaz et al. 2015). In the United States, every state and territory has a State Wildlife Action Plan (SWAP) to guide protection of its unique and at-risk natural resources. These plans are intended to provide a funding roadmap for research, management, restoration, and recovery projects for priority species and important habitats. However, as with most globally available conservation plans, adaptive management and implementation strategies remain largely elusive within SWAPs and the critical initial step related to mapping areas of importance is largely ignored (Fontaine 2011). Furthermore, biogeographic boundaries rarely align with geopolitical boundaries and conservation initiatives developed within states might not identify areas of importance that exist across state boundaries (e.g., Pracheil et al. 2012).

Mapping areas of importance is operationally challenging given the sliding scale of importance across diverse sets of stakeholders. Consequently, the initial mapping stage (conservation assessment) acts as a bottleneck for the remaining conceptual stages (planning, management, and review of progress) necessary for proactive conservation of broad, multispecies landscapes (Knight et al. 2006). In an effort to overcome such bottlenecks, systematic conservation assessments consisting of technical, computer-based evaluation of priority elements, gap analysis, and area design are typically generated to inform and anchor conservation planning and implementation frameworks. Research suggests that successful assessments address three primary objectives: species representation, species resiliency, and the ability to implement sustainable management while accounting for human development (Margules and Pressey 2000; Knight et al. 2006; Williams et al. 2011). The vast majority of conservation assessment research has focused on the first objective of developing plans and algorithms to assess, characterize, and prioritize patterns of biodiversity (Pressey et al. 2007). Assessments that attempt to account for species-specific life history needs and that present the results in a perspective that managers can implement are rare (Knight et al. 2008). Further, efficient broadscale conservation planning and delivery for imperiled species are often confounded by differential jurisdictional management practices and priorities (Soberon and Sarukhan 2009). In a sense, the delineation of areas of importance depend entirely on the viewpoint of a stakeholder(s) or who is asking the question "which areas are important?" (Halpern and Warner 2003). Assessments must address these differences and provide managers perspectives that allow efficiencies in allocating conservation funding and actions within and across jurisdictional units (Hunter and Hutchinson 1994; Vane-Wright 1996; Moilanen and Arponen 2011).

Multiple perspectives can guide delineation of areas of importance. Priority areas might be identified through analysis of spatial patterns in habitat connectivity (e.g., Moilanen et al. 2008), species richness or taxonomic groups (e.g., Stewart et al. 2018), in intact habitats (e.g., Thornbrugh et al. 2018), or along political boundaries (e.g., Moilanen et al. 2013). For freshwater ecosystems in particular, native fish conservation areas (NFCAs; Williams et al. 2011) represent an alternative and innovative conservation perspective that complements traditional, reactive approaches to management of aquatic resources. The NFCAs approach is defined as designating areas that adequately support maintenance of processes that create habitat complexity, protection of all life stages, long-term persistence of priority species, and a framework for sustainable management over time (Williams et al. 2011). For example, Dauwalter et al. (2011) completed an NFCA conservation assessment that identified networks of watersheds large enough to manage for both aquatic species population persistence and compatible human uses in selected watersheds of western North America. Such approaches might be applied to other regions where protection of native fishes and human development are colliding (Limburg et al. 2011). In such regions, preservation of ecosystem goods and services that support both native biodiversity and human populations will become increasingly necessary as water security is altered by human extraction and climate change (Vorosmarty et al. 2010). However, methodologies for the establishment of NFCAs require further development, and to our knowledge, there are no quantitative comparisons for priority areas selected using NFCAs versus conservation perspectives involving surrogate taxa, habitat templates, or political units. These competing prioritizations might now be quantitatively compared with the NFCA perspective using sophisticated statistical frameworks (e.g., Zonation; Lehtomaki and Moilanen 2013) to assess consistencies and inconsistences in spatial priorities (Grantham et al. 2010). Of particular importance is the application of multiple conservation planning perspectives in regions where large numbers of imperiled species inhabit changing landscapes that span multiple jurisdictional boundaries (Kark et al. 2015).

The Great Plains of North America exemplifies a region where NFCAs are needed to ensure the long-term persistence of fishes and water. Agricultural water withdrawals from the High Plains aquifer over the past 60 years have outpaced recharge resulting in depletion of massive portions of the aquifer (Steward et al. 2013; Perkin et al. 2017). These changes in water availability are confounded by mismanagement of surface flows through retention and diversion, at times leading to 90% reductions in flow magnitude downstream of impoundments (Costigan and Daniels 2012). As a result, the majority of fishes endemic to the Great Plains are in need of conservation (Hoagstrom et al. 2011) and are recognized in various categories of imperilment at state, regional, and national levels (Haslouer et al. 2005; Jelks et al. 2008). Causes for fish species declines include habitat fragmentation, dewatering, flow regime alteration, water pollution, and introduction of nonnative species (Gido et al. 2010; Hoagstrom et al. 2011). These factors act in concert to transform fish communities throughout the region and increase the urgency to implement coordinated, multijurisdictional conservation interventions to ensure long-term persistence of native fishes (Perkin et al. 2015b).

Here, we present a workflow, sequence of assessments, and planning tools aimed at providing natural resource managers with a spatial perspective for allocation of conservation actions to help assure persistence of native freshwater fishes of the Great Plains. We utilized species distribution models of 28 priority fishes and a spatial prioritization framework to compare spatial conservation value assignments derived from multiple conservation perspectives. The set of resulting assessments are designed to capture ecological characteristics of priority fishes while accounting for individual species conservation status rankings, connectivity requirements, effects of fragmentation and metapopulation size, habitat condition, and differences between regional and broader jurisdictional spatial management priorities. Additionally, we identified distinct high-priority areas based on distance and compositional similarity of communities, forming a starting point for the delineation of NFCAs for the Great Plains. Finally, we validated our NFCA designations by comparing them to two recent conservation area identification studies (Hoagstrom et al. 2011; Perkin et al. 2015a) and integrated part of the NFCA product into an example framework for implementation of partnership- and stakeholder-led conservation of aquatic resources through regionally based fish habitat partnerships. The three-tier spatial framework (sensu Abell et al. 2007) we used identified broadscale watershed-based management regions, mid-level-scale priority areas consisting of NFCAs, and fine-scale identification of critical management areas for a suite of priority species in the southern Great Plains. This example-tiered framework provides stakeholders with a spatial roadmap and communication tool for coupling science needs and management recommendations to priority areas at relevant scales (Fausch et al. 2002).

Methods

Study area

The Great Plains of central North America extend from the Rocky Mountains eastward to the 95th meridian and from Alberta, Canada south to the Rio Grande. Major river basins include Missouri–White, Platte, Republican–Kansas, Arkansas–Cimarron, Canadian, Red, and Brazos–Colorado (Figure 1) that drain from west to east across Montana, Wyoming, North Dakota, South Dakota, Nebraska, Colorado, Kansas, Oklahoma, New Mexico, and Texas. The interstate nature of basin architectures in the region results in multiple jurisdictional boundaries along water courses. The Great Plains Landscape Conservation Cooperative (GPLCC; www.greatplainslcc.org) was created to coordinate transjurisdictional resource management. The extent chosen for this study was the intersection of the GPLCC's extent with U.S. Geological Survey (USGS) Hydrologic Unit Code 4 (HUC4) boundaries modified to exclude areas far downstream of the GPLCC (Figure 1).

Species data

Species occurrence records were obtained from museum specimen-vouchered occurrences and published data. Museum speci-

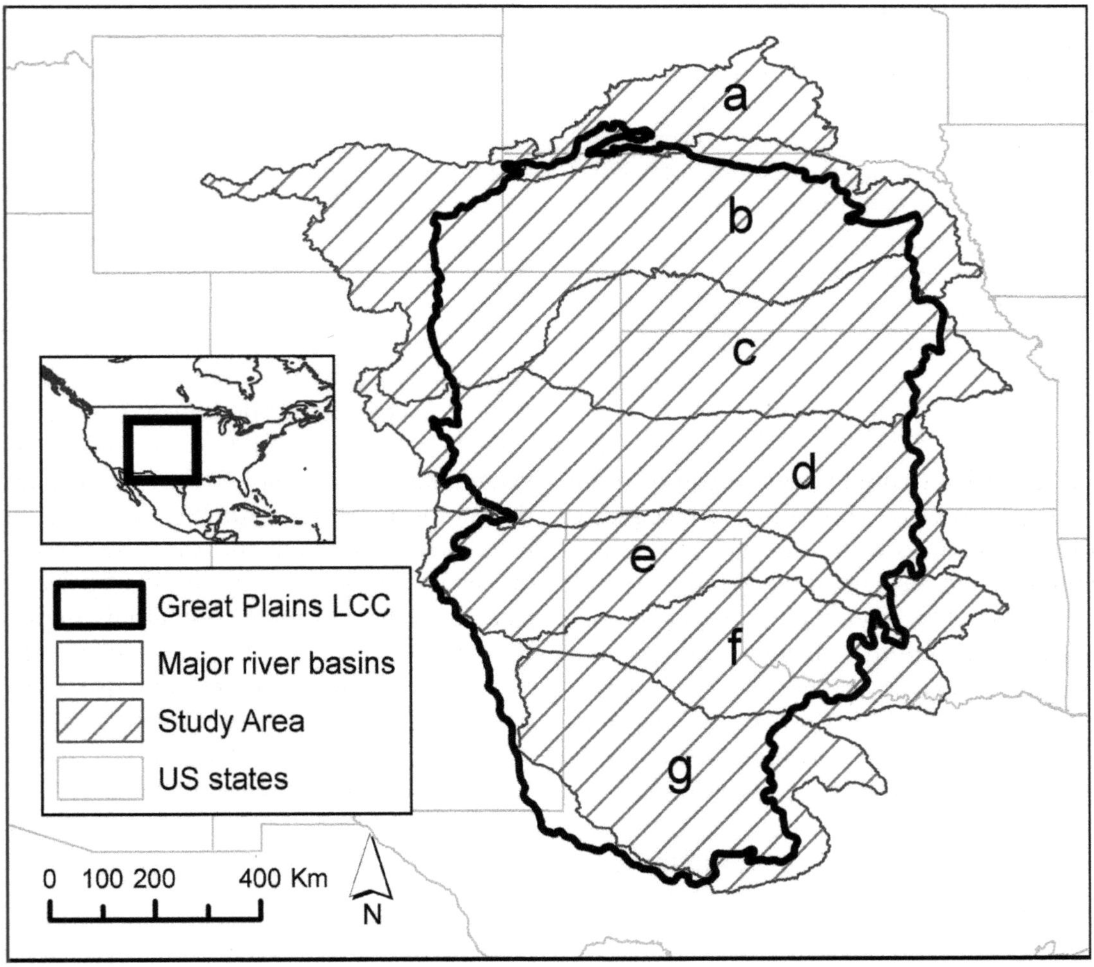

Figure 1. Map of Great Plains Landscape Conservation Cooperative (LCC) extent and study area extent used for analyses. Major river basins indicated and included in the study area include (a) Missouri–White, (b) Platte, (c) Republican–Kansas, (d) Arkansas-Cimarron, (e) Canadian, (f) Red, and (g) Brazos–Colorado.

men-vouchered data sources included two studies (Hendrickson et al. 2010; Cohen et al. 2013) that reviewed and inventoried existing aquatic resource occurrence data from the Global Biodiversity Information Facility (www.gbif.org). Additional museum specimen-voucher records for the entire state of Oklahoma based on collections made by Parham (2009) were obtained from the Sam Noble Oklahoma Museum of Natural History, and data were extracted from the global occurrence database Fishnet2 (http://fishnet2.net). Published data sources included records from the Kansas Aquatic Gap Program, as recently described by Gido et al. (2010), and recent (1993–2013) fish collections made throughout the Great Plains, as described by Perkin et al. (2015b).

These data sets together represent an attempt to use a broad source of primary fish occurrence data, both contemporary and historical. These types of broad, occurrence-based data sets provide a conservative foundation for identifying sets of conservation areas for current and future conservation efforts.

Species distribution models

We used species distribution models (SDMs) to convert point occurrence data into range-wide continuous probability coverages (Guisan et al. 2013). Fish species were chosen on the basis of their state and federal conservation rankings and the authors' expert opinion (Table 1). Environmental variables used in SDM construction (Table 2) were selected in part based on expert evaluation of models created from subsets of variables for a set of species with well-known distributions (see Labay et al. 2011). Selection of functionally relevant environmental predictors is a critical step in species distribution modeling that is ecologically meaningful. The expert opinion-based method to select the set of environmental covariates added an additional model parameterization check in addition to standard statistical validation methods. Climatic, hydrologic, and topographic variables were used to account for broadscale physiological constraints as determinants of distribution (Graham and Hijmans 2006), and two hydrology-based geographic variables controlled for historical zoogeography by categorically constraining predictions of species occurrences to watersheds from which they are documented. Hydrologic variables from the National Hydrology Dataset Plus (NHDPlus Version 1, www.horizon-systems.com/nhdplus/) were converted to 30-arc-second grids using the NHDPlus catchment unit. The spatial extent used for SDM construction was species-specific (Anderson and Raza 2010) and determined by the intersection of input species occurrence data with major watersheds (USGS HUC4). The species-specific extents were divided into a grid of 30-arc-second cells (1 km^2 at equator). The raw environmental grid data were used for modeling, as opposed to a stream-segment based framework, to retain maximum spatial and data resolution.

Species distribution models were constructed using the maximum entropy algorithm encoded in the Maxent software package (version 3.3.4; Phillips et al. 2006). This framework is known to be robust for species distribution modeling with presence-only records (Elith et al. 2006, 2011) and was recently shown to be nearly mathematically equivalent to a Poisson regression model (Renner and Warton 2013). We implemented Maxent following default parameterization recommendations (Phillips and Dudík 2008; Elith et al. 2011) using auto features and 10,000 background points sampled randomly with models cross-validated with 10 replicates to estimate errors around fitted functions. Individual species model performance was evaluated using a receiver operating characteristic (ROC) analysis. The ROC analysis characterizes model performance at all possible thresholds using the area under the curve (AUC). An optimal model with perfect discrimination would have an AUC

Table 1. For each species included in the assessment, number of occurrence records used in model construction and species-specific model and assessment statistics. AUC = area under curve, BQP = boundary quality penalty, and NFCA = Native fish conservation area, determined by similarity of top-ranked river segments based on distance and compositional similarity.

Common name	Genus species	Test AUC	Test AUC SD	No. records in model	Guild group[a]	BQP distance (unit = cells = ~1 km²)	BQP curve[b]	NFCA[c]
Shortnose Gar	*Lepisosteus platostomus*	0.9533	0.0180	336	D	1	1	2–6
Southern Redbelly Dace	*Chrosomus erythrogaster*	0.9309	0.0118	813	A	10	2	3,4,5
Western Silvery Minnow	*Hybognathus argyritis*	0.9637	0.0183	113	B	100	4	2
Brassy Minnow	*H. hankinsoni*	0.9165	0.0231	290	D	1	1	2,3
Plains Minnow	*H. placitus*	0.9529	0.0097	971	B	100	4	3,4,5
Cardinal Shiner	*Luxilus cardinalis*	0.9857	0.0072	266	A	10	2	3,5
Common Shiner	*L. cornutus*	0.9196	0.0127	1,048	A	10	2	2,3
Prairie Chub	*Macrhybopsis australis*	0.9941	0.0038	65	B	100	4	4,6
Sturgeon Chub	*M. gelida*	0.9822	0.0103	56	B	100	4	2,3
Shoal Chub	*M. hyostoma*	0.9626	0.0141	265	B	100	4	2,3
Peppered Chub	*M. tetranema*	0.9753	0.0896	19	B	100	4	3,6
Hornyhead Chub	*Nocomis biguttatus*	0.9593	0.0072	1,044	A	10	2	2,3,5
Emerald Shiner	*N. atherinoides*	0.9503	0.0052	2,910	C	50	3	2,3
River Shiner	*N. blennius*	0.9655	0.0105	591	B	100	4	2,3
Smalleye Shiner	*N. buccula*	0.9765	0.0162	25	B	100	4	7
Arkansas River Shiner	*N. girardi*	0.9712	0.0182	117	B	100	4	3,4
Sharpnose Shiner	*N. oxyrhynchus*	0.9735	0.0178	44	B	100	4	6,7
Chub Shiner	*N. potteri*	0.9766	0.0117	146	B	100	4	6,7
Topeka Shiner	*N. topeka*	0.9844	0.0084	255	A	10	2	3
Suckermouth Minnow	*Phenacobius mirabilis*	0.9598	0.0056	2,067	A	10	2	2,3,6
Slim Minnow	*Pimephales tenellus*	0.9882	0.0040	313	E	1	1	3,4,5
Flathead Chub	*Platygobio gracilis*	0.9682	0.0138	204	C	50	3	2
Blue Sucker	*Cycleptus elongatus*	0.9678	0.0169	209	C	50	3	3
River Redhorse	*Moxostoma carinatum*	0.9639	0.0151	177	C	50	3	2
Neosho Madtom	*Noturus placidus*	0.9934	0.0055	84	E	1	1	3
Northern Plains Killifish	*Fundulus kansae*	0.9607	0.0099	697	A	10	2	3,4
Plains Topminnow	*F. sciadicus*	0.9333	0.0273	106	A	10	2	2,3
Arkansas Darter	*Etheostoma cragini*	0.9891	0.0049	432	A	10	2	3

[a] Guild groupings based on reproductive behaviors as follows: A = lithophilic gravel spawner with adhesive eggs; B = pelagophilic or open substratum or pelagic-broadcast spawning with semibuoyant nonadhesive eggs; C = lithopelagophilic with demersal nonadhesive eggs; D = phytophilic spawner, adhesive eggs onto plants; and E = speleophilic nester, crevices with adhesive eggs.

[b] See Figure 3.

[c] See Figure 4b.

Table 2. Environmental variables used in species distribution models.

Layer category	Description	Source
Topological	Aspect	30-arc second DEM
Topological	Slope	30-arc second DEM
Topological	Compound topological index ln[acc.flow/tan(slope)]	30-arc second DEM
Topological	Altitude	30-arc second DEM
Climate	Annual mean temperature	Wordclim variable 1
Climate	Mean diurnal range [mean of monthly (max temp – min temp)]	Wordclim variable 2
Climate	Isothermality (variable 2/variable 7)(*100)	Wordclim variable 3
Climate	Temperature seasonality (SD*100)	Wordclim variable 4
Climate	Max temperature of warmest month	Wordclim variable 5
Climate	Min temperature of coldest month	Wordclim variable 6
Climate	Temperature annual range (P5-P6)	Wordclim variable 7
Climate	Annual precipitation	Wordclim variable 12
Climate	Precipitation of wettest month	Wordclim variable 13
Climate	Precipitation of driest month	Wordclim variable 14
Climate	Precipitation seasonality (coefficient of variation)	Wordclim variable 15
Climate	Precipitation of wettest quarter	Wordclim variable 16
Climate	Precipitation of driest quarter	Wordclim variable 17
Climate	Precipitation of warmest quarter	Wordclim variable 18
Climate	Precipitation of coldest quarter	Wordclim variable 19
Geographic	Major river basins	U.S. Geologic Survey
Geographic	8-digit hydrologic unit code	U.S. Geologic Survey
Hydrologic	Cumulative drainage	National Hydrology Dataset Plus
Hydrologic	Mean annual flow	National Hydrology Dataset Plus
Hydrologic	Mean annual velocity	National Hydrology Dataset Plus

of 1.0 while a model that predicted species occurrences at random would have an AUC of 0.5 (Hanley and McNeil 1982). We recognize that validation using the AUC statistic has been criticized as misleading as it tends to overestimate model quality, may show spurious high performance with small sample sizes, and may reward overparameterization (Lobo et al. 2008). However, we use AUC (albeit conservatively via a high threshold of acceptability) for its commonality and lack of viable alternative measures and because our intended use of these models was in the context of coarse-scale identification of prime suitable habitat. Species models were considered reliable and included in prioritization analyses if they had an average test AUC over 10 replicates >0.9. Information from rangewide SDMs was restricted to the study area extent for assessment analyses, and all spatial data analyses were conducted at a 30-arc-second (1 km^2 at the equator) grid resolution.

Conservation assessment software

We used the planning software Zonation (Moilanen et al. 2005), which includes many feature options designed to allow sophisticated prioritizations to be performed while accounting for a wide range of parameters. The primary function of the software is to produce a landscape ranking based on conservation value defined by spatially explicit levels of species, habitat, or ecosystem occurrence. It does this in our case by initially considering the entire landscape being analyzed and iteratively removing spatial grid cells that result in the smallest loss of conservation value as defined by SDM estimation of relative probability of occurrence. Zonation allows alternative cell removal rules that emphasize different types of conservation values. The two removal rules used in this study include one that emphasizes species rarity (core-area zonation [CAZ]; Moilanen et al. 2005) and another that emphasizes species richness (additive-benefit zonation [ABZ]; Moilanen 2007). Most analyses performed for this study use the CAZ approach, which aims to identify core priority landscapes for each species regardless of overall species richness. In other words, the CAZ approach can potentially consider two streams equally important, even if one contains substantially more priority species. The intent with the CAZ approach is to identify the set of core areas most relevant for conservation of all priority species. The ABZ approach is used in our final set of assessment analyses that consider differing spatial results from prioritization on the basis of species richness at multiple scales.

Assessment analyses

We produced a series of five analyses to provide differing conservation area perspectives utilizing various Zonation features to "direct" the cell removal process (Figure 2). These five analyses focused on prioritizing areas with (1) intact connectivity where imperiled species persist (primary prioritization), (2) shared occurrences of imperiled species where resources might be focused to promote conservation of native fish assemblages and the habitats they require (NFCA prioritization), (3) least-disturbed habitats (fish habitat quality prioritization), (4) shared political boundaries where jurisdictional lines are frequently drawn (global versus administrative boundary unit prioritization), and (5) high species richness where conservation efforts might be applied to maximize taxonomic benefits (species richness prioritization).

Primary prioritization (CAZ + S + C).— Our primary prioritization utilized the core area removal process (CAZ) and incorporated species-specific weighting (S) and connectivity constraints (C). Species global or subnational (i.e., U.S. state) conservation status ranks set by NatureServe (Faber-Langendoen et al. 2009; Table 3) were used as species-specific weights in the Zonation prioritization algorithm (Moilanen et al. 2005).

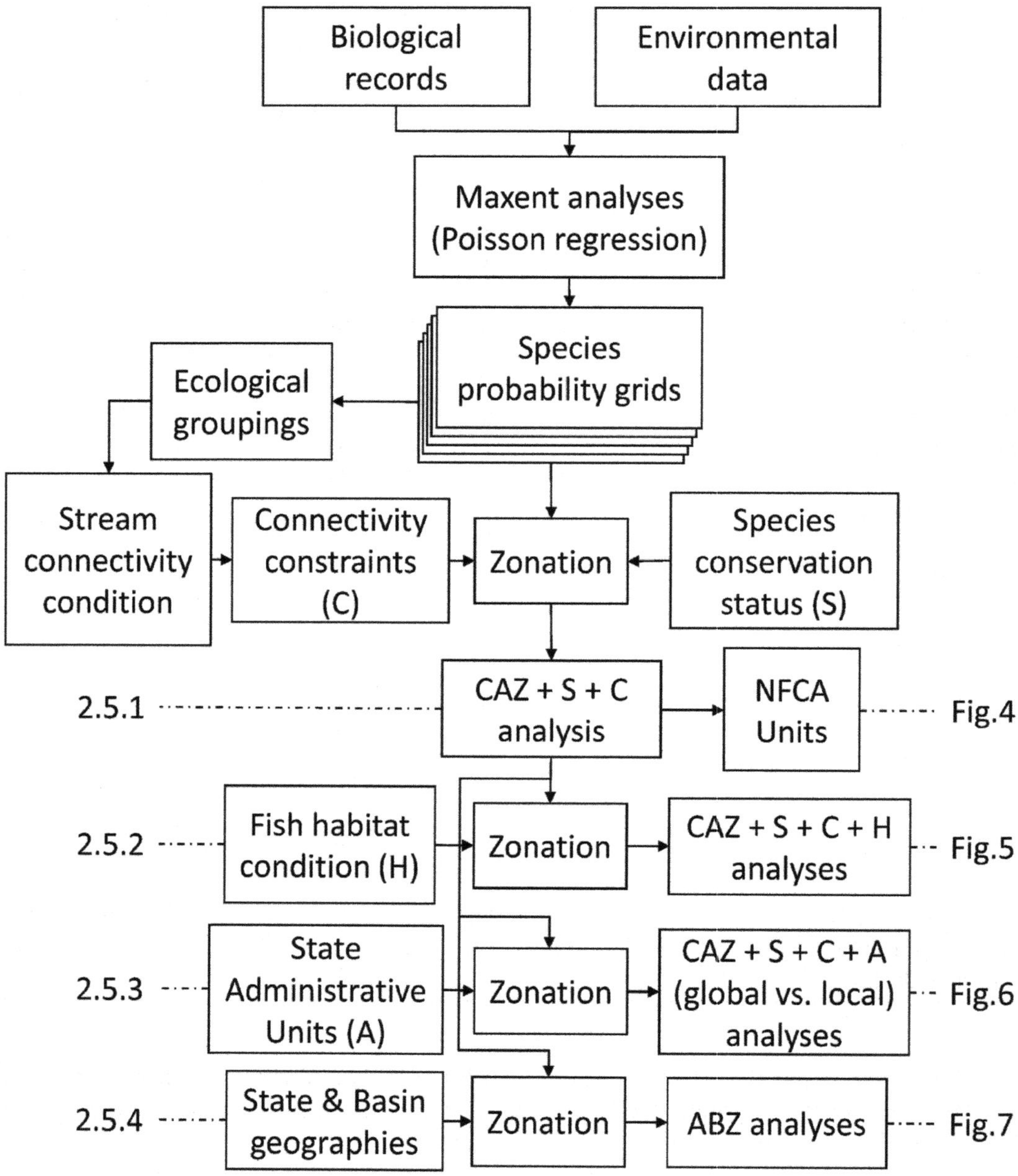

Figure 2. Schematic of the analysis sequence with indication of the sections in methods where each analysis is discussed (left) and the figures that display resultant spatial products (right). CAZ (core-area zonation) and ABZ (additive-benefit zonation) indicate the two removal rules used in this study that emphasizes either species rarity (CAZ, Moilanen et al. 2005) or species richness (ABZ, Moilanen 2007). NFCA = native fish conservation areas.

Species with elevated NatureServe imperilment rankings (e.g., G1 critically imperiled versus G5 demonstrably secure) thus had elevated weights in Zonation, resulting in higher priority for protection of their occurrences (Table 4).

We incorporated habitat connectivity into our analyses via two methods. The first,

Table 3. For each species included in the assessment, NatureServe rank for each state in the study area and globally (Faber-Langendoen et al. 2009). See Table 4 for status code definitions.

Species	SD	WY	NE	CO	KS	OK	NM	TX	Global
Shortnose Gar	S4		S5		S4	S4		S3	G5
Southern Redbelly Dace	S1			S1	S2	S5	S1		G5
Western Silvery Minnow	S5	S2	S5		S2				G4
Brassy Minnow	S5	S5	S4	S3	S1				G5
Plains Minnow	S5	S3	S4	SH	S2	S5	S3	S4	G4
Cardinal Shiner					S3	S4			G4
Common Shiner	S5	S3	S2	S2	S4	S4			G5
Prairie Chub						S4	SNR		G3
Sturgeon Chub	S2	S1	S1		S1				G3
Shoal Chub			S4		S4	S5		SNR	G5
Peppered Chub				SX	S1	S4	S1	S1	G1
Hornyhead Chub	S3	S1	SH	SX	S1				G5
Emerald Shiner	S5		S4		S5	S5		S4	G5
River Shiner	S2		S4		S3	S3		S3	G5
Smalleye Shiner								S2	G2
Arkansas River Shiner					S1	S1	S1	S2	G2
Sharpnose Shiner								S3	G3
Chub Shiner						S4		S4	G4
Topeka Shiner	S2		S1		S2				G3
Suckermouth Minnow	SH	S2	S4	S2	S5	S4	S2	S4	G5
Slim Minnow					S4	S3			G4
Flathead Chub	S5	S5	S5	S3	S1	S1	S4	S2	G5
Slim Minnow	S3		S1		S3	S3	S1	S3	G3
River Redhorse					S1	S1			G4
Neosho Madtom					S2	S1			G2
Northern Plains Killifish		S5	S4	S5	S3	SNR	SNR	SNR	G5
Plains Topminnow	S3	S3	S3	S4	S1	S1			G4
Arkansas Darter				S2	S2	S2			G3

boundary length penalty (BLP; Moilanen and Wintle 2007) is a generalized induction of priority area aggregation based on the structure characteristics (perimeter) of the priority network that produces more compact and less-fragmented solutions. The hierarchy of cell removal is influenced not only by species occurrence probability, but also by the increase/decrease of boundary length that results from cell removal. In our case, this method promotes clustering of priority areas into more continuous segments as opposed to highly fragmented priority reaches along a stream. The second method, boundary quality penalty (BQP; Moilanen and Wintle 2007), is species-specific and accounts for many ecological characteristics of both rivers and species. It is a quantitative way to induce aggregation that accounts for species responses to fragmentation, edge effects, metapopulation size, and connectivity. Here, we used BQP to specifically address species-specific response to fragmentation and thus indirectly account for landscape connectivity and habitat loss. The BQP method was implemented with two expert-defined components, a distance radius determined by when a species is affected by increasing fragmentation or smaller fragment size and a response curve

Table 4. Weights assigned according to NatureServe rank (Faber-Langendoen et al. 2009). Global and state weights were assigned to represent relative priority from one level to the next.

Weight	Status code	Status
0	SX	Presumed extirpated
0	SH	Possibly extirpated
6	S1	Critically imperiled
5	S2	Imperiled
4	S3	Vulnerable
3	S4	Apparently secure
2	S5	Secure
1	SNR	Species not recorded (but present)
0	OR	Out of range
5	G1	Critically imperiled
4	G2	Imperiled
3	G3	Vulnerable
2	G4	Apparently secure
1	G5	Secure

determined by how a species responds to fragmentation and habitat loss. The BQP radius defines the distance (in cell units of the species distribution model grid) across which habitat loss or fragmentation affects a species. For example, a pelagic, broadcast-spawning fish with drifting eggs (e.g., Sharpnose Shiner N) needs long stretches of river (Wilde and Urbanczyk 2013) and thus would have a large radius of fragmentation effect relative to a speleophilic nester that uses crevices and has adhesive eggs (e.g., Slim Minnow; Perkin et al. 2015b). Expert opinion among coauthors (B.J.L., J.S.P.) anchored in fish reproductive guild ecology (Simon 1999; Frimpong and Angermeier 2009) was used to determine radii for each species (Table 1). In absence of life history characteristics that would inform radius size selection for a species, a 10-km radius was used (Hitt and Angermeier 2008, 2011). The second component, BQP response curve, defines the relative magnitude of the effect of fragmentation on each species (Figure 3). For example, for obligate riverine species, we assumed a stronger response to fragmentation and used an extreme penalty curve (see line 4 in Figure 3). We again used coauthor judgment (B.J.L., J.S.P.) anchored in fish reproductive guild ecology to assign response curves (Table 1; Figure 3).

We formally incorporated connectivity constraints into Zonation analyses. Hierarchical prioritization in Zonation can take place on species occurrence while accounting for landscape condition (here, connectivity) and how species differentially respond to that condition. We used the total main-stem distance between upstream and/or downstream dams as a condition layer. This measure was generated by overlaying point locations of large dams on the NHDPlus Version 1 stream network and calculating total main-stem availability (river kilometers) based on distances to main-stem dams (if present) above and below each stream reach (Cooper et al. 2017). Upstream main-stem pathways represent the longest upstream route, whereas downstream main-stem pathways were defined as the shortest route to a stream network outlet (e.g., ocean). Species in guild grouping B (Table 1) were assigned to this connectivity condition factor during select analyses.

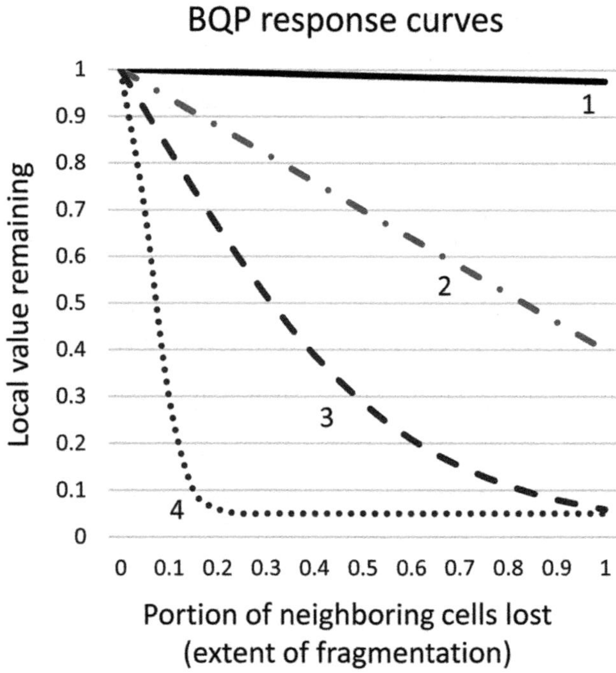

Figure 3. Four species-specific fragmentation penalty curves representing response to fragmentation. BDQ = boundary quality penalty. See Table 1 for species-curve designations.

Native fish conservation areas prioritization.—Zonation allows for the identification of distinct species-based geographic units (here called NFCAs) based on the distance and compositional similarity between priority areas. This is done using four user-specified parameters: (1) percentage of the landscape to consider for inclusion in the management units, (2) minimum inclusion top fraction for each unit (what top fraction must be present in each separate unit), (3) a maximum distance between units, and (4) a maximum difference in compositional similarity between units. To identify NFCAs for Great Plains streams, we used the "CAZ + S + C" primary prioritization as the starting point and chose to consider (1) the top 10% of the landscape ranking, (2) a minimum inclusion of the top 2% (meaning each unit had to contain a portion of the top 2% of the ranking), (3) 25 grid cells (approximately 20 km) for the maximum distance allowed between spatially discrete patches that are included in the same management unit, and (4) a maximum difference in species composition so that two units were determined different if 2 species out of 10 (20%) had a 1-log difference in their probability density (SDM value) between the two areas.

Fish habitat quality prioritization (CAZ + S + C + H).—We also used the index of cumulative disturbance of river fish habitats of the National Fish Habitat Assessment (H), described in Esselman et al. (2011), to rank cells across the landscape as having fish habitat that ranged from undisturbed to highly degraded in relation to occurrence of priority taxa. Incorporating these two condition extremes into separate prioritizations allows for divergent perspectives of stakeholders—those that value and focus on proactive protection of intact habitat versus those that value and focus on restoration of disturbed habitat. The prioritization analyses labeled "CAZ + S + C + H" (see Figure 2) represent two analyses, one in which the fish habitat disturbance condition was used to prioritize selection of low levels of disturbance and another to prioritize areas of high levels of disturbance in need of restoration.

Global versus administrative unit prioritization (CAZ + S + C + A).—We utilized a relative weighting feature in Zonation, also used by Moilanen and Arponen (2011), to provide both an administrative unit prioritization perspective (A) for each state and a broader full study region prioritization. The first is produced by weighting species differentially across states according to each state's NatureServe "S" status (Faber-Langendoen et al. 2009) and the second by weighting species according to their global "G" status.

Species richness prioritization.—We used the additive-benefit cell removal method (Moilanen 2007) to identify richness hot spots within basin, state, and study area geographies. The additive-benefit function calculates a solution where value is additive across biodiversity features (species in this case), and where feature-specific representation is converted to value via concave power functions (Moilanen 2007). Here, the exponent of the power function was set to $z = 0.25$ for all species. The ABZ can be seen as a "bang-for-buck" type of analysis that puts value in areas with more species and downgrades areas that have fewer species. The ABZ approach is not typically used when the analysis features are representing themselves (species); however, we perform these analyses to illustrate the difference in assessment results, and to provide a "bang-for-buck" perspective, particularly for stakeholders and practitioners working at the state level.

Results

Primary prioritization

The primary prioritization (Figure 4a) showed that the majority of the highly ranked, tier 1 (top 2% of landscape) areas were located on unfragmented sections of main-stem streams and, in the southern part of the study area, were bounded downstream or upstream by major impoundments. Examples include the Brazos River upstream of Possum Kingdom Lake, the Red River upstream of Lake Texoma, and the Canadian River downstream of Lake Meredith. This can be attributed to the high numbers of pelagic-broadcast spawning taxa (e.g., Sharpnose Shiner and Arkansas River Shiner) in the analysis and the explicit parameterization concerning their life history requirements related to drifting eggs. However, highly ranked areas also included smaller tributaries, where the core areas of other rare species occurred. These include Brassy Minnow and Plains Topminnow, with core habitat in the tributaries of the Loup River in central Nebraska, and Arkansas Darter and Northern Plains Killifish, with core habitat in western tributaries of the Arkansas River in south-central Kansas. Inclusion of factors to force connectivity and spatial aggregation of resulting rankings in the primary prioritization resulted in a large number of unique areas in the top 10% (1,390). From a mobile organism's perspective, many of these separate areas are connected and should perhaps be managed as combined units.

Native fish conservation area prioritization

The NFCA prioritization grouped priority areas from the primary prioritization according to natural biological communities and proximity. Within the top 10% of the landscape, eight distinct geographic units (i.e., NFCAs) were identified. Each contained a portion of the top 2% of the core areas, and each had sufficient geographic separation and compositional similarity (Figure 4b). Together, these comprised 9.2% of the study area and generally corresponded with major river basins, including (north to south) the White (South Dakota), Platte (Nebraska),

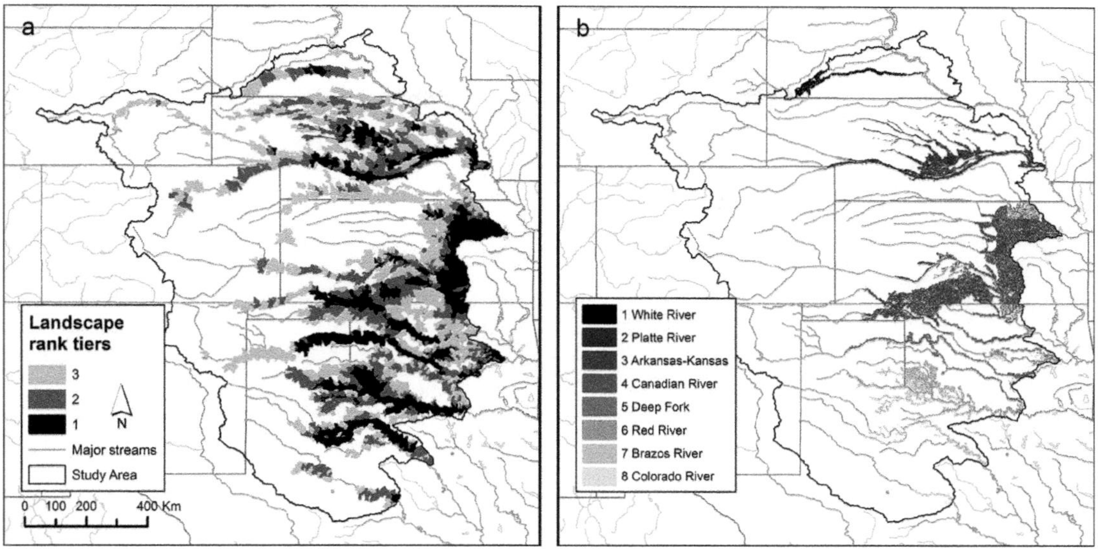

Figure 4. **(a)** Primary prioritization with core-area function zonation analysis, species-specific rankings, connectivity constraints, ecological groupings, and stream connectivity condition factor (CAZ + S + C) for the 28 priority fishes of the Great Plains. Landscape rank tiers represent top fractions of the landscape ranking results and are symbolized by maximum ranking per 12-digit USGS hydrologic unit code: tier 1 = 2% (1–0.98 cell values), tier 2 = 5% (0.98–0.95), and tier 3 = 10% (0.95–0.9). **(b)** Broadscale native fish conservation areas (NFCAs) identified according to distance and compositional similarity among highly ranked regions of the CAZ + S + C analysis.

Arkansas and Kansas (Kansas), Canadian (Texas and Oklahoma), Deep Fork (Oklahoma), Red (Texas and Oklahoma), Brazos (Texas), and Colorado (Texas) River basins.

Fish habitat quality prioritization

Prioritizing "undisturbed" (Figure 5a) and "disturbed" (Figure 5b) habitats each produced top-tier rankings that cover approximately 2% of the study area. These two results overlapped broadly, especially in the Brazos and Red river basins as a result of high-priority core habitats having relatively moderate values of disturbance. Nevertheless, there were notable differences between the two prioritization results. Some of the largest differences occurred in tributaries in the Platte and Arkansas River basins. Within the Platte, relatively undisturbed habitats were identified as core areas in the tributaries of the Loup River (Figure 5a) in Nebraska, whereas disturbed fish habitats were identified along the main-stem Platte River (Figure 5b). In the Arkansas River basin, headwater tributaries of the Salt Fork Arkansas River contained large amounts of core area of relatively undisturbed habitat, whereas the Chikaskia and portions of the Ninnesah contained core areas of disturbed habitat. These results provide guidance for different types of management of habitat for priority fishes (e.g., protection versus rehabilitation).

Global versus administrative unit prioritization

Differences between prioritizations with species weighted according to their global and subnational conservation rankings showed considerable differences. Global weights resulted in less fragmentation of core areas (n = 653) compared to variable

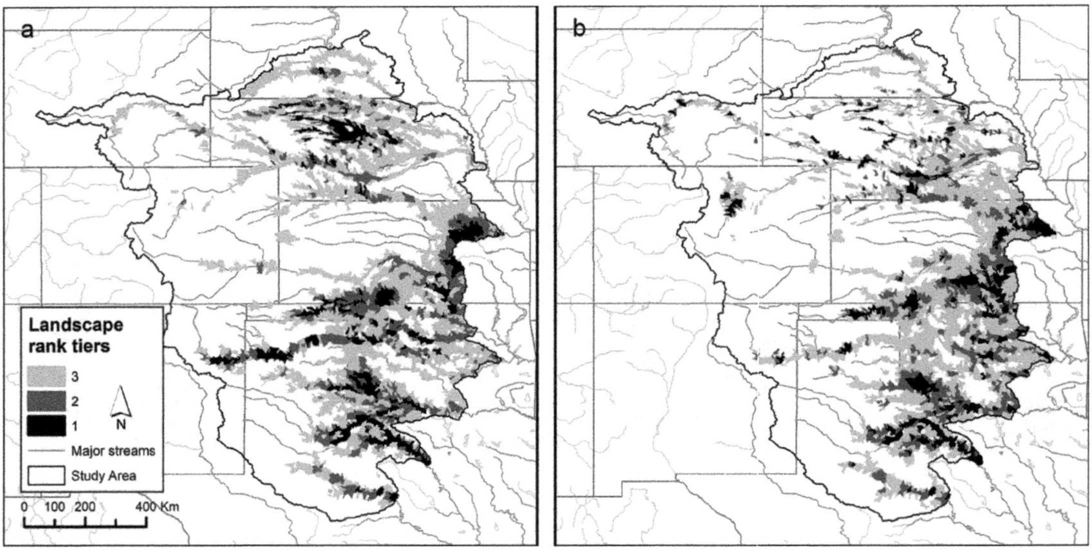

Figure 5. Riverscape prioritization derived from CAZ + S + C setup by adding a fish habitat disturbance condition factor (H) to prioritize areas where the 28 priority fishes of the Great Plains coincide with **(a)** relatively undisturbed habitat and **(b)** relatively disturbed habitat according to the cumulative fish habitat disturbance metric of the National Fish Habitat Action Plan (Esselman et al. 2011). Landscape rank tiers represent top fractions of the landscape ranking results and are symbolized by maximum ranking per 12-digit USGS hydrologic unit code: tier 1 = 2% (1–0.98 cell values), tier 2 = 5% (0.98–0.95), and tier 3 = 10% (0.95–0.9).

state rankings (n = 2,249). Globally weighted rankings generally identified core areas in eastern (Figure 6a) main-stem streams, whereas state-weighted ranking identified more uniform distribution of core areas (Figure 6b). New Mexico, Colorado, and Wyoming completely lacked top-tier (2% of landscape) core areas in the globally weighted prioritization, yet contained 29% of the top-tier core area when prioritization was based on variable state-level conservation weighting for species.

Species richness prioritization

The multiscale prioritization differed dramatically when using the additive benefit cell removal rule (see 2.5.4; providing a "bang-for-buck" perspective) at different spatial extents. Figure 7 shows riverscape prioritizations using the additive benefit cell removal method rule for major river basins (Figure 7a), U.S. states (Figure 7b), and the study area (Figure 7c). The top-tier (2% of landscape) core areas identified vary greatly according to geographic extent of analysis, but in general, core areas are in eastern streams due to higher richness in downstream reaches throughout the study area.

Discussion

Our study provides a comparison of five conservation perspectives useful for identifying areas of conservation interest across the U.S. Great Plains. The primary perspective summarized the distribution of highly connected habitats that are still habited by species in need of conservation, but it identified a large number of small, local areas that were actually nested within broader areas sharing unique assemblages of fishes. Our second perspective based on the NFCA concept (Williams et al. 2011) aggregated these local areas within a smaller number of geographically larger units. The NFCA units tended

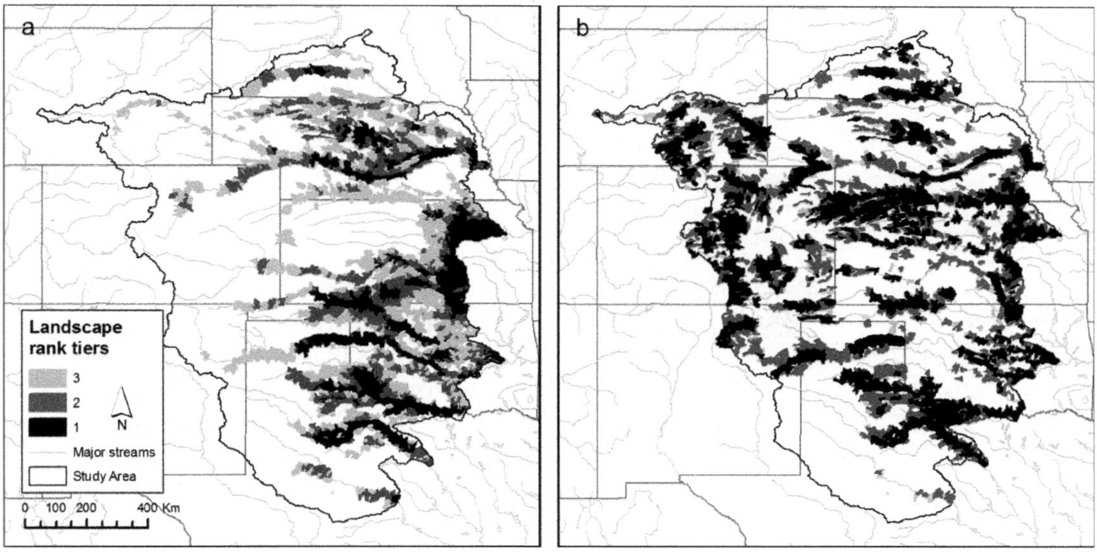

Figure 6. Riverscape prioritization derived from CAZ + S + C setup with **(a)** global species conservation rank weights (Table 3) and **(b)** species weights differentially set in each state based on state-level NatureServe conservation rankings (S rank; Table 3). Landscape rank tiers represent top fractions of the landscape ranking results and are symbolized by maximum ranking per 12-digit USGS hydrologic unit code: tier 1 = 2% (1–0.98 cell values), tier 2 = 5% (0.98–0.95), and tier 3 = 10% (0.95–0.9).

to include the same areas of "undisturbed" habitats identified through the perspective based solely on habitat condition as well as areas identified in the global (Great Plains), but not local (individual states) perspectives. Furthermore, the NFCAs contained the majority of the diversity hotspots identified through the perspective based only on species richness, measured at either the river basin or study area extents (but not state extent). Our results highlight that priorities within geopolitical boundaries (states) were not always spatially consistent with priority areas identified with a broader perspective, meaning broadscale coordination among states with a shared focus on NFCAs is a more efficient conservation methodology compared to state SWAPs operating independently (Neeson et al. 2015). Together, these findings suggest the question "where are areas of interest for fish conservation in the Great Plains?" depends largely on who's asking when viewed from the perspective of individual states, but this subjectivity is reduced considerably when broader geographic perspectives that match the scale of fish biogeography are considered.

Assessment validation with other studies

We assessed the validity of our NFCA framework by comparing it with two previous conservation perspectives for Great Plains fishes (Figure 8). Hoagstrom et al. (2011) suggested that 18 high-quality native fish refuge streams persisted in the Great Plains, each inhabited by at least three Great Plains endemics. Our NFCA model captured 9 of that study's 13 proposed refuge streams that were in our study area (Figure 8). The major area of disagreement between conservation outlooks hinged on inclusion by Hoagstrom et al. (2011) of Plains Sand Shiner *Notropis stramineus missuriensis*, a widely distributed subspecies not included in our NFCA framework because it has a stable conservation status. Endemic fish refuges proposed by

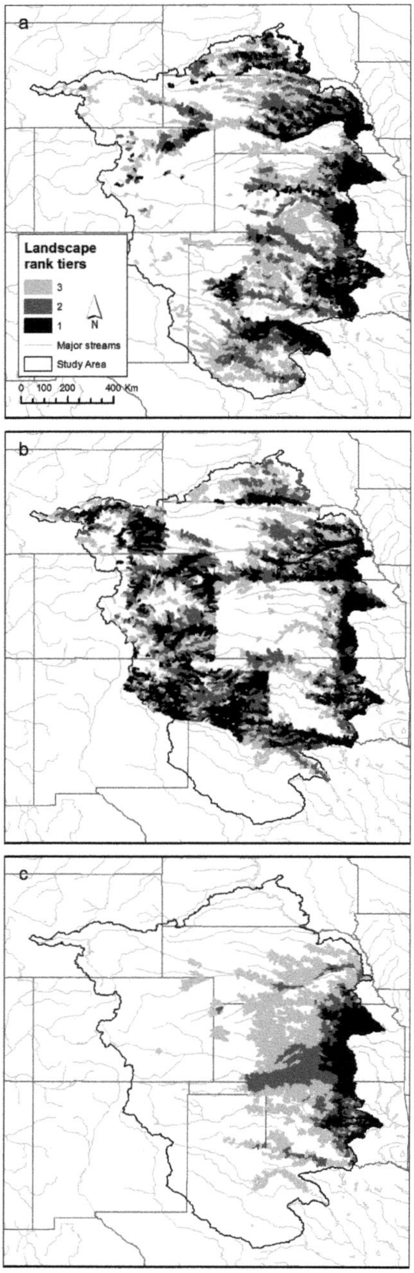

Figure 7. Riverscape prioritization using the additive benefit cell removal prioritization rule for **(a)** major river basin, **(b)** U.S. state, and **(c)** study area geographic extents. Landscape rank tiers represent top fractions of the landscape ranking results and are symbolized by maximum ranking per 12-digit USGS hydrologic unit code: tier 1 = 2% (1–0.98 cell values), tier 2 = 5% (0.98–0.95), and tier 3 = 10% (0.95–0.9).

Hoagstrom et al. (2011) but not identified by our model as having high conservation value tended to be in the western Great Plains where Plains Sand Shiner persists despite declines or extirpations of other native fishes (e.g., Perkin et al. 2015a). Similarly, 13 of 16 stream fragments identified as high conservation priorities by Perkin et al. (2015a) were captured in our NFCA framework, likely because parameterization of our assessment utilized some of the same species-specific life history requirements included by Perkin et al. (2015a). The three not identified in our NFCA framework all occurred along the western edge of the Great Plains where water availability has declined greatly because of depletion of the High Plains aquifer (Steward et al. 2013; Perkin et al. 2017). Although some native species habitat persists in these western extremes of the study area (e.g., the North and South Platte rivers of Wyoming; Steward et al. 2013) and the conservation of these habitats is critical for the maintenance of regional native fish diversity, the ecological future for fish habitat in some portions of the western Great Plains is dire (Falke et al. 2011). Thus, the spatially explicit nature of inclusion and exclusion of previously proposed priority areas for Great Plains fishes is related to the riverscape ranking basis of our modeling framework. That is, excluded stream segments are of high value, albeit at a lower relative ranking compared to the identified NFCA streams. Areas highlighted as having high conservation value in our model are simply the NFCAs in which top-tier priority species' core habitat exists, and we stress that conservation opportunities in the portion of the Great Plains lying outside our proposed NFCAs should not be ignored. Additionally, the NFCA framework identified here is intended to be an initial starting point for local experts to amend as necessary to address stakeholder priorities at finer scales, and consideration of critical exclusions must be made by users of any particular assessment perspective provided here.

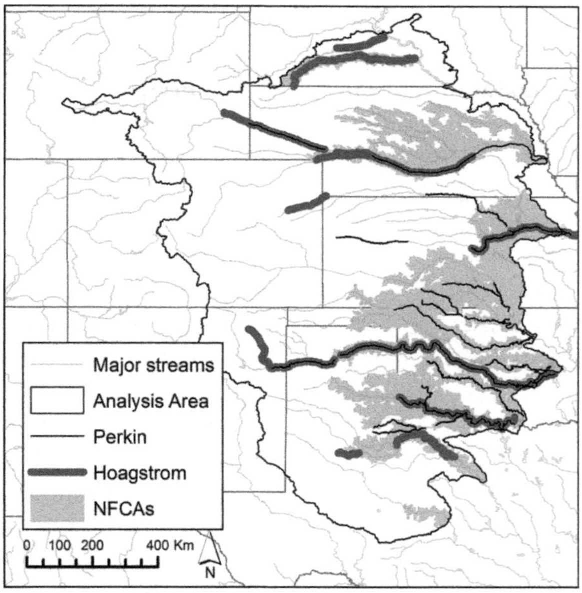

Figure 8. Map showing overlap in between the native fish conservation area (NFCA) framework presented in this study and two previous conservation priority perspectives for Great Plains fishes (Hoagstrom et al. 2011; Perkin et al. 2015a).

Delivery as planning tools for conservation action

To bridge the assessment-implementation gap in conservation planning, conservation assessment products such as those produced here must be complemented with an implementation strategy (Fausch et al. 2002). This can be achieved by mainstreaming of the planning products, coupling their recommendations into communication and collaboration tools used to coordinate among various stakeholders. In practice, this can mean interpreting and redesigning these tools to facilitate a framework for decisions by the diverse personnel and organizations whose work influences natural resource management (e.g., Pierce 2003). As an example, the federally required SWAP of every U.S. state and territory include guidance for interpretive land and stream habitat management (Fontaine 2011). Placing these guidelines into a strategic planning framework, as provided by high-level assessment and planning products such as produced here, facilitates efficiencies for land-use planners and conservation organizations wanting to coordinate and justify decisions and recommendations regarding best management practices.

This mainstreaming of assessment and planning products is not a one-time effort, but requires continuous input and feedback from stakeholders involved. Further, the activity of mainstreaming and purposing these assessment products must be uniquely performed for, and by, individual implementation organizations and researchers, as the information and perspectives presented do not offer the same value across different groups (e.g., federal versus state biologists). That is, the utility of the conservation perspectives provided in our comparative study depends on who is asking for guidance, and our approaches are not intended for fine-scale prioritization requiring local expertise. Resources should be put towards tailoring assessment products into use-case planning tools depending on specific needs of implementation practitioners. At the very minimum, these tools, especially the perspectives of species management units, provide regional managers with organizational frameworks that help them communicate more efficiently with different conservation-implementation and funding groups. However, optimally, individual conservation groups, researchers, and state agencies should be encouraged to come together and actively leverage their unique opportunities to mainstream assessment products, as presented here, to highlight linkages between natural resources and socioeconomic benefits.

Integration into a framework for implementation of the National Fish Habitat Partnership

The National Fish Habitat Action Plan (NFHAP; AFWA 2012; www.fishhabitat.org) provides strategies for cooperative, interjurisdictional, and landscape-scale conservation of fishes and other aquatic resources in the United States and serves as the strategic plan for the National Fish Habitat Partnership and associated network of regionally focused fish habitat partnerships. Within the southern extent of the U.S. Great Plains, the Southeast Aquatic Resources Partnership (SARP; www.southeastaquatics.net) is actively engaged in regional implementation of the NFHAP. The Southeast Aquatic Resources Partnership is a collaborative, multiagency conservation partnership geographically aligned with the 14 member states of the Southeastern Association of Fish and Wildlife Agencies, including the U.S. Great Plains states of Oklahoma and Texas.

Formed in 2001, the mission of SARP is to protect, conserve, and restore aquatic resources, including habitats throughout the region, for the continuing benefit, use, and enjoyment of the American people (SARP 2014). Since the partnership's inception, SARP has served as a regional catalyst and network builder for fish habitat conservation, spearheading regional assessments of flow alteration, riparian condition, and fish passage barriers and supporting on-the-ground delivery of aquatic habitat restoration projects (SARP 2014). The partnership was formally recognized as a fish habitat partnership in 2007, and in 2008, SARP and partners published the Southeast Aquatic Habitat Plan (SARP 2008), which established regional conservation objectives and targets (i.e., 5-, 10-, and 15-year outcomes) used by the partnership to monitor progress and to continually adapt and refine regional fish habitat conservation strategies (SARP 2014).

Critically important to the success of SARP and other fish habitat partnerships is the ability to facilitate communication and cooperative planning among local, state, and federal natural resources management agencies, nongovernmental organizations, and other partners. Cooperative planning allows for identification of shared geographic (e.g., ecoregions, watersheds) and thematic (e.g., dam removal, flow restoration, riparian restoration) priorities and supports strategic investments and leveraging of available technical and financial resources, often allowing for significant expansion of the scope and scale of local conservation projects (e.g., extent of watershed restored, inclusion of project-based monitoring or applied research necessary to evaluate and improve restoration designs). To examine the utility of the conservation assessment products assembled through this study, a pilot conservation planning process was conducted in collaboration with SARP and the GPLCC. The conservation planning process occurred in NFCAs prioritized within the Brazos, Canadian, Colorado, and Red River basins (Figure 4b; Birdsong et al. 2018). Local, state, and federal natural resources management agencies, universities, and nongovernmental organizations were invited to participate in a series of webinars, workshops, and field days. These events were used to profile the conceptual underpinnings of the NFCAs approach (Williams et al. 2011), to review and obtain feedback on the geographic extent of the Great Plains river basins prioritized as NFCAs (Figure 4b), and to identify strategies and project-level conservation actions to restore and preserve focal species and their habitats within the NFCAs. A total of 77 conservation professionals participated in the conservation planning process, collectively identifying 96 project-level conservation actions across the four NFCAs (Birdsong et al. 2018). Actions were prioritized and formulated into multiyear (5–10 years), watershed-based conservation action plans intended to guide and facilitate collaborative conservation within NFCAs of the Brazos, Canadian, Colorado, and Red River basins (Birdsong et al. 2018). Subsequent efforts to implement the conservation action plans assembled for those four NFCAs are described by Birdsong et al. (2019, this volume).

This watershed-based conservation planning within NFCAs of the U.S. Great Plains provided a successful case study that has since been transferred and replicated within other regions of the United States. The Desert Fish Habitat Partnership and Western Native Trout Initiative, two fish habitat partnerships focused on implementation of NFHAP within the southwestern United States, recently cooperated to implement this same conservation assessment and prioritization approach within the Rio Grande basin (Labay et al. 2018), which encompasses the U.S. states of Colorado, New Mexico, and Texas. For the portion of the Rio Grande basin located within the Chihuahuan Desert ecoregion of Texas, six NFCAs were prioritized. The same conservation planning process completed for NFCAs of the U.S. Great Plains was employed by Garrett et al. (2019, this volume) to assemble conservation action plans for those six NFCAs.

Prioritization of NFCAs has served as the impetus for increased and focused investments in native fish conservation (e.g., research, monitoring, habitat restoration, and protection) within select river basins of the U.S. Great Plains and the southwestern United States (Birdsong et al. 2018, 2019; Garrett et al. 2019). These outcomes have demonstrated the utility of assessment products described herein in enhancing communication and fostering collaboration among nongovernmental organizations, universities, and state and federal agencies and in facilitating the leveraging of staff, expertise, project funding, and other resources toward delivery of proactive conservation projects.

Conclusions

Conservation assessment is the technical task of identifying, categorizing, and ranking areas of importance in terms of their relevance to accomplishing conservation goals. This study performed a series of assessments for 28 priority fishes of the U.S. Great Plains that account for various ecological and landscape condition factors that influence fish distributions. The assessment process involved creating a framework for design and implementation of linked components (e.g., species life history traits and stream condition) at a scale appropriate for identification of broad priority areas for streams in the study area.

These assessments constitute a critical component of the broader process of conservation planning. Initial decisions regarding data and assessment design determine how the resultant assessment tools can and should be applied and integrated into the broader planning process, and throughout the setup process, we created an assessment that utilized a NFCA approach to identify important areas for the target fishes while accounting for spatial variability in stakeholder priorities (Williams et al. 2011). Interjurisdictional conservation decisions and solutions require consideration of the priorities held by those who are asking for conservation management direction. Because local protection areas may be insufficient to address regional threats to diversity (e.g., Spurgeon et al. 2014), communication and cooperation among stakeholders are key to successful aquatic resource conservation. Implementation of a broadscale multispecies assessment such as this complements traditional reactive management and restoration by providing tools to facilitate communication, cooperation, and coordination among stakeholders and partners, increasing efficiency of future monitoring and management efforts.

Acknowledgments

This study was supported by the Great Plains Landscape Conservation Cooperative (grant F13AP00523), The University of Texas at Austin, and the Texas Parks and Wildlife Department State Wildlife Grant Program (grant TX T-106-1; CFDA# 15.634). Thanks to the thousands of collectors who contributed to the museum and literature databases of fish occurrence. We also thank Steven Krentz, Adam Cohen, James Broska, and the GPLCC steering committee and science team for discussions and concept reviews, and Jon Paul Pierre for help with geodatabase and metadata provision and management.

References

Abell, R., J. D. Allan, and B. Lehner. 2007. Unlocking the potential of protected areas for freshwaters. Biological Conservation 134:48–63.

AFWA (Association of Fish and Wildlife Agencies). 2012. National fish habitat action plan, 2nd Edition. AFWA, Washington, D.C.

Anderson, R. P., and A. Raza. 2010. The effect of the extent of the study region on GIS models of species geographic distributions and estimates of niche evolution: preliminary tests with montane rodents (genus *Nephelomys*) in Venezuela. Journal of Biogeography 37:1378–1393.

Balmford, A., L. Bennun, B. T. Brink, D. Cooper, I. M. Cote, P. Crane, A. Dobson, N. Dudley, I. Dutton, R. E. Green, R. D. Gregory, J. Harrison, E. T. Kennedy, C. Kremen, N. Leader-Williams, T. E. Lovejpy, G. Mace, R. May, P. Mayaux, P. Morling, J. Phillips, K. Redford, T. H. Ricketts, J. P. Rodriguez, M. Sanjayan, P. J. Schei, A. S. Van Jaarsveld, and B. A. Walther. 2005. Ecology: the convention on biological diversity's 2010 target. Science 307:212–213.

Balmford, A., and T. Whitten. 2003. Who should pay for tropical conservation, and how could the costs be met? Oryx 37:238–250.

Birdsong, T. W., D. C. Dauwalter, G. P. Garrett, B. J. Labay, M. Bean, J. Broska, J. Graham, S. Magnelia, K. B. Mayes, M. McGarrity, K. M. Johnson, S. Robertson, T. Thompson, S. Vail-Muse, and J. B. Whittier. 2018. Native fish conservation areas of the southwestern USA: facilitating landscape-scale conservation of aquatic habitats and freshwater fishes. Southeast Aquatic Resources Partnership, Panama City, Florida.

Birdsong, T. W., G. P. Garrett, B. J. Labay, M. G. Bean, P. T. Bean, J. Botros, M. J. Casarez, A. E. Cohen, T. G. Heger, A. Kalmbach, D. A. Hendrickson, S. J. Magnelia, K. B. Mayes, M. E. McGarrity, R. McGillicuddy, M. M. Parker, and S. Robertson. 2019. Texas native fish conservation areas network. Pages 183–229 *in* D. C. Dauwalter, T. W. Birdsong, and G. P. Garrett, editors. Multispecies and watershed approaches to freshwater fish conservation. American Fisheries Society, Symposium 91, Bethesda, Maryland.

Cohen, A.E., B.J. Labay, D.A. Hendrickson, M. Casarez, S. Sakar. 2013. Final Report: Data provision and projected impact of climate change on fish biodiversity within the Desert LCC. U.S. Department of the Interior, Bureau of Reclamation, Desert Landscape Conservation Cooperative. Available: https://repositories.lib.utexas.edu/handle/2152/22475.

Cooper, A. R., Infante, D. M., Daniel, W. M., Wehrly, K. E., Wang, L., and Brenden, T. O. 2017. Assessment of dam effects on streams and fish assemblages of the conterminous USA. Science of the Total Environment 586:879–889.

Costigan, K. H., and M. D. Daniels. 2012. Damming the prairie: human alteration of Great Plains river regimes. Journal of Hydrology 444:90–99.

Dauwalter, D. C., J. S. Sanderson, J. E. Williams, and J. R. Sedell. 2011. Identification and implementation of native fish conservation areas in the upper Colorado River basin. Fisheries 36:278–288.

Díaz, S., S. Demissew, J. Carabias, C. Joly, M. Lonsdale, N. Ash, A. Larigauderie, J. R. Adhikari, S. Arico, A. Báldi, A. Bartuska, I. A. Baste, A. Bilgin, E. Brondizio, K.M.A. Chan, V. E. Figueroa, A. Duraiappah, M. Fischer, R. Hill, T. Koetz, P. Leadley, P. Lyver, G. M. Mace, B. Martin-Lopez, M. Okumura, D. Pacheco, U. Pascual, E. S. Pérez, B. Reyers, E. Roth, O. Saito, R. J. Scholes, N. Sharma, H. Tallis, R. Thaman, R. Watson, T. Yahara, Z. A. Hamid, C. Akosim, Y. Al-Hafedh, R. Allahverdiyev, E. Amankwah, T. S. Asah, Z. Asfaw, G. Bartus, A. L. Brooks, J. Caillaux, G. Dalle, D. Darnaedi, A. Driver, G. Erpul, P. Escobar-Eyzaguirre, P. Failler, A. M. M. Fouda, B. Fu, H. Gundimeda, S. Hashimoto, F. Homer, S. Lavorel, G. Lichtenstein, W. A. Mala, W. Mandivenyi, P. Matczak, C. Mbizvo, M. Mehrdadi, J. P. Metzger, J. B. Mikissa, H. Moller, H. A. Mooney, P. Mumby, H. Nagendra, C. Nesshover, A. A. Oteng-Yeboah, G. Pataki, M. Roué, J. Rubis, M. Schultz, P. Smith, R. Sumaila, K. Takeuchi, S. Thomas, M. Verma, Y. Yeo-Chang, and D. Zlatanova. 2015. The IPBES conceptual framework—connecting nature and people. Current Opinion in Environmental Sustainability [online serial] 14:1–16.

Dodds, W. K., J. S. Perkin, and J. E. Gerken. 2013. Human impact on freshwater ecosystem services: a global perspective. Environmental Science and Technology 47:9061–9068.

Dudgeon, D., A. H. Arthington, M. O. Gessner, Z. Kawabata, D. J. Knowler, C. Leveque, R. J. Naiman, A. H. Prieur-Richard, D. Soto, M. L. Stiassny, and C. A. Sullivan. 2006. Freshwater biodiversity: importance, threats, status and conservation challenges. Biological Reviews 81:163–182.

Elith, J., C. H. Graham, R. P. Anderson, M. Dudík, S. Ferrier, A. Guisan, R. J. Hijmans, F. Huettmann, J. R. Leathwick, A. Lehmann, J. Li, L. G. Lohmann, B. A. Loiselle, G. Manion, C. Moritz, M. Nakamura, Y. Nakazawa, J. McC. Overton, A. T. Peterson, S. J. Phillips, K. S. Richardson, R. Scachetti-Pereira, R. E. Schapire, J. Soberón, S. Williams, M. S. Wisz, and N. E. Zimmermann. 2006. Novel methods improve prediction of species' distributions from occurrence data. Ecography 29:129–151.

Elith, J., S. J. Phillips, T. Hastie, M. Dudík, Y. E. Chee, and C. Yates. 2011. A statistical explanation of MaxEnt for ecologists. Diversity and Distributions 17:43–57.

Esselman, P. C., D. M. Infante, L. Wang, D. Wu, A. R. Cooper, and W. W. Taylor. 2011. An index of cumulative disturbance to river fish habitats of the conterminous United States from landscape anthropogenic activities. Ecological Restoration 29:133 –151.

Faber-Langendoen, D., L. Master, J. Nichols, L. Master, K. Snow, A. Tomaino, R. Bittman, G. Hammerson, B. Heidel, L. Ramsay, A. Teucher, and B. Young. 2009. NatureServe conservation status assessments: methodology for assigning ranks. NatureServe, Arlington, Virginia.

Falke, J. A., K. D. Fausch, R. Magelky, A. Aldred, D. S. Surnford, L. K. Riley, and R. Oad. 2011. The role of groundwater pumping and drought in shaping ecological futures for stream fishes in a dryland river basin of the western Great Plains, USA. Ecohydrology 4:682–697.

Fausch, K. D., C. E. Torgersen, C. V. Baxter, and H. W. Li. 2002. Landscapes to riverscapes: bridging the gap between research and conservation of stream fishes. BioScience 52:483–498.

Fontaine, J. J. 2011. Improving our legacy: incorporation of adaptive management into state wildlife action plans. Journal of Environmental Management 92:1403–1408.

Frimpong, E. A. and P. L. Angermeier. 2009. Fish traits: a database of ecological and life-history traits of freshwater fishes of the United States. Fisheries 34:487–495.

Garrett, G. P., T. W. Birdsong, M. G. Bean, and B. J. Labay. 2019. Chihuahuan native fish conservation areas: a multispecies and watershed approach to preservation of freshwater fish diversity. Pages 231–252 *in* D. C. Dauwalter, T. W. Birdsong, and G. P. Garrett, editors. Multispecies and watershed approaches to freshwater fish conservation. American Fisheries Society, Symposium 91, Bethesda, Maryland.

Gido, K. B., W. K. Dodds, and M. E. Eberle. 2010. Retrospective analysis of fish community change during a half-century of landuse and streamflow changes. Journal of the North American Benthological Society 29:970–987.

Graham, C. H., and R. J. Hijmans. 2006. A comparison of methods for mapping species ranges and species richness. Global Ecology and Biogeography 15:578–587.

Grantham, H. S., R. L. Pressey, J. A. Wells, and A. J. Beattie. 2010. Effectiveness of biodiversity surrogates for conservation planning: different measures of effectiveness generate a kaleidoscope of variation. PLOS (Public Library of Science) ONE [online serial] 5(7):e11430.

Groves, C. R., D. B. Jensen, L. L. Valutis, K. H. Redford, M. L. Shaffer, J. M. Scott, J. V. Baumgartner, J. V., Higgins, M. W. Beck, and M. G. Anderson. 2002. Planning for biodiversity conservation: putting conservation science into practice. Bioscience 52:499–512.

Guisan, A., R. Tingley, J. B. Baumgartner, I. Naujokaitis-Lewis, and P. R. Sutcliffe. 2013. Predicting species distributions for conservation decisions. Ecology Letters 16:1424–1435.

Halpern, B. S., and R. R. Warner. 2003. Matching marine reserve design to reserve objectives. Proceedings of the Royal Society of London B 270:1871–1878.

Hanley, J. A., and B. J. McNeil. 1982. The meaning and use of the area under a receiver operating characteristic (ROC) curve. Radiology 143:29–36.

Haslouer, S. G., M. E. Eberle, D. R. Edds, K. B. Gido, C. S. Mammoliti, J. R. Triplett, J. T. Collins, D. A. Distler, D. G. Huggins, and W. J. Stark. 2005. Current status of native fish species in Kansas. Transactions of the Kansas Academy of Science 108:32–46.

Hendrickson, D. A., S. Sarkar, and A. Molineux. 2010. Final Report: provision and inventory of diverse aquatic ecosystem-related resources for the Great Plains Landscape Conservation Cooperative (GPLCC). U.S. Fish and Wildlife Service Great Plains Landscape Conservation Cooperative. Available: https://lccnetwork.org/resource/final-report-provision-and-inventory-diverse-aquatic-ecosystem-related-resources-gplcc. (April 2019).

Hitt, N. P., and P. L. Angermeier. 2008. Evidence for fish dispersal from spatial analysis of stream network topology. Journal of the North American Benthological Society 27:304–320.

Hitt, N. P., and P. L. Angermeier. 2011. Fish community and bioassessment responses to stream network position. Journal of the North American Benthological Society 30:296–309.

Hoagstrom, C. W., J. E. Brooks, and S. R. Davenport. 2011. A large-scale conservation perspective considering endemic fishes of the North American plains. Biological Conservation 144:21–34.

Hunter, M. L., Jr., and A. Hutchinson. 1994. The virtues and shortcomings of parochialism: conserving species that are locally rare, but globally common. Conservation Biology 8:1163–1165.

Jelks, H. L., S. J. Walsh, N. M. Burkhead, S. Contreras-Balderas, E. Díaz-Pardo, D. A. Hendrickson, J. Lyons, N. E. Mandrak, F. McCormick, J. S. Nelson, S. P. Platania, B. A. Porter, C. B. Renaud, J. J. Schmitter-Soto, E. B. Taylor, and M. L. Warren. 2008. Conservation status of imperiled North American freshwater and diadromous fishes. Fisheries 33:372–407.

Kark, S., A. Tulloch, A. Gordon, T. Mazor, N. Bunnefeld, and N. Levin. 2015. Cross-boundary collaboration: key to the conservation puzzle. Current Opinion in Environmental Sustainability 12:12–24.

Knight, A. T., R. M. Cowling, and B. M. Campbell. 2006. An operational model for implementing conservation action. Conservation Biology 20:408–419.

Knight, A. T., R. M. Cowling, M. Rouget, A. Balmford, A. T. Lombard, and B. M. Campbell. 2008. Knowing but not doing: selecting priority conservation areas and the research–implementation gap. Conservation Biology 22:610–617.

Labay, B., A. E. Cohen, B. Sissel, D. A. Hendrickson, F. D. Martin, and S. Sarkar. 2011. Assessing historical fish community composition using surveys, historical collection data, and species distribution models. PLOS (Public Library of Science) ONE [online serial] 6(9):e25145.

Labay, B., B. Prince, J. Ogren, and K. Tasker. 2018. Rio Grande fishes conservation assessment and mapping. Siglo Group, Austin, Texas.

Lehtomaki, J., and A. Moilanen. 2013. Methods and workflow for spatial conservation planning using Zonation. Environmental Modelling and Software 47:128–137.

Limburg, K. E., R. M. Hughes, D. C. Jackson, and B. Czech. 2011. Human population increase, economic growth, and fish conservation: collision course or savvy stewardship? Fisheries 36:27–35.

Lobo, J. M., A. Jiménez-Valverde, and R. Real. 2008. AUC: a misleading measure of the performance of predictive distribution models. Global Ecology and Biogeography 17:145–151.

Margules, C. R. and R. L. Pressey. 2000. Systematic conservation planning. Nature (London) 405:243–253.

Martinuzzi, S., S. R. Januchowski-Hartley, B. M. Pracheil, P. B. McIntyre, A. J. Plantinga, D. J. Lewis, and V. C. Radeloff. 2014. Threats and opportunities for freshwater conservation under future land use change scenarios in the United States. Global Change Biology 20:113–124.

Moilanen, A. 2007. Landscape zonation, benefit functions and target-based planning: unifying reserve selection strategies. Biological Conservation 134:571–579.

Moilanen, A., B. J. Anderson, A. Arponen, F. M. Pouzols, and C. D. Thomas. 2013, Edge artefacts and lost performance in national versus continental conservation priority areas. Diversity and Distributions 19:171–183.

Moilanen, A., and A. Arponen. 2011. Administrative regions in conservation: balancing local priorities with regional to global preferences in spatial planning. Biological Conservation 144:1719–1725.

Moilanen, A., A. M. A. Franco, R. I. Early, R. Fox, B. Wintle, and C. D. Thomas. 2005. Prioritizing multiple-use landscapes for conservation: methods for large multispecies planning problems. Proceedings of the Royal Society B 272:1885–1891.

Moilanen, A., and B. A. Wintle. 2007. The boundary-quality penalty: a quantitative method for approximating species responses to fragmentation in reserve selection. Conservation Biology 21:355–364.

Moilanen, A., J. Leathwick, and J. Elith. 2008, A method for spatial freshwater conservation prioritization. Freshwater Biology, 53: 577–592.

Neeson, T. M., M. C. Ferris, M. W. Diebel, P. J. Doran, J. R. O'Hanley, and P. B. McIntyre. 2015. Enhancing ecosystem restoration efficiency through spatial and temporal coordination. Proceedings of the National Academy of Sciences of the United States of America 112:6236–6241.

Palomo, I., C. Montes, B. Martín-López, J. A. González, M. García-Llorente Paloma Alcorlo, and M. R. García Mora. 2014. Incorporating the social–ecological approach in protected areas in the Anthropocene. BioScience 64:181–191.

Parham, R. W. 2009. Structure of assemblages and recent distribution of riverine fishes in Oklahoma. The Southwestern Naturalist 54:382–399.

Perkin, J. S., K. B. Gido, A. R. Cooper, T. F. Turner, M. J. Osborne, E. R. Johnson, and K. B. Mayes. 2015a. Fragmentation and dewatering transform Great Plains stream fish communities. Ecological Monographs 85:73–92.

Perkin, J. S., K. B. Gido, K. H. Costigan, M. D. Daniels, and E. R. Johnson. 2015b. Fragmentation and drying ratchet down Great Plains stream fish diversity. Aquatic Conservation: Marine and Freshwater Ecosystems 25:639–655.

Perkin, J. S., K. B. Gido, J. A. Falke, K. D. Fausch, H. Crockett, E. R. Johnson, and J. Sanderson. 2017. Groundwater declines are linked to changes in Great Plains stream fish assemblages. Proceedings of the National Academy of Sciences of the United States of America 114:7373–7378.

Phillips, S. J., R. P. Anderson, and R. E. Schapire. 2006. Maximum entropy modeling of species geographic distributions. Ecological Modelling 190:231–259.

Phillips, S. J., and M. Dudík. 2008. Modeling of species distributions with Maxent: new extensions and a comprehensive evaluation. Ecography 31:161–175.

Pierce, S. M. 2003. The STEP handbook: integrating the natural environment into land use decisions at the municipal level: towards sustainable development. University of Port Elizabeth, Terrestrial Ecology Research Unit, Report 47, Port Elizabeth, South Africa.

Pracheil, B. M., M. A. Pegg, L. A. Powell, and G. E. Mestl. 2012. Swimways: protecting Paddlefish through movement-centered management. Fisheries 37:449–457.

Pressey, R. L., M. Cabeza, M. E. Watts, R. M. Cowling, and K. A. Wilson. 2007. Conservation planning in a changing world. Trends in Ecology and Evolution 22:583–592.

Renner, I. W., and D. I. Warton. 2013. Equivalence of MAXENT and Poisson point process models for species distribution modeling in ecology. Biometrics 69:274–281.

SARP (Southeast Aquatic Resources Partnership). 2008. Southeast aquatic habitat plan. SARP, Social Circle, Georgia.

SARP (Southeast Aquatic Resources Partnership). 2014. Conserving fish habitat from rivers to the sea: the story of the Southeast Aquatic Resources Partnership. SARP, Social Circle, Georgia.

Simon, T. P. 1999. Assessment of Balon's reproductive guilds with application to Midwestern North American freshwater fishes. Pages 97–121 *in* T. P. Simon, editor. Assessing the sustainability and biological integrity of water resources using fish communities. CRC Press, Boca Raton, Florida.

Soberon, J. M., and J. K. Sarukhan. 2009. A new mechanism for science-policy transfer and biodiversity governance? Environmental Conservation 36:265–267.

Spurgeon, J. J., R. H. Stasiak, G. R. Cunningham, K. L. Pope, and M. A. Pegg. 2014. Status of native stream fishes within selected protected areas of the Niobrara River in western Nebraska. Great Plains Research 24:71–78.

Steward, D. R., P. J. Bruss, X. Yang, S. A. Staggenborg, S. M. Welch, and M. D. Apley. 2013. Tapping unsustainable groundwater stores for agricultural production in the High Plains Aquifer of Kansas, projections to 2110. Proceedings of the National Academy of Sciences of the United States of America 110:E3477–E3486.

Stewart, D. R., Z. E. Underwood, F. J. Rahel, and A. W. Walters. 2018. The effectiveness of surrogate taxa to conserve freshwater biodiversity. Conservation Biology 32:183–194.

Strayer, D. L., and D. Dudgeon. 2010. Freshwater biodiversity conservation: recent progress and future challenges. Journal of the North American Benthological Society 29:344–358.

Thornbrugh, D. J., S. G. Leibowitz, R. A. Hill, M. H. Weber, Z. C. Johnson, A. R. Olsen, and D. V. Peck. 2018. Mapping watershed integrity for the conterminous United States. Ecological indicators 85:1133–1148.

U.S. Fish and Wildlife Service. 2005. Final designation of critical habitat for the Arkansas River basin population of the Arkansas River Shiner (*Notropis girardi*). Federal Register 70:107(13 October 2005):59808–59846.

U.S. Fish and Wildlife Service. 2014. Endangered and threatened wildlife and plants; designation of critical habitat for Sharpnose Shiner and Smalleye Shiner. Federal Register 79:149(4 August 2014):45242–45271.

Vane-Wright, R. I. 1996. Systematics and the conservation of biological diversity. Annals of the Missouri Botanical Garden 83:47–57.

Vorosmarty, C. J., P. B. McIntyre, M. O. Gessner, D. Dudgeon, A. Prusevich, P. Green, S. Glidden, S. E. Bunn, C. A. Sullivan, C. Reidy Liermann, and P. M. Davies. 2010. Global threats to human water security and river biodiversity. Nature (London) 467:555–561.

Wilde, G. R., and A. C. Urbanczyk. 2013. Relationship between river fragment length and persistence of two imperiled Great Plains cyprinids. Journal of Freshwater Ecology 28:445–451.

Williams, J. E., R. N. Williams, R. F. Thurow, L. Elwell, D. P. Philipp, F. A. Harris, J. L. Kershner, P. J. Martinez, D. Miller, G. H. Reeves, C. A. Frissell, and J. R. Sedell. 2011. Native fish conservation areas: a vision for large-scale conservation of native fish communities. Fisheries 36:267–277.

American Fisheries Society Symposium 91:85–108, 2019

Influence of Conservation Planning Methodology on Species Representation and Spatial Distribution of Priority Areas

Nicholas Sievert*
Missouri Cooperative Fish and Wildlife Research Unit
School of Natural Resources, University of Missouri
103 Anheuser-Busch Natural Resources Building, Columbia, Missouri 65211, USA

Joanna B. Whittier
School of Natural Resources, University of Missouri
103 Anheuser-Busch Natural Resources Building, Columbia, Missouri 65211, USA

Craig Paukert
U.S. Geological Survey, Missouri Cooperative Fish and Wildlife Research Unit
School of Natural Resources, University of Missouri
103 Anheuser-Busch Natural Resources Building, Columbia, Missouri 65211, USA

Abstract.—Systematic conservation planning tools offer powerful and flexible means for addressing the protection of biodiversity in freshwater systems. Tools such as the software Zonation can be used to prioritize streams for protection, restoration, and management of aquatic resources. The flexible nature of these tools allow analyses to be tailored to specific objectives but also introduces uncertainty regarding the effects of selected input options on the rankings of stream segments and the representation of fish species within prioritized streams. The objective of our research was to evaluate the effectiveness of several species distribution modeling techniques (generalized additive models, multivariate adaptive regression splines, boosted regression trees, and random forest models, including an ensemble based on these techniques) for characterizing distributions of fish communities and to identify the influence of different prioritization options of Zonation conservation planning software within five input classes (species occurrence data, removal rule, species weighting, connectivity, and protected area masking) on both the resulting stream segment rankings and the representation of species within priority streams. All combinations of input options were compared based on the correlation and congruence of stream rankings and the mean richness of species, minimum level of species representation, and representation of rare species within streams across priority levels. Of the distribution modeling types we evaluated, boosted regression trees performed the best, followed closely by random forest models. The use of an ensemble approach allowed for the largest number of species with robust predicted distributions. Our results also suggested that protected area masking had the largest effects on conservation priority results, followed by choice of removal rule, while

* Corresponding author: nas4tf@mail.missouri.edu

species occurrence data type had limited impacts. The information contained in this chapter is meant to aid planners in understanding how their selection of conservation planning inputs is likely to impact results.

Introduction

Freshwater environments and fish are highly imperiled due to a variety of factors, including habitat degradation, climate change, and invasive species (Dudgeon et al. 2006). In response to these threats, there are calls for conservation action, including the protection and restoration of streams and watersheds (Saunders et al. 2002; Nel et al. 2009; Arthington et al. 2010, 2016; Abell et al. 2016). Systematic conservation planning provides an important tool for the efficient prioritization of stream fish conservation efforts (Linke et al. 2011). Because resources for the creation of new protected areas and management of existing areas are limited, it is vital to make robust, scientifically informed decisions when determining where to implement conservation actions to have the greatest benefits for fish conservation (Thieme et al. 2012; Hermoso et al. 2016; Maire et al. 2017).

The use of systematic conservation planning approaches has been identified as a key to achieving "a strategic, coordinated and landscape-level planning approach to conservation" (Linke et al. 2011). Systematic conservation planning software such as Zonation (Moilanen et al. 2009), Marxan (Watts et al. 2009), and others have been increasingly used to develop freshwater conservation plans. These tools allow users to prioritize streams and watersheds for fish conservation via flexible complementarity-based algorithms that allow planners to use a variety of inputs to meet conservation objectives. Typically, these algorithms seek to maximize complementarity, the contribution of an area to the diversity of focal organisms or features relative to surrounding areas, while minimizing the area or cost required (Margules et al. 2002). Although, these software packages use algorithms that function in different ways (Zonation uses a maximal coverage framework while Marxan uses a minimum set framework), they both incorporate many similar features, including the ability to weight features, account for connectivity, and incorporate existing protected areas (Moilanen et al. 2009; Watts et al. 2009). Plans may be based on achieving a set level of representation for focal features (i.e., minimum five occurrences of each species) or a set level of sites/area (i.e., identify the top 10% of areas). These plans allow decision makers to identify valuable areas for stream fish conservation that could be used to develop networks of high-priority sites (Dauwalter et al. 2011; Strecker et al. 2011; Labay and Hendrickson 2014; Cohen et al. 2018), target areas for the expansion of protected area networks (Nel et al. 2007), identify focal areas within already established protected area networks (Watson et al. 2014; Maire et al. 2017), or determine opportunities for restoration and management actions that would provide the greatest benefits for species conservation (Mantyka-Pringle et al. 2016).

Researchers who use these tools will encounter a variety of different methodological approaches in the literature for setting the algorithm's inputs, including selection of methods for representing species occurrence, whether to weight specific features, choice of prioritization method, and incorporation of stream connectivity or existing protected lands. Understanding how different approaches influence prioritization results may aid researchers in making decisions as to how to set up analyses. Typically, the intent of prioritization is to maximize the representation of focal features such as the occurrence of a suite of freshwater fish species. The representation of species within a prioritization is often accounted for using species distribution models, which are sometimes classified as presence–absence (Strecker et al. 2011; Grantham et al. 2017; Howard et al. 2018) while other times are assigned probabilities of occurrence (Wenger et al. 2009; Dauwalter et al. 2011; Labay and Hendrickson 2014; Stewart et al. 2017; Cohen et al. 2018). Researchers may want to ensure that rare or priority species are well represented and so may provide additional weight to the occurrence of those species (Moilanen et al. 2008; Labay and Hendrickson 2014; Stewart et al. 2017; Cohen et al. 2018). Another factor planners address, via removal rule selection, is whether they are most interested in emphasizing species richness at high-priority sites (examples: Wenger et al. 2008; Dauwalter et al. 2011; Stewart et al. 2017; Howard et al. 2018) or protecting the core habitats of all species (examples: Strecker et al. 2011; Bond et al. 2014; Labay and Hendrickson 2014; Grantham et al. 2017; Cohen et al. 2018; Howard et al. 2018). For species that require connectivity to upstream or downstream areas to complete important aspects of their life history or high-quality upstream watersheds to ensure adequate water quality and habitat characteristics, several methods have been developed to account for the importance of stream network connectivity (Wenger et al. 2008; Dauwalter et al. 2011; Strecker et al. 2011; Bond et al. 2014; Labay and Hendrickson 2014; Grantham et al. 2017; Stewart et al. 2017; Cohen et al. 2018; Howard et al. 2018). In addition, in many areas conservation networks or priority areas already exist, providing a valuable starting point or framework for planning future conservation efforts. By incorporating existing conservation areas into the planning process, planners can develop prioritizations that complement rather than replace ongoing efforts (Grantham et al. 2017; Howard et al. 2018).

The use of these different approaches to setting up a conservation plan allows researchers to create plans that are suited for specific project objectives; however, the impact of different approaches may or may not produce the expected results (Table 1). Altering the options to carry out a prioritization likely results in changes in both the spatial distribution of priority areas and the level of representation of focal features within a set of priorities. Our objective is to determine whether the different approaches for prioritizing stream fish habitat conservation substantially impacts the resulting spatial distribution of priority areas and representation of species using Missouri, USA streams as a case study with Zonation software.

Materials and Methods

Study area

We developed separate conservation prioritizations for the wadeable streams (defined as second–fifth order and classified as perennial) of the Plains and Ozarks aquatic subregions in Missouri, USA (Pflieger 1989; Abbitt et al. 2004; Sowa et al. 2007; Figure 1). These subregions have different geology, landform, soils, dominant land cover types, and levels of groundwater influence that result in unique stream habitats and fish communities in each area (Abbitt et al. 2004). This allowed us to evaluate the influence of different conservation planning inputs on the resulting prioritizations in two relatively unique geographies. The Plains Aquatic Subregion is located in northern and western Missouri and primarily has row crop and pasture land use with relatively low relief (Sowa et al. 2007). Streams are typically low gradient, have high levels of turbidity, and have silt and sand substrates (Pflieger 1989; Sowa et al. 2007). The Ozark Aquatic Subregion, in southern Missouri, has primarily forested land cover with some open pastureland and relatively high relief (Sowa et al. 2007). Streams are typically high-gradient, have clearer water than their Plains counterparts with gravel, cobble, or bedrock substrates, and receive large amounts

Table 1. Expected influence of the use of different inputs on prioritization results.

Input classes	Expected trade-offs
Species occurrence	
Presence–absence versus probability of occurrence:	Increased interpretability but decreased precision of species distribution estimates
Removal rule	
Core area versus additive benefit:	Greater minimum species representation but lower mean species richness across priorities
Species weighting	
Game/listed versus none:	Additional representation of weighted species but decreases in representation of unweighted species across priorities
Connectivity	
Upstream watershed versus none:	Increased connectivity among high priority segments but leads to decreases in several species' representation metrics across priorities
Masking	
Protected areas versus none:	Frames priorities around existing infrastructure but leads to decreases in several species' representation metrics

of groundwater input (Pflieger 1989; Sowa et al. 2007).

Data

Fish species occurrence data were obtained from fish community sampling in wadeable streams conducted by the Missouri Department of Conservation between 1990 and 2011 (Sievert 2014; Sievert et al. 2016). To account for the occurrence of rare species, which were not frequently encountered during community sampling, we obtained additional fish records collected between 1990 and 2011 from the Missouri Natural Heritage Program. The heritage records were only used as supplemental presence-only data for species whose distributions could not be modeled. Twenty-seven variables related to biogeography, stream features, local and upstream landscapes, and anthropogenic impacts associated with each confluence to confluence stream segment were obtained from the Missouri Resource Assessment Partnership (Abbitt et al. 2004). Additional information regarding sampling locations, species, and modeling variables can be found in Sievert (2014).

Species distribution models

We developed species distribution models to predict the occurrence of each species in wadeable stream segments within our study areas. Four different component distribution models were developed for each species using differing approaches: multivariate adaptive regression splines (MARS), generalized additive models (GAMs), boosted regression trees (BRTs), and random forest models (RFMs). Species distribution models were developed for species within each region that had been recorded in at least 40 stream segments. Species presence or absence served as the response variable while predictor variables included both continuous and categorical environmental metrics that were previously linked to fish distributions (Sievert 2014). Models were developed in R using the 'earth' package (MARS; Milborrow 2014), 'gam' package (GAM; Hastie 2014), 'gbm' package (BRT; Ridgeway 2013), and 'randomForest' package (RFM; Liaw and Wiener 2002). A random subset of 70% of the species occurrence data were used for the training data of the models while the remaining 30% was used

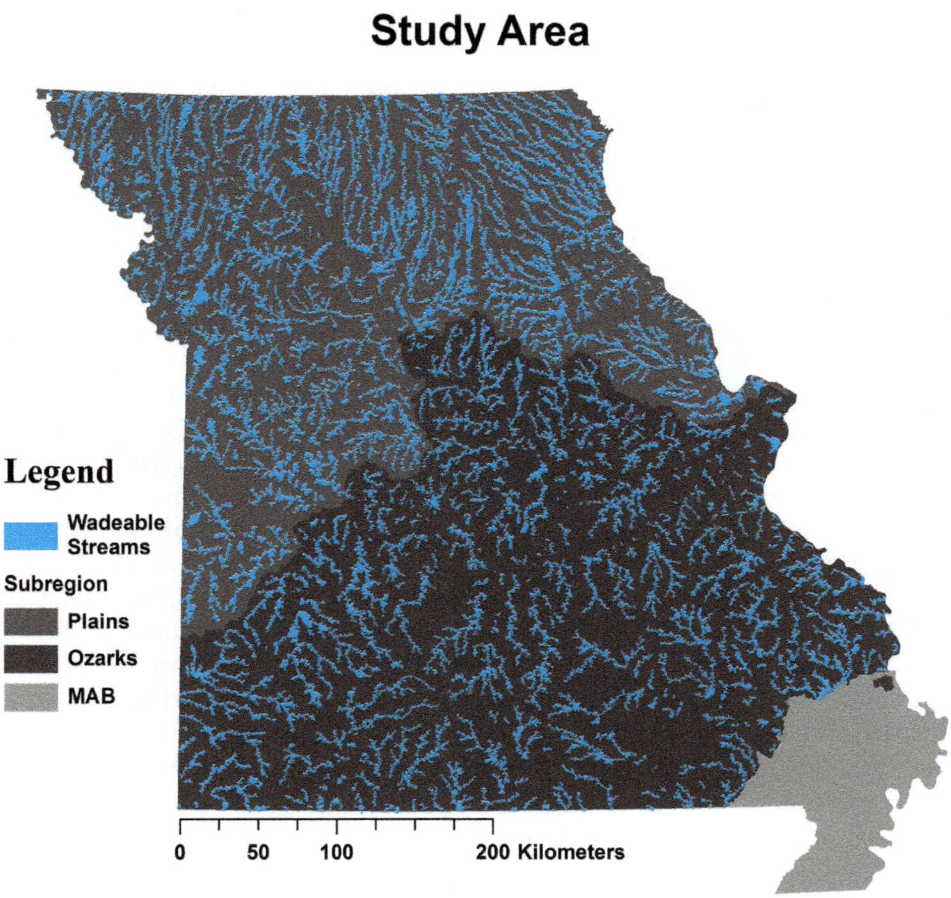

Figure 1. Map showing the study area (Missouri, USA) for this project. MAB: Mississippi alluvial basin (not part of study region).

for model evaluation. Three separate measures were used to evaluate model quality, including the area under the receiver operator characteristic curve (AUC), calculated using the 'ROCR' package in R (Sing et al. 2013), model bias (number of occurrences predicted compared to observed expressed as a percentage), and model fit as the mean absolute error (MAE) of a calibration curve with 10% probability of occurrence bins. An ensemble model was created for each species in each study region by averaging the results of all component models that met our minimum evaluation standards (AUC $\geq$ 0.6, bias of $\pm$25%, and MAE of $\leq$0.125). These ensemble models were used to predict probability of occurrence in all wadeable stream segments. When acceptable models could not be created, a species' occurrence records were used to assign stream segments with an occurrence probability of 1 where the species had been collected while the species' frequency of occurrence within a given watershed was used as the occurrence probability for all other stream segments. Frequency of occurrence was calculated by dividing the number of occurrences by the number of samples within each major watershed.

Prioritization

Conservation planning focuses on creating a spatial prioritization that identifies or ranks areas based on one or several metrics related to the relative conservation value of a site. A variety of software programs, including Zonation, Marxan, and ConsNet, have been used to develop conservation plans with a wide range of objectives in freshwater ecosystems (Ciarleglio et al. 2009; Watts et al. 2009, Moilanen et al. 2014). Each program has different capabilities, strengths, and weaknesses so planners should use the objectives of the project and input data types to guide decisions related to which software to employ. For this comparison, we chose Zonation conservation planning software version 4.0 (Moilanen et al. 2014), as it has been frequently used for freshwater conservation priorities and allows for the development of priority rankings, while the other programs use target-based planning, which identifies a minimum cost solution. Stream segments were prioritized using a suite of different inputs in Zonation conservation planning software version 4.0 following established methods (Sievert 2014; Sievert et al. 2019). Streams were ranked from least to most valuable and assigned a corresponding value from 0 (low) to 1 (high) based on the unique combinations of inputs (hereafter, referred to as variants) for each aquatic subregion. Zonation operates via a removal rule that iteratively removes the least valuable stream segment from the prioritization (this and other key terms defined in Table 2) until only the single most valuable stream reach remains. We ran a total of 48 variants to prioritize streams in each region, representing every possible combination of inputs included in our analyses. These inputs were grouped into five classes, for which a single input from each class was used in each variant: (1) removal rules (core area or additive-benefit function), (2) species occurrence (probability of occurrence or presence–absence), (3) species weighting (all equal, listed species weighting, or game species weighting), (4) connectivity (none or upstream watershed connectivity), and (5) masking (none or protected area masking). With the core-area Zonation removal rule, the most valuable species (species with the smallest proportion of distribution remaining) in each stream segment is identified and the stream segment that results in the smallest decrease in overall value of the remaining streams is removed (Moilanen et al. 2012). The core-area removal rule emphasizes the retention of stream segments that have occurrences of unique species and is expected to result in uncommon species being highly represented among higher ranked streams (Moilanen et al. 2012). The other removal rule, the additive-benefit function, accounts for the value of all species occurring in a segment and removes the stream segment that results in the least decline in conservation value (Moilanen et al. 2012). The use of the additive benefit function removal rule creates priorities that are expected to emphasize species richness and generally result in higher occurrence rates for most species but lower levels of representation for uncommon species (Moilanen et al. 2012).

Species occurrence was incorporated into the prioritization as either probabilities of occurrence from our distribution models or by converting those probabilities to presence (probability of occurrence ≥ 0.5) or absence (probability of occurrence < 0.5). These occurrence data were weighted in three different ways: all species weighted equally, listed species (Missouri species of conservation concern; Missouri Department of Conservation 2017a) weighted 4× unlisted species, game species (Missouri Department of Conservation 2017b) weighted 4× nongame species. We ran prioritizations both with and without a penalty for upstream watershed fragmentation. We developed species-specific penalty curves that represented the response of species to the impacts of upstream watershed fragmentation and penalized the value of segments based on the removal of upstream areas

Table 2. Definitions of key terms used in this manuscript.

Key term	Definition
Congruence	The proportion of segments in which both variants overlap in segments scoring at or above each priority level.
Connectivity	A method of removing value from segments that have decreased connectivity due to the removal of upstream segments from the prioritization, which results in increased connectivity, via clustering, of high-priority segments.
Correlation	A comparison of pairwise variants based on Kendall's rank correlation of the ranking of each segment between prioritizations.
Input classes	Groups of mutually exclusive options for setting up the prioritization algorithm.
Masking	Restricting the removal of segments to certain areas; in the case of this study, masking limits the removal of segments to those that are outside of protected areas before removing any segments within protected areas.
Mean species richness	The average species richness across all sites with scoring at or above each priority level.
Minimum species representation	The minimum number of occurrences by the least well represented species at each priority level.
Pairwise comparisons	A comparison of variants in which all inputs are the same except a single input, which is being evaluated based on a number of metrics (correlation, congruence, mean species richness, minimum species representation, and rare species representation).
Prioritization	The resulting set of stream segment rankings based on the output of a single Zonation run.
Prioritization algorithm	The mathematical formula, typically packaged as a software program such as Zonation, which is used to prioritize or rank stream segments.
Priority level	A measure of the top proportion of stream segments, which ranges from 1 to 0, where 1 is only the most valuable segment, 0 is all segments, and values in between represent a proportion of segments such that a priority level of 0.8 is the top 20% of segments and 0.3 is the top 70% of segments.
Rare species representation	The average number of occurrences by the 10 least well represented species at each priority level.
Removal rule	The method that Zonation uses to determine how to select stream segments for removal in order to create a ranked set of priorities.
Species occurrence	Value that quantifies the predicted likelihood of a species occurring in a given stream segment, which can be summed over a range of segments to predict the total number of occurrences for that range.
Species weighting	The addition of proportional value to occurrences of certain species in order to emphasize their representation in high-priority level segments.
Variant	A unique set of inputs that are used to create a prioritization.

using the directed freshwater connectivity feature (Moilanen et al. 2008; Sievert 2014). This emphasized the maintenance of upstream watershed connectivity in the selection of priority sites for species that are sensitive to watershed degradation. The final input class was masking, which stratifies the removal of stream segments based on whether a segment falls within or outside a specified area. In variants that utilized masking, all segments outside of protected areas (Figg 2011) were removed before any segments within protected areas. This effectively created a ranked set of complementary segments outside of the protected area network and another ranked set of priorities for those within the network. In variants that did not incorporate masking, all segments were equally available for removal without regard as to whether they were within or outside of protected areas.

Comparisons of distribution modeling and prioritization approaches

The effectiveness of different species distribution modeling techniques was assessed for each ecoregion by comparing the average values across all modeled species for each evaluation metric (AUC, bias, and MAE) for each component model (GAMs, MARS, BRTs, and RFMs). We also calculated the percentage of species that met our minimum evaluation standards for each of the four component models. We identified the number of species for which only a single component model was successful and identified the number of additional species distribution models we were able to generate by using the ensemble technique compared to each of the component models individually.

To identify the influence of different input classes on prioritization results (spatial distribution of priorities and species representation within them), pairwise comparisons of variants were constructed in which only a single input differed (Figure 2). The resulting conservation prioritization from each variant was compared to the results from all other combinations for a total of 128 pairwise comparisons in each ecoregion across all priority levels. We used differences in both site rankings and species representation (predicted occurrences of species at a given priority level) to assess the influence of each input class option. Comparisons of the rankings of each segment under each pairwise combination of variants were assessed with Kendall's rank correlation, and congruence (overlap in sites scoring above each priority level). Differences in species representation were assessed across all priority levels by calculating three additional metrics: mean species richness per stream segment (mean species richness), minimum predicted number of occurrences of any species (minimum species representation), and mean predicted number of occurrences of the 10 rarest species at each priority level (rare species representation). To summarize the pairwise comparisons for each input class, we calculated the mean value of each metric across all segments and for high-priority segments (ranked in the top 10%) for each input class (Figure 3). The mean values for each metric (congruence, mean richness, minimum representation, and rare representation) were plotted to visualize these relationships. For each pairwise comparison, we identified the maximum difference for representation metrics and minimum value for congruence and calculated the average for variants in each input class to determine, on average, how far apart results were when using different inputs within an input class.

In addition to testing for input class influence, we also wanted to identify whether certain inputs (namely species weighting, and connectivity) were having their intended effects and if the inefficiencies introduced by masking were outweighed by the increased opportunities afforded by working within already established protected

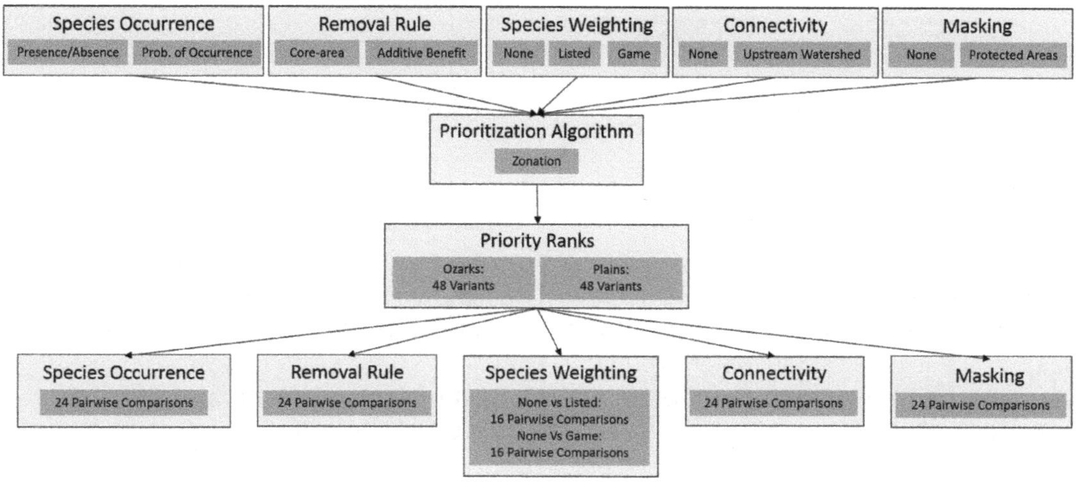

Figure 2. Flowchart of the process used to run a prioritization and compare the results. Top row represents each of the input classes (outlined boxes) and input options (shaded boxes), which are fed into the prioritization algorithm. All combinations of inputs were run, resulting in 48 variants for both the Plains and Ozarks. Pairwise comparisons were made by matching those variants, which shared all but one input. There were 24 pairwise comparisons for each of the input classes (variables of that class were the only ones that differed in each pair), except the weighting comparisons for which there were 16 for each weighting comparison.

area networks. To assess the influence of species weighting on the representation of those species, we calculated the average difference in representation of weighted species in pairwise comparisons where species were weighted to those where they were not. To determine whether the inclusion of the upstream connectivity input resulted in increased clustering, which would demonstrate increased levels of connectivity among high-priority sites, we calculated the standard deviation of rankings within each Hydrologic Unit Code 11 watershed and compared standard deviations of stream rankings within watersheds when upstream fragmentation was and was not included in a variant. To assess the influence of not using a mask on the inclusion of existing protected areas in priorities, we identified the percentage of currently unprotected segments that would be required to create a new network of the same size (total ha) as the current protected network.

We also compared the influence of the input classes on results relative to one another using the mean value of each metric across all priority levels. For correlation and congruence, the lowest value represents the largest difference, while for representation metrics, the highest value represents the largest difference, which necessitates different approaches to calculate relative difference. To calculate the relative influence of an input class using the segment ranking metrics, the following equation was used: relative influence = (input class comparison value − 100)/(absolute value of minimum input class comparison value − 100). To calculate the relative influence of an input class using the representation metrics, the following equation was used: Relative influence = input class comparison value/absolute value of maximum input class comparison value. An input class comparison with a value of 1 (or −1) represents the input class comparison with the largest difference while values between 1 and −1 represent the relative difference for a given input class comparison (i.e., 0.4 means the difference for a given feature comparison is

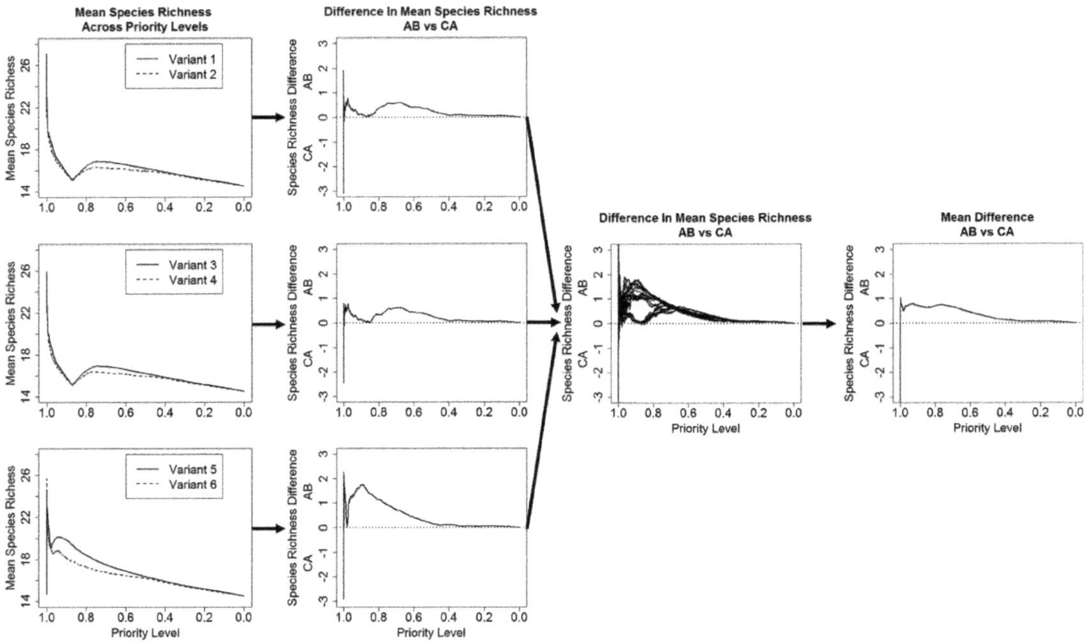

Figure 3. The process by which species representation metrics (pairwise differences in mean species richness, minimum species representation, and rare species representation) were calculated for our comparison of removal rules (additive-benefit function [AB] versus core-area zonation [CA]). In this example, as well as for species occurrence data, connectivity, and masking, there were 24 pairwise comparisons for each, while for species weighting (listed versus no weight and game versus no weight), there were 16 pairwise comparisons for each.

40% of the maximum difference; a negative value simply indicates that the difference was negative).

Results

Distribution modeling techniques

Around half the species occurring within each ecoregion had at least 40 records, which was the minimum number required to attempt to create distribution models (40/84 species in the Plains, 68/132 species in the Ozarks). Random forest models and BRT models were the best at discriminating presence from absence (higher AUC values; Figure 4A). Random forest models and MARS models were the least biased (bias values nearest 1) while BRT models slightly overpredicted the true number of occurrences and GAMs slightly underpredicted occurrences (Figure 4B). Model fit (MAE) was best with BRT models (Figure 4C). All Ozark species had at least one component model that met the minimum evaluation standards (68/68) while in the Plains, 95% of species had at least one acceptable component model (38/40). In both the Plains and Ozarks, the component model types, ranked from most to least number of acceptable models, were RFMs (78% of species in the Plains and 94% in the Ozarks), followed by BRTs (75% and 87%), GAMs (68% and 79%), and MARS (65% and 74%). The BRT method had the most species models that were exclusively created with that approach (two in Plains and two in Ozarks), followed by RFMs (one in Plains and one in Ozarks), MARS (one in Plains),

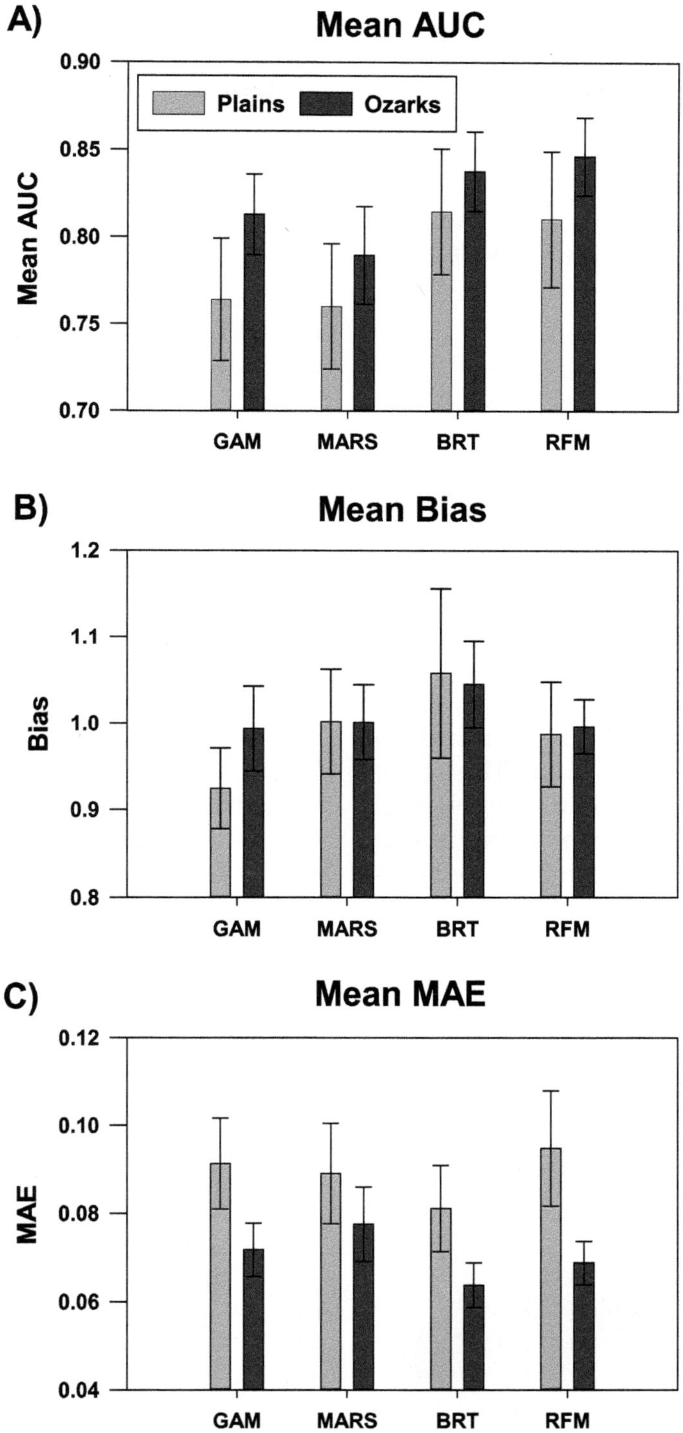

Figure 4. Mean model evaluation metrics with standard error bars for each species distribution model type for the Plains and Ozarks aquatic subregions in Missouri. GAM = generalized additive model, MARS = multivariate adaptive regression splines, BRT = boosted regression trees, RFM = random forest model, AUC = area under the receiver operating characteristic curve, and MAE = mean absolute error.

and GAMs (zero). The ensemble approach increased the number of species represented with occurrence data over any single component model: MARS (32% increase in Plains and 26% in Ozarks), GAMs (29% and 21%), BRTs (21% and 13%), and RFMs (18% and 6%). Further information as well as the species distribution maps can be found in Sievert (2014).

Prioritization

Prioritizations were created for each of the 48 Plains and 48 Ozark variants. These prioritizations ranked streams from most (priority level 1) to least (priority level 0) valuable for each variant. In addition to the individual prioritizations, we also created maps that documented the mean rank and standard deviation of rank for each segment (Figure 5). These maps showed unique patterns in the distribution of high-priority segments and a wide range of standard deviation values (from 0.0 to 0.39). These prioritizations were then utilized to develop pairwise comparisons between variants in which only the inputs representing the input class being considered differed. These pairwise comparisons were then summarized based on input class, with these results described in the next several sessions.

Species occurrence data

The use of presence–absence or probability of occurrence data had minimal impacts on the segment ranking and representation metrics. Segment rankings were highly correlated for the Plains (mean: r = 0.94) and the Ozarks (r = 0.97). Mean congruence was high for both high-priority segments (96% and 95% for the Plains and Ozarks, respectively) and for all priority levels (98% for both Plains and Ozarks), with very little difference in the lowest congruence level (90% and 89% for the Plains and Ozarks, respectively; Figure 6A). The difference in mean species richness between segments using presence–absence and probability of occurrence was negligible across all segments (Plains: 0.00, Ozarks: 0.01) and high-priority segments (–0.04, 0.00), with relatively small maximum differences (–0.25, –0.04; Figure 6A). Mean differences in minimum species representation were relatively small across all segments (–0.12, 0.00) and high-priority segments (–0.01, 0.01), with very low

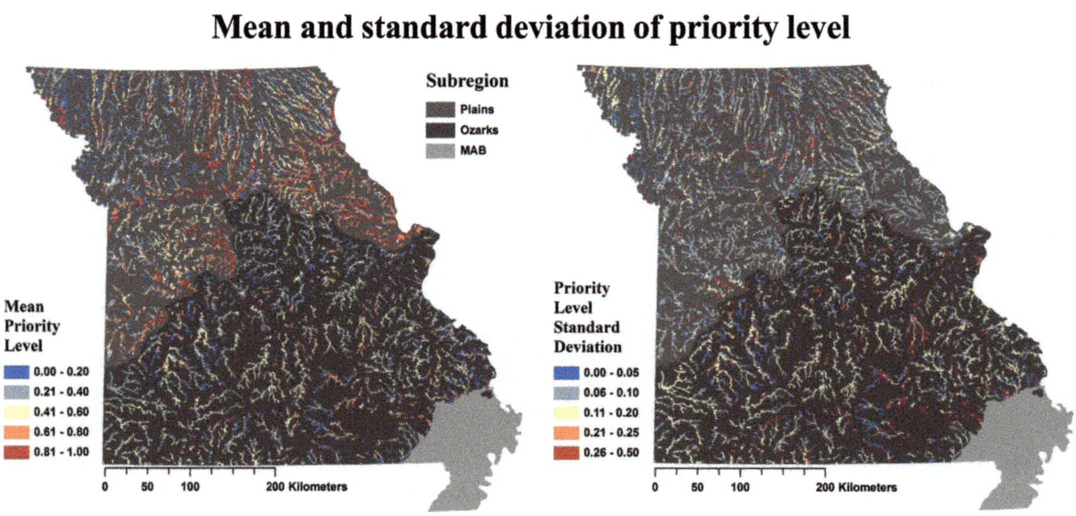

Figure 5. Mean and standard deviation of priority level for each stream segment for all variants. MAB = Mississippi alluvial basin.

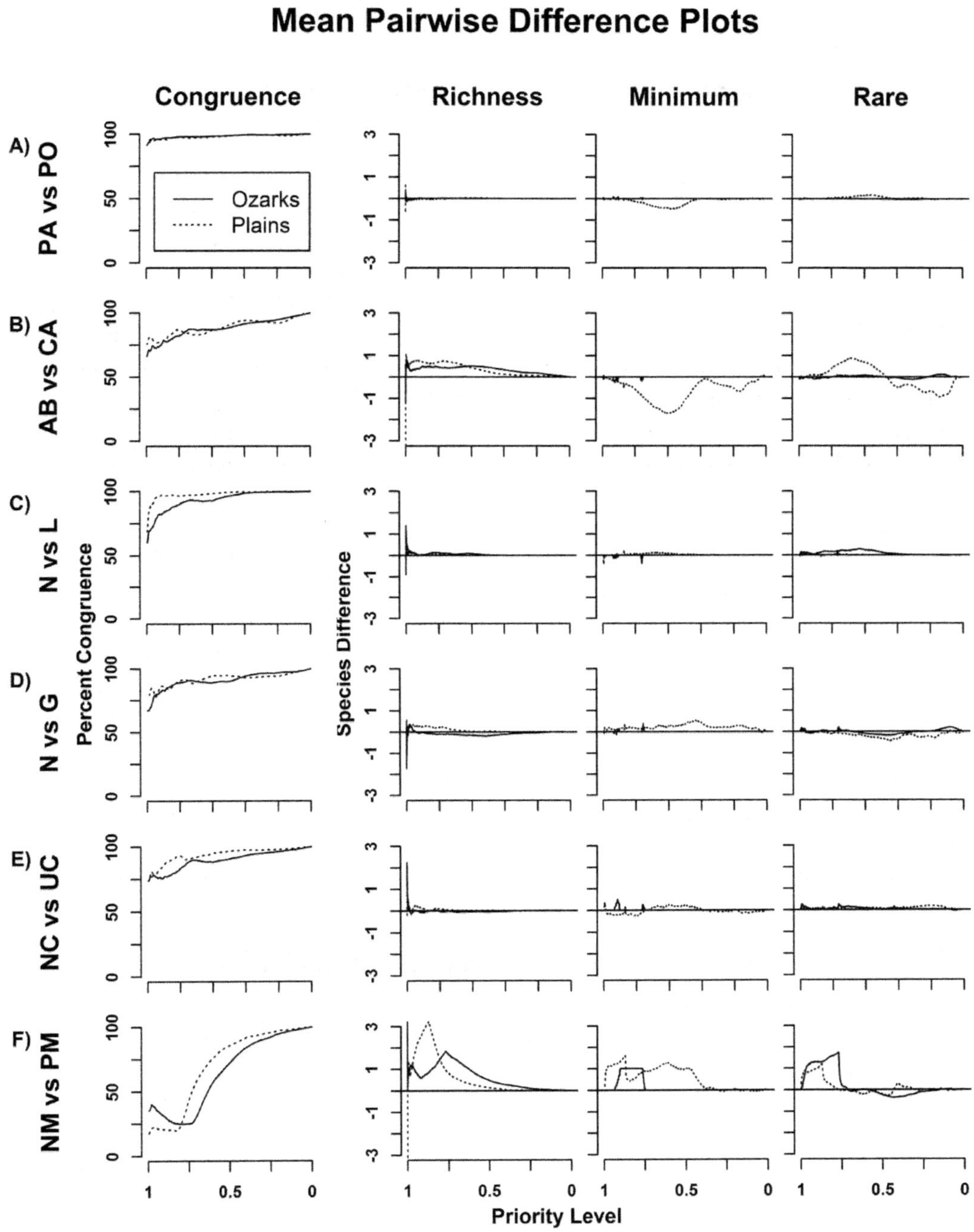

Figure 6. Plots of mean difference in congruence and each of the three representation metrics for the species occurrence data between **(A)** presence–absence (PA) and probability of occurrence (PO), **(B)** additive-benefit function (AB) and core-area zonation (CA), **(C)** no weighting (N) and listed species weighting (L), **(D)** no weighting (N) and game species weighting (G), **(E)** no connectivity (NC) and upstream connectivity (UC), and **(F)** no masking (NM) and protected area masking (PM) variants, across all priority levels in each subregion (Plains and Ozarks). Positive values indicate higher values of a given metric for first-listed variant and negative values indicate higher values for second-listed variant.

maximum differences (−0.84, 0.25; Figure 6A). Mean differences in the mean representation of the 10 rarest species at a given priority level also were minimal across all segments (0.00, 0.00) and high-priority segments (0.00, 0.00), with small maximum differences (0.18, 0.01; Figure 6A).

Removal rule

Segment rankings differed substantially between variants that employed the additive benefit function compared to those using core-area Zonation while mean species representation was consistently higher for variants that employed the additive benefit function compared to core-area zonation, minimum representation was lower and no consistent pattern was observed for rare species. The mean correlation coefficient was 0.79 for both the Plains and Ozarks. Mean congruence for high-priority segments across all pairwise comparisons ranged from 73% (Ozarks) to 78% (Plains) and from 88% (Ozarks) to 89% (Plains) for all priority levels, with a low minimum congruence (67% and 61% in the Plains and Ozarks, respectively; Figure 6B). Additive benefit prioritizations consistently achieved higher mean species richness across all priority levels (Plains: 0.34, Ozarks: 0.33) and high-priority segments (0.70, 0.39), with relatively large maximum differences (−7.9, −1.1; Figure 6B). Core-area variants resulted in higher mean differences in minimum species representation both across all segments (−0.63, −0.01) and high-priority segments (−0.281, −0.05) and with large maximum differences (−1.95, −1; Figure 6B). There was no consistent pattern in the influence of removal rule on the representation of rare species across all priority levels (−0.01, 0.00) or high-priority segments (0.04, −0.04), with relatively high maximum differences (−0.24, −0.09; Figure 6B).

Species weighting

The ranking of stream segments using unweighted species was highly correlated with the corresponding ranking using weighted listed species in the Plains (r = 0.98) and the Ozarks (r = 0.94). Mean congruence for the high-priority segments across all pairwise comparisons of weighted listed to no weighting ranged from 76% (Plains) to 91% (Ozarks) and from 98% (Plains) to 94% (Ozarks) for all priority levels, with minimum congruence levels of 67% and 54% for the Plains and Ozarks, respectively (Figure 6C). Variants without weighting of listed species achieved slightly higher mean differences in species richness in the prioritizations than weighted variants across high-priority segments (0.15, 0.11) and all priority levels (Plains: 0.03, Ozarks: 0.04), with maximum differences of 2.1 and −0.8 in the Plains and Ozarks, respectively (Figure 6C). Weighting listed species also resulted in negligible mean differences for minimum species representation across all segments (0.00, 0.00) and high-priority segments (−0.01, −0.02), with maximum differences of −0.04 and −0.88 in the Plains and Ozarks, respectively (Figure 6C). The mean differences in representation of rare species across all priority levels (0.01, 0.09) and for high-priority segments (0.01, 0.07) were similar when species weighting based on listing status was used, with maximum differences of −0.06 and 0.15 in the Plains and Ozarks, respectively (Figure 6C). Prioritizations where species were weighted based on listing status achieved higher representation of listed species (−0.20, −4.34).

The ranking of stream segments using unweighted species was highly correlated with the corresponding ranking using weighted game species in the Plains (r = 0.84) and Ozarks (r = 0.85). Mean congruence for high-priority segments across all pairwise comparisons of weighted game species to no weighting ranged from 83% (Plains) to 77% (Ozarks) and from 92% (Plains) to 91% (Ozarks) for all priority levels, with minimum congruence values of 72% and 62% in the Plains and Ozarks, respectively (Figure 6D). Prioritizations that did not include game species weighting tended to have slightly higher mean species richness across all priority levels (0.08, −0.07) and high-priority segments (0.26, 0.04), with maximum differences of −1.1 and −2.2 in the Plains and Ozarks, respectively (Figure 6D). Weighting species based on game status resulted in a slight decrease in the mean difference in minimum species representation for all segments (0.22, 0.00) and high-priority segments (0.10, 0.01), with maximum differences of 0.49 and 0.63 in the Plains and Ozarks, respectively (Figure 6D). Prioritizations derived from variants that included game species weighting had negligible differences in the representation of rare species both across all priority levels (−0.16, −0.01) and for high-priority segments (−0.03, 0.00), with maximum differences of −0.68 and 0.20 in the Plains and Ozarks, respectively (Figure 6D). Weighting game species resulted in much higher representation of game species averaged across all priority levels (−13.29, −17.82).

Connectivity

Pairwise comparisons of variants with and without the use of the upstream connectivity input yielded minimal differences in both segment rank and representation metrics. The correlation coefficient was 0.89 for the Plains and 0.81 for the Ozarks. Mean congruence for high priority segments across all pairwise comparisons ranged from 77% (Ozarks) to 81% (Plains) and from 90% (Ozarks) to 94% (Plains) for all priority levels, with minimum congruence levels of 68% and 63% for the Plains and Ozarks, respectively (Figure 6E). Differences in representation metrics were calculated by subtracting the results of prioritizations, which included an upstream connectivity input from those which did not. There was no consistent pattern in the mean difference in mean species richness across all priority levels (Plains: 0.03; Ozarks: −0.01) or high-priority segments (0.17, 0.04), with maximum differences of 3.4 and 4.3 in the Plains and Ozarks, respectively (Figure 6E). The inclusion of upstream connectivity did not result in a consistent change in mean differences in minimum species representation both across all segments (−0.01, 0.01) and high-priority segments (−0.14, 0.09), with maximum differences of 1.0 and 1.0 (Figure 6E). Variants that incorporated a measure of upstream integrity had slightly lower levels of mean differences in rare species representation across all priority levels (0.08, 0.05) and for high-priority segments (0.06, 0.09), with maximum differences of 0.4 and 0.6 in the Plains and Ozarks, respectively (Figure 6E). Our assessment of the clustering of high- and low-value sites, which would be expected to occur with the inclusion of connectivity as a variant, did not indicate that meaningful clustering was occurring. In the Ozarks, the average difference in standard deviation was 0.002, and for the Plains was 0.001.

Protected area masking

Our comparison of the segment rank and representation metrics for variants that included a protected area mask yielded substantial differences. The correlation coefficient was 0.78 for the Plains and 0.62 for the Ozarks. Mean congruence for high-priority segments across all pairwise comparisons ranged from 21% (Plains) to 35% (Ozarks) and from 65% (Ozarks) to 71% (Plains) for all priority levels, with minimum congruence levels of 15% and 23% for the Plains and Ozarks, respectively (Figure 6F). Prioritizations that did not account for protected areas achieved far higher mean species richness across all priority levels (Plains: 0.59, Ozarks: 0.61) and high-priority segments (1.82, 0.86), with maximum differences of −2.8 and 7.0 in the Plains and Ozarks, respectively (Figure 6F). The use of a protected area mask resulted in substantial decreases in minimum species representation both across all segments (0.57, 0.16) and high-priority segments (1.08, 0.15) with maximum differences of 1.8 and 1.0 in the Plains and Ozarks, respectively (Figure 6F). Variants that incorporated protected area masks had lower levels of rare species representation both across all priority levels (0.12, 0.22) and for high-priority segments (0.83, 0.98), with maximum differences of 1.2 and 1.7 in the Plains and Ozarks, respectively (Figure 6F). The development of a new conservation network of equal size to the current network but based on our blank slate prioritizations would require substantial additions of currently unprotected land in both the Ozarks (mean 77%, of segments not already protected) and Plains (77%; Figure 7).

Relative influence of input classes on evaluation metrics

Our assessment of the relative influence of input classes on evaluation metrics showed consistent patterns in which features tended to have large effects across all metrics and which features had limited effects. The correlation of stream segment scores was affected most by masking (lowest correlation), followed by removal rule, while game species weighting and connectivity had intermediate levels of influence and representation input and listed species weighting had relatively low levels of influence (highest correlation; Figure 8A). Protected area masking had a substantially greater influence on congruence than any other factor; no other comparison had greater than 40% of the mean difference achieved by protected area masking for all and high-priority segments in both the Plains and Ozarks (Figure 8B, 8C). In all cases (Plains and Ozarks, high-priority and all segments), representation input had the smallest difference in congruence (<10% of the difference caused by masking; Figure 8B, 8C). Species richness across all and high-priority segments had the greatest differences based on whether or not protected area masking was used; removal rule was the next most influential (~45% of the im-

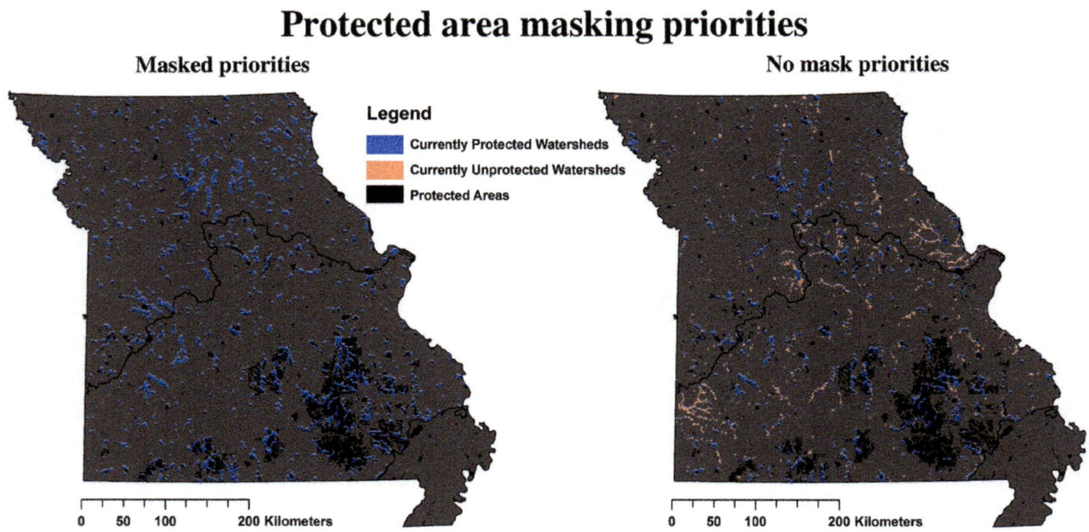

Figure 7. Distribution of the top 1,311 Plains segments and 2,543 ranked Ozark segments (equal to the number of segment watersheds intersecting protected areas in each ecoregion) based on average priority level. In both regions, 100% (1,311 Plains segments and 2,543 Ozark segments) intersect currently protected areas when protected area masking is used; however, in both regions, only 23% (360 Plains segments and 540 Ozark segments) of selected segments are currently protected when masking is not used.

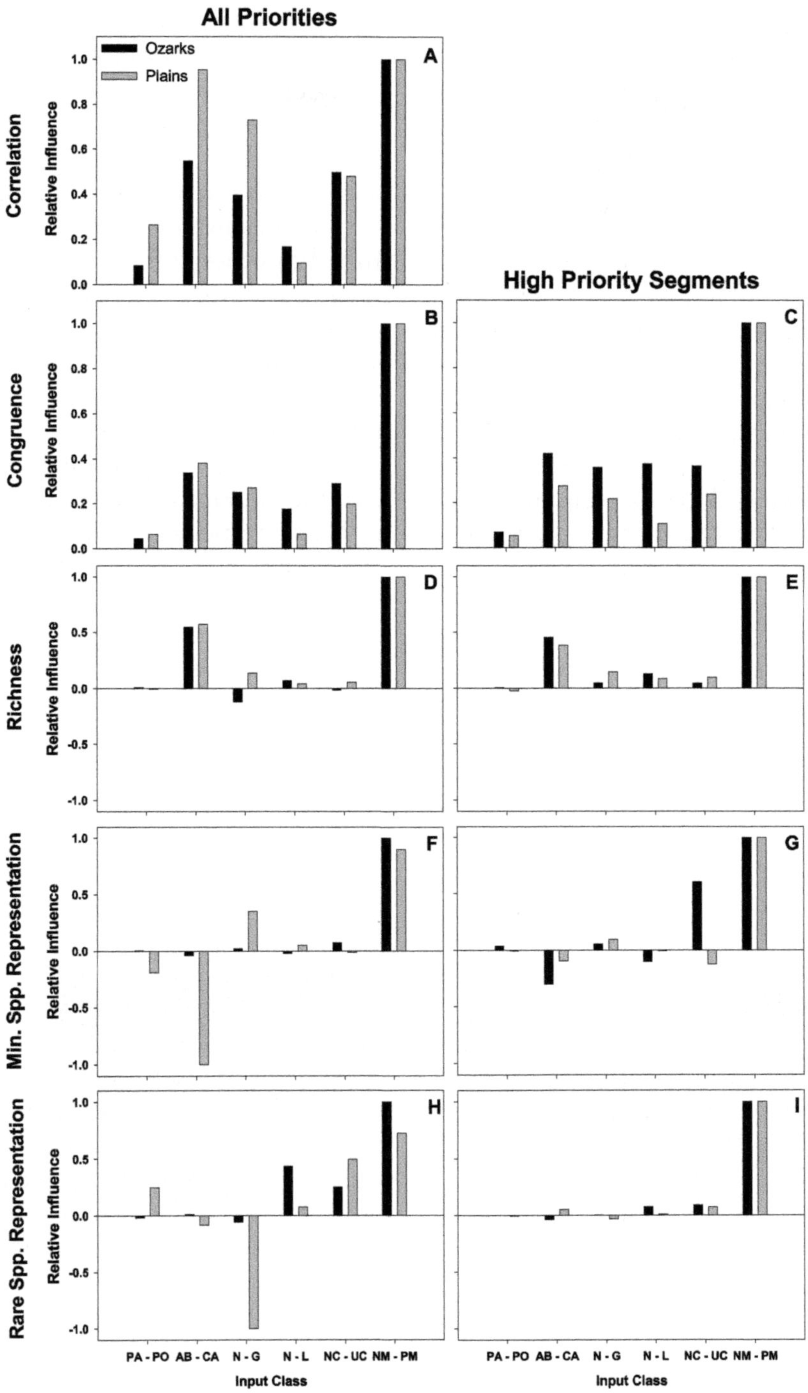

Figure 8. Plots of the relative influence of each input class on each of the evaluation metrics for high-priority (top 10%) and across all priority levels. Abbreviations for inputs include PA = presence–absence, PO = probability of occurrence, AB = additive benefit function, CA = core-area zonation, N = no weight, G = game species weighted, L = listed species weighted, NC = no connectivity, UC = upstream connectivity, NM = no masking, and PM = protected area masking.

pact of masking) while all others had less than 20% of the mean difference caused by masking, with representation input having the least influence on richness (Figure 8D, 8E). For all segments, masking had the largest influence in the Ozarks while removal rule had the largest influence in the Plains (Figure 8F). In the Plains, masking still had a relatively large influence (90%); however, all other comparisons for both the Plains and Ozarks had relative influences of less than 40% (Figure 8F). For high-priority segments, masking had the greatest influence on minimum levels of species representation in both the Plains and Ozarks (Figure 8G). In the Ozarks, connectivity also had a relatively large effect (60% of masking) while all other comparisons for both the Plains and Ozarks had a relatively minor influence (<40%; Figure 8G). Across all priority levels, the Ozarks showed protected area masking having the largest influence and all others having much smaller influences, although listed species weighting had a larger relative influence (43%; Figure 8H). Across all priority levels in the Plains, game species weighting had the highest influence on rare species representation, followed by protected area masking (72%) and connectivity (50%); all others had relatively small influences, less than 25% (Figure 8H). For rare species representation in high-priority segments, priority masking had by far the greatest impact in both the Plains and Ozarks (no other comparison had a relative influence >10%; Figure 8I).

Discussion

Systematic conservation planning is a flexible and powerful approach to aquatic prioritization that can be tailored to a wide variety of situations and objectives. This flexibility allows decision makers to develop customized prioritizations; however, how the input options influence the segment rankings and the representation of species within these areas has not been thoroughly examined in freshwater systems. Zonation is a widely used software for freshwater conservation planning (Moilanen et al. 2008; Dauwalter et al. 2011; Strecker et al. 2011; Grantham et al. 2017; Stewart et al. 2017; Cohen et al. 2018; Frederico et al. 2018), and our findings are likely to be relevant for guiding future efforts in freshwater conservation planning. Although other software programs such as Marxan are based on algorithms that function in a different manner, many shared features such as feature weighting, connectivity, and masking may behave in similar ways to what was observed in our evaluation with Zonation, although further study and confirmation is certainly warranted. We found that species occurrence input type had minimal influence while removal rule and protected area masking had substantial impacts. Understanding how prioritization inputs influence the segment rankings will help natural resource managers craft an approach to attain their conservation goals.

Of the four model types we evaluated, BRT and RFM appear to excel at creating accurate models, as well as effectively creating models for the largest number of different species. We evaluated a variety of methods that have been used to model the distribution of stream fish species, including BRT, MARS, GAM, and RFM modeling techniques. When selecting an approach for species distribution modeling, model accuracy and the ability to successfully develop models for a wide range of species are particularly important to consider. Previous examinations of the topic by Elith et al. (2006) showed BRTs outperforming MARS and GAMs based on measures of model accuracy. Our assessment of the cross-validated results also suggests that BRT models achieve the best discrimination between presence and absence, model fit, and a low level of bias. However, GAM had the poorest discrimination, most bias, and second-worst model fit. Based solely on the number of species for which models could be successfully generated, RFM performed the best, but BRT produced the greatest number of uniquely successful models (species for which only one approach successfully generated an acceptable model). Based on these findings, if a single modeling approach was used, BRT models appear to achieve a balance of creating models that accurately predict species occurrence while also generating robust models for a large number of species. Random forest models performed similarly to BRT models; however, they were slightly less accurate while GAM and MARS models successfully generated fewer species distribution models and were less accurate. While some studies use only a single model type in the construction of species distribution models (Dauwalter et al. 2011; Strecker et al. 2011; Labay and Hendrickson 2014), the use of ensemble modeling techniques that average the results of multiple component models is becoming widely used (Grenouillet et al. 2011; Bouska et al. 2015). The ensemble approach allowed for the inclusion of seven (Plains) and four (Ozarks) additional species compared to the individual component model that was successful for the most species (RFM). These results suggest that using an ensemble modeling approach is likely to allow for the inclusion of more species in a conservation prioritization effort than any single model approach.

Our evaluation revealed consistent responses between two geographically distinct aquatic subregions to the effects of varying conservation planning inputs on segment ranking and species representation metrics. There are a variety of approaches that have been used for developing freshwater prioritizations while using inputs such as species weighting (Moilanen et al. 2008; Stewart et al. 2017; Cohen et al. 2018), measures of connectivity or upstream watershed integrity (Dauwalter et al. 2011; Strecker et al. 2011; Labay and Hendrickson 2014), and/or spatial masking (Grantham et al. 2017; Frederico et al. 2018; Howard et al. 2018). Understanding how different inputs impact results may aid conservation planners in the development of prioritizations that better meet their objectives. Our evaluation revealed consistent responses between two geographically distinct regions relative to the impacts of input class options on the resulting segment rankings and representation of species within priority areas. The results of this analysis can be used to aid planners in other regions in guiding the development of robust aquatic prioritization and decision making for aquatic biodiversity conservation.

Our results suggest that the choice of presence–absence or probability of occurrence data are unlikely to have a substantial impact on either the segment ranking or representation metrics. The higher resolution provided by probability of occurrence data led us to believe that data in that format would perform better in a conservation planning analysis; however, that was not the case. Therefore, prioritizations can be constructed using either presence–absence predictions or with probability of occurrence data without too much concern regarding impacts on the resulting priorities. Presence–absence predictions may be preferred by planners because species distributions can be depicted as presence–absence maps, which can be easier to communicate to stakeholders involved in developing conservation priorities. Although probability of occurrence data lack the clear interpretability of presence–absence predictions, they likely provide a more nuanced and accurate picture of species distributions. The greater resolution provided by probability of occurrence data allows planners and planning software to more fully differentiate between sites across a range of occurrence likelihoods.

The choice of removal rule has a substantial impact on the segment ranking and representation metrics. Core-area zonation emphasizes the representation of uncommon species by prioritizing stream segments in a way that maintains minimum species representation while the additive-benefit function considers the value of all species in a segment rather than only the individual species that has lost the greatest proportion of its occurrences (Moilanen et al. 2009). The additive-benefit approach is expected to result in higher levels of species richness but fewer occurrences for uncommon species across priority levels (Moilanen et al. 2009). Our results corroborated the expectation that the choice of removal rule would show substantial differences in segment ranking and species representation metrics. Our results suggest that for planners who hope to ensure that no species is left out of high-priority segments, core-area zonation appears to be a better approach while for those who want to maximize species richness within high-priority segments, the additive-benefit function is likely the best fit.

In many cases, a planner may have target species they want to emphasize in a prioritization for conservation, ecological, or economic reasons (Arponen et al. 2005). In our analysis, weighting either listed or game species had variable impacts on our comparisons but did result in substantial representation gains for highly weighted species. Weighted game species had greater effects on priority ranks than weighted listed species, potentially because less common species (which were typically listed species) tend to drive the results of parameterization regardless of weighting. Our results suggest that weighting species is likely to substantially increase the representation of those weighted species without having large impacts on species richness, representation of rare species, or minimum representation.

Our results did not indicate that high- and low-priority stream segments were more clustered when upstream watershed connectivity was considered; however, there were impacts on segment ranking and species representation metrics. We expected that variants that included directed freshwater connectivity would result in clustering of high- or low-priority segments within watersheds (rather than having mixtures of high- and low-priority segments within a single watershed). Accounting for upstream watershed fragmentation via stream network connectivity may be crucial to the management and conservation of stream fishes. Several methods have been developed to emphasize the maintenance of upstream watershed connectivity in systematic conservation planning, including directed freshwater connectivity in Zonation (Moilanen et al. 2008), which has been employed in numerous landscapes (Wenger et al. 2009; Dauwalter et al. 2011; Labay and Hendrickson 2014; Stewart et al. 2017; Cohen et al. 2018). Although we failed to detect meaningful levels of clustering, we did observe moderate impacts on segment ranking metrics and variable impacts on species representation metrics, indicating that including upstream connectivity is likely to result in changes to the spatial distribution of high-priority areas and representation of species within them. Our analyses indicate that relatively high penalties for habitat fragmentation are likely needed to cluster priorities within watersheds; however, doing so may result in large changes in segment ranking and representation metrics.

Our study found that spatial masking, in the form of hierarchical removal of streams based on existing protection status, would have substantial impacts on both segment ranking and species representation metrics. Conservation plans that consider the existing conservation network, through the use of spatial masking, may be more efficient (requiring the addition of less currently unprotected lands) and cost-effective. Using this approach, planners are able to target areas that are already under the ownership or control of management agencies or conservation nonprofits. Taking conservation actions on these areas may have less logistical and financial hurdles than attempting to take action on currently unprotected lands;

however, focusing efforts on protected lands may overlook highly valuable streams in unprotected areas, leading to inefficient representation of species (Esselman et al. 2013; Grantham et al. 2017; Frederico et al. 2018). The impacts of masking on segment ranking and species representation metrics were most pronounced for the highest priority sites (those that were contained within protected areas compared to the highest priority sites drawn from anywhere within a study area) with more rare species, higher species richness, and higher minimal representation in variants that did not include masking; however, even with losses in efficiency, the current suite of protected areas does a relatively good job of protecting local fish species, with all but one species (Starhead Topminnow *Fundulus dispar*, a species with a range barely overlapping the Ozarks) occurring within segments intersecting protected areas. The differences in segment ranking and representation metrics between masked and unmasked priorities quickly shrinks for priority ranks outside of the mask extent. This indicates that expanding current networks of protected areas with the highest priority complementary areas outside a network can largely overcome gaps in efficiency with relatively few additions. In our examples, only a relatively small number of high-priority segments identified in prioritizations without masking were found within protected area networks, which indicates that while an approach that does not use masking will lead to higher levels of species representation, it is also likely to require vastly more resources to implement.

Our analysis of the relative influence of each of the input classes was relatively consistent for both subregions, with protected area masking and removal rules accounting for the largest influences while the choice of species occurrence inputs had relatively small impact on all metrics. Masking protected areas and removal rule had by far the largest influence on the correlation of spatial priorities while species occurrence and weighting by listing status had very little impacts on results. Congruence was very high between species occurrence variants and low between masking variants, with intermediate influences for the other input classes. Across the representation metrics, we found that protected area masking consistently had the largest impacts while removal rule also had a high degree of influence. These patterns were largely the same regardless of whether high-priority segments or all segments were considered. This suggests that serious thought should be put into decisions regarding masking protected areas or removal rules, as those decisions are likely to have substantial impacts on results while decisions about which type of species occurrence input is used will not be likely to dramatically change the priorities.

This work highlights some of the costs and benefits to the different approaches of selecting inputs for a conservation prioritization; however, we encourage planners to test the inputs they are considering to see exactly how the results stack up in new settings. Based on our results, decisions regarding the removal rule and the use of feature masking appear to have the largest impacts on results and therefore demand the most consideration when making decisions regarding setting up a project. With a more complete understanding of how different inputs influence both segment rankings and species representation within priorities, planners can develop priorities that better align with a project's objectives. The flexibility of these tools allows planners to customize plans to address a diverse range of conservation challenges; however, it is vital to understand how decisions made regarding the setup of a prioritization influences the outputs. Any change in inputs to a prioritization are likely to influence the results and understanding the trade-offs that result from a choice will allow decision makers to better achieve project objectives. We believe the information provided in this chapter regarding the influence of inputs on resulting conservation priorities can be used to inform future conservation planning efforts.

Acknowledgments

Funding for this work was provided by the U.S. Geological Survey Climate Adaptation Center. Data for this research were provided by the Missouri Department of Conservation and the Missouri Resource Assessment Partnership. Any use of trade, firm, or product names is for descriptive purposes only and does not imply endorsement by the U.S. Government. The Missouri Cooperative Fish and Wildlife Research Unit is sponsored jointly by the U.S. Geological Survey, Missouri Department of Conservation, University of Missouri, Wildlife Management Institute, and U.S. Fish and Wildlife Service.

References

Abbitt, R., G. Annis, and S. Sowa. 2004. Missouri streams with landscape variables. Missouri Resource Assessment Partnership, Columbia.

Abell, R., B. Lehner, M. Thieme, and S. Linke. 2016. Looking beyond the fenceline: assessing protection gaps for the world's rivers. Conservation Letters 10:384–394.

Arponen, A., R. K. Heikkinen, C. D. Tho, and A. Moilanen. 2005. The value of biodiversity in reserve selection: representation, species weighting, and benefit functions. Conservation Biology 19:2009–2014.

Arthington, A. H., N. K. Dulvy, W. Gladstone, and I. J. Winfield. 2016. Fish conservation in freshwater and marine realms: status, threats and management. Aquatic Conservation: Marine and Freshwater Ecosystems 26:838–857.

Arthington, A. H., R. J. Naiman, M. E. Mcclain, and C. Nilsson. 2010. Preserving the biodiversity and ecological services of rivers: new challenges and research opportunities. Freshwater Biology 55:1–16.

Bond, N. R., J. R. Thomson, and P. Reich. 2014. Incorporating climate change in conservation planning for freshwater fishes. Diversity and Distributions 20:931–942.

Bouska, K., G. Whitledge, and C. Lant. 2015. Development and evaluation of species distribution models for fourteen native central U.S. fish species. Hydrobiologia 747:159–176.

Ciarleglio, M., J. Barnes, and S. Sarkar. 2009. ConsNet: new software for the selection of conservation area networks with spatial and multi-criteria analyses. Ecography 32:205–209.

Cohen, A. E., G. P. Garrett, M. J. Casarez, D. A. Hendrickson, B. J. Labay, T. Urban, J. Gentle, D. Wiley, and D. Walling. 2018. Conserving Texas biodiversity: status, trends, and conservation planning for fishes of greatest conservation need. Texas Parks and Wildlife Department, Austin.

Dauwalter, D. C., J. S. Sanderson, J. E. Williams, and J. R. Sedell. 2011. Identification and implementation of native fish conservation areas in the upper Colorado River basin. Fisheries 36:278–288.

Dudgeon, D., A. Arthington, M. Gessner, Z. Kawabata, D. Knowler, C. Lévêque, and C. Sullivan. 2006. Freshwater biodiversity: importance, threats, status and conservation challenges. Biological Reviews 81:163–182.

Elith, J., C. H. Graham, R. P. Anderson, M. Dudík, S. Ferrier, A. Guisan, R. J. Hijmans, F. Huettmann, J. R. Leathwick, A. Lehmann, J. Li, L. G. Lohmann, B. A. Loiselle, G. Manion, C. Moritz, M. Nakamura, Y. Nakazawa, J. M. Overton, T. A. Peterson, S. J. Phillips, K. Richardson, R. Scachetti-Pereira, R. E. Schapire, J. Soberón, S. Williams, M. S. Wisz, and N. E. Zimmermann. 2006. Novel methods improve prediction of species' distributions from occurrence data. Ecography 29:129–151.

Esselman, P., M. Edgar, J. Breck, E. Hay-Chmielewski, and L. Wang. 2013. Riverine connectivity, upstream influences, and multi-taxa representation in a conservation area network for the fishes of Michigan, USA. Aquatic Conservation: Marine and Freshwater Ecosystems 23:7–22.

Figg, D. 2011. Conservation priorities decision

support tool. Missouri Department of Conservation, Jefferson City.

Frederico, R., J. Zuanon, and P. Marco. 2018. Amazon protected areas and its ability to protect stream-dwelling fish fauna. Biological Conservation 219:12–19.

Grantham, T., K. Fesenmyer, R. Peek, E. Holmes, R. Quiñones, A. Bell, N. Santos, J. Howard, J. Viers, and P. Moyle. 2017. Missing the boat on freshwater fish conservation in California. Conservation Letters 10:77–85.

Grenouillet, G., L. Buisson, N. Casajus, and S. Lek. 2011. Ensemble modeling of species distribution: the effects of geographical and environmental ranges. Ecography 34:9–17.

Hastie, T. 2014. gam: generalized additive models. R Package. V. 1.09.1. Available: https://cran.r-project.org/web/packages/gam/gam.pdf.

Hermoso, V., R. Abell, S. Linke, and P. Boon. 2016. The role of protected areas for freshwater biodiversity conservation: challenges and opportunities in a rapidly changing world. Aquatic Conservation: Marine and Freshwater Ecosystems 26(S1):3–11.

Howard, J. K., K. A. Fesenmyer, T. E. Grantham, J. H. Viers, P. R. Ode, P. B. Moyle, S. J. Kupferburg, J. L. Furnish, A. Rehn, J. Slusark, R. D. Mazor, N. R. Santos, R. A. Peek, and A. N. Wright. 2018. A freshwater conservation blueprint for California: prioritizing watersheds for freshwater biodiversity. Freshwater Science 37:417–431.

Linke, S., E. Turak, and J. Nel. 2011. Freshwater conservation planning: the case for systematic approaches. Freshwater Biology 56:6–20.

Maire, A., P. Laffaille, J.-F. Maire, and L. Buisson. 2017. Identification of priority areas for the conservation of stream fish assemblages: implications for river management in France. River Research and Applications 33:524–537.

Mantyka-Pringle, C., T. Martin, D. Moffatt, and J. Udy. 2016. Prioritizing management actions for the conservation of freshwater biodiversity under changing climate and land-cover. Biological Conservation 197:80–89.

Margules, C., R. Pressey, and P. Williams. 2002. Representing biodiversity: data and procedures for identifying priority areas for conservation. Journal of Biosciences 27:309–326.

Milborrow, S. 2014. earth: multivariate adaptive regression spline models. R Package. V. 3.7.

Missouri Department of Conservation. 2017a. Missouri species and communities of conservation concern checklist. Missouri Department of Conservation, Jefferson City.

Missouri Department of Conservation. 2017b. A summary of Missouri fishing regulations. Missouri Department of Conservation, Jefferson City.

Moilanen, A., H. Kujala, and J. Leathwick. 2009. The Zonation framework and software for conservation prioritization. Pages 196–210 in A. M., K. A. Wilson, and H. P. Possingham, editors. Spatial conservation prioritization. Oxford University Press, Oxford, UK.

Moilanen, A., J. Leathwick, and J. Elith. 2008. A method for spatial freshwater conservation prioritization. Freshwater Biology 53:577–592.

Moilanen, A., L. Meller, J. Leppanen, F. Pouzols, A. Arponen, and H. Kujala. 2012. Zonation Conservation Planning Software. Biodiversity Conservation Informatics Group, University of Helsinki, Finland.

Moilanen, A., F. Pouzols, L. Meller, V. Veach, A. Arponen, J. Leppanen, and H. Kujala. 2014 Zonation Conservation Planning Software V4. Biodiversity Conservation Informatics Group, University of Helsinki, Finland.

Nel, J., D. Roux, R. Abell, P. Ashton, R. Cowling, J. Higgins, M. Thieme, and J. Viers. 2009. Progress and challenges in freshwater conservation planning. Aquatic Conservation: Marine and Freshwater Ecosystems 19:474–485.

Nel, J., D. Roux, G. Maree, C. Kleynhans, J. Moolman, B. Reyers, M. Rouget, and R. Cowling. 2007. Rivers in peril inside and outside protected areas: a systematic approach to conservation assessment of river ecosystems. Diversity and Distributions 13:341–352.

Pflieger, W. L. 1989. Aquatic community classification for Missouri. Missouri Department of Conservation, Jefferson City.

Ridgeway, G. 2013. gbm: generalized boosted regression models. R Package. V. 2.1.

Saunders, D., J. Meeuwig, and A. Vincent. 2002. Freshwater protected areas: strategies for conservation. Conservation Biology 16:30–41.

Sievert, N. 2014. An assessment of stream fish vulnerability and an evaluation of conservation networks in Missouri. Master's thesis. University of Missouri, Columbia.

Sievert, N., C. Paukert, Y. Tsang, and D. Infante. 2016. Development and assessment of indices to determine stream fish vulnerability to climate change and habitat alteration. Ecological Indicators 67:403–416.

Sievert, N., C. Paukert, and J. Whittier. 2019. Data supporting a framework to incorporate established conservation networks into freshwater conservation planning. U.S. Geological Survey data release. Available: https://doi.org/10.5066/f7 df6q34.

Sing, T., O. Sander, N. Beerenwinkel, and T. Lengauer. 2013. ROCR: visualizing the performance of scoring classifiers. R Package. V. 1.0–5.

Sowa, S. P., G. Annis, M. E. Morey, and D. D. Diamond. 2007. A gap analysis and comprehensive conservation strategy for riverine ecosystems of Missouri. Ecological Monographs 77:301–334.

Stewart, D. R., Z. E. Underwood, F. J. Rahel, and A. W. Walters. 2017. The effectiveness of surrogate taxa to conserve freshwater biodiversity. Conservation Biology 32:183–194.

Strecker, A. L., J. D. Olden, J. B. Whittier, and C. P. Paukert. 2011. Defining conservation priorities for freshwater fishes according to taxonomic, functional, and phylogenetic diversity. Ecological Applications 21:3002–3013.

Thieme, M. L., J. Rudulph, J. Higgins, and J. A. Takats. 2012. Protected areas and freshwater conservation: a survey of protected area managers in the Tennessee and Cumberland River basins, USA. Journal of Environmental Management 109:189–199.

Watson, J., N. Dudley, D. Segan, and M. Hockings. 2014. The performance and potential of protected areas. Nature (London) 515:67–73.

Watts, M., I. Ball, R. Stewart, and K. C. Software. 2009. Marxan with Zones: software for optimal conservation based land-and sea-use zoning. Environmental Modelling and Software 24:1513–1521.

Wenger, S. J., and M. M. Hagler, and B. J. Freeman. 2009. Prioritizing areas of the Conasauga River sub-basin in Georgia and Tennessee for preservation and restoration. Southeastern Fishes Council Proceedings [online serial] 51:31–38.

Wenger, S. J., J. T. Peterson, M. C. Freeman, B. J. Freeman, and D. D. Homans. 2008. Stream fish occurrence in response to impervious cover, historic land use, and hydrogeomorphic factors. Canadian Journal of Fisheries and Aquatic Sciences 65:1250–1264.

American Fisheries Society Symposium 91:109–133, 2019

Prioritizing Hawaii's Stream Habitats to Conserve Native Species with Changing Climate

Ralph W. Tingley III*[1], Dana M. Infante, and Wesley M. Daniel[2]
and Arthur R. Cooper
Department of Fisheries and Wildlife, Michigan State University
1405 South Harrison Road, Suite 318, East Lansing, Michigan 48823, USA

Gordon C. Smith
U.S. Fish and Wildlife Service, Pacific Islands Office
300 Ala Moana Boulevard, Honolulu, Hawaii 96850, USA

Kyle Herreman
Department of Fisheries and Wildlife, Michigan State University
1405 South Harrison Road, Suite 318, East Lansing, Michigan 48823, USA

Abstract.—Decision makers are increasingly tasked with developing conservation strategies that incorporate information on current human threats along with projected influences of climate change. In this study, we integrated a current stream habitat condition assessment of the five largest Hawaiian Islands, with information on ecological stream characteristics and projected change in rainfall to support proactive stream conservation. We used the software Zonation to identify reaches of high conservation value, given both current conditions and projected changes in climate. The results of the spatial prioritization show that low-elevation stream reaches with high potential for supporting multiple species and limited current human disturbance may be especially valuable to conservation. In addition, leeward-draining streams may lose value due to reductions in rainfall, underscoring the importance of managing stream base flow in these areas to conserve native stream organisms. Our results provide guidance for conservation planning in Hawaii and highlight the relevance of regionwide assessments when planning proactive management strategies.

Introduction

Disturbances to stream habitats have contributed to global declines in stream species biodiversity and increasing numbers of species at risk of extinction (Strayer 2006; Helfman 2007; Jelks et al. 2008). Current human land uses within stream catchments have been linked to degradation of habitats and subsequent declines in species biodiversity (Allan 2004; Helfman 2007). In the contiguous United States, a recent assessment reports that at least 80% of stream fish habitats are at some risk of degradation from human land uses (e.g., urban land use, agriculture, and mining, fragmentation by dams) while nearly 25% of streams are at high risk of degradation (Crawford et al. 2016). Climate change has

* Corresponding author: tingleyr@missouri.edu
[1] Current address: School of Natural Resources, University of Missouri, 302 Anheuser-Busch Natural Resources Building, Columbia, Missouri 65211, USA.
[2] Current address: U.S. Geological Survey, Nonindigenous Aquatic Species Program Wetlands and Aquatic Research Center, 7920 NW 71st Street, Gainesville, Florida 32653, USA.

also had documented effects on stream organisms, including leading to global declines in cool- and coldwater fishes (as reviewed by Comte et al. 2013 and Myers et al. 2017), reductions in Chinook Salmon *Oncorhynchus tshawytscha* populations in the Pacific Northwest (Ward et al. 2015), reduced recruitment of Brook Trout *Salvelinus fontinalis* in the northeastern United States (Bassar et al. 2016), and increases in disease prevalence among Brown Trout *Salmo trutta* in streams of Switzerland (Hari et al. 2006). Given the severity of current human land uses and projected increases in temperature and changes in precipitation, the number of stream species at risk of extinction will continue to increase over the next century.

The deleterious effects of climate change combined with the pervasiveness of current stream habitat degradation resulting from human land uses underscore the need to consider both current and future disturbances when prioritizing stream habitats for conservation. Conservation prioritization efforts that only consider current conditions may not be fully effective at identifying priority habitats to protect at-risk species. Because even pristine systems may change in their ability to support assemblages with changing climate (e.g., Battin et al. 2007), accounting for potential effects of climate change on stream habitats and species can support proactive conservation efforts (Palmer et al. 2009; Zeigler et al. 2012). However, a recent review of more than 300 peer-reviewed spatial prioritization studies targeting biodiversity conservation showed that only 15% incorporated climate change into prioritization planning, and of those, only 10% focused on streams (Jones et al. 2016). This underscores the need for additional efforts that support proactive conservation of stream habitats and organisms in the context of climate change.

Across the Hawaiian Islands, perennial streams have been degraded due to human land use and water diversions, which has, in turn, reduced species abundances and threatened the persistence of native and endemic stream organisms (Brasher 2003; Walter et al. 2012). In addition, declines in base flow over the past century across the island system have been partially attributed to reductions in mean annual rainfall (Bassiouni and Oki 2013; Frazier and Giambelluca 2017). Together, these findings indicate that additional declines in rainfall resulting from climate change may have negative consequences for Hawaiian streams and the species they support. In recent years, conservation decision makers (e.g., the U.S. Fish and Wildlife Service, the Hawaii Division of Aquatic Resources) have used available species and habitat data (e.g., the Atlas of Hawaiian Watersheds and Their Aquatic Resources; Parham et al. 2008) and estimates of current habitat condition to prioritize Hawaiian streams for native species conservation, including fish passage, stream, and estuarine habitat restoration (www.fws.gov/pacificislands/hfhp.html). A lack of projected climate data for the Hawaiian Islands has prevented prioritization approaches from incorporating climate change into stream conservation planning, despite its noted importance as an ecological stressor (Hawaii Fish Habitat Partnership 2010). The release of downscaled, projected rainfall data for Hawaii characterizing average conditions at mid- and late 21st century (e.g., Elison Timm et al. 2015) allows for approaches that can support proactive conservation strategies by allowing decision makers to conceptualize both resilient stream systems and those that are more vulnerable to climate change and that may require immediate conservation action.

The goal of this study is to identify Hawaiian stream habitat of high conservation value given both current conditions and projected changes in climate. Our first objective is to identify areas of high conserva-

tion value across the five largest Hawaiian Islands based solely on current conditions determined by human land uses and natural factors important to stream organisms. To meet this objective, we combine an assessment of current stream habitat condition developed by integrating multiple human disturbance data sets with information on the ecological potential of streams to identify streams of highest conservation value. Our second objective is to identify areas of future conservation value by integrating information on projected climate change into prioritization analyses. We then compare results of the current and future prioritizations to identify areas that are likely to retain high conservation value with climate change and assess where different conservation approaches (i.e., preservation versus restoration) may be most effective. Our results will provide important information to decision makers in Hawaii and will highlight the relevance of quantitative regionwide assessments for proactive conservation planning.

Methods

Study area

The study area included all stream reaches and their catchments from the five largest Hawaiian Islands (Kauai, Oahu, Molokai, Maui, and Hawaii; Figure 1). Interactions between eastern trade winds and topographic characteristics of each island control dominant rainfall patterns and result in generally more precipitation on the windward (eastern) sides of islands and less precipitation on leeward (western) facing slopes (Giambelluca et al. 2013). Groundwater contributions to streamflow vary widely; they are substantial in some catchments and minor in others (Lau and Mink 2006; Izuka et al. 2015). Common human disturbances to stream catchments and channels include agriculture and urban land use, as well as water diversions for irrigation and other purposes (Brasher 2003).

Native Hawaiian stream organisms considered in this study include decapods, fish, and snails (Table 1), all but one of which are endemic to the island chain. All of these organisms are amphidromous; they hatch in streams, spend a short time (weeks to months) developing in the marine environment, and then migrate upstream to mature and reproduce (Nishimoto and Fitzsimons 2006; McDowall 2007). In addition to amphidromous species, coastal fish, most notably Striped Mullet *Mugil cephalus* and flagtails (*Kuhlia xenura* and, less commonly, *K. sandvicensis*), move to and from freshwater stream habitats (Nishimoto et al. 2007; McRae et al. 2011). Several of these organisms are also a traditional food source of Hawaiian people and are regularly harvested for consumption (Brasher 1997). Due to their limited distributions and recent population declines, the International Union for Conservation of Nature lists several of these species as near threatened or vulnerable (World Conservation Monitoring Centre 1996; Cordeiro and Perez 2012; De Grave and Cai 2013).

Study spatial units

The base spatial unit used to assess stream condition and prioritize habitat for conservation was the stream reach, as defined in the Hawaii Fish Habitat Partnership (HFHP) stream layer (see Tingley et al. 2019 for details on development). The HFHP stream layer is a modified version of the 1:24,000 National Hydrography Dataset (NHD 2008; http://nhd.usgs.gov) and includes 11,437 intermittent and perennial stream reaches delineated by hydrologic breaks within the river network (e.g., confluences with other stream reaches, waterfalls, reservoirs, and/or pour points to the marine environment). For this analysis, we reassigned a subset of reaches as perennial (as opposed to intermittent or unclassified in the National Hydrography Dataset) on the islands of Kauai, Oahu, and

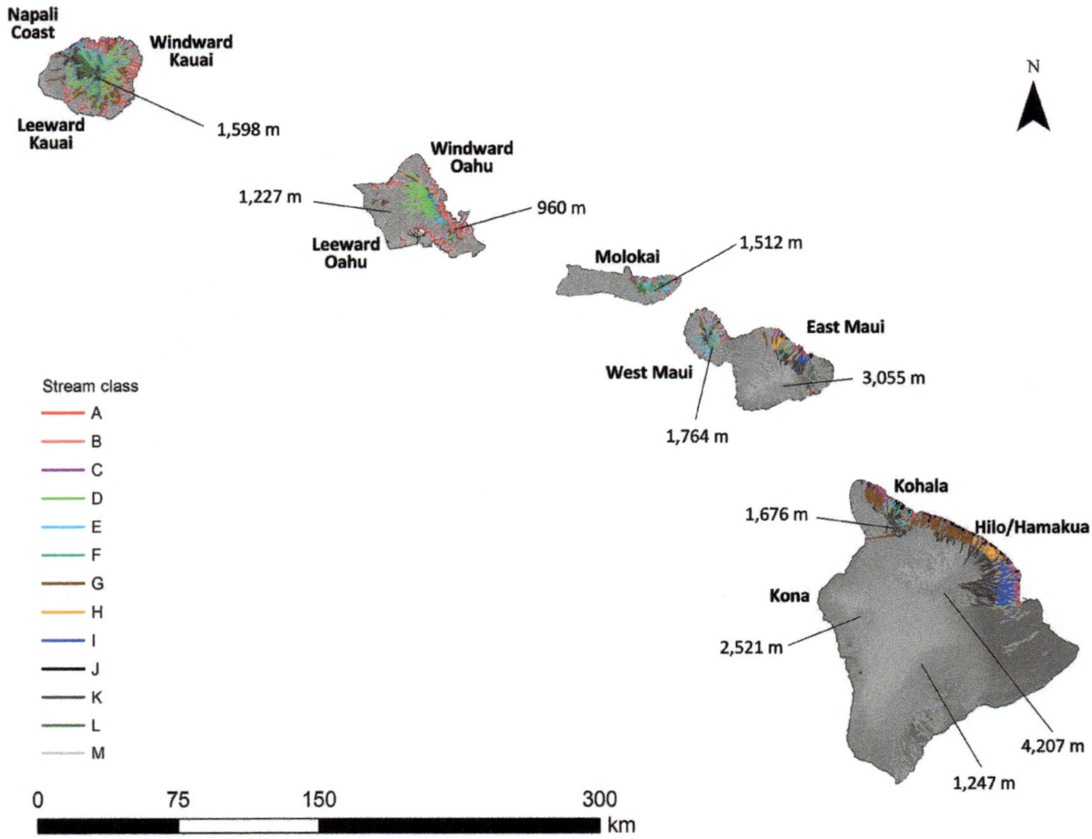

Figure 1. The five Hawaiian islands (from east to west: Kauai, Oahu, Molokai, Maui, and Hawaii Island), with perennial streams included in the spatial prioritization.

Table 1. Native Hawaiian stream taxa and nearshore species considered in the spatial prioritization analysis.

Taxa	Common name	Taxa group
Atyoida bisulcata	ʻopae kalaʻole	Decapod
Lentipes concolor	oʻopu alamoʻo	Fish
Sicyopterus stimpsoni	oʻopu nopili	Fish
Awaous stamineus	oʻopu nakea	Fish
Neritina granosa	hihiwai	Snail
Eleotris sandwicensis	oʻopu ʻakupa	Fish
Stenogobius hawaiiensis	oʻopu nahina	Fish
Macrobrachium grandimanus	opae oehaʻa	Decapod
Kuhlia sp.	āholehole	Fish

Molokai, based on stakeholder feedback. Each stream reach has an associated local catchment generated using a 10-m digital elevation model (see Tingley et al. 2019 for additional details), defined as the portion of the landscape where surface runoff enters the reach directly, as well as a network catchment, defined as the entire area of the landscape draining to a given reach (following Wang et al. 2011). Each reach also has an associated downstream main channel catchment defined as the area of the landscape draining to the portion of the stream that connects a reach to the marine environment and provides a migratory pathway for native species.

Data sets

Stream habitat condition.—We used data sets developed in support of a national assessment of stream fish habitats (2015 National Fish Habitat Assessment; http://assessment.fishhabitat.org) to characterize differences in stream habitat condition across the five islands. An overview of methods used in the Hawaii National Fish Habitat Assessment are described below, with additional details available in Crawford et al. (2016).

The assessment was based on the assumption that more types and greater intensities of human disturbances to stream catchments and channels results in a higher risk of degradation to habitat, which in turn can negatively affect stream organisms (Danz et al. 2007; Esselman et al. 2011; Crawford et al. 2016). Twenty landscape disturbance factors known to influence stream habitat condition in Hawaii were summarized over three spatial extents (local catchment, upstream network catchment, and downstream main channel catchment; Table 2) and were used to generate 19 subindices specific to disturbance types (agricultural land use, urban land use, former plantation lands, point-source pollution, density of ditches associated with water diversions, stream fragmentation from water diversions and dams, and 303(d) listed streams). Subindices were then combined for local catchments, upstream network catchments, and downstream main channel catchments to create three habitat condition indices (HCIs). For this study, we generated an additional index by summing local and upstream network catchment HCI scores to create an upstream HCI (UHCI). We standardized the UHCI from 0 (worst condition) to 1 (best condition) for analysis. For visualization of the risk of habitat degradation across the study area, we used Jenk's natural breaks (De Smith et al. 2015) to assign reaches into five groups based on their UHCI, ranging from very low to very high risk (following Crawford et al. 2016).

Ecological potential.—We used a classification of Hawaiian stream reaches to represent the natural ability of a stream to support native taxa, termed "ecological potential," across the five islands (Tingley et al. 2019; Figure 1). The stream reach classification was developed for perennial reaches in the HFHP stream layer. Stream reaches falling into a single class are defined as having similar ability to support groupings of stream taxa based on important natural landscape features and are broadly defined by prevalent species (Table 3). Stream reach classes were defined using both quantitative and qualitative methods. Eleven reach classes (A–K) were defined based on statistical relationships between natural landscape characteristics (e.g., elevation, channel slope, and mean annual rainfall) and stream taxa. Reach class L, which represents reaches within high-elevation bogs, was also created to further delineate unique Hawaiian stream habitats. Reaches above 750 m with greater than 2,500 mm of mean annual rainfall and more than 5% of a local catchment classified as wetlands were considered to be within high-elevation bogs (Dan Polhemus, U.S. Fish and Wildlife Service, personal communication). We

Table 2. Categories of disturbance and individual landscape disturbance variables used to generate the upstream habitat condition index. Publication dates for landscape data sets developed for the Coastal Change Analysis Program varied by island but represent most recent data available at the time that this analysis was conducted. This table was modified from Crawford et al. (2016).

Category	Variable	Units	Date	Source
Agricultural land use				
	Pasture/hay	%	2005/2010/2011	CCAP[a]
	Cultivated crops	%	2005/2010/2011	CCAP[a]
Urban land use				
	Developed (open)	%	2005/2010/2011	CCAP[a]
	Developed (impervious surface)	%	2005/2010/2011	CCAP[a]
	Population density	#/km²	2010	U.S. Census[b]
	Road density	km/km²	2014	U.S. Census[b]
	Utility pipeline density	m/km²	1983	Hawaii OP[c]
	Percent of catchment covered by golf courses	%	1993	Hawaii OP[c]
Former plantations				
	Percent of catchment that was once used for pineapple production	%	1989	Hawaii OP[c]
	Percent of catchment that was once used for sugarcane production	%	1989	Hawaii OP[c]
Point source pollution				
	Quarry density	#/km²	2003	USGS MRP[d]
	Comprehensive Environmental Response, Compensation, and Liability Information System site density	#/km²	2014	EPA[e]
	Permit Compliance System site density	#/km²	2014	EPA[e]
	Toxic release inventory site density	#/km²	2014	EPA[e]
	Underground injection well density	#/km²	2010	Hawaii DOH[f]
Density of ditches				
	Ditch density	m/km²	2004	Hawaii DAR[g]
Stream fragmentation				
	Stream road crossing density	#/km²	2014	US Census[b]
	Dam density	#/km²	2010	ACOE[h]
	Ditch intersection density	#/km²	2004	Hawaii DAR[g]
303d listed stream				
	Percent of upstream river network classified as 303(d) stream with measured total maximum daily load	%	2012	EPA[i]

Table 2. Continued.

[a] Coastal Change Analysis Program, www.csc.noaa.gov/digitalcoast/data/ccapregional/.
[b] Topologically Integrated Geographic Encoding and Referencing, www.census.gov/geo/maps-data/data/tiger-line.html.
[c] Hawaii Office of Planning, http://planning.hawaii.gov.
[d] U.S. Geological Survey Mineral Resources Program, http://minerals.usgs.gov/)
[e] U.S. Environmental Protection Agency, www.epa.gov/enviro/geospatial-data-download-service.
[f] Hawaii State Department of Health, http://hawaii.gov/health/.
[g] State of Hawaii Division of Aquatic Resources, http://dlnr.hawaii.gov/dar/
[h] U.S. Army Corps of Engineers, www.usace.army.mil.
[i] U.S. Environmental Protection Agency, http://health.hawaii.gov/cwb/site-map/clean-water-branch-home-page.

Table 3. Stream reach classes, associated taxa, and taxa association weightings. Classes that did not support amphidromous species but that represent unique habitats still important for conservation (e.g., reaches within high-elevation bogs) were assigned a taxa association weighting of 1.

Reach type	Description	Taxa association weighting	Associated taxa
A	Coastal	4	o'opu nopili, o'opu nakea, hihiwai, o'opu 'akupa, o'opu nahina, opae oeha'a, āholehole
B	Low-gradient downstream channel at low elevation	3	o'opu nakea, opae oeha'a, o'opu 'akupa, āholehole
C	High-gradient downstream channel at low elevation	4	o'opu alamo'o, o'opu nopili, o'opu nakea, hihiwai, o'opu 'akupa
D	Low gradient, moderate elevation, potential high water table	3	o'opu nopili, o'opu nakea
E	High gradient, moderate elevation, potential high water table	4	'opae kala'ole, o'opu alamo'o, o'opu nopili, o'opu nakea, hihiwai
F	High gradient, high elevation, potential high water table	2	'opae kala'ole
G	Moderate to high elevation, low rainfall	0	No taxa
H	Moderate to high elevation, moderate rainfall	2	'opae kala'ole
I	Moderate to high elevation, high rainfall	3	'opae kala'ole, o'opu alamo'o
J	Terminal falls at low elevation	3	'opae kala'ole, o'opu alamo'o
K	Headwater reaches	1	No taxa
L	Reaches within high elevation bogs	1	No taxa
M	Intermittent reaches	0	No taxa

also included intermittent reaches (class M) from the HFHP stream layer to connect upstream perennial reaches to the marine environment in stream networks that provide temporary migratory pathways that occur during flow events.

Rainfall data.—We used the highest-resolution downscaled projected rainfall data available for the Hawaiian Islands at the time of the study to estimate future annual and dry season rainfall and to characterize how the ecological potential of stream reaches may change with climate. We chose these two variables because climate projections call for drier conditions in many regions of Hawaii, which can, in turn, reduce base flow and decrease available habitat for stream organisms. Statistically downscaled rainfall data generated from 32 general circulation models (250 m resolution; Elison Timm et al. 2015) for the wet (November–April) and dry (May–October) seasons were available from 1978 to 2007 (current period), and the percent change in rainfall from the current period was projected at mid- (2041–2071) and late century (2071–2099) using both moderate- and high-emission scenarios (representative concentration pathway [RCP] 4.5 and 8.5, respectively; IPCC 2013). For each emissions scenario at each time period, we used ArcMap 10.1 to summarize projections of percent change in rainfall for the wet and dry seasons to 10 m raster cells and attributed values to local catchments using the Zonal Statistics as Table tool. We aggregated these values in upstream network catchments using the aggregation procedure described in Tsang et al. (2014). We then calculated changes in projected annual and dry season rainfall totals under both emissions scenarios at both the mid- and late-century time periods relative to the current time period defined in Elison Timm et al. (2015).

We also used current month-year rainfall data from 1992 to 2007 to assess the relative variability in current annual and dry season rainfall for each stream reach. We first attributed monthly rainfall data spatially interpolated from rainfall gauges across the islands by Frazier et al. (2016) to local catchments for each month in each year and added them to calculate mean annual and dry season rainfalls. We then aggregated mean annual and dry season rainfalls in the upstream network catchment for each year. Finally, we calculated the average and standard deviation of annual and dry-season rainfalls for each upstream network catchment to characterize variability in precipitation for Hawaiian streams (for additional details on use within the spatial prioritization analysis, see Identifying areas of future high conservation value).

Identifying areas of current high conservation value

We used the spatial prioritization software program Zonation version 4.0 (Moilanen et al. 2014) to assess and rank current conservation value of stream reaches across the five islands. With Zonation, the conservation value of spatial units is based on the presence of conservation features, which may be important habitats or the presence of species of interest. Zonation uses a complementarity cell-removal approach that generates a balanced representation of conservation features across the study region by iteratively removing the spatial unit that results in the lowest aggregate loss of conservation value (Di Minin et al. 2014). In our study, we ranked conservation value of the spatial units of interest, stream reaches, using five steps (Figure 2).

In step 1, we generated the conservation targets of interest. We generated 13 independent binary raster layers representing each reach class. Raster cells that represented reaches of the class of interest were assigned a value of 1 while all other cells were assigned a value of 0. While Zonation input layers are raster-based, we used the planning unit feature to iteratively remove stream reaches versus individual cells.

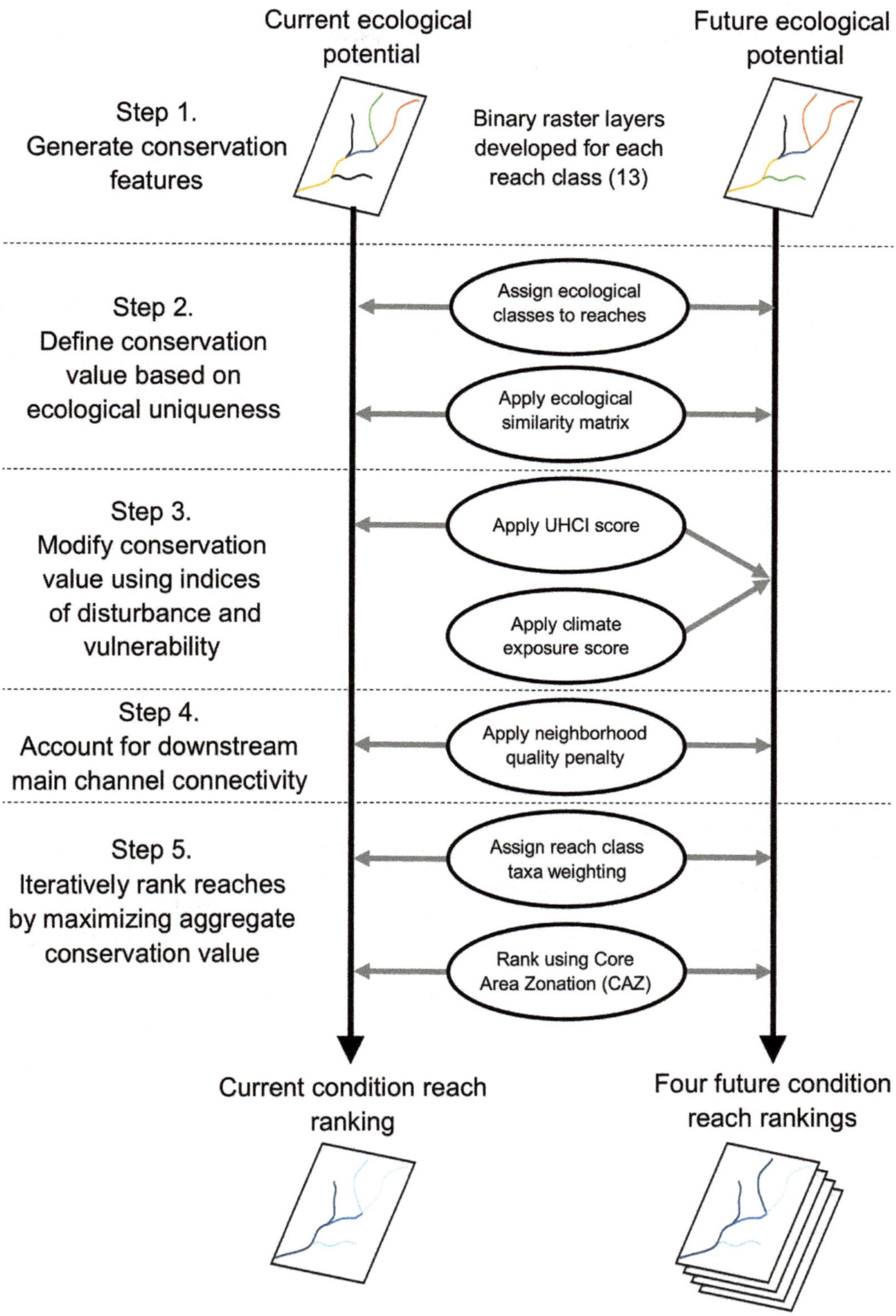

Figure 2. Spatial prioritization approach for the ranking of stream reaches under current conditions and taking into account projected changes in rainfall.

In step 2, we defined conservation feature value for each reach based on its ecological uniqueness. We considered the 13 reach classes described above to be ecologically unique habitats across the islands. Following Leathwick et al. (2010), we then developed a measure of biological similarity for reach classes based on prevalent native stream taxa (Table 3). Similarity values assigned to pairs of stream reach classes were proportions ranging from 0 (fully distinct prevalent taxa) to 1 (all prevalent taxa shared). Using the community analysis option in Zonation (Moilanen et al. 2014), each raster layer was transformed so that all zero values were assigned a value dependent on their similarity to the class of interest. Therefore, reaches assigned to a class that were similar to a given raster layer's class had a higher value than those classes that are less similar to the raster layer's class. This step ensured that rare habitats, based on their potential ability to support unique assemblages, were appropriately represented in our prioritization (Leathwick et al. 2010).

In step 3, we modified conservation feature values based on the relative risk of degradation to the reach from landscape and stream channel disturbance. In the Zonation analysis, a raster layer of UHCI scores was multiplied by conservation values in each of the 13 class layers to achieve "modified" conservation values. For example, if a reach had a conservation value of 0.4 and a UHCI score of 0.5, the modified conservation value would be 0.2. In this way, the conservation value of stream habitat likely to be degraded by disturbance was reduced in the analysis.

In step 4, we accounted for the importance of high-quality habitat within the downstream main channel to the conservation value of each reach. Again, given the amphidromous life histories of most native species, the conservation value of a reach depends on a species' ability to migrate to and from the marine environment, and the suitability of that migration corridor is partially dependent on habitat condition. We used information on stream network connectivity available in the HFHP stream layer to identify the portion of a stream network connecting the reach to the marine environment. We then used the neighborhood quality penalty option (Moilanen and Kujala 2008) to make the value of a reach dependent on the conservation value of reaches in the downstream network. Following Bond et al. (2014), we set the conservation value of a reach to decline linearly with the proportion of the downstream main channel network that has been iteratively removed during the Zonation analysis, essentially increasing the importance of connected migratory pathways. For reaches with downstream channels modified to such an extent that they are not connected via perennial or intermittent streams to the marine environment (e.g., terminating in agricultural ditches that do not drain to the ocean), we assumed that no viable habitat exists for amphidromous species and used the Area Mask option in Zonation to prevent them from receiving a conservation rank.

In step 5, we ranked stream reaches using the core-area selection (CAZ) procedure. Core-area selection is appropriate for our study because it prioritizes stream reaches with the greatest amount of conservation features in the best condition, balancing the uniqueness of an individual reach with the relative condition of all reaches of that class across the study region. Because stream conservation efforts in Hawaii often focus on native amphidromous organisms, we upweighted the value of classes with higher numbers of these taxa (Table 2) prior to running CAZ. In analyses that use reach classes as their conservation features, weighting is an important component of the Zonation algorithm (Moilanen et al. 2014) that allows the planners to prioritize habitats with greater species richness (e.g., Leathwick et al. 2010). We assigned reach classes with five or more prevalent taxa a value of 4, classes with two to four a value of 3, and classes with only one a value of 2. Classes that did not support amphidromous species but that represent unique habitats still important for conservation were assigned a value of 1. Intermittent streams were assigned a value of 0 because we assumed that these streams were least suitable for native Hawaiian stream organisms. Final results included a ranking of each reach, with the reach ranking 100% having the highest current conservation value and the reach with 0% having the lowest current conservation value.

Identifying areas of future high conservation value

We generated four additional rankings that accounted for changes in rainfall under four future projections: moderate emissions at mid-century, high emissions at mid-century, moderate emissions at late century, and high emissions at late century. The same five steps used to rank reaches under current conditions were used for each of the four future projections, with differences described below.

Reaches in four of 13 classes used in our analysis were defined by differences in mean annual rainfall in the upstream catchment. Three are moderate- to high-elevation reaches, likely with limited groundwater contributions to stream flow, differing only in mean annual rainfall (G = low, H = moderate, and I = high; Table 3). Reaches in class L are high-elevation streams currently classified as headwater bogs. We considered these classes most likely to experience changes in ecological potential with changes in rainfall. Therefore, we altered an individual reaches' class based on projected changes in mean annual rainfall in the upstream catchment and used resulting classes as the conservation features for each of the four future rankings.

Because strong deviations from current climate could yield novel conditions that may negatively affect stream organisms, we identified reaches projected to experience substantial declines in rainfall. We determined whether projected mean annual or dry season rainfall was greater than 1 SD from the current (1992–2007) mean generated from the monthly rainfall raster data set. Previous studies have used deviation from current mean rainfall, including absolute change (e.g., Foden et al. 2013) and occurrence outside measures of interannual variability (e.g., 2 SD; Beaumont et al. 2011; Bush et al. 2014) to determine which species or habitats are projected to be exposed to novel or extreme conditions. If a reach was projected to have a mean annual or dry season rainfall more than 1 SD less than the current mean or the reach was projected to have a shift in class associated with declining rainfall, we assigned a "climate exposure score" of 0.5; all other reaches were assigned a value of 1. We then multiplied the climate exposure score by the UHCI score and used the resulting value to modify the conservation value of each reach in step 3. We then completed steps 4 and 5 to develop future projection rankings.

Given that conservation prioritization in Hawaii is often considered in the context of the entire stream network found within an individual watershed (e.g., HFHP 2010), results of our future projection rankings were summarized within existing watersheds used for spatial planning (State of Hawaii Commission of Water Resource Management 2008). We first assessed the overlap in areas of high conservation value under both emissions scenarios for each future time period, considering reaches ranked in the top 10% under both emission scenarios to be "high-priority reaches." We then calculated the percentage of each watershed's stream network that was considered high priority at both the mid- and late century. Finally, to assess how specific actions may differ if proactive approaches are implemented using projected rainfall data, we calculated the overall change in the percentage of high priority reaches from current conditions at the mid-century time period.

Results

Assessment of the current condition of Hawaiian streams

When applied across the study area, the UCHI scores indicate that the relative risk of degradation to Hawaiian streams varies by island (Figure 3). The island of Hawaii has the highest percent of total reaches at high or very high risk of degradation (21.0%), but the majority occur in intermittent stream networks draining the northwest portion of the island. Many perennial streams on the island of Oahu are also severely impacted by human disturbance, with 17.1% of all reaches considered at high or very high risk (Table 4). Coastlines and other areas of lower elevation (i.e., central Oahu and Maui) tend to have high or very high risk of disturbance, a result of urban development and agricultural land use in these areas. Mountainous regions, especially along coastlines that are difficult to access and have sharp gains in elevation (e.g., the Napali coast of Kauai, northeast Molokai, and the Kohala region of Hawaii Island) have very low risks of stream habitat degradation overall.

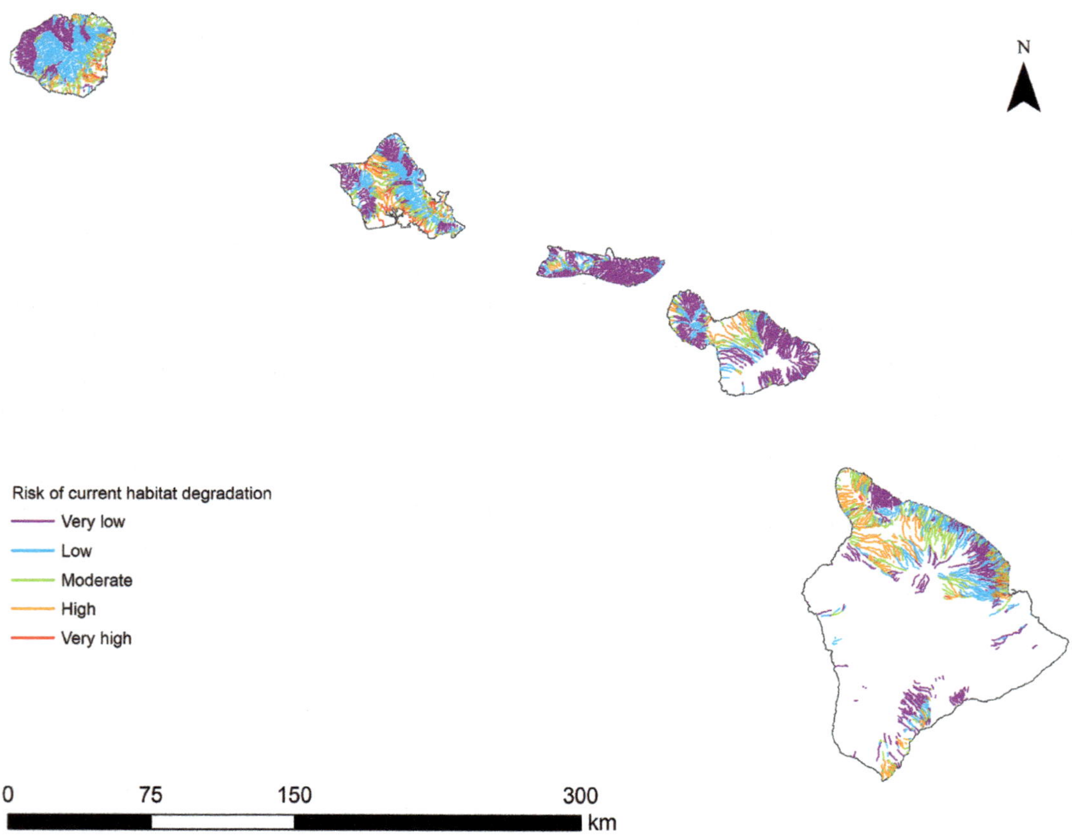

Figure 3. Upstream habitat condition index scores for the five main Hawaiian islands. Continuous values were grouped into five risk of current habitat degradation categories using Jenk's natural breaks.

Table 4. Percent of stream reaches by island in each risk of habitat degradation category. High and very high categories were combined for comparison.

	Risk of degradation			
Island	Very low	Low	Moderate	High/very high
Hawaii	32.6	24.9	21.5	21.0
Maui	63.3	17.7	11.7	7.4
Molokai	82.7	12.1	3.4	1.8
Oahu	29.4	36.9	16.7	17.1
Kauai	29.7	47.8	13.2	9.3

Areas of high conservation value under current conditions

Conservation value of stream reaches based solely on current conditions varies considerably across the study region (Figure 4). However, two trends observed across all islands were the higher ranking of low-elevation reaches and reaches within a short distance to the marine environment, even if they occur in catchments where reaches at higher elevations are ranked relatively low (e.g., windward Kauai). Regions identified in the condition assessment as having a low risk of degradation, such as the Napali coast of Kauai, had many reaches ranked in the top 5% and top 10%. In larger catchments within regions of limited disturbance, reaches ranked in the top 10% occur at high elevations and include the majority of the stream network. The large number of reaches in the top 10% on windward slopes of East and West Maui may result from multiple factors, including a high percentage of perennial reaches within these regions, presence of less common reach classes, and relatively low UHCI scores. Despite a large number of perennial reaches along the southern extent of the Hilo/Hamakua coast, a relatively small number are ranked in the top 10%.

Climate change-driven changes in prioritization

Changes in reach class and climate vulnerability of stream reaches.—The total stream length in each class is anticipated to change under both climate scenarios at each time period (Table 5), suggesting that there may be a shift in the ability of many reaches to support current assemblages of stream organisms. The total length of stream reaches classified as G, H, and L decrease with changes in mean annual rainfall while classes I and K increase for all four future projections. The greatest absolute change in total reach class length occurred to class I under the RCP 4.5 climate scenario (118 km gained), but the greatest change relative to current reach length occurred to class L (bog reaches) under the RCP 8.5 climate scenario (–31%). Changes in average annual or dry season rainfall were at least 1 SD from current estimates of mean annual rainfall for many stream reaches across the study area (Table 5). Declines greater than 1 SD from current annual and dry-season rainfall were more pronounced under at the late-century time period and under the high-emissions scenario (RCP 8.5).

High value conservation areas at the mid- and late-century time periods.—Catchment summaries of high-priority reaches at the mid-century time period show that each of the five islands will continue to support stream networks comprised of reaches of high conservation value and that many of the reaches ranked as high value under current conditions will remain so through mid-century (Figure 5).

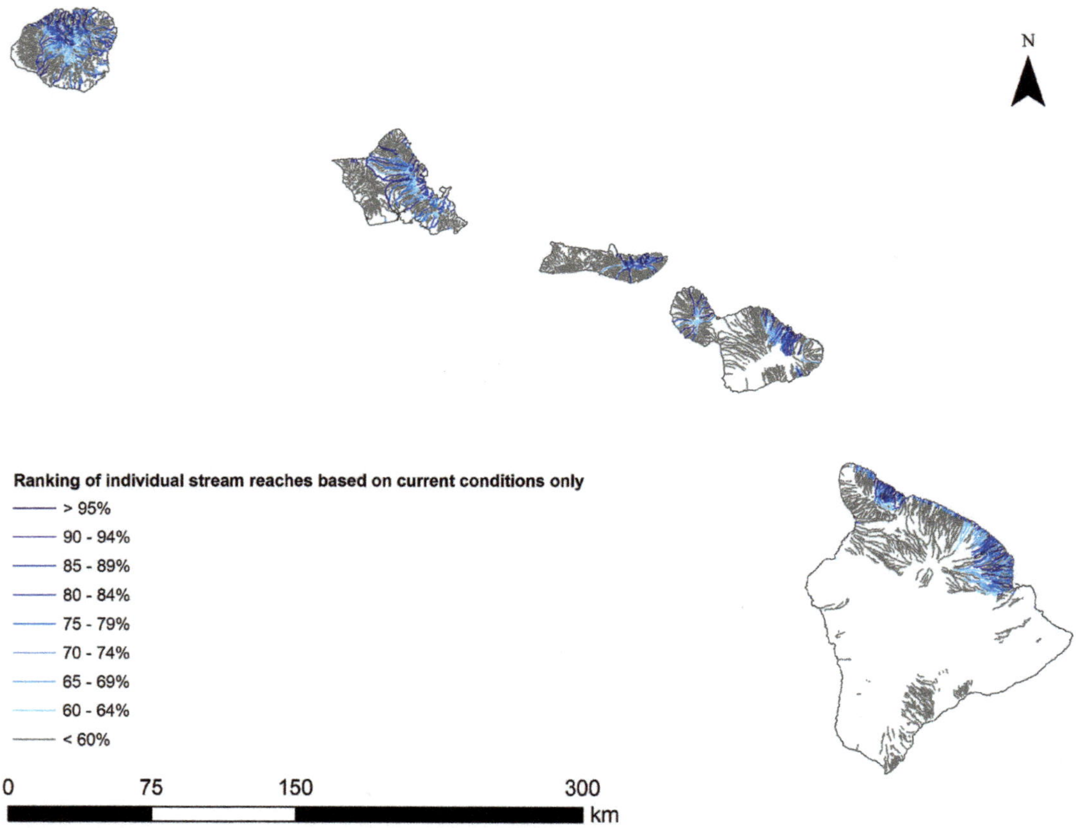

Figure 4. Conservation value of individual stream reaches based on current conditions. Percentages indicate overall value of an individual reach to conservation, with 100% being the most valuable reach and 0% being the least valuable. Priority was based on uniqueness of habitat, higher potential stream biodiversity, less human disturbance to the catchment and channel, and conservation value of reaches in the downstream migration corridor.

Table 5. Changes in stream reach length for each class under each emission scenario and time period. Also shown are total lengths of stream reaches projected to be outside 1 SD of the current mean annual or dry-season rainfall (negative and positive projected changes are shown). RCP = representative concentration pathway.

Time period	Emissions scenario	Class G (km)	Class H (km)	Class I (km)	Class K (km)	Class L (km)	Annual rainfall (−/+ 1 SD; km)	Dry-season rainfall (−/+ 1 SD; km)
Current		849	321	418	783	233		
Mid-century								
	RCP 4.5	−24	−71	+96	25	−25	238/334	521/559
	RCP 8.5	−24	−77	+102	36	−36	477/316	1,334/547
Late century								
	RCP 4.5	−40	−77	+118	41	−41	573/570	909/589
	RCP 8.5	−23	−87	+110	73	−73	2,419/415	3,174/754

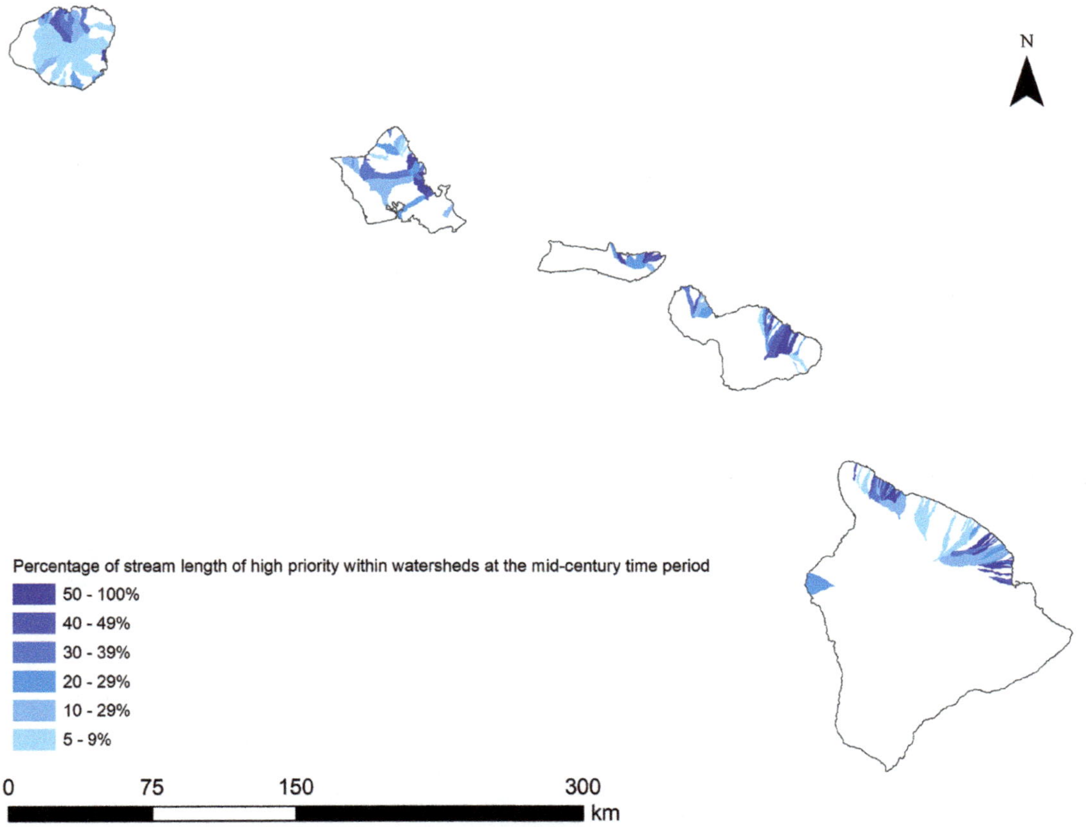

Figure 5. Percent of high-priority reaches (top 10% in both emission scenarios) summarized within watersheds at the mid-century time period. Priority was based on uniqueness of habitat, higher potential stream biodiversity, less human disturbance to the catchment and channel, projected exposure to declines in dry-season or annual rainfall, and conservation value of reaches in the downstream migration corridor.

Regions where perennial streams are dominant and where risk of degradation is low have a large percentage of stream reaches considered high priority. The majority of catchments with greater than 50% of the stream network consisting of high-priority reaches occur in smaller catchments draining slopes with steep elevation gains over short distances, many of which occur on the windward slopes of Oahu. Northern draining streams on Kauai have relatively high numbers of high-priority reaches, as do many larger catchments on windward slopes of East Maui. Other clusters of adjacent catchments with substantial proportions of high-priority reaches in the stream network occur on windward Kauai, East Maui, and the Kohala region of Hawaii. Few catchments draining leeward slopes have greater than 10% of the stream network consisting of high-priority reaches, despite some regions (e.g., West Maui, the Napali coast of Kauai) having reaches ranked in the top 10% under current conditions.

Overall, the ranking of stream reaches at the late-century time period are similar to those of the mid-century, but some regional differences are apparent (Figure 6). For example, catchments with more than 5% of the stream network comprised of

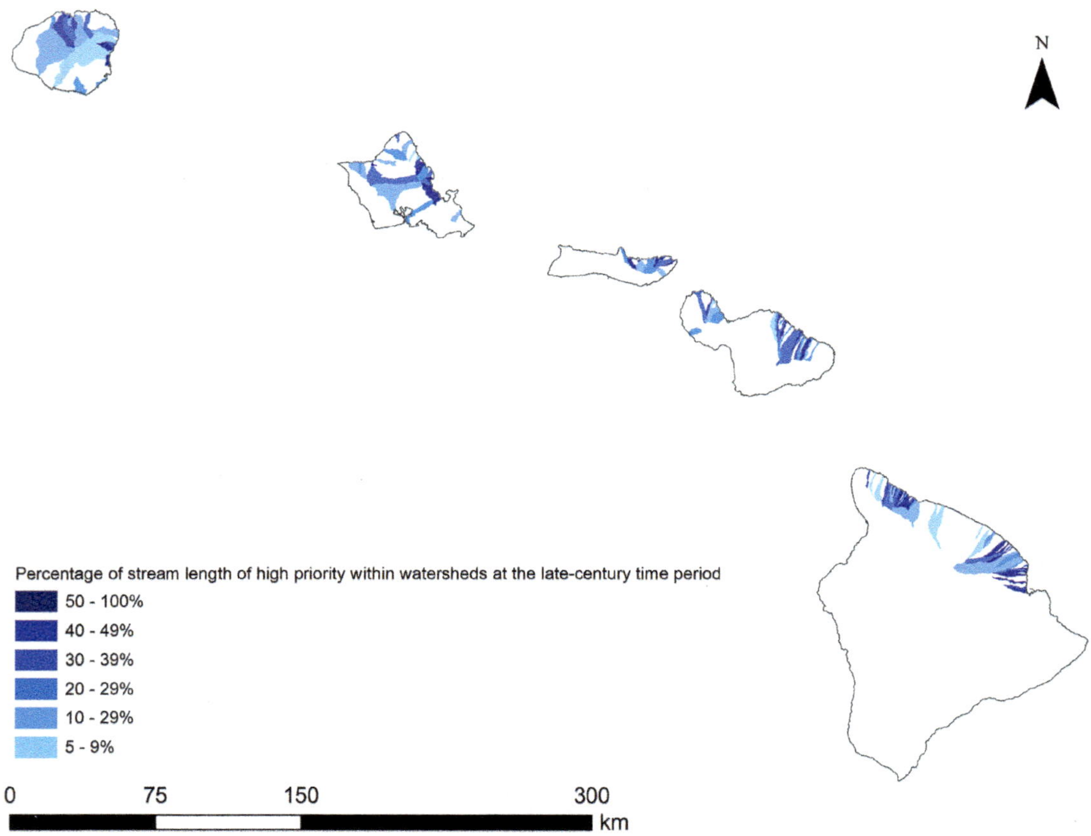

Figure 6. Percent of high-priority reaches (top 10% in both emission scenarios) summarized within watersheds at the late-century time period. Priority was based on uniqueness of habitat, higher potential stream biodiversity, less human disturbance to the catchment and channel, projected exposure to declines in dry season or annual rainfall, and conservation value of reaches in the downstream migration corridor.

high-priority reaches on the Napali coast of Kauai declined from mid- to late century. This is likely the result of total dry season or annual rainfall moving beyond historical norms, given that most reaches in this region do not fall within classes defined by differences in mean annual rainfall (Figure 1). At both mid- and late century, clusters of adjacent catchments on northern Kauai, as well as the windward slopes of Oahu and East Maui, contain high proportions of high-priority reaches. These results suggest these regions may be minimally affected by changes in annual or dry season rainfall by the end of the 21st century.

Changes in areas of high conservation value with changing climate

Comparisons of percentages of high-priority reaches within watersheds between rankings generated using mid-century climate projections and those based solely on current conditions indicate that accounting for changes in climate alters the identification of areas of high conservation value. Both increases (up to 100%) and decreases (as much as –82%) in the total percentage of high-priority reaches within watersheds occur when accounting for mid-century projected changes in rainfall (Figure 7). Most watersheds with increases in high-

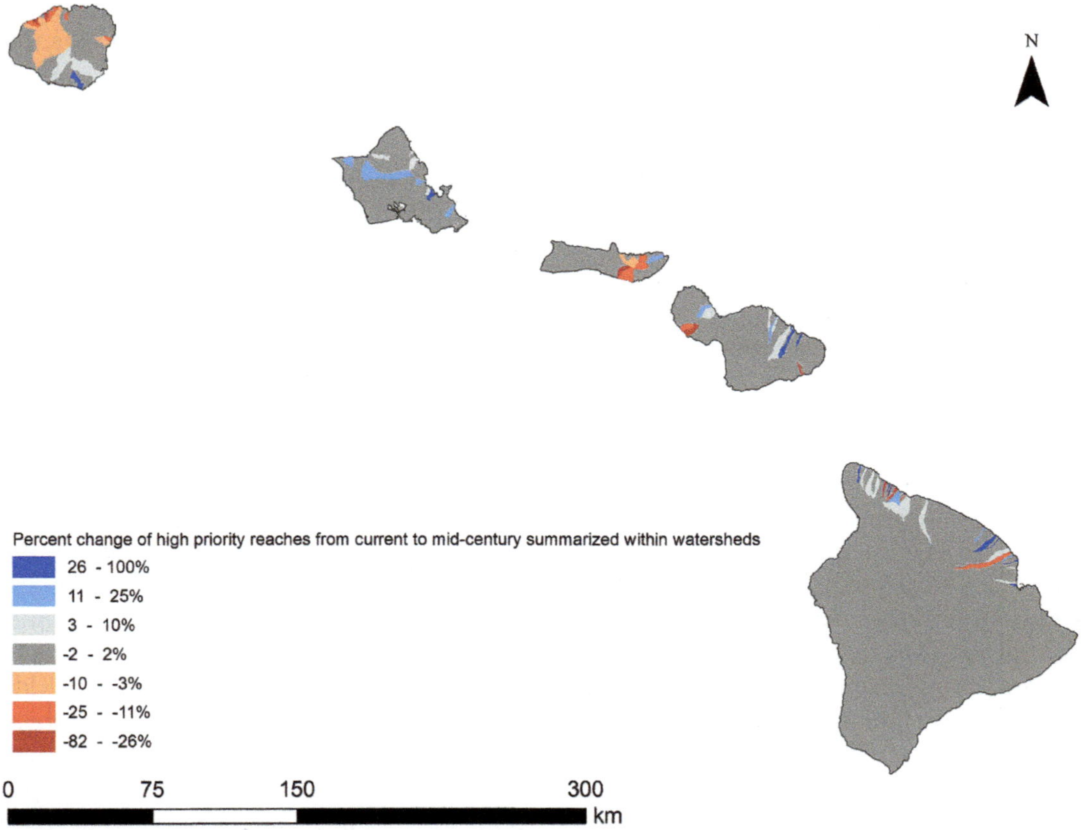

Figure 7. Change in the total percent of high priority reaches within catchments from the current to mid-century time period. Watersheds with negative values contain a lower percentage of high priority reaches when mid-century rainfall projections are incorporated into the analysis, while catchments with positive values contain a higher percentage of high-priority reaches when mid-century rainfall projections are incorporated.

priority reaches are located on windward-facing slopes, with the exception of Oahu where some number of leeward draining streams increase in the percentage of high-priority reaches. Small watersheds on leeward draining slopes of multiple islands have the greatest overall declines in percent high-priority reaches. However, several windward-draining streams on Molokai in areas of very low risk of degradation show declines in the percentage of high-priority stream reaches up to 25%, which is likely the result of both a reduction in the presence of unique stream habitats (i.e., reaches within high-elevation bogs) and mean dry or annual rainfall falling outside historical norms.

Discussion

Proactive strategies that account for both current human disturbance and projected climate change are needed to effectively conserve stream ecosystems (Palmer et al. 2009; Zeigler et al. 2012). In this study, we identified areas of high-conservation value across Hawaiian streams and assessed how prioritization of conservation actions may vary when projected changes in climate are accounted for. Our use of regionwide data sets describing current anthropogenic disturbance and ecological potential allowed for development of a prioritization approach that integrates uniqueness of habitat with overall risk of habitat degradation. By accounting for the quality of habitat connecting a given reach to the marine environment, we also ensured that the prioritization approach has relevance to the conservation of native amphidromous species. The four rankings developed using future rainfall projections indicate that areas of high conservation value will continue to occur across all five islands regardless of the climate scenario or time period considered. While most regions of current high conservation value are not vulnerable to changes in annual or dry season rainfall, several leeward perennial streams of high conservation value may be exposed to substantial declines in rainfall and may experience changes in habitat. Overall, our results help decision makers visualize current and future areas of conservation value and, when paired with local knowledge, can provide guidance on where protection or restoration of stream habitats may be most beneficial to the conservation of Hawaiian streams.

Factors important to conserving stream habitat in Hawaii

Our analysis was based on factors important for managers in Hawaii to consider when prioritizing conservation action in streams: the uniqueness of habitats, the current condition of habitats, and the current condition of migration corridors between stream habitats and the marine environment. Our use of the reach classification allowed for consideration of the native stream species community, which is a priority for managers in Hawaii, who are often focused on the entire species assemblage and the protection or restoration of native stream ecosystems (HFHP 2010). By accounting for similarities in prevalent taxa across classes, we also were able to control for the overrepresentation of similar stream habitats in the prioritization rankings, resulting in a balanced representation of stream conservation value across the five islands.

The use of the UHCI scores in assigning conservation value of individual stream reaches was an important step in our analysis. Integrating stream habitat condition scores ensured that we identified which stream reaches may be in best condition and which may be in most need of conservation action given the relative degradation of other similar habitats. For example, the prioritization rankings indicate that many low-elevation stream classes (i.e., A, B, and C) are of high conservation value. Not only are these classes associated with high numbers of unique taxa that are not likely to migrate far upstream (i.e., *Stenogobius hawaiiensis* and *Macrobrachium grandimanus*; Kido 2008; Parham et al. 2009), but many of these reaches are at high or very high risk of degradation from local and upstream anthropogenic disturbances. This suggests that those low-elevation reaches that have limited disturbance are especially important to Hawaiian stream conservation and are reflected as such in the prioritization rankings.

We accounted for the importance of the condition of migration corridors between stream habitats and the marine environment by making the value of an individual reach dependent on the conservation value of stream reaches in the downstream main channel network. Successful migration to and from the marine environment is critical to the survival of most native Hawaiian stream species (Nishimoto and Fitzsimons 2006; McDowall 2007), but downstream human disturbance can limit migration success through habitat degradation. Reductions in streamflow or complete streambed desiccation (Parham et al. 2009; Gorbach et al. 2012), high water temperatures (Brasher et al. 2004), presence of invasive species (Brasher et al. 2006), and entrainment in stream diversions (Parham et al. 2009) are all associated with increased human disturbance to the catchment and/or channel and may limit the successful migration of Hawaiian stream species. Because we accounted for the conservation value of migratory corridors, final rankings included many connected lengths of streams of high conservation value throughout watersheds that are less likely to be heavily impacted by human disturbance, thus increasing the likelihood that upstream habitats are available to migrating species. This result provides relevant information to decision makers in Hawaii, where conservation efforts are often focused on "Mauka to Makai," or ridge to reef, initiatives.

Planning for the future: proactive conservation strategies

Accounting for climate change in conservation planning increases the likelihood that implemented conservation actions will protect species and habitats into the future. Despite this understanding, it is still uncommon for spatial prioritizations of stream ecosystems to incorporate climate change (Jones et al. 2016). Incorporating potential changes in rainfall into our spatial prioritization effort allowed us to identify reaches and catchments in Hawaii that will likely retain conservation value through the next century. Our results can therefore increase the confidence of decision makers in Hawaii that conservation action taken in the short term will having lasting benefits to stream ecosystems while also allowing them to conceptualize how climate change will impact the Hawaiian Islands. In addition, implementing proactive conservation strategies requires that decisions be made despite uncertainty in the magnitude of future greenhouse gas emissions and environmental change (Hagerman et al. 2010). By including climate projections under two emission scenarios, we accounted for some potential uncertainty associated with future change, as those reaches that are highly ranked under both emission scenarios are the most likely to support native species regardless of differences in the magnitude of future greenhouse gas emissions.

The comparison of prioritization rankings under current conditions to rankings that incorporate future changes in rainfall can provide guidance on which specific conservation actions may be most effective for a given catchment or region (Lawler 2009). For example, if several reaches within a catchment are of high conservation value now but are projected to decline in conservation value in the future, the restriction of further water withdrawals can help retain existing stream habitats and promote climate change resilience. For catchments that contain reaches of high conservation value in both current and future rankings, conservation actions that preserve or protect the upstream catchment and downstream migratory corridor from further anthropogenic disturbance can be prioritized. In instances where streamflow diversions are known to impact catchments with high conservation value in both current and future rankings, for instance in many catchments across West Maui (Cheng 2014), the removal of water diversions can increase available habitat for these systems that are unlikely to experience declines in rainfall. Finally, reaches that have higher conservation value in future rankings compared to current rankings are unlikely to be influenced by reductions in rainfall but may be at current risk from human disturbances. In these systems, including many catchments along windward Oahu, conservation actions that restore or rehabilitate stream habitats can be prioritized.

Data and information needs for future conservation

We used the finest-resolution, statistically downscaled climate data that were available for the Hawaiian Islands to identify areas of high-value conservation under future climate conditions. Statistical downscaling requires the assumption of "stationarity," meaning that current factors that control climate in a region remain the same through time (Timm and Diaz 2009). Dynamic downscaling does not assume stationarity, and projections are available for Hawaii but were not available at fine resolutions (<3 km) across all islands at the time of this study (Zhang et al. 2016). While general patterns of change are in agreement, the magnitude of change between statistical and dynamic downscaled projections vary. For instance, reductions in annual and dry season rainfall in dry regions projected in the statistically downscaled data set are less extreme in the dynamically downscaled data set. In addition, Elison Timm et al. (2015) notes that confidence in dry-season rainfall projections are much lower than wet-season projections, and that caution should be taken when considering the magnitude of change. Conveying such uncertainties associated with projected rainfall in the dry season, especially on leeward facing slopes, should be a priority when results are presented, with a discussion of how such uncertainty may affect final rankings. Despite these uncertainties and differences among downscaled models, decision makers should have confidence that windward-draining reaches of high conservation value have low vulnerability to climate-change driven shifts in mean rainfall given the general agreement in overall regional patterns between the two data sets.

Additional detail on water withdrawals and stream channel modifications would be valuable for refining our analysis and providing additional information for prioritization, as they represent major stressors to many Hawaiian streams. A statewide stream diversion data set that includes the magnitude of water withdrawals, estimated reductions in downstream flow, and the extent to which flow reduction may prevent species movement is not currently available. Such information could be valuable in the development of a new current habitat condition index and could also be used following a landscape-scale prioritization to clearly identify where actions should be implemented to protect areas of high conservation value.

Invasive and nonnative stream species are a major threat to Hawaiian streams (Brasher 2003; Walter et al. 2012), and presence–absence data or distribution models would be valuable for prioritizing stream habitat conservation strategies. We did not include these data in our analysis as they were not made available during the study, but they can still be integrated for strategic planning. For example, in high-value watersheds where most nonnative species are absent but habitat is suitable for invasion, approaches that mitigate the spread of invasive species can be implemented in addition to other conservation actions. Within isolated stream systems or above natural barriers to nonclimbing species, restoration initiatives can include the removal or control of nonnative species in addition to habitat improvement. Because the impacts of nonnative species on native organisms are diverse (e.g., predation, competition, and habitat alteration; as reviewed by Brasher et al. 2003), conservation actions should be tailored to the specific effects of the numerous nonnative species currently in Hawaiian streams. Species distribution models for invasive and nonnative stream organisms (Martin 2018) and a clearer understanding of the role of waterfalls as barriers to nonnative species movement (Martin et al. 2019) have been developed, the results of which have the potential to be used in tandem with our analysis to improve decision making.

Increased study on the effects of changing streamflow regimes on native stream organisms could provide a greater understanding of the vulnerability of individual species to climate change. Our assumption that any substantial reduction in annual or dry-season rainfall will have negative consequences for stream organisms in Hawaii is well founded, as current research suggests that further declines in rainfall will contribute to declines in groundwater recharge and subsequent reductions in stream base flow across the islands (Bassiouni and Oki 2013). However, exposure to climate change is only one component of species vulnerability (Foden et al. 2013), and identifying those species that are most sensitive to change could improve strategic planning.

Conclusion

The effective conservation of stream ecosystems requires that we continue to develop approaches that aid in prioritization of conservation actions in light of projected climate change. In this study, we incorporated the best available information on human disturbance on stream systems in Hawaii to identify areas of conservation value and suggest specific conservation actions. Our results draw attention to potential effects of climate change on Hawaiian streams and provide a template for similar studies in other tropical island systems. We view our work as a building block for future studies that aim to develop a dynamic picture of climate change effects on Hawaiian streams, which can then be used to improve decision making.

Acknowledgments

Financial support for this research was provided through the Hawaii Fish Habitat Partnership. We also acknowledge financial support for travel from the U.S. Geological Survey National Climate Adaptation Science Center (NCASC) through the Science to Action Fellowship and to the National Fish Habitat Partnership for data. We would like to thank members of the Hawaii Fish Habitat Partnership, the Hawaii Division of Aquatic Resources, and the Commission on Water Resources Management's Stream Protection and Management Branch for their participation in project development and the review of results. We would also like to thank Dan Polhemus, Rich MacKenzie, Ayron Strauch, Yin-Phan Tsang, Kim Peyton, and NCASC staff and scientists for insight and advice provided throughout the development of this research, and Abigail Frazier for providing climate data. The findings and conclusions in this article are those of the author(s) and do not necessarily represent the views of the U.S. Fish and Wildlife Service.

References

Allan, J. D. 2004. Landscapes and riverscapes: the influence of land use on stream ecosystems. Annual Review of Ecology Evolution and Systematics 35:257–284.

Bassar, R. D., B. H. Letcher, K. H. Nislow, and A. R. Whiteley. 2016. Changes in seasonal climate outpace compensatory density-dependence in eastern Brook Trout. Global Change Biology 22:577–593.

Bassiouni, M., and D. S. Oki. 2013. Trends and shifts in streamflow in Hawai'i, 1913–2008. Hydrological Processes 27:1484–1500.

Battin, J., M. W. Wiley, M. H. Ruckelshaus, R. N. Palmer, E. Korb, K. K. Bartz, and H. Imaki. 2007. Projected impacts of climate change on salmon habitat restoration. Proceedings of the National Academy of Sciences of the United States of America 104:6720–6725.

Beaumont, L. J., A. Pitman, S. Perkins, N. E. Zimmermann, N. G. Yoccoz, and W. Thuiller. 2011. Impacts of climate change on the world's most exceptional ecoregions. Proceedings of the National Academy of Sciences of the United States of America 108:2366–2311.

Bond, N. R., J. R. Thomson, and P. Reich. 2014. Incorporating climate change in conservation planning for freshwater fishes. Diversity and Distributions 20:931–942.

Brasher, A. M. D. 1997. Habitat use by fish (o'opu), snails (hihiwai), shrimp ('opae) and prawns in two streams on the island of Moloka'i. University of Hawaii, Cooperative National Park Resources Studies Unit, Technical Report 116, Honolulu.

Brasher, A. M. D. 2003. Impacts of human disturbances on biotic communities in Hawaiian streams. Bioscience 53:1052–1060.

Brasher A. M. D., R. H. Wolff, and C. D. Luton. 2004. Associations among land use, habitat characteristics, and invertebrate community structure in nine streams on the island of Oahu, Hawaii. U.S. Geological Survey, Water Resources Investigations Report 03-4256, Reston, Virginia.

Brasher, A. M. D., C. D. Luten, S. L. Goodbred, and R. H. Wolff. 2006. Invasion patterns along elevation and urbanization gradients in Hawaiian streams. Transactions of the American Fisheries Society 135:1109–1129.

Bush, A. A., D. A. Nipperess, D. E. Duursma, G. Theischinger, E. Turak, and L. Hughes. 2014. Continental-scale assessment of risk to the Australian odonata from climate change. PLOS (Public Library of Science) ONE [online serial] 9(2):e88958.

Cheng, C. L. 2014. Low-flow characteristics of streams in the Lahaina District, West Maui, Hawai'i. U.S. Geological Survey, Scientific Investigations Report 2014-5087, Reston, Virginia.

Comte, L., L. Buisson, M. Daufresne, and G. Grenouillet. 2013. Climate-induced changes in the distribution of freshwater fish: observed and predicted trends. Freshwater Biology 58:625–639.

Cordeiro, J., and K. Perez. 2012. Neritina granosa. The IUCN Red List of Threatened Species 2012:e.T189675A1932107. Available: www.iucnredlist.org. (February 2017).

Crawford, S., G. Whelan, D. M. Infante, K. Blackhart, W. M. Daniel, P. L. Fuller, T. Birdsong, D. J. Wieferich, R. McClees-Funinan, S. M. Stedman, K. Herreman, and P. Ruhl. 2016. Through a fish's eye: the status of fish habitats in the United States 2015. National Fish Habitat Partnership. Available: http://assessment.fishhabitat.org/. (February 2017).

Danz, N. P., G. J. Niemi, R. R. Regal, T. Hollenhorst, L. B. Johnson, J. M. Hanowski, and G. E. Host. 2007. Integrated measures of anthropogenic stress in the US Great Lakes basin. Environmental Management 39:631–647.

De Grave, S., and X. Cai. 2013. *Atyoida bisulcata*. The IUCN Red List of Threatened Species 2013:e.T198132A2512737. (January 2017).

De Smith, M. J., M. F. Goodchild, and P. A. Longley. 2015. Geospatial analysis: a comprehensive guide to principles, techniques and software tools. The Winchelsea Press, Winchelsea, UK.

Di Minin, E., V. Veach, J. Lehtomaki, F. M. Pouzols, and A. Moilanen. 2014. A quick introduction to Zonation version 1 (for Zv4). Conservation Biology Informatics Group, Helsinki, Finland. Available: www.researchgate.net/publication/262257846_A_quick_introduction_to_Zonation. (March 2019.)

Elison Timm, O., T. W. Giambelluca, and H. F. Diaz. 2015. Statistical downscaling of rainfall changes in Hawai'i based on the CMIP5 global model projections. Journal of Geophysical Research: Atmospheres 120:92–112.

Esselman, P. C., D. M. Infante, L. Z. Wang, D. Y. Wu, A. R. Cooper, and W. W. Taylor. 2011. An index of cumulative disturbance to river fish habitats of the conterminous United States from landscape anthropogenic activities. Ecological Restoration 29:133–151.

Foden, W. B., S. H. M. Butchart, S. N. Stuart, J.-C. Vié, H. R. Akçakaya, A. Angulo, L. M. DeVantier, A. Gutsche, E. Turak, L. Cao, S. D. Donner, V. Katariya, R. Bernard, R. A. Holland, A. F. Hughes, S. E. O'Hanlon, S. T. Garnett, Ç. H. Şekercioğlu, and G. M. Mace. 2013. Identifying the world's most climate change vulnerable species: a systematic trait-based assessment of all birds, amphibians and corals. PLOS (Public Library of Science) ONE [online serial] 8(6):e65427.

Frazier, A. G., T. W. Giambelluca, H. F. Diaz, and H. L. Needham. 2016. Comparison of geostatistical approaches to spatially interpolate month-year rainfall for the Hawaiian Islands. International Journal of Climatology 12:1459–1470.

Frazier, A. G., and T. W. Giambelluca. 2017. Spatial trend analysis of Hawaiian rainfall from 1920 to 2012. International Journal of Climatology 37:2522–2531.

Giambelluca, T. W., Q. Chen, A. G. Frazier, J. P. Price, Y. L. Chen, P.-S. Chu, J. K. Eischeid, and D. M. Delparte. 2013. Online rainfall atlas of Hawaii. Bulletin of the American Meteorological Society 94: 313–316.

Gorbach, K. R., M. E. Benbow, M. D. McIntosh, and A. J. Burky. 2012. Dispersal and upstream migration of an amphidromous neritid snail: implications for restoring migratory pathways in tropical streams. Freshwater Biology 57:1643–1657.

Hagerman, S., H. Dowlatabadi, T. Satterfield, and T. McDaniels. 2010. Expert views on biodiversity conservation in an era of climate change. Global Environmental Change: Human and Policy Dimensions 20:192–207.

Hari, R. E., D. M. Livingstone, R. Siber, P. Burkhardt-Holm, and H. Guttinger. 2006. Consequences of climatic change for water temperature and Brown Trout populations in Alpine rivers and streams. Global Change Biology 12:10–26.

Hawaii Fish Habitat Partnership. 2010. Hawaii Fish Habitat Partnership Strategic Plan. Available: www.fws.gov/pacificislands/Publications/Hawaii%20FHP%20Strategic%20Plan.pdf. (January 2017).

Helfman, G. S. 2007. Fish conservation: a guide to understanding and restoring global aquatic biodiversity and fishery resources. Island Press, Washington, D.C.

IPCC (Intergovernmental Panel on Climate Change). 2013. Climate change 2013: the physical science basis. Contribution of Working Group I to the fifth assessment report of the Intergovernmental Panel on Climate Change. T. F. Stocker, D. Qin, G. K. Plattner, M. Tignor, S. K. Allen, J. Boschung, A. Nauels, Y. Xia, V. Bex, and P. M. Midgley, editors. Cambridge University Press, Cambridge, UK.

Izuka, S. K., J. A. Engott, M. Bassiouni, A. G. Johnson, L. D. Miller, K. Rotzoll, and A. Mair. 2015. Volcanic aquifers of Hawai'i: hydrogeology, water budgets and conceptual models. U.S. Geological Survey, Scientific Investigations Report 2015-5164, Reston, Virginia.

Jelks, H. L., S. J. Walsh, N. M. Burkhead, S. Contreras-Balderas, E. Diaz-Pardo, D. A. Hendrickson, J. Lyons, N. E. Mandrak, F. McCormick, J. S. Nelson, S. P. Platania, B. A. Porter, C. B. Renuad, J. J. Schmitter-Soto, E. B. Taylor, and M. L. Warren. 2008. Conservation status of imperiled North American freshwater and diadromous fishes. Fisheries 33:372–407.

Jones, K. R., J. E. M. Watson, H. P. Possingham, and C. J. Klein. 2016. Incorporating climate change into spatial conservation prioritisation: a review. Biological Conservation 194:121–130.

Kido, M. H. 2008. A persistent species assemblage structure along a Hawaiian stream from catchment-to-sea. Environmental Biology of Fishes 82:223–235.

Lau, L. S., and J. F. Mink. 2006. Hydrology of the Hawaiian Islands. University of Hawai'i Press, Honolulu.

Lawler, J. J. 2009. Climate change adaptation strategies for resource management and conservation planning. Annals of the New York Academy of Sciences 1162:79–98.

Leathwick, J. R., A. Moilanen, S. Ferrier, and K. Julian. 2010. Complementarity-based conservation prioritization using a community classification, and its application to riverine ecosystems. Biological Conservation 143:984–991.

Martin, B., Y.-P. Tsang, R. W. Tingley, III, H. Clilverd, and D. M. Infante. 2019. Characterizing natural barriers to nonnative stream fauna in Hawai'i. Pages 83–103 *in* R. M. Hughes, D. M. Infante, L. Wang, K. Chen, and B. F. Terra, editors. Advances in understanding landscape influences on freshwater habitats and biological assemblages. American Fisheries Society, Symposium 90, Bethesda, Maryland.

Martin, B. T. 2018. Characterizing the stream environments associated with introduced species across spatial scales in Hawai'i. Master's thesis. University of Hawai'i at Mānoa, Honolulu.

McDowall, R. M. 2007. On amphidromy, a distinct form of diadromy in aquatic organisms. Fish and Fisheries 8:1–13.

McRae, M. G., L. B. McRae, and J. M. Fitzsimons. 2011. Habitats used by juvenile flagtails (*Kuhlia* spp.; Perciformes: Kuhliidae) on the island of Hawai'i. Pacific Science 65:441–450.

Moilanen, A., and H. Kujala. 2008. Zonation spatial conservation planning framework and software v. 2.0, user manual. Available: www.helsinki.fi/bioscience/consplan/. (January 2017).

Moilanen, A, L. Meller, J. Leppanen, F. M. Pouzols, A. Arponen, and H. Kujala. 2014. Zonation spatial conservation planning framework and software v. 4.0, user manual. University of Helsinki, Helsinki, Finland.

Myers, B. J. E., A. J. Lynch, D. B. Bunnell, C. Chu, J. A. Falke, R. P. Kovach, T. J. Krabbenhoft, T. J. Kwak, and C. P. Paukert. 2017. Global synthesis of the documented and projected effects of climate change on inland fishes. Reviews in Fish Biology and Fisheries 27:339–361.

Nishimoto, R. T., and J. M. Fitzsimons. 2006. Status of native Hawaiian stream fishes, a unique amphidromous biota. Pages 59–65 in M. J. Brouder and J. A. Scheurer, editors. Status, distribution, and conservation of native freshwater fishes of western North America: a symposium proceedings. American Fisheries Society, Symposium 53, Bethesda, Maryland.

Nishimoto, R. T., T. E. Shimoda, and L. K. Nishiura. 2007. Mugilids in the Muliwai: a tale of two mullets. Pages 143–156 *in* N. L. Evenhuis and J. M. Fitzsimons, editors. Biology of Hawaiian streams and estuaries. Bishop Museum Bulletin in Cultural and Environmental Studies 3.

Palmer, M. A., D. P. Lettenmaier, N. L. Poff, S. L. Postel, B. Richter, and R. Warner. 2009. Climate change and river ecosystems: protection and adaptation options. Environmental Management 44:1053–1068.

Parham, J. E., G. R. Higashi, E. K. Lapp, D. G. K. Kuamo'o, R. T. Nishimoto, J. M. Fitzsimons, D. A. Polhemus, and W. S. Devick. 2008. Atlas of Hawaiian watersheds and their aquatic resources. Hawaii Division of Aquatic Resources and Bishop Museum, Honolulu.

Parham, J. E., G. R. Higashi, R. T. Nishimoto, S. Hau, D. G. K. Kuamo'o, L. K. Nishiura, T. S. Sakihara, T. E. Shimoda, and T. T. Shindo. 2009. The use of Hawaiian stream habitat evaluation procedure to provide biological resource assessment in support of instream flow standards for East Maui streams. Bishop Museum and the Hawaii Division of Aquatic Resources, Honolulu.

State of Hawaii Commission of Water Resource Management. 2008. Surface water hydrologic unit boundaries (watersheds) for the 8 major Hawaiian islands. (May 2010).

Strayer, D. L. 2006. Challenges for freshwater invertebrate conservation. Journal of the North American Benthological Society 25:271–287.

Timm, O., and H. F. Diaz. 2009. Synoptic-statistical approach to regional downscaling of IPCC twenty-first-century climate projections: seasonal rainfall over the Hawaiian Islands. Journal of Climate 22:4261–4280.

Tingley, R. W., III, D. M. Infante, R. A. MacKenzie, A. R. Cooper, and Y-P. Tsang. 2019. Identifying natural catchment landscape influences on tropical stream organisms: classifying stream reaches of the Hawaiian Islands. Hydrobiologia 826: 67–83.

Tsang, Y-P., D. Wieferich, K. Fang, D. M. Infante, and A. R. Cooper. 2014. An efficient and integrated approach of aggregating upstream catchment information. SpringerPlus [online serial] 3:589.

Wang, L., D. Infante, P. Esselman, A. Cooper, D. Y. Wu, W. Taylor, D. Beard, G. Whelan, and A. Ostrof. 2011. A hierarchical spatial framework and database for the National River Fish Habitat Condition Assessment. Fisheries 36:436–449.

Walter, R. P., J. D. Hogan, M. J. Blum, R. B. Gagne, E. F. Hain, J. F. Gilliam, and P. B. McIntyre. 2012. Climate change and conservation of endemic amphidromous fishes in Hawaiian streams. Endangered Species Research 16:261–272.

Ward, E. J., J. H. Anderson, T. J. Beechie, G. R. Pess, and J. M. Ford. 2015. Increasing hydrologic variability threatens depleted anadromous fish populations. Global Change Biology 21:2500–2509.

World Conservation Monitoring Centre. 1996. Sicyopterus stimpsoni. The IUCN Red List of Threatened Species 1996:e. T44167A10866562.

Zeigler, M. P., A. S. Todd, and C. A. Caldwell. 2012. Evidence of recent climate change within the historic range of Rio Grande Cutthroat Trout: implications for management and future persistence. Transactions of the American Fisheries Society 141:1045–1059.

Zhang, C., Y. Wang, K. Hamilton, and A. Lauer. 2016. Dynamical downscaling of the climate for the Hawaiian Islands part II: projection for the late twenty-first century. Journal of Climate 29:8333–8354.

American Fisheries Society Symposium 91:135–159, 2019

Identification and Utility of Native Fish Conservation Areas in the Upper Snake River Basin, USA

RICHARD N. WILLIAMS*
Department of Biology, The College of Idaho
2112 Cleveland Boulevard, Caldwell, Idaho 83605, USA

DANIEL C. DAUWALTER
Trout Unlimited
910 Main Street, Suite 342, Boise, Idaho 83702, USA

RUSSELL F. THUROW
U.S. Forest Service, Rocky Mountain Research Station
322 East Front Street, Suite 401, Boise, Idaho 83702, USA

DAVID P. PHILIPP
Fisheries Conservation Foundation
302 East Green Street #2102, Champaign, Illinois 61825, USA

JACK E. WILLIAMS
Trout Unlimited
4393 Pioneer Road, Medford, Oregon 97501, USA

CHRIS A. WALSER
Department of Biology, The College of Idaho
2112 Cleveland Boulevard, Caldwell, Idaho 83605, USA

Abstract.—Native fish conservation areas (NFCAs) are watersheds where management emphasizes proactive conservation and restoration for long-term persistence of native fish assemblages while allowing for compatible uses. Native fish conservation areas are intended to complement traditional fisheries management approaches that are often reactive to population stressors and focused on single-species conservation efforts rather than complete assemblages. We identified potential NFCAs in the upper Snake River basin above Hells Canyon Dam using a process that ranked all subwatersheds (Hydrologic Unit Code 12) and used empirical data on distribution, abundance, and genetics for three native trout species (Bull Trout *Salvelinus confluentus*, Columbia River Redband Trout *Oncorhynchus mykiss gairdneri*, and Yellowstone Cutthroat Trout *O. clarkii bouvieri*, including the fine-spotted form) and both known occurrences and modeled potential distributions of native nongame fishes. Rankings also incorporated drainage network connectivity and land-protection status (e.g., national park, wilderness). Clusters of high-ranking subwatersheds were identified as potential NFCAs that were then classified according to the presence of nongame fishes identified as species of greatest conservation need in state wildlife action plans. The Pacific Creek

* Corresponding author: troutdna@gmail.com

and Goose Creek watersheds ranked high in the upper basin (above Shoshone Falls), and Little Jacks Creek and Squaw Creek ranked high in the lower basin. We then contrasted characteristics of a select few potential NFCAs, discuss the practical implementation and benefits of NFCAs for both fishes and other aquatic species in the upper Snake River basin, examined how the NFCA approach could enhance existing conservation partnerships, and discuss how designating select watersheds as NFCAs can create higher public awareness of the value of native fishes and other aquatic species and their habitats.

Introduction

Despite decades of allocating substantial resources to conserve freshwater ecosystems, North American freshwater fishes continue to decline at a much faster rate than their terrestrial counterparts (Master et al. 2000; Jelks et al. 2008). Williams (2019, this volume) discusses how current conservation approaches, such as the National Wildlife Refuge system, have been only moderately successful in protecting riverine ecosystems. Because rivers are linear in nature, approaches based on terrestrial features and land ownership often fail to consider watershed boundaries fundamental to aquatic conservation (Saunders el al. 2002; Roux et al. 2008).

Williams et al. (2011) proposed the concept of native fish conservation areas (NFCAs) to establish entire watersheds cooperatively managed for native fish communities. As a complement to existing conservation approaches (e.g., headwater isolation; Novinger and Rahel 2003), NFCAs emphasize intact and persistent native fish communities and healthy and resilient ecosystems while simultaneously striving to support compatible commercial and recreational uses. Dauwalter et al. (2011) explored the NFCA concept and its application in the upper Colorado River basin in Wyoming through a process that combined known and modeled species distributions, spatial prioritization analysis, and stakeholder discussions. Others have used the NFCA concept as an organizing framework for broad-scale native fish conservation initiatives and associated funding programs (Birdsong et al. 2015 and 2019, this volume).

In this paper, we further explore the utility of the NFCA concept by identifying potential NFCAs and their benefits in the upper Snake River basin upstream from Hells Canyon Dam. As described in detail below, we used known distributions and abundance of native trout species, known and modeled occurrences of nongame fishes, drainage network connectivity interrupted by large dams, and land-protection status (e.g., national parks, wilderness, wild and scenic river) to rank all subwatersheds based on their conservation value and identify potential NFCAs. We then summarize information by potential NFCAs, compare implementation of NFCAs in example watersheds, and discuss the utility of the NFCAs in the upper Snake River basin.

Upper Snake River Basin

The Snake River flows through a large basin draining portions of Wyoming, Idaho, Utah, Nevada, and Oregon. From its headwaters in Yellowstone National Park, Wyoming, the river flows southwest, then west, and then northwest before cascading north through Hells Canyon along the Idaho–Oregon border (Figure 1).

The upper Snake River basin lies above the Hells Canyon Complex of three dams (Brownlee, Oxbow, and Hells Canyon dams), that is, the present upstream limit to anadromous salmon and steelhead *Oncorhynchus mykiss* migrations into the Snake River in Idaho. The upper Snake River basin is naturally divided by Shoshone Falls, a 65-m waterfall near Twin Falls, Idaho that effectively separates the Snake

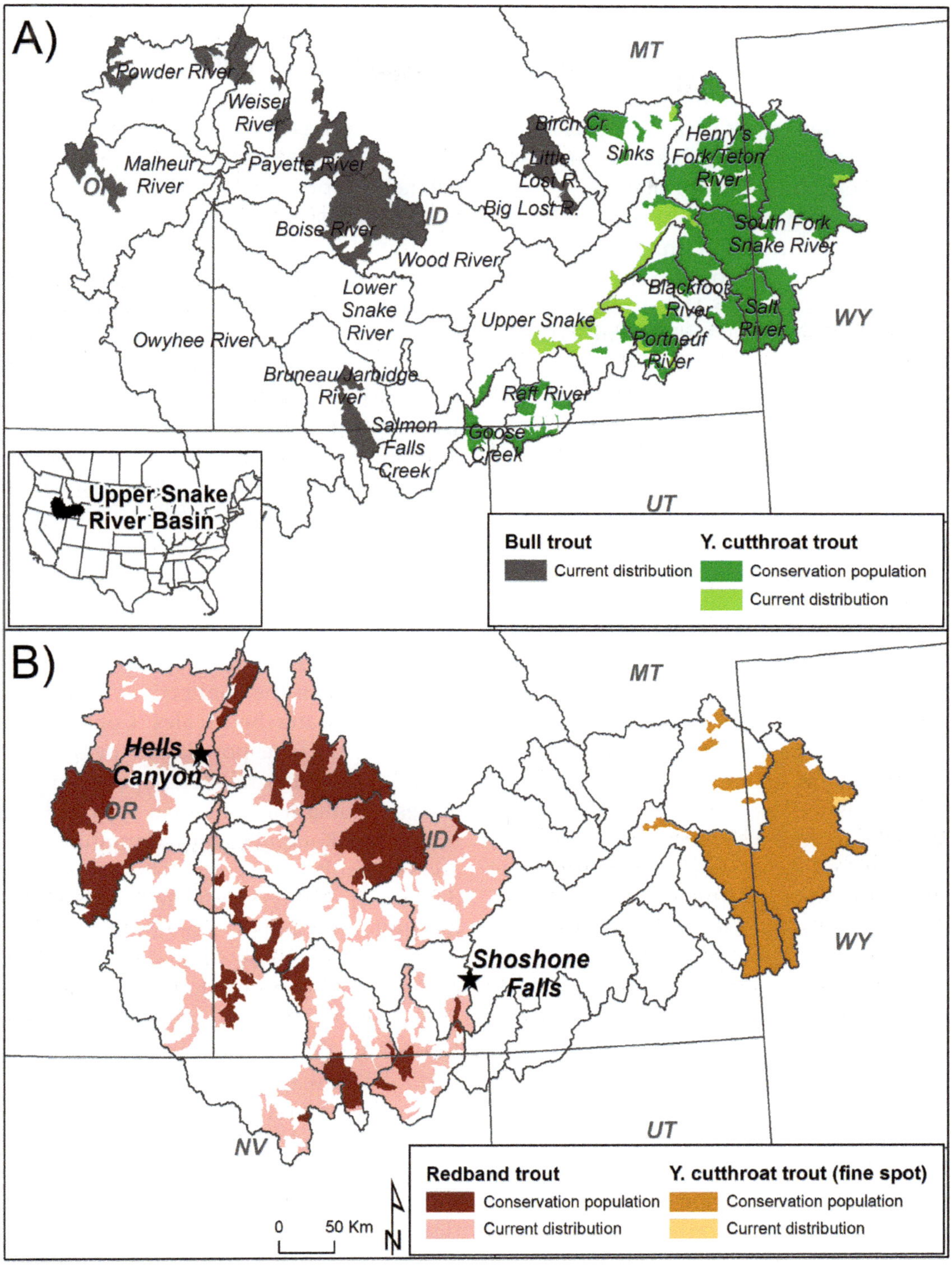

Figure 1. Conservation populations and current distributions of **(A)** Bull Trout *Salvelinus confluentus* and Yellowstone Cutthroat Trout *Oncorhynchus clarkii bouvieri* and **(B)** Columbia River Redband Trout *O. mykiss gairdneri* and the fine-spotted form of Yellowstone Cutthroat Trout in the upper Snake River basin.

River basin into upper and lower basins and is a complete barrier to fish migration (Figure 1; Benke and Cushing 2005). The falls, along with a unique geologic history and past connections with Pleistocene Lake Bonneville, have resulted in a unique history of species colonization that has strongly influenced the biogeography of fishes in the Snake River basin (Smith 1978; Campbell et al. 2011). The lower basin (Shoshone Falls to Hells Canyon Dam) supported 25 native fish species, of which 5 are extirpated (Table 1), compared to the upper basin (Yellowstone Park headwaters to Shoshone Falls), which supports 14 extant native species. Just seven extant species are native to both the upper and lower basins (Table 1; Wallace and Zaroban 2013; Sigler and Zaroban 2018).

Aquatic species management in the Snake River basin has focused primarily on native trout. Yellowstone Cutthroat Trout (including the fine-spotted form) are emphasized above Shoshone Falls, with Columbia River Redband Trout and Bull Trout emphasized below it (Figure 1). All are species of greatest conservation need (SGCN) as well as listed species of special concern in Idaho, Wyoming, Nevada, or Oregon. Bull Trout are also federally listed as a threatened species under the U.S. Endangered Species Act. Nevertheless, eight other SGCN species occur in the lower Snake River basin (such as Leopard Dace, Umatilla Dace, and the Shoshone Sculpin). Four nonsalmonid SGCN species occur in the upper Snake River basin, including Bluehead Sucker and Northern Leatherside Chub (Table 1), with only Mountain Whitefish abundantly occurring in both lower and upper basins. In addition to fishes, the upper Snake River basin provides critical habitat for a plethora of other SGCN species. In addition to the crayfish, fairy shrimp, pond snails, mud snails, mussels, frogs, and toads listed in Table 1, a wide diversity of SGCN birds, mammals, reptiles, and invertebrates persist in the upper Snake River basin (IDFG 2017).

Identification of NFCAs that support at-risk fishes, as well as other aquatic species of concern, has the potential to effectively integrate diverse conservation and management actions within large watersheds across extensive landscapes. This includes the upper Snake River basin, which has a diversity of ecosystems, land ownerships, and uses, as well as at-risk fishes and other aquatic species. For example, in the upper Snake River basin, NFCAs provide an opportunity to link the conservation of headwater Cutthroat Trout populations with downstream habitats supporting nongame fishes such as Bluehead Sucker, Northern Leatherside Chub, as well as conservation opportunities for other at-risk native aquatic species ranging from amphibians to birds to mammals. Consequently, the potential benefits of NFCAs are extremely diverse. For example, they may be managed to sustain the connectivity of critical habitats at watershed scales. Connectivity serves to enhance population persistence by facilitating natural metapopulation processes (Dunham and Rieman 1999; Hilderbrand and Kershner 2000; Compton et al. 2008). Simultaneously, NFCAs may provide discrete hydrologic units in which native fish and other aquatic communities can be isolated, if necessary, from nonnative invasions (Novinger and Rahel 2003; Fausch et al. 2009). To be most effective, proposed NFCAs should be large enough to support natural landscape processes that contribute to the long-term persistence of populations (Haak and Williams 2012). Yet NFCAs may also be small enough to encourage the integration of substantive management actions by diverse entities ranging from federal to tribal to state to private landholders. Consequently, the establishment of NFCAs may facilitate more efficient and effective cooperative actions to benefit a wider assemblage of native aquatic species.

Table 1. Fishes and other taxa native to the upper Snake River basin above Hells Canyon and their general distribution (upper or lower basin relative to Shoshone Falls), species of greatest conservation need (SCGN) status, number of occurrences (and total sites) in survey data, and species weight used to rank subwatersheds.

Taxonomic group	Common name (species abbreviation)	Scientific name	Distribution in basin	SCGN	Occurences (total sites)	Analysis weight
Fishes	Pacific Lamprey (PLY)	*Entosphenus tridentate*	Lower	ID, OR	Extirpated	
	White Sturgeon (WST)	*Acipenser transmontanus*	Lower	ID, OR		
	Chiselmouth (CSM)	*Acrocheilus alutaceus*	Lower		88 (1,452)	0.945
	Utah Chub (UTC)	*Gila atraria*	Upper		23 (1,280)	0.961
	Northern Leatherside Chub (NLC)	*Lepidomeda copei*	Upper	ID, WY	39 (1,210)	0.977
	Peamouth (PMT)	*Mylocheilus caurinus*	Lower		No data	–
	Northern Pikeminnow (PMT)	*Ptychocheilus oregonensis*	Lower		123 (1,448)	0.250
	Redside Shiner (RSS)	*Richardsonius balteatus*	Both		462 (3,045)	0.371
	Longnose Dace (LND)	*Rhinichthys cataractae*	Both		298 (3,002)	0.676
	Speckled Dace (SPD)	*Rhinichthys osculus*	Both		640 (3,047)	0.001
	Kendall Warm Springs Dace (KWD)	*Rhinichthys osculus thermalis*	Upper	WY	No data	–
	Leopard Dace (LPD)	*R. falcatus*	Lower	ID[a]	10 (660)	1.000
	Umatilla Dace (UMD)	*R. umatilla*	Lower	ID[a]	No data	–
	Utah Sucker (UTS)	*Catostomus ardens*	Both		55 (1,310)	0.938
	Bridgelip Sucker (BLS)	*C. columbianus*	Lower		290 (1,609)	0.250
	Wood River Bridgelip Sucker (WBLS)	*C. c. hubbsi*	Lower		No data	–
	Bluehead Sucker (BHS)	*C. discobolus*	Upper	ID[a], UT, WY	46 (1,296)	0.953
	Largescale Sucker (LSS)	*C.macrocheilus*	Lower		101 (1,609)	0.250
	Mountain Sucker (MTS)	*C. platyrhynchus*	Upper		118 (2,736)	0.820
	Snake River Sucker (SRS)	*Chasmistes muriei*	Upper		Extirpated	
	Yellowstone Cutthroat Trout (YCT)	*Oncorhynchus clarkii bouvieri*	Upper	ID, NV, WY		0.991
	fine-spotted form		Upper	ID, WY		0.736
	Columbia River redband trout (Rainbow Trout) (RBT)	*O. mykiss gairdneri*	Lower	ID, OR		0.923
	summer steelhead		Lower	ID, OR	Extirpated	

Table 1. Continued.

Taxonomic group	Common name (species abbreviation)	Scientific name	Distribution in basin	SCGN	Occurences (total sites)	Analysis weight
	Bull Trout (BLT)	*Salvelinus confluentus*	Lower	ID, NV, OR		1.025
	Chinook Salmon (CHS)	*Oncorhynchus tshawytscha*	Lower	ID, NV, OR	Extirpated	
	Sockeye Salmon (SES)	*O. nerka*	Lower	ID, OR	Extirpated	
	Coho Salmon (COS)	*O. kisutch*	Lower	ID, OR	Extirpated	
	Mountain Whitefish (MWF)	*Prosopium williamsoni*	Both	NV, WY	100 (1,354)	0.558
	Mottled Sculpin (MSC)	*Cottus bairdii*	Both		263 (2,628)	0.461
	Paiute Sculpin (PSC)	*C. beldingii*	Both		237 (2,456)	0.340
	Shorthead Sculpin (SSC)	*C. confusus*	Lower		239 (1,609)	0.719
	Shoshone Sculpin (ShSC)	*C. greenei*	Lower	ID	No data	–
	Wood River Sculpin (WSC)	*C. leiopomus*	Lower	ID	44 (154)	0.922
Crayfishes	Snake River pilose crayfish	*Pacifastacus connectens*	Both	ID, WY		
Shrimps	raptor fairy shrimp	*Branchinecta raptor*	Lower	ID		
Snails	Banbury Springs lanx	*Lanx* sp.	Lower	ID		
	pondsnail species group	*Stagnicola* sp.	Both	ID		
	Rocky Mountain dusky snail	*Colligyrus greggi*	Upper	ID		
	Bruneau Hot Spring snail	*Pyrgulopsis bruneauensis*	Lower	ID		
	Bliss Rapids snail	*Taylorconcha serpenticola*	Both	ID		
Mussels	California floater	*Anodonta californiensis*	Both	ID, NV, OR		
	western ridged mussel	*Gonidea angulata*	Both	ID, OR		
	western pearlshell	*Margaritifera falcata*	Both	ID		
Salamander	western tiger salamander	*Ambystoma mavortium*	Upper	WY		
Frogs	northern leopard frog	*Lithobates pipiens*	Both	ID, NV, WY		
	Columbia spotted frog	*Rana luteiventris*	Lower	ID, NV, OR		
Toads	Great Basin spadefoot	*Spea intermontana*	Both	NV, WY		
	western toad	*Anaxyrus boreas*	Both	ID, NV, OR, WY		
	Woodhouse's toad	*A. woodhouseii*	Lower	ID		

[a] Species listed as SGCN but status not assessed and available at time of analysis.

Methods

We identified potential NFCAs through a process that ranked all subwatersheds (Hydrological Unit Code [HUC] 12 watersheds) in the upper Snake River basin based on native trout abundance and distribution, modeled occurrence probabilities for native nongame fishes, differential weighting of species based on their prevalence, drainage network connectivity, and land-protection status. Rankings were intended to identify watershed-scale areas from the headwaters downstream where native trout overlap in distribution with, or occur near, native nongame fishes and where watershed scale conservation would benefit entire fish assemblages. Our analysis did not focus on identifying unique habitats with endemic fishes (e.g., spring habitats with Shoshone Sculpin) or genetically unique subpopulations or subspecies not currently recognized as distinct species (e.g., Wood River Bridgelip Sucker or Big Lost Mountain Whitefish), nor did we focus on large river fishes (e.g., White Sturgeon) or the main-stem Snake River because of the difficulty in managing large rivers to their headwaters per the NFCA concept. Clusters of high-ranking subwatersheds (i.e., the top 25%) were aggregated and characterized based on the native fish assemblage, land ownership and protected status, watershed size, habitat conditions, and future threats.

Fish data

Many different data sources were used to define the distribution and abundance of native trout and the occurrence of native nongame fishes. Distribution and abundance data for native trout were primarily derived from rangewide assessment databases. Yellowstone Cutthroat Trout data (including the fine-spotted form) were based on the 2010 rangewide assessment database (Gresswell 2011), and Columbia River Redband Trout data were based on the 2012 rangewide assessment database (Muhlfeld et al. 2015). For each database, only conservation populations were used to define distributions (Figure 1), and abundance was based on the midpoint of categorical population abundances in the database (e.g., 0–35, 35–100, 101–250, 251–625, 625–1,250, and >1,250 fish/km). Conservation populations were defined as those populations that had less than 10% genetic introgression or had unique genetic, ecological, or behavioral attributes (e.g., adfluvial behavior; UDWR 2000; Gresswell 2011; Muhlfeld et al. 2015). Bull Trout distribution data were obtained from Streamnet, and abundance data from agency databases (see Acknowledgments).

Native nongame fishes data were assembled from fish collections made across the upper Snake River basin, and those data were used to develop species probability of occurrence models. Nongame fishes were sampled primarily by electrofishing, but other methods were also used; for example, nongame species in Idaho were primarily sampled during Idaho Department of Fish and Game (IDFG) and Idaho Department of Environmental Quality electrofishing surveys (Meyer et al. 2013) while both electrofishing and minnow traps were used in other data sets (Blakney 2012). In all, more than 3,047 fish collection records were used to determine species occurrences (see Acknowledgments for sources). From these collections, presence–absence data were used to develop species-specific species distribution models (random forest; Breiman 2001) where probabilities of occurrence were modeled as a function of multiple environmental variables (% converted land, canal density, etc.) (see www.tu.org/USRB-multi-sp-assmt for details). The models for all species, except Leopard Dace, indicated good predictive ability with 10-fold cross-validated AUC values >0.75. The model for Leopard Dace, a species with only 10 occurrences, had the poorest predictive ability (AUC = 0.695). Species-specific models were then used to predict probability of occurrence in perennial stream segments in the National Hydrography Dataset Plus (NHDPlus; version 1; 1:100,000 scale), with drainage areas larger than 159,100 km^2. Predictions were only made in subbasins within the probable native ranges sampled by Meyer et al. (2013) and Gamett (2003).

Watershed rankings

To identify potential NFCAs, every subwatershed in the upper Snake River basin was ranked based on native trout distribution and abundance data, nongame fish probabilities of occurrence data, river network connectivity, and the percent of subwatersheds encumbered in Gap Analysis Program (GAP) Status 1 or 2 protected lands (e.g., wilderness, national parks; USGS 2011). The analysis was constructed to generally give higher ranks to clusters of interconnected subwatersheds with abundant trout populations and high native-fish richness (high biodiversity) within or near protected lands. More specifically, subwatershed rankings were obtained using the additive-benefit function in Zonation v3.0 conservation planning software to account for river system connectivity (Moilanen 2007; Moilanen et al. 2008, 2011). Zonation produces a hierarchical ranking of all subwatersheds (rescaled from 0 to 100) based on the minimum marginal loss across species-specific input values given species weights (minimum biodiversity loss) and offset by a cost function:

$$\delta_i = 1 / c_i w_j \sum_j \Delta V_j ,$$

where δ_i = the marginal loss across all j species for subwatershed i; c_i = 100 − % subwatershed protected, where % protected was based on protected lands identified as GAP status 1 or 2 lands; w_j = the weight for species j based on professional judgment for native trouts or nongame species prevalence, one measure of rarity, defined by Meyer et al. (2013)(see Table 1); and ΔV_j = is the marginal loss of species j values between all remaining subwatersheds minus the value within subwatershed i (see Moilanen et al. 2011). The value of V differed by species. For conservation populations of Yellowstone Cutthroat Trout and Columbia River Redband Trout (see above), V = the midpoint of fish/km ranges reported for those populations, V = subwatershed fish per km average (from fish survey data) for the current distribution of Bull Trout (a value of 1 was used if no abundance data were available), and V = the probability of occurrence (range from 0 to 1) for all nongame species. Rankings were determined by computing the minimum marginal loss across all species for each subwatershed and removing the subwatershed with the lowest marginal loss. This process effectively removed the subwatershed from the landscape that resulted in the minimum biodiversity loss, given species weights and amount of protected lands in the entire upper Snake River basin. The removal process was repeated iteratively until only one subwatershed remained during the last iteration, that is, the subwatershed with the highest marginal loss across all fish species and the most important subwatershed from a biodiversity standpoint. The sequence of removal resulted in the subwatershed rank. Trout data were represented by the spatial hydrography framework for each data source, and the probability of occurrence for nongame species was attributed on NHDPlus. All data were then converted to a 300-m grid for the analysis.

River networks are a nested hierarchy of drainage systems that pose challenges to conservation of fishes residing in riverine environments. Consequently, NFCAs are watersheds managed in their entirety for native aquatic communities. For this reason, river network connectivity was used to impart a penalty on the marginal loss across species based on whether a neighboring subwatershed has already been removed during the ranking process (i.e., has a lower rank). The penalty specifically translated into a reduction in subwatershed value (δ_i) based on the proportion of subwatersheds that had been removed upstream or downstream of the focal subwatershed (see Section 2.4.4 in Moilanen et al. 2011). The same penalties were used for all 21 species and were strong if upstream subwatersheds had already been removed and weak if downstream subwatersheds had been removed. Although connectivity was interrupted by large dams (i.e., those with reservoirs with ≥4 km^2 surface area), smaller dams or other barriers were not used to break connectivity because they are much more capable of being managed for fish passage. Figure 2 illustrates the analytical process.

Potential native fish conservation areas

Clusters of subwatersheds representing independent drainage networks within the top 25% of the landscape (rank > 75) were then aggregated into potential NFCAs. We summarized selected attributes of potential NFCAs: mean subwatershed rank, watershed size, documented native fish occurrences, presence of nongame fishes of greatest conservation need, percent of watershed protected, percent of perennial stream corridor protected, habitat integrity of subwatershed, and future security of subwatershed. The mean rank of subwatersheds in an NFCA was computed as an area-weighted mean from the subwatershed ranking analysis. Documented species occurrences were determined from rangewide assessment databases for native trout and documented presence of native nongame species in fisheries surveys. State wildlife action plans were used to identify species of greatest conservation need in each state (e.g., IDFG 2017). Land protection status was based on GAP Status 1 and 2 lands (USGS 2011). Sub-

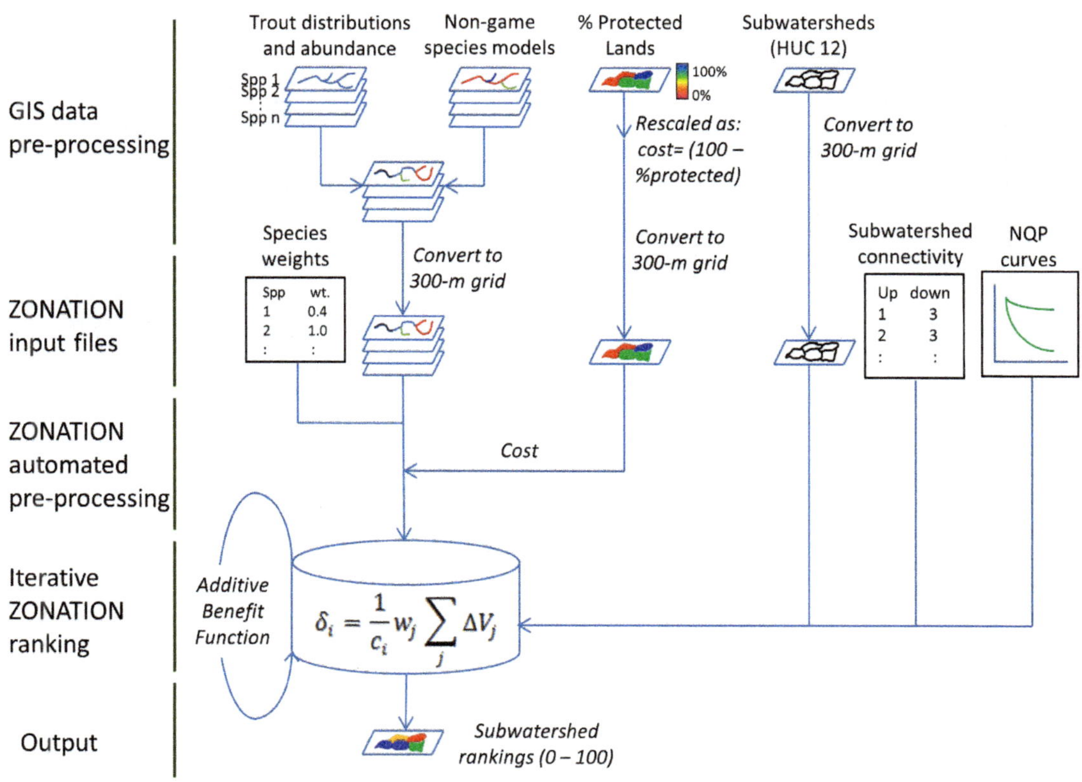

Figure 2. Conceptual model showing data integration for subwatershed ranking analysis.

watershed habitat integrity was based on the Trout Unlimited's conservation success index; a composite index ranging from 5 (low integrity) to 25 (high integrity) based on indicators of land protection, watershed connectivity, watershed condition, water quality, and flow regime, scored each from 1 (low integrity) to 5 (high integrity) (Williams et al. 2007). Future security of subwatersheds was also based on the Trout Unlimited's conservation success index, where future security is an index ranging from 5 (low security) to 25 (high security) based on individuals scores (1–5) for land conversion, resource extraction, energy development, climate change, and introduced species (Williams et al. 2007).

Results

Watershed rankings from the Zonation analysis identified entire subbasins where all subwatersheds ranked high, as well as individual or small clusters of a few high-ranking subwatersheds (Figure 3). Potential NFCA watersheds were primarily clustered in headwater areas, both within individual river systems (e.g., Boise, Payette, Little Jacks, Little Lost, and Goose drainages) and across the Snake River as a whole (e.g., South Fork Snake, Salt, and Blackfoot River subbasins). The Payette, Boise, South Fork Snake, Portneuf, and Blackfoot River subbasins represented large clusters of subwatersheds, with ranks exceeding 75 (of a range from 0 to 100), and representing the top

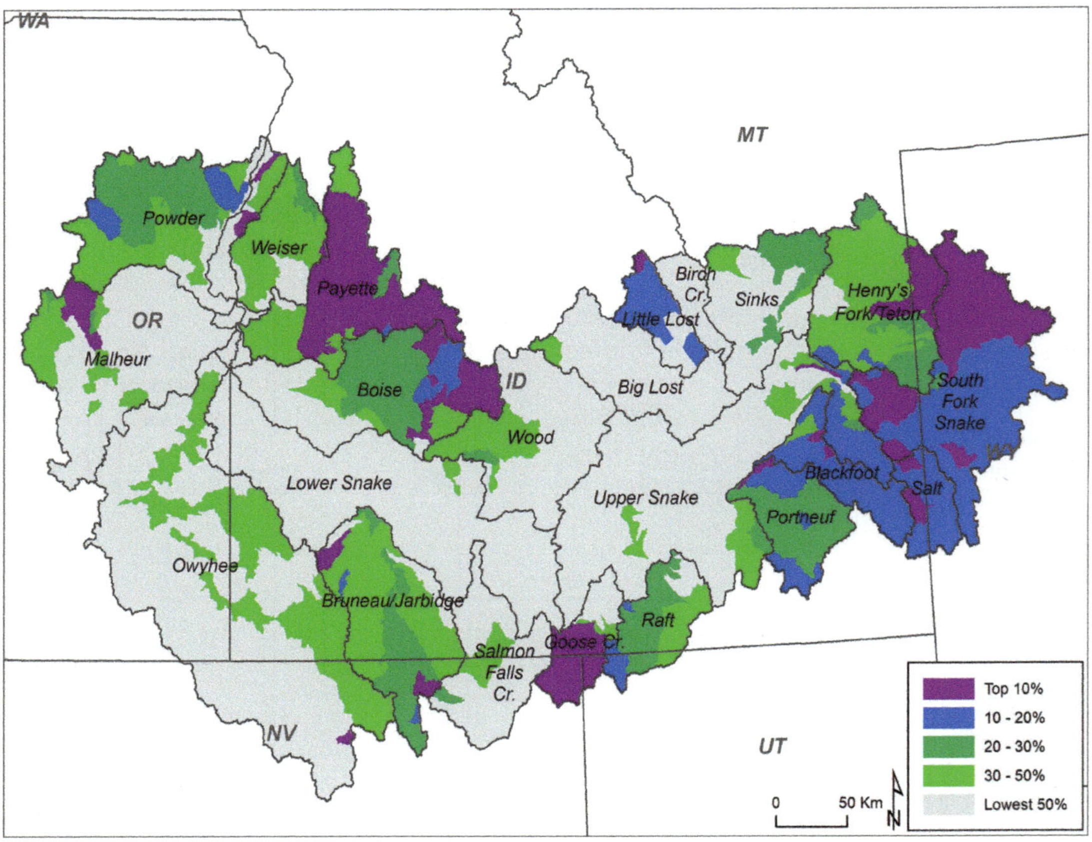

Figure 3. Subwatershed ranks in the upper Snake River basin based on native trout distributions and abundance, native nongame species occurrence probability, drainage network connectivity, and land-protection status.

25% of the entire upper Snake River basin. Smaller aggregations were present in the upper Malheur River, upper Jarbidge River, Goose Creek, Little Lost River, and Fall River/Conant Creek (Henrys Fork watershed). Examples of individual drainages with one or a few high-ranking subwatersheds were Little Jacks Creek, Jack Creek (Nevada), Indian Creek, Cassia Creek, upper Raft River, and Bitch Creek (Figure 4A).

After the clusters of high-ranking subwatersheds (top 25%) were aggregated, 44 watersheds (range from 56 to 4,344 km^2) were identified as potential NFCAs based on their high rank and native-species assemblages (Table 2). All potential NFCAs supported at least one native trout species, and all but three supported at least one native nongame species. Of those, 13 watersheds supported at least one nongame species of greatest state conservation need (Table 2; Figure 4A).

Land status (USGS 2004) within watersheds ranged from 72% private land (Portneuf River) to nearly the entire upper watershed protected in public land and national parks (upper Snake River, Cottonwood Creek, Fall River; Table 2; Figure 4A). Although stream corridors ranged widely in level of protection (Figure 4B), land status was not a reliable predictor of habitat integrity or future security.

Habitat integrity scores ranged from 12 to 25 (Table 2). Habitat integrity in the upper Snake River basin was low in areas with extensive agricultural or urban land use and high in upper elevation mountainous regions (Figure 5A). Watersheds comprised largely of public land (>90%) exhibited habitat integrity scores ranging from the highest possible (25) to 12. In potential NFCAs, habitat integrity was high for Pacific Creek in Grand Teton National Park (score 25 out of 25). In contrast, despite exhibiting species-rich fish assemblages (Table 2), habitat integrity was low for the Portneuf River and Conant Creek (scores of 12 out of 25) due to low scores (1 out of 5) in each of five individual indicators in at least one subwatershed. While increased levels of private land ownership within a watershed generally resulted in lower habitat-integrity scores, of the 11 watersheds with greater than 35% private land ownership, habitat integrity scores ranged from 12 in the Portneuf River and Conant Creek (Henry's Fork subbasin) to 18 in the upper Raft River, where 70% of the watershed is in private ownership.

Likewise, future security of habitats (and fishes) was variable across the basin (Figure 5B). It was typically high in protected areas and in areas with low human development. Within potential NFCAs, future security was high within several watersheds in national parks (scores 21 out of 25), whereas Indian Creek in Hells Canyon had low future security (scores 11 out of 25) because of threats from land conversion, resource extraction, and climate change (Table 2). Bear Creek in the South Fork Snake River subbasin had a future security score of 13, despite being 99% public lands and a habitat integrity score of 22, while Canyon Creek in the Henry's Fork Snake River subbasin, the lower South Fork Snake River, and several Payette River tributaries all had future security scores of 12 and 13 (Table 2).

Identification of potential NFCAs highlighted native-trout distributions and abundance coupled with presence of native nongame species, particularly rare and sensitive species (i.e., species of greatest conservation need; Figure 4). Potential NFCAs in the lower Snake River basin included assemblages of Columbia River Redband Trout and Bull Trout and from 1 to 11 nongame species (except for Indian Creek which contained native trout only). Nongame SGCN species (n = 4) are disproportionally represented in potential NFCAs in the upper basin.

To illustrate differences in the characteristics of potential NFCAs, we contrasted the characteristics of three watersheds: the Jarbidge River, Goose Creek, and the upper Blackfoot River (Figure 6). The Jarbidge

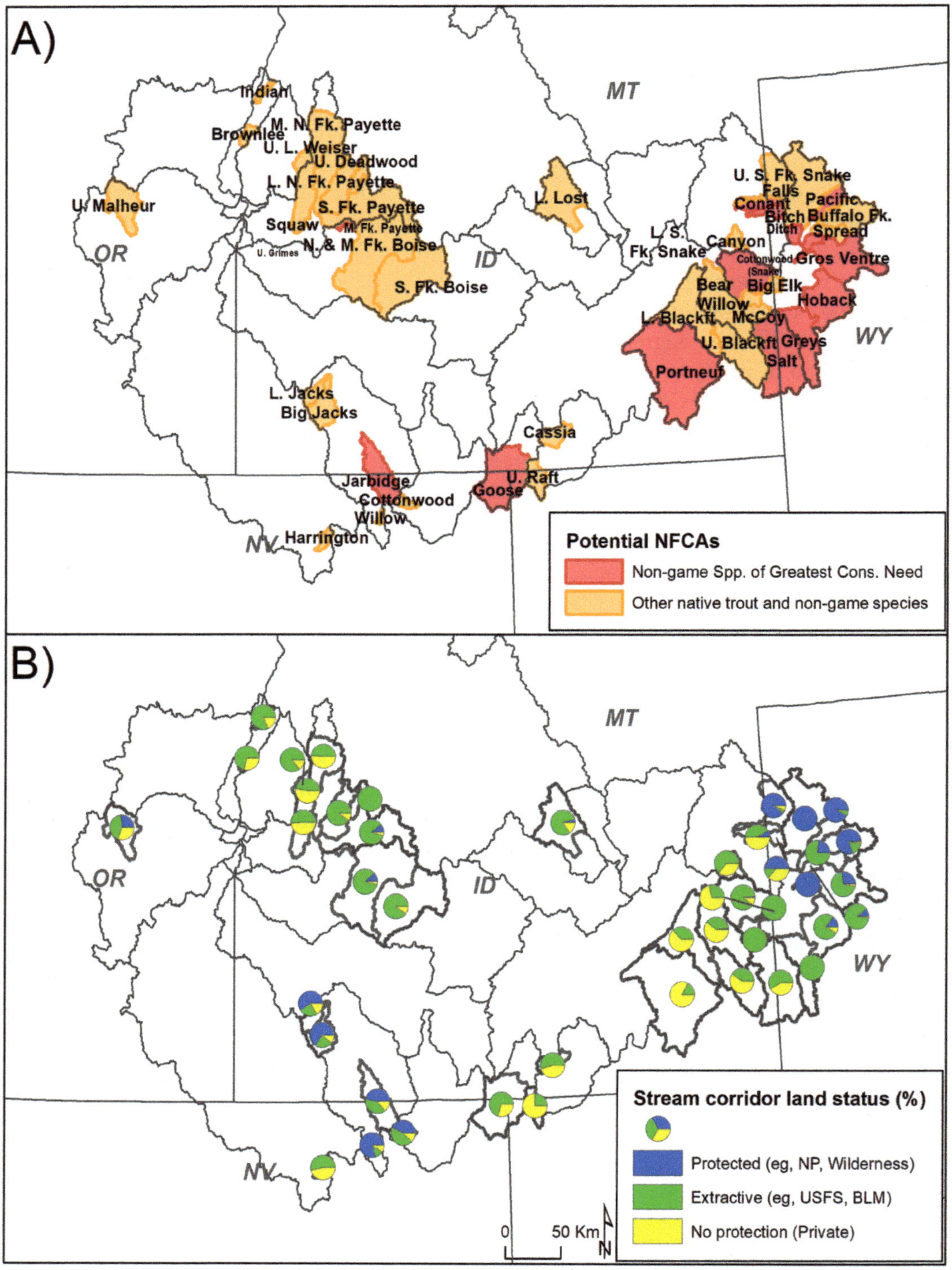

Figure 4. **(A)** Potential native fish conservation areas in the upper Snake River basin with the presence of nongame species of greatest conservation need and **(B)** pie charts illustrating the land status of perennial stream corridors. Details of each watershed are found in Table 2.

Table 2. Characteristics of potential native fish conservation areas in the upper Snake River basin. Species of greatest conservation need shown in bold. R. = River; Cr. = Creek; S. Fk. = South Fork; N. Fk. = North Fork. For Yellowstone Cutthroat Trout (YCT), fine = fine-spotted, lg. = large-spotted, and both = both fine and lg.

Watershed (subbasin)	Mean rank (0–100)	Watershed area (km^2)	Native fishes[a]	Protected (%) (stream corridor)	Public land (%)	Habitat integrity[b] (5–25)	Future security[c] (5–25)
Above Shoshone Falls							
Pacific Cr. (S. Fk. Snake)	98.8	431	**YCT** (both), **NLC**	95.4 (92.3)	97.3	25	21
Upper S. Fk. Snake R.	98.7	2,086	YCT (both), UTS, LND, SPD, RSS, MSC, PSC, UTC	99.9 (99.8)	92.2	24	21
Fall R. (Henrys Fork)	97.6	894	**YCT** (both), MTS, LND, SPD, RSS, MSC, PSC	93.4 (88.8)	97.9	23	16
Buffalo Fork (S. Fk. Snake)	95.8	959	**YCT** (both)	75.4 (80.2)	97.6	23	20
Goose Creek	95.3	1,842	**YCT** (lg.), **BHS**, BLS, MTS, UTS, LND, SPD, RSS, **NLC**, MSC, PSC, SSC, UTC	0.2 (0.1)	85.6	17	16
Conant Cr. (Henrys Fork)	95.0	312	**YCT** (both), **BHS**, MTS, LND, SPD, RSS, MSC, PSC	20.5 (8.8)	51.9	12	14
Cottonwood Cr. (S. Fk. Snake)	94.3	189	**YCT** (both), **MWF**	100 (100)	89.6	23	20
Spread Cr. (S. Fk. Snake)	90.6	344	**YCT** (both), PSC	15.3 (23.7)	90.7	19	19
Lower S. Fk. Snake R.	90.6	1,396	**YCT** (both), **BHS**, LSS, MTS, UTS, LND, SPD, RSS, MSC, PSC, **MWF**	0.7 (1.7)	74.7	16	13
McCoy Cr.(S. Fk. Snake)	89.5	282	**YCT** (both), MTS, UTS, LND, SPD, RSS, MSC, PSC	0.0 (0.0)	99.2	21	14
Ditch Cr. (S. Fk. Snake)	89.2	160	**YCT** (both), **BHS**	41.6 (10.6)	74.4	17	16
Gros Ventre R. (S. Fk. Snake)	88.7	1,617	**YCT** (both), PSC, **MWF**	36.4 (26.8)	97.6	20	18
Hoback R. (S. Fk. Snake)	87.9	1,469	**YCT** (fine), MTS, LND, MSC, **MWF**	19.3 (14.4)	94.1	21	14
Salt R. (S. Fk. Snake)	87.6	2,309	**YCT** (both), **BHS**, MTS, UTS, LND, SPD, RSS, **NLC**, MSC, PSC, **MWF**	0.7 (0.0)	71.3	15	15

Table 2. Continued.

Watershed (subbasin)	Mean rank (0–100)	Watershed area (km^2)	Native fishes[a]	Protected (%) (stream corridor)	Public land (%)	Habitat integrity[b] (5–25)	Future security[c] (5–25)
Above Shoshone Falls (continued)							
Upper Raft R.	87.3	411	**YCT** (lg.), MTS, LND, SPD, RSS, MSC, PSC	4.9 (0.0)	30.2	18	16
Bear Cr. (S. Fk. Snake)	87.0	219	**YCT** (both), SPD, RSS, MSC, PSC	0.0 (0.0)	98.8	22	13
Upper Blackfoot R.	86.0	1,458	**YCT** (lg.), MTS, UTS, LND, SPD, RSS, MSC, PSC, UTC	0.4 (0.1)	56.0	16	15
Greys R. (S. Fk. Snake)	85.2	1,178	**YCT** (both), MTS, LND, MSC, PSC, **MWF**	0.3 (0.1)	99.7	19	16
Little Lost R.	84.3	1,528	**BLT**, MSC, SSC	10.7 (4.0)	94.7	17	21
Lower Blackfoot R.	83.3	1,832	**YCT** (lg.), MTS, UTS, LND, SPD, RSS, MSC, PSC, UTC	0.1 (0.4)	21.6	13	15
Big Elk Cr. (S. Fk. Snake)	81.6	160	**YCT** (both), SPD, MSC, PSC	0.0 (0.0)	97.6	24	14
Portneuf R.	80.0	4,344	**YCT** (lg.), **BHS**, MTS, UTS, LND, SPD, RSS, MSC, PSC, UTC	2.5 (0.1)	27.8	12	17
Willow Cr.	79.6	1,676	**YCT** (lg.), MTS, UTS, LND, SPD, RSS, MSC, PSC	19.0 (1.7)	36.9	15	17
Bitch Cr. (Henrys Fork)	78.7	245	**YCT** (lg.), MSC, PSC, **MWF**	77.2 (50.0)	78.6	22	14
Canyon Cr. (Henrys Fork)	77.4	215	**YCT** (lg.), LND, SPD, PSC, sculpin species	0.0 (0.0)	57.1	13	13
Cassia Cr. (Raft)	74.8	367	**YCT** (lg.), UTS, LND, RSS, MSC, PSC, UTC	0.4 (0.0)	74.8	15	17
Below Shoshone Falls							
Indian Cr. (Hells Canyon)	99.9	104	RBT, BLT	0.1 (0.0)	90.1	14	11
Little Jacks Cr. (Bruneau)	98.3	267	**RBT**, SPD, MSC, SSC	43.5 (59.2)	98.4	18	15
Brownlee Cr. (Hells Canyon)	97.7	162	**RBT**, BLS, SSC, **MWF**	0.0 (0.0)	80.1	16	15
Squaw Cr. (Payette R.)	93.4	880	**RBT**, **BLT**, BLS, LSS, LND, SPD, RSS, NPM, MSC, SSC	0.0 (0.0)	52.8	16	15

Table 2. Continued.

Watershed (subbasin)	Mean rank (0–100)	Watershed area (km^2)	Native fishes[a]	Protected (%) (stream corridor)	Public land (%)	Habitat integrity[b] (5–25)	Future security[c] (5–25)
Below Shoshone Falls (continued)							
Lower N. Fk. Payette R.	93.1	810	**RBT,** LSS, SPD, RSS, NPM, MSC, SSC	0.0 (0.0)	50.3	15	12
Middle N. Fk. Payette R.	92.7	1,166	**RBT**, **BLT**, BLS, LSS, SPD, RSS, NPM, MSC	1.0 (0.4)	40.4	13	13
Middle Fork Payette R.	92.3	878	**RBT**, **BLT**, CSM, LND, SPD, RSS, MSC, SSC, **MWF**	0.2 (0.3)	93.4	18	15
S. Fk. Payette R.	90.0	1,918	**RBT**, **BLT**, LSS, CSM, LND, NPM, SSC	13.5 (10.7)	97.3	20	14
Upper Malheur R.	87.2	900	**RBT**, **BLT**, BLS, LSS, LND, SPD, RSS, NPM, MSC, **MWF**	18.8 (26.4)	76.3	15	18
S. Fk. Boise R.	86.2	2,527	**RBT**, **BLT**, BLS, LSS, CSM, MTS, LND, SPD, RSS, NPM, MSC, SSC, **MWF**	0.0 (0.0)	88.4	17	16
Willow Cr. (Bruneau)	82.8	56	RBT	79.8 (80.5)	95.2	17	15
North/Middle Fork Boise R.	79.7	3,156	**RBT**, **BLT**, BLS, LSS, LND, RSS, MSC, SSC, **MWF**	12.5 (12.3)	93.5	18	16
Jarbidge R. (Bruneau)	78.5	886	**RBT**, **BLT**, BLS, LSS, CSM, MTS, LND, SPD, RSS, NPM, MSC, SSC, **MWF**	47.7 (37.2)	89.6	18	16
Upper Little Weiser R.	78.4	205	**RBT**, **BLT**, LND, MSC, SSC	0.0 (0.0)	87.2	15	14
Upper Deadwood R. (Payette)	77.5	283	**RBT**, **BLT**, SSC, **MWF**	0.0 (0.0)	95.5	21	15
Harrington C. (Owyhee)	69.2	151	**RBT**, SPD, PSC	0.0 (0.0)	61.0	16	16
Big Jacks (Bruneau)	65.3	632	**RBT**, BLS, LND, SPD, RSS, NPM	33.0 (66.1)	97.3	18	15
Cottonwood Cr. (Salmon Falls)	61.2	133	**RBT**, BLS, SPD, RSS, PSC	46.7 (42.5)	96.7	16	18

[a] See Table 1 for species abbreviations. Native fish occurrence based on project database and not expert knowledge.
[b] Habitat integrity from Trout Unlimited's conservation success index. Subwatershed scores range from 5 (poor) to 25 (good). Habitat integrity based in indicators of riparian condition, watershed connectivity, watershed condition, water quality, and flow regime.
[c] Future Security from Trout Unlimited's conservation success index. Subwatershed scores range from 5 (not secure) to 25 (secure). Future security based in indicators of land conversion, resource extraction, energy development, climate change, and introduced species.

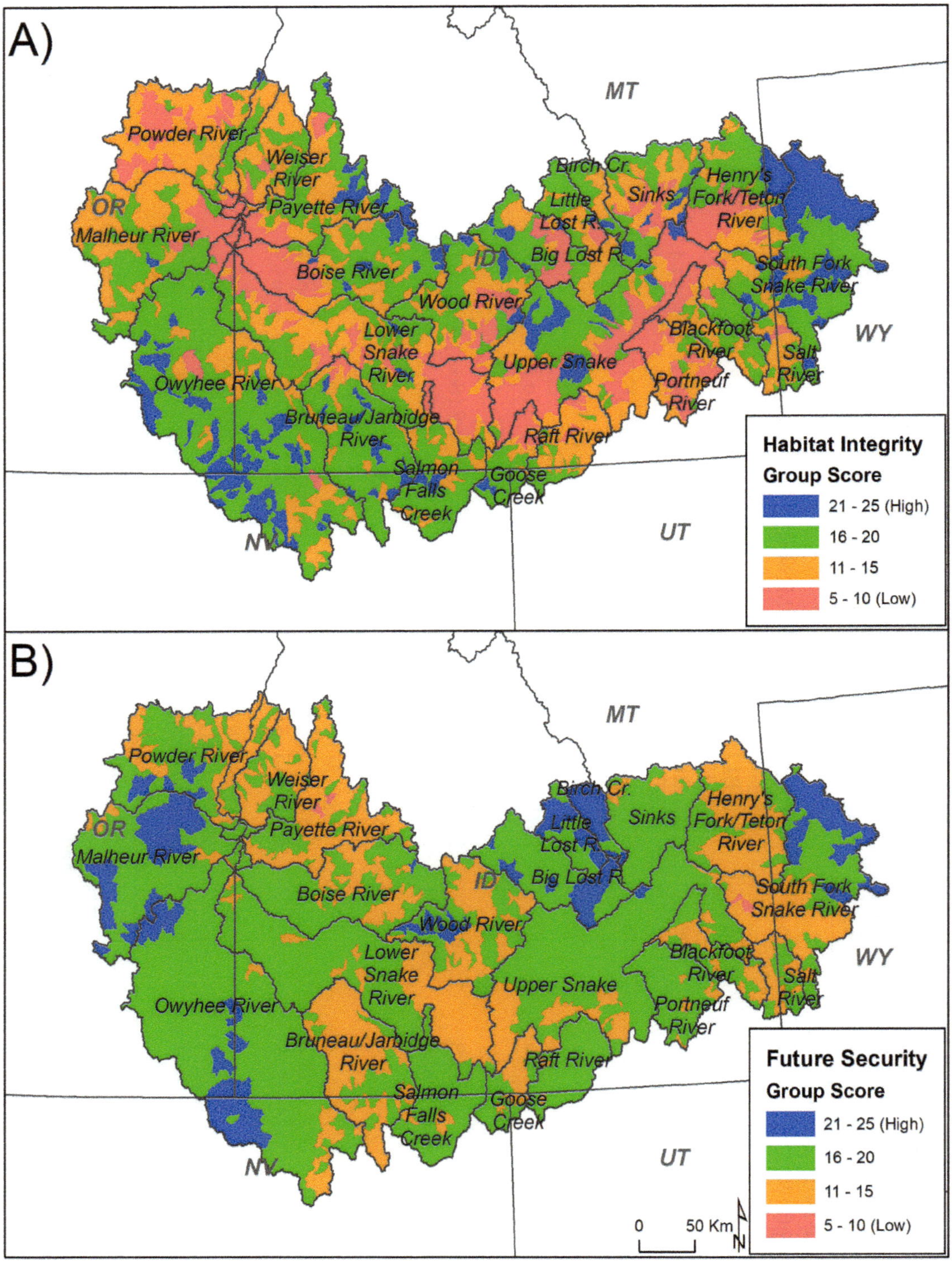

Figure 5. Spatial distribution of **(A)** habitat integrity and **(B)** future security scores for subwatersheds (Hydrologic Unit Code 12: n = 2079) in the upper Snake River basin above Hells Canyon. Habitat integrity and future security indicators were based on Trout Unlimited's conservation success index (Williams et al. 2007).

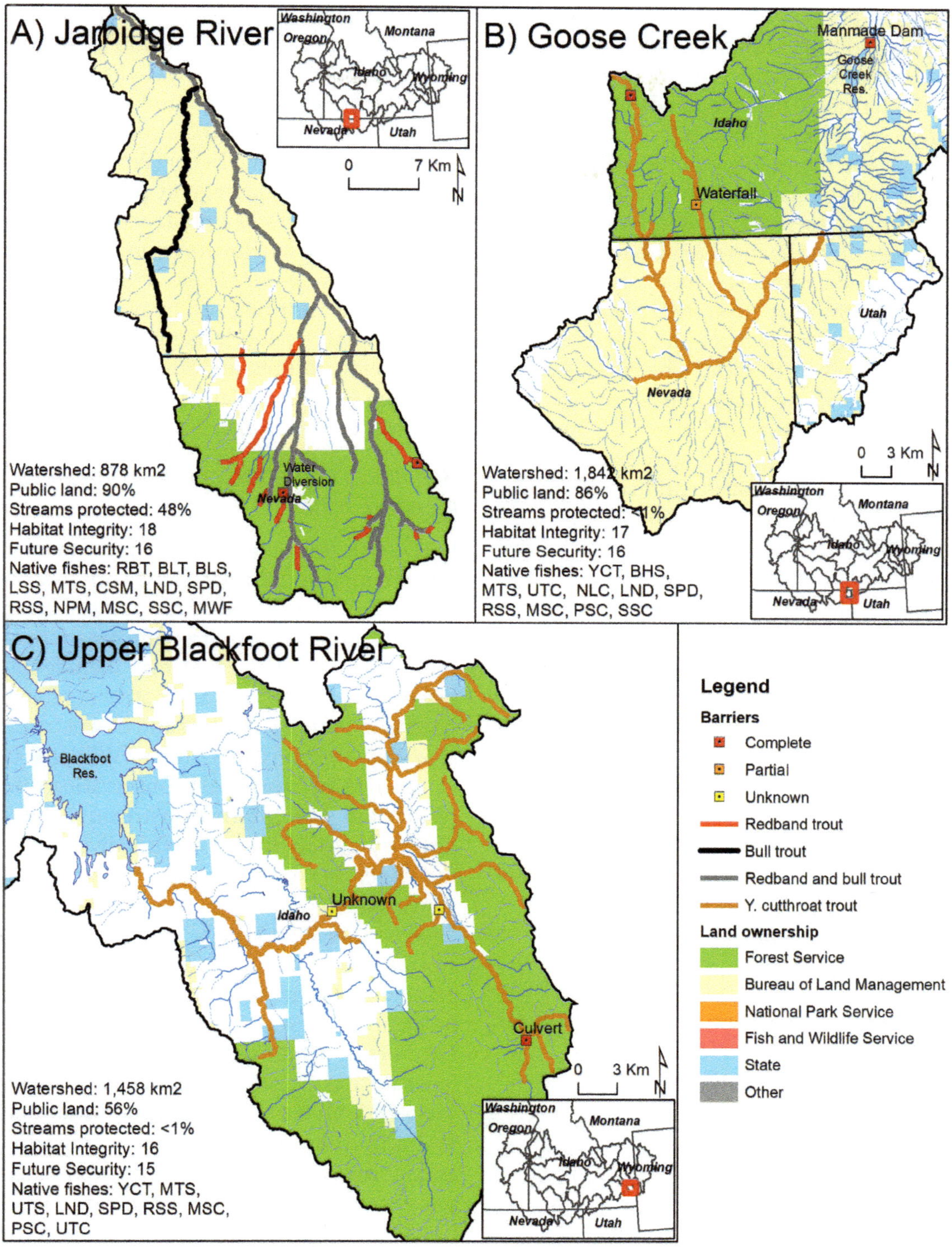

Figure 6. Land ownership and native trout distributions in the **(A)** Jarbidge River, **(B)** Goose Creek, and **(C)** upper Blackfoot River watersheds.

River supports two native trouts, Columbia River Redband Trout and Bull Trout, whereas the others have only Yellowstone Cutthroat Trout. The watersheds ranged in size from 878 to 1,842 km^2, but only two, Goose Creek and upper Blackfoot River, have natural downstream extents defined by dams impounding reservoirs. The Jarbidge River has no definitive downstream boundary and is connected to downstream rivers (a dam isolates the Bruneau River from the Snake River much further downstream). The amount of public land in the watersheds ranges from 56% to 90%, with U.S. Forest Service lands in the headwaters and a mix of Bureau of Land Management and state lands at lower elevations. Perennial stream corridors were 72–87% public land, except in the upper Blackfoot River where 41% of stream kilometers were on public land; however, the main stems of all streams and rivers were largely on private land. Little of the perennial streams had formal land protections, except in the Jarbidge River where 37% of perennial stream kilometers have formal protections in wilderness or wild and scenic river designation. Habitat integrity and future security for these three watersheds were moderate to high, in contrast for example to the future security of the South Fork Snake River and tributaries, which are lower and threatened by introduced species (e.g., Rainbow Trout *Oncorhynchus mykiss*), climate change (high drought risk), and energy development (potential hydropower development).

Discussion

Aquatic habitat in the western United States has been degraded by myriad factors, including overgrazing by domestic livestock (Beschta et al. 2013), water withdrawal (Deacon et al. 2007), oil and gas development (Dauwalter 2013), urbanization (Williams et al. 2007), hydroelectric development (Thurow et al. 2000), and forestry and mining practices (Lee et al. 1997), among other anthropogenic impacts. In turn, the distributions of inland trout have been substantially reduced; for example, all inland Cutthroat Trout subspecies occupy less than half of their historical habitat (Thurow et al. 1997; Haak and Williams 2013). Salmon have been extirpated from some rivers (Praggastis and Williams 2013), and nongame fishes have experienced declines in abundance and distribution in western North America (Brouder and Scheurer 2007). Habitat conditions are expected to continue to change in the future as runoff timing, stream discharge, and water temperatures are altered by climate warming (Wenger et al. 2011; Beschta et al. 2013; Meyer et al. 2014a; Isaak et al. 2015).

Benefits of native fish conservation areas for native fishes

Native fish conservation areas provide both a conceptual and formal framework for collaborative, watershed-scale conservation and restoration efforts targeting broader native fish communities that may include native trout in headwaters and native nongame species in both headwaters and downstream areas in western river systems (Williams et al. 2011). Watershed-scale conservation and restoration can improve instream habitat diversity closely linked to fish diversity in the upper Snake River basin (Walrath et al. 2016). North American Beaver *Castor canadensis* are increasingly important for restoring stream channels incised by livestock grazing and other impacts. Such efforts can be critical for SCGN species like Northern Leatherside Chub that have higher prevalence at sites with complex streamflows associated with beaver dams (Dauwalter and Walrath 2018). Stream habitat restoration and increased connectivity at the watershed scale are expected to improve current conditions for native aquatic assemblages and increase their resiliency to future threats like climate change (Thurow et al. 1997; Zoellick et al. 2005; Isaak et al. 2015).

We used a quantitative, data-driven approach to identify key locations in the upper

Snake River basin where watershed-scale restoration efforts will benefit native fish communities, such as in Goose Creek, the upper Blackfoot River, and Conant Creek (Figure 4A). Watershed-scale restoration might also be a management priority in watersheds that are juxtaposed to lands currently designated for protection as wild and scenic rivers, wilderness areas, or national parks because of the additional future security for the watersheds and aquatic communities. Examples of potential NFCA watersheds in designated status include the upper South Fork Snake, Pacific Creek, and Hoback Creek (Table 2; Figure 4A). Ultimately, adoption of these potential NFCAs as functional NFCAs will require a stakeholder-driven, collaborative process ensuring feasibility assessment and implementation (Dauwalter et al. 2011; Birdsong et al. 2015).

Decades ago, Lee et al. (1997) anticipated the need for NFCAs after examining historical and contemporary distributions of 15 native salmonid taxa in the interior Columbia River basin and portions of the Klamath River and Great Basin. The authors evaluated the native salmonids we examined in this study (Columbia River Redband Trout, two forms of Yellowstone Cutthroat Trout, Bull Trout, and Mountain Whitefish; Lee et al. 1997; Thurow et al. 1997). Their work identified habitat and biodiversity strongholds for native salmonids in the northern Cascades, the central Idaho mountains, and the Snake River headwaters and their connecting river corridors. Thurow et al. (1997) noted that maintaining the integrity of these declining native fishes and aquatic systems for the long term would require a network of well-connected, high-quality habitats (strongholds) that supported a diverse assemblage of native species, including a full expression of their life histories. Our finer-scaled analysis of the upper Snake River basin further illuminates areas of diverse, high-quality, interconnected habitat in the headwaters of the Boise, Payette, and upper Snake River drainages and the potential for NFCAs in several of these same watersheds (Figures 4 and 6).

For the NFCA approach to function and assist both the public interest and management agencies' long-term planning efforts, it is essential that NFCAs be identified at a scale that is practical and manageable. Aggregating smaller subwatersheds with common-species assemblages and management concerns into larger management units (as potential NFCAs) ensures that NFCA designation will be an effective management tool (Birdsong et al. 2019). Although our unit of analysis was a subwatershed (~12,000 ha), our analysis also explicitly accounted for connectivity of adjacent subwatersheds. This facilitated aggregation into larger watersheds where management considerations and actions were likely to be consistent. For example, IDFG manages Yellowstone Cutthroat Trout at the subbasin level (IDFG 2007), a scale representing a natural way to aggregate our results into larger units, as we illustrate (Figure 4).

Watershed scale NFCAs are also compatible with the proposed management strategies of many state and federal natural resource management plans because they rely on identifying management units based on habitat conditions, genetics, and population status. The Idaho Department of Fish and Game's management plan for conservation of Yellowstone Cutthroat Trout (IDFG 2007) defines 13 geographic management units addressing abundance, trends, genetics, and an evaluation of existing threats (Meyer et al. 2006). The status of Columbia River Redband Trout populations is summarized in a similar manner below Shoshone Falls (Meyer et al. 2014b).

The NFCA concept can also be used as an organizational framework for cross-partnership collaborations focusing on strategic restoration of fish habitat. Several multiagency partnerships, such as the Western Native Trout Initiative (WNTI;

www.westernnativetrout.org) and the Desert Fish Habitat Partnership (DFHP; www.desertfhp.org), are collaborative, multi agency and multistate partnerships focused on native trout conservation and restoration across the western United States and conservation and restoration of habitats used by nonsalmonid desert fishes, respectively. The WNTI and DFHP partnerships commonly collaborate on habitat restoration projects (Dauwalter et al. 2019, this volume). The NFCA concept provides an umbrella framework guiding expanded collaboration of watershed-scale restoration efforts that could benefit native trout in headwater streams, as well as native, nongame fishes in downstream, warmer main-stem habitats. For example, our upper Snake River analysis informs identification of focal watersheds for such cross-partnership collaboration, planning, and fundraising (e.g., Dauwalter et al. 2011, 2019; Haak and Williams 2013).

Native sport fishes such as Cutthroat Trout and black bass *Micropterus* spp. have been proposed as key species defining NFCAs because they range widely, are relatively well known compared to many other native stream fishes, and are of primary interest to both recreational anglers and state and federal management agencies, yet they are relatively sensitive to disturbance (Williams et al. 2011; Birdsong et al. 2019). Restoring and reconnecting stream networks has been proposed as a method to conserve native trout in the face of climate change by creating larger stronghold populations and facilitating development of migratory life histories that aid in long-term population persistence (Haak and Williams 2013). Haak and Williams (2013) describe such efforts for Rio Grande Cutthroat Trout *Oncorhynchus clarkii virginalis*, including restoration of 240 km of interconnected habitats in the Rio Costilla area of New Mexico that also creates a refuge for other native fishes of the Rio Grande basin.

The NFCA concept has also been applied to explore watershed level conservation and stakeholder partnerships in Colorado and across the southeastern United States. Dauwalter et al. (2011) compared the distribution of remaining Colorado River Cutthroat Trout *O. c. pleuriticus* with known distributions of three sensitive warmwater fishes (Roundtail Chub *Gila robusta*, Bluehead Sucker, and Flannelmouth Sucker *Catostomus latipinnis*) to identify potential NFCAs in the upper Colorado River drainages of Colorado, Utah, and Wyoming. Several watershed-scale opportunities were identified, including Muddy Creek in the Little Snake River, where multiple agencies and nonprofit groups are working to reconnect headwaters with downstream reaches in order to conserve the entire native fish community (Compton et al. 2008).

In the southeastern United States, the NFCA concept has been proposed as a preferred method to protect native stream fishes. Birdsong et al. (2015) recommended the use of the NFCA concept to protect native black bass, which are keystone species in southeastern stream systems. Their protection in large, functional watersheds would simultaneously help conserve numerous less well-known, native stream fishes in the region. Efforts are underway to identify NFCA watersheds for keystone black bass species, including Guadalupe Bass *Micropterus treculii* in the Llano River watershed in central Texas, Redeye Bass *M. coosae* in South Carolina's upper Savannah River watershed, and Shoal Bass *M. cataractae* in the Chipola River watershed in north-central Florida.

Benefits of native fish conservation areas for other aquatic species

Management focused on resilient watersheds will also likely benefit other imperiled aquatic species such spotted frogs *Rana luteiventris*, western ridged mussel *Gonidea angulata*, other diverse SCGN species (Table 1), as well as riparian obligate species such as the sage grouse *Centrocercus urophasianus* that require wet meadow habitat for early juvenile rearing. Such an approach would engage multiple partners and potentially leverage increased funding sources in a more efficient manner to serve a wide diversity of species (Birdsong et al. 2019).

An example of this multiorganization partnership approach was implemented in 2015 on North Carolina's Little Tennessee River (LTR), which became the nation's first designated NFCA (Harris and Williams 2016; Leslie et al. 2019, this volume). The LTR is an important global biological hotspot and is host to a unique assemblage of fish, amphibians, mollusks, crayfish, and aquatic insects. The North Carolina Wildlife Federation gathered interested organizations, agencies, and business to discuss long-term conservation of the LTR. As a result, the LTR Native Fish Conservation Partnership formed to assist management of the LTR watershed as an NFCA. The LTR Native Fish Conservation Partnership includes 25 collaborative partners comprising nongovernmental organizations, federal and state agencies, the Eastern Band of Cherokee Indians, and several private businesses.

Collaborative benefit of native fish conservation areas

Addressing specific local and regional aquatic conservation issues through the NFCA concept can facilitate important socioecological perspectives that are key to achieving both biological goals and the social and management goals of NFCA partnership members (e.g., the Little Tennessee River NFCA). Establishing the multistakeholder partnership, required to support an NFCA, fosters a collaborative, interdisciplinary approach necessitating active dialogue for sharing values and perspectives. Native fish conservation area collaboration provides the opportunity for effectively discussing and resolving conflict among user groups, managers, and fisheries biologists (Hand et al. 2018; Birdsong et al. 2019). Rieman et al. (2000) used a broadscale and spatially explicit classification of subbasins in the northwest to examine opportunities and conflicts related to terrestrial and aquatic ecosystem management. Birdsong et al. (2019) discuss a 6-year multipartner effort across watersheds in Texas that led to the development of the Texas Native Fish Conservation Areas Network. The Texas NFCA Network protects 91 freshwater fishes considered species of greatest conservation need and their associated watersheds through designation of 20 NFCAs identified to preserve the unique Texas freshwater fish diversity.

Because human disturbance is strongly associated with the condition of both aquatic and terrestrial ecosystems, the commonality in management goals and opportunities challenges managers of multiple disciplines to work together. Spatially explicit classifications such as NFCAs provides a mechanism for better integrating management. Rieman et al. (2000) suggested that such approaches assist integration through (1) communication among disciplines, (2) effective prioritization of limited conservation and restoration resources, and (3) establishing a framework for experimentation and demonstration of restoration techniques. Establishing a multidisciplinary and multiorganization partnership to support NFCAs is a critical element for creating the public and institutional support necessary for more effective, long-term conservation and restoration of aquatic systems, as demonstrated eloquently in the Little Tennessee River (Leslie et al. 2019) and across the Texas NFCA Network (Birdsong et al. 2019).

Acknowledgments

We thank the following individuals for sharing Snake River basin fish population data: K. Meyer, IDFG; B. Gamett, Salmon-Challis National Forest; D. Miller, Wyoming Game and Fish Department; B. Bangs, Oregon

Department of Fish and Wildlife (ODFW); J. Pappani, Idaho Department of Environmental Quality; R. Hughes, U.S. Environmental Protection Agency; E. Keeley, Idaho State University; J. DeRito, Henrys Fork Foundation (HFF)/Trout Unlimited; M. Lien, Friends of the Teton River; and contributors to the HFF database: Bureau of Land Management (BLM), U.S. Forest Service, Wyoming Game and Fish Department (WGFD), IDFG, and Friends of the Teton River (FTR). We thank those who responded to our survey regarding the characterization of NFCAs. We also thank S. Grunder and J. Dillon (IDFG), S. Hoefer and A. Berglund (BLM), R. Perkins (ODFW), M. Lien and A. Vebeten (FTR), L. Mabey (Caribou-Targhee National Forest), and D. Miller (WGFD) for assisting us in reviewing the data and analyses. Funding for this work was provided by the National Fish and Wildlife Foundation and Patagonia's Environmental Fund and World Trout Initiative.

References

Benke, A. C. and C. E. Cushing, editors. 2005. Rivers of North America. Academic Press, Burlington, Massachusetts.

Beschta, R. L., D. L. Donahue, D. A. DellaSala, J. J. Rhodes, J. R. Karr, M. H. O'Brien, T. L. Fleischner, C. Deacon Williams. 2013. Adapting to climate change on western public lands: addressing the ecological effects of domestic, wild, and feral ungulates. Environmental Management 51:474–491.

Birdsong, T. W., M. S. Allen, J. E. Claussen, G. P. Garrett, T. B. Grabowski, J. Graham, F. Harris, A. Hartzog, D. Hendrickson, R. A. Krause, J. Leitner, J. M. Long, C. K. Metcalf, D. P. Phillipp, W. F. Porak, S. Robinson, S. M. Sammons, S. Shaw, J. E. Slaughter, and M. D. Tringali. 2015. Native Black Bass Initiative: implementing watershed-scale approaches to conservation of endemic black bass and other native fishes in the southern United States. Pages 363–378 *in* M. D. Tringali, J. M. Long, T. W. Birdsong, and M. S. Allen, editors. Black bass diversity: multidisciplinary science for conservation. American Fisheries Society, Symposium 82, Bethesda, Maryland.

Birdsong, T. W., G. P. Garrett, B. J. Labay, M. G. Bean, P. T. Bean, J. Botros, M. J. Casarez, A. E. Cohen, T. G. Heger, A. Kalmbach, D. A. Hendrickson, S. J. Magnelia, K. B. Mayes, M. E. McGarrity, R. McGillicuddy, M. M. Parker, and S. Robertson. 2019. Texas Native Fish Conservation Areas Network: strategic investments in restoration and preservation of freshwater fish diversity. Pages 183–229 *in* D. C. Dauwalter, T. W. Birdsong, and G. P. Garrett, editors. Multispecies and watershed approaches to freshwater fish conservation. American Fisheries Society, Symposium 91, Bethesda, Maryland.

Blakney, J. R. 2012. Historical connectivity and contemporary isolation: population genetic structure of a rare high-desert minnow, the Northern Leatherside Chub (*Lepidomeda copei*). Master's thesis. Idaho State University, Pocatello.

Breiman, L. 2001. Random forests. Machine Learning 45:5–32.

Brouder, M. J., and J. A. Scheurer, editors. 2007. Status, distribution, and conservation of native freshwater fishes of western North America. American Fisheries Society, Symposium 53, Bethesda, Maryland.

Campbell, M. R., C. C. Kozfkay, K. A. Meyer, M. S. Powell, and R. N. Williams. 2011. Historical influences of volcanism and glaciation in shaping mitochondrial DNA variation and distribution of Yellowstone Cutthroat Trout across its native range. Transactions of the American Fisheries Society 140:91–107.

Compton, R. I., W. A. Hubert, F. J. Rahel, M. C. Quist, and M. R. Bower. 2008. Influences of fragmentation on three species of native warmwater fishes in a Colorado River basin headwater stream system. North American Journal of Fisheries Management 28:1733–1743.

Dauwalter, D. C. 2013. Fish assemblage associations and thresholds with existing and projected oil and gas development. Fisheries Management and Ecology 20:289–301.

Dauwalter, D. C., J. S. Sanderson, J. E. Williams, and J. R. Sedell. 2011. Identification and implementation of native fish conservation areas in the upper Colorado River basin. Fisheries 36:278–288.

Dauwalter, D. C., S. L. Vail-Muse, T. R. Thompson, J. B. Whittier, K. M. Johnson, and M. G. Bean. 2019. Partnering on multispecies aquatic assessments to inform efficient conservation delivery. Pages 11–32 *in* D. C. Dauwalter, T. W. Birdsong, and G. P. Garrett, editors. Multispecies and watershed approaches to freshwater fish conservation. American Fisheries Society, Symposium 91, Bethesda, Maryland.

Dauwalter, D. C., and J. D. Walrath. 2018. 2018. Beaver dams, streamflow complexity, and the distribution of a rare minnow, *Lepidomeda copei*. Ecology of Freshwater Fish 27:606–616.

Deacon, J. E., A. E. Williams, C. D. Williams, and J. E. Williams. 2007. Fueling population growth in Las Vegas: how large-scale groundwater withdrawal could burn regional biodiversity. BioScience 57:688–698.

Dunham, J. B., and B. E. Rieman. 1999. Metapopulation structure of Bull Trout: influences of physical, biotic and geometrical landscape characteristics. Ecological Applications 9:642–655.

Fausch, K. D., B. E. Rieman, J. B. Dunham, M. K. Young, and D. P. Peterson. 2009. Invasion versus isolation: trade-offs in managing native salmonids with barriers to upstream movement. Conservation Biology 23:859–870.

Gamett, B. L. 2003. A summary: the origin of fishes in the Sinks Drainages of southeastern Idaho. Page 16 *in* R. W. Van Kirk, J. M. Capurso, and B. L. Gamett, editors. The Sinks symposium: exploring the origin and management of fishes in the Sinks Drainages of southeastern Idaho. American Fisheries Society, Idaho Chapter, Bethesda, Maryland.

Gresswell, R. E. 2011. Biology, status, and management of the Yellowstone Cutthroat Trout. North American Journal of Fisheries Management 31:782–812.

Haak, A. L., and J. E. Williams. 2012. Spreading the risk: native trout management in a warmer and less-certain future. North American Journal of Fisheries Management 32:387–401.

Haak, A. L., and J. E. Williams. 2013. Using native trout restoration to jumpstart freshwater conservation planning in the Interior West. Journal of Conservation Planning 9:38–52.

Hand, B. K., C.G. Flint, C. A Frissell, C. C. Muhlfeld, S. P. Devlin, B. P. Kennedy, R. L. Crabtree, W. A. McKee, G. Luikart, and J. A. Stanford. 2018. A social-ecological perspective for riverscape management in the Columbia River basin. Frontiers in Ecology and Environment 16(S1):S23–S33.

Harris, F. and R. N. Williams. 2016. Introducing no. 1: Little Tennessee River basin becomes first native fish conservation area. Flyfisher, spring/summer 2016.

Hilderbrand, R. H., and J. L. Kershner. 2000. Conserving inland Cutthroat Trout in small streams: how much stream is enough? North American Journal of Fisheries Management 20:513–520.

IDFG (Idaho Department of Fish and Game). 2007. Management plan for conservation of Yellowstone Cutthroat Trout in Idaho. Idaho Department of Fish and Game, Technical Report, Boise.

IDFG (Idaho Department of Fish and Game). 2017. Idaho State Wildlife Action Plan, 2015. Prepared for the U.S. Fish and Wildlife Service, Region 1, Office of Migratory Birds & State Programs, Wildlife & Sport Fish Restoration, Portland, Oregon.

Isaak, D. J., M. K. Young, D. E. Nagel, D. L. Horan, and M. C. Groce. 2015. The cold-water climate shield: delineating refugia for preserving salmonid fishes through the 21st century. Global Change Biology 21:2540–2553.

Jelks, H. L., S. J. Walsh, S. Contreras-Balderas, E. Díaz-Pardo, N. M. Burkhead, D. A. Hendrickson, J. Lyons, N. E. Mandrak, F. McCormick, J. S. Nelson, S. P. Platania, B. A. Porter, C. B. Renaud, J. J. Schmitter-Soto, E. B. Taylor, and M. L. Warren. 2008. Conservation status of imperiled North American freshwater and diadromous fishes. Fisheries 33:372–407.

Lee, D. C., J. R. Sedell, B. E. Rieman, R. F. Thurow, and J. E. Williams. 1997. Broadscale assessment of aquatic species and habitats. Pages 1058–1496 *in* T. M. Quigley and S. J. Arbelbide, editors. An assessment of ecosystem components in the interior Columbia basin and portions of the Klamath and Great Basins. U.S. Forest Service, PNW-GTR-405, Portland, Oregon.

Leslie, A., E. McCombs, and F. Harris. 2019. Little Tennessee River Basin Native Fish Conservation Partnership: aquatic conservation on a landscape scale. Pages 415–430 *in* D. C. Dauwalter, T. W. Birdsong, and G. P. Garrett, editors. Multispecies and watershed approaches to freshwater fish conservation. American Fisheries Society, Symposium 91, Bethesda, Maryland.

Master, L. L., B. A. Stein, L. S. Kutner, and G. A. Hammerson. 2000. Vanishing assets: conservation status of U.S. species. Pages 93–118 *in* B. A. Stein, L. S. Kutner, and S. J. Adams, editors. Precious heritage: the status of biodiversity in the United States. Oxford University Press, New York.

Meyer, K. A., D. J. Schill, and J. A. Lamansky, Jr. 2006. Status of Yellowstone Cutthroat Trout in Idaho. Transactions of the American Fisheries Society 135:1329–1347.

Meyer, K. A., J. A. Lamansky, Jr., D. J. Schill, and D. W. Zaroban. 2013. Nongame fish species distribution and habitat associations in the Snake River basin of southern Idaho. Western North American Naturalist 73:20–34.

Meyer, K. A., E. I. Larson, C. L. Sullivan, and B. High. 2014a. Population trends of Yellowstone Cutthroat Trout and nonnative trout in Idaho's upper Snake River basin. Journal of Fish and Wildlife Management 5:227–242.

Meyer, K. A., D. J. Schill, E. R. J. M. Mamer, C. C. Kozfkay, and M. R. Campbell. 2014b. Status of redband trout in the upper Snake River basin of Idaho. North American Journal of Fisheries Management 34:507–523.

Moilanen, A. 2007. Landscape zonation, benefit functions and target-based planning: unifying reserve selection strategies. Biological Conservation 134:571–579.

Moilanen, A., J. Leathwick, and J. Elith. 2008. A method for spatial freshwater conservation prioritization. Freshwater Biology 53:577–592.

Moilanen, A., L. Meller, J. Leppänen, A. Arponen, and H. Kujala. 2011. Zonation: spatial conservation planning framework and software, version 3.0. University of Helsinki, Helsinki, Finland.

Muhlfeld, C. C., S. E. Albeke, S. L. Gunckel, B. J. Writer, B. B. Shepard, and B. E. May. 2015. Status and conservation of interior redband trout in the western United States. North American Journal of Fisheries Management 35:31–53.

Novinger, D. C., and F. J. Rahel. 2003. Isolation management with artificial barriers as a conservation strategy for Cutthroat Trout in headwater streams. Conservation Biology 17:772–781.

Praggastis, A., and J. E. Williams. 2013. Salmon's presence in Nevada's past. Nevada Historical Society Quarterly 56:14–32.

Rieman, B. E., D. C. Lee, R. F. Thurow, P. F. Hessburg, and J. R. Sedell. 2000. Toward an integrated classification of ecosystems: defining opportunities for managing fish and forest health. Environmental Management 25:425–444.

Roux, D. J., J. L. Nel, P. J. Ashton, A. R. Deacon, F. C. de Moor, D. Hardwick, L. Hill, C. J. Kleynhans, G. A. Maree, J. Moolman, and R. J. Scholes. 2008. Designing protected areas to conserve riverine biodiversity: lessons from a hypothetical redesign of the Kruger National Park. Biological Conservation 141:100–117.

Saunders, D. L., J. J. Meeuwig, and A. C. J. Vincent. 2002. Freshwater protected areas: strategies for conservation. Conservation Biology 16:30–41.

Sigler, J., and D. Zaroban. 2018. Fishes of Idaho: a natural history survey. Caxton Press, Caldwell, Idaho.

Smith, G. R. 1978. Biogeography of intermountain fishes. Great Basin Naturalist Memoirs 2(3):17–42.

Thurow, R. F., D. C. Lee, and B. E. Rieman. 1997.

Distribution and status of seven native salmonids in the interior Columbia River basin and portions of the Klamath River and Great basins. North American Journal of Fisheries Management 17:1094–1110.

Thurow, R. F., D. C. Lee, and B. E. Rieman. 2000. Status and distribution of Chinook Salmon and steelhead in the interior Columbia River basin and portions of the Klamath River basin. Pages 133–160 *in* E. Knudsen, C. Steward, D. MacDonald, J. Williams, and D. Reiser, editors. Sustainable fisheries management: Pacific salmon. CRC Press, Boca Raton, Florida.

UDWR (Utah Division of Wildlife Resources). 2000. Cutthroat Trout management: a position paper. Genetic considerations associated with Cutthroat Trout management. UDWR, Publication Number 00-26, Salt Lake City.

USGS (U.S. Geological Survey). 2004. Land ownership in western North America, 180 m. Sage-grouse rangewide conservation assessment. Snake River Field Station, U.S. Geological Survey, Snake River Field Station, Boise, Idaho.

USGS (U.S. Geological Survey). 2011. Protected areas database of the United States (PAD-US), version 1.2. U.S. Geological Survey, Gap Analysis Program (GAP), Moscow, Idaho. Available: https://gapanalysis.usgs.gov/padus/.

Wallace, R. L. and D. W. Zaroban. 2013. Native fishes of Idaho. American Fisheries Society, Bethesda, Maryland.

Walrath, J. D., D. C. Dauwalter, and D. Reinke. 2016. Influence of stream condition on habitat diversity and fish assemblages in an impaired upper Snake River basin watershed. Transactions of the American Fisheries Society 145:821–834.

Wenger, S. J., D. J. Isaak, C. H. Luce, H. M. Neville, K. D. Fausch, J. B. Dunham, D. C. Dauwalter, M. K. Young, M. M. Elsner, B. E. Rieman, A. F. Hamlet, and J. E. Williams. 2011. Flow regime, temperature, and biotic interactions drive differential declines of trout species under climate change. Proceedings of the National Academy of Sciences of the United States of America 108:14175–14180.

Williams, J. E., A. L. Haak, N. G. Gillespie, and W. T. Colyer. 2007. The conservation success index: synthesizing and communicating salmonid condition and management needs. Fisheries 32:477–492.

Williams, J. E., R. N. Williams, R. F. Thurow, L. Elwell, D. P. Philipp, F. A. Harris, J. L. Kershner, P. J. Martinez, D. Miller, G. H. Reeves, C. A. Frissell, and J. R. Sedell. 2011. Native fish conservation areas: a vision for large-scale conservation of native fish communities. Fisheries 36:267–277.

Williams, J. E. 2019. Multispecies and watershed conservation: increasing efficiency and effectiveness for native fish conservation. Pages 1–9 *in* D. C. Dauwalter, T. W. Birdsong, and G. P. Garrett, editors. Multispecies and watershed approaches to freshwater fish conservation. American Fisheries Society, Symposium 91, Bethesda, Maryland.

Zoellick, B. W., D. B. Allen, and B. J. Flatter. 2005. A long-term comparison of redband trout distribution, density, and size structure in southwestern Idaho. North American Journal of Fisheries Management 25:1179–1190.

American Fisheries Society Symposium 91:161–182, 2019

Spatial Conservation Assessment for Balancing Avoidance of Impacts of Tilapia Introduction on Imperiled Fish Biodiversity with Economic Impacts to the Aquaculture Industry

Monica E. McGarrity*
Texas Parks and Wildlife Department
4200 Smith School Road, Austin, Texas 78744, USA

Abstract—Introduction of invasive species, whether deliberate or inadvertent, poses a significant, human-mediated threat to biodiversity that has the potential to exacerbate other anthropogenic impacts and contribute to declines in fish biodiversity. In general, invasive species share life history traits that contribute to their success, such as prolific reproduction and the ability to thrive in highly disturbed habitats. These traits inherently pose difficulties in managing these species. Prevention is widely recognized as the most effective—and often only—management strategy. From a regulatory perspective, prevention often takes the form of a blacklist of prohibited species based on risk assessments, and both the list and associated regulations are developed with input from key stakeholder groups. In the case of prohibited tilapia in Texas, a long-standing regulatory exception allowing possession of Mozambique Tilapia *Oreochromis mossambicus* for pond stocking is believed to have resulted in frequent, unlawful escapes into public waters that may pose a threat to fish species of greatest conservation need and jeopardize the success of conservation initiatives. Spatial conservation prioritization is a method gaining popularity for identifying areas where conservation actions may have the greatest positive benefit, yet the "knowing-doing gap" between science and implementation is commonly recognized as a hurdle that must be overcome. There is a need to develop science to better inform regulatory approaches to prevention and management of invasive species. This is a unique, and successful, case study that demonstrates potential for development and implementation of systematic conservation planning approaches to conservation action that addresses this need. Spatial conservation assessment using Zonation software was a component of the planning framework that facilitated identification of science-based, regulatory zones that consider potential economic impacts to commercial stakeholders. Feedback from these stakeholders, decision makers, and those tasked with enforcement was obtained and incorporated into development of a realistic, balanced regulatory approach to mitigating impacts of invasive tilapia escape on imperiled fishes in Texas.

* Corresponding author: monica.mcgarrity@tpwd.texas.gov

Introduction

Tilapia are invasive cichlids with negative impacts on native fishes in Texas and elsewhere outside of their native range that are well documented in the primary literature (Barnes and Roesler 2017). The deleterious impacts of introduced tilapias on native fishes generally fall into one of three categories—competition, habitat degradation, and predation. Competition for food and nesting space is well documented (Buntz and Manooch 1968; Noble et al. 1975; Germany 1977; Hendricks and Noble 1980; Shafland and Pestrak 1983; Zale 1984; Zale and Gregory 1990; Traxler and Murphy 1995), as is habitat degradation resulting from voracious herbivory and prolific nesting activity (Junor 1969; Courtenay et al. 1974; Germany 1977; Noble and Germany 1986; Starling 1986). Multiple species of tilapia in the genus *Oreochromis*, as well as Redbelly Tilapia *Tilapia zillii*, also prey on other fishes, particularly juveniles (Junor 1969; de Moor et al. 1986; Starling 1986; Fuselier 2001; Moyle 2002; A. A. Schoenherr, Fullerton College, unpublished data, cited in Varela-Romero et al. 2002; Varela-Romero et al. 2002; Martin and Saiki 2005; Scoppettone et al. 2005; Doupé et al. 2009). Nile Tilapia *O. niloticus* populations have demonstrated the capacity to shift from herbivory to omnivory or piscivory over time in response to environmental change (Scoppettone et al. 2005; Bwanika et al. 2006). Introduction of tilapia has also been implicated in increases in nematode and trematode parasite infestations in native fishes (Fuselier 2001; Strecker 2006).

Tilapia are defined in the Texas Administrative Code as "harmful or potentially harmful fish" that may not be introduced into public waters (Harmful or Potentially Harmful Fish, Shellfish, and Aquatic Plants 2017). These rules include an exception that allows stocking of Mozambique Tilapia *O. mossambicus* into private ponds without a permit. In 2016 alone, 1,531 private ponds were stocked with Mozambique Tilapia (McGarrity and Davis 2017). Anecdotal reports suggest that these stockings of tilapia in private ponds may lead to frequent escapes of tilapia into public waters in Texas, and pond escapes are considered a key introduction pathway in Florida (Courtenay et al. 1974). In Texas streams and rivers, springs and wastewater outfalls are ubiquitous, and these thermal refuges pose a significant concern with regards to cold immunity and establishment and persistence of introduced tilapia (Noble and Germany 1986; Howells 1999; USFWS 2002), as well as potential for increased piscivory on juvenile native fishes (Scoppettone et al. 1998).

Decades-old research indicates that hybridization among tilapia species is common in Texas, both in aquaculture and in the wild, and that there may be no genetically distinct Mozambique Tilapia stocks in aquaculture (Howells 1991). Blue Tilapia *O. aureus* and Nile Tilapia, as well as hybrids among the three species, are also commonly cultured and may be sold for consumption by holders of exotic species permits (Harmful or Potentially Harmful Fish, Shellfish, and Aquatic Plants 2017). Neither coloration, nor meristics (which overlap among species), nor morphometrics, nor genetic analysis can reliably distinguish these hybrids from Mozambique Tilapia (Szyper 2000; Dijar Lutz-Carillo, Texas Parks and Wildlife Department, personal communication). In practice, this inability to definitively identify pure Mozambique Tilapia also may result in inadvertent stocking and escape of hybrids and poses difficulties for enforcement of the aforementioned regulatory exception.

Texas statute dictates that no person may introduce nonnative fish species into public waters of the state without an introduction permit issued by the Texas Parks and Wildlife Department (TPWD), stipulating that escape of fish placed in nonpublic water into public water without a permit is also a statutory violation (Introduction of Fish, Shellfish, and Aquatic Plants 2017). Stocking

Policy (2017) charges the TPWD with evaluation and prevention of potential for negative impacts of private-pond stocking on the existing biological ecosystem and state or federally listed threatened or endangered species or their critical habitat. The lack of regulatory oversight of Mozambique Tilapia in private ponds that would allow case-by-case evaluation of potential escape into public waters is problematic as it renders accomplishment of these goals unachievable. The only means of prevention has been through efforts to increase awareness among sellers of the illegality and potential detrimental impacts of tilapia escape.

Consequently, there is a need for science-based assessments to inform larger-scale, regulatory approaches to prevention and management of tilapia introductions in Texas. The first step in achieving this goal was identification of imperiled Texas fishes that may be impacted by inadvertent tilapia introductions. For the purposes of this study, the evaluation considered fish species of greatest conservation need (SGCN)—those rare or declining species identified through expert opinion and public feedback as being in need of conservation action to recover or prevent state or federal regulatory listing.

Potential Tilapia Impacts to Texas Fish Species of Greatest Conservation Need

Identification of Texas fish species of greatest conservation need potentially impacted by tilapia

Necessary first steps in evaluating ecological risks associated with fish stocking include identification of vulnerable species based on potential for negative ecological impacts of the species being introduced (Pearsons and Hopley 1999). A literature review was conducted to identify Texas fish SGCN that may be negatively impacted by tilapia and describe those potential impacts. Literature review generated a list of fish SGCN likely to be impacted by tilapia based on impacts to congeners that have been documented in Texas (Barnes and Roesler 2017). This nontarget list was expanded to consider impacts to congeners observed not only in Texas, but also elsewhere in North America and worldwide, resulting in the addition of genera *Gila* and *Cyprinodon*. Review of supporting literature used by Barnes and Roesler (2017) found that the only studies documenting impacts on *Notropis* were based on Red Shiner *Cyprinella lutrensis* (formerly *Notropis lutrensis*) and predated reclassification of this species (Germany and Noble 1977; Starling 1986). *Cyprinella* were added to the list of vulnerable fish SGCN and *Notropis* removed. Two subspecies recommended by Hendrickson et al. (2015) for SGCN designation in the next revision of the Texas Conservation Action Plan (TPWD 2012)—Rio Grande Largemouth Bass *Micropterus salmoides nuecensis* and Rio Grande Blue Catfish *Ictalurus furcatus* ssp.—were proactively included in the nontarget list for assessment. The Blotched Gambusia *Gambusia senilis* is currently an SGCN (TPWD 2012); however, it is presumed extirpated in Texas (Hubbs et al. 1991) and was not considered further. The final list of vulnerable fish SGCN included in the spatial conservation assessment in the current study included a total of 19 species (Table 1), including two centrarchids, four cyprinids, six cyprinodontids, three ictalurids, and four poeciliids. At least eight of these species are currently listed as critically imperiled at the state level (Table 1), and five are federally listed as endangered. The sections below provide descriptions of the relevant impacts of tilapia to congeners of these species.

Documented impacts of tilapia on Texas fish species of greatest conservation need congeners

Sunfishes (Centrarchidae).—Deleterious impacts of introduced tilapias have been documented for black bass *Micropterus* spp.

Table 1. Texas fish species of greatest conservation need likely to be impacted by tilapia introductions based on documented impacts on congeners. State abundance codes from the most recent Texas Conservation Action Plan were used to assign species weighting ranks (Table 2). Selection for inclusion in the spatial conservation prioritization model was based on availability of adequate geospatial data. State abundance code acronyms are defined in Table 2. Data types are either occurrence (OCC) or species distribution model (SDM; C1—class 1 or C2—class 2).

Species/subspecies by family	State abundance	Data type	Spatial model
Centrarchidae (sunfishes)			
Guadalupe Bass *Micropterus treculii*	S3	C2 SDM	YES
Largemouth Bass *M. salmoides nuecensis*	SNR	NONE	NO
Cyprinidae (minnows)			
Plateau Shiner *Cyprinella lepida*	S1S2	C2 SDM	YES
Proserpine Shiner *C. proserpina*	S2	C2 SDM	YES
Nueces River Shiner *Cyprinella* sp.	S1S2	C2 SDM	YES
Rio Grande Chub *Gila pandora*	S1	OCC	YES
Cyprinodontidae (pupfishes)			
Leon Springs Pupfish *Cyprinodon bovinus*	S1	C2 SDM	YES
Comanche Springs Pupfish *C. elegans*	S1	C2 SDM	YES
Conchos Pupfish *C. eximius*	S1	OCC	YES
Devils River Pupfish *C. eximius* ssp.	SNR	NONE	NO
Pecos Pupfish *C. pecosensis*	S1	C2 SDM	YES
Red River Pupfish *C. rubrofluviatilis*	SNR	C2 SDM	YES
Ictaluridae (North American catfishes)			
Rio Grande Blue Catfish *Ictalurus furcatus* ssp.	SNR	C2 SDM	YES
Headwater Catfish *I. lupus*	S2	C2 SDM	YES
Chihuahua Catfish *Ictalurus* sp.	SNR	OCC	YES
Poeciliidae (topminnows)			
Big Bend Gambusia *Gambusia gaigei*	S1	OCC	YES
Clear Creek Gambusia *G. heterochir*	S1	OCC	YES
Spotfin Gambusia *G. krumholzi*	S1	OCC	YES
Pecos Gambusia *G. nobilis*	S2	C2 SDM	YES

These impacts have been attributed largely to the presence of Mozambique Tilapia and Blue Tilapia, although some studies noted that the co-occurrence of Redbelly Tilapia, Spotted Tilapia *Tilapia mariae*, or Blackchin Tilapia *Sarotherodon melanotheron* may have also contributed to these impacts (Junor 1969; Courtenay et al. 1974). Negative impacts of tilapia on black bass are exerted primarily through competition for food and spawning areas (Buntz and Manooch 1968; Junor 1969; Germany 1977; Zale 1984; Noble and Germany 1986; Starling 1986; Traxler and Murphy 1995). However, impacts from habitat degradation caused by herbivory and nesting activity (Junor 1969; Courtenay et al. 1974; Germany 1977) and direct predation (Junor 1969) have also been documented. Fairfield Lake and Lake Trinidad are Texas power plant cooling reservoirs with year-round warm temperatures that have enabled introduced Blue Tilapia to not only persist, but also proliferate and achieve high biomass (Germany and Noble 1977; Hanifen 1981). In both lakes, Blue Tilapia cold mortality events resulting from power plant shutdowns were followed by an observed decrease in mean catch per unit effort (CPUE) for Blue Tilapia and a subsequent increase in mean CPUE for Largemouth Bass (Germany and Noble 1977; Starling 1986). These studies documented production of a strong Largemouth Bass year-class following Blue Tilapia declines and attributed this finding to reduction in competition with Blue Tilapia for nesting space and harassment that may have resulted in reproductive inhibition.

Minnows (Cyprinidae).—Harmful impacts of introduced tilapias have been documented for the cyprinid genera *Cyprinella* and *Gila*. Following the aforementioned Blue Tilapia cold mortality events at Fairfield Lake and Lake Trinidad, considerable increases in mean CPUE for Red Shiner were also observed (Germany and Noble 1977; Starling 1986). These studies suggest that a reduction in incidental ingestion or disturbance of Red Shiner eggs by Blue Tilapia likely played a key role in increased abundance of Red Shiners. In the Virgin River in Arizona and Nevada, USA, it is believed that Blue Tilapia prey upon Virgin Chub *Gila seminuda* (USFWS 2004; Canonico et al. 2005). Following introduction of Blue Tilapia in the Muddy River, a tributary of the Virgin River in Nevada, declines were observed first in aquatic vegetation and then in native fish abundance (Scoppettone et al. 1998). Analysis of tilapia gut contents as part of this study found that fishes—including the endemic, endangered cyprinids Moapa Dace *Moapa coriacea* and Moapa White River Springfish *Crenichthys baileyi moapae*—comprised 44% of the total volume consumed by tilapia.

Pupfishes (Cyprinodontidae).—Negative impacts of introduced tilapias have been documented for the cyprinodontoid genus *Cyprinodon*. A field study documented broad habitat overlap and agonistic interactions between Blackfin Pupfish *C. beltrani* and the nonindigenous Mozambique Tilapia in Laguna Chichancanab, Mexico (Fuselier 2001). Laboratory experiments conducted as part of the same study found that the presence of Mozambique Tilapia caused a shift in habitat use by Maya Pupfish *C. maya* and Thicklip Pupfish *C. labiosus*, likely resulting in increased behavioral competition with Blackfin Pupfish for nesting space and feeding pits.

In both Mexico and the southwestern United States, introduced tilapias (i.e., Mozambique Tilapia and Redbelly Tilapia considered jointly) are considered one of the greatest threats to the survival of the federally endangered Desert Pupfish *C. macularius* and, in many critical refugia, have largely replaced this imperiled fish (Schoenherr 1981, 1988; Varela-Romero et al. 2002). Abundance of Desert Pupfish was found to be negatively correlated with the combined abundance of Mozambique and Redbelly tilapias within Salt Creek in the Salton Sea basin in California (Martin and Saiki 2005). Although adult tilapia are too large to trespass into the shallow-water spawning areas of Desert Pupfish, juvenile tilapia readily enter these areas and passively affect the pupfish. Dominant males chase juvenile tilapia away, expending energy that would otherwise be allocated to spawning efforts and exposing eggs to greater risk of predation by subdominant males (Schoenherr 1988). Direct predation of Desert Pupfish by tilapia is also a significant concern (Schoenherr 1988). In Laguna Chichancanab, gut contents of Mozambique Tilapia have been found to contain cyprinodontid remains (Fuselier 2001). Predation on Desert Pupfish eggs by Redbelly Tilapia has also been observed (Schoenherr unpublished data, cited in Varela-Romero et al. 2002). The introduction of Mozambique Tilapia has also been found to be associated with substantial increases in ectoparasitic trematode and endoparasitic nematode infestations in the Laguna Chichancanab flock (Fuselier 2001; Strecker 2006).

North American catfishes (Ictaluridae).—The aforementioned Blue Tilapia cold mortality event at Fairfield Lake suggested detrimental impacts of introduced tilapias to the ictalurid genus *Ictalurus*. Mean CPUE for Channel Catfish *I. punctatus* was significantly higher following the Blue Tilapia mortality event as compared to data collected prior to the event (Hanifen 1981). Proportional stock density, young-adult ratio, and total length decreased following the event, suggesting an increase in reproduction and recruitment (Hanifen 1981; Starling 1986). Observed changes in Channel Catfish populations were attributed to reduced competition for food resources and potential reduction in predation of young Channel Catfish by Blue Tilapia. As Blue Tilapia populations recovered, concomitant declines in Channel Catfish CPUE were noted (Starling 1986).

Topminnows (Poeciliidae).—Predation on the Western Mosquitofish *Gambusia affinis* was documented in an analysis of gut contents of Blue Tilapia collected from the Muddy River, a tributary of the Virgin River in Nevada, (Scoppettone et al. 2005). Western Mosquitofish was the predominant fish species consumed and comprised 24% of the total volume consumed.

Systematic Conservation Planning Framework

The systematic conservation planning (SCP) process requires development of a framework for achieving high-level goals in a way that provides a stepwise plan for informing as well as implementing a conservation strategy (Moilanen et al. 2009, 2011b; Pool et al. 2013; Adams et al. 2019). The SCP framework seeks to identify conservation priority areas, find balance between trade-offs (e.g., conservation versus stakeholder needs), and transition from assessment to implementation. The nuances of the problem depend upon the conservation need and goals, the type of conservation action being considered, and the implications thereof.

Context for the SCP framework in the current study is provided by a need to revise current regulations to minimize potential biodiversity impacts on vulnerable Texas fish SGCN of tilapia escapes from ponds. Impetus for this study was generated by the need for a path to move forward with ongoing review and revision of the regulations pertaining to harmful or potentially harmful exotic fish. The overarching SCP framework (Figure 1) for this study had the ultimate, high-level goal of developing and implementing reasonable, enforceable regulations to address this problem and fulfill this need.

The SCP framework for this study (Figure 1) included (1) spatial conservation assessment to identify areas with high conservation potential and low economic impact of regulations-based conservation actions, (2) translation of the assessment model into an implementable form by identifying enforcement-based implementation units as well as management landscapes, and (3) development of a regulations-based conservation action strategy that incorporates stakeholder input. This is a unique, and successful, case study that demonstrates potential for spatial conservation assessment to be used as a tool within a SCP framework to contribute to development and implementation of science-based regulatory approaches to conservation action.

Spatial Conservation Assessment Software and General Setup

The spatial conservation assessment conducted for this study created a conservation prioritization for the entire state of Texas based on balancing potential impacts to fish SGCN with opportunity costs for tilapia sellers to develop a regulations-based conservation action strategy. The iterative process of this assessment incorporated

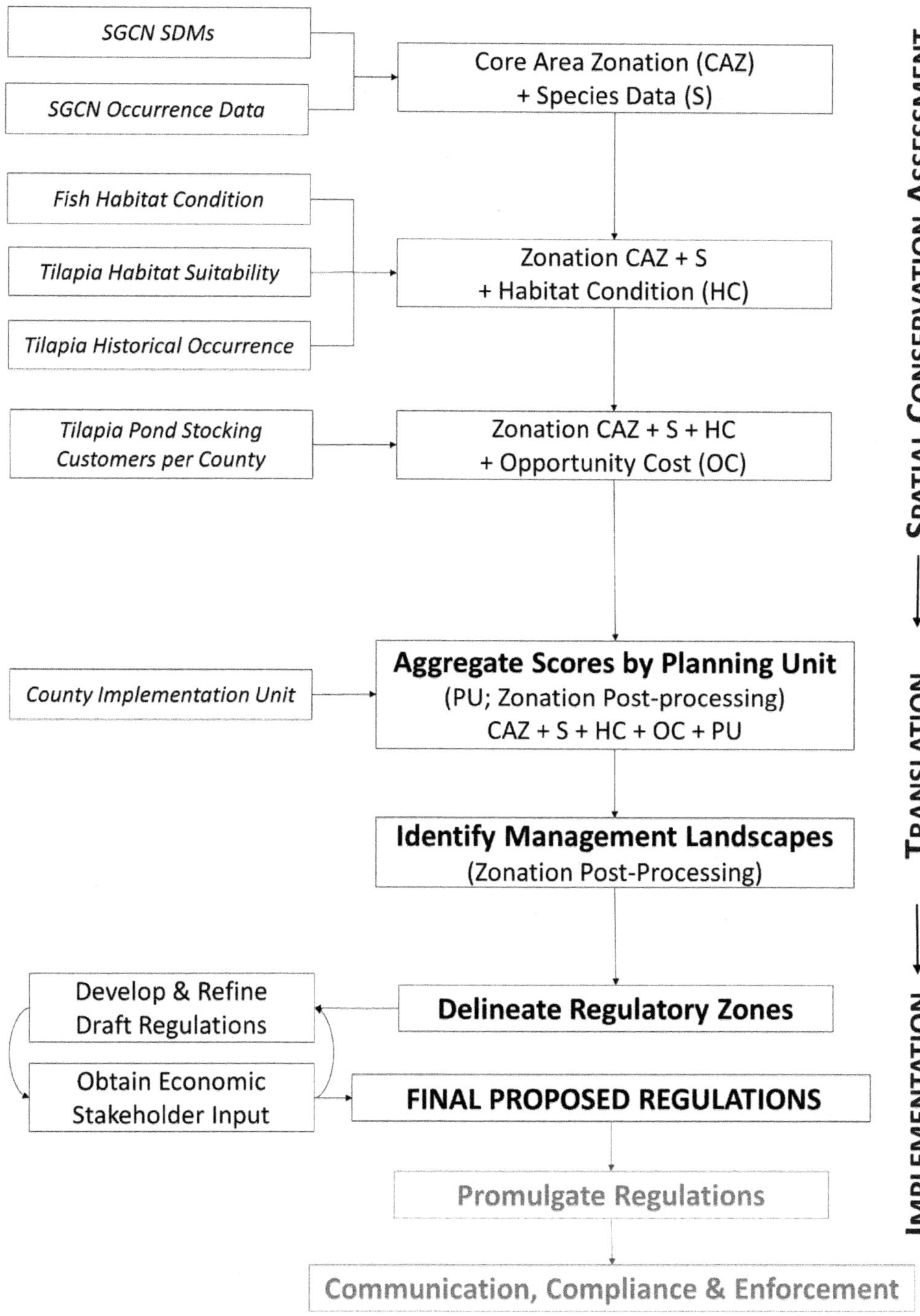

Figure 1. Schematic of the systematic conservation planning (SCP) framework and the spatial conservation assessment components thereof. The SCP framework steps are shown on the right and correspond to the sections in which each step is discussed. In this schematic, species of greatest conservation need (SGCN) refers to those Texas fish SGCN determined likely to be vulnerable to impacts from tilapia escapes from private ponds; SDMs are Fishes of Texas species distribution models.

key biodiversity, habitat condition, and opportunity cost factors. The final product of the assessment is a hierarchical ranking of county implementation units that reflect these trade-offs between protections for imperiled species and avoidance of economic impacts to commercial stakeholders in Texas.

Software selection

The software Zonation (Lehtomäki and Moilanen 2013; Moilanen et al. 2014) was selected for this spatial conservation assessment primarily for the advantages of the flexibility and adaptability that it provides, and it is publicly available free of charge. Zonation inherently assumes that protecting the entire landscape provides the best conservation benefit and removes cells—or planning units—in an iterative ranking process that results in the least loss of biodiversity. The flexibility of this software also facilitates inverse prioritization of features representing lost opportunity costs for stakeholders and translation of landscape-scale prioritization for conservation action to an implementation unit scale (Kareksela et al. 2013).

Zonation models are readily modifiable to enhance or adapt the analysis by incorporating additional components (e.g., new species, fish barriers, and thermal refuges) as data become available or imperiled species lists change. The conservation prioritization concept function, layers included, weights for layers, and other functions and parameters are set in simple text files such that parameter changes and iterative model comparisons can be accomplished easily by creating different versions of, often, a single text file and run consecutively through use of multiple command lines.

Zonation functions

Basic Zonation functions.—The edge removal function of Zonation was employed to reduce computation times and increase the connectivity of high-quality habitats in the landscape structure. The add edge points function was also used as a complement to edge removal that prevents loss of valuable areas without increasing computation time.

Conservation prioritization concept.—The core-area Zonation (CAZ) cell removal function selected for this analysis minimizes loss of biodiversity and considers important species occurring in biodiversity-poor areas (i.e., rarity focus). This prioritization concept seeks to retain the core areas—those locations with highest occurrence—of all species throughout the cell removal process, even the rare ones (Lehtomäki and Moilanen 2013). Core-area Zonation is not a quantitative, target-based approach (e.g., preserve top 10% of habitat), but rather ranks all geographic areas in the analysis landscape from least to most valuable for conservation to facilitate evaluation of trade-offs between conservation and economic losses in an informative and flexible way (Kareksela et al. 2013).

Spatial Conservation Assessment Workflow

Performing a spatial conservation assessment in Zonation is a stepwise modeling process such that each step builds upon previous steps. Many of these steps involve iterative comparisons of alternative model solutions, thus providing significant benefits in terms of model flexibility and adaptability. The current assessment assigned the highest conservation value to areas important for protecting vulnerable fish SGCN biodiversity with high-quality habitat, no records of tilapia, high risk of tilapia establishment and persistence, and low economic value to commercial stakeholders. The final spatial conservation assessment model yielded a prioritization of the Texas landscape across a spectrum that is representative of both conservation and economic interests.

Weighting and analysis of vulnerable fish species of greatest conservation need biodiversity features (CAZ + S)

Spatial distribution data for biodiversity features form the basis of the Zonation spatial conservation assessment model and must represent the best available data for each fish species (S) under consideration. Species distribution models (SDMs; Figure 2a) previously developed by the Fishes of Texas (FoT) project for SGCN consider point occurrence data from museum records along with climate and environmental data to yield a probabilistic model of species occurrence (Hendrickson and Cohen 2015). The FoT class 02 (C2) SDMs incorporate more occurrence and hydrologically relevant environmental data than earlier class 01 (C1) SDMs and account for survey sampling bias (Labay et al. 2012a, 2012b). These SDMs are publicly available from FoT (www.fishesoftexas.org/models) as raster data with 30-arc-second (i.e., 1 km^2 at equator) resolution (Labay et al. 2011) and are publicly available (www.fishesoftexas.org).

Of the 19 Texas fish SGCN for which congeneric impacts were documented, C2 SDMs were available for 11 (Table 1); neither C2 nor C1 SDMs were available for the remaining species. For six of the eight remaining species, occurrence data were publicly available through FoT (www.fishesoftexas.org/data). The FoT occurrence data set is

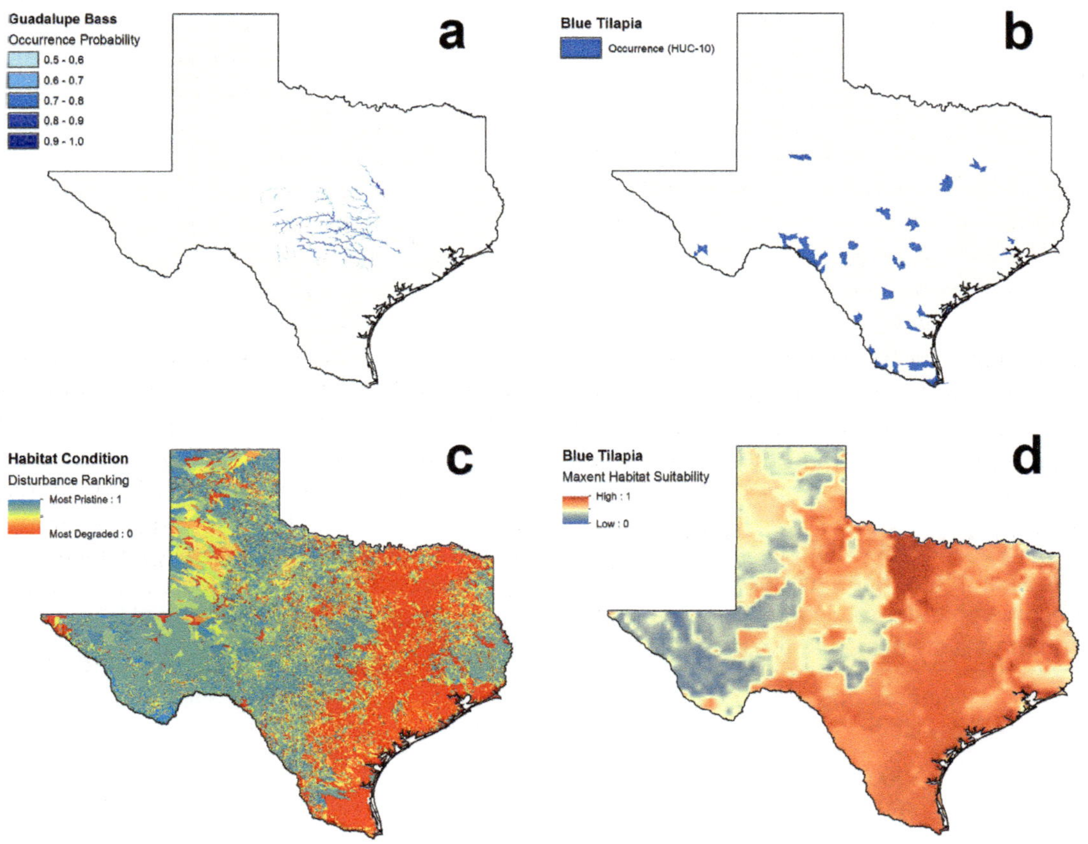

Figure 2. Data used in spatial conservation assessment included **(a)** species distribution models—only values >0.5 shown to represent core areas, **(b)** occurrence data as presence–absence at the Hydrologic Unit Code 10 watershed level, **(c)** National Fish Habitat Partnership habitat condition rankings, and **(d)** Maxent model of tilapia habitat suitability.

the most comprehensive available and includes Global Biodiversity Information Facility data (www.gbif.org). Occurrence data were transformed into binary presence–absence data at the watershed level (i.e., Hydrologic Unit Code 10; USGS-USDA 2012), as shown in Figure 2b. Although aggregation can also be accomplished by summing the number of occurrences within the watershed rather than representing these as binary data (Lehtomäki and Moilanen 2013), doing so could conflate sampling bias and conservation value. Two subspecies—Devils River Pupfish and Rio Grande Largemouth Bass could not be included in the spatial assessment due to lack of sufficient occurrence data. Prior to analysis, occurrence data were converted to rasters, and these and all subsequently described data input layers were resampled to the same resolution as the C2 SDMs and clipped to the same extent.

For each fish SGCN likely to be impacted by tilapia (Table 1), state abundance rankings were translated into positive weighting ranks (Table 2), following a methodology comparable to that of Labay and Hendrickson (2014). Instead of weighting critically threatened and endangered species higher to focus on preventing extinctions, the reverse approach of weighting vulnerable species higher (i.e., SGCN focus) was used to prioritize protection of those species most likely to become endangered in the future. Assignment of a higher weight to a species results in a greater fraction of its distribution, relative to other species with lower weights, being retained throughout the cell removal process (Lehtomäki and Moilanen 2013). The SGCN-focus weighting was selected as the approach most amenable to proactive identification of regulatory conservation actions and minimization of landowner stakeholder concerns regarding regulatory oversight that could result from federal listing of an SGCN. The CAZ + S model (Figure 3a) prioritized geographic areas solely for biodiversity of vulnerable Texas fish SGCN.

Consideration of habitat condition (CAZ + S + HC)

The next step in model development (Figure 3b; CAZ + S + HC) included habitat condition (HC) factors in a manner that assigned higher value to areas that are most pristine, where tilapia presence has been documented previously and where there is a high risk of escaped tilapia becoming established and persisting. Although the Zonation software allows for assignment of groups of species to habitat "condition groups" based on their response to habitat degradation, the response of each fish SGCN to habitat loss and fragmentation was rigorously considered (Labay and Hendrickson 2014) in development

Table 2. Species weights assigned by state abundance code for the species of greatest conservation need focus modeling approach prioritize protection of those species most likely to become endangered in the future (i.e., higher weights). State abundance codes of S1S2 were conservatively assigned the higher S2 weight.

Abundance code	Definition	Species weight
SX	Presumed extirpated	0
SH	Possibly extirpated	0
S1	Critically imperiled	2
S1S2	Critically imperiled/imperiled	3
S2	Imperiled	3
S3	Vulnerable	4
SNR	Unranked	1

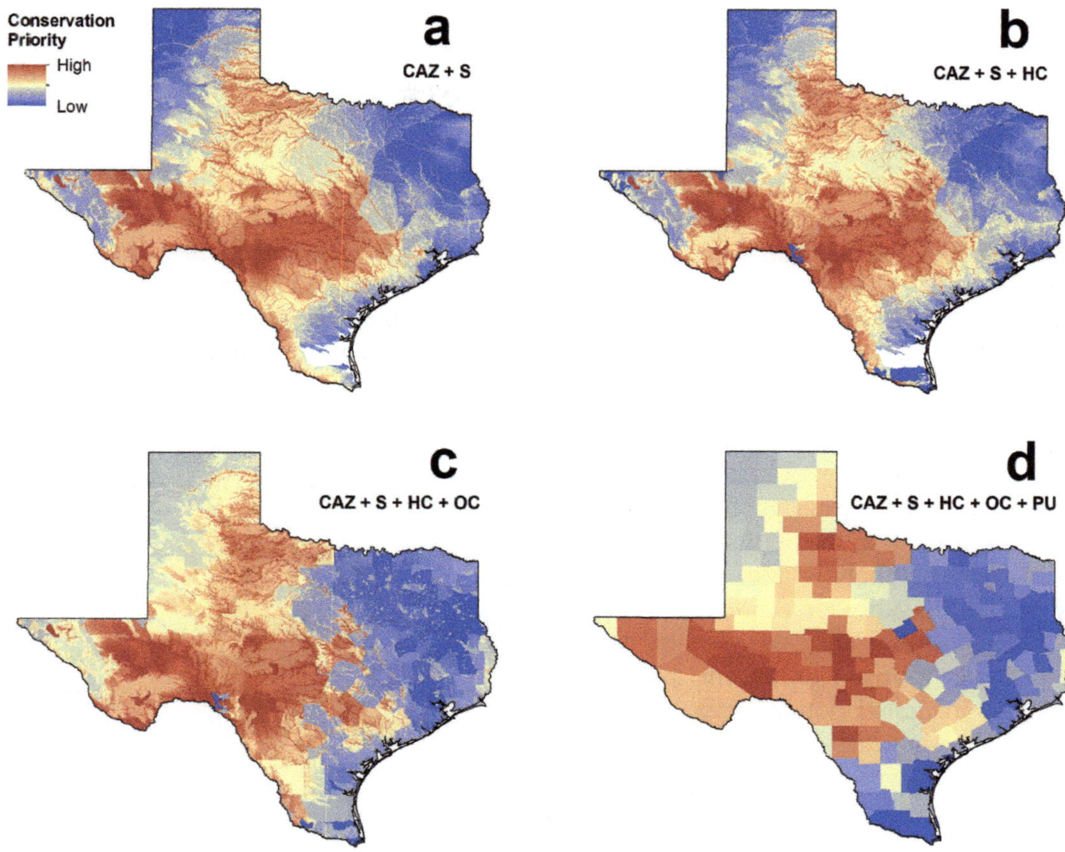

Figure 3. Zonation analysis workflow and intermediary model outputs for conservation prioritization: **(a)** core-area Zonation concept-based prioritization of weighted fish species (S) of greatest conservation need distributions (CAZ + S model), **(b)** addition of habitat condition factors that included inversely weighted tilapia occurrence data (CAZ + S + HC model), and **(c)** consideration of opportunity costs (CAZ + S + HC + OC model). The first step in translation of the final spatial conservation assessment model (c) for implementation involved **(d)** aggregation of spatial prioritization by inclusion of county as a planning unit (CAZ + S + HC + OC + PU model).

of the FoT SDMs, and, for the purposes of this model, assigning species to condition groups would be duplicative. This step in the assessment resulted in a relatively minor refinement of the CAZ + S model.

Fish habitat condition.—The National Fish Habitat Partnership (NFHP) habitat condition index cumulatively scores habitat based on landscape factors that include land use, human population density, roads, dams, mines, and point-source pollution (Esselman et al. 2011). The NFHP scores rank habitat condition from highly degraded to most pristine, such that higher scores represent, presumably, higher conservation value (Figure 2c). The NFHP scores were normalized on a scale of zero (i.e., highly degraded) to one (i.e., most pristine) for incorporation in the current assessment. The assignment of higher NFHP scores to more pristine areas resulted in their retention in the model prioritization further into the removal process of the analysis. Construction of the 11 SDMs used in this analysis in Zonation also incorporated NFHP habitat condition scores (Labay and Hendrickson 2014). The ESRI Band Collection Statistics multivariate analysis tool (ArcGIS 10.5) was used to explore relationships between the NFHP scores and each of the 11 SDMs. The resulting matrix showed no problematic spatial autocorrelations (i.e., all $r \leq 0.21$) that would preclude incorporation of the NFHP scores into the spatial conservation assessment.

Tilapia habitat suitability.—Incorporation of tilapia habitat suitability into the spatial conservation assessment model (Figure 1) ensured that areas with high risk of tilapia establishment and persistence were retained in the prioritization further into the removal process of the analysis. Maxent models of habitat for Mozambique Tilapia, Nile Tilapia, and Blue Tilapia were developed by Barnes and Roesler (2017). The Maxent software is designed to model potential species distributions, scoring raster grid cells on a scale of zero to one, with higher values indicating greater habitat suitability. A Maxent model of habitat suitability for all three species combined was not developed. The Maxent model of Blue Tilapia habitat suitability (Figure 2d) was the most robust and predicted the broadest geographic distribution of suitable habitat. Because physiological tolerances show little variation among these three tilapia species (Barnes and Roesler 2017), the Blue Tilapia Maxent model was deemed appropriate for inclusion in the current assessment to represent overall tilapia habitat suitability.

Historical occurrence of tilapia.—The presence of introduced tilapias in Texas waters has been suggested by commercial stakeholders to be a contraindication for regulatory oversight of tilapias in general. Historical tilapia (i.e., *Oreochromis* spp.) occurrence data were included in the model (i.e., as a biodiversity layer) to account for this supposition. As previously described for fish SGCN occurrence data, tilapia occurrence data were transformed into binary presence–absence data at the watershed level (Figure 2b). Data were negatively weighted such that areas where tilapia presence has been documented were treated by the Zonation software as undesirable and removed from prioritization as soon as possible (Moilanen et al. 2014). There is no hard and fast rule as to the size of negative weights. Relatively large weights confer lower conservation value on areas where tilapia presence has been documented, and smaller negative weights give less importance to the historical occurrence of these invasive species.

Qualitative visual assessment was used to compare CAZ + S + HC model solutions (i.e., alternative models that incorporated all three habitat condition factors) derived using different negative weights for areas with records of tilapia occurrence. These model solutions were compared to core reference areas with high potential economic impact on commercial stakeholders (i.e., sellers of tilapia for pond stocking) or high conservation priority. Areas with high potential economic impact on commercial stakeholders, identified as described in the next section, were used to represent economic interests. Native fish conservation areas (NFCAs) were used to represent areas of high conservation importance for Texas fish SGCN. The NFCAs are watersheds identified through spatial conservation assessment that are critically important to the long-term persistence of imperiled fishes (Birdsong et al. 2019, this volume). The highest negative weight (i.e., −4) for tilapia historical occurrence provided the best balance between economic and conservation priorities and was included in the final CAZ + S + HC model. This weight is comparable to the highest positive weight (i.e., +4) assigned to any fish SGCN.

Avoidance of economic impacts on commercial stakeholders (CAZ + S + HC + OC)

Tilapia are cultured and sold in Texas for three purposes—food, pond stocking, and home aquaponics. Mozambique Tilapia are commonly stocked in private ponds to provide biological control of algae and as prey for sport fish. The aforementioned regulatory exception that allows purchase and stocking of Mozambique Tilapia into private ponds requires that they be purchased from an exotic species permit holder (Harmful or Potentially Harmful Fish, Shellfish, and Aquatic Plants 2017). The number of exotic species permits issued for sale of Mozambique Tilapia for the purposes of pond stocking has shown an upward trend in recent years. As of February 2019, 30 applicants for 2019 exotic species permits, including renewals, have indicated that they intend to sell Mozambique Tilapia for "pond management" (TPWD, unpublished data). Potential conservation actions for minimizing impacts of escape of tilapia stocked in private ponds on Texas fish SGCN must consider the potential for negative economic impacts on commercial stakeholders—those permitted tilapia sellers that might lose customers as a result of conservation action (i.e., experience a lost opportunity cost in terms of lost revenue).

Regulations require that each permitted seller submit Exotic Species Transport Invoices (ESTIs) to the TPWD for each sale of live tilapia, although the purpose of the sale (e.g., private pond stocking) is not required on the ESTIs (Harmful or Potentially Harmful Fish, Shellfish, and Aquatic Plants 2017). These ESTIs must include the address of the destination (i.e., pond location). Because revenue generated by sales of tilapia is not required to be reported, the number of unique destination locations was evaluated as a proxy for lost opportunity cost (OC) to sellers. Sellers' (2016) ESTIs were reviewed and those (1) reporting that only Mozambique Tilapia were sold (i.e., the species that may be sold for stocking in private ponds) and (2) giving no indication that the buyer was a restaurant, market, fish dealer, or aquaponics store (i.e., not sold as food or for aquaponics) were selected for this spatial conservation assessment. Unique 2016 locations were identified based on buyer address; locations with multiple 2016 ESTIs were only counted once.

A total of 1,531 unique, presumptive private pond customer locations (hereafter, "customers") were identified across 141 Texas counties, with up to 58 customers per county. Number of customers was included in the current assessment as a proxy for the potential lost OC to aquaculture stakeholders. Number of customers was aggregated at the county level (Figure 4), and county-level data were normalized prior to assessment on a scale of zero to one, ranked such that higher values equate to higher opportunity cost. The OC was considered in the CAZ + S + HC + OC model by negatively weighting number of customers per county. The negative weighting strategy seeks to avoid economic impacts by reducing the conservation value of areas with high OC (Moilanen et al. 2014). Alternative model solutions were generated for two different negative weights (i.e., −1 and −2) for the OC, with a more negative weight placing higher value on the OC factor.

Qualitative visual assessment of OC weighting strategies was performed, as previously described for CAZ + S + HC, by comparing alternative CAZ + S + HC + OC model solutions to core reference areas with high potential economic impact on commercial stakeholders or high conservation priority. The more negative weight (i.e., −2) strongly favored protection of economic interests and substantially devalued the headwaters of NFCAs in the Edwards Plateau ecoregion of central Texas (Figure 5a, 5b). The less negative weight (i.e., −1) provided a defensible, balanced model for corepresentation of these opposing priorities (Figure 5c, 5d) and was selected for inclusion in the final CAZ + S + HC + OC model (Figure 3c).

Linking Spatial Assessment and Implementation Strategy

The SCP framework required translation of the final spatial conservation assessment

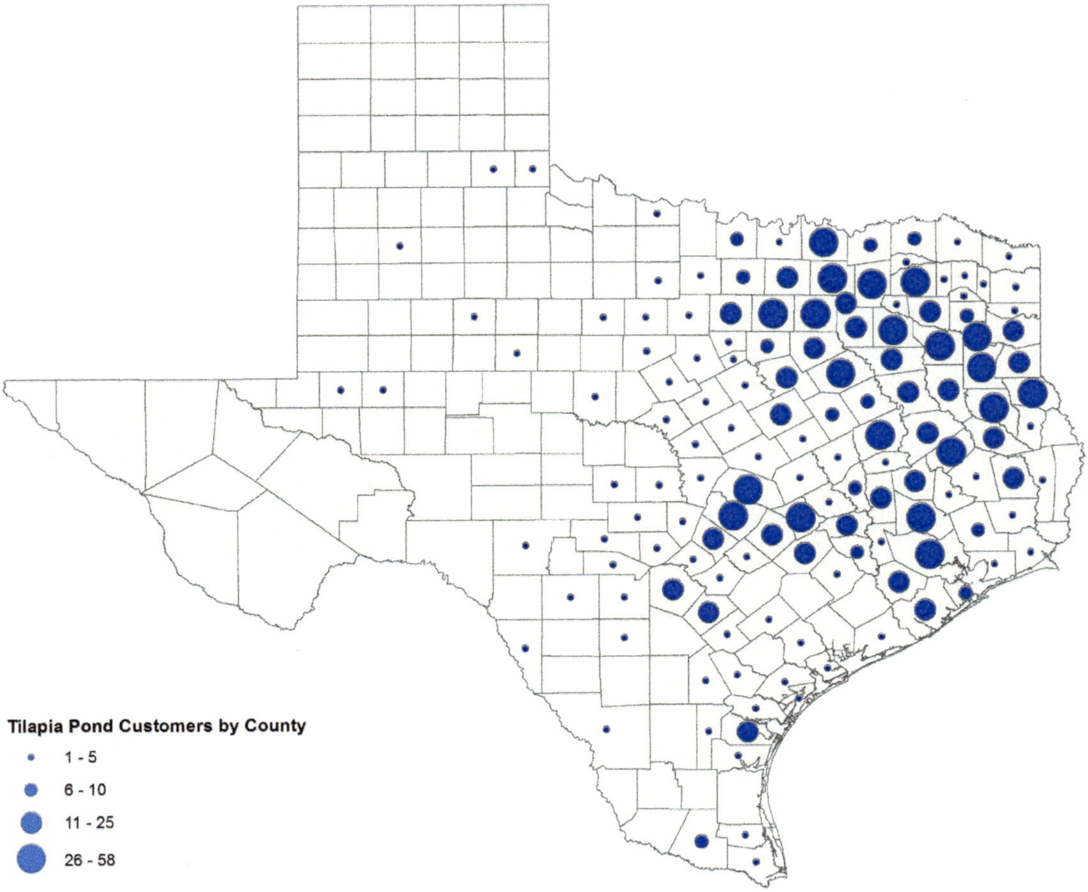

Figure 4. Number of unique 2016 tilapia stocking private-pond customer locations by county.

model (i.e., CAZ + S + HC + OC; Figure 3c) into products that could be used to develop a feasible implementation strategy (Figure 1). This step in the SCP framework was approached collaboratively by incorporation of feedback from an interdisciplinary team of TPWD biologists, law enforcement, permitting staff, and decision makers.

Assessment translation

Aggregate scores by implementation unit (CAZ + S + HC + OC + PU).—Development of a regulatory strategy for conservation action required identification of geographical implementation units that are unequivocal, enforceable, and readily recognizable to both the commercial stakeholders and their customers. County units undoubtedly meet these criteria and have the added benefit of contextual familiarity as many fishing and hunting regulations are enacted, communicated to the public, and enforced at county level. The planning unit function of Zonation is used to translate an assessment model such that grid cells (e.g., individual watersheds) are grouped into distinct units (Moilanen et al. 2014). The final CAZ + S + HC + OC model (Figure 3c) was translated by adding county as a planning unit (PU) layer in model postprocessing (Figure 1). Addition of a planning unit layer yielded an aggregate CAZ + S + HC + OC + PU model with each county ranked and prioritized as a discrete unit (Figure 3d).

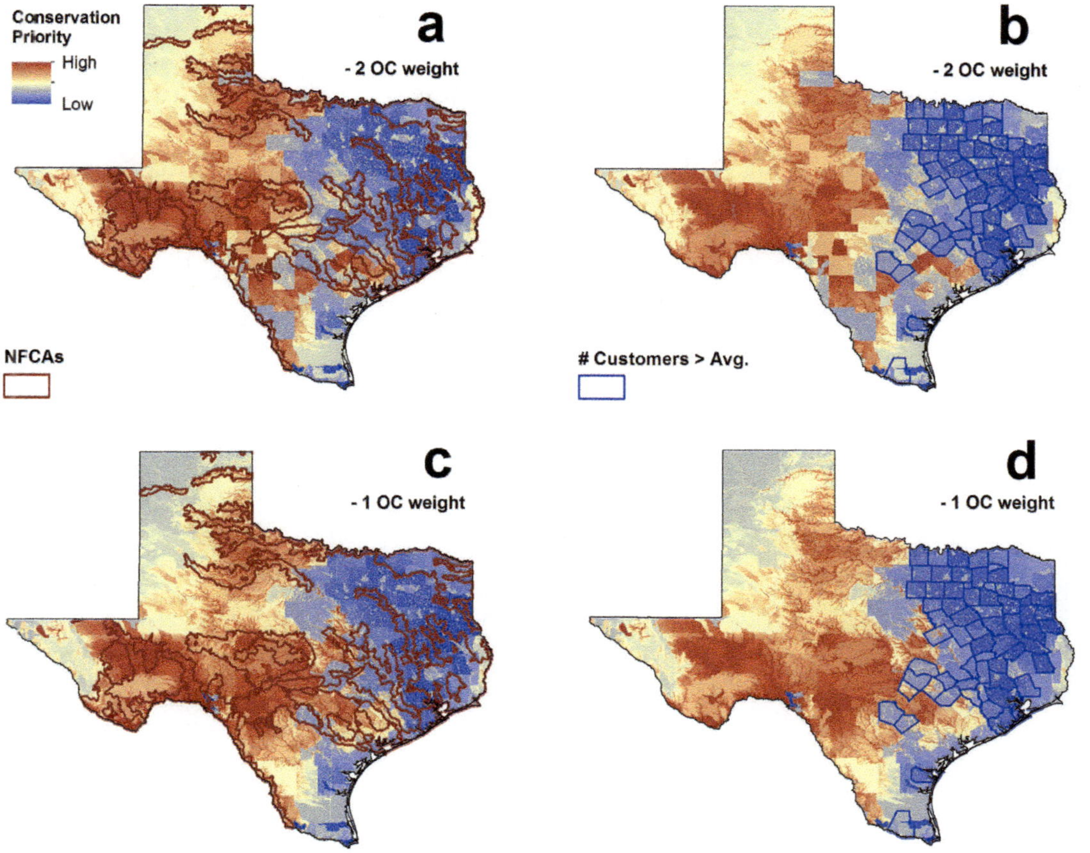

Figure 5. Qualitative visual assessment of two opportunity cost weighting solutions, **(a–b)** –2 and **(c–d)** –1, in comparison to reference areas representing maximal conservation benefits (i.e., native fish conservation areas) versus high opportunity cost (i.e., counties with number of private-pond customer locations greater than the average of six).

Identification of management landscapes.—Landscape identification is a postprocessing tool in Zonation that identifies separate priority landscapes (e.g., NFCAs) based on proximity and similarity of model attributes (Moilanen et al. 2005; Labay and Hendrickson 2014). This postprocessing tool was used to generate an objective basis for delineation of regulatory zones within the current SCP framework (Figure 1). The percentage of the landscape included in the classification was set to 50. Inclusion minimum was set to 10, meaning that each conservation-priority landscape must include at least one county with a conservation value in the top 10% (i.e., rank of ≥0.90 or greater). Nearest neighbor maximum distance allowed between counties joined to the same priority landscape was set to 25 grid cells (approximately 20 km), following the methods of Labay and Hendrickson (2014). Maximum difference in feature composition was set as 0.2, meaning that counties would not be joined to the same priority landscape if ≥20% of all species had a 1-log difference in their probability density. Landscape rankings were created by postprocessing of the CAZ + S + HC + OC + PU model (Figure 6).

Implementation

Delineation of regulatory zones.—Successful implementation necessitates the

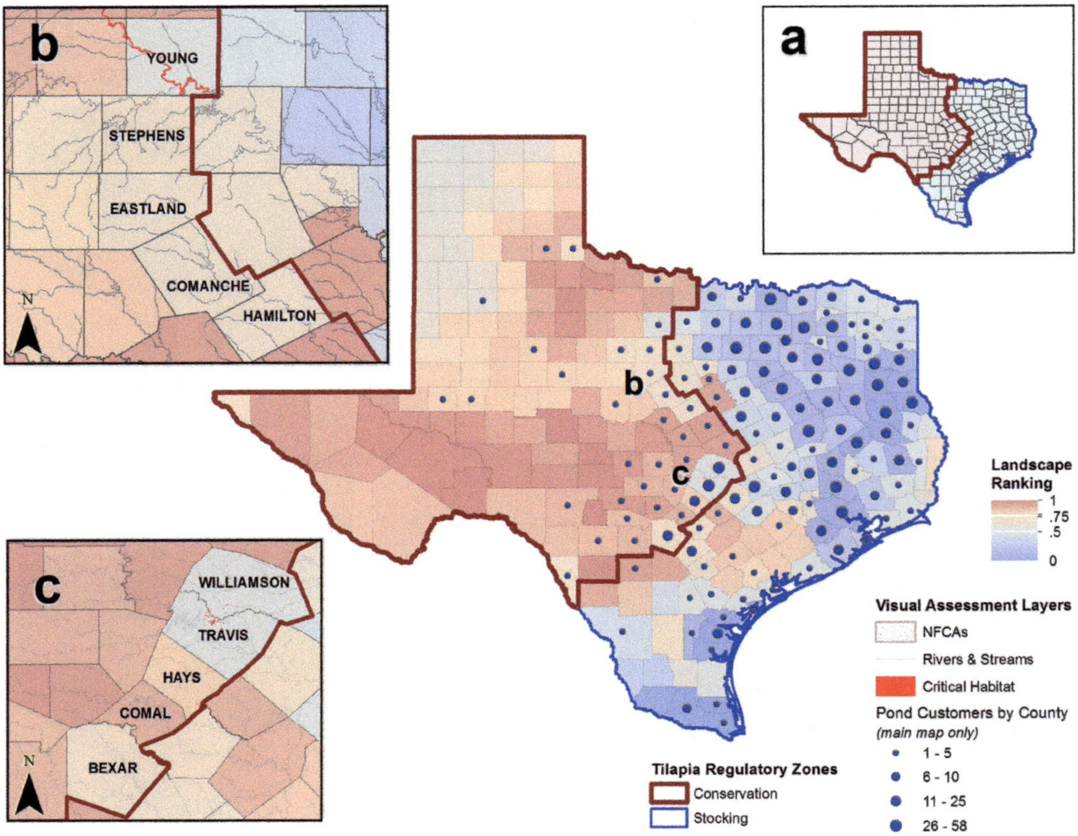

Figure 6. Qualitative visual assessment of conservation landscape rankings in comparison to representations of conservation interests (i.e., native fish conservation areas) versus economic opportunity (i.e., private-pond customer locations by county) used to develop draft regulatory zones. This qualitative assessment, with input from stakeholders and decision makers, yielded **(a)** delineated conservation and stocking zones for regulations-based conservation action. Finer-scale evaluation of potential county-level conflicts near the nexus between identified landscapes also considered **(b, c)** critical habitat designated for any non-cave-dwelling aquatic species and (b) potential for tributary basin discharge into critical habitat or native fish conservation areas in adjacent counties.

ability to demonstrate the linkage between science and regulatory recommendations, communicate proposed regulations to economic stakeholders for the purpose of feedback, and ultimately communicate enacted regulations to the public (i.e., pond owners), commercial stakeholders, and game wardens. Regulatory zones were delineated in the implementation step of the SCP framework based on landscape rankings (Figure 1). Qualitative visual assessment of the conservation landscape rankings generated in model postprocessing (Figure 6) was used to delineate the regulatory zones. The visual assessment involved comparison of landscape rankings with core conservation areas (i.e., NFCAs) and economic opportunity (i.e., raw number of customers by county) to delineate zone boundaries. Two draft zones were identified—a "conservation zone" for vulnerable fish SGCN protection and a "stocking zone" with high potential OC (Figure 6a). Delineation of only two regulatory zones sought to prevent the difficulties in communication, compliance, and enforcement that would likely plague a mosaic of zones.

Zone contiguity was desirable to aid in implementation and enforcement of regulatory-based conservation actions. Therefore, marginal counties between draft-identified zones were carefully evaluated, and each county was placed into one of the two regulatory zones. Seven marginal counties had relatively high conservation priority in the landscape ranking (i.e., >0.7 on a scale of 0–1), but these did not overlap NFCA headwaters or contain critical habitat and were placed in the stocking zone partially for zone contiguity purposes. Ten marginal counties (Figure 6b, 6c) had pond-stocking customers but contained NFCA headwaters, critical habitat, or tributaries draining into these sensitive areas; these were placed in the conservation zone. Comparison of the final zones against vulnerable fish SGCN SDMs found little overlap between core areas (i.e., ≥50% probability of occurrence) and the stocking zone, and no SGCN are likely to occur entirely within the stocking zone.

Development of draft regulations.—Based on the groundwork laid by regulatory zone delineation, draft regulations were developed that would prohibit tilapia stocking in private ponds within the conservation zone to minimize the risk of impacts to vulnerable fish SGCN. Due to a high number of pond-stocking customers in the five marginal counties shown in Figure 6c, an allowance was made for stocking in these counties subject to a determination by the TPWD that the proposed stocking is not likely to result in escape into public waters. Furthermore, all conditionally allowed tilapia species would be regulated equally, alleviating the burden on sellers, buyers, and law enforcement of distinguishing between these species. No permit requirement or associated fee would be instated for private pond stocking of tilapia.

Obtaining and incorporating commercial stakeholder feedback.—Feedback was obtained from the Texas Aquaculture Association (TAA); many members are permitted tilapia sellers and this group was selected to best represent commercial stakeholder interests. The draft-regulations-based strategy for conservation action and the science-based assessment conducted to inform development of regulations were presented to TAA delegates in a stakeholder meeting. Following the meeting, the draft regulations were revised to incorporate their input. The revised rules would allow stocking in all counties of the conservation zone, albeit with the same oversight authority as originally proposed for five marginal counties. No other changes were made to these draft regulations. The final draft regulations were shared with the TAA for review and comment, and their feedback showed support for this approach to development of balanced, science-based regulations and identified no concerns.

Discussion

The vast potential for the use of spatial prioritization to develop regulations-based conservation actions has seldom been explored, yet there are numerous applications for development of conservation-oriented regulations as well as administration of permitting programs in a way that seeks to identify and minimize potential impacts of authorized activities on the existing ecosystem. This study provides a case study of successful application of these methods within a SCP framework to inform development of regulations that provide protection for imperiled fishes. Areas with high conservation priority for fish SGCN vulnerable to impacts from introduced tilapias were identified and refined in the spatial assessment modeling process. These refinements accounted for risk of tilapia establishment and persistence, as well as reduced conservation value of degraded habitats where tilapia are already present.

Potential economic impacts to commercial stakeholders were weighed against conservation priority in the final assessment model. This final model was translated into regulatory zones used to develop a science-based, realistic, balanced regulatory approach to providing protections for imperiled fishes.

The final, future steps in the SCP framework (Figure 1) are to promulgate these regulations, effectively communicate the regulatory requirements to private pond owners, and ensure compliance and enforcement. The zone-based regulations developed in this study are expected to be proposed to the Texas Parks and Wildlife Commission in 2019 for publication in the Texas Register to solicit public comment, after which they will be revised as needed and proposed to the Texas Parks and Wildlife Commission for adoption. Should these rules be enacted, they will enable evaluation of potential for tilapia to escape from ponds before the fish are stocked, with the goal of preventing the harmful impacts of invasive tilapia introductions on Texas' imperiled fishes.

Bridging the gap between the science and ultimate implementation of conservation prioritization may depend upon involving commercial stakeholders in the planning stage and incorporating their needs and potential impacts to these groups into the assessment (Moilanen et al. 2011a; Kareksela et al. 2013). Although there will be many hurdles to overcome in the final steps of the implementation strategy of the current study, working with commercial stakeholders in the regulations development stage may help to facilitate future communication with private pond owners via a seller to buyer pathway and encourage stakeholder (i.e., tilapia seller) compliance. The tilapia sellers engaged in this process as commercial stakeholders are trained fisheries biologists with an interest in conservation of native fishes in Texas through responsible operating procedures that are amenable to implementation of enacted regulations.

Implementation of this regulatory approach would also serve to empower TPWD staff to rise to meet the challenges of the charge to consider potential adverse ecological impacts of private pond stocking and use the best available science to provide leadership in the protection of healthy aquatic ecosystems and recovery of imperiled species (TPWD 2016; Stocking Policy 2017). The proposed regulations-based conservation action could provide added protections for imperiled fishes and their habitats in the southern Great Plains, Edwards Plateau, and Chihuahuan Desert ecoregions. The conservation zone coincided with several Texas NFCAs (Labay and Hendrickson 2014; Birdsong et al. 2019) in the southern Great Plains (i.e., Canadian, Upper Red, and Upper Brazos NFCAs), Edwards Plateau (i.e., Colorado – Hill Country, Guadalupe/San Antonio, and Nueces/Medina NFCAs), and Chihuahuan Desert (i.e., Rio Grande – Big Bend, Pecos, and Devils NFCAs) ecoregions. These NFCAs represent areas of high biological endemism and fish species diversity where the TPWD and other conservation organizations continue to make substantial investments in conservation of native fishes and their habitats. In the southern Great Plains, the majority of the native aquatic fauna is imperiled, and introduction of nonnative species has been implicated in those declines. The Chihuahuan Desert ecoregion supports high biodiversity and endemism and faces threats from multiple invasive species—including tilapias. The Edwards Plateau ecoregion represents a hotspot of biological diversity and endemism yet is rapidly becoming developed and aquatic fauna are threatened by many anthropogenic impacts, including establishment of many invasive species. The regulatory approach developed in this study could support the success of numerous conservation initiatives underway in these ecoregions and NFCAs.

The flexibility and adaptability of the Zonation software used in the spatial conservation assessment in this study lends itself well to revision of the conservation prioritization by incorporating additional data. For example, biodiversity feature data sets representing distributions of Texas fish SGCN identified by new research as vulnerable to tilapia impacts. The role of thermal refuges in spatial variation in tilapia persistence potential warrants research to help inform future refinement of the current assessment and more accurately reflect this risk. Finally, large-scale surveys for tilapia are needed to optimize the assessment and better understand the current distribution of these invasive fishes, which are widely considered to be far more widespread in Texas than has been documented to date.

Acknowledgments

Funding for literature review by Barnes and Roesler (2017) was allocated by the 85th Texas Legislature under Rider 32: Statewide Aquatic Vegetation and Invasive Species Management. Input on the interpretation and implementation of this model in revision of tilapia regulations was provided by T. Birdsong, C. Bonds, K. Kurzawski, and the Texas Aquaculture Association. Review of an earlier draft by two anonymous reviewers helped to significantly strengthen and refine this chapter.

References

Adams, V. M., M. Mills, R. Weeks, D. B. Segan, R. L. Pressey, G. G. Gurney, C. Groves, F. W. Davis, and J. G. Álvarez-Romero. 2019. Implementation strategies for systematic conservation planning. Ambio 48:139–152.

Barnes, M. A., and E. L. Roesler. 2017. Evaluation of research needs related to tilapia in Texas and identification of areas with high potential risks to SGCN fishes. Final report submitted to Texas Parks and Wildlife Department, Austin.

Buntz, J., and C. S. Manooch, III. 1968. Tilapia aurea (Steindachner), a rapidly spreading exotic in south central Florida. Proceedings of the Annual Conference Southeastern Association of Game and Fish Commissioners 22:495–501.

Bwanika, G. N., L. J. Chapman, Y. Kizito, and J. Balirwa. 2006. Cascading effects of introduced Nile Perch (*Lates niloticus*) on the foraging ecology of Nile Tilapia (*Oreochromis niloticus*). Ecology of Freshwater Fish 15:470–48.1.

Canonico, G. C., A. Arthington, J. K. McCrary, and M. L. Thieme. 2005. The effects of introduced tilapias on native biodiversity. Aquatic Conservation: Marine and Freshwater Ecosystems 15:463–483.

Courtenay, W. R., Jr., H. F. Sahlman, W. W. Miley, II, and D. J. Herrema. 1974. Exotic fishes in fresh and brackish waters of Florida. Biological Conservation 6:292–302.

de Moor, F. C., R. C. Wilkinson, and H. M. Herbst. 1986. Food and feeding habits of *Oreochromis mossambicus* (Peters) in hypertrophic Hartbeesport Dam, South Africa. South African Journal of Zoology 21:170–176.

Doupé, R. G., M. J. Knott, J. Schaffer, and D. W. Burrows. 2009. Investigational piscivory of some juvenile Australian freshwater fishes by the introduced Mozambique Tilapia *Oreochromis mossambicus*. Fish Biology 74:2386–2400.

Esselman, P., D., M. Infante, L. Wang, W. Taylor, W. Daniel, R. Tingley, J. Fenner, A. Cooper, D. Wieferich, D. Thornbrugh, and J. Ross. 2011. National Fish Habitat Action Plan (NFHP) 2010 HCI scores and human disturbance data linked to the NHDPLUSV1 for Texas. National Fish Habitat Partnership Data System. Available at https://data.doi.gov/dataset/national-fish-habitat-action-plan-nfhap-2010-hci-scores-and-human-disturbance-data-linked-to-nhd6ea9 .

Fuselier, L. 2001. Impacts of *Oreochromis mossambicus* (Perciformes: Cichlidae) upon habitat segregation among cyprinodontids (Cyprinodontiformes) of a species flock in Mexico. International Journal of Tropical Biology and Conservation 49:647–656.

Germany, R. D. 1977. Population dynamics of the Blue Tilapia and its effects on the fish populations of Trinidad Lake, Texas. Doctoral dissertation. Texas A&M University, College Station.

Germany, R. D., and R. L. Noble. 1977. Population dynamics of Blue Tilapia in Trinidad Lake, Texas. Proceedings of the Annual Conference Southeastern Association of Fish and Wildlife Agencies 31:412–417.

Hanifen, J. G. 1981. Fish populations of Fairfield Reservoir, Texas ten years after impoundment and the effects of the introduction of Tilapia aurea. Master's thesis. Texas A&M University, College Station.

Harmful or Potentially Harmful Fish, Shellfish, and Aquatic Plants. 2017. Texas Administrative Code, section 57.111–57.137.

Hendricks, M. K., and R. L. Noble. 1980. Feeding interactions of three planktivorous fishes in Trinidad Lake, Texas. Proceedings of the Annual Conference Southeastern Association of Fish and Wildlife Agencies 33:324–330.

Hendrickson, D. A., and A. E. Cohen. 2015. Fishes of Texas project database (version 2.0). University of Texas at Austin, Austin. Available: www.fishesoftexas.org/data/ (December 2017).

Hendrickson, D. A., G. P. Garrett, B. J. Labay, A. E. Cohen, and M. Casarez. 2015. Year 1 report for 'Conserving Texas Biodiversity: Status, Trends, and Conservation Planning for Fishes of Greatest Conservation Need.' Annual Report to Texas Parks and Wildlife Department, Austin.

Howells, R. G. 1991. Electrophoretic identification of feral and domestic tilapia in Texas. Texas Parks and Wildlife Department, Fisheries Management Data Series No. 62, Austin.

Howells, R. G. 1999. Guide to identification of harmful and potentially harmful fishes, shellfishes, and aquatic plants prohibited in Texas. Texas Parks and Wildlife Department, Austin.

Hubbs, C., R. J. Edwards, and G. P. Garrett. 1991. An annotated checklist of the freshwater fishes of Texas, with keys to identification of species. The Texas Journal of Science 53(Supplement 4):2–87.

Introduction of Fish, Shellfish, and Aquatic Plants. 1997. Texas Parks and Wildlife Code, section 66.015.

Junor, F. J. R. 1969. Tilapia melanopleura Dum. in artificial lakes and dams in Rhodesia with special reference to its undesirable effects. Rhodesian Journal of Agricultural Research 7:61–69.

Kareksela, S., A. Moilanen, S. Tuominen, and J. S. Kotiaho. 2013. Use of inverse spatial conservation prioritization to avoid biological diversity loss outside protected areas. Conservation Biology 27:1294–1303.

Labay, B. J., A. E. Cohen, B. Sissel, D. A. Hendrickson, F. D. Martin, and S. Sarkar. 2011. Assessing historical fish community composition using surveys, historical collection data, and species distribution models. PLOS (Public Library of Science) ONE [online serial] 6(9):e25145.

Labay, B. J., and D. A. Hendrickson. 2014. Final report: conservation assessment and mapping products for GPLCC priority fish taxa. University of Texas at Austin, Austin.

Labay, B. J., D. A. Hendrickson, and A. E. Cohen. 2012a. Fishes of Texas project class 02 species distributions models (models). Texas Natural History Collection, Texas Natural Science Center, University of Texas at Austin, Austin. Available: www.fishesoftexas.org/models/02/downloads/. (December 2017).

Labay, B. J., D. A. Hendrickson, and A. E. Cohen, A.E. 2012b. Fishes of Texas project class 01 species distributions models (models). Texas Natural History Collection, University of Texas at Austin, Austin, Texas. Available: www.fishesoftexas.org/models/01/downloads/. (December 2017).

Lehtomäki, J., and A. Moilanen. 2013. Methods and workflow for spatial conservation prioritization using Zonation. Environmental Modelling and Software 47:128–137.

Martin, B. A., and M. K. Saiki. 2005. Relation of Desert Pupfish abundance to selected environmental variables in natural and manmade habitats in the Salton Sea basin. Environmental Biology of Fishes 73:97–107.

McGarrity, M. E., and S. Davis. 2017. Geographic distribution and frequency of Mozambique Tilapia stockings in private ponds. Texas Parks and Wildlife Department, Aquatic Invasive Species Prevention and Management Report No. 1, Austin.

Moilanen, A., B. J. Anderson, F. Eigenbrod, A. Heinemeyer, D. B. Roy, S. Gillings, P. R. Armsworth, K. J. Gaston, and C. D. Thomas. 2011a. Balancing alternative land uses in conservation prioritization. Ecological Applications 21:1419–1426.

Moilanen, A., A. M. A. Franco, R. Early, R. Fox, B. Wintle, and C. D. Thomas. 2005. Prioritising multiple-use landscapes for conservation: methods for large multi-species planning problems. Proceedings of the Royal Society of London Series B 272:1885–1891.

Moilanen, A., J. R. Leathwick, and J. M. Quinn. 2011b. Spatial prioritization of conservation management. Conservation Letters 4:383–393.

Moilanen, A., H. Possingham, and K. A. Wilson. 2009. Spatial conservation prioritization: past, present, and future. Pages 260–268 *in* A. Moilanen, H. K. Wilson, and Possingham, editors. Spatial conservation prioritization quantitative methods and computational tools. Oxford University Press, Oxford, UK.

Moilanen, A., F. M. Pouzols, L. Meller, V. Veach, A. Arponen, J. Leppänen, and H. Kujala. 2014. Zonation: spatial conservation planning methods and software, version 4 user manual. Conservation Biology Informatics Group, Department of Biosciences, University of Helsinki, Helsinki, Finland.

Moyle, P. B. 2002. Inland fishes of California. University of California Press, Berkeley.

Noble, R. L., and R. D. Germany. 1986. Changes in fish populations of Trinidad Lake, Texas, in response to abundance of Blue Tilapia. Fish Culture in Fisheries Management 455–461.

Noble, R. L., R. D. Germany, and C. R. Hall. 1975. Interactions of Blue Tilapia and Largemouth Bass in a power plant cooling reservoir. Proceedings of the Annual Conference Southeastern Association Game and Fish Commissioners 29:247–251.

Pearsons, T. N., and C. W. Hopley. 1999. A practical approach for assessing ecological risks associated with fish stocking programs. Fisheries 24:16–23.

TPWD (Texas Parks and Wildlife Department). 2012. Texas Conservation Action Plan 2012–2016: overview. W. Connally, editor. TPWD, Austin.

TPWD (Texas Parks and Wildlife Department). 2016. A strategic plan for Texas Parks and Wildlife Department: natural agenda, fiscal years 2017–2021. TPWD, Austin.

Pool, T. K., A. L. Strecker, and J. D. Olden. 2013. Identifying preservation and restoration priority areas for desert fishes in an increasingly invaded world. Environmental Management 51:631–641.

Schoenherr, A. A. 1981. Replacement of *Cyprinodon macularius* by *Tilapia zillii* in an irrigation drain near the Salton Sea. Proceedings of the Desert Fishes Council 13:65–66.

Schoenherr, A. A. 1988. A review of the life history and status of the Desert Pupfish, *Cyprinodon macularius*. Bulletin of the Southern California Academy of Sciences 87:104–134.

Scoppettone, G. G., P. H. Rissler, M. B. Nielsen, and J. E. Harvey. 1998. The status of *Moapa coriacea* and *Gila seminuda* and status information on other fishes of the Muddy River, Clark County, Nevada. Southwestern Association of Naturalists 43:115–122.

Scoppettone, G. G., J. A. Salgado, and M. B. Nielsen. 2005. Blue Tilapia (*Oreochromis aureus*) predation on fishes in the Muddy River system, Clark County, Nevada. Western North American Naturalist 65:410–415.

Shafland, P. L., and J. M. Pestrak. 1983. Suppression of Largemouth Bass production by Blue Tilapia in ponds. Proceedings of the Annual Conference Southeastern Association Fish and Wildlife Agencies 37:441–446.

Starling, S. M. 1986. Effects of a reduction of Blue Tilapia *Oreochromis aureus* on the ichthyofauna of a power-plant reservoir. Master's thesis. Texas A&M University, College Station.

Stocking Policy. 2017. Texas Administrative Code, section 52.101–52.401.

Strecker, U. 2006. The impact of invasive fish

on an endemic *Cyprinodon* species flock (Teleostei) from Laguna Chichancanab, Yucatan, Mexico. Ecology of Freshwater Fish 15:408–418.

Szyper, J. P. 2000. History and prospects of tilapia stocks in Hawaii, USA. Pages 663–672 *in* K. Fitzsimmons and J. C. Filho, editors. Tilapia culture in the 21st century. American Tilapia Association, Charles Town, West Virginia and ICLARM, Penang, Malaysia.

Traxler, S. L., and B. Murphy. 1995. Experimental trophic ecology of juvenile Largemouth Bass, *Micropterus salmoides*, and Blue Tilapia, *Oreochromis aureus*. Experimental Biology of Fishes 42:201–211.

USFWS (U.S. Fish and Wildlife Service). 2004. Tilapia removal program on the Virgin River, Clark County, Nevada and Mohave County, Arizona. USFWS, Southern Nevada Field Office, Las Vegas, Nevada.

USGS-USDA (U.S. Geological Survey and U.S. Department of Agriculture, Natural Resources Conservation Service). 2012. Federal standards and procedures for the National Watershed Boundary Dataset (WBD), 3rd edition. U.S. Geological Survey, Techniques and Methods 11-A3, Reston, Virginia.

Varela-Romero, A., G. Ruiz-Campos, L. M. Yépiz-Velázquez, and J. Alaníz-García. 2002. Distribution, habitat, and conservation status of Desert Pupfish, (*Cyprinodon macularius*) in the lower Colorado River basin, Mexico. Reviews in Fish Biology and Fisheries 12:157–165.

Zale, A. V. 1984. Applied aspects of the thermal biology, ecology, and life history of the Blue Tilapia, *Tilapia aurea* (Pisces: Cichlidae). Doctoral dissertation. University of Florida, Gainesville.

Zale, A. V., and R. W. Gregory. 1990. Food selection by early life stages of Blue Tilapia, *Oreochromis aureus*, in Lake George, Florida: overlap with sympatric shad larvae. Florida Scientist 53:123–129.

American Fisheries Society Symposium 91:183–229, 2019

Texas Native Fish Conservation Areas Network: Strategic Investments in Restoration and Preservation of Freshwater Fish Diversity

Timothy W. Birdsong*
Texas Parks and Wildlife Department, Inland Fisheries Division
4200 Smith School Road, Austin, Texas 78744, USA

Gary P. Garrett
Biodiversity Center, Department of Integrative Biology, University of Texas at Austin
10100 Burnet Road, PRC176 East/R4000, Austin, Texas 78758, USA

Ben J. Labay
Siglo Group
1619 West 6th Street, Austin, Texas 78703, USA

Megan G. Bean and Preston T. Bean
Texas Parks and Wildlife Department, Inland Fisheries Division
5103 Junction Highway, Mountain Home, Texas 78058, USA

John Botros
Texas Parks and Wildlife Department, Inland Fisheries Division
505 Staples Road, Building 1, San Marcos, Texas 78666, USA

Melissa J. Casarez and Adam E. Cohen
Biodiversity Center, Department of Integrative Biology, University of Texas at Austin
10100 Burnet Road, PRC176 East/R4000, Austin, Texas 78758, USA

Thomas G. Heger
Texas Parks and Wildlife Department, Inland Fisheries Division
4200 Smith School Road, Austin, Texas 78744, USA

Arlene Kalmbach
Texas Parks and Wildlife Department, Wildlife Division
1887 County Road 324, Lincoln, Texas 78948, USA

Dean A. Hendrickson
Biodiversity Center, Department of Integrative Biology, University of Texas at Austin
10100 Burnet Road, PRC176 East/R4000, Austin, Texas 78758, USA

Stephan J. Magnelia and Kevin B. Mayes
Texas Parks and Wildlife Department, Inland Fisheries Division
505 Staples Road, Building 1, San Marcos, Texas 78666, USA

* Corresponding author: timothy.birdsong@tpwd.texas.gov

Monica E. McGarrity
Texas Parks and Wildlife Department, Inland Fisheries Division
4200 Smith School Road, Austin, Texas 78744, USA

Ryan McGillicuddy, M. Melissa Parker, and Sarah Robertson
Texas Parks and Wildlife Department, Inland Fisheries Division
505 Staples Road, Building 1, San Marcos, Texas 78666, USA

Abstract.—Texas harbors 191 species of native freshwater fishes, 48% of which are considered imperiled. The primary cause of fish species imperilment in Texas is anthropogenic alteration of freshwater systems, which continues to occur at rates and scales that threaten the long-term resiliency of freshwater habitats, species, and ecosystems. Innovative conservation approaches are needed to restore and maintain functional watershed processes, restore freshwater habitats, and conserve native species while simultaneously supporting human needs, such as flood control, municipal and agricultural water supply, water quality protection, and water-based recreation. The need for an integrated and holistic approach to conservation of freshwater systems has been the impetus for development of the Texas Native Fish Conservation Areas Network (hereafter "Texas NFCAs Network"). The Texas NFCAs Network consists of springs, ciénegas, creeks, rivers, and associated watersheds uniquely valued in preservation of Texas freshwater fish diversity. Twenty native fish conservation areas have been designated throughout the state. These were selected based on a spatial prioritization focused on identification of freshwater systems critically important to the long-term persistence of 91 freshwater fishes considered species of greatest conservation need. Through a shared vision of collaborative stewardship, conservation partnerships have formed among nongovernmental organizations, universities, and state and federal agencies to plan and deliver actions within the Texas NFCAs Network to restore and preserve native fishes and their habitats. Furthermore, the Texas NFCAs Network has increased awareness of the ecological, recreational, and economic values of Texas freshwater systems and helped increase interest and capacity of local landowners, communities, and recreational users (e.g., paddlers, anglers) to act as advocates and local stewards of these systems. By facilitating partnership development, coordinating broad-based conservation planning, and leveraging technical and financial resources toward strategic conservation investments, the Texas NFCAs Network has served as a catalyst for collaborative, science-based stewardship of native freshwater fishes and their habitats in Texas. The Texas NFCAs Network offers a successful case study in multispecies and watershed approaches to freshwater fish conservation transferrable to other states in the United States, with particular relevance to those states that, similar to Texas, consist predominately of privately owned landscapes.

Conservation Needs of Texas Freshwater Fishes

Texas harbors 191 species of native freshwater fishes, 48% of which are considered imperiled (Table 1). An additional 67 species of native estuarine fishes have also been documented to occur in Texas freshwater systems, with 9 of those species considered imperiled (Table 1). The primary cause of fish species imperilment in Texas is anthropogenic alteration of freshwater systems, which continues to occur at rates and scales that threaten the long-term re-

Table 1. Fishes of the Texas Native Fish Conservation Areas Network. F = NFCA focal species and listed as a Texas species of greatest conservation need; FR = NFCA focal species and a recommended addition to the list of Texas species of greatest conservation need (Cohen et al. 2018); FL = NFCA focal species that is a Texas species of greatest conservation need and is listed as state or federally threatened or endangered; NN = species that occurs in and is non-native to the NFCA; C = species that is considered common that occurs in the NFCA; CR = species that is currently listed as a Texas species of greatest conservation need, but that is recommended for removal from that list by Cohen et al. (2018).

	Native fish conservation area																			
Taxa	Upper Brazos River	Upper Red River	Upper Canadian River	Central Edwards Plateau Rivers	Guadalupe and San Antonio Rivers	Southern Edwards Plateau Rivers	Upper Big Bend	Lower Big Bend	Guadalupe Mountains Streams	Davis Mountains Streams	Pecos River	Devils River	Lower Rio Grande	Central Coast Rivers and Streams	Lower Colorado River	Middle Brazos River	San Gabriel River	Lower Brazos River	Southeast Texas Rivers	Northeast Texas Rivers
Chestnut Lamprey *Ichthyomyzon castaneus*																			C	C
Southern Brook Lamprey *I. gagei*																		C	C	
Atlantic Stingray *Dasyatis sabina*														C					C	
Shovelnose Sturgeon *Scaphirhynchus platorynchus*							FL	FL			FL									FL
Paddlefish *Polyodon spathula*																			FL	FL
Alligator Gar *Atractosteus spatula*					F	F	F	F			F	F	F	F	F	F		F	F	F
Spotted Gar *Lepisosteus oculatus*	C	C			C	C	C	C			C	C	C	C	C	C	C	C	C	C
Longnose Gar *L. osseus*	C	C		C	C	C	C	C			C	C	C	C	C	C	C	C	C	C
Shornose Gar *L. platostomus*		C																		C
Bowfin *Amia calva*													C	C			C	C	C	
Goldeye *Hiodon alosoides*		F																		F
Ladyfish *Elops saurus*														C					C	
Tarpon *Megalops atlanticus*														F					F	
American Eel *Anguilla rostrata*	F	F	F	F	F	F	F	F			F	F	F	F	F	F	F	F	F	F
Speckled Worm Eel *Myrophis punctatus*																			C	
Striped Anchovy *Anchoa hepsetus*														C					C	

Table 1. Continued.

Taxa	Native fish conservation area																			
	Upper Brazos River	Upper Red River	Upper Canadian River	Central Edwards Plateau Rivers	Guadalupe and San Antonio Rivers	Southern Edwards Plateau Rivers	Upper Big Bend	Lower Big Bend	Guadalupe Mountains Streams	Davis Mountains Streams	Pecos River	Devils River	Lower Rio Grande	Central Coast Rivers and Streams	Lower Colorado River	Middle Brazos River	San Gabriel River	Lower Brazos River	Southeast Texas Rivers	Northeast Texas Rivers
Bay Anchovy *A. mitchilli*														C				C	C	
Skipjack Herring *Alosa chrysochloris*													C				C	C		
Finescale Menhadden *Brevoortia gunteri*														C					C	
Gulf Menhaden *B. patronus*														C				C	C	
Gizzard Shad *Dorosoma cepedianum*	C	C		C	C	C	C	C		C	C	C	C	C	C	C	C	C	C	C
Threadfin Shad *D. petenense*	C			C	C	C	NN	NN		NN		NN	C	C	C		C	C	C	C
Scaled Sardine *Harengula jaguana*														C						
Central Stoneroller *Campostoma anomalum*	C	C	C	C	C	C					C	C	C	C	C	C	C	C	C	C
Mexican Stoneroller *C. ornatum*							FL	FL												
Highland Stoneroller *C. spadiceum*																				FR
Goldfish *Carassius auratus*	NN			NN			NN	NN		NN	NN	NN			NN	NN	NN	NN	NN	NN
Grass Carp *Ctenopharyngodon idella*													NN	NN	NN					NN
Plateau Shiner *Cyprinella lepida*						F														
Red Shiner *C. lutrensis*	C	C	C	C	C	C	C	C		C	C	C	C	C	C	C	C	C	C	C
Maravillas Red Shiner *C. l. blairi*								F												
Conchos Shiner C. panarcys							FL													
Proserpine Shiner *C. proserpina*											FL	FL								
Nueces River Shiner *Cyprinella* sp.						F														
Blacktail Shiner *C. venusta*	C	C		C	C	C	NN	NN		NN	NN	C	C	C	C	C	C	C	C	C
Steelcolor Shiner *C. whipplei*																				C
Common Carp *Cyprinus carpio*	NN	NN	NN	NN	NN	NN	NN	NN		NN	NN	NN	NN	NN	NN	NN	NN	NN	NN	NN
Manatial Roundhouse Minnow *Dionda argentosa*											F	F								

Table 1. Continued.

Taxa	Native fish conservation area																			
	Upper Brazos River	Upper Red River	Upper Canadian River	Central Edwards Plateau Rivers	Guadalupe and San Antonio Rivers	Southern Edwards Plateau Rivers	Upper Big Bend	Lower Big Bend	Guadalupe Mountains Streams	Davis Mountains Streams	Pecos River	Devils River	Lower Rio Grande	Central Coast Rivers and Streams	Lower Colorado River	Middle Brazos River	San Gabriel River	Lower Brazos River	Southeast Texas Rivers	Northeast Texas Rivers
Devils River Minnopw *D. diaboli*												FL								
Roundnose Minnow *D. episcopa*										F	F									
Guadalupe Roundnose Minnow *D. flavipinnis*				FR	FR															
Medina Roundnose Minnow *D. nigrotaeniata*						F														
Frio Roundnose Minnow *D. serena*						F														
Conchos Roundnose Minnow *Dionda* sp. 1						FR														
Colorado Roundnose Minnow *Dionda* sp. 3				FR																
Nueces Roundnose Minnow *D. texensis*						FR														
Rio Grande Chub *Gila pandora*									FL	FL										
Rio Grande Silvery Minnow *Hybognathus amarus*							FL	FL			FL		FL							
Cypress Minnow *H. hayi*																				C
Mississippi Silvery Minnow *H. nuchalis*																FR		FR	FR	FR
Plains Minnow *H. placitus*	FR	FR	FR	FR	NN							NN				FR	FR	FR		
Pallid Shiner *Hybopsis amnis*					FR									FR	FR	FR	FR	FR	FR	FR
Striped Shiner *Luxilus chrysocephalus*																				C
Ribbon Shiner *Lythrurus fumeus*					C									C	C			C	C	C
Redfin Shiner *L. umbratilis*																			C	C
Speckled Chub *Macrhybopsis aestivalis*							F	F			F	F	F							
Prairie Chub *M. australis*		F																		
Shoal Chub *M. hyostoma*	FR	FR		FR										FR	FR	FR	FR	FR	FR	FR
Burrhead Chub *M. marconis*				FR	FR	FR									FR					

Table 1. Continued.

Taxa	Upper Brazos River	Upper Red River	Upper Canadian River	Central Edwards Plateau Rivers	Guadalupe and San Antonio Rivers	Southern Edwards Plateau Rivers	Upper Big Bend	Lower Big Bend	Guadalupe Mountains Streams	Davis Mountains Streams	Pecos River	Devils River	Lower Rio Grande	Central Coast Rivers and Streams	Lower Colorado River	Middle Brazos River	San Gabriel River	Lower Brazos River	Southeast Texas Rivers	Northeast Texas Rivers
	Native fish conservation area																			
Silver Chub *M. storeriana*	F	F														F		F		F
Peppered Chub *M. tetranema*			FR																	
Golden Shiner *Notemigonus crysoleucas*	NN	NN	NN	NN	C	C	NN	NN		NN	NN		NN	C	C	NN	NN	C	C	C
Texas Shiner *Notropis amabilis*				CR	CR	CR					CR	CR	CR		CR					
Emerald Shiner *N. atherinoides*		C	C												NN			NN	C	C
Blackspot Shiner *N. atrocaudalis*																		F	F	F
Red River Shiner *N. bairdi*		F																		
River Shiner *N. blennius*		F																		F
Bigeye Shiner *N. boops*		C																		
Tamaulipas Shiner *N. braytoni*							F	F			F	F	F							
Smalleye Shiner *N. buccula*	FL			FL											FL	FL	FL	FL		
Ghost Shiner *N. buchanani*	C	C		C	C	C					C	C	C	C	C	C	C	C	C	C
Ironcolor Shiner *N. chalybaeus*					F														F	F
Chihuahua Shiner *N. chihuahua*							FL	FL												
Arkansas River Shiner *N. girardi*			FL																	
Rio Grande Shiner *N. jemezanus*							F	F			F	F	F							
Taillight Shiner *N. maculatus*																				F
West Texas Shiner *N. megalops*											FR	FR								
Phantom Shiner *N. orca*							F	F			F		F							
Sharpnose Shiner *N. oxyrhynchus*	FL			FL											FL	FL	FL	FL		
Chub Shiner *N. potteri*	F	F												F		F	F	F	F	F
Sabine Shiner *N. sabinae*														F					F	F

Table 1. Continued.

Taxa	Upper Brazos River	Upper Red River	Upper Canadian River	Central Edwards Plateau Rivers	Guadalupe and San Antonio Rivers	Southern Edwards Plateau Rivers	Upper Big Bend	Lower Big Bend	Guadalupe Mountains Streams	Davis Mountains Streams	Pecos River	Devils River	Lower Rio Grande	Central Coast Rivers and Streams	Lower Colorado River	Middle Brazos River	San Gabriel River	Lower Brazos River	Southeast Texas Rivers	Northeast Texas Rivers
	Native fish conservation area																			
Silverband Shiner *N. shumardi*	F	F												F	F	F	F	F	F	F
Pecos Bluntnose Shiner *N. simus pecosensis*											FL									
Rio Grande Bluntnose Shiner *N. s. simus*							F	F												
Sand Shiner *N. stramineus*	C	C	C	C	C	C	C	C			C	C			C	NN	NN	NN	NN	
Weed Shiner *N. texanus*				C	C	C								C	C	NN				C
Mimic Shiner *N. volucellus*	C			C	C	C								C	C	C	C	C	C	C
Pugnose Minnow *Opsopoeodus emiliae*				C	C	C								C	C	C	C	C	C	C
Suckermouth Minnow *Phenacobius mirabilis*		FR	FR	FR											FR				FR	FR
Fathead Minnow *Pimephales promelas*	C	C	C	C	C	C	C	C			C			C	C	C	C	C	C	C
Bullhead Minnow *P. vigilax*	C	C	NN	C	C	C	NN	NN		NN	NN	C	C	C	C	C	C	C	C	C
Flathead Chub *Platygobio gracilis*			FR																	
Bluehead Shiner *Pteronotropis hubbsi*																				FL
Longnose Dace *Rhinichthys cataractae*							F	F			F		F							
Rudd *Scardinius erythrophthalmus*													NN							
Creek Chub *Semotilus atromaculatus*																			C	C
River Carpsucker *Carpiodes carpio*	C	C		C	C	C	C	C		C	C	C	C	C	C	C	C	C	C	C
Quillback *C. cyprinus*																				C
Llano River Carpsucker *Carpiodes* sp.				FR																
Blue Sucker *Cycleptus elongatus*		FL		FL											FL	FL	FL	FL	FL	FL
Rio Grande Blue Sucker *Cycleptus* sp.							FL	FL			FL	FL	FL							
Western Creek Chubsucker *Erimyzon claviformis*														FL				FL	FL	FL
Lake Chubsucker *E. sucetta*					C									C				C	C	C

Table 1. Continued.

Taxa	Upper Brazos River	Upper Red River	Upper Canadian River	Central Edwards Plateau Rivers	Guadalupe and San Antonio Rivers	Southern Edwards Plateau Rivers	Upper Big Bend	Lower Big Bend	Guadalupe Mountains Streams	Davis Mountains Streams	Pecos River	Devils River	Lower Rio Grande	Central Coast Rivers and Streams	Lower Colorado River	Middle Brazos River	San Gabriel River	Lower Brazos River	Southeast Texas Rivers	Northeast Texas Rivers
	Native fish conservation area																			
Smallmouth Buffalo *Ictiobus bubalus*	C	C		C			C	C		C	C		C	C	C	C	C	C	C	C
Bigmouth Buffalo *I. cyprinellus*																			C	C
Black Buffalo *I. niger*		NN					NN	NN			NN	NN							NN	
Spotted Sucker *Minytrema melanops*				F										C		C	C	C	C	C
Longlip Jumprock *Moxostoma albidum*											FR	FR								
Mexican Redhorse *M. austrinum*							F	F			F		F							
Gray Redhorse *M. congestum*	C			C	C	C	C	C		C	C	C	C		C	C	C	C		
Black Redhorse *M. duquesnii*																				C
Golden Redhorse *M. erythrurum*																				C
Blacktail Redhorse *M. poecilurum*														C					C	C
Mexican Tetra *Astyanax mexicanus*				NN	NN	C	C	C		C	C	C	C		NN		NN	NN	NN	
Black Bullhead *Ameiurus melas*	C	C	C	C	C	C			C	C	C	C	C	C	C	C	C	C	C	C
Yellow Bullhead *A. natalis*	C	C		C	C	C	NN	NN						C	C	C	C	C	C	C
Blue Catfish *Ictalurus furcatus*	C						C	C			C	C	C	C		C	C	C	C	C
Headwater Catfish *I. lupus*				F	F	F	F	F		F	F	F	F							
Channel Catfish *I. punctatus*	C	C	C	C	C	C	C	C		C	C	C	C	C	C	C	C	C	C	C
Chihuahua Catfish *Ictalurus* sp.							F	F	F	F	F	F								
Rio Grande Blue Catfish *Ictalurus* sp.							FR	FR			FR	FR								
Tadpole Madtom *Noturus gyrinus*	C				C								NN	C	C	C	C	C	C	C
Freckled Madtom *N. nocturnus*	C													C		C	C	C	C	C
Mexican Blindcat *Prietella phreatophila*											FL	FL								
Flathead Catfish *Pylodictis olivaris*	C	C	C	C	C	C	C	C			C	C	C	C	C	C	C	C	C	C

Table 1. Continued.

Taxa	Native fish conservation area																			
	Upper Brazos River	Upper Red River	Upper Canadian River	Central Edwards Plateau Rivers	Guadalupe and San Antonio Rivers	Southern Edwards Plateau Rivers	Upper Big Bend	Lower Big Bend	Guadalupe Mountains Streams	Davis Mountains Streams	Pecos River	Devils River	Lower Rio Grande	Central Coast Rivers and Streams	Lower Colorado River	Middle Brazos River	San Gabriel River	Lower Brazos River	Southeast Texas Rivers	Northeast Texas Rivers
Widemouth Blindcat *Satan eurystomus*					FL	FL														
Toothless Blindcat *Trogloglanis pattersoni*					FL	FL														
Hardhead Catfish *Ariopsis felis*														C					C	
Gafftopsail Catfish *Bagre marinus*														C					C	
Armadillo Del Rio *Hypostomus* sp.												NN	NN							
Southern Sailfin Catfish *Pterygoplichthys anisitsi*														NN						
Vermiculated Sailfin Catfish *P. disjunctivus*					NN															
Orinoco Sailfin Catfish *P. multiradiatus*														NN						
Redfin Pickerel *Esox americanus*														C				C	C	C
Chain Pickerel *E. niger*																				C
Rio Grande Cutthroat Trout *Oncorhynchus clarkii virginalis*									F	F										
Rainbow Trout *O. mykiss*						NN			NN										NN	
Pirate Perch *Aphredoderus sayanus*														C			C	C	C	C
Mountain Mullet *Agonostomus monticola*													FR	FR	FR	FR	FR	FR	FR	
Striped Mullet *Mugil cephalus*													C	C	C			C	C	
White Mullet *M. curema*														C					C	
Brook Silverside *Labidesthes sicculus*	NN	NN														NN	NN		C	C
Rough Silverside *Membras martinica*														C			NN	C	C	
Inland Silverside *Menidia beryllina*	NN	NN	NN	NN	C	NN	NN	NN		NN	NN	NN	NN	C	C	NN	NN	C	C	NN
Texas Silverside *M. clarkhubbsi*														F					F	
Tidewater Silverside *M. peninsulae*														C					C	

Table 1. Continued.

Taxa	Native fish conservation area																			
	Upper Brazos River	Upper Red River	Upper Canadian River	Central Edwards Plateau Rivers	Guadalupe and San Antonio Rivers	Southern Edwards Plateau Rivers	Upper Big Bend	Lower Big Bend	Guadalupe Mountains Streams	Davis Mountains Streams	Pecos River	Devils River	Lower Rio Grande	Central Coast Rivers and Streams	Lower Colorado River	Middle Brazos River	San Gabriel River	Lower Brazos River	Southeast Texas Rivers	Northeast Texas Rivers
Atlantic Needlefish *Strongylura marina*														C		C	C	C	C	
Diamond Killifish *Adinia xenica*														C					C	
Western Starhead Topminnow *Fundulus blairae*														C				C	C	C
Golden Topminnow *F. chrysotus*														C				C	C	C
Gulf Killifish *F. grandis*	NN	NN		NN	NN	NN	NN	NN		NN	NN		NN	C	C	NN	NN	C	C	
Saltmarsh Topminnow *F. jenkinsi*														F					F	
Northern Plains Killifish *F. kansae*			C																	
Blackstripe Topminnow *F. notatus*	C	C		C	C	NN								C	C	C	C	C	C	C
Blackspotted Topminnow *F. olivaceous*														C		C	C	C	C	C
Bayou Killifish *F. pulvereus*														C					C	
Longnose Killifish *F. similis*														C						
Plains Killifish *F. zebrinus*	C	C		C			NN	NN		C	C	C			C	C	C	C	C	
Bluefin Killifish *Lucania goodei*														NN						
Rainwater Killifish *L. parva*					NN				C	C	C		C	C					C	
Pike Killifish *Belonesox belizanus*						NN														
Western Mosquitofish *Gambusia affinis*	C	C	C	C	C	C	C	C	C	C	C	C	C	C	C	C	C	C	C	C
Amistad Gambusia *G. amistadensis*											F									
Big Bend Gambusia *G. gaigei*								FL												
Largespring Gambusia *G. geiseri*				NN	C	NN				NN	NN	NN								
San Marcos Gambusia *G. georgei*					FL															
Clear Creek Gambusia *G. heterochir*				FL																
Spotfin Gambusia *G. krumholzi*												FL								

Table 1. Continued.

Taxa	Native fish conservation area																			
	Upper Brazos River	Upper Red River	Upper Canadian River	Central Edwards Plateau Rivers	Guadalupe and San Antonio Rivers	Southern Edwards Plateau Rivers	Upper Big Bend	Lower Big Bend	Guadalupe Mountains Streams	Davis Mountains Streams	Pecos River	Devils River	Lower Rio Grande	Central Coast Rivers and Streams	Lower Colorado River	Middle Brazos River	San Gabriel River	Lower Brazos River	Southeast Texas Rivers	Northeast Texas Rivers
Pecos Gambusia *G. nobilis*										FL	FL									
Blotched Gambusia *G. senilis*												FL								
Tex-Mex Gambusia *G. speciosa*											C	C								
Least Killifish *Heterandria formosa*																			C	
Amazon Molly *Poecilia formosa*					NN	NN							C					NN		
Sailfin Molly *P. latipinna*				NN	NN	NN						NN	C	C	C	NN		C	C	
Guppy *P. reticulata*						NN														
Green Swordtail *Xiphophorus hellerii*						NN														
Variable Platyfish *X. variatus*															NN					
Leon Springs Pupfish *Cyprinodon bovinus*											FL									
Comanche Springs Pupfish *C. elegans*										FL	FL									
Conchos Pupfish *C. eximius*							FL					FL								
Pecos Pupfish *C. pecosensis*											FL									
Red River Pupfish *C. rubrofluviatilis*	F	F	F	NN																
Sheepshead Minnow *C. variegatus*	NN			NN	NN					NN	NN	NN	C	C	C			C	C	
Opossum Pipefish *Microphis brachyurus*														FL					FL	
Dusky Pipefish *Syngnathus floridae*														C					C	
Chain Pipefish *S. louisianae*														C					C	
Gulf Pipefish *S. scovelli*														C					C	
White Bass *Morone chrysops*	NN	C	C	NN			NN	NN		NN	NN	NN	NN	NN	NN	NN	NN	NN	NN	C
Yellow Bass *M. mississippiensis*														C					C	C
Striped Bass *M. saxatilis*	NN			NN										NN				NN	NN	NN

Table 1. Continued.

Taxa	Native fish conservation area																			
	Upper Brazos River	Upper Red River	Upper Canadian River	Central Edwards Plateau Rivers	Guadalupe and San Antonio Rivers	Southern Edwards Plateau Rivers	Upper Big Bend	Lower Big Bend	Guadalupe Mountains Streams	Davis Mountains Streams	Pecos River	Devils River	Lower Rio Grande	Central Coast Rivers and Streams	Lower Colorado River	Middle Brazos River	San Gabriel River	Lower Brazos River	Southeast Texas Rivers	Northeast Texas Rivers
Rock Bass *Ambloplites rupestris*					NN															
Flier *Centrarchus macropterus*														C					C	C
Redbreast Sunfish *Lepomis auritus*	NN	NN		NN	NN	NN	NN	NN		NN	NN	NN	NN	NN	NN	NN	NN	NN	NN	NN
Green Sunfish *L. cyanellus*	C	C	C	C	C	C	C	C	C	C	C	C	C	C	C	C	C	C	C	C
Warmouth *L. gulosus*	C	C		C	C	C	C	C		C	C	C	C	C	C	C	C	C	C	C
Orangespotted Sunfish *L. humilis*	C	C	C	C	NN									C	C	C	C	C	C	C
Bluegill *L. macrochirus*	C	C	C	C	C	C	C	C	C	C	C	C	C	C	C	C	C	C	C	C
Dollar Sunfish *L. marginatus*														C				C	C	C
Longear Sunfish *L. megalotis*	C	C	C	C	C	C	C	C	C	C	C	C	C	C	C	C	C	C	C	C
Redear Sunfish *L. microlophus*	C	C	C	C	C	C	NN	NN		NN	NN	NN	NN	C	C	C	C	C	C	C
Redspotted Sunfish *L. miniatus*	C			C	C	C						C		C	C	C	C	C	C	C
Bantam Sunfish *L. symmetricus*														C	C			C	C	C
Smallmouth Bass *Micropterus dolomieu*				NN	NN	NN						NN			NN	NN	NN	NN	NN	
Spotted Bass *M. punctulatus*	C	C			C									C	C	C	C	C	C	C
Largemouth Bass *M. salmoides*	C	C	C	C	C									C	C	C	C	C	C	C
Florida Largemouth Bass *M. salmoides floridanus*	NN	NN	NN	NN	NN	NN	NN	NN		NN	NN	NN	NN	NN	NN	NN	NN	NN	NN	NN
Rio Grande Largemouth Bass *M. salmoides nuecensis*						FR	FR	FR		FR	FR	FR	FR							
Guadalupe Bass *M. treculii*				F	F	F									F	F	F			
White Crappie *Pomoxis annularis*	NN	C		C	NN	NN	NN	NN					NN	C	C	C	C	C	C	C
Black Crappie *P. nigromaculatus*	NN			NN	NN								NN	C	C	C	C	C	C	C
Western Sand Darter *Ammocrypta clara*																			F	F
Scaly Sand Darter *A. vivax*																			C	C

Table 1. Continued.

Taxa	Upper Brazos River	Upper Red River	Upper Canadian River	Central Edwards Plateau Rivers	Guadalupe and San Antonio Rivers	Southern Edwards Plateau Rivers	Upper Big Bend	Lower Big Bend	Guadalupe Mountains Streams	Davis Mountains Streams	Pecos River	Devils River	Lower Rio Grande	Central Coast Rivers and Streams	Lower Colorado River	Middle Brazos River	San Gabriel River	Lower Brazos River	Southeast Texas Rivers	Northeast Texas Rivers
	Native fish conservation area																			
Redspot Darter *Etheostoma artesiae*																			C	C
Mud Darter *E. asprigene*																				C
Bluntnose Darter *E. chlorosoma*					C									C	C			C	C	C
Fountain Darter *E. fonticola*					FL															
Swamp Darter *E. fusiforme*																			C	
Slough Darter *E. gracile*					C									C	C	C	C	C	C	C
Rio Grande Darter *E. grahami*											FL	FL	FL							
Harlequin Darter *E. histrio*																			C	C
Greenthroat Darter *E. lepidum*				C	C	C														
Goldstripe Darter *E. parvipinne*														C	C			C	C	C
Cypress Darter *E. proeliare*														C	C				C	C
Plains Orangethroat Darter *E. pulchellum*	C	C		C	C	C									C	C	C	C		C
Orangebelly Darter *E. radiosum*																				F
Speckled Darter *E. stigmaeum*																			C	
Gumbo Darter *E. thompsoni*																			FR	FR
Yellow Perch *Perca flavescens*			NN																	
Guadalupe Darter *Percina apristis*					F															
Logperch *P. caprodes*		C																		C
Texas Logperch *P. carbonaria*	C			C	C	C									C	C	C	C	C	
Bigscale Logperch *P. macrolepida*	C	C		C								C	C	C	C	C	C	C	C	C
Blackside Darter *P. maculata*														FL						FL
Slenderhead Darter *P. phoxocephala*																				C

Table 1. Continued.

Taxa	Native fish conservation area																			
	Upper Brazos River	Upper Red River	Upper Canadian River	Central Edwards Plateau Rivers	Guadalupe and San Antonio Rivers	Southern Edwards Plateau Rivers	Upper Big Bend	Lower Big Bend	Guadalupe Mountains Streams	Davis Mountains Streams	Pecos River	Devils River	Lower Rio Grande	Central Coast Rivers and Streams	Lower Colorado River	Middle Brazos River	San Gabriel River	Lower Brazos River	Southeast Texas Rivers	Northeast Texas Rivers
Dusky Darter *P. sciera*	C			C										C	C	C	C	C	C	C
River Darter *P. shumardi*		FR			FR														FR	FR
Sauger *Sander canadensis*		NN																		NN
Walleye *S. vitreus*	NN		NN		NN															
Crevalle Jack *Caranx hippos*																			C	
Leatherjacket *Oligoplites saurus*														C					C	
Spotfin Mojarra *Eucinostomus argenteus*																			C	
Silver Jenny *E. gula*																			C	
Tidewater Mojarra *E. harengulus*														C						
Flagfin Mojarra *E. melanopterus*														C					C	
Sheepshead *Archosargus probatocephalus*																			C	
Pinfish *Lagodon rhomboides*																			C	
Atlantic Threadfin *Polydactylus octonemus*																			C	
Freshwater Drum *Aplodinotus grunniens*	C	C		C	C	C	C	C		C	C	C	C	C	C	C	C	C	C	C
Silver Perch *Bairdiella chrysoura*														C						
Sand Seatrout *Cynoscion arenarius*														C					C	
Spotted Seatrout *C. nebulosus*														C					C	
Silver Seatrout *C. nothus*																			C	
Spot *Leiostomus xanthurus*														C					C	
Atlantic Croaker *Micropogonias undulatus*														C					C	
Black Drum *Pogonias cromis*														C					C	
Red Drum *Sciaenops ocellatus*														C					C	

Table 1. Continued.

Taxa	Native fish conservation area																			
	Upper Brazos River	Upper Red River	Upper Canadian River	Central Edwards Plateau Rivers	Guadalupe and San Antonio Rivers	Southern Edwards Plateau Rivers	Upper Big Bend	Lower Big Bend	Guadalupe Mountains Streams	Davis Mountains Streams	Pecos River	Devils River	Lower Rio Grande	Central Coast Rivers and Streams	Lower Colorado River	Middle Brazos River	San Gabriel River	Lower Brazos River	Southeast Texas Rivers	Northeast Texas Rivers
Star Drum *Stellifer lanceolatus*																			C	
Banded Pygmy Sunfish *Elassoma zonatum*														C				C	C	C
Rio Grande Cichlid *Herichthys cyanoguttatus*	NN			NN	NN	NN	C	C		C	C	C	C	NN	NN	NN	NN	NN		
Blue Tilapia *Oreochromis aureus*	NN			NN	NN	NN	NN	NN		NN	NN	NN	NN	NN	NN	NN	NN	NN	NN	
Mozambique Tilapia *O. mossambicus*					NN						NN	NN								
Redbelly Tilapia *Tilapia zillii*					NN															
Fat Sleeper *Dormitator maculatus*														C				C	C	
Largescaled Spinycheek Sleeper *Eleotris amblyopsis*														C					C	
Bigmouth Sleeper *Gobiomorus dormitor*														C						
Darter Goby *Ctenogobius boleosoma*														C					C	
Freshwater Goby *C. shufeldti*														C					C	
Lyre Goby *Evorthodus lyricus*																			C	
Violet Goby *Gobioides broussonetii*														C					C	
Highfin Goby *Gobionellus oceanicus*														C					C	
Naked Goby *Gobiosoma bosc*														C					C	
Code Goby *G. robustum*														C					C	
Clown Goby *Microgobius gulosus*														C					C	
Spotted Whiff *Citharichthys macrops*																			C	
Bay Whiff *C. spilopterus*														C					C	
Southern Flounder *Paralichthys lethostigma*														F					F	
Lined Sole *Achirus lineatus*														C					C	
Hogchoker *Trinectes maculatus*														C					C	
Least Puffer *Sphoeroides parvus*																			C	

siliency of freshwater habitats, species, and ecosystems (Dodds et al. 2013). Extraction of groundwater for agricultural irrigation, energy development, and municipal water supply has substantially altered groundwater levels and resulted in concomitant reductions in spring discharge and instream flows (Costigan and Daniels 2012; Steward et al. 2013; Garrett et al., in press). The construction of dams has fragmented rivers, altered natural flow patterns, and reduced the availability of suitable habitats for native fishes (Costigan et al. 2012; Wilde and Urbanczyk 2013; Perkin et al. 2014; Worthington et al. 2014; Perkin et al. 2015; Mayes et al. 2019, this volume; Smith et al. 2019, this volume). The cumulative impacts of urbanization and other land use changes have substantially altered natural watershed processes. These and a myriad of other interrelated challenges—degradation of water quality, instream habitat degradation, and the negative effects of nonindigenous species (e.g., predation, competition, and hybridization with native species)—threaten freshwater fish diversity (Gido et al. 2010; Hoagstrom et al. 2011). Left unchecked, these issues will likely continue to contribute to the imperilment and loss of native fishes and other freshwater species (Gido et al. 2010; Hoagstrom et al. 2011). Coordinated conservation intervention is urgently needed to ensure the preservation of native freshwater fish diversity (Hoagstrom et al. 2011; Perkin et al. 2015).

Declining freshwater fish diversity is a conservation issue not unique to Texas (Haslouer et al. 2005; Jelks et al. 2008). Freshwater fish diversity is threatened globally (Dodds et al. 2013; Dudgeon et al. 2006; Strayer and Dudgeon 2010). Innovative and systematic conservation approaches are needed that can be effective at restoration and maintenance of the functional watershed processes necessary to sustain freshwater systems, including native fishes and the habitats they need to thrive (Margules and Pressey 2000; Groves et al. 2002; Balmford and Whitten 2003; Abell et al. 2007; Martinuzzi et al. 2014; Donlan 2015). For those approaches to be supported sociopolitically, they must simultaneously address human needs (e.g., flood abatement, municipal and agricultural water supply, water quality protection, and recreation; Limburg et al. 2011). Collaborative, integrated, and holistic approaches will become increasingly necessary as conflicts and competition for dwindling freshwater resources become unavoidable (Vorosmarty et al. 2010; Dodds et al. 2013; Martinuzzi et al. 2014; Palomo et al. 2014). Lynch et al. (2016) summarized climate trends that have the potential to influence freshwater fishes and compiled findings of 31 peer-reviewed studies from throughout North America that documented climate change effects on freshwater fishes. Of particular concern for native freshwater fishes of Texas is the forecast for droughts to increase in frequency and severity, especially in arid rivers, affecting the timing and frequency of flows and water levels necessary to support spawning and other habitat requirements of freshwater fishes. Complex interactions are also expected to occur between climate change and existing anthropogenic stressors (e.g., invasive species, watershed alteration, river fragmentation, and pumping of groundwater and surface water), making separating and understanding the relative magnitude of climate effects challenging. Paukert et al. (2016) recommended that to support the ability of native freshwater fishes to adapt to climate change, resource managers will need to increasingly focus available resources on actions that enhance the resiliency of watersheds, habitats, and ecosystems.

Although conflicts and competition for available freshwater resources seem inevitable, case studies are available that demonstrate innovative and effective approaches involving cooperation among diverse and seemingly competing interests to deliver beneficial and shared solutions (Garrett 2003; Birdsong et al. 2015; Garrett et al. 2015; Magnelia et al. 2019a, this volume; Robertson et al. 2019, this volume; Smith et al. 2019). These case studies have utilized the native fish conservation areas (NFCAs) approach (Dauwalter et al. 2011; Williams et al. 2011; Garrett et al. 2019, this volume; Harris et al. 2019, this volume; Labay et al. 2019, this volume), which ensures consideration of the conservation needs of native fishes and their habitats in river and watershed management. Furthermore, the NFCAs approach emphasizes management of freshwater systems toward an explicit goal of sustaining native fish diversity while encouraging compatible human uses. This concept is especially relevant to the conservation of freshwater systems in Texas, where many fish species have experienced population declines associated with anthropogenic changes in watersheds where public lands are scarce and large-scale preservation opportunities are limited.

Spatial Prioritization of Native Fish Conservation Areas in Texas

The 2012 Texas Conservation Action Plan (i.e., the State Wildlife Action Plan for Texas; TPWD 2012) identified the need to prioritize areas of the state critically important to the restoration and preservation of freshwater, marine, and terrestrial species of greatest conservation need (SGCN), with the intent of informing future investments of project funding, staff, and other resources toward strategic conservation actions (Margules and Pressey 2000; Fausch et al. 2002; Knight et al. 2006; Knight et al. 2008; Fontaine 2011). To address this need for freshwater fishes, known and modeled fish species distributions, hydrologic and environmental variables, and spatial prioritization analysis were combined to identify high-priority freshwater systems throughout Texas that meet the four critical elements of an NFCA, as described by Williams et al. (2011): (1) natural physical processes remain intact (or have the potential to be restored) within the watershed that support the maintenance of freshwater habitat complexity, diversity and connectivity; (2) habitats are contained within the watershed that support all life history stages of the fish species being preserved; (3) the watershed or fragmented river segment is large enough to provide for long-term persistence of native fish populations (e.g., effective population size); and (4) management plans and other agreements can be developed that will allow the watershed or river segment to be managed in a manner that sustains aquatic and riparian habitat integrity over time and across management jurisdictions and land ownerships. This effort resulted in development of conservation planning products, including species distribution models for freshwater fish SGCN, a landscape ranking and prioritization that identifies focal areas for conservation of freshwater fish SGCN, and a spatial framework for conservation planning and delivery via identification of high-priority freshwater systems considered NFCAs. Detailed descriptions of the concepts and methods used in creation of these products are discussed by Labay et al. (2019), who used consistent methods to prioritize freshwater systems for conservation of fish diversity in the U.S. Great Plains. Thus, methods will be only briefly discussed here.

Fish species chosen for distribution modeling and subsequent analyses were selected on the basis of their inclusion in a recommended list of Texas freshwater fish SGCN assembled by Cohen et al. (2018). The list identifies 91 species of freshwater fishes, each with a conservation status that warrants listing as SGCN (Table 1). It is anticipated that those 91 species will be listed as SGCN in the forthcoming update of the Texas Conservation Action Plan in 2023. Species distribution models (SDMs) were assembled for 85 of the 90 species; SDMs were not assembled for four species and subspecies that are likely extinct (Maravillas Red Shiner, San Marcos Gambusia, Phantom Shiner, Rio Grande Bluntnose Shiner) or for one additional species considered a unique, disjunct population (Spotted Sucker).

The SDMs converted point occurrence data into rangewide probabilities of occurrence (Guisan et al. 2013). Fish occurrence data used in development of SDMs consisted of museum-vouchered specimens available from the University of Texas at Austin Fishes of Texas database and data available from the Global Biodiversity Information Facility that were compiled, reviewed, and partially normalized by Hendrickson et al. (2010) and Cohen et al. (2013). Specific hydrologic, climatic, and topographic variables included in SDM development are described by Labay et al. (2019). Individual SDMs in geographic information system-ready formats and detailed information on the model production methodology can be accessed through the University of Texas at Austin Fishes of Texas Project model download portal (www.fishesoftexas.org/models/).

The SDMs were used within Zonation (Moilanen et al. 2005), a conservation planning software, to spatially rank and prioritize freshwater systems statewide based on their value in conservation of the diversity of Texas freshwater fish SGCN. Conservation value was assessed based on spatially explicit levels of species, habitat, or ecosystem occurrence, as defined by SDM estimation of the relative probability of occurrence. The prioritization emphasized species rarity as opposed to species richness (Moilanen et al. 2005). This approach resulted in prioritization of freshwater systems important in preservation of the diversity of freshwater fish SGCN (Figure 1). Zonation was then used to identify species-based geographic management units, here referred to as NFCAs, based on distance and compositional similarity among the priority freshwater systems. This analysis resulted in identification of 20 NFCAs for inclusion in the Texas Native Fish Conservation Areas Network (hereafter, "Texas NFCAs Network"; Figure 2), which represents a selection of springs, ciénegas, creeks, rivers, and associated watersheds that serve as native fish strongholds and that are now considered priority landscapes for conservation investments by the Texas Parks and Wildlife Department (TPWD) and partners. The diversity of native and nonnative fishes known to occur within each of the 20 NFCAs are outlined in Table 1.

Initial efforts to prioritize freshwater systems for inclusion in the Texas NFCAs Network were conducted in 2013. That pilot phase focused on prioritization of freshwater systems for conservation of 71 fish SGCN that occur within six ecoregions located in the northwestern portion of the state (i.e., High Plains, Southwestern Tablelands, Central Great Plains, Arizona/New Mexico Mountains, Chihuahuan Desert, and Edwards Plateau ecoregions; Griffith et al. 2007). Prioritization of freshwater systems in that geography identified the following 11 conservation priority areas: Upper Canadian River NFCA, Upper Red River NFCA, Upper Brazos River NFCA, Central Edwards Plateau Rivers NFCA, Southern Edwards Plateau Rivers NFCA, Devils River NFCA, Pecos River NFCA, Guadalupe Mountains Streams NFCA, Davis Mountains Streams NFCA, Upper Big Bend NFCA, and Lower Big Bend NFCA (Figure 2). The statewide NFCA prioritization was subsequently completed in 2015, identifying nine additional conservation priority areas located in the eastern and southern portions of the state: Northeast Texas Rivers NFCA, Southeast Texas Rivers NFCA, San Gabriel River NFCA, Middle Brazos River NFCA, Lower Brazos River NFCA, Lower Colorado River NFCA, Guadalupe and San Antonio Rivers NFCA, Central Coast Rivers and Streams NFCA, and Lower Rio Grande NFCA (Figure 2).

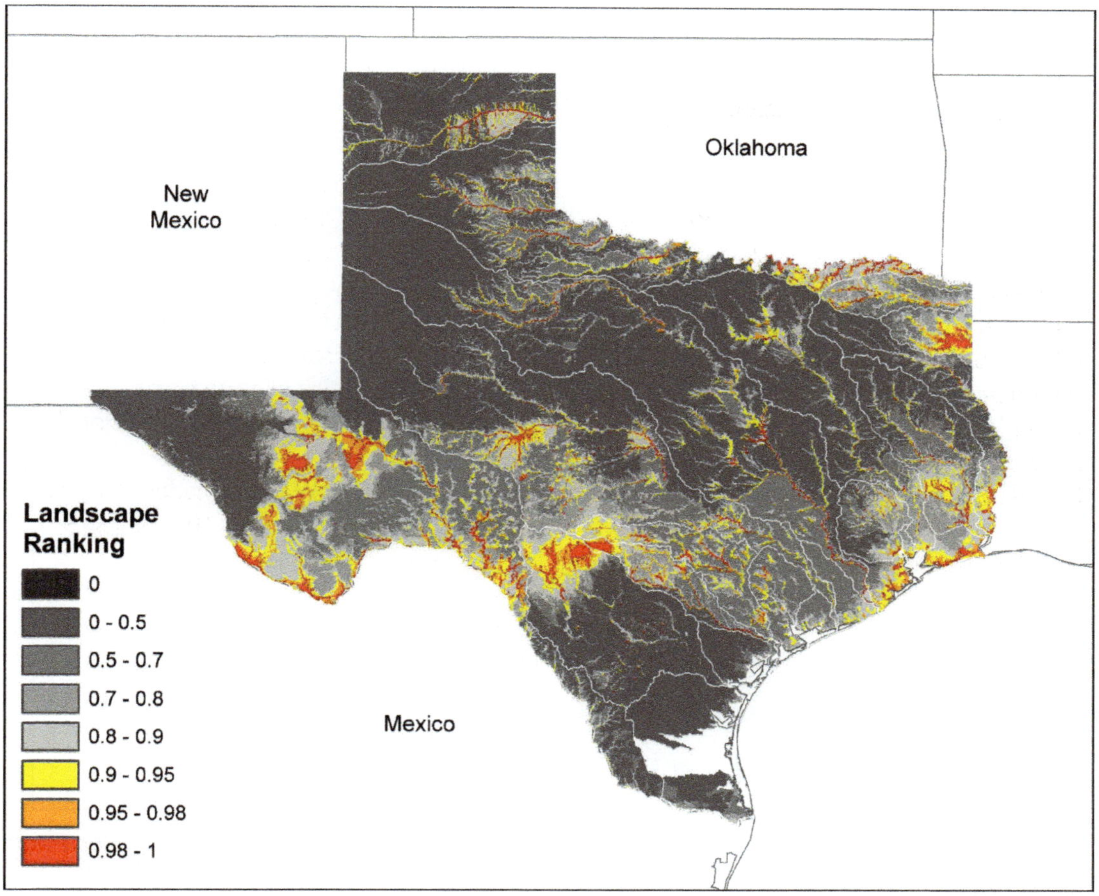

Figure 1. Raw spatial prioritization of freshwater systems for conservation of Texas freshwater fish species of greatest conservation need.

Conservation Planning and Delivery within the Texas Native Fish Conservation Areas Network

Since the initial NFCA prioritization in 2013, the TPWD and partners have focused available technical and financial resources toward conservation planning and delivery within the Texas NFCAs Network. This has included investments in habitat restoration, habitat preservation, research, and surveys. Those investments have contributed toward achieving the vision for the Texas NFCAs Network of restoring and preserving freshwater systems to the level that native fishes thrive as stable components of diverse ecological communities, simultaneously providing clean water, outstanding outdoor recreation, and a stable economic base for present and future citizens. This vision was adapted from the vision established for the Little Tennessee River NFCA (Harris et al. 2019), which is considered the first NFCA officially designated in the United States.

Conservation planning and delivery within the Texas NFCAs Network has involved a diverse group of conservation partners from nongovernmental organizations, state and federal agencies, and universities with a desire to participate in collaborative conservation and a shared mission to restore and preserve wild and native fishes and the habitats they need to thrive. Partners have focused on guiding strategic investments

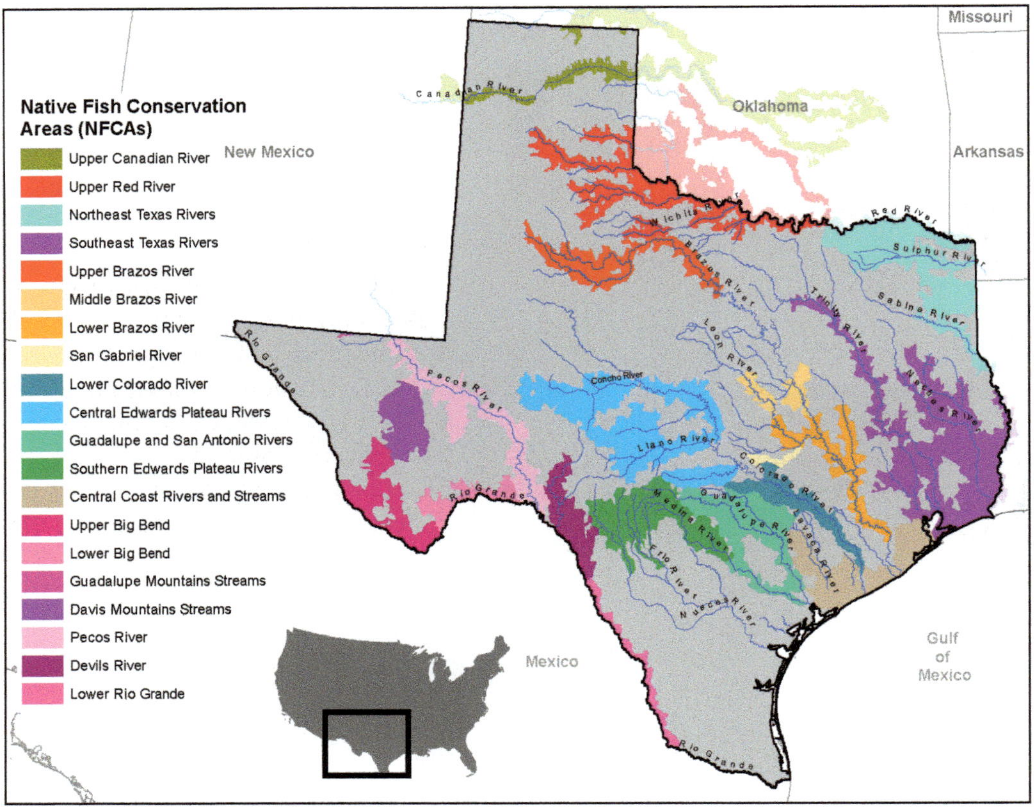

Figure 2. Freshwater systems adopted as native fish conservation areas by the Texas Parks and Wildlife Department.

and leveraging available technical and financial resources to achieve scale-appropriate and transformative actions for conservation of native fishes, their habitats, and other freshwater resources. The critical elements of NFCAs described by Williams et al. (2011), outlined previously in this chapter, have been adopted as the core principles of the Texas NFCAs Network. To facilitate conservation planning and align specific conservation actions undertaken within the Texas NFCAs Network with those core principles, the following eight goals were established for the Texas NFCAs Network:

1. Protect and maintain intact habitats
2. Restore altered habitats
3. Restore instream and floodplain connectivity
4. Mitigate effects of invasive species
5. Organize and facilitate conservation partnership networks
6. Establish conservation demonstration areas
7. Conduct research to fill critical science needs
8. Monitor conservation outcomes and perform adaptive management

Adoption of these eight goals was intended to promote the restoration of watershed functions, emphasizing actions that curtail or eliminate activities on the landscape that degrade water quality, reduce water quantity, degrade riparian systems, favor nonnative species, or fragment river systems while encouraging a wide array of sustainable land-use and water-based recreational activities that are compatible with freshwater fish conservation. Furthermore, these goals have served as thematic topics used to facilitate cooperative planning and identification of NFCA-specific conservation needs, related conservation strategies, project-level conservation actions, and research and monitoring needs. Summaries of conservation investments toward achieving each of these eight goals from 2013 to 2018 are provided below.

Goal 1: protect and maintain intact habitats

Effective January 2016, the Texas Legislature authorized and provided funding for the TPWD to administer the Texas Farm and Ranch Lands Conservation Program (TFRLCP), a grant program designed to provide cost-share funding to land trusts for the purchase of conservation easements on private lands in support of the following objectives: (1) conserve water or protect water quality, (2) conserve native wildlife species through protection of their habitat, (3) conserve rare or sensitive species, (4) demonstrably contribute to preservation of a landscape of conservation lands, or (5) protect productive open-space land threated by fragmentation or development. Specific project scoring, ranking, and selection criteria were assembled by the TPWD and approved by the governor-appointed Texas Farm and Ranch Lands Council, which provides leadership and oversight of the TFRLCP.

The 100 possible points awarded to grant applications submitted to the TFRLCP consider the following scoring criteria: (1) threat of development (20 points), (2) value and cost-effectiveness (20 points), (3) value in protection of watershed processes and aquatic habitats (20 points), (4) value in protection of habitats for SGCN (20 points), (5) contributions to protection of a conservation landscape such as a wildlife migration route or riparian corridor (10 points), and (6) terms of the conservation easement (10 points). Application scoring criteria 3–5 for the TFRLCP directly relate to variables considered in selection of freshwater systems included in the Texas NFCAs Network. Grant applications to the TFRLCP that proposed conservation easements on private lands located within the Texas NFCAs Network scored considerably higher than those located elsewhere in the state. As such, the Texas NFCAs Network prioritization substantially influenced scoring of individual applications, and 13 of the 14 grants awarded by the TFRLCP in 2016–2018 supported preservation of private lands within the Texas NFCAs Network (Table 2).

Those 13 conservation easements funded by the TFRLCP protected 10,563 ha of springs and instream, riparian, and upland habitats within the Central Coast Rivers and Streams, Central Edwards Plateau Rivers, Guadalupe Mountains Streams, Guadalupe and San Antonio Rivers (Figure 3), Lower Colorado River (Figure 4), San Gabriel River, Southeast Texas Rivers, Southern Edwards Plateau Rivers, and Upper Red River NFCAs. Conservation biologists from the TPWD Inland Fisheries Division conducted site visits to the private properties selected for funding and consulted with landowners and partnering land trusts on terms and conditions of the conservation easements to maximize their benefit and value in long-term protection of native fishes, their habitats, and other freshwater resources within those nine NFCAs.

The Western Association of Fish and Wildlife Agencies Crucial Habitat Assessment Tool (CHAT; www.wafwachat.org) identifies important fish and wildlife habitats and corridors across the 17 western states of the United States, including Texas. The purpose of the tool is to incorporate and inform consideration of fish and wildlife habitats in land-use planning, zoning, and development decisions, such as planning of new energy or transportation corridors. Habitats identified as priorities within CHAT were selected by cooperating state fish and wildlife

Table 2. Conservation easements secured within the Texas Native Fish Conservation Areas Network through the Texas Farm and Ranch Lands Conservation Program (2016–2018).

Native fish conservation area	Number of conservatgion easements	Area protected (ha)	Habitat types protected
Central Coast Rivers and Streams	2	2,276	Freshwater wetlands in the Colorado River and Tres Palacios River
Central Edwards Plateau Rivers	1	554	Riparian buffer of the Colorado River
Guadalupe Mountains Streams	1	2,925	Tributaries and riparian buffer of McKittrick Creek
Guadalupe and San Antonio Rivers	1	85	Tributaries and riparian buffer of the Blanco River
Lower Colorado River	3	635	Tributaries and riparian buffer of Barton and Onion creeks
San Gabriel River	1	248	Tributaries and riparian buffer of the San Gabriel River
Southeast Texas Rivers	1	2,230	Tributary Streams and riparian buffer of the Neches River
Southern Edwards Plateau Rivers	2	953	Tributaries and riparian buffer of the Nueces River
Upper Red River	1	656	Riparian buffer of the Red River

Figure 3. Riparian corridor of Wanslow Creek, a stream located within the Guadalupe and San Antonio Rivers Native Fish Conservation Area that was protected through a conservation easement supported through the Texas Farm and Ranch Lands Conservation Program.

Figure 4. Riparian corridor of Onion Creek, a stream located within the Lower Colorado River Native Fish Conservation Area that was protected through multiple conservation easements supported through the Texas Farm and Ranch Lands Conservation Program.

agencies. Each state agency utilized a standard set of definitions, guidelines, and criteria to achieve a consistent regional approach. In 2018, the TPWD finalized CHAT input layers for Texas, including the Aquatic CHAT layer (Figure 5), which incorporated priority freshwater systems of the Texas NFCAs Network as a top-tier priority for conservation of native fishes and other freshwater SGCN. As is the intent of CHAT, this is expected to encourage land developers toward increased consideration, avoidance, and protection of freshwater systems contained within the Texas NFCAs Network.

Goal 2: restore altered habitats

Actions to restore altered habitats within the Texas NFCAs Network have primarily centered on restoration of habitats for Guadalupe Bass, a SGCN and the official state fish of Texas (Figure 6; Birdsong et al. 2015; Garrett et al. 2015; Magnelia et al. 2019a, 2019b). Supported through funding provided by the Southeast Aquatic Resources Partnership, National Fish and Wildlife Foundation, and U.S. Fish and Wildlife Service (USFWS) Texas Partners for Fish and Wildlife Program, the TPWD and partners restored 3,199 ha of springs, creeks, and riparian buffers within the Central Edwards Plateau Rivers NFCA (Table 3), improving habitat conditions for native fishes in approximately 89 km of the James, Llano, and Pedernales rivers and their tributaries (Figures 7 and 8). Additionally, the TPWD provided technical guidance to approximately 850 landowners and other local stakeholders on recommended stewardship practices for management of instream and riparian habitats. Stewardship practices to maintain or restore physical watershed processes have been implemented on approximately 42,389 ha of ranchlands. Habitat restoration was conducted in con-

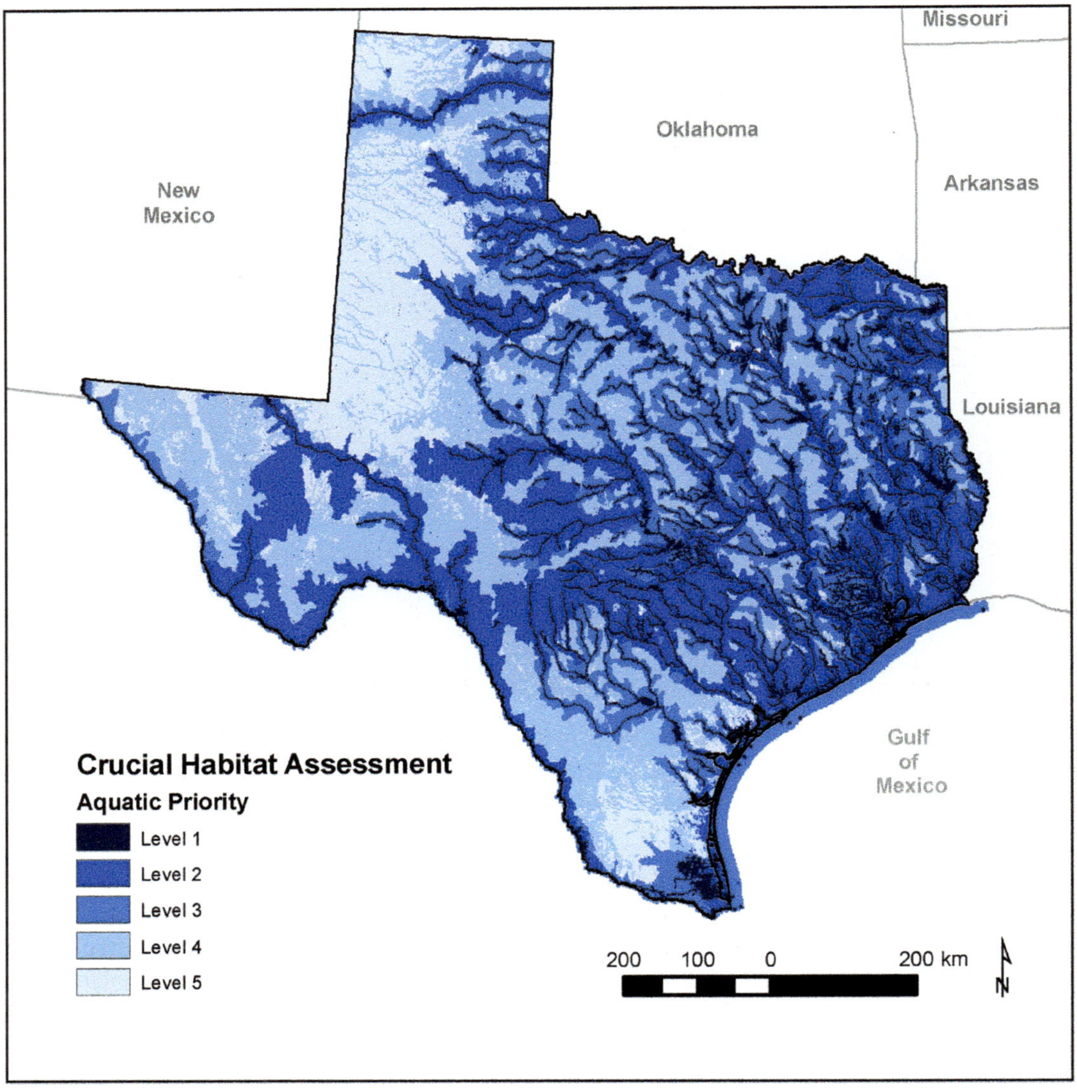

Figure 5. Spatial priorities for conservation of freshwater fish species of greatest conservation need identified within the Western Association of Fish and Wildlife Agencies Crucial Habitat Assessment Tool (Tier 1 = perennial streams within the native range of fish species of greatest conservation need; Tier 2 = Texas Native Fish Conservation Areas Network; Tier 3 = Hydrologic Unit Code 12s that contain perennial streams within the native ranges of fish species of greatest conservation need; Tier 5 = all other areas of the state).

junction with Guadalupe Bass genetic restoration efforts, which involved the production and stocking of more than 793,629 genetically pure Guadalupe Bass fingerlings to ameliorate hybridization between Guadalupe Bass and nonnative, introduced Smallmouth Bass (Lutz-Carrillo et al. 2015).

Within the Southern Edwards Plateau Rivers NFCA, the TPWD collaborated with private landowners and nongovernmental organizations to restore 209 ha of grasslands, riparian buffers, and instream habitats (Table 3). Biologists from the TPWD cooperated with landowners to implement

Figure 6. Guadalupe Bass, the official state fish of Texas and a species of greatest conservation need.

Table 3. Landowner technical guidance and habitat restoration completed within the Texas Native Fish Conservation Areas Network (2013–2018).

Native fish conservation area	Area of ranchlands that received prescriptive guidance on best practices for watershed management (ha)	Quantity of habitats restored (ha)	Habitat types restored
Central Edwards Plateau Rivers	42,389	3,199	Grasslands, springs, riparian, and instream habitats
Guadalupe and San Antonio Rivers	472	142	Riparian and instream habitats
Pecos River	5,036	0.4	Springs and instream habitats
Southern Edwards Plateau Rivers	9,930	209	Grasslands, riparian, and instream habitats
Upper Big Bend	120,343	16,596	Grasslands, springs, riparian, and instream habitats
Upper Brazos River	2,711	2,711	Riparian and instream habitats

Figure 7. Llano River, located within the Central Edwards Plateau Rivers Native Fish Conservation Area.

Figure 8. Pedernales River, located within the Central Edwards Plateau Rivers Native Fish Conservation Area.

conservation best management practices on 9,930 ha within the watersheds of the Frio, Medina, Nueces, and upper Guadalupe rivers. Additionally, a partnership was formed among the TPWD, The Nature Conservancy, the USFWS Partners for Fish and Wildlife Program, and the Texas Master Naturalists Program to provide landowners and citizen scientists with technical guidance on watershed stewardship practices, aquatic species identification, and aquatic resources monitoring strategies.

In the Guadalupe and San Antonio Rivers NFCA, the TPWD and partners organized 13 riparian habitat restoration workshops attended by 544 riparian landowners and other local stakeholders. A landowner restoration manual was assembled that identifies strategies for accommodating recreational access to rivers while maintaining riparian functions (Asher et al. 2017). Similar to conservation efforts in the Central Edwards Plateau Rivers NFCA, the Southeast Aquatic Resources Partnership provided funding to the TPWD and partners to deliver restoration of 142 ha of riparian buffer along the Blanco River (Table 3). Restoration included planting of 3,300 native riparian saplings and 15,000 riparian sedges to support revegetation of erosional riverbanks denuded of vegetation following catastrophic flooding. Habitat restoration was completed in conjunction with a nonnative Smallmouth Bass removal and Guadalupe Bass stocking program, which successfully repatriated Guadalupe Bass to a fragmented reach of the Blanco River (Magnelia et al. 2019b).

Freshwater systems of the Texas NFCAs Network have also been adopted as geographic priorities for investments by multiple conservation funding programs administered by federal agencies and foundations. The most recent 5-year strategic plan of the USFWS Texas Partners for Fish and Wildlife Program (2017–2021) adopted the Davis Mountains Streams, Pecos River, Upper Big Bend, Lower Big Bend, Devils River, Central Edwards Plateau Rivers, Southern Edwards Plateau Rivers, Guadalupe and San Antonio Rivers, and Lower Colorado River NFCAs as geographic focus areas for investments in habitat restoration and species conservation. The USFWS Texas Partners for Fish and Wildlife Program has an active partnership with the TPWD Landowner Incentive Program focused on restoration of grasslands, riparian buffers, and instream habitats. From 2013 to 2018, the two organizations cooperated on restoration of 1,793 ha of grasslands and 11 km of instream habitats and riparian buffers within the Upper Big Bend NFCA (Figures 2 and 9). Comparable investments are expected to continue across the Texas NFCAs Network.

The Texas Parks and Wildlife Foundation (TPWF) is the official nonprofit partner of the TPWD, and in 2013, the TPWD and TPWF partnered to establish the Conserving Texas Rivers Initiative (CTRI), a fundraising program that has supported habitat restoration, native fish conservation, river access, and conservation demonstration within the Texas NFCAs Network. The CTRI represents a public–private partnership, in which private donations have been leveraged with public funding available to the TPWD (e.g., state fishing license revenues, federal grants through the Wildlife and Sport Fish Restoration Program, and federal grants through the National Fish Habitat Partnership). During 2013–2018, private donors contributed US$190,000, which was leveraged against approximately $1.2 million in state and federal funding to implement conservation projects in the Central Edwards Plateau Rivers, Guadalupe and San Antonio Rivers, and Devils River NFCAs. In 2017, the TPWD designed a new vehicle license plate with artwork that features a Texas river (Figure 10). An annual fee of $30 is paid by Texas drivers to display the plate, with $22 allocated to the CTRI. Nearly 1,000 plates were sold in the initial 12 months that the plate was available for purchase, and efforts to market and raise pub-

Figure 9. Rio Grande at the Terlingua Creek confluence within the Upper Big Bend Native Fish Conservation Area.

Figure 10. Texas rivers-themed vehicle license plate, which is sold to raise funding for habitat restoration, native fish conservation, river access, and conservation demonstration implemented within the Texas Native Fish Conservation Areas Network.

lic awareness of the plate are ongoing. The CTRI continues to address a critical need of providing nonfederal funds to meet the cost-share requirements of grants that support conservation projects within the Texas NFCAs Network.

The National Fish and Wildlife Foundation Southwest Rivers Program was established in 2017 to "fund effective conservation projects that achieve measurable outcomes and fill knowledge gaps where they exist, reinvigorating habitats throughout this unique American landscape" (www.nfwf.org/swrivers). The NFCAs located in the Chihuahuan Desert ecoregion of Texas (i.e., Davis Mountains Streams, Guadalupe Mountains Streams, Pecos River, Devils River, Upper Big Bend, and Lower Big Bend NFCAs) were adopted as focal watersheds in the initial request for proposals of the Southwest Rivers Program. Grants totaling $1,535,755 were subsequently awarded by the National Fish and Wildlife Foundation in 2018 to the TPWD, the TPWF, and Sul Ross State University for restoration of streams, riparian buffers, and grasslands in those six NFCAs. A 10-year business plan is currently being assembled for the Southwest Rivers Program. The NFCAs of the Chihuahuan Desert ecoregion of Texas and the associated freshwater fish SGCN are being considered by the National Fish and Wildlife Foundation as strategic priorities are formalized within the plan.

State Technical Advisory Committees (STACs) serve in an advisory capacity to the Natural Resources Conservation Service (NRCS) and other agencies within the U.S. Department of Agriculture on the implementation of the natural resources conservation provisions of the U.S. Farm Bill legislation. The Texas STAC includes an active Wildlife Sub-Committee that informs consideration of fish and wildlife conservation needs and that recommends geographic (e.g., focal watersheds, species ranges) and thematic priorities (e.g., riparian restoration, instream habitat improvements) for conservation initiatives supported through the U.S. Farm Bill in Texas. Since the statewide Texas NFCAs Network prioritization was completed in 2015, the TPWD has recommended that the Texas STAC adopt the Texas NFCAs Network as geographic priorities for a variety of programs, including the Conservation Stewardship Program, Environmental Quality Incentives Program, and Agricultural Conservation Easement Program. Additionally, the Texas STAC Wildlife Sub-Committee initiated establishment of an ad-hoc working group in 2018 to identify riparian habitat conservation priorities in Texas, and the TPWD has encouraged the working group to consider inclusion of the Texas NFCAs Network within that prioritization.

Also occurring in 2018, the NRCS awarded $5,150,000 through the Regional Conservation Partnership Program for habitat restoration and conservation easements in portions of the Central Edwards Plateau Rivers, Lower Colorado River (Figure 11), and Guadalupe and San Antonio Rivers NFCAs. The NRCS also selected the Lower Colorado River NFCA as a 2018 strategic priority for aquatic species conservation through the Working Lands for Wildlife Program. This is expected to increase funding available for restoration and preservation of instream and riparian habitats, potentially benefiting the 12 freshwater fish SGCN that occur in the Lower Colorado River NFCA (Table 1).

Goal 3: restore instream and floodplain connectivity

Efforts to restore instream connectivity (i.e., longitudinal connectivity) within rivers and streams of the Texas NFCAs Network has primarily centered on the removal of low-head dams and the redesign or removal of culverted stream crossings. In 2014, the TPWD and the Texas Commission on Environmental Quality cooperated on removal of a 1.2-m-tall low-head dam spanning a 55-m-wide reach of the North Fork Guadalupe River (Figures 12 and 13), which is located within the Southern

Figure 11. Lower Colorado River Native Fish Conservation Area.

Edwards Plateau Rivers NFCA. In 2016, the USFWS, the TPWD, and local partners cooperated on the removal of Ottine Dam, a 4-m-tall and 30-m-wide low-head dam on the San Marcos River, located within the Guadalupe and San Antonio Rivers NFCA. Removal of Ottine Dam restored instream connectivity in 63 km of the San Marcos River. Also occurring in 2016, the TPWD cooperated with the USFWS on removal of a 2.4-m-tall and 30-m-wide culverted stream crossing in the Upper Brazos River NFCA on the Double Mountain Fork Brazos River. Removal of the crossing restored instream connectivity for the last remaining populations of Sharpnose Shiner and Smalleye Shiner, two highly migratory, pelagic spawning prairie minnows currently listed as federally endangered. Additionally, the TPWD has consulted on the redesign of several culverted stream crossings planned for renovation in the Central Edwards Plateau Rivers and Southern Edwards Plateau Rivers NFCAs.

To undertake a more proactive, strategic approach to restoration of instream connectivity, the TPWD is currently partnering with the Southeast Aquatic Resources Partnership to complete a barrier inventory and prioritization for a portion of the Southern Edwards Plateau Rivers NFCA. Initiated in 2017, this project is expected to serve as a pilot program for possible expansion of the Southeast Aquatic Connectivity Program into Texas (Graham et al. 2019, this volume). The mission of the program is to restore connectivity, habitat, and ecological functions to streams by identifying and removing dams as well as other barriers to aquatic species passage.

Restoration of floodplain connectivity (i.e., lateral connectivity) in rivers and streams of the Texas NFCAs Network has been primarily limited to the Northeast Texas Rivers NFCA (Smith et al. 2019), where The Nature Conservancy, the U.S. Army Corps of Engineers, the Northeast Texas Municipal

Figure 12. Dam removed from the North Fork Guadalupe River to restore instream connectivity and fish passage.

Figure 13. Active removal of a dam on the North Fork Guadalupe River to restore instream connectivity and fish passage.

Water District, the TPWD, the Caddo Lake Institute, and numerous other local conservation partners cooperated on a flow agreement to restore a more natural flow regime in Big Cypress Bayou downstream of Lake O' the Pines. The flow regime included prescriptions for high-flow pulses and overbank flows intended to reconnect the river to its natural floodplain and benefit floodplain spawning fish SGCN, including Ironcolor Shiner and Taillight Shiner. Instream flow recommendations for high-flow pulses and overbank flows to support longitudinal and lateral connectivity within the Texas NFCAs Network are also expected to result from research described within the summary for Goal 7.

Goal 4: mitigate effects of invasive species

Efforts to address the negative effects of invasive species within the Texas NFCAs Network have focused on identification and implementation of regulatory and permitting measures to mitigate impacts of invasive tilapia *Oreochromis* spp. (McGarrity 2019, this volume) and control of invasive riparian plants that form dense, monotypic stands and degrade riparian habitat quality (Bell 1997; Di Tomaso 1998). Efforts to control invasive riparian plants have primarily focused on management of saltcedar *Tamarix* spp. and river cane *Arundo donax*. These species have been shown to accumulate sediment, narrow stream channels, isolate floodplains, reduce instream flow, degrade water quality, increase erosion, and alter instream habitats (Blackburn et al. 1982; Shafroth et al. 2002; Birken and Cooper 2006; Stromberg et al. 2007; Merritt and Poff 2010; Dean and Schmidt 2011; Dean et al. 2011).

In the Guadalupe and San Antonio Rivers, Central Edwards Plateau Rivers, and Southern Edwards Plateau Rivers NFCAs, the TPWD has partnered with The Nature Conservancy, the Hill Country Alliance, the Texas Department of Transportation, river authorities, local municipalities, and approximately 400 cooperating riparian landowners to implement large-scale management of river cane along 200 km of the Blanco, Guadalupe, Medina, Nueces, and Pedernales rivers and their tributaries. In the Blanco and Pedernales rivers, the scope of these efforts was expanded to include mapping of other invasive plants and restoration plantings to augment passive recolonization by native species. Biological monitoring sites were also established along Barons Creek, a tributary of the Pedernales River, to evaluate effects of control efforts on riparian plant communities, fish and invertebrate communities, water quality and quantity, and channel morphology. Similar efforts to implement large-scale control of river cane and to re-establish native riparian vegetation are being implemented by the National Park Service, World Wildlife Fund, Rio Grande Joint Venture, and TPWD in the Upper Big Bend and Lower Big Bend NFCAs (Garrett et al. 2019).

In the Upper Brazos River NFCA, the TPWD has partnered with the USFWS Partners for Fish and Wildlife Program, Texas A&M AgriLife, Texas Tech University, University of Texas at Austin, and more than 60 riparian landowners to manage 4,209 ha of saltcedar, focusing initial efforts along 286 km of the Double Mountain Fork Brazos River. Aerial surveys of saltcedar were completed throughout the entire Upper Brazos River NFCA, and control efforts were expanded to the Salt Fork Brazos River in 2018, with restoration planting of cottonwood *Salix populus* currently in the planning stages. Research is being conducted in partnership with the University of Texas at Austin Bureau of Economic Geology to evaluate the effects of saltcedar control on water budget, water quality, river channel morphology, and riparian plant communities (Mayes et al. 2019).

In the Central Edwards Plateau Rivers NFCA, the TPWD has partnered with the Texas Tech University Llano River Field Station, the Llano River Watershed Alliance, cooperating landowners, and volunteers to implement management of invasive elephant ear *Colocasia esculenta* along more than 80 km of the Llano River and Gorman Creek. Partners have also implemented management of river cane at the South Llano River State Park. Restoration plantings and changes to stewardship practices implemented at the South Llano River and Colorado Bend state parks will be used to provide demonstration sites for outreach to increase awareness of the negative impacts of invasive riparian plants.

Goal 5: organize and facilitate conservation partnership networks

To plan and coordinate conservation delivery within the Texas NFCAs Network, the TPWD, University of Texas at Austin, and Siglo Group facilitated development of regional conservation partnerships. To initiate the partnerships, local, state, and federal natural resources management agencies, universities, nongovernmental organizations, and other local conservation partners were invited to participate in conservation planning webinars, workshops, and field days. Webinars were used to present an overview of the Texas NFCAs Network prioritization and to invite technical review and input into conservation plans assembled for each of the NFCAs (accessible for download at www.nativefishconservation.org). The conservation plans identify desired outcomes for populations of freshwater fish SGCN and their habitats, with an emphasis on maintenance of watershed processes, restoration of degraded habitats, preservation of intact habitats, and long-term strategies for ensuring that conservation actions are sustainable. Each plan contains a brief description of the NFCA, a checklist of fish species, descriptions of the biology and life history of each fish SGCN, a description of the current and desired biological status of each fish SGCN, and a summary of conservation goals for the NFCA.

Facilitated workshops were subsequently organized with partners to refine the conservation plans. During 2015–2018, workshops were conducted for the Upper Canadian River, Upper Red River, Upper Brazos River, Lower Colorado River, Central Edwards Plateau Rivers, Upper Big Bend, Lower Big Bend, Guadalupe Mountains Streams, Davis Mountains Streams, Pecos River, and Devils River NFCAs (Birdsong et al. 2018). Those workshops were attended by 132 fish and wildlife conservation professionals tasked with identification and prioritization of specific conservation projects that could be conducted to conserve native fishes within the NFCAs (e.g., improved land management practices within associated watersheds, barrier removal, water rights acquisition, flow agreements, and research). Workshop participants identified and prioritized 172 individual conservation projects, which are identified in Figure 14 by location and by the corresponding goal of the Texas NFCAs Network addressed by the project.

Additional webinars and conservation planning workshops are scheduled to occur during 2019–2020 in the Northeast Texas Rivers, Southeast Texas Rivers, Guadalupe and San Antonio Rivers, Central Coast Rivers and Streams, San Gabriel River, Middle Brazos River, Lower Brazos River, Southern Edwards Plateau Rivers, and Lower Rio Grande NFCAs. Detailed outcomes of the conservation planning workshops, including descriptions of the planning process and priority projects, will continue to be made accessible at www.nativefishconservation.org.

Goal 6: establish conservation demonstration areas

Through a partnership among the TPWD, the Texas Council of Fly Fishers International, Keep Texas Beautiful, the Devils River Conservancy, the Llano River Watershed Alliance, the Hill Country Alliance, All Water Guides, the Colorado River Alliance, and other local partners, an extensive list of service-oriented river stewardship projects has been

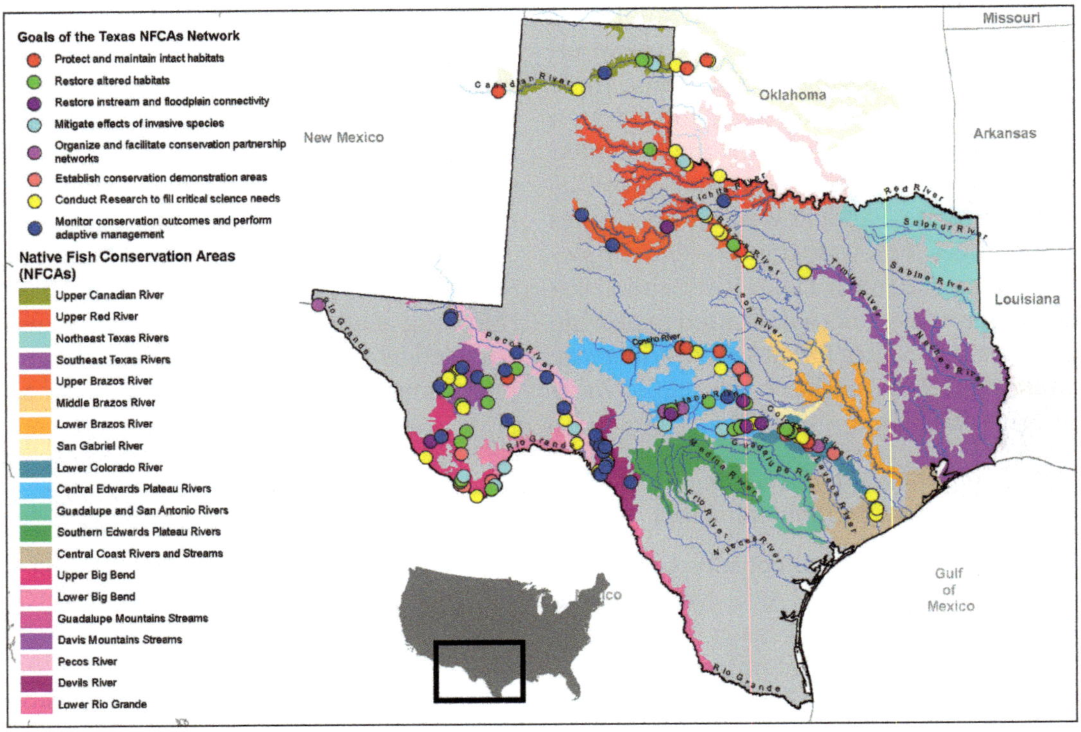

Figure 14. Proposed conservation projects prioritized at conservation planning workshops, with projects identified by location and by the corresponding goal of the Texas Native Fish Conservation Areas Network that would be addressed by each proposed project.

organized and conducted within the Texas NFCAs Network. River stewardship projects have consisted of river-wide trash cleanups, invasive fish and plant removal, planting of native trees and reseeding of erosional banks, establishment of nature trails, installation of educational kiosks, and creation of paddler manuals and other educational resources for river users. Partners have hosted river stewardship workshops for landowners and local communities in order to demonstrate and promote best management practices for conservation of riparian and instream habitats.

River stewardship projects have primarily been conducted in reaches of river where public river access is supported through partnerships with local communities or through lease agreements with willing riparian landowners (Figure 15). The intent of the TPWD-supported river access areas is to facilitate nature-oriented recreation on Texas rivers (e.g., paddling, fishing, and wildlife viewing) and to demonstrate and encourage best practices in the management and conservation of instream and riparian habitats. During 2016–2018, lease payments and habitat improvements were funded through a grant provided by the U.S. Department of Agriculture's Voluntary Public Access and Habitat Incentive Program. Lease agreements between the TPWD and cooperating landowners supported public river access and conservation demonstration at 15 riparian properties within the Texas NFCAs Network. Public river access at these properties provided enhanced paddling and river fishing within approximately 274 km of the NFCAs and served as a catalyst for grassroots involvement in river stewardship activities (Birdsong et al. 2019).

Partnering landowners cooperated with the TPWD and local conservation organizations to assemble resource conservation

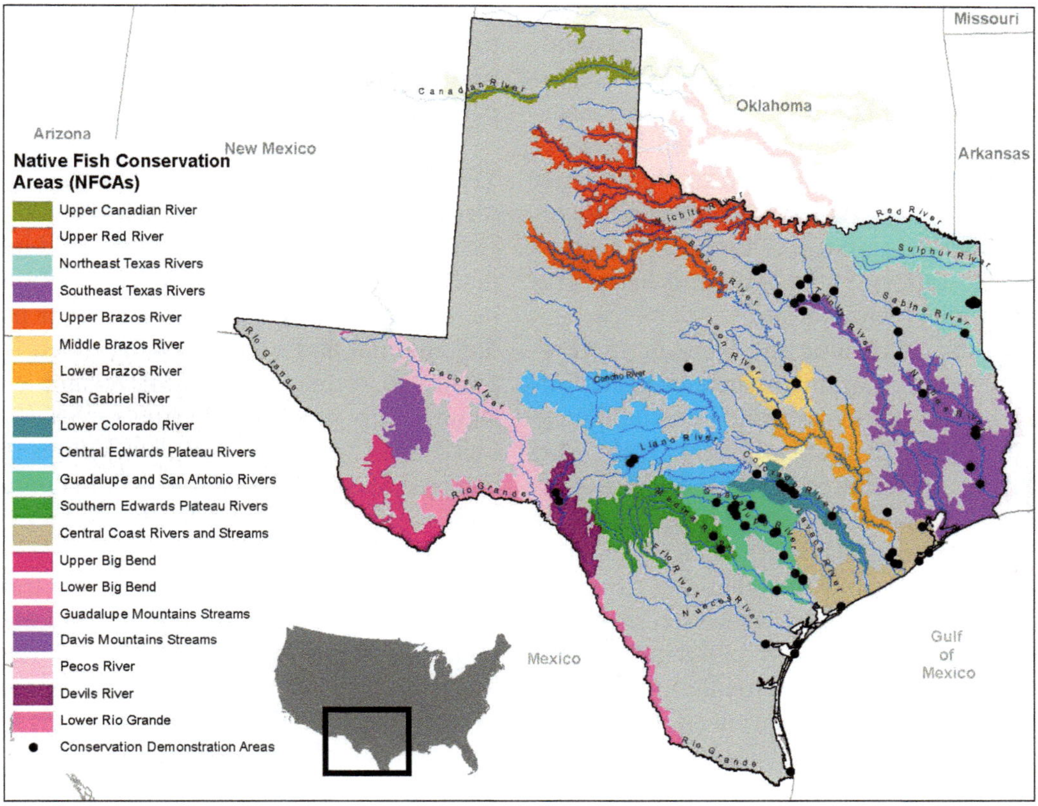

Figure 15. Location of public river access areas utilized by Texas Parks and Wildlife Department as conservation demonstration areas.

plans and deliver habitat improvements and recreational enhancements (e.g., trail maintenance, development of primitive campsites for river users) at the riparian properties. In partnership with Keep Texas Beautiful, a state-based nonprofit organization with numerous local chapters, 19 community outreach and service projects were conducted at the river access areas located within the Lower Colorado River, Guadalupe and San Antonio Rivers, and Southeast Texas Rivers NFCAs. Outreach events promoted river stewardship to 611 attendees. Service projects were supported by 329 volunteers and resulted in removal of 1,411 kg of litter, 280 kg of recyclables, and 675 tires.

In the Central Edwards Plateau Rivers NFCA, the TPWD partnered with the Hill Country Alliance, Llano River Watershed Alliance, and Texas Council of Fly Fishers International to conduct river-wide trash cleanups, install protective exclosures to support recruitment of native riparian seedlings and saplings (decimated by herbivory by nonnative ungulates), and broadcast native riparian seed mix on bare, erosional riverbanks. A series of public "town hall" conversations were also organized in the Central Edwards Plateau Rivers NFCA. These aired on Texas Public Radio and involved a dialogue among the general public, landowners, elected officials, and subject matter experts on river conservation topics such as groundwater management, invasive species management, riparian restoration, conservation easements, and ecosystem services. Partners also produced a 14-min video on the importance of effective riparian management in promoting river resilience; hosted workshops on riparian management, rotational grazing of livestock, and preservation of natural landscapes (through conservation easements); and organized a series of Wild and Scenic Film Festivals that further promoted river conservation through inspirational videos on topics such as protection and restoration of wild lands and waters and biodiversity conservation.

In the Guadalupe and San Antonio Rivers NFCA, a riparian conservation demonstration area was established on the Blanco River in partnership with the Lady Bird Johnson Wildflower Center and cooperating landowners. Stewardship practices implemented at the site included native plant seeding and installation, extensive tree plantings, invasive species control, assimilation of large woody debris into site design, soil compaction remediation for seep restoration, and installation of native turf grasses for access areas. Guided tours of the site began in spring 2018 and are expected to continue, with 50 land managers, to date, having received hands-on instruction in riparian stewardship practices.

In the Devils River NFCA (Figure 16), the TPWD partnered with the Devils River Conservancy to conduct four river-wide trash cleanups, invasive fish removal, and outreach to paddlers and landowners. Outreach included production of a Devils River paddler manual and accompanying video that promotes recreational etiquette and river stewardship practices. Partners also organized two river stewardship workshops that engaged Devils River landowners in demonstration of riparian and land management practices that support healthy rivers.

The Texas Council of Fly Fishers International played an active role in supporting identification of specific reaches of rivers where anglers desired improved access and in establishing and maintaining positive relationships with cooperating landowners of

Figure 16. Devils River Native Fish Conservation Area.

the leasing program. The organization has also distributed 5,000 citrus fruit bags (Figure 17; used for river trash cleanups) to their network of 20 local fly-fishing clubs located throughout the state. The clubs conducted river trash cleanups in conjunction with routine club fishing trips to the access areas.

In 2018, the TPWD was awarded a grant from the USFWS Sport Fish Restoration Recreational Boating Access Grant Program in the amount of $240,000 dollars that will allow for establishment of additional river access and conservation demonstration areas on rivers throughout the state. The grant is expected to support 20 lease agreements with private riparian landowners, opening approximately 322 km of rivers for paddling and kayak fishing. The grant will also add 20 new and maintain 133 existing river access areas supported through partnerships with local communities, providing opportunities for paddling and fishing on 1,081 km of Texas rivers. More than half of those river access areas occur within the Texas NFCAs Network, and through cooperation with local communities, the TPWD intends to utilize these access areas (riparian properties) and the recreationally accessible reaches of river as conservation demonstration areas. In addition to the variety of service-oriented

Figure 17. A youth volunteer holds a mesh bag used to collect trash during a river cleanup on Onion Creek, located within the Lower Colorado River Native Fish Conservation Area.

stewardship projects referenced above, these reaches of river were recently prioritized by the TPWD for management of invasive riparian plants, with an emphasis on management of the problematic species referenced within the summary for Goal 4.

Goal 7: conduct research to fill critical science needs

Since completion of the initial Texas NFCAs Network prioritization in 2013, the TPWD has invested approximately $3 million in State Wildlife Grant funding to fill critical science needs for conservation of freshwater fish SGCN in the Texas NFCAs Network. A primary emphasis of this research has been to quantify flow-ecology relationships for flow-dependent fishes, freshwater mussels, and riparian productivity within highly managed and regulated river reaches (i.e., downstream of reservoirs). This research is ongoing in the Guadalupe and San Antonio Rivers, Middle Brazos River, Lower Brazos River, Lower Colorado River, and Southeast Texas Rivers NFCAs. Results (see TIFP and SARA 2017) are expected to inform strategies for environmental flow restoration and protection (Valente et al. 2019, this volume), adaptive management of environmental flow standards, and related management of river flows and reservoir water levels.

In the Devils River NFCA, similar investments of State Wildlife Grant funding have been made to improve understanding of the relationships among groundwater levels, spring discharge, river flows, and habitat conditions for fish and freshwater mussel SGCN (Robertson et al. 2019, this volume). Potential establishment of a groundwater management district for the portion of the Edwards-Trinity Aquifer that is the source of spring discharge and base flows in the Devils River NFCA has been contemplated by the Texas State Legislature. Meanwhile, the comprehensive science needed to inform such actions has historically been lacking. Over the past 5 years, the TPWD and partners have prioritized investments of State Wildlife Grant funding within the Devils River NFCA to provide the science needed to ensure consideration of the instream flow needs of native fishes and mussels, their habitats, and river recreation in water management decisions for this spring-dominated system.

In the Guadalupe Mountains Streams NFCA, the TPWD partnered with Trout Unlimited, the National Park Service, and the U.S. Geological Survey New Mexico Cooperative Fish and Wildlife Research Unit in 2013–2014 to assess the potential for repatriation of Rio Grande Cutthroat Trout and Rio Grande Chub into McKittrick Creek (Zeigler and Caldwell 2017), a stream that currently hosts a nonnative, introduced population of Rainbow Trout (Garrett and Matlock 1991).

Additional research supported by State Wildlife Grants within the Texas NFCAs Network has centered on filling critical science needs in the life history, distribution, and status of freshwater fish SGCN. This included ongoing research to address the status and distribution of American Eel in the Central Coast Rivers and Streams NFCA and the status of the undescribed Chihuahua Catfish and other regionally endemic fish SGCN in the Central Edwards Plateau Rivers, Southern Edwards Plateau Rivers, Devils River, Pecos River, Davis Mountains Streams, Upper Big Bend, and Lower Big Bend NFCAs.

Priority research needs identified at NFCA conservation planning workshops have also been communicated to other science funding programs. For example, priority research needs identified during the conservation planning workshops held for the Central Edwards Plateau Rivers, Upper Brazos River, Upper Canadian River, and Upper Red River NFCAs during 2015–2016 were presented to the Steering Committee of the Great Plains Landscape Conservation Cooperative in spring 2016. The Great Plains Landscape Conservation Cooperative adopted a subset of those priority research needs for their spring 2016 request for proposals, emphasizing the desire to receive proposals for projects that would examine opportunities for water leases, water rights acquisition, and voluntary incentive-based programs to achieve flow restoration targets within those four NFCAs. One of the projects selected for funding through that request for proposals is described by Valente et al. (2019).

Goal 8: monitor conservation outcomes and perform adaptive management

To fill data gaps and monitor status and trends of fish SGCN within the Texas NFCAs Network, the TPWD and the University of Texas at Austin collaborated on development and implementation of an aquatic gap sampling program. Initiated in 2013 in conjunction with the pilot phase of the Texas NFCAs Network prioritization, the partnership has surveyed 187 locations within the Texas NFCAs Network (Figures 18 and 19). This has resulted in 46,617 museum-vouchered fish specimens and 316 corresponding tissue samples deposited and permanently housed at the University of Texas at Austin Biodiversity Collections. Surveys resulted in the addition of one new species to the state's faunal list (i.e., Bigeye Shiner), additional records of rarely collected fishes (e.g., Pallid Shiner, Cypress Minnow, Emerald Shiner, and Arkansas River Shiner), and records representing range expansions for native (i.e., Least Killifish) and invasive species (i.e., Sheepshead Minnow and Gulf Killifish in the Red River). Surveys have also

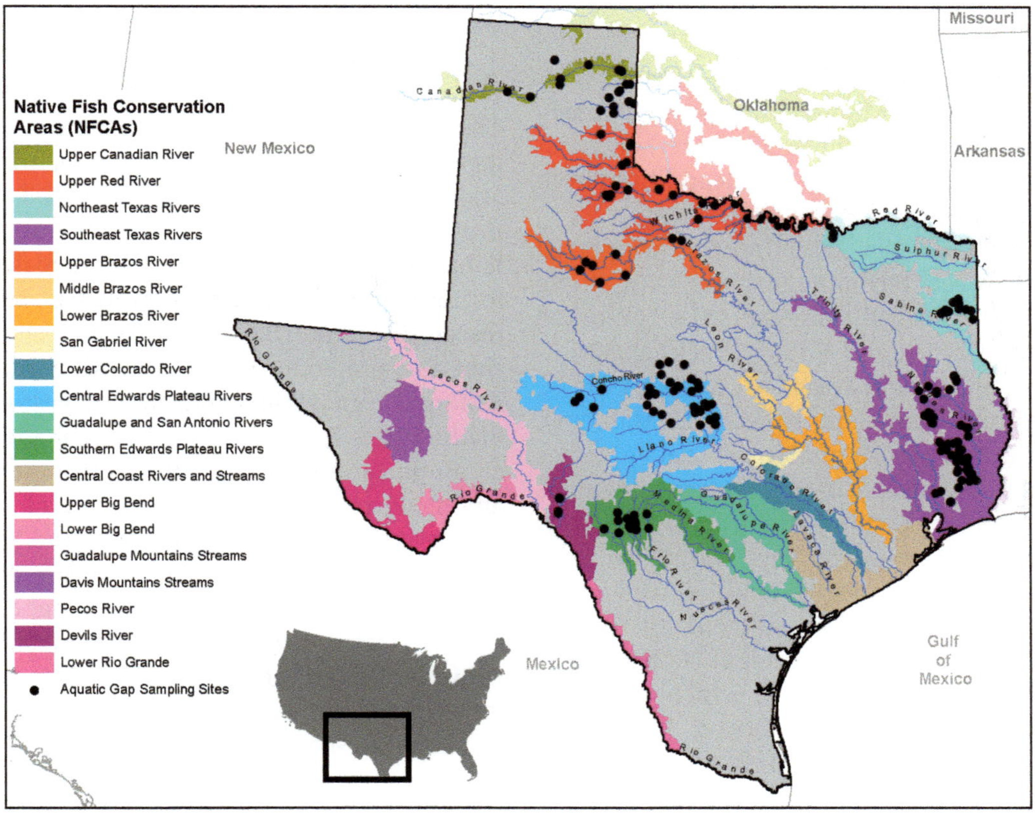

Figure 18. Location of surveys conducted through the joint Texas Parks and Wildlife Department and University of Texas at Austin Aquatic Gap Sampling Program (2013–2018).

Figure 19. Biologists from Texas Parks and Wildlife Department and the University of Texas at Austin conduct surveys within the Upper Canadian River Native Fish Conservation Area.

provided evidence of range reductions (e.g., Suckermouth Minnow, Red River Pupfish, Pallid Shiner, and Emerald Shiner) and possible extirpations (e.g., Flathead Chub) from the state.

Aquatic gap sampling has supported data-driven recommendations to multiple TPWD-managed state parks (i.e., Colorado Bend, Garner, and Village Creek) and wildlife management areas (i.e., Alabama Creek, Alazan Bayou, Gene Howe, and Matador), increasing consideration of native fishes and their habitats in site management plans (Robertson 2015; Robertson et al. 2016, 2017; Cohen et al. 2018). Surveys conducted in the Northeast Texas Rivers NFCA served as the baseline biological assessment for the Cypress Basin Flows Agreement (Robertson et al. 2016; Smith et al. 2019), which was previously referenced under Goal 3. Restoration of the natural flow regime in the Cypress basin is expected to improve instream and riparian habitats, support repatriation of Paddlefish, and benefit conservation of other fish SGCN.

Discussion

The Texas NFCAs Network represents a strategic, science-based approach to planning and delivery of multispecies, watershed approaches to freshwater fish conservation. It has served as the impetus for increased, focused, and sustained investments (e.g., research, monitoring, and habitat restoration and protection) in native fish conservation within priority freshwater systems of the state. Furthermore, it has enhanced communication and fostered collaboration among Texas-based nongovernmental organizations, universities, and state and federal agencies and has facilitated the leveraging of staff, expertise, project funding, and other resources toward delivery of proactive, voluntary conservation projects. These voluntary investments have complemented traditional state-based native fish conservation efforts in Texas, which historically involved reactive, regulatory activities (e.g., state listing as threatened or endangered, permitting of nongame fish collection, permitting of fish stocking in public waters, and permitting of dredging and other instream habitat disturbances) and consultation with other agencies on water management and watershed development projects seeking state or federal permits through requirements of the U.S. Clean Water Act, the U.S. National Environmental Policy Act, or other state and federal laws (e.g., water rights permitting, hydropower relicensing, dam construction, and urban development).

The majority of the technical and financial resources that have supported the conservation activities referenced in this chapter are accessible to fish and wildlife agencies throughout the United States. As such, this chapter offers a case study transferrable to other state fish and wildlife agencies, and that is particularly relevant to those states that, similar to Texas, consist predominately of privately owned lands. This holistic, landscape-scale approach has been adopted in response to the multitude of interrelated natural and anthropogenic stressors affecting native freshwater fishes in Texas, including climate change. As fish and wildlife agencies plan and prepare for current and anticipated effects of climate change, the Texas NFCAs Network offers a case study in engagement of the broader community of conservation practitioners to scale up investments in conservation planning and delivery to enhance the resiliency of freshwater systems. Directly aligned with the goals of the National Fish, Wildlife and Plants Climate Adaptation Strategy (NFWPCAP 2012), efforts to implement the Texas NFCAs Network have emphasized conservation and management of natural landscapes, watershed processes, habitats, species, and ecosystems in a manner that enhances their resiliency and adaptive capacity,

The greatest challenge and opportunity for successful implementation of a statewide network of NFCAs has been the ability to build and sustain grassroots interest, capacity and leadership for river conservation. An important step taken by the TPWD to facilitate grassroots engagement was investment of modest financial support in the operations of local nonprofit organizations, including the Devils River Conservancy, Llano River Watershed Alliance, Hill Country Alliance, and numerous local chapters of the Texas Council of Fly Fishers International and Keep Texas Beautiful. Small capacity grants provided project-based funding to organize and host local outreach events and community service projects, such as riparian stewardship workshops, preparation of seed mix for revegetation projects, tagging of invasive riparian plants at parks and greenspace (for follow up treatment by the TPWD), installation of educational kiosks at river access areas, river trash cleanups, and work days at conservation demonstration areas. In many cases, those projects effectively served as gateway activities that increased stakeholder interest and awareness of broader conservation challenges facing their hometown rivers. River trash cleanups have since given way to grassroots engagement on arguably more significant river conservation issues, such as advocating the need for establishment of more holistic and integrated frameworks for management of aquifer levels, spring discharge, and instream flows, especially for the karst, spring-fed streams that occur in central and west Texas.

An additional lesson learned from implementation of the Texas NFCAs Network was recognition of the importance of building relationships with local thought leaders, early adopters, and trendsetters willing to cooperate on a pilot project. Furthermore, by selecting an initial set of conservation projects that could be completed in a relatively short amount of time, local cooperators were able to quickly demonstrate the feasibility, value, and benefits of the NFCAs approach. The resulting proof of concept was then showcased and promoted to neighboring communities and landowners, supporting expansion throughout the watershed and offering a case study transferable to other NFCAs. Initial successes in delivery of the NFCAs approach in the Central Edwards Plateau Rivers NFCA, notably within the watersheds of the Llano and Pedernales rivers, generated considerable momentum and facilitated rapid expansion into other NFCAs of the state. Although significant conservation challenges remain for Texas freshwater systems, the Texas NFCAs Network offers a holistic and innovative approach that has undeniably increased the scope and scale of efforts to restore and preserve Texas freshwater fish diversity.

Acknowledgments

We thank the numerous individuals and partner organizations who have actively contributed to the conservation outcomes described in this chapter. Your interest, enthusiasm, dedication, and substantive contributions are appreciated and we look forward to building upon our shared vision and partnership to continue to advance conservation goals within the Texas NFCAs Network. Conservation outcomes described in this chapter were financially supported by an extensive list of individual donors, foundations, and state and federal agencies, including the Desert Fish Habitat Partnership, National Fish and Wildlife Foundation, Texas Parks and Wildlife Foundation, Southeast Aquatic Resources Partnership, Texas Parks and Wildlife Department, U.S. Fish and Wildlife Service, and Wildlife and Sport Fish Restoration Program.

References

Abell, R., J. D. Allan, and B. Lehner. 2007. Unlocking the potential of protected areas for freshwaters. Biological Conservation 134:48–63.

Asher, H. A., M. Bertelsen, and M. O'Toole. 2017. Blanco River design guidelines. Lady Bird Johnson Wildflower Center, Austin, Texas.

Balmford, A., and T. Whitten. 2003. Who should pay for tropical conservation, and how could the costs be met? Oryx 37:238–250.

Bell, G. P. 1997. Ecology and management of Arundo donax, and approaches to riparian habitat restoration in Southern California. Pages 103–111 *in* J. H. Brock, M. Wade, P. Pysek, and D. Green, editors. Plant invasions: studies from North America and Europe. Backhuys Publishers, Leiden, Netherlands.

Birdsong, T. W., M. S. Allen, J. E. Claussen, G. P. Garrett, T. B. Grabowski, J. Graham, F. Harris, A. Hartzog, A., D. Hendrickson, R. A. Krause, J. K. Leitner, J. M. Long, C. K. Metcalf, D. P. Phillipp, W. F. Porak, S. Robinson, S. M. Sammons, S. L. Shaw, J. E. Slaughter, IV, and M. D. Tringali. 2015. Native black bass initiative: implementing watershed-scale approaches to conservation of endemic black bass and other native fishes in the southern United States. Pages 363–378 *in* M. D. Tringali, J. M. Long, T. W. Birdsong, and M. S. Allen, editors. Black bass diversity: multidisciplinary science for conservation. American Fisheries Society, Symposium 82, Bethesda, Maryland.

Birdsong, T. W., J. Botros, S. Magnelia, J. Anderson, M. Bean, T. Broad, D. Cortez, T. Grabowski, C. Kowaleski, C. Chute-Canal, J. East, K. Glenewinkel, B. Hester, R. Husted, J. Joplin, J. Lewey, S. Nichols, D. Oppenheimer, M. Parker, S. Robertson, and A. Stevens. 2019. Texas River Access and Conservation Areas Program: partnering with private landowners to expand paddling and fishing opportunities on Texas rivers. Texas Parks and Wildlife Department, Austin.

Birdsong, T. W., D. C. Dauwalter, G. P. Garrett, B. J. Labay, M. Bean, J. Broska, J. Graham, S. Magnelia, K. B. Mayes, M. McGarrity, K. M. Johnson, S. Robertson, T. Thompson, S. Vail-Muse, and J. B. Whittier. 2018. Native fish conservation areas of the southwestern USA: facilitating landscape-scale conservation of aquatic habitats and freshwater fishes. Southeast Aquatic Resources Partnership, Panama City, Florida.

Birken, A. S., and D. J. Cooper. 2006. Processes of Tamarix invasion and floodplain development along the lower Green River, Utah. Ecological Applications 16:1103–1120.

Blackburn, W. H., R. W. Knight, and J. L. Schuster. 1982. Saltcedar influence on sedimentation in the Brazos River. Journal of Soil and Water Conservation 37:298–301.

Cohen, A. E., B. J. Labay, D. A. Hendrickson, M. Casarez, and S. Sarkar. 2013. Data provision and projected impact of climate change on fish biodiversity within the Desert LCC. Final Report submitted to U.S. Department of Interior, Bureau of Reclamation, Desert Landscape Conservation Cooperative, Austin, Texas.

Cohen, A. E., G. P. Garrett, M. J. Casarez, D. A. Hendrickson, B. J. Labay, T. Urban, J. Gentle, D. Wylie, and D. Walling. 2018. Conserving Texas biodiversity: status, trends, and conservation planning for fishes of greatest conservation need. Texas Parks and Wildlife Department, Austin.

Costigan, K. H., and M. D. Daniels. 2012. Damming the prairie: human alteration of Great Plains river regimes. Journal of Hydrology 444:90–99.

Dauwalter, D. C., J. S. Sanderson, J. E. Williams, and J. R. Sedell. 2011. Identification and implementation of native fish conservation areas in the upper Colorado River basin. Fisheries 36:278–288.

Dean, D. J., and J. C. Schmidt. 2011. The role of feedback mechanisms in historic channel changes of the lower Rio Grande in the Big Bend region. Geomorphology 126:333–349.

Dean, D. J., M. L. Scott, P. B. Shafroth, and J. C. Schmidt. 2011. Stratigraphic, sedimentologic, and dendrogeomorphic analyses of rapid floodplain formation along the Rio Grande in Big Bend National Park, Texas. Geological Society of America Bulletin 123:1908–1925.

Di Tomaso, J. M. 1998. Impact, biology, and ecology of saltcedar (*Tamarix* spp.) in the southwestern United States. Weed Technology 12:326–336.

Dodds, W. K., J. S. Perkin, and J. E. Gerken. 2013. Human impact on freshwater ecosystem services: a global perspective. Environmental Science and Technology 47:9061–9068.

Donlan, C. J., 2015. Proactive strategies for protecting species: pre-listing conservation and the Endangered Species Act. University of California Press, Berkeley.

Dudgeon, D., A. H. Arthington, M. O. Gessner, Z. I. Kawabata, D. J. Knowler, C. Lévêque, R. J. Naiman, A. H. Prieur-Richard, D. Soto, M. L. J. Stiassny, and C. A. Sullivan. 2006. Freshwater biodiversity: importance, threats, status and conservation challenges. Biological Reviews 81:163–182.

Fausch, K. D., C. E. Torgersen, C. V. Baxter, and H. W. Li. 2002. Landscapes to riverscapes: bridging the gap between research and conservation of stream fishes. BioScience 52:483–498.

Fleming, B. P., G. P. Garrett, and N. G. Smith. 2015. Reducing hybridization and introgression in wild populations of Guadalupe Bass through supplemental stocking. Pages 537–547 *in* M. D. Tringali, J. M. Long, T. W. Birdsong, and M. S. Allen, editors. Black bass diversity: multidisciplinary science for conservation. American Fisheries Society, Symposium 82, Bethesda, Maryland.

Fontaine, J. J. 2011. Improving our legacy: incorporation of adaptive management into state wildlife action plans. Journal of Environmental Management 92:1403–1408.

Garrett, G. P. 2003. Innovative approaches to recover endangered species. Pages 151–160 *in* G. P. Garrett and N. L. Allan, editors. Aquatic fauna of the northern Chihuahuan Desert. Museum of Texas Tech University, Special Publication 46, Lubbock.

Garrett, G. P., T. W. Birdsong, M. G. Bean, and R. McGillicuddy. 2015. Guadalupe Bass restoration initiative. Pages 635–657 *in* M. D. Tringali, J. M. Long, T. W. Birdsong, and M. S. Allen, editors. Black bass diversity: multidisciplinary science for conservation. American Fisheries Society, Symposium 82, Bethesda, Maryland.

Garrett, G. P., and G. C. Matlock. 1991. Rio Grande Cutthroat Trout in Texas. Texas Journal of Science 43:405–410.

Garrett, G. P., M. G. Bean, R. J. Edwards, and D. A. Hendrickson. In press. Mining hidden waters: groundwater depletion, aquatic habitat degradation and loss of fish diversity in the Chihuahuan Desert ecoregion of Texas. In D. L. Propst, J. E. Williams, K. R. Bestgen and C. W. Hoagstrom, editors. Standing between life and extinction: ethics and ecology of conserving aquatic species in the American Southwest. University of Chicago Press, Chicago.

Garrett, G. P., T. W. Birdsong, M. G. Bean, and B. J. Labay. 2019. Chihuahuan Desert native fish conservation areas: a multispecies and watershed approach to preservation of freshwater fish diversity. Pages 231–252 *in* D. C. Dauwalter, T. W. Birdsong, and G. P. Garrett, editors. Multispecies and watershed approaches to freshwater fish conservation. American Fisheries Society, Symposium 91, Bethesda, Maryland.

Gido, K. B., W. K. Dodds, and M. E. Eberle. 2010. Retrospective analysis of fish community change during a half-century of landuse and streamflow changes. Journal of the North American Benthological Society 29:970–987.

Griffith, G., S. Bryce, J. Omernik, and A. Rogers. 2007. Ecoregions of Texas. Texas Commission on Environmental Quality, Austin.

Groves, C. R., D. B. Jensen, L. L. Valutis, K. H. Redford, M. L. Shaffer, J. M. Scott, J. V. Baumgartner, J. V. Higgins, M. W. Beck, and M. G. Anderson. 2002. Planning for biodiversity conservation: Putting conservation science into practice: a seven-step framework for developing regional plans to conserve biological diversity, based upon principles of conservation biology and ecology, is being used extensively by the Nature Conservancy to identify priority areas for conservation. BioScience 52:499–512.

Guisan, A., R. Tingley, J. B. Baumgartner, I. Naujokaitis-Lewis, P. R. Sutcliffe, A. I. T. Tulloch, T. J. Regan, L. Brotons, E. McDonald-Madden, C. Mantyka-Pringle, T. G. Martin, J. R. Rhodes, R. Maggini, S. A. Setterfield, J. Elith, M. W. Schwartz, B. A. Wintle, O. Broennimann, M. Austin, S. Ferrier, M. R. Kearney, H. P. Possingham, and Y. M. Buckley. 2013. Predicting species distributions for conservation decisions. Ecology Letters 16:1424–1435.

Haslouer, S. G., M. E. Eberle, D. R. Edds, K. B. Gido, C. S. Mammoliti, J. R. Triplett, J. T. Collins, D. A. Distler, D. G. Huggins, and W. J. Stark. 2005. Current status of native fish species in Kansas. Transactions of the Kansas Academy of Science 108:32–46.

Hendrickson, D. A., S. Sarkar, A. Molineux. 2010. Provision and inventory of diverse aquatic ecosystem-related resources for the Great Plains Landscape Conservation Cooperative (GPLCC). Final report on Grant Agreement Number 20181AG91. Great Plains Conservation Cooperative. University of Texas, Austin.

Hoagstrom, C. W., J. E. Brooks, and S. R. Davenport. 2011. A large-scale conservation perspective considering endemic fishes of the North American plains. Biological Conservation 144:21–34.

Jelks, H. L., S. J. Walsh, N. M. Burkhead, S. Contreras-Balderas, E. Diaz-Pardo, D. A. Hendrickson, J. Lyons, N. E. Mandrak, F. McCormick, J. S. Nelson, and S. P. Platania. 2008. Conservation status of imperiled North American freshwater and diadromous fishes. Fisheries 33:372–407.

Knight, A. T., R. M. Cowling, and B. M. Campbell. 2006. An operational model for implementing conservation action. Conservation Biology 20:408–419.

Knight, A. T., R. M. Cowling, M. Rouget, A. Balmford, A. T. Lombard, and B. M. Campbell. 2008. Knowing but not doing: selecting priority conservation areas and the research-implementation gap. Conservation Biology 22:610–617.

Labay, B. J., J. S. Perkin, D. A. Hendrickson, A. R. Cooper, G. P. Garrett, and T. W. Birdsong. 2019. Who's asking? Interjurisdictional conservation assessment and planning for Great Plains fishes. Pages 57–83 *in* D. C. Dauwalter, T. W. Birdsong, and G. P. Garrett, editors. Multispecies and watershed approaches to freshwater fish conservation. American Fisheries Society, Symposium 91, Bethesda, Maryland.

Leslie, A., E. McCombs, and F. Harris. 2019. Little Tennessee Native Fish Conservation Partnership: aquatic conservation on a landscape scale. Pages 415–430 *in* D. C.

Dauwalter, T. W. Birdsong, and G. P. Garrett, editors. Multispecies and watershed approaches to freshwater fish conservation. American Fisheries Society, Symposium 91, Bethesda, Maryland.

Limburg, K. E., R. M. Hughes, D. C. Jackson, and B. Czech. 2011. Human population increase, economic growth, and fish conservation: collision course or savvy stewardship? Fisheries 36:27–35.

Lutz-Carrillo, D., C. Thibodeaux, M. Elliott, N. A. Rathjen, C. Kittel, L. T. Fries, and G. P. Garrett. 2015. Inferred reproductive behavior of captive Guadalupe Bass. Pages 549–583 *in* M. D. Tringali, J. M. Long, T. W. Birdsong, and M. S. Allen, editors. Black bass diversity: multidisciplinary science for conservation. American Fisheries Society, Symposium 82, Bethesda, Maryland.

Lynch, A.J., Bonnie J. E. Myers, Cindy Chu, Lisa A. Eby, Jeffrey A. Falke, Ryan P. Kovach, Trevor J. Krabbenhoft, Thomas J. Kwak, John Lyons, Craig P. Paukert, and James E. Whitney. 2016. Climate change effects on North American inland fish populations and assemblages. Fisheries 41:346–361.

Magnelia, S., G. Linam, R. McGillicuddy, K. Saunders, M. Parker, T. Birdsong, D. Lutz-Carillo, J. Williamson, R. Ranft, and T. Bonner. 2019b. Repatriation of Guadalupe Bass in the Blanco River, Texas: a case study in the opportunistic use of drought as a fisheries management tool. Pages 213–230 *in* M. J. Siepker and J. W. Quinn, editors. Managing centrarchid fisheries in rivers and streams. American Fisheries Society, Symposium 87, Bethesda, Maryland.

Magnelia, S. J., K. B. Mayes, M. G. Bean, C. L. Loeffler. and D. D. Bradsby. 2019a. Four decades of conserving native fish in the Colorado River watershed, Texas. Pages 269–292 *in* D. C. Dauwalter, T. W. Birdsong, and G. P. Garrett, editors. Multispecies and watershed approaches to freshwater fish conservation. American Fisheries Society, Symposium 91, Bethesda, Maryland.

Margules, C. R., and R. L. Pressey. 2000. Systematic conservation planning. Nature (London) 405:243–253.

Martinuzzi, S., S. R. Januchowski-Hartley, B. M. Pracheil, P. B. McIntyre, A. J. Plantinga, D. J. Lewis, and V. C. Radeloff. 2014. Threats and opportunities for freshwater conservation under future land use change scenarios in the United States. Global Change Biology 20:113–124.

Mayes, K. B., G. R. Wilde, M. E. McGarrity, B. D. Wolaver, and T. G. Caldwell. 2019. Watershed-scale conservation of native fishes in the Brazos River basin, Texas. Pages 315–343 *in* D. C. Dauwalter, T. W. Birdsong, and G. P. Garrett, editors. Multispecies and watershed approaches to freshwater fish conservation. American Fisheries Society, Symposium 91, Bethesda, Maryland.

Merritt, D. M., and N. L. Poff. 2010. Shifting dominance of riparian Populus and Tamarix along gradients of flow alteration in western North American rivers. Ecological Applications 20:135–152.

Moilanen, A., A. M. A. Franco, R. I. Early, R. Fox, B. Wintle, and C. D. Thomas. 2005. Prioritizing multiple-use landscapes for conservation: methods for large multi-species planning problems. Proceedings of the Royal Society of London B 272:1885–1891.

NFWPCAP (National Fish, Wildlife and Plants Climate Adaptation Partnership). 2012. National fish, wildlife and plants climate adaptation strategy. Association of Fish and Wildlife Agencies, Council on Environmental Quality, Great Lakes Indian Fish and Wildlife Commission, National Oceanic and Atmospheric Administration, and U.S. Fish and Wildlife Service, Washington, DC.

Paukert, C. P., B. A. Glazer, G. J. A. Hansen, B. J. Irwin, P. C. Jacobson, J. L. Kershner, B. J. Shuter, J. E. Whitney, and A. J. Lynch. 2016. Adapting inland fisheries management to a changing climate. Fisheries 41:374–384.

Palomo, I., C. Montes, B. Martín-López, J. A. González, M. García-Llorente, P. Alcorlo, and M. R. G. Mora. 2014. Incorporating the social-ecological approach in protected areas in the Anthropocene. BioScience 64(3):181–191.

Perkin, J. S., K. B. Gido, A. R. Cooper, T. F. Turner, M. J. Osborne, E. R. Johnson, and K. B. Mayes. 2015. Fragmentation and dewatering transform Great Plains stream fish communities. Ecological Monographs 85:73–92.

Perkin, J. S., K. B. Gido, K. H. Costigan, M. D. Daniels, and E. R. Johnson. 2014. Fragmentation and drying ratchet down Great Plains stream fish diversity. Aquatic Conservation: Marine and Freshwater Ecosystems 25:639–655.

Robertson, S. 2015. Upper Frio River basin bioassessment: Dry Frio and Frio rivers in Real and Uvalde counties, Texas. Texas Parks and Wildlife Department, River Studies Report No. 23, Austin.

Robertson, S., M. Parker, G. Linam, C. Robertson, A. Grubh, and M. Casarez. 2016. Village Creek watershed bioassessment. Texas Parks and Wildlife Department, River Studies Report No. 25, Austin.

Robertson, S., M. Parker, G. Linam, C. Robertson, A. Grubh, and M. Casarez. 2017. Canadian River basin bioassessment. Texas Parks and Wildlife Department, River Studies Report No. 26, Austin.

Robertson, S., B. D. Wolaver, T. G. Caldwell, T. W. Birdsong, R. Smith, T. Hardy, J. Lewey, and J. Joplin. 2019. Developing the science and public support needed to preserve the Devils River: a case study in collaborative conservation. Pages 293–314 *in* D. C. Dauwalter, T. W. Birdsong, and G. P. Garrett, editors. Multispecies and watershed approaches to freshwater fish conservation. American Fisheries Society, Symposium 91, Bethesda, Maryland.

Shafroth, P. B., J. C. Stromberg, and D. T. Patten. 2002. Riparian vegetation response to altered disturbance and stress regimes. Ecological Applications 12:107–123.

Smith, R., J. Trungale, R. Lowerre, T. Hayes, M. Montagne, T. Bister, L.-A. Overdyke, and M. Hackett. 2019. Instream flow restoration and watershed conservation in the Cypress basin, Texas. Pages 345–366 *in* D. C. Dauwalter, T. W. Birdsong, and G. P. Garrett, editors. Multispecies and watershed approaches to freshwater fish conservation. American Fisheries Society, Symposium 91, Bethesda, Maryland.

Steward, D. R., P. J. Bruss, X. Yang, S. A. Staggenborg, S. M. Welch, and M. D. Apley. 2013. Tapping unsustainable groundwater stores for agricultural production in the High Plains Aquifer of Kansas, projections to 2110. Proceedings of the National Academy of Sciences of the United States of America 110:E3477–E3486.

Strayer, D. L., and D. Dudgeon. 2010. Freshwater biodiversity conservation: recent progress and future challenges. Freshwater Science 29:344–358.

Stromberg, J. C., S. J. Lite, R. Marler, C. Paradzick, P. B. Shafroth, D. Shorrock, J. M. White, and M. S. White. 2007. Altered stream-flow regimes and invasive plant species: the Tamarix case. Global Ecology and Biogeography 16:381–393.

TIFP (Texas Instream Flow Program) and SARA (San Antonio River Authority). 2017. Instream flow recommendations for the lower San Antonio River and lower Cibolo Creek. Available: www.twdb.texas.gov/surfacewater/flows/instream/lower_san_antonio/doc/instream_flow_study_of_the_lower_san_antonio_river_and_lower_cibolo_creek.pdf?d=20692.785. (April 2019).

TPWD (Texas Parks and Wildlife Department). 2012. Texas conservation action plan. TPWD, Austin, Texas. Available: https://tpwd.texas.gov/landwater/land/tcap/. (April 2019).

Valente, J., D. Bradsby, K. B. Mayes, C. Loeffler, L. Hamlin, D. Geeslin, K. Horndeski, D. Young, J. Trungale, R. Smith, K. Garmany, and T. Hayes. 2019. Developing a geospatial decision support tool for protecting and restoring environmental flows in Texas rivers and streams. Pages 253–267 *in* D. C. Dauwalter, T. W. Birdsong, and G. P. Garrett, editors. Multispecies and watershed approaches to freshwater fish conservation. American Fisheries Society, Symposium 91, Bethesda, Maryland.

Vorosmarty, C. J., P. B. McIntyre, M. O. Gessner, D. Dudgeon, A. Prusevich, P. Green, S. Glidden, S. E. Bunn, C. A. Sullivan,

C. R. Liermann, and P. M. Davies. 2010. Global threats to human water security and river biodiversity. Nature (London) 467:555–561.

Wilde, G. R., and A. C. Urbanczyk. 2013. Relationship between river fragment length and persistence of two imperiled Great Plains cyprinids. Journal of Freshwater Ecology 28:445–451.

Williams, J. E., R. N. Williams, R. F. Thurow, L. Elwell, D. P. Philipp, F. A. Harris, J. L. Kershner, P. J. Martinez, D. Miller, G. H. Reeves, C. A. Frissell, and J. R. Sedell. 2011. Native fish conservation areas: a vision for large-scale conservation of native fish communities. Fisheries 36:267–277.

Worthington, T. A., S. K. Brewer, T. B. Grabowski, and J. Mueller. 2014. Backcasting the decline of a vulnerable Great Plains reproductive ecotype: identifying threats and conservation priorities. Global Change Biology 20:89–102.

Zeigler, M. P., and C. A. Caldwell. 2017. Feasibility study: establishing native fish fauna to McKittrick Creek, Guadalupe Mountains National Park, Texas. National Park Service, Guadalupe Mountains National Park, Final Report, Salt Flat, Texas.

American Fisheries Society Symposium 91:231–252, 2019

Chihuahuan Desert Native Fish Conservation Areas: A Multispecies and Watershed Approach to Preservation of Freshwater Fish Diversity

Gary P. Garrett*
Texas Natural History Collections, Department of Integrative Biology
University of Texas at Austin
10100 Burnet Road, Austin, Texas 78758, USA

Timothy W. Birdsong
Texas Parks and Wildlife Department, Habitat Conservation Branch, Inland Fisheries
4200 Smith School Road, Austin, Texas 78744, USA

Megan G. Bean
Texas Parks and Wildlife Department, Habitat Conservation Branch, Inland Fisheries
5103 Junction Highway, Mountain Home, Texas 78058, USA

Ben J. Labay
Siglo Group
1619 West 6th Street, Austin, Texas 78703, USA

Abstract.—Native fish conservation areas in the Chihuahuan Desert of Texas were identified and designated as part of a statewide network of focal watersheds uniquely valued in preservation of Texas freshwater fish diversity. Native fish conservation areas represent a holistic, multispecies, and habitat-based approach to native fish conservation that encourages and facilitates coordination among landowners, nongovernmental organizations, state and federal agencies, universities, and local governments to achieve landscape-scale conservation within focal watersheds. This approach to native fish conservation provides an effective method for addressing the common nature and magnitude of threats facing species and their habitats in freshwater systems. Desert fishes and their habitats are particularly susceptible to habitat alteration, especially anthropogenic land use and water consumption patterns, which continue to create conservation challenges. The strategic and science-based conservation strategies embodied by the native fish conservation areas approach represent an innovative path forward for addressing the conservation needs of native fishes and their habitats in the Chihuahuan Desert of Texas. In this chapter, we describe six native fish conservation areas designated within the Chihuahuan Desert of Texas and profile multiagency conservation planning and delivery that has substantially increased the scope and scale of conservation investments for restoration and protection of native fishes and their habitats in the region.

* Corresponding author: garygarrett@utexas.edu

Introduction

The Texas Parks and Wildlife Department (TPWD), in partnership with the University of Texas at Austin and Siglo Group, has prioritized and established a statewide network of watersheds uniquely valued in preservation of Texas freshwater fish diversity (Figure 1; Cohen et al. 2018; Birdsong et al. 2019, this volume). Termed "native fish conservation areas" (NFCAs), these focal watersheds were identified through a spatial prioritization that considered distributions of focal fishes, watershed characteristics, and habitat conditions necessary to fulfill their life histories (Dauwalter et al. 2011; Labay et al. 2019, this volume). Furthermore, the watersheds selected fulfilled the following criteria established by Williams et al. (2011) for watersheds being considered for designation as NFCAs:

1. Protect and, if necessary, restore watershed-scale processes that create and maintain freshwater habitat complexity, diversity, and connectivity.
    a. Aquatic habitats are dynamic over space and time and are shaped by processes in the immediate stream valley and upstream watershed.
    b. Native fish conservation areas should encompass sufficient upslope lands and upstream waters to provide for the continuation of these processes over time.
2. Support habitats necessary to complete the life cycle of native species.
    a. Native fish conservation areas must include habitats that are necessary to support various life stages for established and reintroduced native species.
    b. For wide-ranging species, conservation and restoration of migratory corridors as well as spawning and rearing areas are essential.
3. Include a large-enough watershed to provide for long-term persistence of native fish populations.
    a. Native fish conservation areas should be large enough not only to support life history stages of the native fishes, but also to support sufficiently large populations that have a high likelihood of long-term persistence.
    b. If the goal is to support a metapopulation or to provide interconnected habitat patches, a larger amount of habitat would be needed.
4. Provide management that is sustainable over time.
    a. Management plans or other agreements should be in place to ensure that the NFCA can be managed in a manner that sustains aquatic and riparian habitat integrity over time and across management jurisdictions and land ownerships.
    b. Depending on the size and complexity of the watershed, this may require local community commitments, landowner agreements, and local and state government support.

Native fish conservation areas represent a holistic, habitat-based approach to restoration and protection of instream, riparian, and upland habitats, with overarching goals centered on achieving environmentally and economically healthy watersheds. The approach taken in Texas to designate NFCAs (Birdsong et al. 2019) enables a strategic focus on delivery of habitat conservation projects within targeted watersheds considered strongholds for preservation of freshwater fish diversity. The approach encourages and facilitates coordination among landowners, nongovernmental organizations, Texas and federal agencies, universities, and local governments toward delivery of shared conservation objectives.

Over the years, there have been numerous groups and initiatives (e.g., Big Bend Conservation Cooperative, Desert Landscape Conservation Cooperative, Chihuahuan Desert ecoregion plan, Big Bend/Río

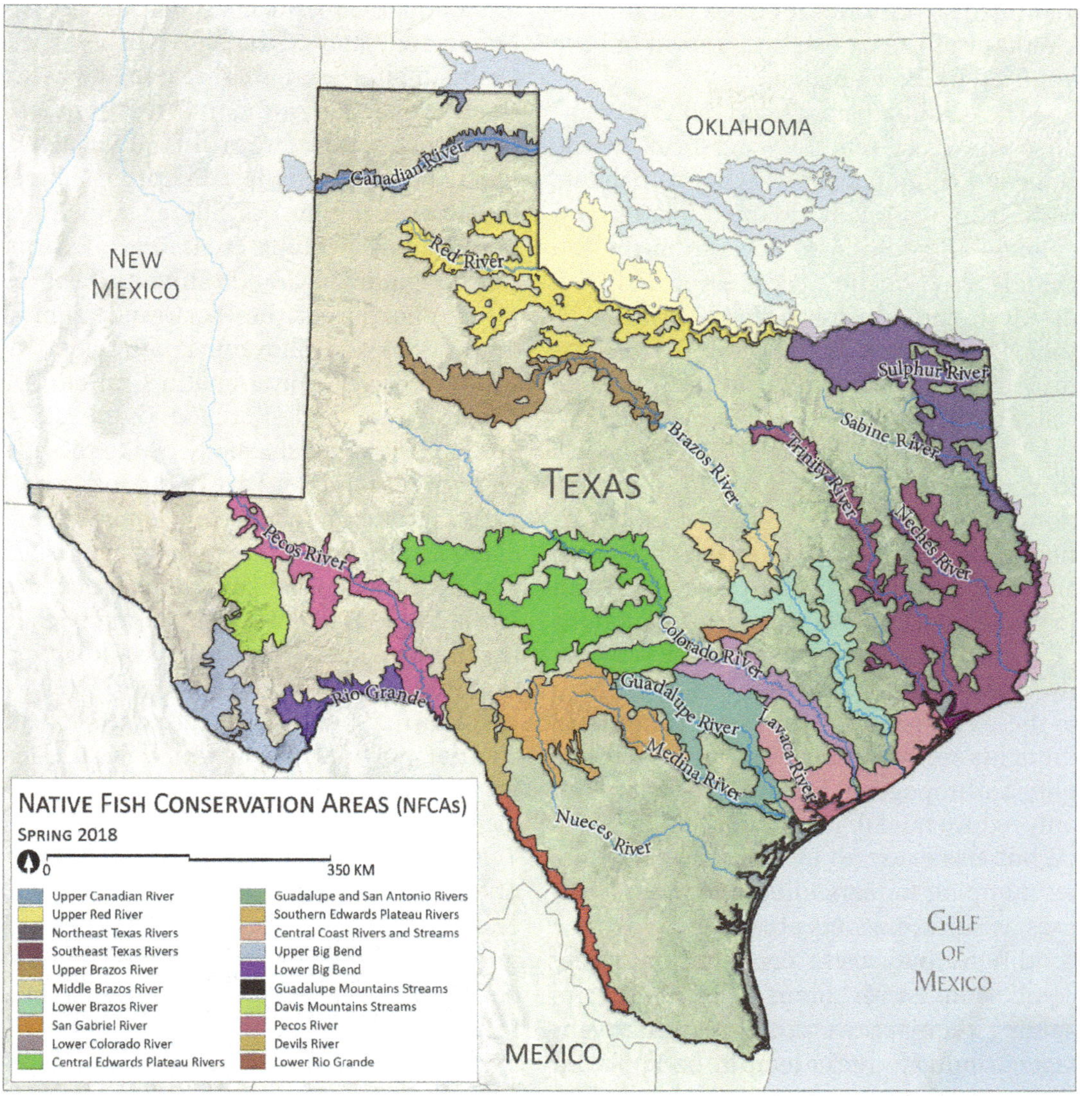

Figure 1. Native fish conservation areas of Texas.

Bravo Transboundary Landscape Conservation Project, Conservation Assessment for the Big Bend-Río Bravo Region, Desert Fish Habitat Partnership, Big Bend Conservation Alliance; Pronatura Noreste 2004; Commission for Environmental Cooperation 2014) that have sought to catalyze conservation in the Chihuahuan Desert. Many have had a variety of successes, but some have not. The objective of the TPWD network of NFCAs is not to appropriate, replace, or compete with other efforts, but to provide a forum for support and synergistic interaction among all interests.

Characterization of Chihuahuan Desert Native Fish Conservation Areas

In the Chihuahuan Desert ecoregion of Texas, six NFCAs were delineated in the Rio Grande, Pecos River, and Devils River basins (Figure 1), and 40 focal fish species were identified as priorities for conservation (Table 1) out of the 93 fish taxa known to have occurred in the ecoregion.

The Chihuahuan Desert NFCAs comprise the most remote watersheds in the state and contain a wide variety of habitats with many uniquely adapted plants and animals. In addition to flows from the Rio Grande, Río Conchos, Pecos River, and Devils River, three major (Hueco-Mesilla Bolsons, Pecos Valley, and Edwards-Trinity Plateau) and six or more minor aquifers provide water to the region.

Desert ecosystems are fragile and slow to recover from perturbations. Some disruptions may not be recoverable. Deep downcutting of streams by erosion from overgrazing and deforestation (Ohmart and Anderson 1982), reduced instream flow, introductions of nonnative species, and extinction of native species may cause irreversible damage to these ecosystems. Under such conditions, droughts are even more devastating and amplify anthropogenic impacts. Droughts not only reduce rainfall magnitude and frequency, but also cause an increase in groundwater pumping for agricultural and municipal uses as surface waters abate. Such extreme conditions put stress on fish community equilibrium, with more tolerant species gaining a competitive and numerical advantage. Tributary creeks tend to be impacted more severely yet are critical to the breeding and rearing of young of many of the indigenous species (e.g., Mexican Stoneroller, Chihuahua Shiner, Big Bend Gambusia, and Conchos Pupfish; Hubbs et al. 1977). These changes have been gradual and long-term, taking place since the mid-1800s (Miller 1961), but their effects have been compounded over time and are now becoming dramatic. While perturbations such as pollution, reduced groundwater, and dam construction are theoretically fixable, recovery to a pristine state is unlikely.

Although many data gaps exist, what is known about Chihuahuan Desert fish and other aquatic organisms is distressing. Approximately half of the native fishes of the Chihuahuan Desert are threatened with extinction or already are extinct (Hubbs 1990). Likely extinctions from this area include Maravillas Red Shiner, Phantom Shiner, Rio Grande Bluntnose Shiner, and Amistad Gambusia (Miller et al. 1989). Extirpations include Rio Grande Shiner in the New Mexico portion of the Rio Grande (Propst et al. 1987) and Shovelnose Sturgeon, Rio Grande Silvery Minnow, Rio Grande Cutthroat Trout, and Blotched Gambusia in Texas (Hubbs et al. 2008). Endemic species, such as mussels, are being lost as well (Rio Grande monkeyface *Quadrula couchiana*, false spike *Quincuncina mitchelli*, and Mexican fawnsfoot *Truncilla cognata*; Howells and Garrett 1995). Left unchecked, this trend of species extirpation and extinction is likely to continue.

Another significant threat to a substantial portion of the Chihuahuan Desert NFCAs is the establishment of invasive giant reed *Arundo donax* and saltcedar *Tamarix* spp. These nonnative plants have effectively channelized stream segments and the resulting constricted flow has reduced shallow, backwater habitat, and changed bottom sediments from a mixture of sand and gravels to one of primarily larger gravels and cobble. The dense stands have also armored and stabilized the riverbanks, thus preventing natural sediments and sand from being available within the river itself, impacting important habitat for many species (Upper Rio Grande Basin and Bay Expert Science Team 2012; Garrett and Edwards 2014).

The streams of the Chihuahuan Desert NFCAs hardly resemble their natural state where many of the original water courses were once lined with gallery forests and diverse riparian zones. Of the more than 100 moderate (2.8–28 L/s) and major (>28 L/s) historical springs, 50% are no longer extant (Brune 1981; Garrett et al., in press). Early

Table 1. Fishes of the Chihuahuan Desert native fish conservation areas (NFCAs). F = NFCA focal species and listed as a Texas species of greatest conservation need; FR = NFCA focal species and a recommended addition to the list of Texas species of greatest conservation need (Cohen et al. 2018); FL = NFCA focal species that is a Texas species of greatest conservation need and is listed as state or federally threatened or endangered; NN = species that is nonnative to the NFCA; C = species that is considered common in the NFCA; and CR = species that is currently listed as a Texas species of greatest conservation need, but is recommended for removal from that list by Cohen et al. (2018).

Taxa	Upper Big Bend	Lower Big Bend	Guadalupe Mountains Streams	Davis Mountains Streams	Pecos River	Devils River
Shovelnose Sturgeon *Scaphirhynchus platorynchus*	FL	FL			FL	
Alligator Gar *Atractosteus spatula*	F	F			F	F
Spotted Gar *Lepisosteus oculatus*	C	C			C	C
Longnose Gar *L. osseus*	C	C			C	C
American Eel *Anguilla rostrata*	F	F			F	F
Gizzard Shad *Dorosoma cepedianum*	C	C		C	C	C
Threadfin Shad *D. petenense*	NN	NN		NN		NN
Central Stoneroller *Campostoma anomalum*					C	C
Mexican Stoneroller *C. ornatum*	FL	FL				
Goldfish *Carassius auratus*	NN	NN		NN	NN	NN
Red Shiner *Cyprinella lutrensis*	C	C		C	C	C
Maravillas Red Shiner *C. l. blairi*		F				
Conchos Shiner *C. panarcys*	FL					
Proserpine Shiner *C. proserpina*					FL	FL
Blacktail Shiner *C. venusta*	NN	NN		NN	NN	C
Common Carp *Cyprinus carpio*	NN	NN		NN	NN	NN
Manantial Roundnose Minnow *Dionda argentosa*					F	F
Devils River Minnow *D. diaboli*						FL
Roundnose Minnow *D. episcopa*				F	F	
Conchos Roundnose Minnow *Dionda* sp. 1	FR					
Rio Grande Chub *Gila pandora*			FL	FL		
Rio Grande Silvery Minnow *Hybognathus amarus*	FL	FL			FL	
Plains Minnow *H. placitus*						NN
Speckled Chub *Macrhybopsis aestivalis*	F	F			F	F
Golden Shiner *Notemigonus crysoleucas*	NN	NN		NN	NN	
Texas Shiner *Notropis amabilis*					CR	CR
Tamaulipas Shiner *N. braytoni*	F	F			F	F
Ghost Shiner *N. buchanani*					C	C
Chihuahua Shiner *N. chihuahua*	FL	FL				

Table 1. Continued.

Taxa	Upper Big Bend	Lower Big Bend	Guadalupe Mountains Streams	Davis Mountains Streams	Pecos River	Devils River
Rio Grande Shiner *N. jemezanus*	F	F			F	F
West Texas Shiner *N. megalops*					FR	FR
Phantom Shiner *N. orca*	F	F			F	
Pecos Bluntnose Shiner *N. simus pecosensis*					FL	
Rio Grande Bluntnose Shiner *N. s. simus*	F	F				
Sand Shiner *N. stramineus*	C	C			C	C
Fathead Minnow *Pimephales promelas*	C	C			C	
Bullhead Minnow *P. vigilax*	NN	NN		NN	NN	C
Longnose Dace *Rhinichthys cataractae*	F	F			F	
River Carpsucker *Carpiodes carpio*	C	C		C	C	C
Rio Grande Blue Sucker *Cycleptus* sp.	FL	FL			FL	FL
Smallmouth Buffalo *Ictiobus bubalus*	C	C		C	C	
Black Buffalo *I. niger*	NN	NN			NN	NN
Longlip Jumprock *Moxostoma albidum*					FR	FR
Mexican Redhorse *M. austrinum*	F	F			F	
Gray Redhorse *M. congestum*	C	C		C	C	C
Mexican Tetra *Astyanax mexicanus*	C	C		C	C	C
Black Bullhead *Ameiurus melas*			C	C	C	C
Yellow Bullhead *A. natalis*	NN	NN				
Blue Catfish *Ictalurus furcatus*	C	C			C	C
Headwater Catfish *I. lupus*	F	F		F	F	F
Channel Catfish *I. punctatus*	C	C		C	C	C
Chihuahua Catfish *Ictalurus* sp. 1	F	F	F	F	F	F
Rio Grande Blue Catfish *Ictalurus* sp. 2	FR	FR			FR	FR
Mexican Blindcat *Prietella phreatophila*					FL	FL
Flathead Catfish *Pylodictis olivaris*	C	C			C	C
Armadillo Del Rio *Hypostomus* sp.						NN
Rio Grande Cutthroat Trout *Oncorhynchus clarkii virginalis*			F	F		
Rainbow Trout *O. mykiss*			NN			
Inland Silverside *Menidia beryllina*	NN	NN		NN	NN	NN
Gulf Killifish *Fundulus grandis*	NN	NN		NN		
Plains Killifish *F. zebrinus*	NN	NN		C	C	C
Rainwater Killifish *Lucania parva*			C	C	C	
Western Mosquitofish *Gambusia affinis*	C	C	C	C	C	C
Amistad Gambusia *G. amistadensis*					F	
Big Bend Gambusia *G. gaigei*		FL				

Table 1. Continued.

Taxa	Upper Big Bend	Lower Big Bend	Guadalupe Mountains Streams	Davis Mountains Streams	Pecos River	Devils River
Largespring Gambusia *G. geiseri*				NN	NN	NN
Spotfin Gambusia *G. krumholzi*						FL
Pecos Gambusia *G. nobilis*				FL	FL	
Blotched Gambusia *G. senilis*						FL
Tex-Mex Gambusia *G. speciosa*					C	C
Sailfin Molly *Poecilia latipinna*						NN
Leon Springs Pupfish *Cyprinodon bovinus*					FL	
Comanche Springs Pupfish *C. elegans*				FL	FL	
Conchos Pupfish *C. eximius*	FL					FL
Pecos Pupfish *C. pecosensis*					FL	
Sheepshead Minnow *C. variegatus*				NN	NN	NN
White Bass *Morone chrysops*	NN	NN		NN	NN	NN
Redbreast Sunfish *Lepomis auritus*	NN	NN		NN	NN	NN
Green Sunfish *L. cyanellus*	C	C	C	C	C	C
Warmouth *L. gulosus*	C	C		C	C	C
Bluegill *L. macrochirus*	C	C	C	C	C	C
Longear Sunfish *L. megalotis*	C	C	C	C	C	C
Redear Sunfish *L. microlophus*	NN	NN		NN	NN	NN
Redspotted Sunfish *L. miniatus*						C
Smallmouth Bass *Micropterus dolomieu*						NN
Florida Largemouth Bass *M. salmoides floridanus*	NN	NN		NN	NN	NN
Rio Grande Largemouth Bass *M. s. nuecensis*	FR	FR		FR	FR	FR
White Crappie *Pomoxis annularis*	NN	NN				
Rio Grande Darter *Etheostoma grahami*					FL	FL
Bigscale Logperch *Percina macrolepida*						C
Freshwater Drum *Aplodinotus grunniens*	C	C		C	C	C
Rio Grande Cichlid *Herichthys cyanoguttatus*	C	C		C	C	C
Blue Tilapia *Oreochromis aureus*	NN	NN		NN	NN	NN
Mozambique Tilapia *O. mossambicus*					NN	NN

records, some as far back as 1583, mention expansive ciénegas (desert wetlands) and abundant fishes (Brune 1981). Exploitation of limited resources, particularly groundwater pumping, has degraded that environment, caused extirpation and extinction of species, and ultimately perpetuated the loss of habitats and ecosystems (Smith and Miller 1985). The few relatively natural faunas and reasonably intact ecosystems that remain need careful management if they are to be preserved.

Native Fish Conservation Areas of Big Bend

The Upper Big Bend NFCA and Lower Big Bend NFCA are located along the international border in Presidio, Brewster, and Terrell counties (Figures 1 and 2). They represent two contiguous, but very diverse stream segments (upstream and downstream of Mariscal Canyon) with distinct differences in base flow, sediment movement, and water quality. These differences are primarily due to reduced base flow and water quality in the Upper Big Bend NFCA and considerable spring flow inputs and improved water quality in the Lower Big Bend NFCA (Bennett et al. 2014). As a result, the Lower Big Bend NFCA remains somewhat ecologically intact and still supports a high diversity of native aquatic species (Bennett and Urbanczyk 2014). However, more than half of the 42 native fish species in the two Big Bend NFCAs are imperiled. In addition, 29% of the fish species that currently occur in these two NFCAs are nonnative (Table 1).

Bennett et al. (2014) noted that numerous threats to aquatic resources have been documented within the two Big Bend NFCAs, including channel narrowing and sediment accumulation (Dean and Schmidt 2011; Dean et al. 2011), deteriorating aquatic habitat (Heard et al. 2012), invasive and nonnative species (Everitt 1998), water-quality deterioration (Sandoval-Solis et al. 2010; Bennett et al. 2012), elevated concentrations of mercury, arsenic, and selenium in fishes (Schmitt et al. 2005), groundwater extraction (Donnelly 2007), and climate change (Ingol-Blanco 2011). The combination of regional water management and invasive, nonnative

Figure 2. Big Bend Ranch State Park, part of the Upper Big Bend Native Fish Conservation Area. Photo by Chase A. Fountain, ©Texas Parks and Wildlife Department.

riparian species has changed streamflow, sediment dynamics, and riparian vegetation cover (Everitt 1998; Schmidt et al. 2003; Dean and Schmidt 2011). The once wide and shallow channel of the Rio Grande has become narrow and deep. Nonnative riparian plants (primarily giant reed and saltcedar) affect channel sediment dynamics, aquatic habitat, and riparian communities by covering up and eliminating backwaters and side channels, diminishing channel conveyance capacity, and increasing flooding frequency (Dean and Schmidt 2011; Garrett and Edwards 2014). Feral pigs *Sus scrofa*, burros *Equus africanus asinus*, horses *E. ferus caballus*, and cattle *Bos* spp. occur in the area and have negatively impacted natural resources within the river corridor by further degrading riparian habitats.

The focal fish species of the Big Bend NFCAs include four undescribed taxa: Conchos Roundnose Minnow, Rio Grande Blue Sucker, Chihuahua Catfish, and Rio Grande Blue Catfish. In addition to the focal fish species, other aquatic species of concern include Salina mucket *Potamilus metnecktayi* (also known as *Disconaias salinasensis*), Tampico pearlymussel *Cyrtonaias tampicoensis*, Texas hornshell *Popenaias popeii*, Big Bend rough-footed mud turtle *Kinosternon hirtipes murrayi*, Big Bend slider *Trachemys gaigeae*, North American beaver *Castor canadensis*, and Pecos River muskrat *Ondatra zibethicus ripensis*.

Areas being actively managed for conservation of species and habitats include Big Bend National Park, Rio Grande Wild and Scenic River, Big Bend Ranch State Park, Black Gap Wildlife Management Area, Chinati Mountains State Natural Area, Terlingua Creek watershed, and Alamito Creek watershed.

Native Fish Conservation Areas of the Pecos River Basin

The Pecos River NFCA (Figures 1 and 3) encompasses a wide variety of habitat quality. Agricultural and municipal water diversions have greatly diminished water quantity in the upper reaches and increased salinity (as high as 12,000 mg/L; Miyamoto et al. 2008). High salinity has resulted in the loss of many fish species (Hoagstrom 2009) and the repeated occurrence of toxic golden alga *Prymnesium parvum* blooms (Patiño et al. 2014). Freshwater inputs from Independence Creek and other springs greatly improve water quality and quantity in the lower segment. Other threats include groundwater extraction, oil and gas development, and invasive species. In addition to the focal fish species (Table 1), other aquatic species of concern include the Rio Grande cooter *Pseudemys gorzugi*, North American beaver, and Pecos River muskrat. An 8,000-ha preserve along Independence Creek is owned by The Nature Conservancy, with an adjacent 280 ha under conservation easement (Karges 2014a).

The Davis Mountains Streams NFCA includes desert habitats such as the Balmorhea Springs Complex, as well as streams in the Davis Mountains (Figure 1). The mountain streams harbor at least two species of conservation concern: Rio Grande Chub and Chihuahua Catfish. The Balmorhea Springs Complex is considered one of the largest and most important of the remaining desert spring systems in West Texas (Karges 2014b). The main springs include Phantom Lake, San Solomon, Giffin, Saragosa, Toyah Creek, East Sandia, and West Sandia. This was once a massive, interconnected network of ciénegas fed by cumulative spring discharges of approximately 130,000,000 L/d (White et al. 1941). Groundwater pumping and draining of ciénegas for agriculture has reduced flow by more than one-third and disconnected the ciénega network (Garrett et al., in press). All that remains for aquatic habitat are three small springs (East Sandia, West Sandia, and Giffin), artificial refuge ciénegas, and irrigation canals (Garrett 2003). These remaining habitats are essential for the survival of the two federally endangered fishes, Pecos Gam-

Figure 3. Pecos River, part of the Pecos River Native Fish Conservation Area. Photo by Earl Nottingham, ©Texas Parks and Wildlife Department.

busia and the endemic Comanche Springs Pupfish.

In addition to the focal fish species (Table 1), other aquatic species of concern in this NFCA include the diminutive amphipod *Gammarus hyalelloides*, phantom cave snail *Pyrgulopsis texana*, phantom springsnail (also known as phantom tryonia) *Tryonia cheatumi*, Rio Grande cooter, and Pecos sunflower *Helianthus paradoxus*. Areas being actively managed for species and habitat conservation include Davis Mountains State Park, Balmorhea State Park, Guadalupe Mountains National Park, Phantom Lake, and The Nature Conservancy's Sandia Springs and Davis Mountains preserves.

The Pecos River NFCA also includes Comanche, Leon, and Diamond-Y springs in Pecos County near Fort Stockton. Comanche Springs no longer flow, and as Gunnar Brune (1981) noted, "failure of Comanche Springs was probably the most spectacular example in Texas of man's abuse of nature." Flowing at 1,200–1,900 L/s, this spring was one of the largest in Texas, but flow completely ceased in 1962 due to aquifer pumping for irrigation from a well field upgradient of the springs during the drought of the 1950s (Mace 2001). The outflow from the springs, Comanche Creek, supported a vast ciénega of approximately 25 km in length. The drying of the springs extirpated the Pecos Gambusia and Comanche Springs Pupfish and was not only an ecological disaster, but also had severe impacts on the more than 100 farmers who had, since the 1860s, depended on waters flowing from Comanche Springs and the ciénega for irrigation of approximately 2,500 ha of cropland (Brune 1981).

Leon Springs, upgradient and in the same aquifer as Comanche Springs, were also modified to provide irrigation for farming. Originally the springs were deep and up to 30 m in diameter and supported a large ciénega that extended for many kilometers downstream (Brune 1981). During the 1920s, a stone and earth dam created Lake Leon (Scudday 2003) that backed water up to, or over, Leon Springs. Unfortunately, this modification likely led to the extirpation of both Pecos Gambusia and Leon Springs Pupfish as none were collected by Carl Hubbs in his 1938 survey of this type locality for the Leon Springs Pupfish (Hubbs 1980; Minckley et al. 1991). The same groundwater pumping that led to the demise of Comanche Springs also dried Leon Springs in 1958 (Brune 1981).

Although Diamond-Y Springs were not as large as Comanche and Leon springs, they continue to flow, so far, and provide habitat for the federally endangered Pecos Gambusia and Leon Springs Pupfish, as well as several other species; however, the flow is greatly reduced from historical levels (Scudday 2003). Fortunately, the Diamond-Y Springs ecosystem does not derive all of its flow from the same aquifer as Comanche and Leon springs (Sharp et al. 2003). Other rare species in this system include federally endangered invertebrates: Diamond tryonia *Pseudotryonia* (also *Tryonia*) *adamantina*, Gonzales tryonia *Tryonia circumstriata*, Pecos amphipod *Gammarus pecos*, and Pecos assiminea *Assiminea pecos*, and the federally threatened Pecos sunflower. Some degree of protection is afforded the inhabitants of the ciénega at Diamond-Y Springs in that The Nature Conservancy owns 1,600 ha that encompass it and is committed to its maintenance and perpetuation. Although state water law recognizes that The Nature Conservancy owns the water beneath their land, it does not allow them to protect this owned water from pumping. Additionally, the Diamond-Y Springs is adjacent to an active oil and gas extraction field (with some of the active pumps located on the preserve). Working wells are within 100 m of surface water, a natural gas refinery is 30 m upslope from the spring, and old brine pits are just a few meters away (Garrett et al., in press).

The Guadalupe Mountains Streams NFCA (Figures 1 and 4) is unique in that it supports a population of introduced Rainbow Trout and might have had a native population of Rio Grande Cutthroat Trout (Garrett and Matlock 1991; Petersen 2002). The stream could provide a viable option for creating sustainable populations of Rio Grande Cutthroat Trout and Rio Grande Chub.

Devils River Native Fish Conservation Area

The Devils River NFCA is situated in an ecological transition zone at the confluence of three biotic provinces (Chihuahuan, Balconian, and Tamaulipan) and, as a result, supports a high level of aquatic biodiversity and endemism. This NFCA includes the Devils River (Figures 1 and 5), which extends 100 km from its headwaters at Pecan Springs to Amistad International Reservoir, and San Felipe Creek in Del Rio. The springs of the Devils River and surrounding area are fed by the Edwards-Trinity Plateau Aquifer, which produces the largest number of springs in Texas, with 46 occurring in Val Verde County alone, as well as the third (Goodenough Springs) and fourth (San Felipe Springs) largest springs in the state (Brune 1981). Goodenough Springs, now covered by Amistad Reservoir, still maintain a significant discharge under the lake surface (Ashworth and Stein 2005). Amistad Gambusia was endemic to the headsprings and the 1.3-km spring run downstream to its confluence with the Rio Grande (Peden 1973), but inundation by the reservoir resulted in its extinction.

The primary threat in this NFCA is groundwater extraction and the resultant reduction or loss of spring flows. The biodiversity is well represented by the fishes of this region, including the federally endangered Mexican Blindcat; federally threatened Devils River Minnow; and state threatened Proserpine Shiner, Spotfin

Figure 4. McKittrick Creek in the Guadalupe Mountains Native Fish Conservation Area. Photo by Gary Garrett.

Figure 5. Devils River, part of the Devils River Native Fish Conservation Area. Photo by Chase A. Fountain, ©Texas Parks and Wildlife Department.

Gambusia, Blotched Gambusia, Conchos Pupfish, and Rio Grande Darter, as well as other focal species, including the Manantial Roundnose Minnow, Tamaulipas Shiner, Rio Grande Shiner, West Texas Shiner, Longlip Jumprock, Headwater Catfish, and Rio Grande Largemouth Bass. In addition to the focal fish species (Table 1), other species of concern include the Texas hornshell, Rio Grande cooter, spring salamander *Eurycea* spp., and endemic spring invertebrates. North American beaver also inhabit the river but suffer from habitat loss, changes to the natural hydrological regime, competition with nonnative nutria *Myocastor coypus*, decreased food supply, and the presence of the invasive and nonnative giant reed and saltcedar (Garrett et al. 2014). Although most of the Devils River flows adjacent to private property, several conservation areas and initiatives exist within the watershed. The TPWD currently protects 15,000 ha in the Del Norte Unit and Dan A. Hughes Unit of the Devils River State Natural Area. In addition, The Nature Conservancy owns and manages the 1,900-ha Dolan Falls Preserve and a total of 63,000 ha of private and public lands are currently under conservation easements (Garrett et al. 2014).

Conservation Planning

In order to restore and protect these NFCAs, the TPWD and partners intend to employ a multispecies, habitat-based approach to species conservation that provides an improved method for addressing the common nature and magnitude of threats facing these ecosystems and their component species. It also improves efficiency and cost-effectiveness and is more likely to be implemented (Knight et al. 2006). This approach is designed to coordinate projects to improve water quality, increase water quantity, restore natural habitats, reduce impacts of nonnative species, diminish stream system fragmentation, and restore proper function of springs, ciénegas, creeks, rivers, and riparian areas. It will only be effective if it is able to inform and influence water management, land-use planning and zoning, and land-management decisions that will determine current and future conditions of rivers and streams and the associated habitat quality for native fishes. Additionally, to provide long-term benefits to focal species populations, conservation actions must be coordinated at sufficient scales to meet all life-history stages of these species and must adopt conservation approaches that are cost-effective and sustainable over time.

To accomplish this goal, it is necessary to develop a holistic, habitat-oriented approach to conservation of focal species, restore and protect habitat, restore habitat connectivity, and reduce deleterious effects of nonnative species. Threat factors need to be delineated and prioritized based on threat level and what can be managed. Currently known threats in the Chihuahuan Desert NFCAs include

1. habitat fragmentation,
2. barriers to migration,
3. loss of natural flow regime,
4. reduced streamflow,
5. spring flow declines and aquifer depletion,
6. channel narrowing and sediment accumulation,
7. groundwater pollution,
8. habitat loss, and
9. nonnative species—habitat modification, hybridization, competition, and predation.

In order to develop and refine conservation plans for the NFCAs in the Chihuahuan Desert, an interdisciplinary team of 55 individuals, representing the TPWD Inland Fisheries Division, TPWD Wildlife Division, TPWD Parks Division, U.S. Fish and Wildlife Service (USFWS) Partners for Fish and Wildlife Program, USFWS Texas Fish and Wildlife Conservation Office, National Park Service, University of Texas, Texas Tech University, Fort Worth Zoo, The Nature Conservancy of Texas, World Wildlife Fund, Desert Fish Habitat Partnership, Big Bend Conservation Alliance, and Devils River Conservancy, formed the Chihuahuan Desert Native Fish Conservation Network (NFCN) Advisory Council. They met by webinar in fall 2016 and then workshops in spring 2017 and spring 2018. Members were tasked with

1. Identifying priority research, monitoring, and restoration actions for preservation of native fishes, their habitats, and other aquatic resources in the Chihuahuan Desert of Texas;
2. Catalyzing cooperation, collaboration, and leveraging of technical and financial resources among local, state and federal natural resource management agencies, universities, nongovernmental organizations, and other local partners that contribute to the conservation of native fishes and other aquatic resources in the watersheds of the Chihuahuan Desert; and
3. Facilitating local implementation of the National Fish Habitat Action Plan (National Fish Habitat Partnership 2012) and Texas Conservation Action Plan in the Chihuahuan Desert NFCAs (Cohen et al. 2018).

Priorities for research, monitoring, and restoration were focused on the eight primary objectives developed for the NFCAs in Texas (Cohen et al. 2018).

Objective 1: Protect and maintain intact, healthy habitats

- Determine locations and extent of healthy habitats.
- Assess degree of threats and limiting factors present in healthy habitats.
- Develop a priority list of stream segments for protection actions.
- Organize technical advisory teams for individual stream segments to analyze current data, define challenges, determine conservation methods, and engage public support.
- Develop action plans for addressing the objectives, select the best watershed management alternatives, list strategies for implementing alternatives, and determine appropriate milestones for measuring progress.
- Maintain floodplain functions such as aquifer recharge, natural flow regime, base flows, spring flows, water quality, soil moistening, habitat diversity, and sediment transport.
- Maintain appropriate sediment transport and avoid channel narrowing.
- Maintain native vegetation throughout stream segments, including riparian corridors, floodplains, and upland areas.
- Develop voluntary, nonregulatory tools such as financial incentives, conservation easements, landowner agreements, and targeted acquisition.
- Seek appropriate easements, water-rights acquisitions, and flow agreements to maintain appropriate hydrologic conditions.
- Adopt conservation approaches that are cost-effective and sustainable over time.
- Convene stakeholder groups to foster support of action plans.
- Monitor conservation efforts and assess benefits to focal species populations.

Objective 2: Restore impacted habitats

- Determine locations, extent, and type of impacted habitats.
- Assess degree of threats and limiting factors present in impacted habitats.
- Develop a priority list of stream segments for restoration actions.
- Organize technical advisory teams for individual stream segments to analyze data, define challenges, determine restoration methods, and engage public support.
- Develop action plans for addressing

the objectives, select the best watershed management alternatives, list strategies for implementing alternatives, and determine appropriate milestones for measuring progress.

- Where feasible, restore floodplain functions such as aquifer recharge, natural flow regime, base flows, spring flows, water quality, soil moistening, habitat diversity, and sediment transport.
  - o Restore appropriate sediment transport and reduce channel narrowing.
  - o Restore native vegetation throughout stream segments, including riparian corridors, floodplains, and upland areas.
- Develop voluntary, nonregulatory tools such as financial incentives, conservation easements, landowner agreements, and targeted acquisition.
- Seek appropriate easements, water-rights acquisitions, and flow agreements to improve appropriate hydrologic conditions.
- Adopt conservation approaches that are cost-effective and sustainable over time.
- Convene stakeholder groups to foster support of action plans.
- Monitor restoration efforts and assess benefits to focal species populations.

Objective 3: Restore stream and habitat connectivity

- Inventory fish passage barriers and delineate impacts on ecology of focal species.
- Where feasible, diminish or remove fish passage barriers and restore aquatic connectivity.

Objective 4: Mitigate effects of invasive species

- Assess current status of focal species affected by invasive species.
- Develop methods for reducing nonnative species in targeted areas.
- Develop methods to prevent introductions of invasive species and minimize impacts of existing invasive species.
- Restore or improve the ecological balance in habitats negatively affected by nonnative, invasive, or problem species.
- Re-establish genetic integrity of hybridized populations in targeted areas.

Objective 5: Organize networks of public and private landowners

- Provide technical guidance workshops, newsletters, social media, and so forth to facilitate development and expansion of local citizen-based partnerships.
- Landowner networks should be committed to the cooperative conservation of land and water resources within the watershed.
- Landowner networks should promote values of functional upland, riparian, and stream systems and emphasize the conservation of native fish communities and supporting habitats.
- Landowner networks should work to reduce or eliminate activities on the landscape that degrade water quality, reduce water quantity, degrade riparian systems, favor nonnative species, or fragment stream systems.
- Landowner networks should encourage an array of sustainable land-use activities that are compatible with aquatic resource conservation.
- Landowner networks should promote collaboration across jurisdictional and landownership boundaries.

Objective 6: Develop conservation demonstration areas

- Provide fishing, paddling, and hiking opportunities.
- Promote sustainable public use of rivers.
- Describe benefits to other native species.
- Demonstrate best management practices.
- Highlight restoration actions through educational kiosks.

Objective 7: Conduct research to fill critical information gaps

- Identify knowledge gaps critical to restoration and conservation of the focal species.
- Design and conduct research as needed to enhance conservation efforts outlined in Objectives 1–4.
- Initial sampling at representative locations within each NFCA should be quarterly and include
  - o Biological characteristics of focal species: population size, population structure (genetics and demographics), fecundity, food habits, habitat selectivity, flow–ecology relationships, and associated species
  - o Habitat structure: flow and discharge rates, channel width, channel morphology, substrate types, depth, cover, and trends in surrounding land use
  - o Water quality: temperature, pH, dissolved oxygen, conductivity, total dissolved solids, alkalinity, hardness, and chemical and biological oxygen demand
- Threats and limiting factors for the focal species will determine the scale at which the monitoring is designed. As baseline data are developed, monitoring parameters can be modified and streamlined to address critical issues and needs for the focal species.

Objective 8: Adaptive management and reporting

- Develop annual and long-term reporting requirements to document acquired data, departures from plan, and evaluations necessary for adaptive management.
- Determine research needs for refining restoration and management actions.
- Periodically modify strategies based on monitoring, evaluation, and research results.
- Share information with the public in an easy to use and understandable format.

The Chihuahuan Desert NFCN Advisory Council met initially by webinar to familiarize participants on the rationale and development of the NFCAs and the process for developing and ranking conservation needs and data gaps. We then held in-person workshops to identify and prioritize research, monitoring and restoration projects, or actions that need to be initiated within the six NFCAs. These discussions resulted in the identification of proposed conservation actions (approximately 80), funding needs, and a list of top-priority projects. Funding and other means of support for the priority projects are now being developed by the TPWD and the team. The highest priority projects were to

- Develop ciénega research and management recommendations.
  - o Increase understanding of flow alteration, marsh encroachment, and other changes to the system.
  - o Establish benchmarks based on historical conditions.
  - o Develop strategies to inform effective management.
- Support science to increase understanding of groundwater–surface water interactions, establish environmental flows targets, and develop strategies to effectively manage groundwater to achieve specific targets for spring discharge and instream flows.
  - o Connect desired future conditions to environmental flows for the Rio Grande and tributaries.
  - o Independence Creek and the lower Pecos River are recommended priorities for this study.
- Complete studies of temperature tolerances, distribution, and fish hosts for Mexican fawnsfoot and Salina mucket.
- Refine Devils River habitat models, including additional monitoring wells and data collection.

- Conduct a geographic information system-based regional assessment of water tables in ephemeral streams of the region to determine restoration potential.
- Explore establishment of an incentives-based program for Davis Mountains fuel reduction/thinning and riparian livestock fencing.
- Work with Sul Ross State University to explore programmatic partnership to increase involvement in biological monitoring and riparian restoration in the region.
- Build capacity for riparian restoration in Big Bend tributaries.
    - o Provide support to increase supply of plant materials by establishing a nursery or supporting a contract with an existing greenhouse.
    - o Explore partnerships with nonprofits to administer a large-scale riparian restoration program, including riparian restoration workshops, volunteer coordination, and planning/delivery of service projects.
- Conduct a data and information mining/sharing project on relevant research, monitoring, and restoration actions that have occurred within the Chihuahuan Desert over the past 10 years and assemble an historical summary of regional conservation efforts.
    - o Work with the Devils River Conservancy to conduct a pilot project on data mining for the Devils River watershed.
    - o The assembled proof of concept and lessons learned will guide a work plan for the entire Chihuahuan Desert.

Conservation Delivery

The partnerships and collaborations forged in the workshop process will enable a leveraging of federal and nonfederal dollars to initiate and complete many of the projects identified in the workshops. Conservation efforts currently underway include collaborating with the USFWS Partners for Fish and Wildlife Program through the TPWD Landowner Incentive Program (LIP; https://tpwd.texas.gov/landwater/land/private/lip/) to work with private landowners on habitat restoration and enhancement projects. Field staff work one on one with landowners and ranch managers to develop management plans that address ranch objectives (many of these large properties are still working cattle ranches) and habitat needs for upland, riparian, and aquatic species, as well as to design potential habitat restoration and enhancement projects. Field staff cooperate with landowners to submit applications for proposed habitat restoration and enhancement projects to the competitive LIP request for proposals. Applications are reviewed, ranked, and selected for funding by a panel of conservation practitioners based on expected benefits of the proposed projects to focal watersheds and species. The Chihuahuan Desert NFCAs were adopted as priorities of LIP and the USFWS Partners for Fish and Wildlife Program in order to more effectively target areas in need of restoration that would have the highest ecological value.

Based on input from the Chihuahuan Desert NFCN Advisory Council, restoration efforts have focused on tributary streams in the Upper Big Bend NFCA. Riparian areas of these streams historically consisted of large cottonwood and willow gallery forests, but the trees were harvested for local mining operations and very few remnant stands remain. Mining records show that one company harvested 6 km of cord wood in just one 3-year period from the Terlingua Creek watershed (Jeff Bennett, American Bird Conservancy/Rio Grande Joint Venture, personal communication). This is equivalent to an amount of wood stacked 1.2 m by 1.2 m by 6.5 km—a significant harvest (one of many) from a tributary stream that flows 135 km. Recovery of the riparian area was hindered by poor land management and grazing practices, which prevented recruitment and growth of these woody riparian species. A primary focus of habitat restoration investments through LIP has been to cooperate with private landowners to reestablish colonies of riparian vegetation. Riparian vegetation stabilizes stream banks, filters sediment from overland flow, provides inputs for aquatic food web nutrient cycling, and provides stream shading (and therefore thermal refuges). These colonies will serve as nursery areas to provide material for future restoration sites.

The Chihuahuan Desert NFCN Advisory Council also identified the need for long-term monitoring of restoration sites, so the TPWD directed State Wildlife Grant funds into a project to establish aquatic monitoring stations at these restoration sites. Information from this study will also be used to build predictive models for recovery and species recolonization and guide future restoration projects.

Another focus of LIP project funds has been to restore riparian areas through the eradication of saltcedar. Disturbed project sites are easily colonized by invasive, non-native species, so project partners are establishing long-term treatment plans to prevent invasive species from taking these restoration sites.

The TPWD was awarded a grant in spring 2018 through the National Fish and Wildlife Foundation through their Pecos Watershed Conservation Initiative to continue work with partners to plant riparian vegetation colonies, fence streams from feral livestock, continue to build partnerships with landowners, conserve focal species, and provide technical guidance and conservation best management practices to promote habitat conservation and restoration.

The Chihuahuan Desert NFCN Advisory Council will hold annual or biannual meetings to continue refining priorities and leverage new funding to complete projects. The priority actions developed by the interdisciplinary advisory council will continue to provide the foundation to guide strategies and partner investments over the next 5–10 years in cooperative, watershed-scale conservation of native fishes and other aquatic resources in each of the NFCAs. Additional information on development of NFCAs, workshop products, and project details can be found at http://nativefishconservation.org/.

Unfortunately, ongoing and impending land use and water consumption patterns continue to create challenges to conservation. Innovative, strategic, and science-based conservation strategies are needed, as well as a coordinated approach by natural resource professionals and private landowners. What happens next will be determined by numerous factors ranging from hard work and commitment to appropriate timing and good luck.

Acknowledgments

We thank the many collaborators and partners representing nongovernmental organizations, local stakeholder groups, universities, and state and federal agencies who joined together and participated in the Chihuahuan Desert NFCN Advisory Council. Their interest, dedication, and substantive contributions at the workshop are appreciated and we look forward to building upon our shared vision and partnership to move conservation and restoration goals forward in these NFCAs. We thank the TPWD LIP for their willingness to work in this remote region of the state for holistic watershed conservation of upland, riparian, and instream habitats. We also appreciate and thank the USFWS Partners for Fish and Wildlife Program in Texas; their partnership and vested interest in developing, implementing, and promoting projects in the region are invaluable. We especially thank our partners who have brought forth funding to implement the vision and goals outlined by the Chihuahuan

Desert NFCN. The USFWS Partners for Fish and Wildlife Program and the State of Texas, through legislative appropriated Statewide Aquatic Vegetation and Invasive Species Management funds, have provided support for habitat restoration projects and invasive species treatment. The USFWS State Wildlife Grant program funded the modeling and development of NFCAs in Texas and provided funds to partner with universities and other organizations to fill critical research gaps. The Western Native Trout Initiative and Desert Fish Habitat Partnership are collaborating through a USFWS Multistate Conservation Grant to fund modeling and development of NFCAs for the remainder of the Rio Grande and Pecos River watersheds outside of Texas. Finally, we thank the National Fish and Wildlife Foundation for their support and shared vision in restoration of aquatic resources in the Chihuahuan Desert ecoregion of Texas. We look forward to implementing many conservation projects and outcomes with their long-term support.

References

Ashworth, J. B., and W. G. Stein. 2005. Springs of Kinney and Val Verde counties. LBG-Guyton Associates, Austin, Texas.

Bennett, J., B. Brauch, and K. Urbanczyk. 2012. Estimating ground water contribution from the Edwards-Trinity Plateau Aquifer to the Big Bend reach of the Rio Grande, Texas. Geological Society of America Abstracts with Programs 44:2.

Bennett, J., M. Briggs, and S. Sandoval Soliz. 2014. Rio Grande—Río Bravo river corridor. Pages 21–23 *in* M. D. Wesson, C. Hallmich, J. Bennett, C. Sifuentes Lugo, A. Garcia, A. M. Roberson, J. Karges, and G. P. Garrett, editors. Conservation assessment for the Big Bend-Río Bravo region: a binational collaborative approach to conservation. Commission for Environmental Cooperation, Montreal, Quebec.

Bennett, J., and K. Urbanczyk. 2014. Springs of the Lower Canyons. Pages 23–24 *in* M. D. Wesson, C. Hallmich, J. Bennett, C. Sifuentes Lugo, A. Garcia, A. M. Roberson, J. Karges, and G. P. Garrett, editors. Conservation assessment for the Big Bend-Río Bravo region: a binational collaborative approach to conservation. Commission for Environmental Cooperation, Montreal, Quebec.

Birdsong, T. W., G. P. Garrett, B. J. Labay, M. G. Bean, P. T. Bean, J. Botros, M. J. Casarez, A. E. Cohen, T. G. Heger, A. Kalmbach, D. A. Hendrickson, S. J. Magnelia, K. B. Mayes, M. E. McGarrity, R. McGillicuddy, M. M. Parker, and S. Robertson. 2019. Texas Native Fish Conservation Areas Network: strategic investments in restoration and preservation of freshwater fish diversity. Pages 183–229 *in* D. C. Dauwalter, T. W. Birdsong, and G. P. Garrett, editors. Multispecies and watershed approaches to freshwater fish conservation. American Fisheries Society, Symposium 91. Bethesda, Maryland.

Brune, G. 1981. Springs of Texas, volume I. Branch-Smith, Fort Worth, Texas.

Cohen, A. E., G. P. Garrett, M. J. Casarez, D. A. Hendrickson, B. J. Labay, T. Urban, J. Gentle, D. Wiley, and D. Walling. 2018. Conserving Texas biodiversity: status, trends, and conservation planning for fishes of greatest conservation need. Texas Parks and Wildlife Department, Austin.

Commission for Environmental Cooperation. 2014. Conservation assessment for the Big Bend-Río Bravo region: a binational collaborative approach to conservation. M. D. Wesson, C. Hallmich, J. Bennett, C. Sifuentes Lugo, A. Garcia, A. M. Roberson, J. Karges and G. P. Garrett, editors. Commission for Environmental Cooperation, Montreal, Quebec.

Dauwalter, D. C., J. S. Sanderson, J. E. Williams, and J. R. Sedell. 2011. Identification and implementation of native fish conservation areas in the upper Colorado River basin. Fisheries 36:278–288.

Dean, D. J., and J. C. Schmidt. 2011. The role of feedback mechanisms in historic channel changes of the lower Rio Grande in the Big Bend region. Geomorphology 126:333–349.

Dean, D. J., M. L. Scott, P. B. Shafroth, and J. C. Schmidt. 2011. Stratigraphic, sedimentologic, and dendrogeomorphic analyses of rapid floodplain formation along the Rio Grande in Big Bend National Park, Texas. Geological Society of America Bulletin 123:1908–1925.

Donnelly, A. C. A. 2007. Groundwater availability run. Texas Water Development Board, Report No. 06-16, Austin.

Everitt, B. L. 1998. Chronology of the spread of tamarisk in the central Rio Grande. Wetlands 18:658–668.

Garrett, G. P. 2003. Innovative approaches to recover endangered species. Pages 151–160 *in* G. P. Garrett and N. L. Allan, editors. Aquatic fauna of the northern Chihuahuan Desert. Museum of Texas Tech University Special Publication 46.

Garrett, G. P., M. G. Bean, R. J. Edwards, and D. A. Hendrickson. In press. Mining hidden waters: groundwater depletion, aquatic habitat degradation and loss of fish diversity in the Chihuahuan Desert ecoregion of Texas. *In* D. L. Propst, J. E. Williams, K. R. Bestgen and C. W. Hoagstrom, editors. Standing between life and extinction: ethics and ecology of conserving aquatic species in the American Southwest. University of Chicago Press, Chicago.

Garrett, G. P., and R. J. Edwards. 2014. Changes in fish populations in the Lower Canyons of the Rio Grande. Pages 396–408 *in* C. A. Hoyt and J. Karges, editors. Proceedings of the sixth symposium on the natural resources of the Chihuahuan Desert region. Chihuahuan Desert Research Institute, Fort Davis, Texas.

Garrett, G. P., J. Karges, and E. Verdecchia. 2014. Devils River. Pages 29–30 *in* M. D. Wesson, C. Hallmich, J. Bennett, C. Sifuentes Lugo, A. Garcia, A. M. Roberson, J. Karges, and G. P. Garrett, editors. Conservation assessment for the Big Bend-Río Bravo region: a binational collaborative approach to conservation. Commission for Environmental Cooperation, Montreal, Quebec.

Garrett, G. P., and G. C. Matlock. 1991. Rio Grande Cutthroat Trout in Texas. Texas Journal of Science 43:405–410.

Heard, T. C., J. S. Perkin, and T. H. Bonner. 2012. Intra-annual variation in fish communities and habitat associations in a Chihuahuan Desert reach of the Rio Grande/Río Bravo del Norte. Western North American Naturalist 72:1–15.

Hoagstrom, C. W. 2009. Causes and impacts of salinization in the lower Pecos River. Great Plains Research 19:27–44.

Howells, R. G., and G. P. Garrett. 1995. Freshwater mussel surveys of Rio Grande tributaries in Chihuahua, Mexico. Triannual Unionid Report 8:10.

Hubbs, C. 1980. The solution to the *Cyprinodon bovinus* problem: eradication of a pupfish genome. Proceedings of the Desert Fishes Council 10:9–18.

Hubbs, C. 1990. Declining fishes of the Chihuahuan Desert. Pages 89–96 *in* A. M. Powell, R. R. Hollander, J. C. Barlow, W. B. McGillivray, and D. J. Schmidly, editors. Third symposium on resources of the Chihuahuan Desert region, United States and Mexico. Chihuahuan Desert Research Institute, Alpine, Texas.

Hubbs, C., R. J. Edwards, and G. P. Garrett. 2008. An annotated checklist of the freshwater fishes of Texas, with keys to identification of species, 2nd edition. Texas Journal of Science 43(4).

Hubbs, C., R. R. Miller, R. J. Edwards, K. W. Thompson, E. Marsh, G. P. Garrett, G. L. Powell, D. J. Morris, and R. W. Zerr. 1977. Fishes inhabiting the Rio Grande, Texas-Mexico, between El Paso and the Pecos confluence. Pages 91–97 *in* R. R. Johnson and D. A. Jones, editors. Importance, preservation and management of riparian habitat: a symposium. U.S. Forest Service, General Technical Report RM-43, Washington, D.C.

Ingol-Blanco, E. 2011. Modeling climate change impacts on hydrology and water resources: case study Río Conchos basin. Doctoral dissertation. University of Texas at Austin, Austin.

Karges, J. 2014a. Pecos River. Pages 32–33 *in* M. D. Wesson, C. Hallmich, J. Bennett, C. Sifuentes Lugo, A. Garcia, A. M. Roberson, J. Karges, and G. P. Garrett, editors. Conservation assessment for the Big Bend-Río Bravo region: a binational collaborative approach to conservation. Commission for Environmental Cooperation, Montreal, Quebec.

Karges, J. 2014b. Balmorhea Springs Complex. Pages 34–35 *in* M. D. Wesson, C. Hallmich, J. Bennett, C. Sifuentes Lugo, A. Garcia, A. M. Roberson, J. Karges, and G. P. Garrett, editors. Conservation assessment for the Big Bend-Río Bravo region: A binational collaborative approach to conservation. Commission for Environmental Cooperation, Montreal, Quebec.

Knight, A. T., R. M. Cowling, and B. M. Campbell. 2006. An operational model for implementing conservation action. Conservation Biology 20:408–419.

Labay, B. J., J. S. Perkin, D. A. Hendrickson, A. R. Cooper, G. P. Garrett, and T. W. Birdsong. 2019. Who's asking? Interjurisdictional conservation assessment and planning for Great Plains fishes. Pages 57–83 *in* D. C. Dauwalter, T. W. Birdsong, and G. P. Garrett, editors. Multispecies and watershed approaches to freshwater fish conservation. American Fisheries Society, Symposium 91, Bethesda, Maryland.

Mace, R. E. 2001. Aquifers of West Texas: an overview. Pages 1–16 *in* R. E. Mace, W. F. Mullican, III, and E. S. Angle, editors. Aquifers of West Texas. Texas Water Development Board, Report 356, Austin.

Miller, R. R. 1961. Man and the changing fish fauna of the American Southwest. Papers of the Michigan Academy of Science, Arts, and Letters 46:365–404.

Miller, R. R., J. D. Williams, and J. E. Williams. 1989. Extinctions of North American fishes during the past century. Fisheries 14(6):22–38.

Minckley, W. L., G. K. Meffe, and D. L. Soltz. 1991. Conservation and management of short-lived fishes: the cyprinodontoids. Pages 247–282 *in* W. L. Minckley and J. E. Deacon, editors. Battle against extinction. The University of Arizona Press, Tucson.

Miyamoto, S., S. Anand, and W. Hatler. 2008. Hydrology, salinity, and salinity control possibilities of the Middle Pecos River: a reconnaissance report. Texas Water Resources Institute, Technical Report TR-315, College Station.

National Fish Habitat Partnership. 2012. National Fish Habitat Action Plan, 2nd edition. Association of Fish and Wildlife Agencies, Washington, D.C.

Ohmart, R. D., and B. W. Anderson. 1982. North American desert riparian ecosystems. Pages 433–466 *in* G. L. Bender, editor. Reference handbook on the deserts of North America. Greenwood Press, Westport, Connecticut.

Patiño, R., D. Dawson, and M. M. VanLandeghem. 2014. Retrospective analysis of associations between water quality and toxic blooms of golden alga (*Prymnesium parvum*) in Texas reservoirs: implications for understanding dispersal mechanisms and impacts of climate change. Harmful Algae 33:1–11.

Peden, A. E. 1973. Virtual extinction of *Gambusia amistadensis* n. sp., a poeciliid fish from Texas. Copeia 1973:210–221.

Petersen, A. M. 2002. The ecology of fishes in McKittrick Creek, Guadalupe Mountains National Park, Texas. Master's thesis. Colorado State University, Fort Collins.

Probst, D. L., G. L. Burton, and B. H. Pridgeon. 1987. Fishes of the Rio Grande between Elephant Butte and Caballo reservoirs, New Mexico. Southwestern Naturalist 32:408–411.

Pronatura Noreste. 2004. Ecoregional conservation assessment of the Chihuahuan Desert, 2nd edition. Pronatura Noreste, Monterrey, Mexico; The Nature Conservancy, Arlington, Virginia; and World Wildlife Fund, Gland Switzerland.

Sandoval-Solis, S., B. Reith, and D. C. McKinney. 2010. Hydrologic analysis before and after reservoir alteration at the Big Bend reach, Rio Grande/Río Bravo. University of Texas at Austin, Center for Research in Water Resources Online Report 10-06, Austin.

Schmidt, J. C., B. L. Everitt, and G. A. Richard. 2003. Hydrology and geomorphology of the Rio Grande and implications for river rehabilitation. Pages 25–45 *in* G. P. Garrett and N. L. Allan, editors. Aquatic fauna of the northern Chihuahuan Desert. Museum of Texas Tech University Special Publications 46.

Schmitt, C. J., J. E. Hinck, V. S. Blazer, N. D. Denslow, G. M. Dethloff, T. M. Bartish, J. J. Coyle, and D. E. Tillitt. 2005. Environmental contaminants and biomarker responses in fish from the Rio Grande and its U.S. tributaries: spatial and temporal trends. Science of the Total Environment 350:161–193.

Scudday, J. F. 2003. My favorite old fishing holes in West Texas: where did they go? Pages 135–140 *in* G. P. Garrett and N. L. Allan, editors. Aquatic fauna of the northern Chihuahuan Desert. Museum of Texas Tech University Special Publication 46.

Sharp, J. M. Jr., R. Boghici, and M. Uliana. 2003. Groundwater systems feeding the springs of West Texas. Pages 1–11 *in* G. P. Garrett and N. L. Allan, editors. Aquatic fauna of the northern Chihuahuan Desert. Museum of Texas Tech University Special Publications 46.

Smith, M. L., and R. R. Miller. 1985. Conservation of desert spring habitats and their endemic fauna in northern Chihuahua, Mexico. Proceedings of the Desert Fishes Council 13:54–63.

Upper Rio Grande Basin and Bay Expert Science Team. 2012. Environmental flows recommendations report. Final submission to the Environmental Flows Advisory Group, Rio Grande Basin and Bay Area Stakeholders Committee, and Texas Commission on Environmental Quality, Austin.

White, W. N., H. S. Gale, and S. S. Nye. 1941. Geology and ground-water resources of the Balmorhea area, western Texas. U.S. Geological Survey Water-Supply Paper 849-C.

Williams, J. E., R. N. Williams, R. F. Thurow, L. Elwell, D. P. Philipp, F. A. Harris, J. L. Kershner, P. J. Martinez, D. Miller, G. H. Reeves, C. A. Frissell, and J. R. Sedell. 2011. Native fish conservation areas: a vision for large-scale conservation of native fish communities. Fisheries 36:267–277.

American Fisheries Society Symposium 91:253–267, 2019

Developing a Geospatial Decision Support Tool for Protecting and Restoring Environmental Flows in Texas Rivers and Streams

JOHANNA VALENTE* AND DAVID BRADSBY
Texas Parks and Wildlife Department
4200 Smith School Road, Austin, Texas 78744, USA

KEVIN B. MAYES
Texas Parks and Wildlife Department, Inland Fisheries Division
505 Staples Road, Building 1, San Marcos, Texas 78666, USA

CINDY LOEFFLER, LYNNE HAMLIN, DAKUS GEESLIN, KIMBERLY HORNDESKI, AND DAVID YOUNG
Texas Parks and Wildlife Department
4200 Smith School Road, Austin, Texas 78744, USA

JOE TRUNGALE
Texas Conservation Science, Inc.
Post Office Box 150894, Austin, Texas, 78715, USA

RYAN SMITH AND KYLE GARMANY
The Nature Conservancy of Texas
Post Office Box 1440, San Antonio, Texas, 78295, USA

TOM HAYES
Texas Conservation Science, Inc.
Post Office Box 150894, Austin, Texas, 78715, USA

Abstract.—Alteration of environmental flows is a growing concern within the conservation community as human demand for water increases. Alteration of the natural flow regime of rivers and streams in Texas has degraded aquatic and riparian habitats and contributed to declines in native fishes and other aquatic taxa. To inform identification and implementation of voluntary environmental flow protection and restoration strategies, conservation partners initiated development of the Environmental Flow Information Toolkit. The Environmental Flow Information Toolkit is being developed as a Web-based geospatial platform that will provide data and information on water use, hydrologic alteration, and environmental flow targets to meet conservation objectives and offer other data layers supported by end-user recommendations and feedback. The Environmental Flow Information Toolkit is expected to increase efficiencies and effectiveness of environmental flow conservation efforts, providing a valuable decision support tool for resource managers, water providers, and conservation organizations facing the challenges of meeting human water demand while conserving Texas' rich natural heritage.

* Corresponding author: johanna.valente@tpwd.texas.gov

Introduction

Texas comprises 15 major river basins and more than 191,000 mi (307,384 km) of natural waterways within 12 ecoregions (Griffith et al. 2004), supporting 191 native freshwater fish species (Birdsong et al. 2019, this volume), more than 50 freshwater mussel species (Howells et al. 1996), and numerous macroinvertebrate taxa. In addition to rivers and streams, Texas' aquatic resources include seven major estuaries, more than 1,100 public lakes, and at least 200 major springs (Loeffler 2015). Aquatic and riparian ecosystems and the species they support are dependent on natural flow regimes—consisting of subsistence (low flows occurring during dry times), base flows, high flow pulses, and overbanking flows (see TIFP 2008). The hydrologic characteristics of each of these flow components play important roles in maintaining ecological, chemical, and physical processes that support native biodiversity, water quality, channel maintenance and sediment transport, and connectivity in vertical, longitudinal, and lateral dimensions (Annear et al. 2004). Sources of environmental flow include alluvial and artesian aquifers that contribute to springs and base flows and storm-driven rainfall events that lead to high flow pulses and overbanking flows.

Forty-eight percent of Texas' freshwater fishes are imperiled (Birdsong et al. 2019). Hydrologic alteration of river systems contributes to imperilment of these fishes and other river species. The natural timing, frequency, and magnitude of streamflows direct the evolutionary adaptations of river biota (Bunn and Arthington 2002) and influence physical and chemical processes (Poff et al. 2010); therefore, anthropogenic alterations of streamflows can have significant effects on ecosystem structure and function (Carlisle et al. 2011). Surface water diversions, groundwater pumping, and invasive riparian vegetation cause hydrologic alterations by reducing flows, especially during periods of drought. Operations of dams alter natural flow regimes by modifying hydrologic characteristics (e.g., magnitudes, seasonal timing, rate of change, duration, and frequency) of base flows, high flow pulses and overbanking flows and by extending dry and no-flow periods. Temporal changes in flow regimes and barriers to movement can disrupt completion of fish life cycles through impaired migration, spawning, and recruitment. Furthermore, extended duration and frequencies of zero-flow days and modified base flows alter the quantity and quality of instream habitat conditions necessary for aquatic life. All these effects, including fragmentation and reduced longitudinal and lateral connectivity, contribute to losses of aquatic biodiversity (Bunn and Arthington 2002; Perkin et al. 2014).

In Texas, the water right process depends on whether the source is groundwater or surface water. Surface water flowing in a clearly defined water course (i.e., river or stream) is considered state property; groundwater is water found beneath the land's surface and is considered the property of the landowner. Groundwater is governed by the rule of capture, sometimes referred to as the "law of the biggest straw," which allows landowners the right to capture the water beneath their property, regardless of the impacts of that pumping on neighboring wells, as long as the water is not wasted. Surface water, on the other hand, belongs to the state of Texas, and landowners must obtain a water right from the state to impound, divert, or use state water. Water right permits specify the amount of water to be diverted or stored, as well as the location, diversion rate, type use, and priority date of the right. Texas operates primarily on the prior appropriation system in which the oldest water rights have the first claim on available water—first in time, first in right. Prior to 1985, the state generally did not consider impacts on fish and wildlife when granting surface water right permits. With more than

90% issued before 1985, most Texas water rights do not contain environmental provisions. Since 1985, the Texas Commission on Environmental Quality (TCEQ) and its predecessor agencies have included some level of environmental flow protections on new water rights; however, the level of protection varies widely from permit to permit. These environmental conditions normally involve a restriction on when, or how much, water can be taken from a river or stream to ensure some water still flows downstream.

More recently, new surface water right permits and major amendments include special conditions related to the state's environmental flow standards. These standards represent the outcome of the legislatively mandated Senate Bill 3 (SB 3) environmental flows process created by the 80th Texas Legislature in 2007. The process set out a comprehensive, statewide approach for reserving unappropriated surface water necessary to protect environmental flow regimes in Texas rivers and bays by establishing environmental flow standards to be applied to newly issued or amended water rights (Loeffler 2015). By 2014, environmental flow standards had been adopted for most of the state (with the exception of the Canadian, Red, Cypress, and Sulfur River basins). The process for identifying environmental flows to benefit river, bay, and estuary ecosystems while balancing human needs relied upon input from local stakeholder groups, with membership ranging from municipal and agricultural water users to commercial anglers, along with basin-specific expert science teams. Although the TCEQ adopted flow standards, in many cases, existing unappropriated flows are not sufficient to meet environmental flow needs. A critical step in the SB 3 process is to identify voluntary strategies to meet environmental flow standards, especially in fully appropriated basins. Texas Water Code, section 11.0235(d-3)(2) instructs that in those basins where the environmental flow standards are not met, "a variety of market approaches, both public and private, for filling the gap must be explored and pursued."

Entities including federal and state agencies, universities, nongovernmental organizations, and conservation partnerships have identified restoration and protection of environmental flows as a critical conservation issue, given the increasing human demand for water (TWDB 2016). Efforts to secure water for the environment would be more efficient and better informed if decision support tools were available to practitioners, regulators, and conservation organizations. For example, the Texas Environmental Flow Initiative (TEFI) represents a stakeholder collaboration involving the Meadows Center for Water and the Environment at Texas State University, The Nature Conservancy, Ducks Unlimited, the Harte Research Institute for Gulf of Mexico Studies at Texas A&M University–Corpus Christi, and the National Wildlife Federation focused on developing environmental water transactions and water markets in Texas. The TEFI set out in 2015 with the goal of identifying water rights that could be acquired through voluntary transactions to meet freshwater inflow needs of Matagorda Bay, Guadalupe/San Antonio Bay, and Galveston Bay and associated estuaries. An evaluation framework for water rights based on their proximity to the estuary and their respective characteristics (e.g., priority date, authorized diversion volume, and modeled reliability) was developed. To assist with the characterization and evaluation of the framework, the group contracted the development of the Texas Environmental Flows Project Geospatial Water Rights Tool (EFP Tool). The EFP Tool was a significant advancement in integrating the TCEQ's geographic information system (GIS) data sets by spatially connecting those within the river networks and included information on historical water use and water reliability based on water availability models (WAMs). The EFP Tool assisted TEFI practitioners in spatial assessment of water rights, determining how much water had been authorized, and what their respective acquisition and dedication would mean for freshwater inflows into those bay and estuary systems. Using the EFP Tool, the TEFI was able to identify multiple water rights and prioritize them for acquisition. Subsequent decisions on the preferred transaction scenarios were developed and their relative costs assessed.

Also in 2015, the Texas Parks and Wildlife Department (TPWD) initiated development of a pilot geospatial framework with the science needed to inform implementation of voluntary water right transactions as a flow restoration strategy. The Llano River basin was selected as a pilot watershed based on its status as a priority conservation area for Guadalupe Bass *Micropterus treculii* (Birdsong et al. 2015; Garrett et al. 2015), access to long-term streamflow records, and presence of hydrologic alteration. Using a combination of U.S. Geological Survey (USGS) streamflow data, The Nature Conservancy's Indicators of Hydrologic Alteration (TNC 2009), Hydrology-based Environmental Flow Regime (HEFR) software developed specifically for the SB 3 process (Opdyke et al. 2014), science team flow recommendations developed during the SB 3 environmental flows process (Colorado River BBEST 2011), and conservation goals for Guadalupe Bass in the Llano River basin, the TPWD developed a flow restoration target for the North Llano River. Flow-ecology hypotheses established by Grabowski (2014) associated negative impacts to Guadalupe Bass from hydrological alterations that reduce productivity, availability, connectivity, or accessibility of riffles and runs. Groeschel (2013) found a negative correlation between low discharges and the growth of age-0 Guadalupe Bass and lower recruitment during reduced streamflow conditions. Although the Groeschel (2013) study was performed on the South Llano River, the TPWD was able to develop a streamflow target for the North Llano River, an adjacent tributary to the Llano River, due to long-term streamflow records indicating a pattern of seasonal flow alteration and a high proportion of zero-flow days. The hypotheses and data indicated that low base flow or subsistence flow conditions were limiting Guadalupe Bass growth and recruitment. Therefore, the TPWD calculated the summer Q95 (or flow exceeded 95% of the time) for July–September using USGS streamflow gage #08148500 (North Llano River near Junction, Texas), which was 0.0113 m^3/s. With a goal to double the summer Q95, flow restoration targets were then set at 0.0227 m^3/s, which amounts to 113,346 m^3 of water for the summer months.

Although conceptual, the flow target was intended to provide habitat connectivity and maintain water quality within the river during the summer season. Hydrologic data were then compiled for an assessment of water rights for hypothetical environmental water transactions. Four separate TCEQ data sets were used in a Llano River pilot flow restoration project, which included a water right permit database, self-reported water use data, the geographic locations of water right permits, and the basin's WAM model. The permit priority date, authorized volume, geographic position of the diversion point, historical water use, seasonal period of interest, and volume reliability were calculated using the basin WAM to identify a set of water rights that, through voluntary transactions, could meet the flow restoration target and volume. The total number of water rights, their collective permitted volume, and pattern of use were used to determine the feasibility of reaching the flow target. Singular and combined collections of water rights were then identified to form the basis of potential voluntary flow restoration strategies to meet the target.

The TPWD and TEFI pilots provided significant advances in the methods and tool development to support environmental flow management within Texas' legal and institutional setting and highlighted the need for an advanced decision support tool. While both pilots confirmed the primary data sets necessary for build-out of an environmental flow decision support tool, limitations of these initial tools were evident. For example, both had limited geographic scope, few target flow options, and limited end-user or stakeholder input in their development. In addition, the pilot tools were not developed to be publicly available, further restricting their audience, and user options were narrow.

Environmental Flow Information Toolkit

Building upon the TPWD and TEFI pilots, partners initiated development of a decision support tool for environmental flow management, the Environmental Flow Information Toolkit (EFIT). The objective of EFIT is to identify opportunity areas for the protection and restoration of environmental flows and align potential water management strategies, including voluntary acquisition, easement, leases, and other water market-based approaches, reservoir reoperations (e.g., Smith et al. 2019, this volume), and watershed management plans (e.g., saltcedar management; see Mayes et al. 2019, this volume) to meet flow restoration goals. The four primary components of EFIT are (1) a hydrologic foundation, including information on surface and groundwater use and assessments of hydrologic alteration; (2) environmental streamflow targets (flow targets) based on existing environmental flow standards, hydrologic analyses, and flow-ecology relationships that provide goals for conserving native fishes and sustaining aquatic and riparian ecosystems; (3) a Web-based geospatial platform for including these and other regionally relevant data layers important for aligning restoration and protection opportunity areas with water management strategies; and (4) end-user input.

Hydrologic foundation

Understanding surface water use and availability is a necessary foundation for determining the impact and outcomes of environmental flow protection and restoration strategies. For example, determining the impact of the exercise of a particular water right on an aquatic ecosystem depends on a number of factors, including the structure of the aquatic community and the ecological functions provided by instream flows and freshwater inflows. From a water management perspective, resource managers need to know the amount of current and future water use associated with a given water right or, conversely, the benefits to flow conditions if a specific water right was not fully used. The hydrologic foundation of EFIT includes the National Hydrography Dataset (NHDPlus, Horizon Systems), historical surface water use data, related water right permit information, and data developed through the application of basin-specific WAMs. Water availability analyses consist of simulating current conditions in each basin and potential future scenarios with water rights fully utilized with no return flows. Given that the data sets were not previously associated spatially, GIS tools were applied to ensure water right permit locations were linked to the correct river segments in georeferenced project files, shapefiles, and maps summarizing water right and subbasin data.

Another important element of the hydrologic foundation is the assessment of hydrologic alteration to identify time periods of natural and altered flow conditions. This step is critical for development of hydrology-based flow targets using a natural period of record minimally altered by human uses of water (e.g., water storage projects, major diversions, and groundwater pumping). We analyzed data from 784 USGS streamflow gages and developed summaries of upstream reservoirs (i.e., the construction date, size, and drainage area of reservoirs upstream of each gage). The Pettitt test (Pettitt 1979) for change-point detection was applied to low, medium, and high flow annual flow statistics to quantitatively identify the onset of a shift in the underlying distribution of flows. Results were compared with the date at which upstream reservoir development began. Notably, different parts of the flow regime appear to have been affected differently; for example, high flow pulse regimes are more significantly impacted by reservoirs, which included significant flow control storage, whereas the ratio of reservoir-controlled drainage area versus the total drainage area for the gage coincided more consistently with change point detection results for the medium and low flow statistics. Given the statewide scope of EFIT, a simple rule was developed to define the altered date as the date at which 20% of the upstream area was regulated by upstream reservoir development. This is a conservative estimate to ensure that the flow records are consistent with natural conditions but does not account for long-term groundwater pumping at this time. Next, we needed to identify the minimum period of record, in number of years, sufficient to support development of hydrology-based environmental flow targets (see Flow targets). A 20-year minimum period of record was recommended by The Nature Conservancy (TNC 2009), and that period should be sufficient to capture the full range of Texas' climatic conditions. However, this threshold eliminates many aforementioned USGS gages for development of target flows. To potentially expand the pool of gages, we used a Kolmogorov–Smirnov test (R Development Core Team 2012) to identify all *n*-year lengths of discharge data that contained the range of variability (as defined by indices of climatic condition) observed within the entire 1900–2018 period of record. Based on this analysis, gages with preimpact periods of greater than 10 years but less than 20 years, which contained variability consistent with the long-term record, were also included.

Flow targets

An important step in environmental flow management is determining flow targets to achieve restoration or protection goals, such as native fish conservation. Target flows may take the form of prescribed flow regimes, specific flow components for use in flow standards (e.g., Opdyke et al. 2014), or dam reoperation scenarios (Smith et al. 2019, this volume). Additionally, streamflow targets may take the form of specific ranges of flow metrics or other flow descriptors that are important to sustain conservation objectives (e.g., focal fish species, guilds, or assemblages). Flow target application and objectives may differ substantially among user groups; therefore, EFIT will include several options for flow targets.

Senate Bill 3 flow targets.—The SB 3 environmental flow process included environmental flow regime recommendations produced by basin-specific expert science teams, as well as environmental flow standards adopted by the TCEQ and contained in the Texas Administrative Code, title 30, chapter 298 (Figure 1). The science team flow recommendations were based solely on best available data, with the goal of maintaining a sound ecological environment. In most cases, the science teams relied upon a desktop evaluation of USGS streamflow records overlain with any known flow–ecology relationships or instream flow study results. The desktop hydrology-based approach utilized Indicators of Hydrologic Alteration and HEFR to produce a matrix of flow components that included subsistence flows, base flows, high flow pulses, and overbanking flows (Figure 2). Science teams then used any additional information on biology, water quality, geomorphology or other factors to refine recommended flow regime components. On the other hand, flow standards, as adopted by TCEQ, reflect a balancing of environmental needs with future water development needs and may not reflect environ-

United States Geological Survey Gage 08082500, Brazos River at Seymour

Season	Subsistence	Hydrologic Condition	Base	Dry Condition Seasonal Pulse	Average Condition Seasonal Pulse	Wet Condition Seasonal Pulse
Winter	1 cfs	Dry	10 cfs	N/A	N/A	N/A
		Average	25 cfs			
		Wet	46 cfs			
Spring	1 cfs	Dry	7 cfs	1 per season Trigger: 560 cfs Volume: 2,960 af Duration: 10 days	2 per season Trigger: 560 cfs Volume: 2,960 af Duration: 10 days	1 per season Trigger: 1,040 cfs Volume: 5,870 af Duration: 12 days
		Average	19 cfs			
		Wet	35 cfs			
Summer	1 cfs	Dry	4 cfs	1 per season Trigger: 370 cfs Volume: 1,870 afDuration: 8 days	2 per season Trigger: 370 cfs Volume: 1,870 af Duration: 8 days	1 per season Trigger: 800 cfs Volume: 4,290 af Duration: 11 days
		Average	13 cfs			
		Wet	32 cfs			

cfs = cubic feet per second
af = acre-feet
N/A = not applicable

Figure 1. Environmental flow regime standard for the Brazos River at Seymour, Texas adopted by the Texas Commission on Environmental Quality (Texas Administrative Code, title 30, section 298.480(3)).

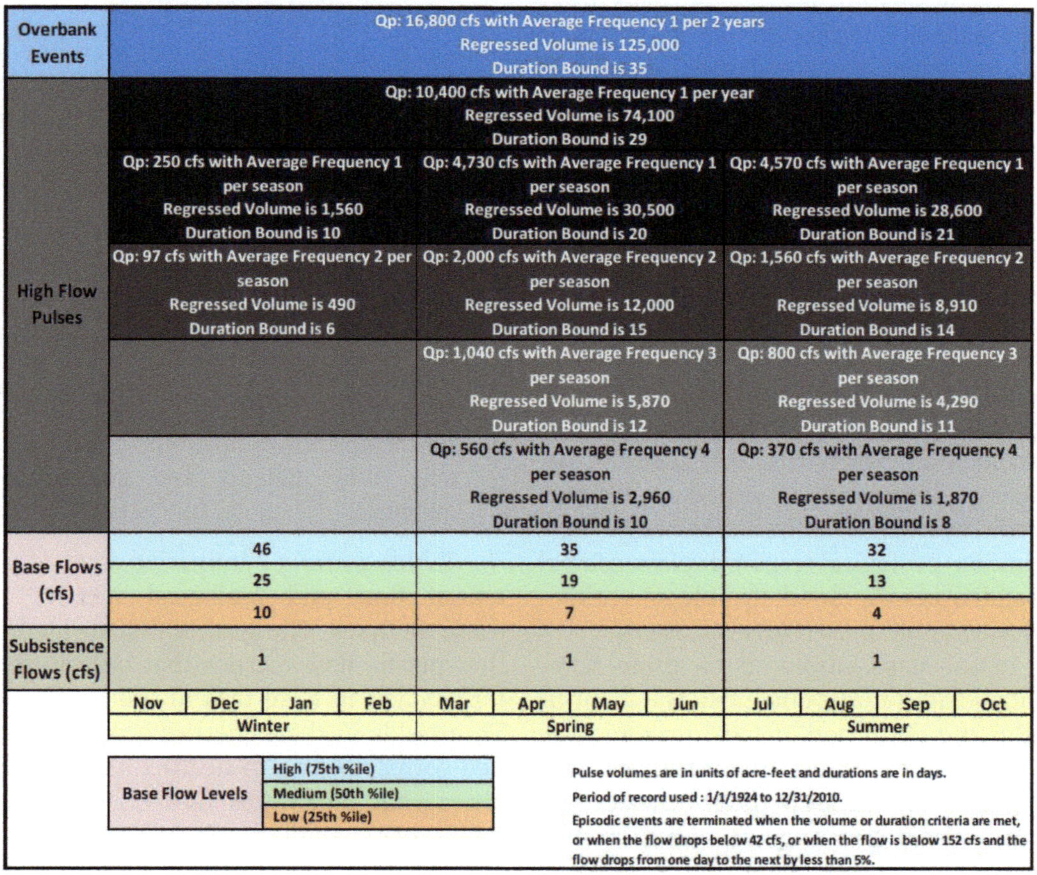

Figure 2. Environmental flow regime matrix for the Brazos River at Seymour, Texas recommended by the Brazos River Basin and Bay Expert Science Team (Brazos River BBEST 2012).

mental flow amounts necessary to maintain or restore aquatic ecosystems. Thus, to meet the needs of the most end users, EFIT will include both expert science team recommendations and flow standards.

Hydrology-based flow targets.—Hydrology-based flow targets will also be developed as an option to address two issues with SB 3 flow targets. First, SB 3 flow targets are only available for a small fraction of USGS gages in Texas, and several basins have no recommendations or flow standards. Second, full periods of record were used in most HEFR applications by expert science teams, and those included altered flow conditions. We will generate flow targets using natural periods of record (as described in the hydrologic foundation section) for the 316 USGS gages in Texas, with sufficient predevelopment data. As above, the HEFR software will be used to derive flow targets using a consistent framework.

Flow–ecology-based flow targets.—Flow targets based on flow–ecology relationships derived from published literature and statistical analyses will also be included in EFIT. These flow targets are defined by flow needs of focal taxa, guilds, and assemblages (focal taxa for short) and may be further defined using methodologies such as ecological limits of hydrologic alteration (ELOHA). The ELOHA process synthesizes ecological data and site-specific hydrologic analysis to generate "flow alteration–ecological response relationships" expanded across varied hydrologic regimes and geographic areas (Poff et al. 2010).

Initial steps employed to guide flow–ecology-based targets included the identification and definition of focal taxa and the execution of a thorough literature review focused on native fishes in the Southern Great Plains, generally defined as the portion of the Great Plains Landscape Conservation Cooperative geography contained within Texas. We selected a group of focal taxa (i.e., fish species and reproductive guilds; Table 1) that represent iconic and environmentally sensitive components of Great Plains fish communities. Initial selection of these species and guilds was guided by Great Plains Landscape Conservation Cooperative priority fish taxa (Labay et al. 2019, this volume). Concurrently, a regionally focused literature review was performed, with emphasis on descriptive and quantitative relationships connecting ecological measures to flow metrics. While a variety of methods exist to define flow targets (e.g., Annear et al. 2004; Acreman et al. 2014; Stewart-Kostner et al. 2014; Poff et al. 2016; Wheeler et al. 2018), we will initially focus on flow or flow-alteration metrics that describe ecological conditions that are also simple enough to be usable in flow management and decision support tools (e.g., Cartwright et al. 2017).

Findings from the regional literature review were summarized and compiled into a database and used to define flow–ecology hypotheses and an analytical framework for additional flow–ecology analyses to address the principal research questions:

- What are the trends in changes of focal taxa over time and how do they relate to variation in flow metrics and groundwater levels?
- How can flow–ecology relationships describe current conditions among focal taxa, flow metrics, and groundwater metrics?

Tables 2 and 3 summarize a set of potential flow and biological metrics for these analyses. The goal is first to identify the specific flow metrics that best account for changes in focal taxa in the study area. Thresholds in these metrics can be identified corresponding to healthy fish communities that constitute flow–ecology-based flow targets. While work currently focuses on the Southern Great Plains, efforts to expand the analytical framework and literature review to the rest of the river basins in Texas are underway.

Table 1. Focal guilds and species for the Southern Great Plains. Federal listing status designated as endangered (E) and threatened (T).

Focal guild	Species	River basin(s) inhabited
Pelagic broadcast-spawning minnows	Smalleye Shiner *Notropis buccula* (E)	Brazos
	Sharpnose Shiner *N. oxyrhynchus* (E)	Brazos
	Arkansas River Shiner *N. girardi* (T)	Arkansas (Canadian River basin in TX)
	Plains Minnow *Hybognathus placitus*	Arkansas (Canadian in TX), Red, Brazos
Potential pelagic broadcast-spawning minnows	Peppered Chub *Macrhybopsis tetranema*	Arkansas (Canadian in TX)
	Shoal Chub *M. hyostoma*	Brazos, Colorado
	Prairie Chub *M. australis*	Red
	Burrhead Chub *M. marconis*	Colorado
	Red River Shiner *Notropis bairdi*	Red
	Chub Shiner *N. potteri*	Red, Brazos
Lithopelagic benthic-spawning minnows	Silver Chub *Macrhybopsis storeriana*	Red, Brazos
	Flathead Chub *Platygobio gracilis*	Arkansas (Canadian in TX)
	Chub Shiner *N. potteri*	Red, Brazos
	Emerald Shiner *Notropis atherinoides*	Red
Benthic-spawning minnows	Suckermouth Minnow *Phenacobius mirabilis*	Arkansas (Canadian in TX), Red, Colorado
Other fishes	Red River Pupfish *Cyprinodon rubrofluviatilis*	Red, Brazos
	Shovelnose Sturgeon *Scaphirhynchus platorynchus*	Red
	Paddlefish *Polyodon spathula*	Red
	American Eel *Anguilla rostrata*	All
	Silverband Shiner *Notropis shumardi*	Red, Brazos, Colorado
	River Shiner *N. blennius*	Red
	Arkansas Darter *Etheostoma cragini*	Arkansas (not in TX)
	Topeka Shiner *N. topeka*	Arkansas (not in TX)

Web-based geospatial platform

The identification of protection and restoration opportunity areas using a geospatial platform is an essential element of EFIT's project goals. A geospatial platform pulls from geographical imagery with the intent of defining information spatially based on global position and scale. The Environmental Systems Research Institute (ESRI) develops and maintains a suite of GIS software and geodatabase management applications. As ESRI has evolved, the company has transitioned from a focus on desktop software and applications, such as ArcMap and ArcCatalog, to Web-based GIS. ESRI's ArcGIS Online is a Web-based platform in which all data and analytic processing functionality is stored in the cloud. Data and geospatial services hosted in the cloud are linked to a portal allowing users to share information with the general public or a target audience.

Table 2. Flow attributes and associated environmental flow metrics for the initial focus of the flow-ecology analysis in the Southern Great Plains.

Flow attribute	Flow metrics
Base flow during spawning season	Monthly mean daily flows Mean daily flow Apr–Sep Monthly base flows Apr–Sep Change in mean daily flow Apr–Sep Attainment frequency of base flows
High flow pulses during spawning season	Frequency of high flow pulses Magnitude of high flow pulses Attainment frequency of subsistence flows
Continuous flowing water	Number of zero-flow days Attainment frequency of subsistence flows

A valuable component of ArcGIS Online is the ability to build intuitive two-dimensional and three-dimensional Web applications without writing code. Users are able to engage powerful tools via ArcGIS Online's Web AppBuilder to pick and choose the functionality and design components controlling the configuration of HTML/JavaScript-based applications. The ArcGIS Online platform using Web AppBuilder was chosen for EFIT based on its compatibility with TPWD's current geospatial licensing, coding skill levels, and the necessity to design a decision support tool that is publicly available. Once complete, EFIT will enable end users to access the EFIT application online, free of charge.

Another aspect of Web AppBuilder is the capability to further customize the tool by including supporting data and tailored functionality. The Environmental Flow Information Toolkit has been developed to include a variety of regional data layers including fish data and ranges, native fish conservation areas, critical habitat, in-stream barriers, groundwater, reservoirs, water quality, playa lakes, and riparian areas with nonnative, invasive saltcedar. Supporting geographic information, such as county lines, cities, regional boundaries and water planning group locations, was obtained from the Texas Natural Resource Information System.

Table 3. Selected fish community metrics to be the initial focus of the flow-ecology analysis.

Ecological attribute	Ecology metric	Trend, current or both
Presence or occurrence	Probability of occurrence/capture of focal taxa (guilds/species)	Trend
Abundance	Relative abundance of focal taxa (guilds/species)	Both
Reproductive success	Relative abundance of focal guilds and species (versus 1- or 2-year antecedent flows)	Both
	Relative abundance of age-0 fish (if available)	Current
Traits	To be determined	Both

End-user input

Critical to the development of EFIT is the integration of end-user feedback into tool design. By soliciting input prior to tool development, the finished product is more likely to include data, relationships, and functionality that addresses end-user objectives. Texas Parks and Wildlife Department EFIT team members attended meetings and conferences to present information about the tool and solicit input from peers, colleagues, and flow practitioners. Community engagement was used to communicate the importance of streamflow protection and restoration, inform the public of EFIT, and gather stakeholder input for project design.

Stakeholder participation.—During the early development phase of EFIT, an initial list of more than 200 names, addresses, and organizations was gathered and categorized by type and location. The preliminary list of potential end users included staff from state and federal natural resource agencies, academics, regional water planning group members, landowners, agriculture community representatives, conservation group members, and others with a perceived interest in environmental flow and water management issues. The list of end users was used to target individuals and audiences for obtaining input on EFIT development.

Several surveys were developed with an emphasis on understanding the viewpoints of specific end users and stakeholders. One-on-one and small-group interviews were conducted using formatted questions within an informal setting designed to engage the end user and receive unbiased feedback. Survey results and end-user feedback were then compiled and quantified (Table 4), to be incorporated into the design of the tool.

The importance of stakeholder input and feedback cannot be overstated. For example, most riparian lands throughout the United States are under private ownership. Considering Texas' longstanding tradition of robust private property rights, the percentage of private riparian lands is significantly high. For fish and other biota, access to adjacent riparian resources, especially during overbank flows, is critical, given that almost all biomass within riverine systems is produced within riparian habitats (Junk et al. 1989). Riparian sustainability depends upon connectivity with the river, a naturally variable flow regime that maintains both low and high flows, and enough area to foster geomorphic processes and significant ecosystem services (Opperman et al. 2010).

The landscape scale necessary for the long-term sustainability of riverine ecosystems places landowner and public input at the forefront of a successful decision process for environmental flows. The rapid decline of riparian habitats on private lands across Texas poses serious threats to downstream resources, including aquatic and terrestrial habitats, and the quantity and quality of streamflow (King et al. 2009; Hayes 2016). Voluntary conservation and management of private riparian lands is considered ideal, as it is generally more effective than a regulatory approach. End users of EFIT will be able to use the tool to identify opportunity areas and develop appropriate landowner incentives to benefit riparian health (Hardy and Davis 2014).

Incorporating feedback.—Agile project management is an iterative approach to planning and guiding project processes. A modified Agile methodology was employed to effectively integrate end-user response data into EFIT's development. A component of Agile project management is the incorporation of user stories and user scenarios. User stories help describe the type of user, what they want, and why. User stories are short, simple descriptions of a feature told from the perspective of the person who desires the new capability. User stories identify end-user needs and objectives and can aid in creation of a simplified description of a requirement of the final tool.

User scenarios, on the other hand, are the stories that predict how certain types of end users will interact with a Web site or tool in a given situation in order to complete a given goal. Scenarios describe the stories and context behind why a specific end user or user group visits a Web site or accesses a tool. They note the goals and questions to be achieved and sometimes define the possibilities of how the user(s) can achieve them through an interactive interface. Scenarios are critical both for designing an interface and for usability testing. Throughout the development of EFIT, multiple user scenarios have been discussed and will be incorporated in the final tool. User options within the tool could provide pathways to further investigate aspects of flow-ecology, hydrologic change, or water rights, for example.

Discussion

Restoration and protection of environmental flows is a critical conservation issue for native freshwater fishes, and the development of EFIT will provide a decision support tool that is expected to increase efficiencies and effectiveness of environmental flow conservation efforts. The Environmental Flow Information Toolkit is being designed to use a broad, watershed-scale conservation framework, a number of interdisciplinary metrics, and a spatial framework for querying diverse underlying data layers to support the decision-making process for multiple user groups. Based on individual selected problem solving, end users will be able to explore various conservation scenarios and gain information and recommendations catered specifically towards their unique inquiries and analyses. The Environmental Flow Information Toolkit is expected to be a "living" tool, with the ability to ingest new data, science, and technologies as they become available. Once built, there will be a regular process of review, updates, and end-user communication to ensure that EFIT information is current and reliable. The Environmental Flow Information Toolkit will enable environmental flow practitioners and stakeholders to communicate and collaborate more effectively to achieve voluntary environmental flow protection and restoration strategies. The potential of such a tool developed to scale a range of strategies from communities to watersheds to the state of Texas is exciting and endless. Pioneering the public distribution of strategic conservation science within a Web-based geospatial platform may be the key to successful water resource management and native fish conservation in the future.

Acknowledgments

Funding for development of EFIT was provided by the U.S. Fish and Wildlife Service Science Applications Program and U.S. Fish and Wildlife Service State Wildlife Grant Program. The views and conclusions contained in this document are those of the authors and should not be interpreted as representing the opinions or policies of the U.S. Government.

References

Acreman, M. C., I. C. Overton, J. King, P. J. Wood, I. G. Cowx, M. J. Dunbar, E. Kendy, and W. J. Young. 2014. The changing role of ecohydrological science in guiding environmental flows, Hydrological Sciences Journal 59:433–450.

Annear, T., I. Chisholm, H. Beecher, A. Locke, P. Aarrestad, C. Coomer, C. Estes, J. Hunt, R. Jacobson, G. Jöbsis, J. Kauffman, J. Marshall, K. Mayes, G. Smith, R. Wentworth, and C. Stalnaker. 2004. Instream flows for riverine resource stewardship, revised edition. Instream Flow Council, Cheyenne, Wyoming.

Birdsong, T. W., M. S. Allen, J. E. Claussen, G. P. Garrett, T. B. Grabowski, J. Graham, F. Harris, A. Hartzog, D. Hendrickson, R. A. Krause, J. K. Leitner, J. M. Long, C. K. Metcalf, D. P. Philipp, W. F. Porak, S. Robinson, S. M. Sammons, S. Shaw, J. E. Slaughter, IV, and M. D. Tringali. 2015. Native Black Bass Initiative: implementing watershed-scale approaches to conservation of endemic black bass and other native fishes in the southern United States. Pages 363–378 *in* M. D. Tringali, J. M. Long, T. W. Birdsong, and M. S. Allen, editors. Black bass diversity: multidisciplinary science for conservation. American Fisheries Society, Symposium 82, Bethesda, Maryland.

Birdsong, T. W., G. P. Garrett, B. J. Labay, M. G. Bean, P. T. Bean, J. Botros, M. J. Casarez, A. E. Cohen, T. G. Heger, A. Kalmbach, D. A. Hendrickson, S. J. Magnelia, K. B. Mayes, M. E. McGarrity, R. McGillicuddy, M. M. Parker, and S. Robertson. 2019. Texas Native Fish Conservation Areas Network: strategic investments in restoration and preservation of freshwater fish diversity. Pages 183–229 *in* D. C. Dauwalter, T. W. Birdsong, and G. P. Garrett, editors. Multispecies and watershed approaches to freshwater fish conservation. American Fisheries Society, Symposium 91, Bethesda, Maryland.

Brazos River BBEST (Basin and Bay Expert Science Team). 2012. Brazos River Basin and Bay Expert Science Team environmental flow regime recommendations report. Final submission to the Brazos River Basin and Bay Area Stakeholder Committee, Environmental Flows Advisory Group, and the Texas Commission on Environmental Quality, Austin, Texas.

Bunn, S. E., and A. H. Arthington. 2002. Basic principles and ecological consequences of altered flow regimes for aquatic biodiversity. Environmental Management 30:492–507.

Carlisle, D. M., D. M. Wolock, and M. R. Meador. 2011. Alteration of streamflow magnitudes and potential ecological consequences: a multiregional assessment. Frontiers in Ecology and the Environment 9:264–270.

Cartwright, J., C. Caldwell, S. Nebiker, and R. Knight. 2017. Putting flow–ecology relationships into practice: a decision-support system to assess fish community response to water-management scenarios. Water [online serial] 9(3):196.

Colorado River BBEST (Colorado and Lavaca Rivers and Matagorda and Lavaca Bays Basin and Bay Expert Science Team). 2011. Environmental flow regime recommendations report. Final submission to the Colorado and Lavaca Rivers and Matagorda and Lavaca Bays Basin and Bay Area Stakeholder Committee, Environmental Flows Advisory Group, and Texas Commission on Environmental Quality, Austin, Texas.

Garrett, G. P., T. W. Birdsong, M. G. Bean and R. McGillicuddy. 2015. Guadalupe Bass Restoration Initiative. Pages 379–385 *in* M. D. Tringali, J. M. Long, T. M. Birdsong and M. S. Allen, editors. Black bass diversity: multidisciplinary science for conservation. American Fisheries Society, Symposium 82, Bethesda, Maryland.

Grabowski, T. 2014. Guadalupe Bass (*Micropterus treculii*). Pages 42–48 *in* M. Davis and S. Brewer, editors. Gulf Coast Prairie Landscape Conservation Cooperative regional hypotheses of ecological responses to flow alteration. A report by the GCP LCC Flow-Ecology Hypotheses Committee to the Southeast Aquatic Resources Partnership. Wildlife Management Institute, Washington, D.C.

Griffith, G. E., S. A. Bryce, J. M. Omernik, J. A. Comstock, A. C. Rogers, B. Harrison, S. L. Hatch, and D. Bezanson. 2004. Ecoregions of Texas. U.S. Environmental Protection Agency, Corvallis, Oregon.

Groeschel, J. 2013. Evaluations of growth and habitat use by Guadalupe Bass at the riverscape scale in the South Llano River, Texas. Master's thesis. Texas Tech University, Lubbock.

Hardy, T., and N. Davis. 2014. Texas riparian areas. Texas A&M University Press, College Station.

Hayes, T. 2016. Riparian assessment on the Guadalupe and Brazos rivers. Final report submitted to Texas Parks and Wildlife Department and Texas Water Development Board, Austin.

Howells, R. G., R. W. Neck, and H. D. Murray. 1996. Freshwater mussels of Texas. Texas Parks and Wildlife Press, Austin.

Junk, W. J., P. B. Bayley, and R. E. Sparks. 1989.

The flood pulse concept in river-floodplain systems. Pages 110–127 *in* D. P. Dodge, editor. Proceedings of the international large river symposium. Canadian Special Publication of Fisheries and Aquatic Sciences 106.

King, S. L., R. R. Sharitz, J. W. Groninger, and L. L Battaglia. 2009. The ecology, restoration, and management of southeastern floodplain ecosystems: a synthesis. Wetlands 29:624–634.

Labay, B. J., J. S. Perkin, D. A. Hendrickson, A. R. Cooper, G. P. Garrett, and T. W. Birdsong. 2019. Who's asking? Interjurisdictional conservation assessment and planning for Great Plains fishes. Pages 57–83*in* D. C. Dauwalter, T. W. Birdsong, and G. P. Garrett, editors. Multispecies and watershed approaches to freshwater fish conservation. American Fisheries Society, Symposium 91, Bethesda, Maryland.

Loeffler, C. 2015. A brief history of environmental flows in Texas. Pages 2350–2359 *in* K. Karvazy and V. L. Webster, editors. Proceedings of the world environmental and water resources congress 2015: floods, droughts, and ecosystems. American Society of Civil Engineers, Reston, Virginia.

Mayes, K. B., G. R. Wilde, M. E. McGarrity, B. D. Wolaver, and T. G. Caldwell. 2019. Watershed-scale conservation of native fishes in the Brazos River basin, Texas. Pages 315–343 *in* D. C. Dauwalter, T. W. Birdsong, and G. P. Garrett, editors. Multispecies and watershed approaches to freshwater fish conservation. American Fisheries Society, Symposium 91, Bethesda, Maryland.

Opdyke, D. R., E. L. Oborny, S. K. Vaugh, and K. B. Mayes. 2014. Texas environmental flow standards and the hydrology-based environmental flow regime methodology. Hydrological Sciences Journal 59:820–830.

Opperman, J. J., R. Luster, B. A. McKenney, M. Roberts, and A. W. Meadows. 2010. Ecologically functional floodplains: connectivity, flow regime, and scale. Journal of the American Water Resources Association 46:211–226.

Perkin, J. S., K. B. Gido, K. H. Costigan, M. D. Daniels, and E. R. Johnson. 2014. Fragmentation and drying ratchet down Great Plains stream fish diversity. Aquatic Conservation: Marine and Freshwater Ecosystems 25:639–655.

Pettitt, A. N. 1979. A non-parametric approach to the change-point problem. Applied Statistics 28:126–135.

Poff, N. L., C. M. Brown, T. E. Grantham, J. H. Matthews, M. A. Palmer, C. M. Spence, R. L. Wilby, M. Haasnoot, G. F. Mendoza, K. C. Dominique, and A. Baeza. 2016. Sustainable water management under future uncertainty with eco-engineering decision scaling. Nature Climate Change 6:25–34.

Poff, N. L., B. D. Richter, A. H. Arthington, S. E. Bunn, R. J. Naiman, E. Kendy, M. Acreman, C. Apse, B. P. Bledsoe, M. C. Freeman, J. Henriksen, R. B. Jacobson, J. G. Kennen, D. M. Merritt, J. H. O'Keeffe, J. D. Olden, K. Rogers, R. E. Tharme, and A. Warner. 2010. The ecological limits of hydrologic alteration (ELOHA): a new framework for developing regional environmental flow standards. Freshwater Biology 55:147–170.

R Development Core Team. 2012. R: a language and environment for statistical computing [online]. R Foundation for Statistical Computing, Vienna.

Smith, R., J. Trungale, R. Lowerre, T. Hayes, M. Montagne, T. Bister, L.-A. Overdyke, and M. Hackett. 2019. Instream flow restoration and watershed conservation in the Cypress basin, Texas. Pages 345–366 *in* D. C. Dauwalter, T. W. Birdsong, and G. P. Garrett, editors. Multispecies and watershed approaches to freshwater fish conservation. American Fisheries Society, Symposium 91, Bethesda, Maryland.

Stewart-Koster, B., J. D. Olden, and K. B. Gido. 2014. Quantifying flow–ecology relationships with functional linear models. Hydrological Sciences Journal 59:629–644.

TWDB (Texas Water Development Board). 2016. Water for Texas: 2017 state water plan. TWDB, Austin.

TIFP (Texas Instream Flow Program). 2008. Texas instream flow studies: technical overview. Texas Commission on Environmental Quality, Texas Parks and Wildlife

Department, and Texas Water Development Board, Austin.

TNC (The Nature Conservancy). 2009. Indicators of hydrologic alteration, version 7.1 user's manual. TNC, Arlington, Virginia.

Wheeler, K., S. J. Wenger, and M. C. Freeman. 2018. States and rates: complementary approaches to developing flow–ecology relationships. Freshwater Biology 63:906–916.

American Fisheries Society Symposium 91:269–292, 2019

Four Decades of Conserving Native Fish in the Colorado River Watershed, Texas

STEPHAN J. MAGNELIA* AND KEVIN B. MAYES
Texas Parks and Wildlife Department, Inland Fisheries Division
505 Staples Road, Building 1, San Marcos, Texas 78666, USA

MEGAN G. BEAN
Texas Parks and Wildlife Department
5103 Junction Highway, Mountain Home, Texas 78058, USA

CINDY L. LOEFFLER AND DAVID D. BRADSBY
Texas Parks and Wildlife Department
4200 Smith School Road, Austin, Texas 78744, USA

Abstract.—The Colorado River of Texas originates in the northwest portion of the state and runs southeasterly across the environmentally sensitive Edwards Plateau ecoregion, eventually emptying into Matagorda Bay on the Gulf of Mexico. The Texas Parks and Wildlife Department (TPWD) and other conservation partners have worked for decades to conserve native fishes in the river and its associated estuary. The river and its tributaries are also the major water source for millions of people, and portions of the watershed are undergoing unprecedented human population growth and a concomitant increased demand for water. Drought, which is a frequent occurrence in Texas, exacerbates these effects. Providing adequate streamflow and water quality to conserve aquatic species while still fulfilling obligations for municipal, industrial, agricultural, hydropower, and recreational water use is increasingly challenging. Since the 1980s, the TPWD has promoted water management policies in the watershed to benefit aquatic life, conducted research to maintain a sound ecological environment for fish and other aquatic taxa, provided technical guidance and financial incentives to private landowners to improve land practices that benefit water quality and quantity, and improved public access to foster increased stewardship of the river. The TPWD has also collaborated with researchers to study the potential impacts of altered hydrology on two endemic, flow-dependent fish species, Guadalupe Bass *Micropterus treculii* and Blue Sucker *Cycleptus elongatus*. Two portions of the Colorado River watershed have been identified as native fish conservation areas in part for persistence of populations of these native fishes. Outcomes from these activities are now being leveraged using federally funded research and conservation delivery programs to further advance the conservation of native fish populations and other aquatic life in native fish conservation areas of the Colorado River watershed.

* Corresponding author: stephan.magnelia@tpwd.texas.gov

Introduction

The Texas Parks and Wildlife Department (TPWD) and its conservation partners have been actively working to conserve aquatic species in the Colorado River watershed since the early 1980s. Staff routinely provides input into water management policies and processes to ensure that aquatic species needs are considered. This input relies on collaborative research to fill data gaps and inform adaptive management decisions for aquatic species. In addition, the TPWD works directly with landowners to improve land practices to enhance habitat conditions for fish and wildlife and uses education and outreach to increase public appreciation and stewardship of aquatic resources. This chapter provides a historical summary of the TPWD's collaborative partnership efforts to protect and conserve native fish populations in the Colorado River watershed.

Colorado River watershed

The Colorado River originates in northwest Texas and flows in a southeasterly direction for approximately 965 km, eventually flowing into Matagorda Bay and the Gulf of Mexico (Figure 1). The watershed drains 103,341 km^2 and runs through six ecoregions (Clay and Kleiner 2010), including the Edwards Plateau ecoregion, an environmentally sensitive area in central Texas that harbors at least 14 endemic freshwater fishes (Bowles and Arsuffi 1993; Hubbs et al. 2008). Instream flows have been impacted by human activity for more than 100 years, as water from the river is used for flood control, municipal, agricultural and industrial supply, hydropower generation, and recreation. A chain of six main-stem reservoirs (the Highland Lakes), starting northwest of the city of Austin and continuing through the city, inundates approximately one-third (322 km) of the main-stem river. The Lower Colorado River Authority (LCRA) has jurisdiction for managing the portion of the Colorado River from Lake Buchanan, the northernmost of the Highland Lakes, downstream to the Gulf of Mexico. Flows downstream from Austin are largely regulated by releases from Lake Buchanan and Lake Travis, the two largest of the Highland Lakes. Water released from these reservoirs to meet water demands has substantially altered the natural flow regime and water quality in downstream reaches of the river (Mosier and Ray 1992; Acre 2019).

Austin is the 11th largest city in the United States, and the city and surrounding area have undergone unprecedented population growth. For example, according to data from the 2015 U.S. Census, the population in Austin's five-county metropolitan area increased 37.7% over the preceding decade, exceeding 2 million people in 2015. Just west of Austin, the population in the 17-county Texas Hill Country portion of the environmentally sensitive Edwards Plateau ecoregion grew from approximately 800,000 in 1950 to 2.6 million in 2000 and is projected to grow to 4.3 million by 2030 (Hill Country Alliance 2008). Associated increased demand for surface water from the Highland Lakes and groundwater from the Edwards Aquifer, and effects those withdrawals may have on water quantity in the Colorado River watershed, has become a major concern for aquatic resource managers. Severe drought, such as that experienced in central Texas from 2011 to 2015 (Winters 2013), and projected increasing frequency of severe drought for Texas as a result of climate change (U.S. Global Change Research Program 2018), has and will likely exacerbate negative effects on aquatic species from human water demand.

Land use has also changed in the Hill Country as a result of population growth. While the area was historically farm and ranch land, single-family home developments are now commonplace (Hill Country Alliance 2008). The associated increase in impervious surfaces from this change has altered the natural flow regime in portions of the Colorado River watershed (Pease et al. 2017, 2018). With few exceptions, local land-

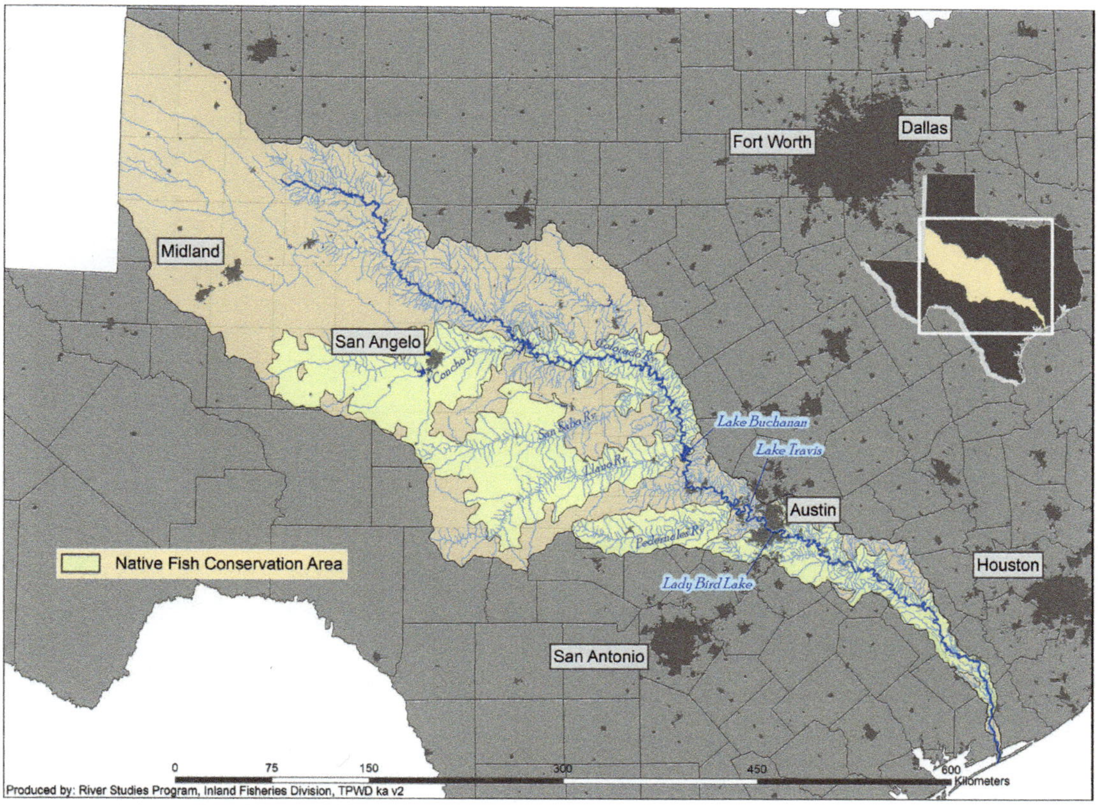

Figure 1. The Colorado River watershed in Texas, including the Lower Colorado River and Central Edwards Plateau Rivers native fish conservation areas.

development regulations and ordinances for protecting water quantity and quality are generally lacking, and further land development in sensitive areas (e.g., riparian areas or aquifer recharge zones) will likely have deleterious effects on groundwater and surface water systems. The ability to effectively mitigate impacts to aquatic resources from land-use change continues to be a challenge given that the vast majority of property in the watershed is privately owned.

Guadalupe Bass *Micropterus treculii*, the Texas state fish, is a central Texas endemic, flow-dependent, black bass species (Figure 2). In addition to being a highly sought after sport fish for river anglers (Thomas et al. 2015), it is also a species of conservation concern (Hubbs et al. 2008; Bean 2017) and a species of greatest conservation need (SGCN; TPWD 2012). Decreased streamflow (Hurst et al. 1975), reservoir construction (Edwards 1978), habitat degradation, and hybridization with nonnative Smallmouth Bass *M. dolomieu* (Edwards 1980) have all been identified as threats to the persistence of Guadalupe Bass populations. As the human population increased in central Texas, habitat degradation was identified as a chronic threat to the conservation of Guadalupe Bass (Bean et al. 2013; Curtis et al. 2015). The clear and fast-flowing Colorado River tributary streams of the Edwards Plateau ecoregion (Figure 3) generally provide high-quality Guadalupe Bass habitat, but base flows are largely dependent upon contributions from groundwater. In addition to land-use-related habitat degradation, groundwater depletion from increased pumping has also become a major threat to aquatic resources in this area. Many small

Figure 2. A trophy-size Guadalupe Bass caught by an angler in the lower Colorado River.

Figure 3. Example of the clear and fast-flowing waters of a typical Hill Country stream.

springs and streams throughout Texas have experienced general declines in annual flows or ceased to flow permanently as a result of increased demands on groundwater (Bowles and Arsuffi 1993). Changes to Guadalupe Bass population structure resulting from altered hydrology have been hypothesized (Grabowski 2014a) but are largely unstudied, with a few exceptions (e.g., Edwards 1978; Groeschel 2013; Pease et al. 2017). Filling this knowledge gap would be critical to effectively managing populations. Long-term reduction in base flow could decrease growth of Guadalupe Bass (Groeschel 2013) and ultimately change population structure (Pease et al. 2017). Reduction of spring flows would also likely negatively impact the 13 endemic species of fish in the Edwards Plateau ecoregion considered spring-associated obligates (Craig et al. 2016).

Blue Sucker *Cycleptus elongatus* (Figure 4), once common in the lower Colorado River, is now considered a SGCN and are considered imperiled or vulnerable in 19 of the 22 states in which they historically occurred (NatureServe 2019). In Texas, they were listed on the first protected species list in 1977 and have been a species of conservation con-

Figure 4. A Blue Sucker collected during the Texas Tech University study of the lower Colorado River Blue Sucker population.

cern since their original listing. Blue Sucker is a fluvial specialist associated with big river ecosystems (Mettee 2000). Early life history stages are hypothesized to be sensitive to flow alteration (Adams et al. 2006) and are considered vulnerable throughout their range (Jelks et al. 2008) as the hydrology of many big river systems, similar to the lower Colorado River, has been highly altered. Spawning requirements and factors affecting juvenile survival in the lower Colorado River remain poorly understood (Mosier and Ray 1992; Acre 2019).

Resource protection

To address the increasing threats to fish and wildlife resources, the Texas Legislature granted to the TPWD resource protection authority, which the agency began exercising in 1987, providing recommendations on scheduling of instream flows for Texas rivers and freshwater inflows to Texas estuaries and bays for the management of fish and wildlife resources. Special conditions to protect these flows have been incorporated into water right permits, including those managed by the LCRA. The LCRA's current Water Management Plan (WMP) provides guidance for releases from the Highland Lakes system, in part to maintain a healthy aquatic community in the lower Colorado River, Matagorda Bay, and associated estuaries (LCRA 2015).

Colorado River watershed native fish conservation areas

In 2015, the University of Texas at Austin and the TPWD completed a statewide conservation prioritization that identified a network of 20 subwatersheds considered priorities for conservation of freshwater fish SGCN (Birdsong et al. 2019b, this volume). Termed "native fish conservation areas" (NFCAs; Williams et al. 2011), two subwatersheds were identified as priorities in the Colorado River watershed (Figure 1): the Central Edwards Plateau Rivers (CEPR) NFCA and the Lower Colorado River (LCR) NFCA. The CEPR NFCA consists of a portion of the main-stem Colorado River and rocky, spring-fed tributaries that flow across the Edwards Plateau ecoregion, including the Concho, San Saba, Llano, and Pedernales rivers. The Central Edwards Plateau Rivers NFCA hosts 54 species of native freshwater fishes, including 15 SGCN (Cohen et al. 2018; Birdsong et al. 2019b). The LCR NFCA consists of the lower main stem and its tributaries from the city of Austin to almost Matagorda Bay, a distance of approximately 470 km. Guadalupe Bass and Blue Sucker were identified as focal species of the NFCAs, defined as species that are indicators of habitat quality or in need of conservation action (Cohen et al. 2018; Birdsong et al. 2019b). The LCR NFCA is home to 64 native species of fish, including 12 SGCN (Birdsong et al. 2019b). Unlike the fast-flowing and narrow headwater streams located in the CEPR NFCA, the river here is relatively wide and slow-moving with intermittent pools, riffles, occasional rapids, and intermittent boulder fields (Mosier and Ray 1992; BIO-WEST, Inc. 2008; Magnelia 2018). Chute and rapid habitats with bedrock substrate strewn with boulders provide good habitat for Blue Sucker (Mosier and Ray 1992; Figure 5). While spawning requirements and factors affecting juvenile survival in the lower Colorado River remain poorly understood (Mosier and Ray 1992; Acre 2019), instream flows for providing Blue Sucker spawning habitat were recommended and implemented in the LCRA WMP (Mosier and Ray 1992; BIO-WEST, Inc. 2008). In addition to Blue Sucker, the LCR NFCA also supports a unique Guadalupe Bass population, with a much higher density than populations found in more typical habitat on the Edwards Plateau and exhibiting higher growth rates (Pease et al. 2017). The fast rate of growth allows individuals to attain a larger maximum size. The world-record Guadalupe Bass (1.68 kg, 43 cm) was caught here in 2014, and the reach

Figure 5. Texas Parks and Wildlife Department staff electrofishing a rapid and boulder field in the lower Colorado River for Blue Sucker.

from Austin to Columbus, Texas is considered the state's premier sport fishery for trophy-size (≥380 mm total length) Guadalupe Bass (Bean 2017; Cummings 2018).

In 2015, the Southeast Aquatics Resource Partnership, the TPWD, and the University of Texas at Austin organized and facilitated a series of conservation planning workshops within the CEPR and LCR NFCAs in which participants were asked to serve as catalysts for local implementation of the National Fish Habitat Action Plan and Texas' State Wildlife Action Plan (Birdsong et al. 2018). The workshops centered on identifying science gaps and priority research, monitoring needs, potential conservation delivery projects, and public outreach efforts that would further protect or enhance native fish populations. Results from the conservation planning workshops (Birdsong et al. 2018) provided a basic framework for mitigating threats and enhancing the watershed's native fish populations into the future.

Lower Colorado River Authority Water Management Plan and the Environmental Flows Process

Prior to 1985, there was no explicit recognition of environmental flows (instream flows and freshwater inflows to estuaries) under Texas water law. Water rights were issued, often in perpetuity, without regard to environmental flow needs. The 69th Texas Legislature amended both the TPWD Code and the Texas Water Code, creating for the first time in Texas an opportunity to address environmental flow issues within the context of water regulation and management. Section 12.0011 was added to the TPWD Code, explicitly stating that the TPWD's resource protection activities include providing recommendations to the Texas Department of Water Resources (predecessor agency of the Texas Water Development Board and Texas Water Commission) on scheduling of instream flows and freshwater inflows to Texas estuaries for the management of fish and wildlife resources. These changes led to influence by the TPWD in development of the LCRA WMP, which included recommending environmental flows for the lower Colorado River and Matagorda Bay.

Since 1987, special conditions to protect instream flows in rivers and freshwater inflows to estuaries have been incorporated into water right permits, including those managed by the LCRA. The state of Texas approved the first lower Colorado River WMP in 1989, and updates were made in 1991, 1992, 1999, 2010, and 2015 (Table 1). In 2018, the LCRA began the process of updating the 2015 WMP. As part of the WMP process, the LCRA convenes an advisory committee consisting of members representing diverse interests to provide input to the LCRA related to key aspects of the plan. In the past, committee members represented municipal and industrial customers, Highland Lakes' area businesses and residents, agriculture, and the environment. For the 1999, 2010, 2015, and 2018 revisions, the TPWD served on the advisory committee. Involvement in the WMP process is the result of the TPWD's recognized expertise in instream flow and freshwater inflow science, as well as the TPWD's legal status as a full party in any hearing on an application to store, take, or divert water. The TPWD has also been a party in several contested case hearings related to LCRA water right applications and WMP revisions. Through hearings and, perhaps more importantly, the negotiation process, the TPWD, the LCRA, and others have promoted advancements in the use of science and policy regarding environmental flows in the lower Colorado River.

Water management plan triggers and flow targets

Lake Buchanan and Lake Travis water supplies managed by the LCRA are divided into firm and interruptible water. Firm water is available even during a repeat of the worst drought on record. During water shortages, the supply of interruptible water, which is mostly used for agriculture, is reduced or cut off completely. Based upon the amount of water stored in Lakes Travis and Buchanan, the WMP establishes trigger levels for determining the amount of water provided to meet interruptible contracts and environmental flow needs. Identification of trigger levels are the subject of stakeholder and LCRA negotiations and have been adjusted with WMP revisions. The WMP also requires the LCRA to provide interruptible and firm water from the lakes to help meet the environmental needs of the LCR NFCA and Matagorda Bay. Actual water demands from Lakes Buchanan and Travis for firm water users continue to increase, and the amount of interruptible stored water available from the lakes is expected to continue to decrease into the future, which will likely further change the hydrology in the LCR NFCA.

The environmental flow science supporting the WMP evolved over time. An initial study of instream flow needs of the lower Colorado River by the TPWD and LCRA in the 1980s focused on identifying critical and target flows for the native fish community at several downstream locations (Mosier and Ray 1992). Critical minimum flows to protect water quality included seasonally adjusted target flows to provide suitable habitat conditions for Blue Sucker spawning. Subsistence and critical flows were classified as a noninterruptible demand to be met or exceeded at all times. Target flows, a schedule of flows that provides an optimal range of habitat complexity to support a well-balanced, native aquatic community, were identified by the study as streamflows to be maintained whenever water supplies were adequate. Target flows were adjusted monthly to emulate normal seasonal variations (Mosier and Ray 1992).

Instream flow targets for the current WMP (2015) are based on the most recent

Table 1. Changes in the Lower Colorado River Authority Water Management Plan for the lower Colorado River, 1989–2015.

Year	Instream flow targets	Bay and estuary freshwater inflow targets	Number of storage triggers/evaluation date(s)	Stored water for the environment	Science supporting environmental flow targets
1989[a]	5.66 m^3/s minimum mean monthly flow placeholder until instream flow studies completed	335,656,313 m^3 minimal annual flow, 10.62 m^3/s minimum mean monthly flow placeholder until freshwater inflow studies completed	None	30,837,946 m^3/year	Placeholder environmental flows used while studies conducted
1991[b]	No change from 1989	No change from 1989	1/Jan 1	No change from 1989	Ongoing instream flow and freshwater inflow studies
1992[c]	Monthly subsistence/critical flows at two locations; target flows at three locations; springtime Blue Sucker spawning flows	No change from 1989	1/Jan 1	39,224,723 m^3/year	Instream flow targets: Mosier and Ray 1992
1999[d]	No change from 1992	Monthly freshwater inflow target flows totaling 1,270,486,299 m^3/year, monthly critical flows	1/Jan 1	No change from 1992	Freshwater inflow targets: LCRA 1997
2010[e]	Minimum flow requirements at one location modified, attainment frequencies added	Intermediate freshwater inflow level = 150% critical freshwater inflow target added	1/Jan 1	41,198,294 m^3/year	Freshwater inflow targets: LCRA 2006, 2008 Instream flow targets: BIO-WEST, Inc. 2008

Table 1. Continued.

Year	Instream flow targets	Bay and estuary freshwater inflow targets	Number of storage triggers/evaluation date(s)	Stored water for the environment	Science supporting environmental flow targets
2015[f]	Subsistence, base-dry and base-average instream flow criteria at four locations; Blue Sucker springtime spawning flows to protect 90% of spawning habitat	2010 criteria replaced by levels of seasonal freshwater inflows as determined by 2008 Matagorda Bay Health Evaluation Study, ranging from 222,026,732 m^3/year to 1,406,169,301 m^3/year	2/Mar 1 and Jul 1	No change from 2010	Freshwater inflow targets: LCRA 2008 Instream flow targets: BIO-WEST 2008

[a] September 20, 1989 agreed order approving LCRA's Water Management Plan and amending certificate of adjudication Nos. 14-5478 and 14-5482.
[b] December 23, 1991 agreed order approving LCRA's Drought Management Plan.
[c] December 18, 1992 agreed order approving amendments to LCRA's Water Management Plan and Drought Management Plan.
[d] March 1, 1999 agreed order approving amendments to LCRA's Water Management Plan and Drought Management Plan.
[e] January 27, 2010 agreed order approving amendments to LCRA's Water Management Plan.
[f] November 18, 2015 agreed order approving amendments to LCRA's Water Management Plan.

instream flow studies available (BIO-WEST, Inc. 2008). Biological and physical data were collected and used to develop hydraulic, habitat, water quality, and sediment transport models to identify subsistence and base flow targets. The WMP includes three categories for instream flow targets (subsistence, base-dry, and base-average) for U.S. Geological Survey (USGS) streamflow gage locations at Austin, Bastrop, Columbus, and Wharton. Flow targets vary by month to emulate the natural seasonal variations of flow in the river. Subsistence flows are intended to maintain acceptable water quality conditions (e.g., dissolved oxygen at or above 5.0 mg/L) and aquatic habitats during drought conditions. Special consideration for Blue Sucker is reflected in the February and March instream flow targets at certain locations to help ensure that 90% of spawning habitat is maintained. The base flow targets provide habitat conditions and year-to-year variability in habitat type, amount, and distribution needed to support a sound ecological environment, with base-dry targets met about 80% of the time and base-average targets met about 60% of the time.

The WMP also includes five categories for freshwater inflows to Matagorda Bay. Freshwater inflow targets and threshold values were determined using the latest data and science (LCRA 2008). Computer models and data analysis were used to develop relationships between freshwater inflow, bay salinity, habitat, species abundance, and nutrient supply. Freshwater inflow targets were designed to cover the full range of inflow conditions into Matagorda Bay, with a regime that incorporates five levels of inflow, each with an associated desired achievement guideline. The lowest level is a fixed monthly threshold value that would ideally be achieved 100% of the time to provide refuge habitat conditions during periods of extreme low flow. The remaining levels represent different inflow targets with varying levels of achievement intended to mimic natural patterns of inflows and include seasonal variability. The WMP environmental flow requirements in place from March through June are based on the amount of water stored in Lakes Buchanan and Travis on March 1, and the environmental flow requirements in place from July through the following February are based on storage on July 1.

Environmental flows process

Another important water policy process for the Colorado River watershed occurred when the 80th Texas Legislature in 2007 passed Senate Bill 3 (SB3), which set out a comprehensive, statewide approach for reserving unappropriated surface water necessary to protect environmental flow regimes in Texas rivers and bays by establishing environmental flow standards to be applied to newly issued or amended water rights (Loeffler 2015). Environmental flow regimes were developed and recommended by expert science teams working with technical support from state agencies. The legislation specifically mandated that expert science team recommendations were to be based solely on best available science. For this process, an environmental flow regime was defined as "a schedule of flow quantities that reflects seasonal and yearly fluctuations that typically would vary geographically, by a specific location in a watershed, and that are shown to be adequate to support a sound ecological environment and to maintain the productivity, extent, and persistence of key aquatic habitats in and along the affected water bodies." Given the relatively short timelines provided and the directive to the science teams to base their work on existing science, reliance upon USGS gaged streamflow data became critical given these data are the most widely available and systematically measured data available for rivers. The Hydrology-based Environmental Flow Regime tool was developed during the SB3 process by a team of scientists that included the TPWD, with guidance from a science advisory committee (Opdyke et al. 2014). It provided an efficient method to use USGS daily streamflow data to quickly calculate relevant statistics and populate an environmental flow regime matrix for multiple control points in a watershed.

Once the expert science team recommendations were presented, each basin stakeholder committee reviewed the findings and then recommended its own set of environmental flow regimes to the Texas Commission on Environmental Quality (TCEQ). In some cases, the science committee recommendations were forwarded with only some modification to the recommendation of the stakeholder committees, but in other cases, the stakeholders did make significant changes before submitting their final recommendations. After receiving recommendations from both the science committees and stakeholder committees, the TCEQ drafted proposed environmental flow standards rule packages that were released for review and comment through a formal rulemaking process. The end result of the formal rulemaking process was a set of adopted environmental flow standards for select locations. The TCEQ may not issue new water right permits or amend water right permits if the issuance of the permit will impair environmental flows as established by rule. Adopted environmental flow standards and implementation strategies are subject to "adaptive management," meaning that the success and/or failure of management measures will be assessed and adjusted as new science and information becomes available. TPWD has been heavily involved in all phases of the process, developing tools, providing technical assistance, and offering review comments during the development of environmental flow standards.

In crafting the environmental flow standards for the lower Colorado River and Matagorda Bay, the same studies that were used in developing the environmental flow requirements for the LCRA WMP were used for making flow recommendations. Not surprisingly, the recommendations and resulting flow standards for the most part mirror the WMP flow requirements. Having been involved in discussions and recommendations during the planning and data analyses for the latest instream flow study (BIO-WEST, Inc. 2008) and freshwater inflow study (LCRA 2008), the TPWD supported the use of the reports and their background science in the creation of the environmental flow standards for the lower Colorado River and Matagorda Bay.

Adaptive Management and Filling Science Gaps for Focal Fish Species

Even though instream flow science has been evolving since the 1980s, the need to gain a better understanding of effects of altered flows became more apparent in the early to mid-2000s. During the most recent severe drought from 2011 to 2015 (Winters 2013), the two water supply reservoirs in the Highland Lakes chain, Lakes Buchanan and Travis, dropped to 34% of their storage capacity, a level not seen since the 1950s drought of record for Texas. In 2014 as the drought lingered, the LCRA sought, and was granted by the TCEQ, emergency relief regarding environmental flow requirements for Blue Sucker. Prescriptive flows for Blue Sucker were reduced to 8.49 m^3/s, which decreased spawning habitat (bedrock outcrops, boulders, and cobble riffles) by 7–14% (TCEQ 2015). Given the severity of this drought and the potential for more frequent severe droughts as a result of climate change (U.S. Global Change Research Program 2018), the TPWD was interested in providing science-based recommendations for maximizing the efficiency of prescribed flows for Blue Sucker, including the timing, duration, and magnitude of flows needed for successful spawning and recruitment. This information could then be provided to the LCRA when developing environmental flow recommendations for future versions of the WMP.

Gaining a better understanding of effects of hydrology on reproduction and recruitment became an area of emphasis for both Guadalupe Bass and Blue Sucker research by the TPWD and partners starting in 2011. For both species, the TPWD funded collaborative studies using State Wildlife Grants with the USGS Texas Cooperative Fish and Wildlife Research Unit and Texas Tech University. For Guadalupe Bass, assessing the effects of anthropogenic disturbance throughout the watershed and how changing land use and hydrology might affect population dynamics were important deliverables (Pease et al. 2017). Key findings included the following: (1) growth and recruitment showed a positive correlation with increased base flows and mean monthly flows, but long-term monitoring was recommended as a means to confirm or refine the relationship; (2) reductions in streamflow in the Colorado River watershed would likely result in lower Guadalupe Bass growth rates with the potential to impact the size-structure of populations; and (3) conservation initiatives solely focused on improving physical instream or riparian habitat were unlikely to be as beneficial as those focused on restoring or maintaining adequate streamflow (Pease et al. 2017). Although a specific flow rate was not provided, the authors recommended water managers consider increasing releases to the LCR NFCA immediately prior to and following Guadalupe Bass spawning as a means to preserve the population's trophy-size structure. Future research to document long-term Guadalupe Bass recruitment trends under a wider range of flows to confirm and refine recommendations from the study is a priority for the TPWD.

Another study assessed the timing and extent of streamflow on spawning habitat use and recruitment in the LCR NFCA for Blue Sucker. Specific objectives for the study were to (1) estimate population size, (2) identify the effects of streamflow on population demographics with an emphasis on recruitment, (3) describe habitat use and selection, and (4) elucidate movement, including spawning migrations, and spawning habitat use under various releases. Results could be used to refine flow release strategies from the Highland Lakes or for recommending other conservation strategies for Blue Sucker. Important findings included the following: (1) a small population size within the 292 river kilometer study area, estimated within a range of only 449–1,089 individuals; (2) the frequency of extreme low summer discharges (<4.3 m^3/s) had significantly increased since the first Highland Lakes dams were built, and modeled results suggested higher summer discharges and frequency of high flow pulses (≥80.99 m^3/s as calculated at the Bastrop USGS gage) could improve year-class strength; (3) spawning habitat appeared limited in the study area and a potentially vital spawning area upstream of the city of Bastrop was identified which could be targeted for protection; and (4) the population has been decreasing since 2009 (Acre 2019). Given the population trend since 2009, continued monitoring of the population size was recommended to determine if conservation efforts, such as a captive propagation program and stocking or construction of additional spawning areas, would be needed to sustain the population (Acre 2019).

Recognizing the need to further inform adaptive management of flow recommendations, additional study of flow-ecology relationships for Guadalupe Bass and Blue Sucker was one of three high-priority science needs identified during the Colorado River watershed NFCA workshops held in 2015 (Birdsong et al. 2018). A better understanding of the potential effects of altered hydrology on these species was needed to recommend and implement management practices (e.g., recommended flows in future versions of the LCRA WMP) to mitigate for these effects. Documenting a population dynamics response to altered hydrology, and recommendations to mitigate for the alteration, was a desired outcome of research projects. It was also surmised that responses of these focal species from an altered flow regime might apply to other native fish species with similar life-history strategies (i.e., a guild-level response) (Grabowski 2014a, 2014b).

For Guadalupe Bass, increased extreme low flows and increased frequency of high flow pulses during their spawning period were hypothesized to have potential negative effects on growth, and reproductive success, respectively (Grabowski 2014a). In the CEPR NFCA, decreased annual base flow resulting from groundwater depletion and increased high flow pulses during the spawning season as a result of increased impervious surfaces in the watershed (Pease et al. 2017; Pease et al. 2018) could decrease growth (Groeschel 2013) and reproductive success (Grabowski 2014a). In the LCR NFCA, where a sport fishery for trophy Guadalupe Bass exists, extended low flows could decrease growth rates (Groeschel 2013; Pease et al. 2017). Guadalupe Bass are relatively short-lived (<6 years; Pease et al. 2017) and reduced growth rates could potentially shift the population structure towards smaller individuals, reducing the trophy size-structure of the fishery. Gaining a thorough understanding of Guadalupe Bass response to the effect of anthropogenic disturbances, including increased urbanization on streamflow in the watershed, would further assist in implementing compensatory on-the-ground conservation actions.

The fall 2015 NFCA workshops also identified threats, knowledge gaps, and research needs for catostomids in the lower Colorado River (Birdsong et al. 2018). For example, low flows during the spawning season were thought to reduce availability or quality of spawning habitats (Grabowski and Isely 2007a, 2007b) and depress growth rates or recruitment due to a resultant lack of access to off-channel habitats (Dutterer et al. 2012; Grabowski et al. 2012) while high flows resulting from hydropower peaking could lead to mortality of early life history stages (Weyers et al. 2003). Though adult Blue Suckers were commonly collected, few juveniles had historically been collected from the lower Colorado River (e.g., see Acre 2019). Little was known about the specific early life history needs of the species (Mosier and Ray 1992) and under what conditions strong year-classes were produced (Acre 2019). Blue Sucker is a relatively long-lived species (up to 22 years; Vokoun et al. 2003), making populations potentially resilient to year-class failure, but it was hypothesized that chronic year-class failure could eventually lead to their extirpation from the LCR NFCA. While a specific instream flow recommendation (i.e., 14.1 m^3/s for a 6-week period between March and May; see Mosier and Ray 1992; BIOWEST, Inc. 2008) to maintain spawning habitat was incorporated into the LCRA WMP, there had been no evaluation to determine if the recommendation had any effect on producing strong year-classes. While the recent research project conducted by the USGS Texas Cooperative Fish and Wildlife Research Unit and Texas Tech University on the Blue Sucker population in the lower Colorado River (Acre 2019) provided some recommendations for potentially improving the population, streamflow and water temperature only partially explained when, and possibly if, Blue Sucker make spawning migrations. Gaining a more thorough understanding of how a long-lived species such as Blue Sucker respond to flow conditions will require additional research over a longer time frame. Continuing to gather Blue Sucker spawning and recruitment data in the lower Colorado River under varying flow regimes, and investigating the importance of tributaries for spawning are future priority research needs.

Central Edwards Plateau Rivers Native Fish Conservation Area Watershed Enhancement

Through the Guadalupe Bass Restoration Initiative (GBRI; Garrett et al. 2015), the TPWD has been working with landowners, local communities, and conservation partners in the CEPR NFCA to address anthropogenic threats on a landscape scale to benefit riparian and aquatic resources. Healthy riparian areas are needed for functioning and healthy headwater stream systems as they provide bank stabilization, erosion control, sediment filtration, stream shading, and inputs for nutrient and food web cycling. Riparian areas became a focus of private landowner education and on-the-ground restoration efforts. With more than 95% of the land in Texas privately owned, it was important for the TPWD and partners to build relationships and work with landowners to implement upland and riparian conservation and restoration initiatives. Fish and wildlife biologists at the TPWD identified landowners with riparian, spring, creek, and river frontage to provide technical guidance on land management practices, work with adjoining landowners to implement watershed-scale conservation goals, and develop specific recommendations for restoration and enhancement projects.

Through funds provided by the National Fish and Wildlife Foundation Native Black Bass Initiative, the TPWD was able to offer financial incentives to landowners for implementing restoration, enhancement, and preservation projects to benefit aquatic resources. Common threats to upland, riparian, and instream habitat encountered during site visits included poor land management practices (i.e., continuous grazing or mowing of riparian areas), alteration of habitat by invasive species, and no riparian vegetation recruitment due to excessive browsing by high-density populations of native and exotic ungulates.

To achieve conservation and restoration goals for aquatic systems, biologists also had to address habitat problems arising in the uplands, as well as targeting issues in riparian and aquatic areas. Specific restoration projects included restoring grasslands immediately adjacent to important springs and spring complexes, implementing rotational grazing practices in upland habitats, constructing riparian fencing to exclude cattle and feral hogs, building small exclosures to protect high-priority riparian species from browse, improving water quantity and quality by improving and increasing riparian buffers, and planting a diversity of riparian plant species (many that have been grazed or browsed out of the area). Landowners agreed to long-term monitoring of restoration sites through surveys, fixed photo points, and time-lapse cameras. Time-lapse photography was especially valuable for talking about land management strategies and demonstrating the effects of restoration projects to landowners. Momentum and experience gained through the GBRI was used to continue work with private landowners in other portions of the CEPR and LCR NFCAs.

Another critical component to engaging communities and landowners through the GBRI has been working with local, grassroots, stakeholder-driven watershed groups. The Llano River Watershed Alliance is one such group that was started by local landowners and community partners. Their mission is to "preserve and enhance the Llano River and adjoining watersheds by encouraging land and water stewardship through collaboration, education, and community participation." Two reports were created in 2008 to provide a biological overview of the South Llano River, which initiated engagement of local landowners and stakeholders in an organized way to discuss impacts and future needs of the watershed. A community-driven group was formed from workshops used in this process and they started the South Llano Watershed Alliance. The group gained nonprofit status and has since expanded their footprint; the group was renamed the Llano River Watershed Alliance to reflect expansion of the geography covered. Another organization active in the CEPR NFCA is the Hill Country Alliance. The Hill Country Alliance covers a much broader geography in the Edwards Plateau ecoregion connecting communities and landowners of many of the Colorado River headwater streams. Their mission statement is "to raise public awareness and build community support around the need to preserve the natural resources and heritage of the Central Texas Hill Country." Both the Llano River Watershed Alliance and Hill Country Alliance have been active partners with the TPWD for engaging landowners on a range of issues impacting land management, land conservation, river and stream restoration and conservation, and public engagement and outreach. Community outreach was an emphasis of both organizations as they hosted workshops to engage landowners and stakeholders on a wide range of issues deemed important by the TPWD, including land management, wildfire recovery, riparian management, streambank stabilization, and river clean-up work days. Continuing to work with the Llano River Watershed Alliance, the Hill Country Alliance and other groups to engage landowners build landowner and community groups to enhance land conservation goals, host workshops to provide technical guidance, and engage local groups interested in conservation continues to be a priority for the TPWD in both Colorado River NFCAs.

As a result of one-on-one technical guidance for landowners, and implementation of restoration projects, during the GBRI, the TPWD and partners delivered the following conservation outcomes: (1) 25 springs were restored, preserved or protected; (2) 87 km of stream and riparian areas were conserved; (3) 3,138 ha of spring, stream, riparian, and upland habitats were restored; (4) conservation best management practices were provided for ranches on 31,565 ha in the Llano River watershed; (5) 747 landowners were engaged through technical site visits, habitat restoration projects, workshops, and Llano River Watershed Alliance and Watershed Protection Plan events; (6) 64 articles, scientific publications, reports, presentations, and research projects were completed; and (7) 618,108 fingerlings were stocked in the South Llano and Llano rivers for genetic restoration of Guadalupe Bass populations (Fleming et al. 2015), resulting in a decrease in hybridization levels from 3.9% to approximately 1% (Bean 2017). In addition, an invasive species treatment program targeting giant cane *Arundo donax* was also implemented, with a long-term study initiated to evaluate the effects of herbicide treatments on fish, aquatic invertebrates, and stream geomorphology.

Increasing Public Stewardship

Increasing public stewardship for the Colorado River watershed by providing increased opportunity for recreational use by improving public access was also part of the overall strategy for conserving native fishes in both NFCAs. Availability and ease of access have consistently been identified as an impediment to the public's river recreational use of rivers in Texas (Baker 1998; Thomas and Arsuffi 2015; Thomas et al. 2015). With more than 95% of the land in Texas privately owned and the state's population expected to potentially more than double by 2050 (Hoque et al. 2014), the public's access to land and waters for recreational use, especially areas close to major urban areas where demand is greatest, is in increasingly short supply. Increasing access opportunities for the public to make a personal connection with the river might increase the likelihood these individuals would take the initiative to contact legislators about policies aimed at conserving it. Developing a nonregulatory approach for the public to learn about and explore a natural resource, rather than dictating behavior (e.g., government-mandated water restrictions), may be far more effective at inspiring public concern for protecting the environment (Kaplan 2000). In addition, the environmental concerns people develop are associated with the degree in which they view their interconnectedness and inclusion with the natural world (Schultz 2000), and there is a positive correlation between connecting with nature and endorsement of pro-environmental attitudes (Nisbet et al. 2009) and greater willingness to engage in environmentally sustainable behavior (Zelenski et al. 2015).

Improved access also had the potential to provide direct economic impact to communities along the river as nonlocal anglers and paddlers visited the area. Economic impact studies of recreational users are often initiated by the TPWD as a means to help better inform community leaders of economic benefits of that use when there are multiple uses of a natural resource. For example, both Colorado River NFCAs have popular Guadalupe Bass fisheries. Fishing for this species generated US$30 million in economic impact and supported 325 full-time jobs in a 24-county region of Texas, including counties in both Colorado River NFCAs (Thomas et al. 2015).

To improve river access, the TPWD developed the Texas Paddling Trails Program in 1998 and by 2018 had established 54 trails on inland waters. The trails provide well-mapped and accessible day trips for individuals with all levels of paddling experience. Each trail requires a local community partner and put-in and take-out points at already established public access points (e.g., boat ramps or public parks). Four paddling trails were developed in the Colorado River NFCAs, which opened up 53 km of the river for public use, but there were still reaches in both NFCAs that had no public access or were too long to paddle in one day.

Beyond the Paddling Trails Program, the ability to expand the amount of low-cost outdoor recreational opportunity on the Colorado River centered on efforts to secure public-access rights across private lands. In 2011, the TPWD developed the River Access and Conservation Area Program (RACA; Birdsong et al. 2019a). The program focused on expanding recreational access for the public to rivers surrounded by private lands, as well as providing landowners with information about their resources and prescriptive guidelines to help them sustainably manage them. Landowners entered into lease contracts with the TPWD that allowed the public to use their property for gaining access to rivers. In return for this access, the TPWD provided a monthly payment to the landowner. Leases were partially funded by the U.S. Department of Agriculture Volunteer Public Access-Habitat Incentive Program, which is a competitive grants program that helps state and tribal governments increase public access to private lands for wildlife-dependent recreation such as hunting, fishing, nature watching, or hiking. During the course of these agreements, a variety of improvements were also completed on the landowner's property, including development of improved parking areas, trash clean-ups, installation of informational kiosks that promoted riparian conservation, and installation of fencing and gated entries to prevent trespassing and separate cattle and other livestock from fishing areas and sensitive habitats. Resource conservation plans were developed for most of the RACA sites, which included information on site utilization, basic creel statistics, landowner satisfaction with the program, a baseline assessment of fish and the riparian area, and recommendations for access and habitat improvements. Lease agreements were initiated in 2016 with three private landowners to provide additional public access to the LCR NFCA between Austin and Smithville, Texas. In the CEP NFCA, lease agreements were initiated with landowners on the South Llano and Llano rivers. Three of the access sites had good fisheries for Largemouth Bass *Micropterus salmoides*, Guadalupe Bass, and catfishes *Ictalurus* spp. (Magnelia et al. 2003; Cummings and De Jesus 2017). The fourth site located on the main-stem Llano River upstream of one of the Highland Lakes, Lake Lyndon Baines Johnson, allowed bank and wade anglers access to a popular spawning run of White Bass *Morone chrysops*. These five leases opened up approximately 66 km of additional public access for anglers and paddlers.

Expansion of the Texas Paddling Trails and RACA programs in both Colorado River NFCAs and throughout the state continues to be a high priority for the TPWD. Given the general lack of access to rivers in Texas, the public's opportunity for hands-on interaction with these natural resources has been limited. Beyond recreation, rivers and streams provide many other vital ecosystem services to society, including the supply of water for drinking, irrigation, and hydropower generation (Postel and Thompson 2005), of which the public may not be aware of. For example, in a 2014 survey commissioned by the Texas Water Foundation found that only 28% of Texans could identify the natural source of their drinking water (Baselice & Associates Inc. 2014). While the TPWD makes the assumption that improved access will increase stewardship for the river and awareness of the benefits of healthy river ecosystems to society, no evaluation of these efforts has been undertaken. As both the Texas Paddling Trails and RACA programs continue to expand, an evaluation to validate this assumption is needed.

Leveraging Funding

Mitigating threats to native fish species in the Colorado River watershed will likely become more difficult in the future as more pressure is placed on water resources. Obtaining additional funding for conservation is a key component for mitigating those effects. Realizing this, a strategy employed by the TPWD was to align TPWD efforts with federal land and wildlife conservation funding programs. Being directly involved with these federal programs or making federal staff aware of state aquatics initiatives or geographical focus areas such as NFCAs has often successfully led to additional or enhanced funding for research or conservation delivery. For example, in 2010, TPWD staff promoted Guadalupe Bass as the focal species for headwater streams in the Gulf Coast Prairie Landscape Conservation Cooperative (GCP LCC) geography. A research project to further study Guadalupe Bass flow-ecology relationships with an emphasis on the impact of flow on recruitment was developed and eventually funded by the GCP LCC. This project supplemented similar flow-ecology research administered by the TPWD using State Wildlife Grant funds (e.g., Pease et al. 2017). Recognizing the importance of the Colorado River watershed to fish and wildlife conservation, the GCP LCC also funded a conservation planning project that included the LCR NFCA, which provided guidance for conservation delivery that might benefit both terrestrial and aquatic species. Another research project funded through a State Wildlife Grant, building on a similar project funded by the Great Plains Landscape Conservation Cooperative, explored the development of a decision support tool to identify priority opportunity areas and voluntary strategies such as water leases, water rights acquisition, and incentive-based programs to achieve flow restoration targets for conservation of native fishes. Protection of riparian buffers using landowner incentives was identified as a high priority in the Colorado River NFCA workshop (Birdsong et al. 2018). While the Landowner Incentive Program was the funding mechanism in which the TPWD provided financial incentive for landowners to enact land conservation best management practices (BMPs) for enhancing watershed habitat in the CEPR NFCA, more robust financial assistance became available through the U.S. Department of Agriculture Regional Conservation Partnership Program (RCPP). This federal program encourages partnerships to increase restoration and sustainable use of natural resources on regional or watershed scales. In 2018, the TPWD provided partner support for a RCPP project initiated by the Hill Country Conservancy. The Hill Country Headwaters Conservation Initiative will provide up to $5.15 million in direct payments for habitat restoration and conservation easements to landowners in the Llano and Pedernales watersheds of the CEPR NFCA and the Onion Creek watershed in the LCR NFCA. Under the initiative, U.S. Department of Agriculture Natural Resources Conservation Services staff will work with landowners to implement Environmental Quality Incentives, Agricultural Conservation Easement, and Conservation Stewardship program funding for conservation easements, as well as the installation of land management BMPs, such as brush control and enhancement of riparian buffers, for improving water quantity and quality. The leveraging of TPWD resources with federal landowner assistance funding programs, which often have much greater financial capacity relative to state programs, and alignment of the NFCAs with federal land management program focus areas, is an ongoing priority for the TPWD.

The use of conservation easements in the Colorado River NFCAs was also identified as a high priority for protecting riparian areas and spring systems. Conservation easements restrict future use of private property to protect its natural, productive, or cultural features, including limiting the right to develop or subdivide the property. While use of this conservation tool is not a new concept in Texas, the emphasis on conserving lands for the purpose of protecting water resources seems to have only recently gained traction in many areas of the state. The burgeoning demand for water as the population grows in Texas and the severe drought from 2011 to 2015 (Winters 2013) may have played a role in increasing awareness of the importance of water resources. In 2016, the TPWD was provided increased capacity to fund conservation easements with development of the Texas Farm and Ranch Lands Conservation Program (https://tpwd.texas.gov/landwater/land/private/farm-and-ranch/). The program complements the TPWD's mission to conserve natural resources by protecting working lands from fragmentation and development. As part of the program, the TPWD aquatics staff modified scoring criteria for these grants to include consideration for protecting headwater systems, springs, creeks, wetlands, riparian buffers, and aquifer recharge features (i.e., watershed value). In 2018, the TPWD awarded four conservation easements in the LCR NFCA totaling 720 ha, in part for their contribution to enhancing water quality, water quantity, and overall habitat conditions for native fishes.

Conclusions

Water-resource decision makers are faced with the need to balance competing water needs, which becomes increasingly difficult with growing demands and a changing climate. Science-based recommendations are critical for informing those decisions. There is a critical need to continue ongoing research to better understand and quantify instream flow needs, especially for focal species such as Guadalupe Bass and the state threatened Blue Sucker. Developing innovative strategies for maintaining water quantity in unregulated streams, which in many cases are already overappropriated, such as leasing or buying or voluntary nonuse of water rights will likely be needed to meet flow requirements for many native fish species. Species that are fluvial specialists will be especially vulnerable, and extirpation or extinction is certainly possible. Given the increasing future demand predicted for water from the Highland Lakes, innovative strategies to supplement environmental flow releases in the LCR NFCA will likely be needed. Releases that are designed to maintain human water needs, support positive ecological outcomes, and promote native fish species rather than mimic the natural flow regime (Chen and Olden 2017) could also be considered. For example, when water is available, releases from the Highland Lakes that provides a higher frequency of high-flow pulses during the summer may fulfill downstream water obligations as well as promote successful recruitment of Blue Sucker. Given the low population size estimated for this species in the LCR NFCA and based on the decreasing population trend from 2009 to 2014, this species may reach extinction in 25 years (Acre 2019) without changes in releases from the Highland Lakes.

Since the TPWD has no regulatory authority to set land management policies that often directly affect water quantity and quality, the need exists for continued and expanded promotion of voluntary land-use strategies as urban areas in central Texas continue to grow. For example, recommending building practices that minimize impervious surfaces, locating green space that maximizes riparian buffers, and promoting voluntary water conservation could help ameliorate effects of increased population growth and urbanization on aquatic species. A concerted outreach effort by the TPWD to communicate to urban planners recommendations that protect aquatic species will likely be required.

While the TPWD's Wildlife Division has been working with private landowners for decades to manage terrestrial game species, aquatic resources were rarely considered in land management plans. Collaboration between terrestrial and aquatics agency biologists to develop a more holistic approach to land management within the TPWD was one of the most valuable outcomes of the GBRI. The CEP NFCA was the first area where the TPWD focused on and implemented a watershed conservation approach. Lessons learned are now being expanded to other NFCAs across the state to implement conservation on a watershed scale. A continued focus by the TPWD on nonregulatory approaches to watershed management, such as providing science on the effects of flows on native aquatic species to water authorities, providing land management BMPs and technical support to municipalities and landowners, and promoting voluntary and/or incentive-based land conservation practices, will likely continue to be the path forward for conservation of native fishes in the Colorado River watershed and elsewhere in Texas.

Acknowledgments

Funding to support completion of this manuscript was provided through the Federal Aid in Sport Fish Restoration Act. The authors acknowledge TPWD River Studies Program staff cartographer Karim Aziz for producing the manuscript figure, two anonymous reviewers who provided valuable input for improving early drafts, and TPWD staff members Dakus Geeslin and Timothy Birdsong who provided photos of the lower Colorado River.

References

Acre, M. R. 2019. Assessing demography, habitat use, and flow regime effects on spawning migrations of Blue Sucker in the lower Colorado River, Texas. Doctoral dissertation. Texas Tech University, Lubbock.

Adams, S. R., M. B. Flinn, B. M. Burr, M. R. Whiles, and J. E. Garvey. 2006. Ecology of larval Blue Sucker (*Cycleptus elongatus*) in the Mississippi River. Ecology of Freshwater Fish 15:291–300.

Baker, T. L. 1998. A study of Texas rivers with attention to river access and recreational fisheries. Undergraduate honors thesis. Texas A&M University, College Station.

Baselice & Associates Inc. 2014. Texas statewide water conservation survey. Available: www.texaswater.org/wp-content/uploads/2014/09/Texas-Statewide-Water-Conservation-Survey.pdf. (August 2018).

Bean, P. T., D. J. Lutz-Carrillo, and T. H. Bonner. 2013. Rangewide survey of the introgressive status of Guadalupe Bass: implications for conservation and management. Transactions of the American Fisheries Society 142:681–689.

Bean, P. 2017. Guadalupe Bass conservation plan: a ten-year plan for restoring and preserving the state fish of Texas 2017–2026. Texas Parks and Wildlife Department, Austin.

BIO-WEST, Inc. 2008. Lower Colorado River, Texas instream flow guidelines. Colorado River flow relationships to aquatic habitat and state threatened species: Blue Sucker. Available: http://environmental-stewardship.org/wp-content/uploads/2012/04/Exhibit12-BIO_LSWP_IFguidelines_FINAL.pdf. (May 2018).

Birdsong, T. W., J. Botros, S. Magnelia, J. Anderson, M. Bean, T. Broad, D. Cortez, T. Grabowski, C. Kowaleski, C. Chute-Canal, J. East, K. Glenewinkel, B. Hester, R. Husted, J. Joplin, J. Lewey, S. Nichols, D. Oppenheimer, M. Parker, S. Robertson, and A. Stevens. 2019a. Texas River Access and Conservation Areas Program: partnering with private landowners to expand paddling and fishing opportunities on Texas rivers. Texas Parks and Wildlife Department, Austin.

Birdsong, T. W., D. C. Dauwalter, G. P. Garrett, B. J. Labay, M. Bean, J. Broska, J. Graham, S. Magnelia, K. B. Mayes, M. McGarrity, K. M. Johnson, S. Robertson, T. Thompson, S. Vail-Muse, and J. B. Whittier. 2018. Native fish conservation areas of the southwestern USA: facilitating landscape-scale conservation of aquatic habitats and freshwater fishes. Southeast Aquatic Resources Partnership, Panama City, Florida.

Birdsong, T. W., G. P. Garrett, B. J. Labay, M. G. Bean, P. T. Bean, J. Botros, M. J. Casarez, A. E. Cohen, T. G. Heger, A. Kalmbach, D. A. Hendrickson, S. J. Magnelia, K. B. Mayes, M. E. McGarrity, R. McGillicuddy, M. M. Parker, and S. Robertson. 2019b. Texas Native Fish Conservation Areas Network: strategic investments in restoration and preservation of freshwater fish diversity. Pages 183–229 *in* D. C. Dauwalter, T. W. Birdsong, and G. P. Garrett, editors. Multispecies and watershed approaches to freshwater fish conservation. American Fisheries Society, Symposium 91, Bethesda, Maryland.

Bowles, D. E., and T. L. Arsuffi. 1993. Karst aquatic ecosystems of the Edwards Plateau region of central Texas, USA: a consideration of their importance, threats to their existence, and efforts for their conservation. Aquatic Conservation: Marine and Freshwater Ecosystems 3:317–328.

Chen, W., and J. D. Olden. 2017. Designing flows to resolve human and environmental water needs in a dam regulated river. Nature Communications 8:2185.

Clay, C., and D. Kleiner. 2010. Colorado River handbook of Texas online. Texas State Historical Association, Austin. Available: https://tshaonline.org/handbook/online/articles/rnc10. (May 2018).

Cohen, A. E., G. P. Garrett, M. J. Casarez, D. A. Hendrickson, B. J. Labay, T. Urban, J. Gentle, D. Wiley, and D. Walling. 2018. Conserving Texas biodiversity: status, trends, and conservation planning for fishes of greatest conservation need. Texas Parks and Wildlife Department, Final Report, Austin.

Craig, C. A., K. A. Kollaus, K. P. K. Behen, and T. H Bonner. 2016. Relationships among spring flow, habitats, and fishes within evolutionary refugia of the Edwards Plateau. Ecosphere [online serial] 7(2):e01205.

Cummings, G. 2018. Proposed standard length categories for Guadalupe Bass. Texas Parks and Wildlife Department, Management Data Series No. 293, Austin.

Cummings, G., and M. De Jesus. 2017. Survey report for the Colorado River Hwy 183 bridge crossing, Austin to SH 71 bridge crossing, La Grange Texas, 2003. Texas Parks and Wildlife Department, Austin.

Curtis, S. G., J. S. Perkin, P. T. Bean, M. L. Sullivan, and T. H. Bonner. 2015. Guadalupe

Bass *Micropterus treculii* (Vaillant & Bocourt, 1874). Pages 55–60 *in* M. D. Tringali, J. M. Long, T. W. Birdsong, and M. S. Allen, editors. Black bass diversity: multidisciplinary science for conservation. American Fisheries Society, Symposium 82, Bethesda, Maryland.

Dutterer, A. C., C. Mesing, R. Cailteux, M. S. Allen, W.E. Pine, and P. A. Strickland. 2012. Fish recruitment is influenced by river flows and floodplain inundation at Apalachicola River, Florida. River Research and Applications 29:1110–1118.

Edwards, R. J. 1978. The effect of hypolimnion reservoir releases on fish distribution and species diversity. Transactions of the American Fisheries Society 107:71–77.

Edwards, R. J. 1980. The ecology and geographic variation of the Guadalupe Bass, *Micropterus treculi*. Doctoral dissertation. University of Texas, Austin.

Fleming, B. P., G. P. Garrett, and N. G. Smith. 2015. Reducing hybridization and introgression in wild populations of Guadalupe Bass through supplemental stocking. Pages 537–548 *in* M. D. Tringali, J. M. Long, T. M. Birdsong, and M. S. Allen, editors. Black bass diversity: multidisciplinary science for conservation. American Fisheries Society, Symposium 82, Bethesda, Maryland.

Garrett, G. P., T. W. Birdsong, M. G. Bean and R. McGillicuddy. 2015. Guadalupe Bass restoration initiative. Pages 379–385 *in* M. D. Tringali, J. M. Long, T. M. Birdsong and M. S. Allen, editors. Black bass diversity: multidisciplinary science for conservation. American Fisheries Society, Symposium 82, Bethesda, Maryland.

Grabowski, T. B., and J. J. Isely. 2007a. Effects of flow fluctuations on riverine fish spawning habitat. Southeastern Naturalist 6:471–478.

Grabowski, T. B., and J. J. Isely. 2007b. Spatial and temporal segregation of spawning habitat by catostomids in the Savannah River, Georgia and South Carolina, U.S.A. Journal of Fish Biology 70:782–798.

Grabowski, T. B., S. P. Young, J. J. Isely, and P. C. Ely. 2012. Age, growth, and reproductive biology of catostomid species from the Apalachicola River, Florida. Journal of Fish and Wildlife Management 3:223–237.

Grabowski, T. 2014a. Guadalupe Bass (*Micropterus treculii*). Pages 42–48 *in* M. Davis and S. Brewer, editors. Gulf Coast Prairie Landscape Conservation Cooperative regional hypotheses of ecological responses to flow alteration. A report by the GCP LCC Flow-Ecology Hypotheses Committee to the Southeast Aquatic Resources Partnership. Wildlife Management Institute, Washington, D.C.

Grabowski, T. 2014b. Suckers (particularly Moxostoma spp.). Pages 56–63 *in* M. Davis and S. Brewer, editors. Gulf Coast Prairie Landscape Conservation Cooperative regional hypotheses of ecological responses to flow alteration. A report by the GCP LCC Flow-Ecology Hypotheses Committee to the Southeast Aquatic Resources Partnership. Wildlife Management Institute, Washington, D.C.

Groeschel, J. 2013. Evaluations of growth and habitat use by Guadalupe Bass at the riverscape scale in the South Llano River, Texas. Master's thesis. Texas Tech University, Lubbock.

Hill Country Alliance. 2008. A look at the Texas Hill Country: following the path we are on today through 2030. Available: www.hillcountryalliance.org/uploads/HCA/GrowthScenario2.pdf. (May 2018).

Hoque, N., C. McNeill, and J. Granato. 2014. Projections of the population of Texas and counties in Texas by age, sex, and race/ethnicity from 2010 to 2050. University of Houston Hobby Center for Public Policy, Houston. Available: www.uh.edu/class/hobby/_docs/research/population/2014%20PPRLE-SV2.pdf. (May 2018).

Hubbs, C., R. J. Edwards and G. P. Garrett. 2008. An annotated checklist of the freshwater fishes of Texas, with keys to identification of species. Texas Academy of Science. Available: www.texasacademyofscience.org (December 2018).

Hurst, H., G. Bass, and C. Hubbs. 1975. The biology of the Guadalupe, Suwanee, and Redeye basses. Pages 47–53 *in* R. Stroud and H. Clepper, editors. Black bass biology and management. Sport Fishing Institute, Washington, D.C.

Jelks, H. L., S. J. Walsh, N. M. Burkhead, S. Conteras-Balderas, E. Diaz-Pardo, D. A. Hendrickson, J. Lyons, N. E. Mandrak, F. McCormick, J. S. Nelson, S. P. Platania, B. A. Porter, C. B. Renaud, J. J. Schmitter-Soto, E. B. Taylor, and M. L. Warren, Jr. 2008. Conservation status of imperiled North American freshwater and diadromous fishes. Fisheries 33:372–407.

Kaplan, S. 2000. Human nature and environmentally responsible behavior. Journal of Social Issues 56:372–508.

LCRA (Lower Colorado River Authority). 1997. Freshwater inflow needs of the Matagorda Bay system. LCRA, Austin, Texas.

LCRA (Lower Colorado River Authority). 2006. Matagorda Bay freshwater inflow needs study. LCRA, Texas Commission on Environmental Quality, Texas Parks and Wildlife Department, Texas Water Development Board, Austin.

LCRA (Lower Colorado River Authority). 2008. Matagorda Bay health evaluation final report, Matagorda Bay inflow criteria (Colorado River). LCRA, Austin, Texas.

LCRA (Lower Colorado River Authority). 2015. Lakes Buchanan and Travis water management plan and drought contingency plans. LCRA, Austin, Texas. Available: www.lcra.org/water/water-supply-planning/water-management-plan-for-lower-colorado-river-basin/Documents/FINAL-WMP-AsApprovedbyTCEQ-Nov-2015.pdf. (April 2019).

Loeffler, C. 2015. A brief history of environmental flows in Texas. Pages 2350–2359 *in* K. Karvazy and V. L. Webster, editors. Proceedings of the World Environmental and Water Resources Congress 2015: floods, droughts, and ecosystems. American Society of Civil Engineers, Reston, Virginia.

Magnelia, S. J., C. C. Bonds, and J. L. Duty. 2003. Survey report for the Colorado River FM 973 bridge crossing, Austin to State Highway 71 bridge crossing, La Grange Texas, 2003. Texas Parks and Wildlife Department, Austin.

Magnelia, S. J. 2018. Resource conservation plan for the lower Colorado River, Travis and Bastrop counties, Texas. Texas Parks and Wildlife Department, Austin.

Mettee, M. F. 2000. Blue Sucker research. River Crossings 9:9.

Mosier, D., and R. Ray. 1992. Instream flows for the lower Colorado River: reconciling traditional beneficial uses with the ecological requirements of the native aquatic community. Lower Colorado River Authority, Austin, Texas.

NatureServe. 2018. NatureServe Explorer: an online encyclopedia of life [web application]. Version 7.1. NatureServe, Arlington, Virginia. Available: http://explorer.natureserve.org. (February 2019).

Nisbet, E. K., J. M. Zelenski, and S. A. Murphy. 2009. The nature relatedness scale: linking individuals' connection with nature to environmental concern and behavior. Environment and Behavior 41:715–740.

Opdyke, D. R., E. L. Oborny, S. K. Vaugh, and K. B. Mayes. 2014. Texas environmental flow standards and the hydrology-based environmental flow regime methodology. Hydrological Sciences Journal 59:820–830.

Pease, J. E., T. B. Grabowski, and A. A. Pease. 2017. Variation and plasticity and their interaction with urbanization in Guadalupe Bass populations on and off the Edwards Plateau. U.S. Fish and Wildlife Service, Cooperator Science Series FWS/CSS-125-2017, Washington, D.C.

Pease, J. E., T. B. Grabowski, A. A. Pease, and P. T. Bean. 2018. Changing environmental gradients over forty years alter ecomorphological variation in Guadalupe Bass *Micropterus treculii* throughout a river basin. Ecology and Evolution 8:8508–8522.

Postel, S. L., and B. H. Thompson, Jr. 2005. Watershed protection: capturing the benefits of nature's water supply services. Natural Resources Forum 29:98–108.

Schultz, P. W. 2000. Empathizing with nature: the effects of perspective taking on concern for environmental issues. Journal of Social Issues 57:391–406.

TCEQ (Texas Commission on Environmental Quality). 2015. An order affirming, with modification, an emergency order granted by the executive director to the Lower Colorado River Authority reducing instream flow requirements for the Blue Sucker. TCEQ, Austin.

Thomas, Z. A. and T. L. Arsuffi. 2015. Assessing landowner attitudes toward the development of a paddling trail on the South Llano River and recreationist attitudes toward recreational paddling. Texas Tech University Llano River Field Station, Junction.

Thomas, Z. A., T. L. Arsuffi, and S. J. Magnelia. 2015. Fishing warmwater streams with limited public access: angling behavior, economic impact, and the role of Guadalupe Bass in a twenty-four county region of Texas. Pages 123–137 *in* M. D. Tringali, J. M. Long, T. M. Birdsong, and M. S. Allen, editors. Black bass diversity: multidisciplinary science for conservation. American Fisheries Society, Symposium 82, Bethesda, Maryland.

TPWD (Texas Parks and Wildlife Department). 2012. Texas conservation action plan 2012–2016: overview. Wendy Connally, editor. TPWD, Austin.

U.S. Global Change Research Program. 2018. Southern Great Plains. Chapter 23 *in* D. R. Reidmiller, C. W. Avery, D. R. Easterling, K. E. Kunkel, K. L. M. Lewis, T. K. Maycock, and B. C. Stewart, editors. Impacts, risks, and adaptation in the United States: fourth national climate assessment, volume II. U.S. Global Change Research Program, Washington, D.C.

Vokoun, J. C., T. L. Guerrant, and C. F. Rabeni. 2003. Demographics and chronology of a spawning aggregation of Blue Sucker (*Cycleptus elongatus*) in the Grand River, Missouri, USA. Journal of Freshwater Ecology 18:567–575.

Weyers, R. S., C. A. Jennings, and M. C. Freeman. 2003. Effects of pulsed, high-velocity water flow on larval Robust Redhorse and V-lip Redhorse. Transactions of the American Fisheries Society 132:84–91.

Williams, J. E. , R. N. Williams, R. F. Thurow, L. Elwell, D. P. Philipp, F. A. Harris, J. L. Kershner, P. J. Martinez, D. Miller, G. H. Reeves, C. A. Frissell, and J. R. Sedell. 2011. Native fish conservation areas: a vision for large-scale conservation of native fish communities. Fisheries 36:267–277.

Winters, K. E. 2013. A historical perspective on precipitation, drought severity, and streamflow in Texas during 1951–56 and 2011. Prepared in cooperation with the Texas Water Development Board. U.S. Geological Survey, Scientific Investigations Report 2013-5113, Reston, Virginia.

Zelenski, J. M., R. L. Dopko, and C. A. Capaldi. 2015. Cooperation is in our nature: nature exposure may promote cooperative and environmentally sustainable behavior. Journal of Environmental Psychology 42:24–31.

American Fisheries Society Symposium 91:293–314, 2019

Developing the Science and Public Support Needed to Preserve the Devils River: A Case Study in Collaborative Conservation

Sarah Robertson*
Texas Parks and Wildlife Department, Inland Fisheries Division
505 Staples Road, Building 1, San Marcos, Texas 78666, USA

Brad D. Wolaver and Todd G. Caldwell
Bureau of Economic Geology, Jackson School of Geosciences
The University of Texas at Austin
University Station, Box X, Austin, Texas 78713, USA

Timothy W. Birdsong
Texas Parks and Wildlife Department
4200 Smith School Road, Austin, Texas 78744, USA

Ryan Smith
The Nature Conservancy of Texas
200 East Grayson Street, Suite 202, San Antonio, Texas 78215, USA

Thomas Hardy
Texas State University at San Marcos
601 University Drive, San Marcos, Texas 78666, USA

Julie Lewey
Devils River Conservancy
201 West 5th Street, Austin, Texas 78701, USA

Joe Joplin
Texas Parks and Wildlife Department
HC 01, Box 513, Del Rio, Texas 78840, USA

Abstract.—The Devils River is a groundwater-dominated, semiarid river in southwest Texas and considered one of the most pristine rivers in the state. It is one of the last strongholds for multiple species of regionally endemic freshwater fishes and mussels. However, groundwater pumping in the watershed poses an imminent threat to the river and its fragile ecosystem. Reductions in groundwater availability have the potential to result in concomitant reductions in spring discharge and thus instream flows. Base flow reductions would negatively impact many already imperiled aquatic species and degrade one of the state's most remote and scenic paddling and angling destinations. Development of a comprehensive basinwide fish and mussel conservation plan is ideal due to the relatively small size of the watershed. However, challenges include

* Corresponding author: sarah.robertson@tpwd.texas.gov

the isolated location of the river and the low proportion of publicly held lands for implementing on-the-ground conservation measures. To best determine science needs, focus resources, and increase informed stewardship of the river, the Texas Parks and Wildlife Department has partnered with governmental agencies, universities, nonprofit organizations, and landowners interested in preserving this unique resource. Through collaborative research aimed at a better understanding of groundwater–surface water interactions and instream flow needs of endemic species, and by building cooperative relationships with landowners and nonprofit conservation organizations, steps are underway to preserve the esthetic, ecological, and recreational values of the Devils River.

The Devils River, Texas

The Devils River, a spring-fed river in southwest Texas, is often described as the most pristine river in the state (Figure 1). It is also considered one of the last true wilderness paddling experiences in Texas and is revered for its biological, esthetic, cultural, and recreational values (El-Hage and Moulton 2001; NPS 2018). It has been nominated as a national wild and scenic river and is recognized as an ecologically significant stream segment and a native fish conservation area by the Texas Parks and Wildlife Department (TPWD; El-Hage and Moulton 2001; Birdsong et al. 2019, this volume). The Devils River watershed spans five counties (Crockett, Edwards, Schleicher, Sutton, and Val Verde; Figure 2); however, the perennially flowing reach of the Devils River is entirely contained within Val Verde County, Texas. Pecan Springs, located in central Val Verde County, is considered the headwaters of the river due to its consistent flow (USFWS 1999). While the Devils River is typically characterized as 145 km in length, the perennial reach of the river flows approximately 97 km to its confluence with the Rio Grande at Amistad Reservoir (Figure 2). The river is a gaining stream over its entire length (Green et al. 2014), receiving groundwater discharge from numerous springs (Brune 1981; Abolt et al. 2018).

The Devils River occurs at the juncture of the Chihuahuan Desert, South Texas Plains, and Edwards Plateau ecoregions (Figure 2; Griffith et al. 2004). This unique intersection of arid desert, brushland, and karst topography provides diverse habitat types that support numerous aquatic and terrestrial species, including several regional endemics classified as threatened or endangered by the TPWD and U.S. Fish and Wildlife Service (USFWS). The Devils River and its major tributary, Dolan Creek (Figure 2), are home to four state-threatened (ST) fish species: Proserpine Shiner *Cyprinella proserpina*, Devils River Minnow *Dionda diaboli*, Conchos Pupfish *Cyprinodon eximius*, and Rio Grande Darter *Etheostoma grahami* (Figure 3; El-Hage and Moulton 2001). The Devils River Minnow was also listed as federally threatened (FT) in 1999 based on documented population declines attributed to a loss of habitat within the species' range due to the construction of Amistad Reservoir, spring dewatering, and stream modifications (Garrett et al. 1992; USFWS 1999). The Devils River is also home to Texas hornshell *Popenaias popeii* (Figure 3), the only native mussel species known to occur in the river (Howells 2014; Randklev et al. 2018). Texas hornshell was listed as federally endangered (FE) in 2018 due to the threat imposed by habitat degradation in the form of hydrologic alteration, sedimentation, predation, instream fish passage barriers, and water quality impairment (USFWS 2018a).

The river and riparian corridor are utilized by several state and federally listed bird species, including the black-capped vireo *Vireo atricapilla* (state-endangered [SE]), tropical parula *Parula pitiayumi* (ST),

Figure 1. The Devils River in Val Verde County, Texas. Photograph by Clinton Robertson, Texas Parks and Wildlife Department.

interior least tern *Sterna antillarum athalassos* (SE, FE), and zone-tailed hawk *Buteo albonotatus* (ST) (El-Hage and Moulton 2001). The Devils River corridor is also home to the Texas snowbell *Styrax platanifolius texanus* (SE, FE), a riparian shrub (USFWS 2008) and the Tobusch fishhook cactus *Sclerocactus brevihamatus tobuschii* (SE, FT), a species known to occur in flood-prone riparian areas (TPWD 1995; USFWS 2018b). Additionally, the Devils River watershed is located along a major migratory path for the monarch butterfly *Danaus plexippus* (Reppert et al. 2010), which is under review for federal listing (USFWS 2014). The watershed is home to Fern Cave, which serves as a maternity roost for approximately 10 million Mexican free-tailed bats *Tadarida brasiliensis* from May through October each year (TNC 2008). Springs and streams in the watershed are also home to an undescribed species of salamander *Eurycea* sp. and several rare, regionally endemic invertebrates (TNC 2004; USFWS 2017).

Arguably the largest threat currently facing the river and the many species that depend on it is the potential for declining groundwater levels, which in turn impact spring discharge and river flows (TNC 2004). Water quality in the Devils River has historically ranked as excellent when compared to standards established by the Texas Commission on Environmental Quality (TCEQ; TCEQ 2004). This high-quality surface water can in part be attributed to the many springs along the river's length (Brune 1981) and the rural, undeveloped nature of the watershed (Anderson et al. 2014). Springs supply the river with the majority of its base flow and, via Amistad Reservoir, provide approximately 15% of the water needed for municipal and agricultural water supplies in the lower Rio

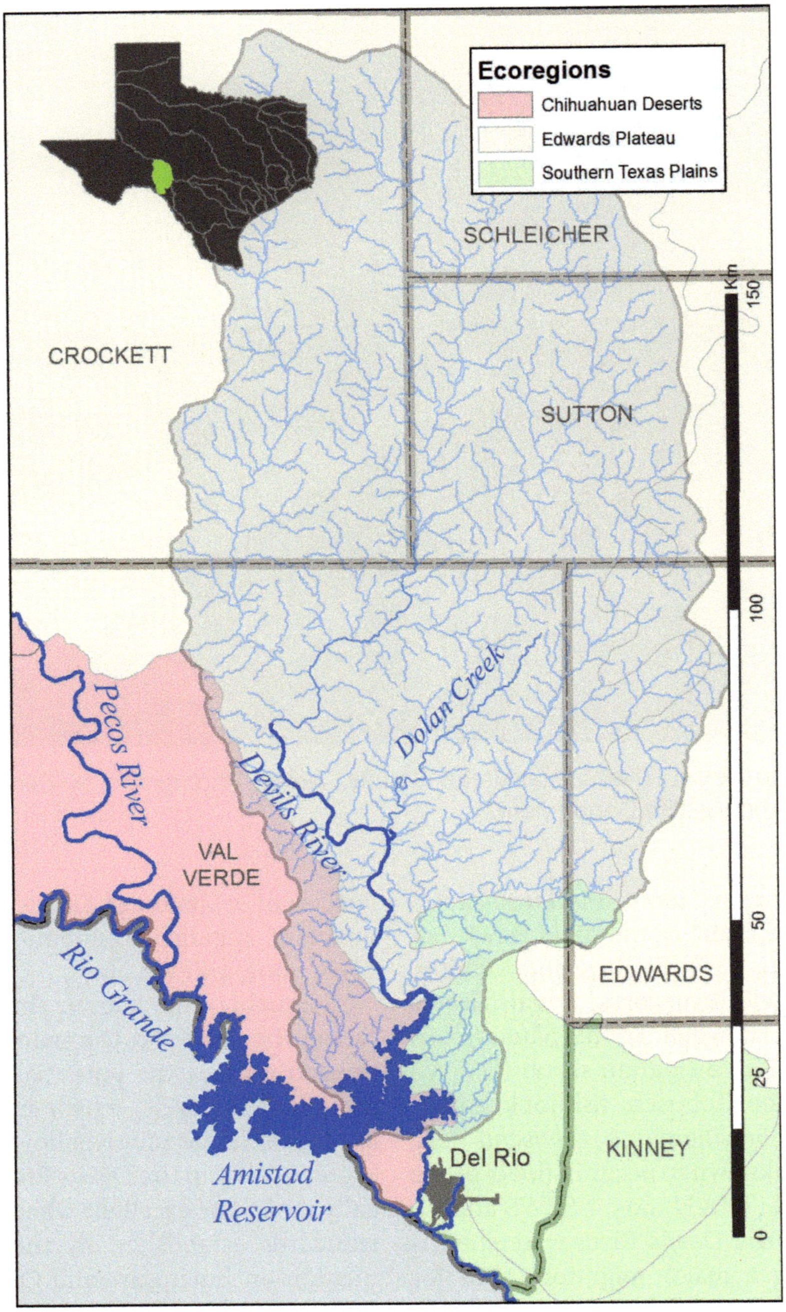

Figure 2. Watershed area, ecoregions, and perennially flowing reaches (darker blue line) of the Devils River, Texas.

Grande valley (Green et al. 2014). While discharge from these springs provides an oasis of high-quality water in an otherwise arid environment, it also makes the river vulnerable to reductions in base flow from reduced groundwater availability. In addition to a predicted 73% increase in human population between 2020 and 2070, Texas is projected to suffer a 24% decrease in groundwater availability statewide (TWDB 2017). This

Figure 3. Listed fish and mussel species inhabiting the Devils River from top left to right are Rio Grande Darter (state-threatened [ST]), Devils River Minnow (ST, federal-threatened [FT]), and Proserpine Shiner (ST); and from bottom left to right are Conchos Pupfish (ST), Texas hornshell (ST, federal-endangered [FE]), and Texas hornshell in situ. Photographs by Sarah Robertson, Texas Parks and Wildlife Department.

decline is largely due to reduced supply from the Ogallala and Gulf Coast aquifers; however, when coupled with increasing municipal, manufacturing, and energy water demands across the state (TWDB 2017), the chance for increased pressure on other aquifers, such as the Edwards-Trinity Plateau Aquifer that feeds the Devils River, becomes reasonably foreseeable—especially if groundwater were pumped and exported out of the basin.

Additional threats include watershed alteration, introduction and expansion of invasive species, and increased recreational use. Land use within the watershed has historically been composed of large-acreage cattle, sheep, and goat ranches. Chronic overgrazing in portions of the watershed has led to changes in terrestrial ecosystems, with a shift from native grassland prairie habitats to bare soil and shrubland (Brune 1981; TNC 2004; TPWD 2010). Increases in bare ground has the potential to increase surface runoff and decrease groundwater recharge (Brune 1981).

Fragmentation and changes in land use on some of the historically large ranches in the watershed have had an impact on the river. Some of these lands have been sold for housing developments, which, when overseen with limited regulation, pose potential threats to the river through point-source pollution from faulty septic systems (McQuillan 2004; TNC 2004). Degradation of groundwater quality from contamination may be exacerbated by the high-permeability karstic nature of the Edwards-Trinity Plateau Aquifer (Katz et al. 2010). Other ranches have shifted from domestic livestock to exotic game. Escapement of exotic ungulates from these ranches has resulted in the establishment of feral populations of, most notably, axis deer *Cervus axis* and aoudad or Barbary sheep *Ammotragus lervia*. The addition of exotic ungulates to the Devils River landscape has increased the abundance of foraging species in an already sensitive and overgrazed system. The proliferation of these exotic grazing species reduces forage diversity for native species and increases runoff with associated potential for increased suspended sediment loadings to the river (TPWD 2010). Increased runoff and lack of vegetation contribute to further reductions in groundwater infiltration. Several exotic aquatic species have also become established, including Common Carp *Cyprinus carpio*, Blue Tilapia *Oreochromis aureus*, Asian clam *Corbicula fluminea*, and red-rim melania *Melanoides tuberculatus*.

In the face of these challenges, there is an urgent need for a holistic river and watershed management plan supported by multidisciplinary studies (Stoffels et al. 2018). Stakeholder involvement is paramount for the successful adoption and application of conservation measures (Reed 2008; Ulibarri 2018), especially in development of instream flow recommendations (Poff et al. 2003; Richter et al. 2006; TIFP 2008; Conallin et al. 2018). This chapter presents collaborative efforts by stakeholders, consisting of state agencies, nongovernmental organizations, universities, and landowners, to preserve the unique esthetic, ecological, recreational, and economic values of the Devils River. Conservation efforts in the Devils River watershed have focused on development of watershed-based partnerships, implementation of best practices in watershed stewardship, management of river recreation for sustainable use, and advancement of scientific understanding of river and watershed processes, with an emphasis on quantification of hydrologic relationships among groundwater levels, spring discharge, and instream flows.

Building Partnerships

The terrestrial landscape of Texas is more than 95% privately owned (Anderson et al. 2014). Private land ownership can make access to publicly owned rivers for on-the-ground research challenging; however, it also presents an opportunity to engage landowners on a topic in which they are personally invested: river conservation. Resource managers and researchers can build mutually beneficial partnerships with property owners who are personally vested in river stewardship, although building these connections is not without challenges. In the Devils River watershed, stakeholders often have conflicting views on river and groundwater management and recreational access and use (e.g., paddling and camping). Along with these conflicting views, there is also often an inherent distrust among landowners of government agencies, conservation organizations, and even other landowners. However, if all vested parties can come together in support of a common goal, such as the belief that the Devils River warrants protection for future generations, then steps can be made to advance collaborative efforts toward that goal.

Prior to 1988, the Devils River riparian corridor was entirely privately owned. In 1988, the TPWD purchased an 80-km^2 ranch now known as the Del Norte Unit of the Devils River State Natural Area (DRSNA; Figures 4 and 5). This land acquisition provided protection for approximately 1.6 km of riparian corridor and one of the largest spring complexes on the river, Finegan Springs (Figure 5; TPWD 2012a). The purchase provided public access to this unique resource for the first time.

The Nature Conservancy (TNC) began further protection efforts along the Devils River in 1991 with acquisition of the 19-km^2 Dolan Falls Preserve, located adjacent to the DRSNA–Del Norte Unit (Figure 4). The Nature Conservancy continued these efforts from 1990 until 2005 through protection of more than 70 km^2 of land in fee title and nearly 500 km^2 of land in conservation easements or other protections. This included further protection for the DRSNA in the form of a conservation easement that provided permanent protection to upland, riparian, and riverine habitats. Another property that TNC protected through use

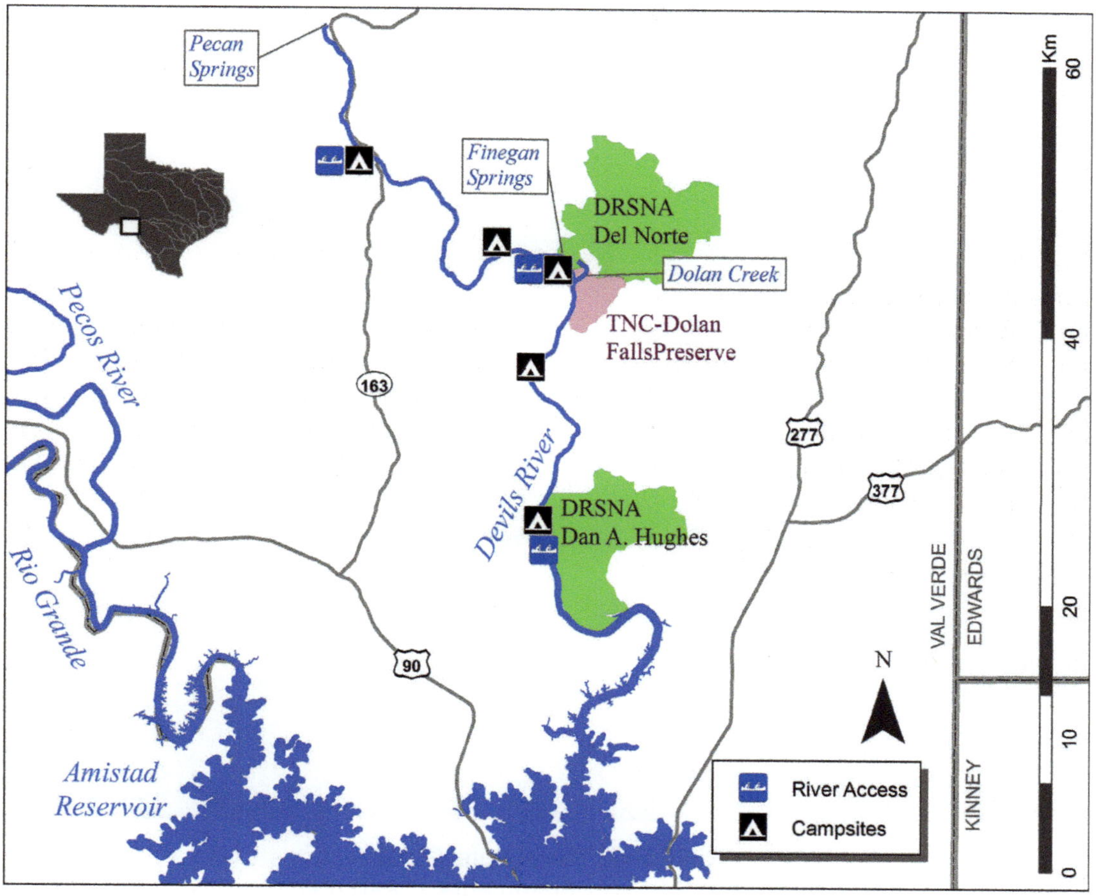

Figure 4. Perennial portion of the Devils River with public access points, paddler camps, and points of interest in Val Verde County, Texas

of a conservation easement was a 73-km² ranch that bordered 16.5 km of the Devils River just upstream of Lake Amistad. In 2011, through the help of many partners, the TPWD acquired this property and named it the Dan A. Hughes Unit. The property is located approximately 21 km downstream of the Del Norte Unit and together the two properties make up the DRSNA (Figures 4 and 6; TPWD 2012a).

The acquisition of the Dan A. Hughes Unit made the TPWD one of the largest land holders in the watershed. The agency has a vested interested in conservation of the river as a landowner and as a state agency entrusted with the management and conservation of natural and cultural resources. Furthermore, the TPWD sought to author a plan for the integrated management of both units, keeping in mind the vision for holistic management of the entire watershed (TPWD 2012a). Success for such a plan was contingent on the input and support from stakeholders invested in the long-term health of the river. With this in mind, the TPWD chartered the Devils River Working Group (DRWG) in 2010 to provide input and feedback to the agency regarding strategies for protection of the river and surrounding watershed, information valuable to the TPWD to inform management of the two units of the DRSNA (TPWD 2012a).

To represent the broad interests and perspectives of the stakeholder communi-

Figure 5. View of the Devils River with The Nature Conservancy's Dolan Falls Preserve on the left and Texas Parks and Wildlife Department's (TPWD) Devils River State Natural Area–Del Norte Unit on the right with an inset of an individual spring in the Finegan Springs complex, which is one of the largest contributors to base flows on the river. Photographs by Earl Nottingham, TPWD.

Figure 6. The Devils River at the Dan A. Hughes Unit of Texas Parks and Wildlife Department's (TPWD) Devils River State Natural Area. Photograph by Clinton Robertson, TPWD.

ty, landowners, paddlers, anglers, DRSNA staff, researchers, conservation-oriented nonprofits, and recreational outfitters were appointed. The official charge of the working group was to identify issues of concern regarding sustainability of the resource, outline goals to address concerns, and identify specific management actions to achieve those goals.

The DRWG outlined the following conservation goals (reprinted from TPWD 2012a):

1. Protection of the river's water quality and quantity,
2. Protection of the river's ecological integrity and biota,
3. Protection of cultural resources along the river corridor,
4. Management of public access to the river,
5. Management to keep the river free of trash and waste,
6. Maintenance of the river's wilderness experience, and
7. Increased public outreach and education.

The DRWG also documented several challenges to achieving these goals, including land fragmentation, incomplete science, and nonnative species. Last, the group provided an outline of recommended actions to the TPWD (reprinted from TPWD 2012a):

1. Continue to seek input from a coalition of vested stakeholders,
2. Develop a comprehensive Devils River use management plan,
3. Develop a river access permit system,
4. Explore the feasibility of acquiring management control over Baker's Crossing (a put-in access point located along a public roadway),
5. Develop a comprehensive education and outreach campaign to inform the public of good recreational river stewardship practices,
6. Increase law enforcement presence on the river,
7. Have TPWD Commission members visit the state natural area complex,
8. Develop a volunteer river patrol program to monitor river use and health,
9. Support consolidation and gathering of scientific data on the river,
10. Continue working with partners to minimize land fragmentation,
11. Increase penalties for damage to cultural and natural resources, and
12. Implement fisheries management policies that support sustainable populations of fish.

Of these 12 recommendations, at the top of the list was the recommendation for continued dialogue among the TPWD and vested stakeholders. This recommendation was arguably the most critical, as without input and support from the varied interests within the watershed, many of the other goals set forth could not be achieved. For example, without the support of landowners, science conducted in the watershed would be primarily limited to the DRSNA, which is not necessarily an accurate representation of the entirety of the river or watershed. Additionally, landowners provide invaluable first-hand knowledge of the river, both historical and present day, giving resource managers an improved basis for recommending conservation actions.

Following this recommendation, the TPWD formed the Devils River Working Group II (DRWG II; TPWD 2014). The DRWG II was tasked with further detailing recommendations of the DRWG and identifying additional strategies to support the long-term health of the river. The DRWG II was similarly composed of watershed landowners, recreational users, land and river conservation leaders, and TPWD personnel. The DRWG II presented an updated list of recommendations to the TPWD in 2014 (TPWD 2014), including establishing regulations, campsites, and paddler education materials to reduce recreational impacts; supporting hydrology, recreation, and fisheries studies; compiling all historical and current Devils River-related research; encouraging landowners to employ best conservation practices; encouraging conservation easements within the watershed; and promoting catch and release of native fishes.

These working groups set the stage for increased communication and understanding among groups of stakeholders, who in the past have had opposing viewpoints on management of the river. Increased dialogue among landowners, agency staff, recreational users, and conservation-based nonprofit organizations has led to the formation of new partnerships focused around the desire to carry out recommended actions identified by the working groups. This includes a partnership between the TPWD and the Devils River Conservancy (DRC), a landowner-driven conservation advocacy nonprofit organization. In an effort to continue the dialogue among stakeholders, the TPWD and DRC initiated educational outreach workshops (Figure 7), production of a paddler manual promoting good river stewardship practices and ongoing scientific research. Historically, much of the research conducted on the Devils River did not involve private property and landowners; however, the DRC is now connecting the TPWD and other agency and university researchers with property owners to discuss and conduct ongoing science in the watershed.

One way the DRC is making these connections is by holding joint DRC, TPWD, and landowner paddle trips. These 48-km paddle trips occur twice a year with the objectives of evaluating recreational impacts and engaging landowners (DRC 2016, 2017a, 2017b). Over a 4-d period, participants pick up litter, document illegal fires and campsites, document and remove invasive species,

Figure 7. The Devils River Conservancy and Texas Parks and Wildlife Department have partnered together to put on workshops and organize paddle trips for landowners, agency staff, and nonprofit staff. Workshop topics promote implementation of best management practices and a better understanding of the resource and have included best rural road building practices (top left) and the aquatic life of the Devils River (top right, bottom left). Biannual paddle trips (bottom right) focus on trash removal, documentation of recreational impacts, and increased dialog between partners.

qualitatively document the general health of the river, and discuss river conservation issues (Figure 7). Bringing together resource managers and landowners on the river is arguably the most important part of these trips and has fostered open discussions and facilitated opportunities for future collaboration. Furthermore, it directly aligns with the vision outlined in the working group reports (TPWD 2012a, 2014).

Another recommendation outlined in the DRWG II report is that the TPWD continue to support science needs in the watershed by compiling existing data and conduct and fund new research to fill data gaps (TPWD 2014). A central repository does not currently exist for data, manuscripts, and technical reports generated through those research projects, making it difficult to outline data gaps and prioritize research needs. This situation has also made it difficult to assess population trends due to the inability to compare species data over time. To help remedy this, the TPWD is working with DRC to compile published manuscripts, agency publications, gray literature, technical presentations, raw data, historical photographs, and other pertinent data and information. The goal of this project is to house all compiled materials in a searchable database platform with a Web-based interface that will allow it to be utilized by partners to inform research priorities and conservation actions and provide a better overall understanding of the resource.

The need to collect more data to inform conservation efforts in the watershed is crucial. To determine the most urgent research needs, the TPWD and TNC pooled knowledge of existing research within the Devils River watershed and have worked to collaboratively fund research to fill priority data needs. Like the TPWD, TNC has a dual interest in the Devils River, both as a conservation advocate and a landowner, and has long been involved in research and conservation in the watershed.

In the early 2000s, TNC coordinated early conservation planning efforts with an array of partners, including the TPWD, other state and federal agencies, and landowners. The resulting plan (TNC 2004) guided many initial conservation efforts in the watershed. It sought to complement land protection with the following primary recommendations: enhance black-capped vireo populations and Texas snowbell populations, reduce nonnative fish species, maintain or enhance the condition and beauty of riparian gallery woodlands, and promote public recreation that protects natural resources and quality of life for residents.

In 2006, TNC, the TPWD, and the Environmental Defense Fund updated the original conservation plan to prioritize aquatic conservation and research efforts. Priorities included a springs inventory and biological assessment; establishment of groundwater monitoring wells; characterization of the instream flow regime; and a watershed-wide aquifer characterization, water budget, and groundwater model. This plan led to the establishment of several ongoing monitoring efforts that have supported collaborations among TNC, the TPWD, Texas A&M University (TAMU), Texas State University (TSU), the USFWS, the U.S. Geological Survey, the Texas Water Development Board (TWDB), the TCEQ, and The University of Texas at Austin (UT Austin) Bureau of Economic Geology (Valdes Cantu and Winemiller 1997; TCEQ 2004; Kollaus and Bonner 2012; Robertson et al. 2016; USFWS 2017; Abolt et al. 2018).

Partnerships have been vital to furthering conservation and guiding investments in science in the Devils River watershed. While statewide agencies such as the TPWD often bring scientific expertise, funding, and other resources to fulfill research, restoration, or education objectives, they often lack the time necessary to successfully accomplish all the goals set forth within a targeted watershed. Partnerships between state agencies and local nonprofit conservation organizations are an effective way to accomplish localized goals. Additionally, opening dialogues with other entities that are also funding research in the watershed is critical to ensure that research efforts are not duplicative and that the timeliest needs are prioritized and addressed.

Managing Sustainable Recreation

Two of the major concerns expressed by landowners have been increasing litter and trespassing along the river, which have been attributed to an increase in recreational access and use by paddlers. These concerns were documented in the DRWG I and DRWG II reports and led the TPWD to create the Devils River Access Permit (DRAP) as a mechanism to manage recreational use in a sustainable manner (TPWD 2012a, 2014). The DRAP is required of any paddler accessing or camping along the river at either unit of the DRSNA. This allows the TPWD to monitor and limit usage to a sustainable level and educate paddlers on river use etiquette and river stewardship practices prior to their trip, through educational materials provided to them with their permits.

Data collected in association with the DRAP confirms that recreation on the Devils River is increasing. From 2013 to 2017, issuance of permits increased 300% for single-day permits (n = 67 in 2013; n = 200 in 2017) and 200% for overnight permits (n = 767 in 2013; n = 1,464 in 2017; TPWD 2018). These numbers only include paddlers utilizing the TPWD's DRAP system. It is unknown how many paddlers utilized the river over that time through an unlicensed outfitter or by self-shuttling without a permit, both of which are known to occur. Increased recreational use can be attributed to several factors, including increased public access, increased media and advertising, and an overall increase in paddle sports participation in Texas (TPWD 2018).

Beyond the DRAP, the TPWD has employed several methods to manage recreational use and encourage river stewardship. Increased law enforcement allows officers to stop undesirable behavior, such as trespassing, littering, unauthorized ground fires, and other illegal activities that threaten the health of the river, as well as disseminate information on the resource and how paddlers can protect the river while still enjoying it (TPWD 2018).

A second measure employed by the TPWD to reduce recreational impacts was the addition of two new paddler campsites, bringing the total number of public campsites along the river to five (Figure 4; TPWD 2018). Prior to the addition of those two new paddler campsites, exit survey results from permit holders indicated that the average paddler traveled 8–16 km/d during a multiday canoe or kayak trip. Meanwhile, the average distance between public river access points was 25 km (TPWD 2018). This meant that most paddlers were finding alternative camping locations to complete their trip, and given that most of the properties adjacent to the river are privately owned, paddlers were limited to camping on islands or within the gradient boundary (i.e., the dividing line between the public streambed and private riparian property) to avoid trespassing. There are few islands large enough for camping, and determining the location of the gradient boundary can be difficult; therefore, this lack of clearly defined legal campsites led to conflicts between paddlers and riparian landowners.

As trespassing complaints increased, the TPWD recognized the need to develop two additional paddler campsites to reduce the distance between public access points. The goal of the campsites was to provide paddlers with additional legal camping options, leading to fewer landowner–paddler conflicts and a more enjoyable paddler experience. The challenge of establishing new campsites was the fact that the river is bordered by private lands in these reaches and the TPWD did not have the resources to purchase and manage new properties. Rather than purchase additional public lands in the watershed, in 2016 the TPWD's Inland Fisheries and State Parks divisions partnered to utilize the TPWD's River Access and Conservation Areas (RACA) program to lease private property for the establishment of two new campsites. The RACA program utilizes funding from the Texas Parks and Wildlife Foundation, U.S. Department of Agriculture's Voluntary Public Access and Habitat Incentive Program, and Sport Fish Restoration Recreational Boating Access Grant Program to provide anglers and paddlers with public river access across private lands. Given the goals of the RACA-supported campsites on the Devils River, the TPWD restricted access to paddle-up camping with a one-night limit.

Another challenge was landowner perceptions that the TPWD was opening the new sites to encourage increased recreation rather than to better manage existing recreation. Anticipating potential backlash, the TPWD sought recommendations and input from the DRC prior to securing the lease agreements with cooperating private landowners. Although somewhat hesitant that additional paddler campsites would encourage increased visitation, the DRC recognized that these new campsites had the potential to condense paddler overnight locations, potentially reducing trespassing and related riparian degradation on private lands. The DRC endorsed the new sites on a 1-year trial basis, and in return, the TPWD agreed to follow up and seek the DRC's input in evaluation of the campsites after the initial year, prior to lease renewals (TPWD 2018).

During the first year of use, more than 500 paddlers (37% of all permitted paddlers) utilized the new paddler campsites (TPWD 2018). In a report by the TPWD evaluating the pilot project, recreational usage was compared with river citations issued during the initial 10 months of site usage in 2017 versus the same 10-month period the prior year. Despite increased river usage and increased river patrols, there was a decrease in the number of citations issued. This included a decrease in the number of trespassing and littering violations, issues of special concern to landowners. The DRC provided a letter of support for continuation of the paddler campsites noting that

> they have served to consolidate human impacts, alleviate trespassing and encourage a high standard of river ethics by visitors.

The DRC went on to say they are grateful for the DRAP and new paddler camps and

> It is crucial to the long-term protection of the Devils River from recreational abuse and over-use that stakeholders continue to collaborate in promoting the highest standard of river ethics.

Based on the data and feedback compiled, the TPWD renewed the lease of the two new paddler campsites. In addition to positive feedback from landowners, the TPWD received positive feedback from paddlers utilizing the new campsites. The feedback reaffirmed that these sites have provided an enjoyable paddler experience and reduced conflicts between paddlers and landowners. The TPWD and DRC intend to continue to work together to monitor and evaluate the effectiveness of the paddler campsites in meeting their intended goals and to determine whether additional paddler campsites are warranted.

Science as the Foundation for Conservation

The time spent building trust and developing partnerships with landholders in the watershed has allowed for expanded opportunities to conduct the science needed to guide conservation. Due to the richness of endemic species that occur in the Devils River watershed, the area has long been a focal point for ecological research. River sustainability requires an understanding of the instream flow regime necessary to maintain ecological functions and processes (Annear et al. 2004). Instream flow science requires a multidisciplinary approach. To identify environmental flow needs for the Devils River, hydrology, biology, geomorphology, water quality, and connectivity must all be considered (Annear et al. 2004). The Devils River flows over and interacts with a karst aquifer system; thus, instream flows are intrinsically linked with groundwater availability and the interaction between groundwater and surface water.

Current threats to groundwater availability in the Devils River watershed are primarily centered on lack of regulation and increasing water demands. High-quality groundwater can be taken from one part of the state and exported to other parts where it is needed. Much of Texas is arid or semiarid, and urban areas in these regions will likely need additional water supply given forecasted increases in their populations. Municipal water demand in Texas is anticipated to increase from 6.4×10^{12} m^3 per year in 2020 to 10.4×10^{12} m^3 in 2070 (TWDB 2017). Texas groundwater law states that, under rule of capture, any individual has the right to pump an unlimited amount of groundwater from their property (Bath 1999), which can hinder conservation efforts of groundwater-dependent streams. Exceptions exist in areas where groundwater conservation districts (GCDs) have been established to provide regulatory authority over the management of groundwater; however, Val Verde County is not currently governed by a district. In recent years, there have been proposals to pump substantial volumes of groundwater from the Devils River watershed to support municipal water demands in urban centers such as the cities of San Antonio, Texas and San Angelo, Texas (Satija 2014). While this would be expected to have detrimental effects on the Devils River ecosystem, the science is lacking to model the quantity of groundwater pumped before impacts occur and how those impacts would be manifested. Groundwater–surface water interaction is just one facet of the science still needed to inform protection of the Devils River, although it is currently considered the most pressing need. State agencies, nonprofit conservation organizations, and university researchers have had an increased level of cooperation and collaboration over the past 5 years to coordinate and align research in the watershed with the goal of addressing the question of how much water the Devils River needs to sustain a functioning ecological system.

Biological monitoring

Annual longitudinal fish surveys were conducted by the TPWD from 2000 to 2004 (TPWD, unpublished data). This monitoring, along with research by TAMU (Valdes Cantu and Winemiller 1997; Robertson and Winemiller 2001, 2003), TSU (Kollaus and Bonner 2012; Robertson et al. 2016), and the USFWS (Phillips et al. 2011), provide an important baseline of fish data prior to the 2011–2012 drought. In 2011, the TPWD compiled a report summarizing historical fisheries surveys in the Devils River watershed and outlining recommended fisheries management strategies (TPWD 2011). Hindrances to effective fisheries management included sport fish regulations that varied by river reach, nonstandardized fisheries data collection, limited data on aquatic habitat and riparian condition, and the threat exotic species pose to the listed fish species in the Devils River (TPWD 2011). Recommendations to address those issues included standardization of the frequency, timing, location, and gear employed in the collection of fisheries data in the Devils River. In response, a Devils River standardized aquatic monitoring plan was developed (TPWD 2012b). Components of this plan included mapping of aquatic habitats, annual fish assemblage sampling, genetic monitoring for target species, black bass monitoring, and riparian evaluations.

In 2012, the TPWD and TNC began standardized monitoring of the fish communities in the Devils River and Dolan Creek. Sites were selected to encompass available habitats and collect species representative of the Devils River. Fish were collected using seines from representative mesohabitats (i.e., riffle, run, pool, and backwater), and aquatic habitat parameters were measured within each sampled mesohabitat (TPWD, unpublished data).

Annual fish community monitoring has continued with some minor changes, including the inclusion of gill nets to better document larger species such as Longnose Gar *Lepisosteus osseus*, Headwater Catfish *Ictalurus lupus*, and Largemouth Bass *Micropterus salmoides*. Additionally, methods have been refined to include collection of the various habitat parameters at each seine haul and gill net location to collect data at a fine enough resolution to allow for development of habitat suitability criteria.

In 2017, the TPWD began longitudinal surveys, which, when completed, will document species distribution and habitat use in 85 km of river for three focal species: Devils River Minnow, Conchos Pupfish, and Texas hornshell (Figure 3). The Devils River Minnow (ST, FT) is vulnerable to potential impacts due to its narrow distribution and reliance on imperiled spring-fed streams (Garrett et al. 2002; Jelks et al. 2008). It is assumed that this species has already been extirpated from several streams and likely only remains in the Devils River, Dolan Creek, San Felipe Creek, and Pinto Creek in Texas, with an unknown distribution in Mexico (Garrett et al. 2004).

The Conchos Pupfish (ST) was selected also based on its narrow distribution, as well as its unique habitat use. It is only known to occur in Alamito Creek (Presidio County), the Devils River, and Dolan Creek in Texas and in the Rio Conchos watershed in Mexico (Garrett et al. 2005; Hubbs et al. 2008). In the Devils River and Dolan Creek, Conchos Pupfish primarily utilize shallow (25–150 mm) margin habitats on inundated bedrock shelves (TPWD, unpublished data). The primary conservation concern for this species is dewatering of shelf habitats in times of reduced flow. It is currently unknown what habitats this species utilizes during times in which these shallow-margin shelf habitats are unavailable and if other habitats can support the species long term.

The distribution of Texas hornshell (ST, FE) is limited to the Rio Grande basin of Texas, New Mexico, and northern Mexico (USFWS 2018a), and until recently, little was known about its status in the Devils River (Howells 1996; Karatayev et al. 2012). Recent collections have found that habitat preference of Texas hornshell in the Devils River (shallow riffles and runs; TPWD, unpublished data) makes them particularly vulnerable during periods of reduced flows, which can lead to stranding or increased predation by raccoon *Procyon lotor* (USFWS 2018a). Because Texas hornshell utilizes shallow water habitats, it is likely that developing flow recommendations for this species will in turn serve as protection for most, if not all, fish species in the river.

The TPWD and TNC recognize the importance of collecting a long-term monitoring data set using standardized sampling methodologies and plan to continue annual fish community monitoring on the Devils River and Dolan Creek and longitudinal surveys for focal species every 3 to 5 years. The development of long-term data sets is an important tool for monitoring species abundance and can be used to evaluate potential impacts to the ecosystem, especially given future water availability concerns.

Understanding the relationship between habitat and surface water

The first effort to model habitat availability as a function of Devils River discharge was made as part of Texas' environmental flows process in the Rio Grande basin. The Upper Rio Grande Bay and Basin Expert Science Team (URGBBEST) developed habitat suitability criteria for 10 Devils River fishes and, with assistance from the TPWD and TWDB, coordinated development of a one-dimensional hydraulics model (TES 2012) to estimate the amount of instream habitat maintained by the recommended flow regime (URGBBEST 2012). The habitat model was developed for the Devils River adjacent to the Dolan Falls Preserve, and model simulations concluded that the range of base flows recommended by the URGBBEST would provide adequate habitats for the fishes studied. This work was an important component of the flow recommendations (URGBBEST 2012) that were adopted by the TCEQ as flow standards for the Devils River.

In an effort to refine and expand upon URGBBEST recommendations, the TPWD partnered with TSU in 2012 to develop a two-dimensional habitat model for Devils River Minnow and nine other priority native fishes in two additional reaches of the Devils River at DRSNA–Del Norte Unit and one reach of Dolan Creek at the Dolan Falls Preserve (Hardy 2014).

Extreme drought during the study period (2013) resulted in relatively stable discharges in the Devils River (upstream reach: 0.54–0.94 m^3/s; downstream reach: 1.14–1.63 m^3/s). In combination with flat instream bathymetry within the study reaches, these historically low flows limited inferences regarding changes in available suitable habitat under variable flow conditions. Moderate declines in available suitable habitat for Devils River Minnow as instream flow increased were predicted by the model at the downstream Devils River study site. However, changes in depth or velocity predicted by the model were narrow under the observed discharges, and results at this site may not be indicative of actual changes. Changes in available suitable habitat at the upstream Devils River site were inconsequential. At the Dolan Creek site, habitat could not be modeled because variability in flow observed during data collection trips was insufficient to develop the stage-discharge relationship needed to calibrate the hydraulic model. It was recommended that additional physical habitat data be collected under a wider range of discharges for the model to more effectively predict changes in instream habitat availability.

Based on recommendations from Hardy (2014), and foreseeing future applications of the data, the TPWD partnered with UT Austin to collect airborne Light Detection and Ranging (Lidar) bathymetry data for the entire perennial reach of the Devils River and most of Dolan Creek. These data will produce a digital elevation model of both the surface and subsurface (to depths of 3 m), as well as riparian vegetation densities that will allow researchers to develop better hydraulic models for longer contiguous reaches of river. This combined bathymetric, topographic, and visual imagery survey of approximately 70 km of the stream can be used as a primary input for instream habitat models, assessing instream flow targets and mapping aquatic vegetation, which provides important habitat for Devils River Minnow (Robertson et al. 2016).

Understanding groundwater–surface water interactions

Essential to developing effective instream flow recommendations in groundwater-dependent streams is an understanding of how variations in groundwater level affect spring discharge and how groundwater inputs maintain instream flows. Hydrogeologic research efforts (e.g., Veni 1995, 1996; Mace et al. 2004; LBG-Guyton Associates 2005, 2009) have resulted in a conceptual understanding of aquifer connections in the watershed and provided basis for recent advances. Simulation models of groundwater flow and aquifer–surface water connections are an important tool in understanding aquifer systems and evaluating groundwater management scenarios. The Devils River has been the focus of several recent modeling efforts (Ecokai Environmental Inc. 2014; Green et al. 2014) with the goal of better understanding potential effects of proposed pumping scenarios on the Devils River and downstream water supply in the Rio Grande. A coupled groundwater–surface water model developed by the Southwest Research Institute (Green et al. 2015; Toll et al. 2017) that includes conceptualization of preferential flow paths (Green et al. 2014) provides the best current tool for relating aquifer dynamics to instream flows. Simulations of three pumping scenarios (Toll et al. 2017) suggest that current low levels of pumping may have already had significant effects on the river, especially considering low recharge rates (Green et al. 2012). Thus, expanded pumping in sensitive areas above the river's headwaters could significantly reduce river flows.

To strengthen understanding of these relationships and further refine groundwater models, the TPWD partnered with UT Austin in 2015 to collect hydrologic data needed to understand the relationships among groundwater, spring discharge, surface water, and aquatic habitats of Devils River Minnow, Texas hornshell, and other species of conservation interest. Specifically, the goals of this study are to measure groundwater levels to understand baseline groundwater variability; monitor spring and stream water salinity (e.g., conductivity) and temperature to understand rainfall-runoff and spring flow relationships affecting streamflow; record temporal water temperature changes in Devils River Minnow habitat; measure diurnal thalweg water temperature variations; develop a stage-discharge relationship, enabling calculation of spring discharge; understand the influence of climate variability on aquatic habitat by monitoring meteorological parameters; measure substrate temperature and stream stage at three riffles where Texas hornshell has been detected; and collect thermal infrared imagery of surface water temperature around spring discharge sites (see Abolt et al. 2018).

The hydrologic data generated by this study will be directly used by the TPWD and USFWS to implement key components of a recovery plan for Texas hornshell (upon completion by the USFWS) and the Devils River Minnow recovery plan (USFWS 2005), including development of a groundwater management plan for streamflow protection that ensures protection for the Devils River. The results of the hydrology study will inform management of threats to Texas hornshell identified in the Texas hornshell species status assessment (USFWS 2018c). To this end, study results will be conveyed to state, federal, and local stakeholders to increase opportunities for successful conservation of aquatic habitats for the Devils River Minnow, Texas hornshell, and other species of interest (TPWD 2012c; Cohen et al. 2018) in the Devils River.

Discussion and Future Conservation Steps

While much progress has been made over the past decade to build partnerships, strengthen scientific research efforts in the watershed, align research priorities, and promote river stewardship by recreational users of the Devils River, more work is needed to conserve this special resource. Partners should continue to cooperate on delivery of conservation goals that have been established through collaborative stakeholder processes, such as the DRWG and DRWG II. Reconvening a DRWG every 5 to 10 years would support timely reevaluation of conservation strategies, priorities, and threats.

Continuing to develop working relationships with landowners is of the utmost importance in any watershed that is predominately privately owned. The Devils River watershed, with its many large ranches and relatively few property owners, offers a unique opportunity to reach a large portion of the watershed through educational events and landowner meetings. These relationships are mutually beneficial and help ensure that conservation objectives are relevant and achievable.

The issue regarding creation of a GCD to protect the Devils River continues to be discussed. This is especially important as various entities continue to draft proposals to develop groundwater resources in the county. In Texas, the authority to establish GCDs resides with the state legislature or TCEQ. To ensure that the state can make an informed decision about creation of a GCD, partners must work together to focus research on understanding groundwater and surface water relationships and effects on instream habitat. More importantly, they must be able to communicate the science to decision makers effectively. The use of instream flow science tools, such as hydraulic habitat models in conjunction with groundwater–surface water simulation models, is necessary to inform groundwater management.

In the past decade, communication and cooperation among state agencies, conservation-based nonprofit organizations, universities, and landowners has increased substantially to the benefit of the resource. As river conservation efforts in Texas have shifted toward a holistic, watershed approach in recent years, conservation efforts in the Devils River have also shifted toward endeavors inclusive of all vested stakeholders.

Acknowledgments

Funding to support completion of this manuscript was provided by U.S. Fish and Wildlife Service Sport Fish Restoration Act Grant TX-F-139-T7. The authors thank Stephan Magnelia, Clinton Robertson, and John McMillan for their review of early versions of this manuscript. We also thank the numerous partners that continue to support the research and collaborative efforts to conserve the Devils River watershed.

References

Abolt, C. J., T. G. Caldwell, B. D. Wolaver, and H. Pai. 2018. Unmanned aerial vehicle-based monitoring of groundwater inputs to surface waters using an economical thermal infrared camera. Optical Engineering 57:1–9.

Anderson, R., A. Engeling, A. Grones, R. Lopez, B. Pierce, K. Skow, and T. Snelgrove. 2014. Status update and trends of Texas rural working lands. Texas A&M Institute of Renewable Natural Resources, College Station.

Annear, T., I. Chisholm, H. Beecher, A. Locke, P. Aarrestad, C. Coomer, C. Estes, J. Hunt, R. Jacobson, G. Jöbsis, J. Kauffman, J. Marshall, K. Mayes, G. Smith, R. Wentworth, and C. Stalnaker. 2004. Instream flows for riverine resource stewardship, revised edition. Instream Flow Council, Cheyenne, Wyoming.

Bath, C. R. 1999. A commentary on Texas water law and policy. Natural Resources Journal 39:121–128.

Birdsong, T. W., G. P. Garrett, B. J. Labay, M. G. Bean, P. T. Bean, M. J. Casarez, A. E. Cohen, T. G. Heger, A. Kalmbach, D. A. Hendrickson, S. J. Magnelia, K. B. Mayes, M. E. McGarrity, R. McGillicuddy, M. M. Parker, and S. Robertson. 2019. Texas Native Fish Conservation Areas Network: strategic investments in restoration and preservation of freshwater fish diversity. Pages 183–229 *in* T. W. Birdsong, D. C. Dauwalter, and G. P. Garrett, editors. Multispecies and watershed approaches to freshwater fish conservation: science, planning, and implementation. American Fisheries Society, Symposium 91, Bethesda, Maryland.

Brune, G. M. 1981. Springs of Texas, volume 1. Branch-Smith, Inc., Fort Worth, Texas.

Cohen, A. E., G. P. Garrett, M. C. Casarez, D. A. Hendrickson, B. J. Labay, T. Urban, J. Gentle, D. Wylie, and D. Walling. 2018. Conserving Texas biodiversity: status, trends, and conservation planning for fishes of greatest conservation need. Texas Parks and Wildlife Department, Austin.

Conallin, J., C. A. McLoughlin, J. Campbell, R. Knight, T. Bright, and I. Fisher. 2018. Stakeholder participation in freshwater monitoring and evaluation programs: applying thresholds of potential concern within environmental flows. Environmental Management 61:408–420.

DRC (Devils River Conservancy). 2016. Devils River working days report: August 2016. DRC, Austin, Texas.

DRC (Devils River Conservancy). 2017a. Devils River working days I report: March 2017. DRC, Austin, Texas.

DRC (Devils River Conservancy). 2017b. Devils River working days II report: September 2017. DRC, Austin, Texas.

Ecokai Environmental Inc. 2014. Hydrogeological study for Val Verde County and City of Del Rio. Final draft report to Val Verde County and City of Del Rio, Del Rio, Texas.

El-Hage, A., and D. W. Moulton. 2001. Ecologically significant river and stream segments of Region J (Plateau), regional water planning area. Texas Parks and Wildlife Department, Austin.

Garrett, G. P., R. J. Edwards, and C. Hubbs. 2004. Discovery of a new population of Devils River Minnow (*Dionda diaboli*), with implications for conservation of the species. Southwestern Naturalist 49:435–441.

Garrett, G. P., R. J. Edwards, and A. H. Price. 1992. Distribution and status of the Devils River Minnow, *Dionda diaboli*. Southwestern Naturalist 37:259–267.

Garrett, G. P., C. Hubbs, and R. J. Edwards. 2002. Threatened fishes of the world: *Dionda diaboli* Hubbs and Brown 1956 (Cyprinidae). Environmental Biology of Fishes 65:478.

Garrett, G. P., C. Hubbs, and R. J. Edwards. 2005. Threatened fishes of the world: *Cyprinodon eximius* Girard 1859 (Cyprinodontidae). Environmental Biology of Fishes 72:98.

Green, R., F. Bertetti, and M. Hernandez. 2012. Recharge variability in semi-arid climates. Nature Education Knowledge 3(10):34.

Green, R. T., F. P. Bertetti, and M. S. Miller. 2014. Focused groundwater flow in a carbonate aquifer in a semi-arid environment. Journal of Hydrology 517:284–297.

Green, R. T., N. Toll, and F. P. Bertetti. 2015. Sustainable yield of a carbonate aquifer dominated by preferential flow: Devils River watershed, South Texas, USA. Final Report to Nueces River Authority, Uvalde, Texas.

Griffith, G. E., S. A. Bryce, J. M. Omernik, J. A. Comstock, A. C. Rogers, B. Harrison, S. L. Hatch, and D. Bezanson. 2004. Ecoregions of Texas. U.S. Geological Survey, Reston, Virginia.

Hardy, T. B. 2014. Relationship between stream discharge and habitat availability for the Devils River Minnow (*Dionda diaboli*) and other native fishes in portions of the Devils River and Dolan Creek, Val Verde County. Final Report to Texas Parks and Wildlife Department, Austin.

Howells, R. G. 1996. Distributional surveys of freshwater bivalves in Texas: progress report for 1995. Texas Parks and Wildlife Department, Management Data Series 125, Kerrville.

Howells, R. G. 2014. Field guide to Texas freshwater mussels, 2nd edition. BioStudies, Kerrville, Texas.

Hubbs, C., R. J. Edwards, and G. P. Garrett. 2008. An annotated checklist of the freshwater fishes of Texas, with keys to identification of species, 2nd edition. Texas Journal of Science Supplement 53(4):2–87.

Jelks, H. L., S. J. Walsh, N. M. Burkhead, S. Contreras-Balderas, E. Diaz-Pardo, D. A. Hendrickson, J. Lyons, N. E. Mandrak, F. McCormick, J. S. Nelson, S. P. Platania, B. A. Porter, C. B. Renaud, J. J. Schmitter-Soto, E. B. Taylor, and M. J. Warren, Jr. 2008. Conservation status of imperiled North American freshwater and diadromous fishes. Fisheries 33:372–407.

Karatayev, A. Y., T. D. Miller, and L. E. Burlakova. 2012. Long-term changes in unionid assemblages in the Rio Grande, one of the world's top 10 rivers at risk. Aquatic Conservation: Marine and Freshwater Ecosystems 22:206–219.

Katz, B. G., D. W. Griffin, P. B. McMahon, H. Harden, E. Wade, R. W. Hicks, and J. Chanton. 2010. Fate of effluent-borne contaminants beneath septic tank drainfields overlying a karst aquifer. Journal of Environmental Quality 39:1181–1195.

Kollaus, K. A., and T. H. Bonner. 2012. Habitat associations of a semi-arid fish community in a karst spring-fed stream. Journal of Arid Environments 76:72–79.

LBG-Guyton Associates. 2005. Springs of Kinney and Val Verde counties. Final Report prepared for the Plateau Regional Water Planning Group by LBG-Guyton Associates, Austin, Texas.

LBG-Guyton Associates. 2009. Groundwater data acquisition in Edwards, Kinney and Val Verde counties, Texas. Final Report to the Plateau Region Water Planning Group and Texas Water Development Board, Austin.

Mace, R. E., E. S. Angle, and W. F. Mullican, III, editors. 2004. Aquifers of the Edwards Plateau. Texas Water Development Board, Austin, Texas.

McQuillan, D. 2004. Ground-water quality impacts from on-site septic systems. In Proceedings National Onsite Wastewater Recycling Association, 13th annual conference, Albuquerque, NM, November 7–10, 2004. National Onsite Wastewater Recycling Association, Tacoma, Washington.

NPS (National Parks Service). 2018. Nationwide rivers inventory [online database]. Available: www.nps.gov/subjects/rivers/nationwide-rivers-inventory.htm. (May 2018.)

Phillips, C. T., J. R. Gibson, and J. N. Fries. 2011. Spawning behavior and nest association by *Dionda diaboli* in the Devils River, Texas. Southwestern Naturalist 56:108–112.

Poff, N. L., J. D. Allan, M. A. Palmer, D. D. Hart, B. D. Richter, A. H. Arthington, K. H. Rogers, J. L. Meyer, and J. A. Stanford. 2003. River flows and water wars: emerging science for environmental decision making. Frontiers in Ecology and the Environment 1:298–306.

Randklev, C. R., T. Miller, M. Hart, J. Morton, N. A. Johnson, K. Skow, K. Inoue, E. T. Tsakiris, S. Oetker, R. Smith, C. Robertson, and R. Lopez. 2018. A semi-arid river in distress: contributing factors and recovery solutions for three imperiled freshwater mussels (family Unionidae) endemic to the Rio Grande basin in North America. Science of the Total Environment 631–632:733–744.

Reed, M. S. 2008. Stakeholder participation for environmental management: a literature review. Biological Conservation 141:2417–2431.

Reppert, S. M., R. J. Gegear, and C. Merlin. 2010. Navigational mechanisms of migrating monarch butterflies. Trends in Neurosciences 33:399–406.

Richter, B. D., A. T. Warner, J. L. Meyer, and K. Lutz. 2006. A collaborative and adaptive process for developing instream flow recommendations. River Research and Applications 22:297–318.

Robertson, M. S., and K. O. Winemiller. 2001. Diet and growth of Smallmouth Bass in the Devils River, Texas. Southwestern Naturalist 46:216–221.

Robertson, M. S., and K. O. Winemiller. 2003. Habitat associations of fishes in the Devils River. Journal of Freshwater Ecology 18:115–127.

Robertson, S. M., T. H. Bonner, and J. N. Fries. 2016. Effects of habitat utilization on the life histories of two imperiled, sympatric *Dionda* (Cyprinidae) in the Rio Grande basin, Texas. American Midland Naturalist 175:222–232.

Satija, N. 2014. Rio Grande water users fear groundwater pumping project. Texas Tribune, January 29, 2014, Austin, Texas.

Stoffels, R. J., N. R. Bond, and S. Nicol. 2018. Science to support the management of riverine flows. Freshwater Biology 63:996–1010.

TCEQ (Texas Commission on Environmental Quality). 2004. An assessment of water quality of Segment 2309 (Devils River). Texas Commission on Environmental Quality, Austin.

TES (Trungale Engineering & Science). 2012. Instream flow-habitat relationships in the upper Rio Grande river basin. Prepared for Senate Bill 3 Upper Rio Grande Basin and Bay Expert Science Team. Texas Water Development Board, Austin.

TIFP (Texas Instream Flow Program). 2008. Texas instream flow studies: technical overview. Texas Water Development Board, Report 369, Austin.

TNC (The Nature Conservancy). 2004. Conservation area plan for the Devils River. TNC, Austin, Texas.

TNC (The Nature Conservancy). 2008. Devils River. TNC, Austin, Texas.

TPWD (Texas Parks and Wildlife Department). 1995. Project No. 30: a study of the reproductive biology of the Tobusch fishhook cactus (*Ancistrocactus tobuschii*). TPWD, Austin.

TPWD (Texas Parks and Wildlife Department). 2010. Wildlife management activities and practices: comprehensive wildlife management planning guidelines for the Trans-Pecos ecological region. TPWD, Austin.

TPWD (Texas Parks and Wildlife Department). 2011. Inland fisheries report on Devils River fish community. TPWD, Austin.

TPWD (Texas Parks and Wildlife Department). 2012a. Devils River working group: final report and recommendations. TPWD, Austin.

TPWD (Texas Parks and Wildlife Department). 2012b. Devils River standardized aquatic monitoring plan. TPWD, Austin.

TPWD (Texas Parks and Wildlife Department). 2012c. Texas conservation action plan 2012–2016: Edwards Plateau handbook. TPWD, Austin.

TPWD (Texas Parks and Wildlife Department). 2014. Devils River working group action plan. TPWD, Austin.

TPWD (Texas Parks and Wildlife Department). 2018. Devils River recreational use assessment: evaluation of river stewardship outcomes achieved through establishment of the mile 12 and mile 20 paddler camps. TPWD, Austin.

TWDB (Texas Water Development Board). 2017. 2017 State water plan. TWDB, Austin.

Toll, N., S. B. Fratesi, R. T. Green, F. P. Bertetti, and R. Nunu. 2017. Water-resource management of the Devils River watershed. Final Report to Southwest Research Institute, San Antonio, Texas.

Ulibarri, N. 2018. Collaborative model development increases trust in and use of scientific information in environmental decision-making. Environmental Science and Policy 82:136–142.

URGBBEST (Upper Rio Grande Basin and Bay Expert Science Team). 2012. Environmental flows recommendations report. Final Submission to the Environmental Flows Advisory Group, Rio Grande Basin and Bay Area Stakeholders Committee and Texas Commission on Environmental Quality by the Texas Commission on Environmental Quality, Austin.

USFWS (U.S. Fish and Wildlife Service). 1999. Endangered and threatened wildlife and plants; final rule to list the Devils River Minnow as threatened. Federal Register 64:202(20 October 1999):56596–56609.

USFWS (U.S. Fish and Wildlife Service). 2005. Devils River Minnow *Dionda diaboli* recovery plan. USFWS, Albuquerque, New Mexico.

USFWS (U.S. Fish and Wildlife Service). 2008. Texas snowbells (*Styrax platanifolius* ssp. *texanus*) 5-year review: summary and evaluation. USFWS, Austin, Texas.

USFWS (U.S. Fish and Wildlife Service). 2014. Endangered and threatened wildlife and plants; 90-day findings on two petitions. Federal Register 79:250(31 December 2014):78775–78778.

USFWS (U.S. Fish and Wildlife Service). 2017. Dolan Springs assessment. Final Report to The Nature Conservancy, Austin, Texas.

USFWS (U.S. Fish and Wildlife Service). 2018a. Endangered and threatened wildlife and plants; endangered species status for Texas hornshell. Federal Register 83:28(9 February 2018):5720–5735.

USFWS (U.S. Fish and Wildlife Service). 2018b. Endangered and threatened wildlife and plants; reclassifying Tobusch fishhook cactus from endangered to threatened and adopting a new scientific name. Federal Register 83:94(15 May 2018):22392–22401.

USFWS (U.S. Fish and Wildlife Service). 2018c. Species status assessment report for the Texas hornshell (*Popenaias popeii*), version 1.2. USFWS, Albuquerque, New Mexico.

Valdes Cantu, N. E., and K. O. Winemiller. 1997. Structure and habitat associations of Devils River fish assemblages. Southwestern Naturalist 42:265–278.

Veni, G. 1995. Hydrogeologic characterization of Devils River State Natural Area, Val Verde County, Texas. Final Report to Texas Parks and Wildlife Department, Austin.

Veni, G. 1996. Drainage basin delineation and preliminary hydrogeologic assessment of the Dolan Springs, Val Verde County, Texas. Final Report to The Nature Conservancy, Austin, Texas.

American Fisheries Society Symposium 91:315–343, 2019

Watershed-Scale Conservation of Native Fishes in the Brazos River Basin, Texas

Kevin B. Mayes*
Texas Parks and Wildlife Department, Inland Fisheries Division
505 Staples Road, Building 1, San Marcos, Texas 78666, USA

Gene R. Wilde
Department of Biological Sciences, Texas Tech University
Box 43131, Lubbock, Texas 79409, USA

Monica E. McGarrity
Texas Parks and Wildlife Department
4200 Smith School Road, Austin, Texas 78744, USA

Brad D. Wolaver and Todd G. Caldwell
Bureau of Economic Geology, Jackson School of Geosciences
The University of Texas at Austin
10100 Burnet Road, Austin, Texas 78758, USA

Abstract.—The Brazos River crosses eight ecoregions on its journey from New Mexico through the heart of Texas to the Gulf of Mexico. This diverse stream ecosystem supports at least 85 fish species, many of which—including two endangered, migratory, pelagic broadcast-spawning cyprinids, Smalleye Shiner *Notropis buccula* and Sharpnose Shiner *N. oxyrhynchus*—have life histories that track the natural flow regime. These two shiners were listed as endangered in part because of severe range reductions that left each with one viable population in the upper Brazos River. Given their short life span, a single adverse event, such as a persistent drought of two consecutive years, could lead to extinction. This concern was nearly realized in 2011 when a record drought and heatwave resulted in complete reproductive failure of these species, which led to rescue efforts for imperiled shiners confined to drying pools. Seventeen major reservoirs control streamflow and create distinct, disconnected fragments in the Brazos River basin. Long-term ecological studies have provided a strong science foundation for guiding water and environmental flow management and watershed conservation. Implementation of both upland and riparian best management practices in the upper Brazos River watershed, including management of invasive saltcedar *Tamarix* spp., seeks to improve habitat for fish and wildlife. Hydrological monitoring and modeling is being conducted to evaluate the potential for saltcedar control to improve base flows. Identification of stream reaches most threatened by drying and where aquifer pumping may reduce groundwater inflows to streams is the focus of ongoing research on groundwater–surface water relationships. Fish passage barriers hinder successful recruitment, migration, and recolo-

* Corresponding author: kevin.mayes@tpwd.texas.gov

nization of prairie fishes. Removal and mitigation of barriers, as appropriate, will be critical to restoring ecological functions and connectivity required for migratory fishes. Research on propagation and repatriation of prairie fishes is needed to inform conservation and recovery efforts. A watershed-scale, multidisciplinary approach coordinated across borders and among entities is critical to ensure conservation efforts result in the persistence of native fishes in the Great Plains, including the Brazos River.

Introduction

Cycles of drought and flood foretell the fate and the storyline of the Brazos River in Texas. People and cities face this reality daily as they plan for adequate water supplies to meet social, environmental, and economic needs and watch the sky for storms that may cause the Brazos River to spill out of its banks. Like many streams of the U.S. Great Plains (Dodds et al. 2004), the Brazos River is part of a diverse and rich prairie ecosystem adapted to extremes. The rhythms of reproduction and recruitment of river species and floodplain communities track the natural flow regime (Poff et al. 1997; Lytle and Poff 2004).

Native Americans and Spanish settlers used the river for agriculture and built numerous irrigation canals and small dams in the 17th and 18th centuries (Kimmel 2011). In the early 19th century, European settlers began to construct larger dams and locks to harness the river for transportation of agricultural goods, control swelling floodwaters, and tame the parching effects of drought (Archer 2015). In 1915 and 1916, the U.S. Army Corps of Engineers (USACE) constructed two of eight proposed lock-and-dam structures to extend boat navigation from the Gulf of Mexico upstream to Waco, Texas to take full advantage of the fertile Brazos valley soils. Completion of the remaining six structures was abandoned because of rising costs and growing importance of railways, as well as the shifting course of the river and natural floods. Recurrent floods diverted attention and money to construction of flood control and multiuse reservoirs. In 1929, the Brazos River Conservation and Reclamation District, later renamed Brazos River Authority (BRA), was formed to control flood waters of the Brazos River. The district laid out a plan to construct 13 dams, including five on the Brazos River main stem, and it was this plan that inspired John Graves to canoe the Brazos River and narrate an intimate farewell to his stretch of the river in the 1960 classic *Goodbye to a River* (Graves 1960). The first reservoir on the Brazos River, Possum Kingdom Lake, was constructed by BRA in 1941. Since then, 16 additional major reservoirs (storage capacities >6.2 million m^3 [50,000 acre-feet]) were constructed within the Brazos River basin, with the last one, Lake Alan Henry, completed in 1994.

The Brazos River spans eight ecoregions on its 1,350-km journey from New Mexico, across the High Plains, down through the prairies of Texas, out onto the Coastal Plain and into the Gulf of Mexico (Figure 1). Its watershed area, more than 11 million ha, is second only in Texas to the Rio Grande. Streamflow is generated through groundwater sources and precipitation events from spring and summer thunderstorms, which frequently turn the waters muddy and red. The Brazos River emerges from the Llano Estacado in west Texas and New Mexico in the formation of several watersheds: the Permian brine spring-fed Salt Fork Brazos River and its fresher tributary, White River, and the Double Mountain Fork Brazos River and its tributary North Fork Double Mountain Fork Brazos River (Figure 1). The Brazos River proper forms at the confluence of the Salt and Double Mountain forks and is impounded by three major reservoirs on its way to the Gulf of Mexico. The character of

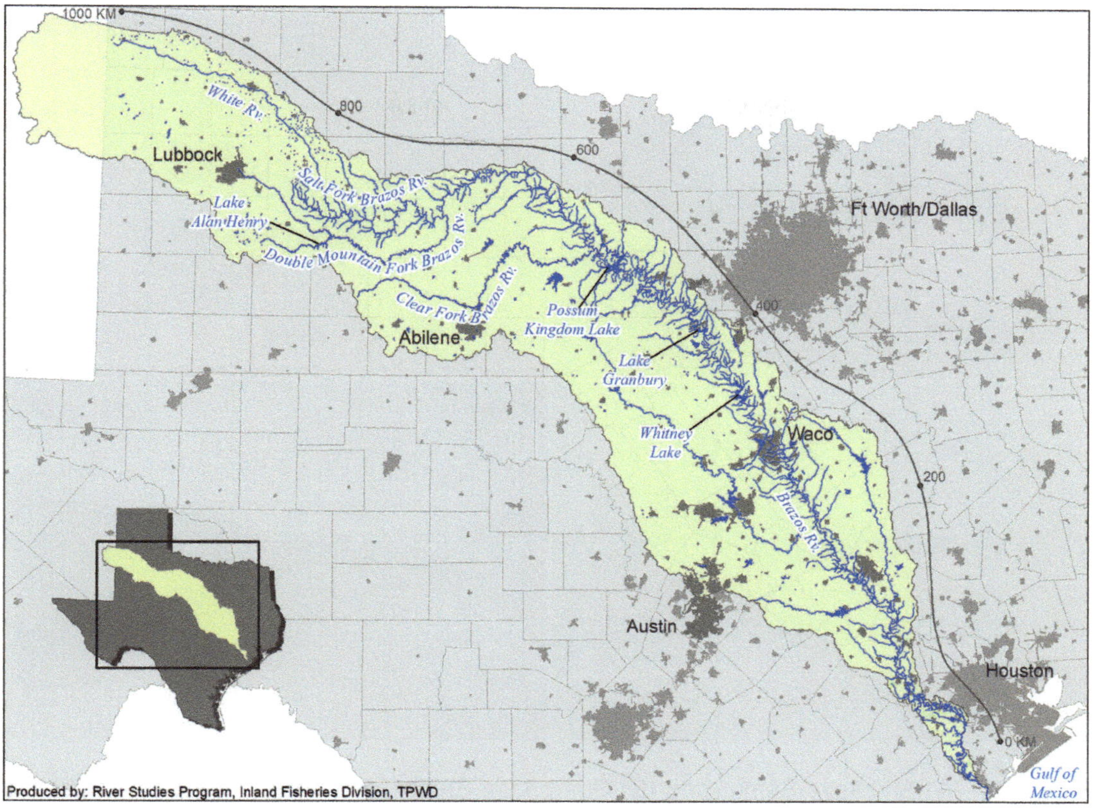

Figure 1. The Brazos River basin spans from the Southern Great Plains to the Gulf of Mexico.

the Brazos River transitions from a braided, sandy prairie stream in the upstream reaches, through the clearer and coarser impounded reaches, into a sinuous floodplain river marked by limestone outcrops, bottomlands, and oxbows.

Native Fish Conservation Areas

The Brazos River ecosystem supports at least 85 native freshwater fish species, including 17 identified as species of greatest conservation need (TPWD 2012); two of these, Smalleye Shiner *Notropis buccula* (Figure 2A) and Sharpnose Shiner *N. oxyrhynchus* (Figure 2B), are federally listed as endangered. These two species are highly migratory and belong to the reproductive guild of pelagic, broadcast-spawning cyprinids that is imperiled across the Great Plains (Worthington et al. 2018). Fifteen nonnative fishes are also reported. Four native fish conservation areas (NFCAs) have been proposed for the Brazos River basin (Figure 3; Cohen et al. 2018; see Birdsong et al. 2019, this volume).

The Lower Brazos River NFCA (Figure 3) begins downstream from Lake Brazos in Waco, Texas and is fed by upstream reservoirs and several major tributaries, including the Little and Navasota rivers. The lower Brazos River is characterized by a relatively wide channel dominated by sand substrates, historically flood-dependent riparian areas, and high turbidity. Run and pool habitats are common, bedrock shoals punctuate the reach in several places, and riffles are rarely seen under most flow conditions but become more exposed under low base flows (TIFP 2018). Unlike most other Texas rivers, sediment-laden water flows through a tidal channel directly connected to the Gulf of Mexico, as the river's estuary has filled

Figure 2. (A) Smalleye Shiner and **(B)** Sharpnose Shiner, endangered species and native Texans. These specimens were rescued **(C)** from drying pools **(D)** during a severe drought in September 2011. Credit: Texas Parks and Wildlife Department, C. Robertson.

with sediment and formed a bulge on the Texas coastline (BEG 1996). The lower Brazos River is unimpounded, although operations of upstream and tributary reservoirs alter flow regimes by shifting the seasonal timing and reducing variability of base flows (TIFP 2018). Constructed between 1954 and 1983, tributary reservoirs (eight in total) are operated for flood control, recreation, and water supply. Sand and gravel mining in the lower river contributes to channel degradation (TIFP 2018). The Lower Brazos River NFCA is the most diverse in the basin, supporting large river fishes such as floodplain-spawning Alligator Gar *Atractosteus spatula*; migratory, broadcast-spawning minnows including Mississippi Silvery Minnow *Hybognathus nuchalis*, Shoal Chub *Macrhybopsis hyostoma*, and Silverband Shiner *Notropis shumardi*; and diadromous American Eel *Anguilla rostrata* and Mountain Mullet *Agonostomus monticola*. In total, 83 native fishes (see Table 1 in Birdsong et al. 2019) have been collected from the Lower Brazos River NFCA, representing Edwards Plateau, Coastal Plains, and Mississippian faunal elements (Conner and Suttkus 1986). Bonner and Runyan (2007) reported significant shifts, following reservoir construction, in fish assemblage structure, including increases in generalist fishes such as Red Shiner *Cyprinella lutrensis* and Bullhead Minnow *Pimephales vigilax* and decreases in fluvial specialists, including Chub Shiner *N. potteri*, Sharpnose Shiner, and Smalleye Shiner; the latter two are likely extirpated from the Lower Brazos River NFCA (Moss and Mayes 1993; USFWS 2014b).

Emanating from Edwards Plateau springs, the San Gabriel River NFCA (Figure

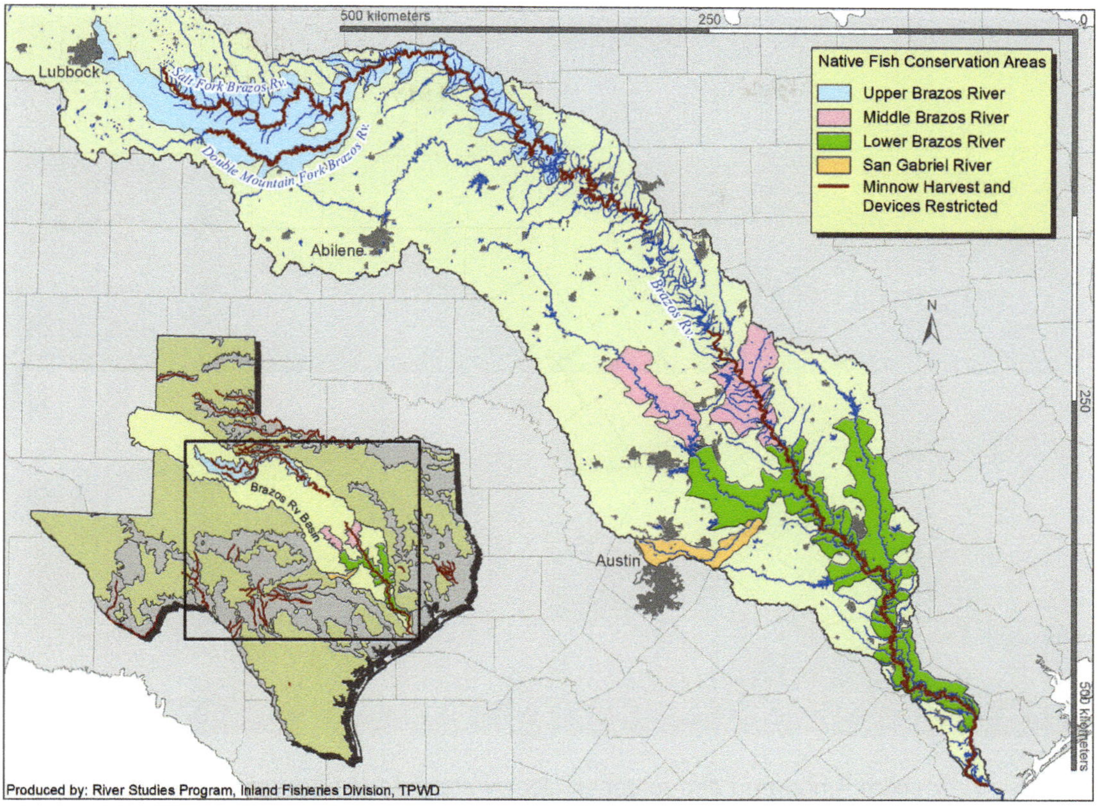

Figure 3. Native fish conservation areas (NFCAs) of the Brazos River basin. Dark red lines indicate areas where minnow harvest and devices are restricted to minimize impacts of commercial nongame fish harvest permitted by the Texas Parks and Wildlife Department on imperiled native fishes within the Brazos River basin. The inset shows those areas across the state overlain with all Texas NFCAs (Birdsong et al. 2019).

3) watersheds comprise Brushy Creek and all forks of the San Gabriel River—one of three major tributaries to the Little River watershed, which feeds the lower Brazos River. The USACE operates two major reservoirs: Lake Georgetown on the North Fork San Gabriel River (constructed in 1979) and Lake Granger on the San Gabriel River (1977). The San Gabriel River watershed provides habitat for Guadalupe Bass *Micropterus treculii* and is a priority area for conservation measures outlined in the Texas Parks and Wildlife Department (TPWD) Guadalupe Bass Conservation Plan (Bean 2017). Seventy-two fishes (57 native and 15 nonnative) are known from the San Gabriel River NFCA (see Table 1 in Birdsong et al. 2019).

The middle Brazos River extends from Possum Kingdom Lake downstream to Lake Brazos in Waco, Texas (Figure 1). The reach downstream from Possum Kingdom Lake (Figure 4A) was designated by the 79th Texas Legislature as the John Graves Scenic Riverway to protect the river by regulating quarries (TCEQ 2008). It is, indeed, a scenic river, lined with magnificent pecan trees and towering limestone bluffs. Following completion of Possum Kingdom Lake (in 1941 by BRA), Lake Whitney (USACE) began impounding waters in 1951, followed by Lake Granbury (BRA) in 1969 in the stretch between the two (Figure 4B). These three reservoirs, built for flood control, water supply, and hydropower generation, altered the

Figure 4. Possum Kingdom Lake releases clear water into **(A)** the middle Brazos River. In 2012, biologists seined the middle Brazos River downstream from **(B)** Lake Granbury looking for Smalleye Shiner and Sharpnose Shiner. The study (Labay et al. 2013) confirmed earlier findings that imperiled shiners have been extirpated from the middle Brazos River and found that the nonnative fishes **(C)** Sheepshead Minnow and **(D)** Gulf Killifish are common. Credits for (B), (C), and (D): University of Texas at Austin.

character of the Brazos River by releasing clear, cool water, modifying the flow regime, reducing meander migration rates (Gillespie and Giardino 1997), and blocking the downstream transport of sand and fine silt—in Graves' (1960) parlance, "filtering out the West Texas drainings." Responding to these changes, fish assemblages in the middle Brazos River have shifted from fluvial specialists to habitat generalists (Moss and Mayes 1993), and the nonnative fishes, Sheepshead Minnow *Cyprinodon variegatus* (Figure 4C) and Gulf Killifish *Fundulus grandis* (Figure 4D), are now common (Labay et al. 2013). While the Lake Whitney hydropower project continues to generate daily peaking hydropower, the Federal Energy Regulatory Commission license for the Possum Kingdom hydropower project was surrendered by BRA in 2014. As part of the surrender process, BRA agreed to continue downstream flow releases as stipulated in the original license order and now releases water from higher reservoir elevations to improve dissolved oxygen levels and temperatures that support aquatic life. The Brazos River downstream from Lake Whitney to Waco and the Bosque River watershed are proposed as the Middle Brazos River NFCA (Figure 3), which supports 58 native and 14 nonnative fishes (see Table 1 in Birdsong et al. 2019).

The Upper Brazos River NFCA (Figure 3) comprises the Brazos River and two of its tributaries, Double Mountain Fork Brazos

River (Figure 5A) and Salt Fork Brazos River (Figure 5B), upstream from Possum Kingdom Lake. The Upper Brazos River NFCA has 47 native and 16 nonnative fishes and will be the focus of the remainder of this chapter because it is where the majority of research and other actions directed at conservation of native freshwater fishes have occurred, to date, and coincides with critical habitat for the two endangered shiners (USFWS 2014a). This watershed is largely free-flowing and characterized by shallow, sandy, and braided stream channels (Figure 5C). The Salt Fork Brazos River is fed by numerous Permian brine springs and, at times, results in salinities greater than the Gulf of Mexico (Baker et al. 1964; Brune 2002) and very clear waters under low flow conditions (Figure 5D). The Salt Fork contributes more than 85% of the chloride load (770 metric tons daily average) to the Brazos River (Baker et al. 1964), and the volume of these brine springs created a great saline lake evident on 19th century maps of Texas (e.g., Mitchell 1846). The White River, a tributary of the Salt Fork, was once fed by the fresher Ogallala Aquifer at a rate of 49,210 m^3/d (Evermann and Kendall 1894) and was impounded in 1963 with the construction of White River Reservoir. Leakage from the dam forms a small trickle, and groundwater pumping for irrigation has dried up nearly all springs fed by the Ogallala Aquifer in the Salt Fork watershed (Brune 2002). Lake Alan Henry, construct-

Figure 5. The **(A)** Double Mountain Fork Brazos River and **(B)** Salt Fork Brazos River are two major tributaries of the upper Brazos River. **(C)** Shallow, sandy, braided river channels provide habitat for native fishes, including two imperiled shiners. To support water quality studies, Texas Park and Wildlife Department staff sampled **(D)** the Salt Fork, which provides habitat for salt-tolerant, native fishes such as Red River Pupfish *Cyprinodon rubrofluviatilis* and Plains Killifish *Fundulus zebrinus*.

ed by the city of Lubbock, impounds the Double Mountain Fork Brazos River before its confluence with the North Fork Double Mountain Fork Brazos River. Portions of the Double Mountain Fork are also underlain by the Ogallala Aquifer (Brune 2002).

Ongoing Impacts and Potential Threats to Upper Brazos River Native Fish Conservation Area

Reservoir construction and operation, groundwater pumping and surface water diversions, invasion of the Brazos River riverscape by saltcedar *Tamarix* spp., fish passage barriers, and introduction of nonnative fishes have altered the Brazos River riverscape and its riparian and aquatic communities. Demand for water is expected to increase (TWDB 2017), and water availability may decline as global warming is predicted to increase average temperatures and the frequency, duration, and intensity of extreme heat events and drought in the Southern Great Plains (Kloesel et al. 2018). Both of these factors create more need for water storage and appropriation for human use.

Reservoirs fragment prairie stream ecosystems, precluding upstream and downstream passage of migratory fishes, and can reduce stream reaches to lengths that no longer support some native fishes (e.g., Perkin and Gido 2011). Reservoirs with large surface areas also increase evaporation volumes, reducing streamflows downstream. Hydropower operations and downstream delivery schedules can change flow regimes on daily, seasonal, and annual scales. Flood-control reservoirs also change surface water-groundwater interactions; in semiarid rivers like the upper Brazos River, regular floods recharge bank storage, which can provide sustained base flows for years (Doble et al. 2012; Simpson et al. 2013).

River fragmentation and hydrologic alteration associated with reservoir development have led to substantial changes in fish communities, including extirpations and assemblage shifts across the Great Plains (Perkin et al. 2015a, 2015b). Water extraction reduces base flows, altering instream habitat and water quality and contributing to increased duration and frequency of drying/intermittency; during drought, the effects of reduced base flows are intensified. While some fishes are more tolerant of the harsh environmental conditions in intermittent pools, others perish as temperatures and salinities increase and dissolved oxygen levels decline (Ostrand and Wilde 2004).

Since its introduction to the upper Brazos River watershed in the first half of the 1900s, nonnative saltcedar has expanded and is present in high-density stands along the river's edge (Figure 6). New growth is often found in the river channel and in lines along high-water marks. Saltcedar colonizes stream floodplains and terraces, armoring river terraces and reducing the ability of a stream to meander. As the channel narrows, stream depth increases and temperature decreases (Dean et al. 2011), altering habitat for fishes adapted to the stream's natural geomorphology and water chemistry. Saltcedar may be more drought-tolerant than native species (Glenn and Nagler 2005) and often becomes the dominant woody species in riparian areas in the western United States (Dudley et al. 2000; Doody et al. 2011). In the Brazos River main stem upstream from Possum Kingdom Lake, phreatophytic, woody vegetation in the floodplain increased from 39% in 1969 to 57% by 1979 (Blackburn et al. 1982). Saltcedar dominated areas where the floodplain was narrow and the stream channel was straight, providing optimum water table conditions for its growth and regeneration (Busby and Schuster 1973). Saltcedar invasion caused 3 m of sediment accumulation, reduced Brazos River width by approximately 90 m, reduced sediment input to Possum Kingdom Lake, and resulted in higher flood stages.

All riparian vegetation transpires large amounts of water, and saltcedar is no excep-

Figure 6. Saltcedar is an invasive, nonnative woody plant invading the Upper Brazos River Native Fish Conservation Area. High-density stands along the river's edge may consume precious water and alter habitat for native fish; wildlife such as turkey, quail, and deer; and people and their livestock.

tion. Saltcedar evapotranspiration has been measured and found to be moderate, albeit higher than native, riparian, woody species such as cottonwood (Shafroth et al. 2005). However, evapotranspiration and water use by saltcedar vary greatly with tree density, stand size, stand location with respect to the water table, season, hydrological conditions, climatic conditions, and other factors (Shafroth et al. 2005; Nagler et al. 2008; Cleverly 2013). Although potential exists for modest water gains to result from saltcedar control, such gains are dependent upon replacement of saltcedar with xeric shrubs and grasses. Detection of gains could be confounded by concomitant increases in groundwater pumping (Cleverly 2013).

The supposition that saltcedar management will result in water salvage has led to management efforts in west Texas and other areas of the southwestern United States. Several studies have found only modest water yields following saltcedar abatement, and substantial yields have not yet been observed at watershed scales (Wilcox 2002; Wilcox et al. 2006; Doody et al. 2011; Cleverly 2013; McDonald et al. 2015). On the Brazos River upstream from Possum Kingdom Lake, water use by saltcedar was estimated to be high—in excess of 5.4 million m^3/year (Busby and Schuster 1973). Long-term hydrological monitoring will be required to determine whether ongoing saltcedar control on the upper Brazos River will be followed by detectable water gains. Potential impacts of posttreatment water chemistry changes (e.g., increased salinity pulses) on fish communities is largely unknown (Hart et al. 2005). Furthermore, evaluation of posttreatment changes in geomorphology and impacts to native fishes and instream habitat remain a key science need for management of invasive plants in riparian areas.

Introduction of nonnative fishes—such as Sheepshead Minnow and Gulf Killifish—through bait bucket releases (Hillis et al. 1980; Ross 2001; Hubbs et al. 2008) or other pathways presents a threat to endemic fishes. Sheepshead Minnow, native to coastal waters in Texas, was found in the upper Brazos River during the 2011 drought, as were putative hybrids between that species and the native Red River Pupfish (Wilde 2015). Hybridization has the potential to result in genetic swamping of the native form, as happened in the Pecos River (Wilde and Echelle 1992; Childs et al. 1996). While not presently in the upper Brazos River, Gulf Killifish (also native to coastal waters in Texas) is found in high numbers in the middle Brazos River (Labay et al. 2013) and poses a serious predation threat to native fishes, especially during dry periods when fish are confined to isolated pools. In the Pecos River, Gulf Killifish was found to be primarily piscivorous (Vaughan et al. 2016).

Upper Brazos River Fish Assemblages and Life Histories

Native fishes of the Upper Brazos River NFCA (n = 47; see Table 1 in Birdsong et al. 2019) are structured by spatial and temporal variations in environmental conditions such as salinity and hydraulic habitat (Echelle et al. 1972; Ostrand and Wilde 2001; Higgins and Wilde 2005), as well as emigration and immigration constraints imposed by downstream dams. In the saline reaches, such as most of the Salt Fork Brazos River, salt-tolerant Red River Pupfish and Plains Killifish dominate assemblages, while in fresher reaches (North Fork Double Mountain Fork, Double Mountain Fork, and the Brazos River), fish species richness is greater due to the presence of numerous cyprinids (Moss and Mayes 1993; Ostrand and Wilde 2002). Using data from Moss and Mayes (1993) and recent fish collections from the upper Brazos River basin, Wilde (2015) assessed temporal changes in fish abundance (Figure 7). Cyprinids in the pelagic, broadcast-spawning reproductive guild (i.e., Smalleye Shiner, Sharpnose Shiner, Chub Shiner, and Plains Minnow *Hybognathus placitus*) declined in relative abundance, whereas crevice-spawning Red Shiner, nest-building Red River Pupfish, and live-bearing Western Mosquitofish *Gambusia affinis* increased, indicating substantial fish assemblage shifts (Wilde 2015). Silver Chub *Macrhybopsis storeriana* (a pelagic broadcast-spawning cyprinid) was historically present in the upper Brazos River (Moss and Mayes 1993) but has not been collected there since the early 1990s (Wilde and Urbanczyk 2017). Fishes that showed the greatest declines in the upper Brazos River are pelagic, broadcast-spawning cyprinids (Figure 7). These migratory cyprinids are iconic fishes of prairie streams such as the Brazos River and are highly vulnerable due to their complex reproductive strategies and life history requirements (Durham and Wilde 2014; Worthington et al. 2018).

Flowing water is essential to the successful reproduction of pelagic, broadcast-spawning cyprinids and the survival of eggs and larval fish (Durham and Wilde 2009a; Worthington et al. 2018). Spawning migrations are made by adults to upstream reaches. Females release semibuoyant eggs into the water column as males wrap them and simultaneously release sperm to ensure fertilization (see Platania and Altenbach 1998 for spawning behavior of Plains Minnow). Relatively long (e.g., 275 river kilometers for the two Brazos River shiners; Wilde and Urbanczyk 2013; USFWS 2015), unobstructed river reaches are required to allow fertilized eggs enough distance to drift with the river's current. Once hatched, larval fish must stay suspended in the current (to prevent larval settlement) until they become

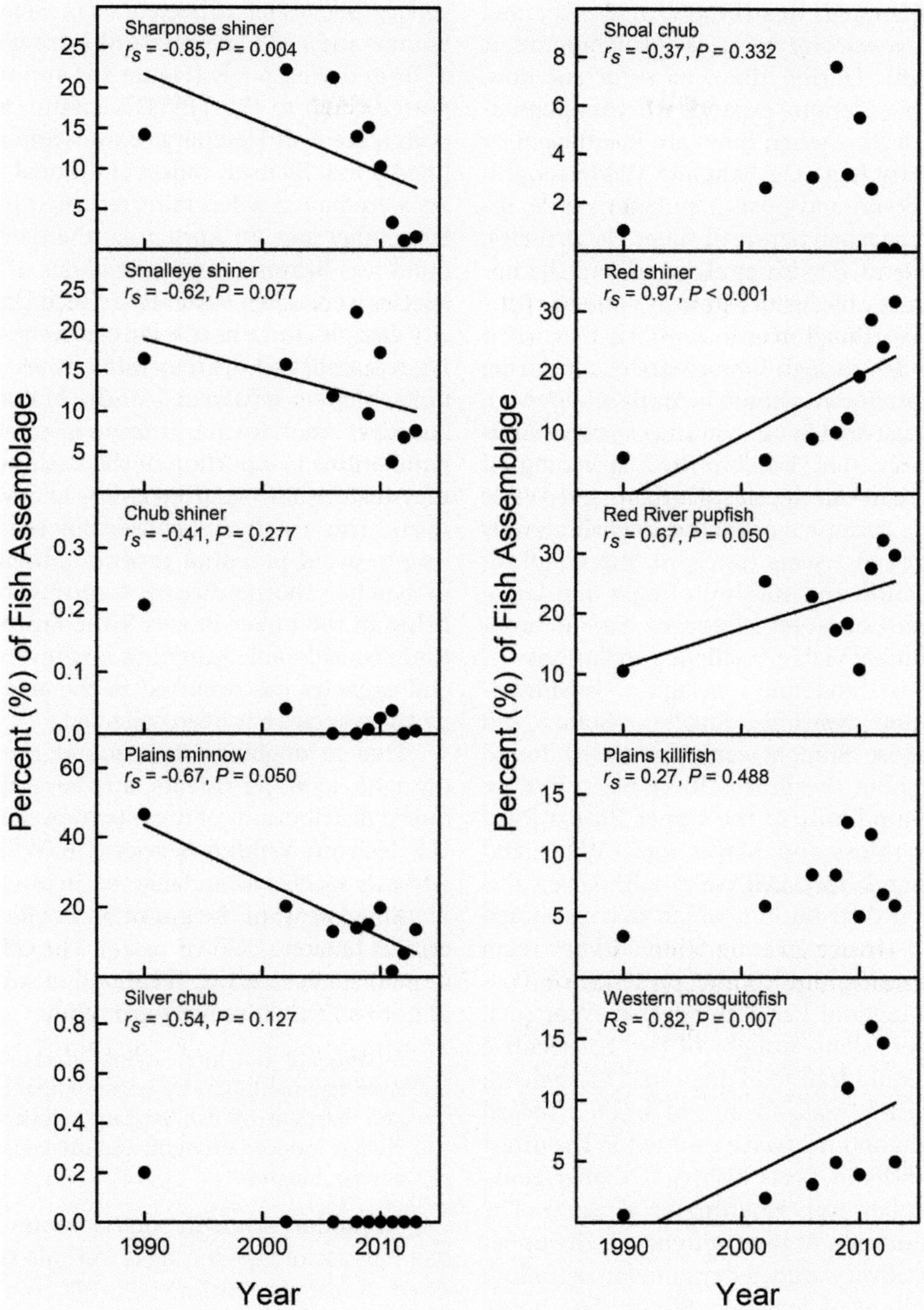

Figure 7. Trends in the relative abundance of native fishes in the Upper Brazos River Native Fish Conservation Area for nine 1-year periods spanning 1988–2014. Spearman's rank correlation (r_s) was used to assess the presence of a trend for increasing, or decreasing, relative abundance of each species. For species that showed a significant ($P \leq 0.05$) trend in relative abundance through time, a least-squares line was plotted to illustrate the general trend. Source: Modified from Wilde 2015.

strong enough to swim out of the channel into low-velocity habitats (Worthington et al. 2018). During times of sufficient flow, their populations expand, whereas populations decline when flows are insufficient or rivers dry (e.g., Durham and Wilde 2009b). This boom-and-bust population cycle reflects the harsh nature of Great Plains prairie ecosystems (Dodds et al. 2004) and the opportunistic life history strategy of these cyprinids (Worthington et al. 2018). Further, when storm-driven high-flow events occur during the reproductive season (~April–September), broadcast-spawning cyprinids spawn simultaneously (i.e., synchronized spawning of more than one species; Durham and Wilde 2009a). Because most of these cyprinids only live for 1 to 2 years (rarely 3), successful reproduction and survival of eggs and larval fish must occur on a frequent basis in order to maintain viable, resilient populations.

Two endemic pelagic, broadcast-spawning cyprinids, Smalleye Shiner and Sharpnose Shiner, were historically found throughout the Brazos River basin but are now found only in the Upper Brazos River NFCA (Moss and Mayes 1993; Wilde and Urbanczyk 2013; USFWS 2014b). Given this reduced distribution, which is constrained from downstream emigration and upstream colonization due to the presence of Possum Kingdom Lake, an adverse event such as a persistent drought of two consecutive years could lead to extinction. This concern was nearly realized in 2011 when a record drought and heatwave resulted in the driest 12 months in Texas history (October 2010–September 2011; Hoerling et al. 2013). The long duration of intermittency in the upper Brazos River watershed resulted in reproductive failure of Smalleye Shiner, Sharpnose Shiner, and other broadcast-spawning species (Wilde 2015). The combination of extended intermittency, high temperatures, and reproductive failure led to a coordinated rescue effort (Figure 2c) in mid-September 2011 in which the TPWD and Texas Tech University personnel collected 1,725 Smalleye Shiners and 1,388 Sharpnose Shiners (Figure 2) from drying pools (Figure 2d) and transported them to the TPWD's Possum Kingdom State Fish Hatchery to overwinter in a lined pond. Plans to collect additional shiners were halted when rains returned in late September 2011. In April 2012, the hatchery pond was harvested; 50 individuals of each species were taken to the Texas Tech University campus in Lubbock for captive spawning research, and the remaining fishes (~370 of each) were repatriated to the lower Brazos River near Hearne, Texas to re-establish populations in a portion of the basin where they historically occurred (Moss and Mayes 1993). This reach was chosen for repatriation to avoid potential genetic bottlenecks or hatchery-borne disease transmission to fishes in the upper Brazos River. Since that time, considerable sampling by universities and agencies has occurred in the area, but neither species has been collected.

Due to ongoing impacts and the 2011 drought, looming threats, and severely reduced distributions of these two species, the U.S. Fish and Wildlife Service (USFWS) listed both species as endangered in 2014 and identified most of the upper Brazos River as critical habitat (USFWS 2014a). The USFWS (2014a) states that the likelihood of survival of both shiners is reduced primarily

> by ongoing and intensifying effects of river fragmentation by impoundments and barriers to fish passage, climate change induced drought, and salt cedar encroachment.

Secondary stressors such as commercial bait harvest and golden algae were also identified. A recovery outline (USFWS 2015) was generated to guide conservation planning and delivery prior to a formal recovery process. The recovery outline addresses what both species need to avoid extinction (in terms of redundancy, resiliency, and representation), primary and secondary stressors, and conservation efforts to support recovery.

Conservation efforts include improved water management and conservation, restoring instream flows, improving habitat and connectivity, and captive propagation (USFWS 2015).

Conservation Opportunities

Basic research on the life history, environmental and habitat requirements, and reproductive strategies of many Great Plains fishes is needed, especially for broadcast-spawning cyprinids (see Worthington et al. 2018). Such biological information is needed to better inform impact assessments for water development and discharge proposals; processes for listing, recovery, and delisting; and conservation/recovery efforts and regulatory measures.

Measures to address commercial baitfish harvest

Regulations and permitting procedures implemented using a science-based approach can play a key role in conservation of native fishes. Take and sale of nongame fish, including commercial baitfish harvest, is regulated by the TPWD. Whether harvest of a species may be permitted is contingent on said species being designated by the Texas Parks and Wildlife Commission as in need of management (Nongame Species 1997). Currently, these permits may be issued for minnows of the genera *Campostoma*, *Cyprinella*, *Hybognathus*, *Macrhybopsis*, *Notemigonus*, *Notropis*, *Opsopoeodus*, *Phenacobius*, *Pimephales*, *Rhinichthys*, and *Semotilus* (Permits to Collect or Sell Nongame Fish Taken from Public Fresh Water 2017). Several of these genera include listed species and species of greatest conservation need (SGCN) identified in the 2012 Texas Conservation Action Plan (TPWD 2012).

Historically, bait dealers targeted prairie rivers and streams across the Great Plains because of the abundance of large minnows such as Plains Minnow. The ability, at that time, to drive in the riverbed to target minnows found in drying, intermittent pools during drought likely exacerbated impacts of drought on imperiled fishes (Ostrand and Wilde 2004; Hoerling et al. 2013). Enactment of state laws restricting motor vehicle use in navigable rivers below the gradient boundary (Access to Protected Freshwater Areas 2003) was followed by a decline in the number of permits requested for commercial bait harvest in the rivers of the Great Plains in Texas.

In 2014, as part of a multifaceted strategy for addressing the potential impacts of permitted commercial nongame fish harvest activities on SGCN fishes, the TPWD revised the application form for this permit to require lists of proposed devices and species requested for harvest to be water-body-specific. This update played a pivotal role in enabling the TPWD to evaluate potential impacts prior to permit issuance or renewal. Concomitantly, areas were identified by TPWD biologists where it was recommended that, for the take of minnows, use of certain devices (e.g., minnow seines) be disallowed due to potential bycatch of selected fish SGCN. A geospatial decision support tool was developed (see Figure 3) to facilitate implementation of enhanced permitting procedures based on these guidelines. To further support these efforts, regulations regarding these permits (Permits to Collect or Sell Nongame Fish Taken from Public Fresh Water 2017) were revised in 2017 to more clearly stipulate that the TPWD may not only deny a permit, but also impose conservation restrictions on permitted species, water bodies, and devices when the prospective take may be detrimental to any aquatic species.

In 2014, the TPWD conducted a review of all existing permits for commercial bait harvest. Only one remaining permit in the upper Brazos River basin was found. In 2015, renewal of said permit was denied based on the potential for the activity to be detrimental to fish SGCN. Furthermore, several permits for other river basins were denied or restricted on similar grounds. The decision support tool provides TPWD staff with readily available, species-specific justifications for permit denial or conservation restrictions intended to minimize impacts to SGCN fishes of the Brazos River and other sensitive areas.

Captive spawning and repatriation research

The species status assessment report for the two listed Brazos River shiners (USFWS 2014b) highlights the need to increase redundancy for these species, given that only single populations of each remain. Captive propagation and rearing will be critical for a successful refugia program, repatriation efforts, and basic research (e.g., environmental tolerances and requirements for all life stages, and conservation genetics). The Prairie Stream Fish Conservation and Propagation Laboratory located at Texas Tech University developed protocols for spawning and rearing 10 species. To support repatriation efforts, the laboratory raised two surrogate species, Plains Minnow and Shoal Chub (Wilde and Urbanczyk 2017). A total of 20,080 Plains Minnows and 2,840 Shoal Chub were batch-marked and released in the Brazos River downstream from Possum Kingdom Lake from 2014 to 2016 to assess that reach's viability for repatriation of prairie stream fishes. The middle Brazos River was chosen because the fishes historically occurred in this reach, daily hydropower peaking from Possum Kingdom was no longer expected, and the reach was possibly long enough to support reproduction and recruitment of imperiled shiners (Wilde and Urbanczyk 2013). Follow-up monitoring resulted in recapture of six fish (four Plains Minnows and two Shoal Chub); most were recaptured near release sites and one Shoal Chub was recaptured more than 62 km upstream. Additional monitoring is needed to determine if recruitment of either species has occurred in that reach of the river. If populations of either species are restored, then the process to repatriate listed fishes to the middle Brazos River could begin. Using wild-caught and propagated Smalleye Shiners and Sharpnose Shiners, an assessment of conservation genetics was performed to inform guidelines for maintaining genetic diversity in future refugia populations and repatriation efforts (Vazquez and Wilde 2019). Propagated fish are also being used to test salinity tolerances of larval and juvenile shiners to inform discharge permitting decisions and to determine hydraulic preferences related to instream habitat (Wilde 2018). While the Prairie Stream Fish Conservation and Propagation Laboratory can produce and rear sufficient fish for research and repatriation experiments, larger facilities that include spawning and hatching systems and grow-out ponds will be needed for large-scale repatriation efforts and captive refugia.

Barrier mitigation to improve connectivity

The three major dams and their impoundments in the Upper Brazos River NFCA effectively disrupt upstream fish migrations, precluding recolonization of reaches following local extirpation events. Large impoundments also act as barriers to downstream dispersal during times of severe drought and drying and, because of low current velocities in impounded headwaters, can also act as sinks for drifting early life stages. Wilde and Ostrand (1999) documented changes in fish assemblages in the Double Mountain Fork Brazos River upstream from Lake Alan Henry, including reductions in cyprinids and increases in Red River Pupfish, Plains Killifish, and other species normally found in impoundments. Plains Minnow was not collected 5 years postimpoundment, and Smalleye Shiner abundances appeared low (Wilde and Ostrand 1999). Neither species has been collected from this area since 1998 (Wilde 2015). Likewise, Possum Kingdom Lake precludes upstream recolonization from the middle and lower Brazos River and provides a terminus for other small-bodied native fishes—and any other fish regardless of body size. Several new reservoirs are proposed for development in the Upper Brazos River NFCA, including Post Reservoir on the North Fork Double Mountain Fork Brazos River and Cedar Ridge Reservoir on the Clear Fork, a tributary that feeds the Brazos River upstream from Possum Kingdom Lake. Life history requirements of native fishes could be considered in the engineering of dam infrastructure. For example, fish passage structures engineered and built upfront rather than after the dam is constructed could be more economical in the long term. Gated release-structures with multilevel outlets can be engineered to purposefully release high-flow pulses of sufficient hydrologic character (magnitude, rate of change, timing, duration, and frequency) and quality (e.g., temperature, dissolved oxygen) to support reproduction, downstream transport of early life stages, and successful recruitment.

Smaller dams and impoundments and roads (e.g., low water or culverted crossings) can also affect movement, recolonization, and recruitment. Improving connectivity at these types of structures can be accomplished through complete removal and restoration of river and riparian areas or through creation of fish passage structures that enable upstream migrations (see Pennock et al. 2017). In 2016, the TPWD, the USFWS, and other agencies oversaw the removal of a 2.4-m-tall and 30-m-wide culverted road crossing that had been illegally constructed by an oil and gas company on the Double Mountain Fork Brazos River (Figure 8A). Removal of the crossing restored access to more than 27 river kilometers for riverine fishes, including Sharpnose Shiner and Smalleye Shiner. Following removal, the river channel was reshaped and riparian vegetation restored in the affected area. Other structures such as pipeline coverings (Figure 8B) can create barriers to upstream migration, especially under lower flow conditions; poor design or construction can lead to undercutting and decrease connectivity. Pennock et al. (2017) documented movement of small-bodied Great Plains fishes through a fishway at a dam on the Arkansas River in Kansas. Similar fishways and rock ramps could mitigate small-scale barriers to upstream and downstream movements of small-bodied prairie stream fishes.

Environmental flows management

Natural flow regimes drive the ecological processes that support fish assemblages in the Upper Brazos River NFCA. Flow regimes include subsistence flows that occur during droughts and dry periods, base flows derived from spring inputs and bank storage, storm-driven high-flow pulses that cue and synchronize spawning, and overbanking flows that promote lateral connectivity and geomorphic processes (Annear et al. 2004; TIFP 2008). Several frameworks exist to help navigate practitioners through the complex, challenging process of environmental flow management (e.g., ecological limits of hydrologic alteration in Poff et al. 2010). Key elements include determining flow requirements/targets, identifying sources of hydrologic alteration, and aligning water management strategies to protect and restore environmental flow regimes.

Flow requirements for broadcast-spawning cyprinids include "flowing waters" (Durham and Wilde 2009a; USFWS 2014b), which can be indicated by a minimum flow value or by the number of zero-flow days in a given season or year. Flowing water is critical for preventing semibuoyant eggs and larvae of broadcast-spawning fishes from settling (Worthington et al. 2018). Additionally, literature-based flow targets include mean summer (May–September spawning season) discharge of 6.43 m^3/s for Smalleye Shiner (Durham and Wilde 2009b) and 2.61 m^3/s for Sharpnose Shiner (Durham 2007) at the Brazos River at Seymour, Texas U.S. Geological Survey (USGS) discharge gage

Figure 8. **(A)** A culverted road crossing on the Double Mountain Fork Brazos River being removed to restore connectivity for migratory fishes. Small barriers, such as **(B)** this pipeline covering on the Brazos River, can restrict fish passage under low-flow conditions.

(#08082500). These two studies used population dynamics models and validated results with field collections to identify flow conditions that support survival of age-0 fishes and growing populations. Following completion of Lake Alan Henry on the Double Mountain Fork Brazos River in 1994, the mean summer discharge in the Brazos River at Seymour decreased approximately 14% during the May–September spawning season (Durham and Wilde 2009b). The USFWS (2014a) also noted the decreased frequency with which mean summer discharge met the flow volumes for Sharpnose Shiner and Smalleye Shiner. High-flow pulses are also a critical flow component for broadcast-spawning cyprinids because the volumes in peak flows contribute to the mean summer discharge in addition to serving as a cue to synchronize spawning among these fishes (Durham and Wilde 2009a).

Sources of hydrologic alteration, including human water use (e.g., surface water diversions, water rights, groundwater pumping, and reservoir operations) and changes in land use and cover (e.g., distribution of invasive riparian vegetation), also need to be identified and quantified. In addition to reservoir construction, groundwater development in the Great Plains has depleted or substantially reduced streamflows in groundwater-dependent streams, including those in Texas (Brune 2002), impacting riverine habitat and prairie stream fishes (Hoagstrom et al. 2011; Perkin et al. 2017). For example, pumping from the Ogallala Aquifer has lowered the water table as much as 50 m (Konikow 2013), depleted springs that fed streams (Brune 2002), and caused streams flowing across depleted aquifers to lose water (Parsons Engineering Science Inc. 1999). An ongoing research project seeks to quantify the relationship between the groundwater in the Seymour Aquifer, the alluvial aquifer of the Brazos River, and surface water flows in the upper Brazos River to identify stream reaches most threatened by drying where aquifer pumping may reduce groundwater inflows to streams. Given the complexity of groundwater–surface water relationships in the upper Brazos River, several approaches will be used, including a conductivity-based hydrograph separation (Matsubayashi et al. 1993; Cox et al. 2007; Stewart et al. 2007) and assessment of long-term stream discharge records and calculation of trends in base flow index, which is the percent of streamflow consisting of groundwater inflows (Maidment 1993; Richter et al. 1996; Fenelon and Moreo 2002). Using a complementary approach, gaining reaches at risk of reduced streamflow or drying from nearby groundwater pumping will be identified by examining groundwater monitoring well records. Wells in the Texas Water Development Board's groundwater database (TWDB 2014; some of which were analyzed in Jones et al. [2012] and Jigmond et al. [2014]) with long-term groundwater level records located in the Seymour Aquifer or Brazos River alluvium will be assessed to identify declining groundwater trends. Following the general approach of Reynolds et al. (2015), long-term stream gage records will be used to calculate metrics that include changes in zero-flow days, zero-flow months, minimum 90-d streamflow, and mean summer spawning season discharge. In addition, the effects of the 2011 drought on streamflow will be assessed, along with evaluating antecedent conditions, which may inform future rescue efforts. High-flow pulses will also be assessed to identify changes in magnitude, frequency, and duration, with a focus on spawning flows and if a characteristic high-flow pulse threshold may recharge alluvial aquifer bank storage and maintain subsequent base flows. Finally, minimum- and high-flow metrics will be related to a suite of environmental variables (e.g., precipitation, land use, and drainage area) using regression approaches to understand how these factors may affect streamflows.

Identifying and implementing environmental flow strategies presents many challenges, including scientific uncertainty, lack of political and public support, and limitations in resources (Opperman et al. 2018). However, opportunities exist to achieve flow restoration and protection and minimize impacts of future water management strategies. For example, using a geospatial tool called the Environmental Flow Information Toolkit (EFIT), the TPWD and collaborators are identifying flow restoration and protection strategies that provide the greatest conservation value in the Great Plains region of Texas (see Valente et al. 2019, this volume). Potential restoration strategies include flow and water level agreements with reservoir operators (where appropriate and feasible), dry-year water leases with irrigators, and improved land or riparian management (e.g., saltcedar control). Protection strategies would be appropriate in areas where no or limited human water uses exists or where instream and riparian habitats are preserved in natural states.

Saltcedar management, monitoring, and modeling

Saltcedar control strategies.—Texas, unlike many other western states where saltcedar has become invasive, is approximately 95% privately owned and thus significant landowner involvement is required to implement watershed-scale saltcedar management. In 2015, the TPWD began a partnership with landowners as well as other agencies (e.g., USFWS Partners for Fish and Wildlife Program and Texas A&M AgriLife Extension) to manage saltcedar in the upper Brazos River. In 2016–2018, 4,209 ha along the major forks of the upper Brazos River were treated with herbicide using aerial application (Figure 9). Helicopter-based, foliar application of Imazapyr has been shown to be highly effective for achieving control of saltcedar (Duncan and McDaniel 1998). Helicopter-based treatment is more precise and the added maneuverability (as compared to fixed-wing aircraft) enables the pilot to avoid eastern cottonwood *Populus deltoides*—which are highly cherished along the upper Brazos River—and other desirable native trees (McDaniel and Taylor 2003).

Imazapyr is a member of the imidazolinone family of herbicides and inhibits the synthesis of enzymes that play a key role in amino acid production found only in plants. Aquatic formulations of Imazapyr used for saltcedar control have been shown to be "practically nontoxic" to fish and wildlife and break down quickly in water (Shaner and O'Connor 1991). However, the use of this chemical treatment is not without consequence. Much of the Imazapyr binds to the soil and degrades primarily because of microbial activity. As a result, Imazapyr can have long-lasting, residual activity in the soil (Shaner and O'Connor 1991). This can be prolonged during periods of drought and hinder re-establishment of native plants (Rod Goodwin, Canadian River Municipal Water District, personal communication).

Saltcedar beetles *Diorhabda* spp., a biological control agent found in the tree's native range, have been introduced across the western United States, including Texas, to aid in saltcedar management (Bean et al. 2013). At least one species is present in the upper Brazos River watershed. Anecdotal reports suggest that Brazos populations of the beetles are transient and often found in low numbers. The transient nature of saltcedar beetles is often attributed to their preference for feeding on new growth, yet several years of consistent defoliation by the beetles is required to result in saltcedar mortality (Bean et al. 2013). Complete eradication of an invasive plant such as saltcedar is virtually impossible. Rather than expend significant cost and manpower to implement ground-based treatment of small creeks and draws that are often inaccessible or hazardous to helicopters (e.g., crossed by power lines), these small areas were left untreated in the upper Brazos River watershed to serve as refuges for beetles as part of an integrated pest management strategy. Following herbicide treatment, fresh re-sprouts may be more appetizing to beetles than older, dense stands that dominated prior to treatment. Active defoliation in these areas, where beetles previously had little or no impact, could enhance their effectiveness.

The goal of saltcedar management is "designed disturbance," essentially an effort to mimic, to some degree, the natural processes that regulate riparian plant communities and river channel migration within the floodplain. This strategy can help to create a mosaic of areas in which habitat can be restored to provide ecosystem services and benefit fish and wildlife. Implementation of an adaptive, integrated management strategy for saltcedar must begin with identification of achievable goals (e.g., treatment of 75% of the saltcedar on the North Fork Double Mountain Fork and Double Mountain Fork Brazos River). Management of saltcedar alone without high-flow events to serve as a reset for the system will not suffice to restore these habitats. However, saltcedar management can help to maximize the benefits of flood events and promote improvements in river function, instream habitat quality, and diversity of plants and wildlife (Keller et al. 2014).

Instream habitat and riparian vegetation monitoring.—Saltcedar management in the upper Brazos River has the potential to restore instream habitat through channel widening and support reestablishment of native riparian flora (Auerbach et al. 2013). To assess potential long-term changes, monitoring sites were established on the Double Mountain Fork Brazos River and Salt Fork Brazos River (see Figure 9) prior to saltcedar treatments. At each site, 11 instream habitat transects were established in all mesohabitats (e.g., riffles, runs, and pools) along a reach length of 40 channel widths. Transects were surveyed prior to saltcedar treatment and will be resurveyed following treatment and after sufficient response time and discharge events (i.e., high-flow pulses and floods) have occurred. Instream habitat data from each transect includes cross-section profile (depth and water surface elevation), substrate types (e.g., sand, silt, sand gravel), and current velocities. Discharge is also taken at each site if a USGS discharge gage is not located nearby. Vegetation surveys associated with five randomly selected instream habitat transects were also conducted. Percent cover of dominant woody species, grasses and forbs, leaf litter/woody debris, and gravel/rock/bare earth in ground cover/shrub, subcanopy, and canopy layers along 1-m by 50-m belt transects on both riverbanks. Vegetation transects were surveyed prior to and 1 year after treatment and will be resurveyed each year.

Post-saltcedar treatment hydrological monitoring and modeling.—Monitoring the hydrological impacts of post-saltcedar abatement requires multiscale knowledge of soil moisture and groundwater and surface water fluxes (Figure 10). In 2017, six long-term sites (see Figure 9) were established, each with three monitoring locations in treated riparian areas along the Double Mountain Fork ($n = 3$), Salt Fork ($n = 2$), and the Brazos River main stem ($n = 1$). Each location has equipment installed to monitor groundwater level, temperature, and salinity at three landscape positions. Soil moisture sensors installed at five depths track storage of water in the root zone.

Monitored, in situ data provide quantification of hydrological alterations following saltcedar abatement along the Brazos River. There are several limitations to this research: (1) there is no control (e.g., untreated) or comparable watershed, (2) duration of data acquisition is contingent upon continued funding, and (3) the logistics of saltcedar control and landowner agreements can limit the spatial extent of control efforts. However, the potential upper limit of hydro-

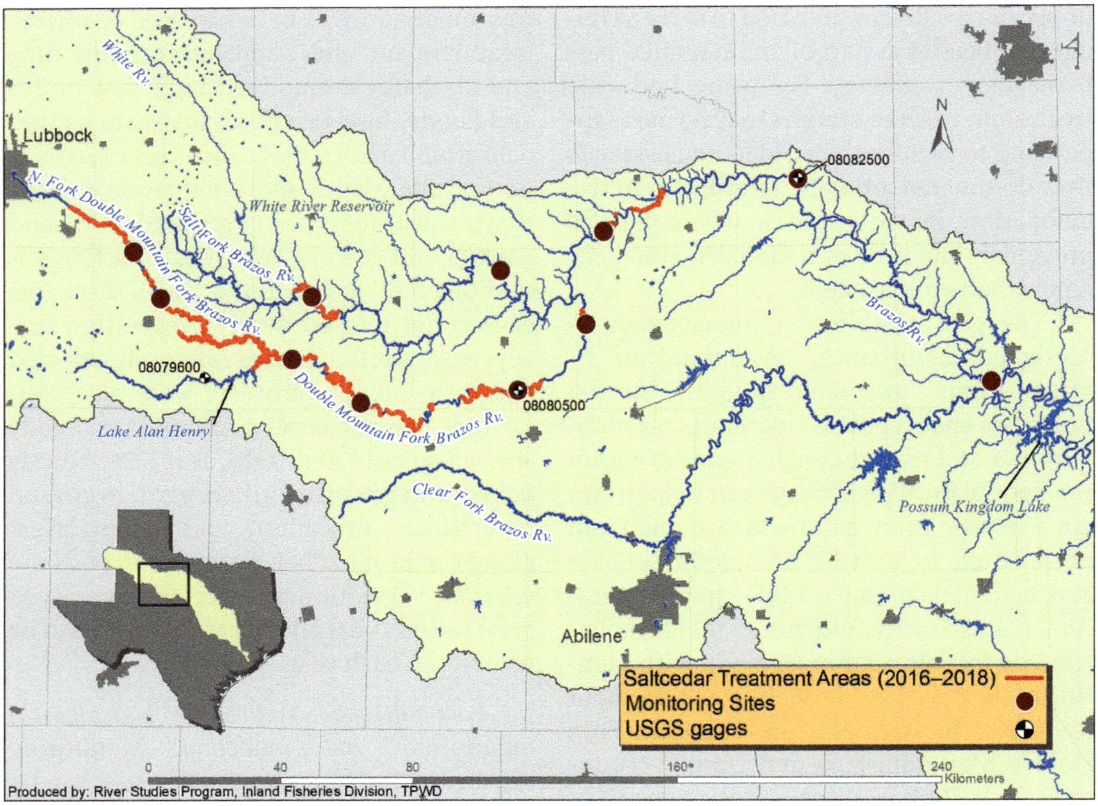

Figure 9. Saltcedar herbicide treatments facilitated by the Texas Parks and Wildlife Department from 2016 to 2018 covered 4,209 ha. Sites (dark red circles) for instream and riparian habitat and hydrological studies show the geographic scope of monitoring activities. U.S. Geological Survey discharge gage numbers are shown for the three gages referenced in the chapter: Brazos River at Seymour, Texas (#08082500); Double Mountain Fork Brazos River near Aspermont, Texas (#08080500); and Double Mountain Fork Brazos River at Justiceburg, Texas (#08079600).

logic impact from saltcedar management can be estimated using numerical modeling. We focused efforts on the upper reaches of the Double Mountain Fork Brazos River and a parsimonious numerical model: the coupled routing and excess storage (CREST) model (Wang et al. 2011). CREST is a distributed hydrological model developed to simulate the spatial and temporal variation of land surface, and subsurface water fluxes and storages.

The primary inputs to CREST are spatially and temporally distributed rainfall and potential evapotranspiration (PET). CREST evaluates water fluxes by routing surface and subsurface runoff between model cells based upon watershed topography using a digital elevation model. Our objective was to simulate historical discharge on the Double Mountain Fork Brazos River near Aspermont, Texas (USGS discharge gage #08080500), then model scenarios of different levels of saltcedar abatement by decreasing leaf area and PET along the stream. The model was calibrated in two steps beginning upstream from Lake Alan Henry on the Double Mountain Fork Brazos River at Justiceburg, Texas (USGS discharge gage #08079600), then calibrating the remaining watershed at the Aspermont gage (see Fig-

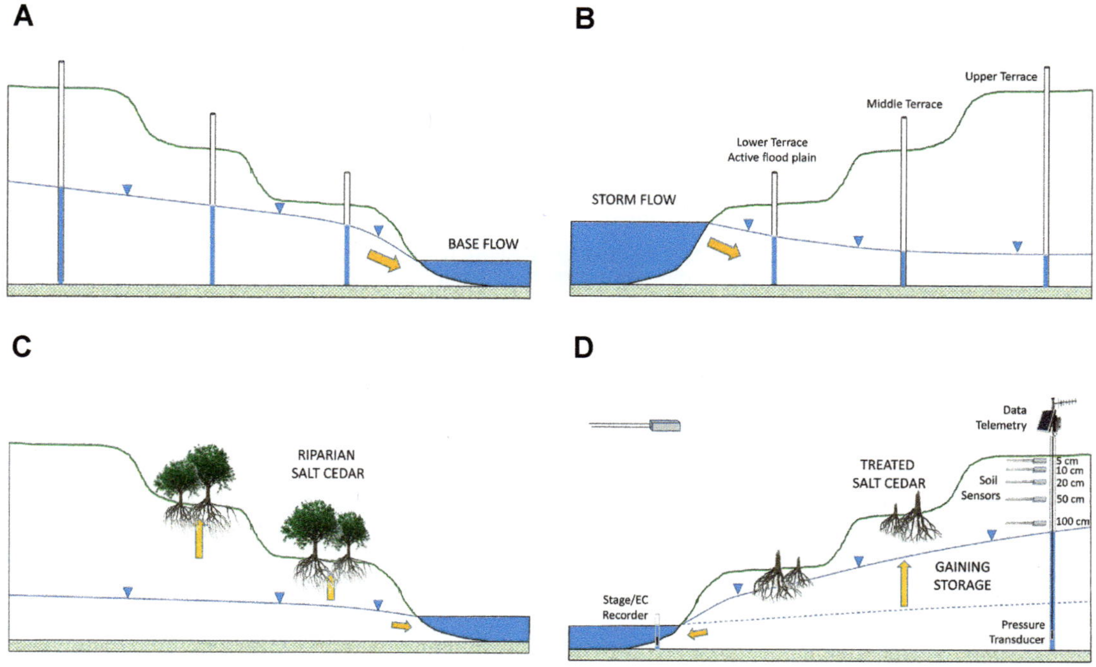

Figure 10. Surface water–groundwater interaction in alluvial semiarid basins like the upper Brazos River fluctuate between **(A)** base flow conditions when the hydraulic gradient in bank sediments is higher than river stage and groundwater discharges and **(B)** storm flow when surface waters replenish bank storage, reversing the hydraulic gradient. Similar to a pumping well, **(C)** riparian vegetation (e.g., saltcedar) alters this hydraulic gradient through evapotranspiration, lowering the alluvial water table. **(D)** Saltcedar control will reduce this gradient leaving more water in the river, but bank storage must be replenished before any increase in discharge is realized. To monitor these dynamics, six locations are equipped to measure groundwater levels and soil moisture at lower, middle, and upper river terrace positions. Along with river stage, this real-time data will improve our understanding of the complex interaction between surface water, groundwater, and saltcedar control.

ure 9). Once the model was calibrated, PET was manipulated along the riparian corridor to simulate reductions in saltcedar following herbicide application to a 1-km (model grid size) buffer within the treated area. Riparian consumptive water use is numerically governed by leaf area index and PET. Thus, we modeled the effects of a 10, 25, 50, and 75% reduction in PET within the treatment corridor to assess resulting changes in stream discharge at the downstream discharge gage.

Figure 11 shows monthly simulated and observed discharge for the study period. The correlation coefficients for monthly flows were 0.83 and 0.72, respectfully, with root mean square errors of 3.09 m^3/s and 1.54 m^3/s at the Aspermont (#08080500) and Justiceburg (#08079600) streamflow gages, respectively. The model was successful in reproducing base flow conditions at monthly time steps but was unable to reproduce storm events at daily time steps. However, it was adequate to evaluate decreases in leaf area index and PET from abatement scenarios, which are very liberal estimates of both the treatment area, efficiency, and the impact on PET.

Results suggest average annual increase in water yield from saltcedar treatment would be 5,630 and 10,800 L/ha treated within the

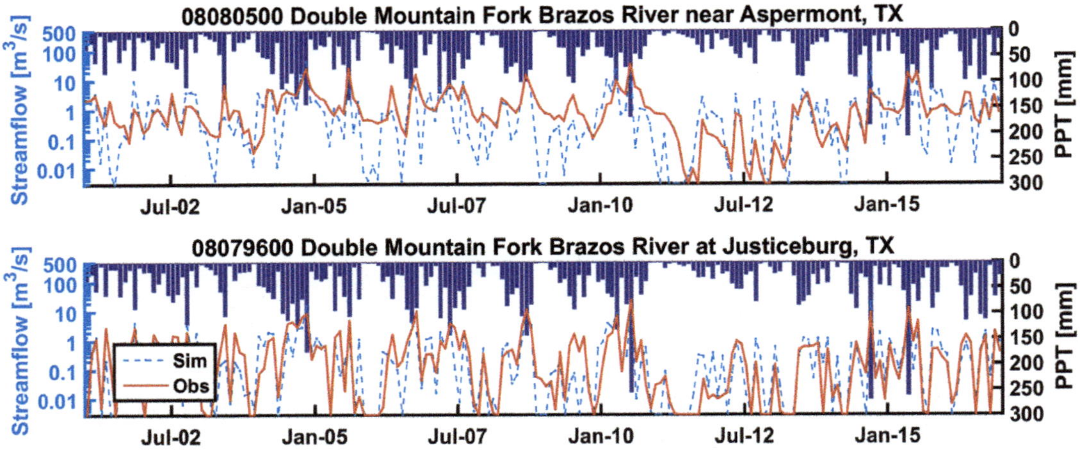

Figure 11. Observed streamflow (orange solid line) on the Double Mountain Fork Brazos River and simulated monthly streamflow (blue dashed line) using the calibrated CREST model. Annual precipitation (PPT) shown as blue bars.

watershed upstream of the Aspermont and Justiceburg gages, respectively, when a 10% reduction was modeled. A 75% reduction in saltcedar would increase water yield to 175,000 and 344,000 L/ha. Figure 12 shows the annual mean streamflow volume difference compared to no treatment. During wet years, the highest change was observed with little to no change during dry years. Effective implementation of environmental flows requires a clear scientific connection between the potential gains in streamflow during wet years and its improvement to habitat conditions (e.g., channel morphology, water temperature, and water chemistry), and fish abundance and richness. Similar results were obtained by Harwell et al. (2016) along the Double Mountain Fork upstream from Lake Alan Henry using the Soil and Water Assessment Tool (SWAT) model. Simulated average annual water-yield was 55,000–322,000 L/ha of shrubland replaced by grassland. Although at least one field study has suggested only modest benefits to water resources following saltcedar abatement (McDonald et al. 2015), there were key differences between that study and saltcedar management in the upper Brazos River watershed. Differences in width and density of saltcedar infestation and groundwater dynamics (i.e., losing versus gaining reaches) suggest that implementation of this larger-scale monitoring effort could yield beneficial insights to help guide future saltcedar management decisions.

Conclusions

Over the past 80 years, human activities have greatly modified the riverscape of the Brazos River, yet the natural cycles of flood and drought continue. Historically, native fish populations would rise and fall in response to long-term climatic conditions. As upstream reaches began to dry, fishes could move downstream, following the water, and return upstream when wetter flow conditions prevailed. Today, native fishes face impediments to downstream movement as barriers and impoundments block downstream migration routes and access to downstream refugia habitat. Three major dams on the Brazos River have effectively fragmented the river into three zones that disrupt connectivity between populations. Water supply and flood control reservoirs have been built or are proposed to be built on all major tributaries in the Brazos River watershed, including those in the Upper Brazos River

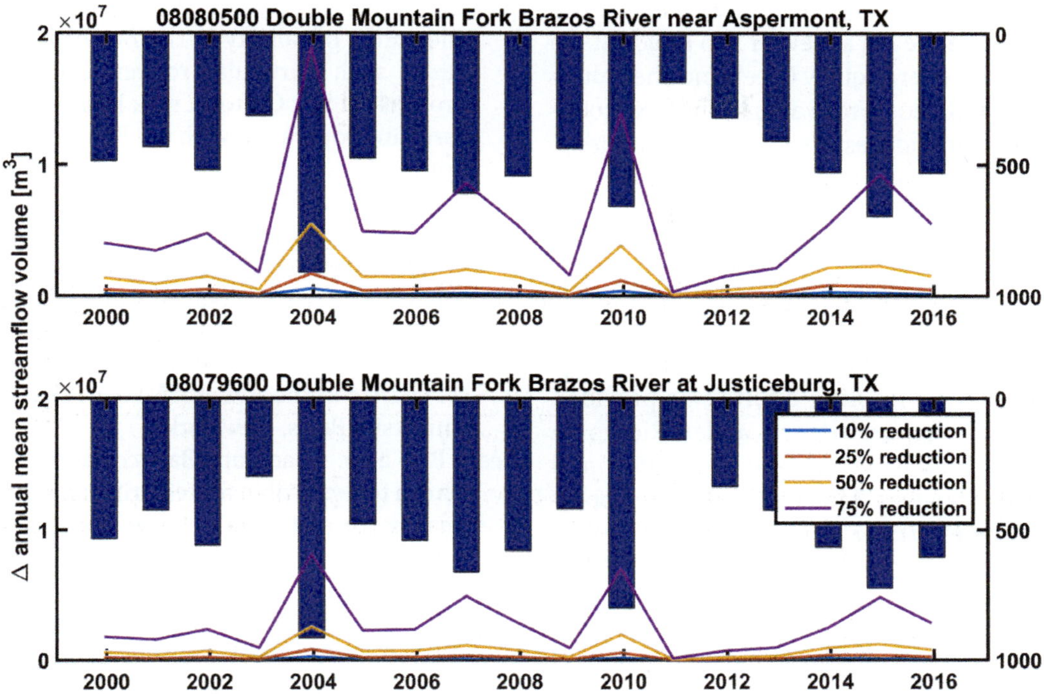

Figure 12. Simulated change in mean annual volume ($m^3 \times 10^7$) for a 10, 25, 50, and 75% reduction in saltcedar evapotranspiration when compared to no treatment at U.S. Geological Survey discharge gages #08080500 Double Mountain Fork Brazos River near Aspermont, Texas and #08079600 Double Mountain Fork Brazos River at Justiceburg, Texas for 2000–2016. Annual precipitation (PPT) shown as blue bars. See Figure 9 for gage locations.

NFCA. Aside from the changes in hydrology, water quality, and sediment transport imposed by reservoirs and operations, the span of river reaches available for these unique and highly migratory species to fulfill their life history is shrinking (see Perkin and Gido 2011). Reduced base flows and drying from drought, groundwater pumping, surface water extractions, and invasive plants that alter hydrology and instream and riparian habitat compound the effects of fragmentation, leading to compromised biodiversity (Perkin et al. 2015b) and potentially altering the availability of water for human uses and values. Among Great Plains states, Texas boasts the richest assemblage of prairie minnows but also the greatest number of imperiled fishes (Worthington et al. 2018). Our science and research programs—as well as those of many scientists that have stained their seines in the turbid rivers of the Great Plains—can inform decisions about water and river management and policy. However, conservation delivery targeted at flow restoration and protection, fish passage, and fish repatriation needs to be coordinated across boundaries and entities before more prairie stream fishes become imperiled or lost from the Great Plains.

Dedication

This chapter is dedicated to Ed Lowe (1949–2018), who passed away on a canoe trip while scouting the banks of the Rio Grande. Ed was a river advocate, conservationist, and activist and founded the Friends of the Brazos to keep clean, flowing water in the Brazos River. He was an educator, storyteller, and adventurist who inspired the young and old to enjoy, explore, and care about all rivers. In 2012, he led a crew of fish biologists to Tres Rios, where Squaw Creek and the Paluxy meet the Brazos. From a high bluff overlooking the confluences, we slid down the steep, sandy slope with our seines and buckets. I wondered how the heck I was going to get back up. Follow Ed! ...that's what I did. K.B.M.

Acknowledgments

Funding for saltcedar management and hydrological monitoring efforts was provided by the Texas Legislature under Riders 32 and 34: Statewide Aquatic Vegetation and Invasive Species Management. We appreciate the efforts of Karim Aziz (TPWD) for producing maps, Clint Robertson (TPWD) for editing photographs, Stephan Magnelia (TPWD) for helpful comments on an earlier draft, and Tara Bongiovanni (University of Texas Bureau of Economic Geology) for producing water budget graphs.

References

Access to Protected Freshwater Areas. 2003. Texas Parks and Wildlife Code, sections 90.001–90.011. Available: https://statutes.capitol.texas.gov/Docs/PW/htm/PW.90.htm. (April 2019).

Annear, T., I. Chisholm, H. Beecher, A. Locke, P. Aarrestad, C. Coomer, C. Estes, J. Hunt, R. Jacobson, G. Jöbsis, J. Kauffman, J. Marshall, K. Mayes, G. Smith, R. Wentworth, and C. Stalnaker. 2004. Instream flows for riverine resource stewardship, revised edition. Instream Flow Council, Cheyenne, Wyoming.

Archer, K. L. 2015. Unruly waters: a social and environmental history of the Brazos River. University of New Mexico Press, Albuquerque.

Auerbach D. A., D. M. Merritt, and P. B. Shafroth. 2013. *Tamarix*, hydrology, and fluvial geomorphology. Pages 99–122 *in* A. Sher and M. F. Quigley, editors. Tamarix: a case study of ecological change in the American West. Oxford Press, New York.

Baker, R., L. S. Hughes, and I. Yost. 1964. Natural sources of salinity in the Brazos River, Texas, with particular reference to the Croton and Salt Croton Creek basins. Prepared in cooperation with the Brazos River Authority. U.S. Geological Survey Water-Supply Paper 1669-CC.

Bean, D. W., T. Dudley, and K. Hultine. 2013. Bring on the beetles. The history and impact of tamarisk biological control. Pages 377–403 *in* A. Sher and M. F. Quigley, editors. Tamarix: a case study of ecological change in the American West. Oxford University Press, New York.

Bean, P. T. 2017. Guadalupe Bass conservation plan: a ten-year plan for restoring and preserving the state fish of Texas 2017–2026. Texas Parks and Wildlife Department, PWD PL T3200-2078, Austin.

BEG (Bureau of Economic Geology). 1996. River basin map of Texas. University of Texas, Austin, Texas. Available: www.beg.utexas.edu/UTopia/images/pagesizemaps/river_basin.pdf.

Birdsong,T. W., G. P. Garrett, B. J. Labay, M. G. Bean, P. T. Bean, J. Botros, M. J. Casarez, A. E. Cohen, T. G. Heger, A. Kalmbach, D. A. Hendrickson, S. J. Magnelia, K. B. Mayes, M. E. McGarrity, R. McGillicuddy, M. M. Parker, and S. Robertson. 2019. Texas Native Fish Conservation Areas Network: strategic investments in restoration and preservation of freshwater fish diversity. Pages 183–229 *in* D. C. Dauwalter, T. W. Birdsong, and G. P. Garrett, editors. Multispecies and watershed approaches to freshwater fish conservation. American Fisheries Society, Symposium 91. Bethesda, Maryland.

Blackburn, W. H., R. W. Knight, and J. L. Schuster. 1982. Saltcedar influence on sedimentation in the Brazos River. Journal of Soil and Water Conservation 37:298–301.

Bonner, T. H., and D. T. Runyan. 2007. Fish assemblage changes in three western Gulf slope drainages. Texas Water Development Board, Final Report, Austin.

Brune, G. M. 2002. Springs of Texas, volume 1. Texas A&M University Press, College Station.

Busby, F. E., Jr., and J. L. Schuster. 1973. Woody phreatophytes along the Brazos River and selected tributaries above Possum Kingdom Lake. Texas Water Development Board, Report 168, Austin.

Childs, M. R., A. A. Echelle, and T. E. Dowling. 1996. Development of the hybrid swarm between Pecos Pupfish (Cyprinodontidae: *Cyprinodon pecosensis*) and Sheepshead Minnow (*Cyprinodon variegatus*): a perspective from allozymes and mtDNA. Evolution 50:2014–2022.

Cleverly, J. R. 2013. Water use by *Tamarix*. Pages 85–98 *in* A. Sher and M. F. Quigley, editors. Tamarix: a case study of ecological change in the American West. Oxford University Press, New York.

Cohen, A. E., G. P. Garrett, M. J. Casarez, D. A. Hendrickson, B. J. Labay, T. Urban, J. Gentle, D. Wiley, and D. Walling. 2018. Conserving Texas biodiversity: status, trends, and conservation planning for fishes of greatest conservation need. Texas Parks and Wildlife Department, Austin.

Conner, J. V., and R. D. Suttkus. 1986. Zoogeography of freshwater fishes of the western gulf slope of North America. Pages 413–456 *in* C. H. Hocutt and E. O. Wiley, editors. The zoogeography of North American freshwater fishes. Wiley, New York.

Cox, M. H., G. W. Su, and J. Constantz. 2007. Heat, chloride, and specific conductance as ground water tracers near streams. Groundwater 45:187–195.

Dean, D., M. Scott, P. Shafroth, and J. Schmidt. 2011. Stratigraphic, sedimentologic, and dendrogeomorphic analyses of rapid floodplain formation along the Rio Grande in Big Bend National Park, Texas. Geological Society of America Bulletin 123:1908–1925.

Doble, R., P. Brunner, J. McCallum, and P. G. Cook. 2012. An analysis of river bank slope and unsaturated flow effects on bank storage. Groundwater 50:77–86.

Dodds, W. K., K. Gido, M. R. Whiles, K. M. Fritz, and W. J. Matthews. 2004. Life on the edge: the ecology of Great Plains prairie streams. BioScience 54:205–216.

Doody, T. M., P. L. Nagler, E. P Glenn., G. W. Moore, K. Morino, K. R. Hultine, and R. G. Benyon. 2011. Potential for water salvage by removal of non-native woody vegetation from dryland river systems. Hydrological Processes 25:4117–4131.

Dudley, T. L., C. J. DeLoach, J. E. Lovich, and R. I. Carruthers. 2000. Saltcedar invasion of western riparian areas: impacts and new prospects for control. Pages 345–381 *in* E. McCabe and S. E. Loos, editors. Transactions of the 65th North American wildlife and natural resources conference. Wildlife Management Institute, Washington, D.C.

Duncan, K. W., and K. C. McDaniel. 1998. Saltcedar (*Tamarix* spp.) management with Imazapyr. Weed Technology 12:337–344.

Durham, B. W. 2007. Reproductive ecology, habitat associations, and population dynamics of two imperiled cyprinids in a Great Plains river. Doctoral dissertation. Texas Tech University, Lubbock.

Durham, B. W., and G. R. Wilde. 2009a. Effects of streamflow and intermittency on the reproductive success of two broadcast-spawning cyprinid fishes. Copeia 2009:21–28.

Durham, B. W., and G. R. Wilde. 2009b. Population dynamics of the Smalleye Shiner, an imperiled cyprinid fish endemic to the Brazos River, Texas. Transactions of the American Fisheries Society 138:666–674.

Durham, B. W., and G. R. Wilde. 2014. Understanding complex reproductive ecology in fishes: the importance of individual and population-scale information. Aquatic Ecology 48:91–106.

Echelle, A. A., A. F. Echelle, and L. G. Hill. 1972. Interspecific interactions and limiting factors of abundance and distribution in the Red River Pupfish, *Cyprinodon rubrofluviatilis*. American Midland Naturalist 88:109–130.

Evermann, B. W., and W. C. Kendall. 1894. The fishes of Texas and Rio Grande basin, considered chiefly with reference to their geographic distribution. U.S. Fish Commission Bulletin, Washington, D.C.

Fenelon, J., and M. Moreo. 2002. Trend analysis of ground-water levels and spring discharge in the Yucca Mountain region,

Nevada and California 1960–2000. U.S. Geological Survey, Water-Resources Investigations Report 02-4178, Reston, Virginia.

Gillespie, B. M., and J. R. Giardino. 1997. The nature of channel planform change: Brazos River, Texas. Texas Journal of Science 49:109–142.

Glenn, E. P., and P. L. Nagler. 2005. Comparative ecophysiology of *Tamarix ramosissima* and native trees in western US riparian zones. Journal of Arid Environments 61:419–446.

Graves, J. 1960. Goodbye to a river. Knopf, Inc., New York.

Hart, C. R., L. D. White, A. McDonald, and Z. P. Sheng. 2005. Saltcedar control and water salvage on the Pecos River, Texas, 1999–2003. Journal of Environmental Management 75:399–409.

Harwell, G. R., V. G. Stengel, and J. R. Bumgarner. 2016. Simulation of streamflow and the effects of brush management on water yields in the Double Mountain Fork Brazos River watershed, western Texas 1994–2013. U.S. Geological Survey, Scientific Investigations Report 2016-5032, Reston, Virginia.

Higgins, C. L., and G. R. Wilde. 2005. The role of salinity in structuring fish assemblages in a prairie stream system. Hydrobiologia 549:197–203.

Hillis, D. M., E. Milstead, and S. L. Campbell. 1980. Inland records of *Fundulus grandis* (Cyprinodontidae) in Texas. The Southwestern Naturalist 25:271–272.

Hoagstrom, C. W., J. E. Brooks, and S. R. Davenport. 2011. A large-scale conservation perspective considering endemic fishes of the North American plains. Biological Conservation 144:21–34.

Hoerling, M., A. Kumar, R. Dole, J. W. Nielsen-Gammon, J. Eischeid, J. Perlwitz, X.-W. Quan, T. Zhang, P. Pegion, and M. Chen. 2013. Anatomy of an extreme event. Journal of Climate 26:2811–2832.

Hubbs, C., R. J. Edwards and G. P. Garrett. 2008. An annotated checklist of the freshwater fishes of Texas, with keys to identification of species, 2nd edition. Texas Journal of Science Supplement 43(4).

Jigmond, M., W. R. Hutchison, and J. J. Shi. 2014. Final report: groundwater availability model of the Seymour Aquifer in Haskell, Knox, and Baylor counties. Texas Water Development Board, Austin.

Jones, T. L., J. E. Ewing, T. Yan, J. F. Pickens, B. R. Scanlon, J. Olyphant, and A. Chastain-Howley. 2012. Final report: conceptual model for the refined Seymour Aquifer groundwater availability model: Haskell, Knox, and Baylor counties. Prepared for the Texas Water Development Board, Austin.

Keller, D. L., B. G. Laub, P. Birdsey, and D. J. Dean. 2014. Effects of flooding and tamarisk removal on habitat for sensitive fish species in the San Rafael River, Utah: implications for fish habitat enhancement and future restoration efforts Environmental Management 4:465–478.

Kimmel, J. 2011. Exploring the Brazos River: from beginning to end. Texas A&M University Press, College Station.

Kloesel, K., B. Bartush, J. Banner, D. Brown, J. Lemory, X. Lin, G. McManus, E. Mullens, J. Nielsen-Gammon, M. Shafer, C. Sorenson, S. Sperry, D. Wildcat, and J. Ziolkowska. 2018. Southern Great Plains. Pages 987–1035 *in* D. R. Reidmiller, C. W Avery, D. R. Easterling, K. E. Kunkel, K. L. M. Lewis, T. K. Maycock, and B. C. Stewart, editors. Impacts, risks, and adaptation in the United States: fourth national climate assessment, volume II. U.S. Global Change Research Program, Washington, D.C.

Konikow, L. F. 2013. Groundwater depletion in the United States (1900–2008). U.S. Geological Survey, Scientific Investigations Report 2013-5079, Reston, Virginia.

Labay, B. J., A. E. Cohen, D. A. Hendrickson, B. Sissel, S. Sarkar, and M. Casarez. 2013. Data compilation, distribution models, conservation planning, and status survey for selected fishes of concern in Texas and region. University of Texas at Austin, Final Report, Austin.

Lytle, D. A., and N. L. Poff. 2004. Adaptation to natural flow regimes. Trends in Ecology and Evolution 19:94–100.

Maidment, D. R. 1993. Handbook of hydrology. McGraw-Hill, New York.

Matsubayashi, U., G. T. Velasquez, and F. Takagi. 1993. Hydrograph separation and flow analysis by specific electrical conductance of water. Journal of Hydrology 152:179–199.

McDaniel, K. C., and J. P. Taylor. 2003. Aerial spraying and mechanical saltcedar control. Pages 100–105 *in* Proceedings of the saltcedar and water resources in the West symposium. Texas Agricultural Experiment Station and Texas Cooperative Extension, College Station.

McDonald, A. K., B. P. Wilcox, G. W. Moore, C. R. Hart, Z. P. Sheng, and M. K. Owens. 2015. Tamarix transpiration along a semiarid river has negligible impact on water resources. Water Resources Research 51:5117–5127.

Mitchell, S. A. 1846. Map of the state of Texas: engraved to illustrate Mitchell's, school and family geography. Available: texashistory.unt.edu/ark:/67531/metapth192714/. (August 2018).

Moss, R. E., and K. B. Mayes. 1993. Current status of *Notropis buccula* and *Notropis oxyrhynchus* in Texas. Texas Parks and Wildlife Department, River Studies Report No. 8, Austin.

Nagler, P. L., E. P. Glenn, K. Didan, J. Osterberg, F. Jordan, and J. Cunningham. 2008. Wide-area estimates of stand structure and use of *Tamarix* spp. on the lower Colorado River: implications for restoration and water management projects. Restoration Ecology 16:136–145.

Nongame Species. 1997. Texas Parks & Wildlife Code, section 67.004. Available: https://statutes.capitol.texas.gov/Docs/PW/htm/PW.67.htm. (April 2019).

Opperman, J. J., E. Kendy, R. E. Tharme, A. T. Warner, E. Barrios, and B. D. Richter. 2018. A three-level framework for assessing and implementing environmental flows. Frontiers in Environmental Science 6:1–13.

Ostrand, K. G., and G. R. Wilde. 2001. Temperature, dissolved oxygen, and salinity tolerances of five prairie stream fishes and their role in explaining fish assemblage patterns. Transactions of the American Fisheries Society 130:742–749.

Ostrand, K. G., and G. R. Wilde. 2002. Seasonal and spatial variation in a prairie stream-fish assemblage. Ecology of Freshwater Fishes 11:137–149.

Ostrand, K. G., and G. R. Wilde. 2004. Changes in prairie stream fish assemblages restricted to isolated streambed pools. Transactions of the American Fisheries Society 133:1329–1338.

Parsons Engineering Science Inc. 1999. Surface water/groundwater interaction evaluation for the 22 Texas river basins. Prepared for the Texas National Resource Conservation Commission, Austin.

Pennock, C. A., D. Bender, J. Hofmeier, J. A. Mounts, R. Waters, V. D. Weaver, and K. B. Gido. 2017. Can fishways mitigate fragmentation effects on Great Plains fish communities? Canadian Journal of Fisheries and Aquatic Sciences 75:121–130.

Perkin, J. S., and K. B. Gido. 2011. Stream fragmentation thresholds for a reproductive guild of Great Plains fishes. Fisheries 36:371–383.

Perkin, J. S., K. B. Gido, A. R. Cooper, T. F. Turner, M. J. Osborne, E. R. Johnson, and K. B. Mayes. 2015a. Fragmentation and dewatering transform Great Plains stream fish communities. Ecological Monographs 85:73–92.

Perkin, J. S., K. B. Gido, K. H. Costigan, M. D. Daniels, and E. R. Johnson. 2015b. Fragmentation and drying ratchet down Great Plains stream fish diversity. Aquatic Conservation: Marine and Freshwater Ecosystems 25:639–655.

Perkin, J. S., K. B. Gido, J. A. Falke, K. D. Fausch, H. Crockett, E. R. Johnson, and J. Sanderson. 2017. Groundwater declines are linked to changes in Great Plains stream fish assemblages. Proceedings of the National Academy of Sciences 114:7373–7378.

Permits to Collect or Sell Nongame Fish Taken from Public Fresh Water. 2017 (revised). Texas Administrative Code, sections 57.377–57.386. Available: https://texreg.sos.state.tx.us/public/readtac$ext.ViewTAC?tac_view=5&ti=31&pt=2&ch=57&sch=E&rl=Y. (April 2019).

Platania, S. P., and C. S. Altenbach. 1998. Reproductive strategies and egg types of seven Rio Grande cyprinids. Copeia 1998:559–569.

Poff, N. L., J. D. Allan, M. B. Bain, J. R. Karr, K. L. Prestegaard, B. D. Richter, R. Sparks, and J. C. Stromberg. 1997. The natural flow regime: a paradigm for river conservation and restoration. BioScience 47:769–784.

Poff, N. L., B. D. Richter, A. H. Arthington, S. E. Bunn, R. J. Naiman, E. Kendy, M. Acreman, C. Apse, B. P. Bledsoe, M. C. Freeman, J. Henriksen, R. B. Jacobson, J. G. Kennen, D. M. Merritt, J. H. O'Keeffe, J. D. Olden, K. Rogers, R. E. Tharme, and A. Warner. 2010. The ecological limits of hydrologic alteration (ELOHA): a new framework for developing regional environmental flow standards. Freshwater Biology 55:147–170.

Reynolds, L. V., P. B. Shafroth, and N. L. Poff. 2015. Modeled intermittency risk for small streams in the upper Colorado River basin under climate change. Journal of Hydrology 523:768–780.

Richter, B. D., J. V. Baumgartner, J. Powell, and D. P. Braun. 1996. A method for assessing hydrologic alteration within ecosystems. Conservation Biology 10:1163–1174.

Ross, S. T. 2001. The inland fishes of Mississippi. University Press of Mississippi, Jackson.

Shafroth, P. B., J. R. Cleverly, T. L. Dudley, J. P. Taylor, C. V. Riper, III, E. P. Weeks, and J. N. Stuart. 2005. Control of *Tamarix* in the western United States: implications for water salvage, wildlife use, and riparian restoration. Environmental Management 35:231–246.

Shaner, D. L., and S. L. O'Connor. 1991. The imidazolinone herbicides. CRC Press, Boca Rotan, Florida.

Simpson, S. C., T. Meixner, and J. F. Hogan. 2013. The role of flood size and duration on streamflow and riparian groundwater composition in a semi-arid basin. Journal of Hydrology 488:126–135.

Stewart, M., J. Cimino, and M. Ross. 2007. Calibration of base flow separation methods with streamflow conductivity. Groundwater 45:17–27.

TCEQ (Texas Commission on Environmental Quality). 2008. The John Graves Scenic Riverway: a report to the 81st Legislature. TCEQ, SFR-087/08, Austin.

TIFP (Texas Instream Flow Program). 2008. Texas instream flow studies: technical overview. Texas Commission on Environmental Quality, Texas Parks and Wildlife Department, and Texas Water Development Board, Report 369, Austin.

TIFP (Texas Instream Flow Program). 2018. Instream flow study of the middle and lower Brazos River. Texas Commission on Environmental Quality, Texas Parks and Wildlife Department, and Texas Water Development Board, Austin.

TPWD (Texas Parks and Wildlife Department). 2012. Texas conservation action plan. TPWD, Austin.

TWDB (Texas Water Development Board). 2014. Groundwater database. TWDB, Austin.

TWDB (Texas Water Development Board). 2017. 2017 state water plan: water for Texas. TWDB, Austin.

USFWS (U.S. Fish and Wildlife Service). 2014a. Endangered and threatened wildlife and plants; designation of critical habitat for the Sharpnose Shiner and Smalleye Shiner; final rule. Federal Register 79:149(4 August 2014):45242–45271.

USFWS (U.S. Fish and Wildlife Service). 2014b. Species status assessment report for the Sharpnose Shiner (*Notropis oxyrhynchus*) and Smalleye Shiner (*N. buccula*). USFWS, Ecological Services Field Office, Arlington, Texas.

USFWS (U.S. Fish and Wildlife Service). 2015. Recovery outline for Sharpnose Shiner and Smalleye Shiner. USFWS, Ecological Services Field Office, Arlington, Texas.

Valente, J., D. Bradsby, K. B. Mayes, C. Loeffler, L. Hamlin, D. Geeslin, K. Horndeski, D. Young, J. Trungale, R. Smith, K. Garmany, and T. Hayes. 2019. Developing a geospatial decision support tool for protecting and restoring environmental flows in Texas rivers and streams. Pages 253–267 *in* D. C. Dauwalter, T. W. Birdsong, and G. P. Garrett, editors. Multispecies and watershed approaches to freshwater fish conservation. American Fisheries Society, Symposium 91. Bethesda, Maryland.

Vaughan, C. M., J. H. Breaux, J. L. East, and A. A. Pease. 2016. Feeding ecology of nonnative, inland *Fundulus grandis* in the lower Pecos River. The Southwestern Naturalist 61:74–78.

Vazquez, R., and G. R. Wilde. 2019. Conservation genetics of wild and propagated Sharpnose Shiners and Smalleye Shiners. Texas Tech University, Lubbock. Final performance report submitted to Texas Parks and Wildlife Department, Austin.

Wang, J. H., H. Yang, L. Li, J. J. Gourley, I. K. Sadiq, K. K. Yilmaz, R. F. Adler, F. S. Policelli, S. Habib, D. Irwn, A. S. Limaye, T. Korme, and L. Okello. 2011. The coupled routing and excess storage (CREST) distributed hydrological model. Hydrological Sciences Journal 56:84–98.

Wilcox, B. P. 2002. Shrub control and streamflow on rangelands: a process based viewpoint. Journal of Range Management 55:318–326.

Wilcox, B. P., M. K. Owens, W. A. Dugas, D. N. Ueckert, and C. R. Hart. 2006. Shrubs, streamflow, and the paradox of scale. Hydrological Processes 20:3245–3259.

Wilde, G. R. 2015. Reproductive ecology and population dynamics of fishes in the upper Brazos River. Texas Tech University, Lubbock. Final performance report submitted to Texas Parks and Wildlife Department, Austin.

Wilde, G. R. 2018. Effects of saltcedar encroachment on hydrology and instream habitat availability for Sharpnose Shiner and Smalleye Shiner in the upper Brazos River. Texas Tech University, Lubbock. Final performance report submitted to Texas Parks and Widlife Department, Austin.

Wilde, G. R., and A. A. Echelle. 1992. Genetic status of Pecos Pupfish populations after establishment of a hybrid swarm involving an introduced congener. Transactions of the American Fisheries Society 121:277–286.

Wilde, G. R., and K. G. Ostrand. 1999. Changes in the fish assemblage of an intermittent prairie stream upstream from a Texas impoundment. Texas Journal of Science 51:203–210.

Wilde, G. R., and A. C. Urbanczyk. 2013. Relationship between river fragment length and persistence of two imperiled Great Plains cyprinids. Journal of Freshwater Ecology 28:445–451.

Wilde, G. R., and A. C. Urbanczyk. 2017. Propagation and repatriation of native prairie stream minnows in the middle Brazos River. Texas Tech University, Lubbock. Final performance report submitted to Texas Parks and Wildlife Department, Austin.

Worthington, T. A., A. A. Echelle, J. S. Perkin, R. Mollenhauer, N. Farless, J. J. Dyer, D. Logue, and S. K. Brewer. 2018. The emblematic minnows of the North American Great Plains: a synthesis of threats and conservation opportunities. Fish and Fisheries 19:271–307.

American Fisheries Society Symposium 91:345–366, 2019

Instream Flow Restoration and Watershed Conservation in the Cypress Basin, Texas

RYAN SMITH*
The Nature Conservancy of Texas
200 East Grayson Street, Suite 202, San Antonio, Texas 78215, USA

JOE TRUNGALE
Trungale Engineering & Science
1206 San Antonio Street, Austin, Texas 78701, USA

RICK LOWERRE
Caddo Lake Institute
1206 San Antonio Street, Austin, Texas 78701, USA

TOM HAYES
Texas Conservation Science, Inc.
Post Office Box 150894, Austin, Texas 78715, USA

MIKE MONTAGNE
U.S. Fish and Wildlife Service, Texas Fish and Wildlife Conservation Office
500 East McCarty Lane, San Marcos, Texas 78666, USA

TIM BISTER
Texas Parks and Wildlife Department
3802 East End Boulevard South, Marshall, Texas 75672, USA

LAURA-ASHLEY OVERDYKE
Caddo Lake Institute
400 Edwards Street, Shreveport, Louisiana 71101, USA

MARCIA HACKETT
U.S. Army Corps of Engineers
819 Taylor Street, Fort Worth, Texas 76102, USA

Abstract.—Caddo Lake, along with its swamps and tributary bayous, supports a diversity of aquatic ecosystems and has been designated a wetland of global significance by the Ramsar Convention. The life blood of Caddo Lake is the network of tributary creeks and bayous that drain into the wetland complex of the lake's upper reaches. The ecology of the main tributary, Big Cypress Bayou, however, has been altered by flow regulation by Lake O' the Pines dam. Additional threats from giant salvinia *Salvinia molesta* and other invasive plants, water quality impacts, and land uses have added stress to the ecosystem. Several conservation organizations, led by the Caddo Lake Institute, have formed partnerships to address these threats to the watershed. The Sustainable Rivers Program, a partnership of The Nature Conservancy

* Corresponding author: ryan_smith@tnc.org

and the U.S. Army Corps of Engineers (Corps), has managed dam operations to enhance the natural ecology of Big Cypress Bayou and Caddo Lake downstream. The Corps has been releasing recommended flows to allow researchers to gather more information to evaluate the success of restoration efforts. Early monitoring results indicate a potential positive response of the fish community to these flow releases. We present results of flow restoration work and associated ecological monitoring. We also summarize floodplain vegetation monitoring, paddlefish restoration and invasive species management projects in Caddo Lake and the Cypress River basin.

Introduction

The Cypress River basin is situated on the border of northeast Texas and northwest Louisiana (Figure 1) and covers approximately 11,000 ha. Caddo Lake, the most prominent natural and cultural feature of the basin, is an impoundment of Twelve Mile Bayou in the lower Red River basin. Major tributary streams in the Cypress basin are Big Cypress Bayou, Black Cypress Bayou, Little Cypress Bayou, and James Bayou. Caddo Lake is one of few large naturally formed lakes in Texas and is thought to have been formed by the Great Raft, a logjam on the Red River that was more than 160 km long at its height in the 1800s (Bagur 2001). A majority of the lake is a complex of shallow cypress swamps and channels.

Caddo Lake is an ecological treasure recognized by the Convention on Wetlands (i.e., Ramsar Convention) as a wetland of international importance, with particular significance for migratory waterfowl, colonial waterbirds, and other ecological elements (Ramsar 1987). The lake and its wetlands support a diverse array of species, including more than 200 bird species, 47 mammals, 90 reptiles and amphibians, and more than 80 fishes. These include many species of conservation concern, including the Paddlefish *Polyodon spathula*, Bluehead Shiner *Pteronotropis hubbsi*, and alligator snapping turtle *Macrochelys temminckii*. The watershed formerly supported a reproducing population of Paddlefish that was able to connect to downstream populations in the Red River, but the population has been greatly reduced or eliminated in the watershed due to dam construction, flow alteration, and floodplain habitat impacts (USFWS 2015).

The ecology of Caddo Lake and its watershed has been highly impacted by several modifications, most notably reduced river network connectivity and altered flow regimes by a series of dams. Water levels in Caddo Lake were raised and stabilized in 1914 with the addition of a concrete weir on Twelve Mile Bayou, which was later replaced in 1971. The weir functionally isolated the Cypress basin from the Red River. Construction of large reservoirs in the watershed, particularly, in 1959, Lake O' the Pines on Big Cypress Bayou upstream from Caddo Lake, further fragmented the watershed and altered the hydrologic regime. Operation of Lake O' the Pines dam, including the cessation of high-flow events greater than 3,000 ft^3/s (1 ft^s/s = 0.3 m^s/s), has had a profound effect on Big Cypress Bayou and Caddo Lake. The channel of Big Cypress Bayou and the upper wetland complex of Caddo Lake have been impacted by dredging for boat navigation.

Invasive species are another significant threat to Caddo Lake, with giant salvinia *Salvinia molesta*, water hyacinth *Eichhornia crassipes*, and hydrilla *Hydrilla verticillata* being extensive throughout. Dense stands of these plants choke many areas of the lake during the growing season, with associated impacts to recreation, esthetics, water quality, and wildlife habitat. Water quality has also been impacted by land-use changes and several pollution sources. Water-quality impairments in the watershed include low

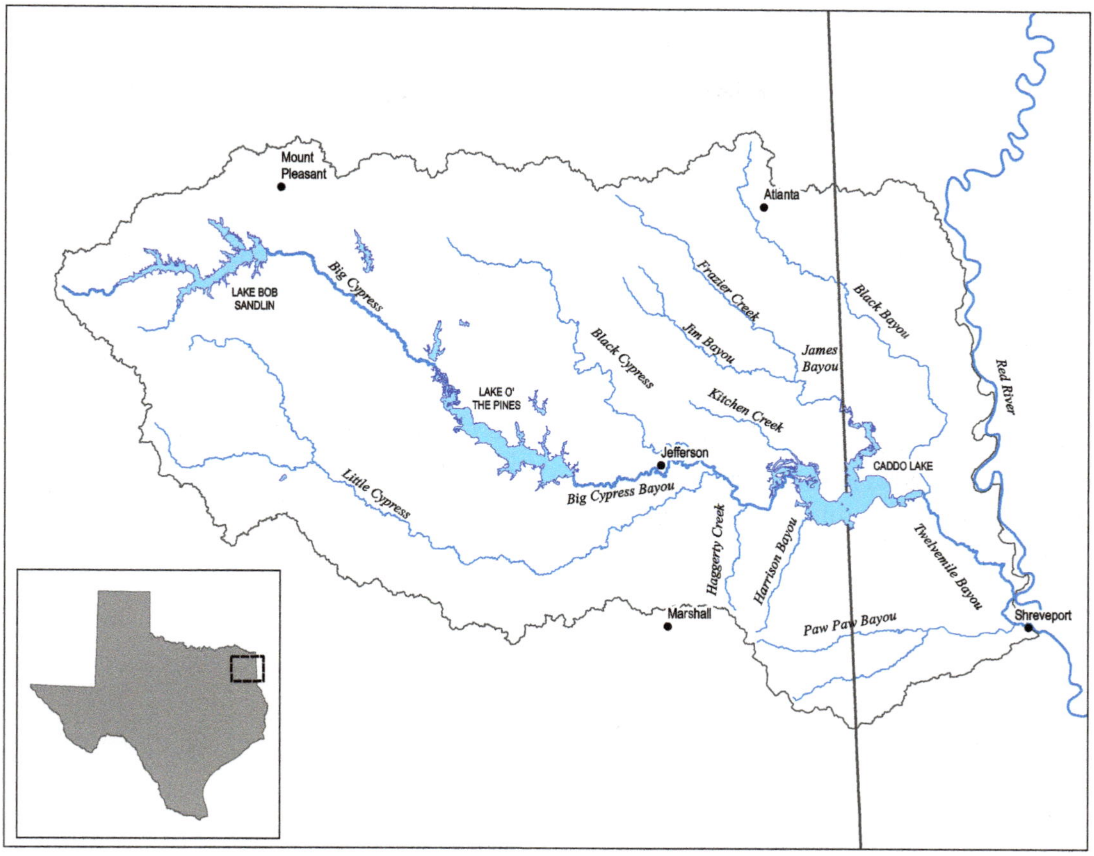

Figure 1. Location of the Cypress River basin and Caddo Lake in northeast Texas and northwest Louisiana.

dissolved oxygen, low pH, and elevated mercury (TCEQ 2014), and high levels of nutrients are also a concern. Despite the existence of many of these impairments for long periods of time, only the dissolved oxygen impairment in Lake O' the Pines has been addressed through a total maximum daily load process (TCEQ 2006). A watershed protection plan (TCEQ 2011) was begun to address bacteria and dissolved oxygen in Caddo Lake and its watershed but was never completed or implemented. Riparian and floodplain forests have also been highly impacted by clearing for grazing and row crops.

Caddo Lake is also a vital economic, recreational, and cultural resource for the region, and local interest in preserving the lake is high (Dahmer 1995). A broad array of stakeholders, including state and federal agencies, local nongovernment organizations, municipalities, and industry, have mobilized to engage in various conservation and management activities to maintain the lake and its waterways and vital wetlands. The goals of this multifaceted conservation effort include

1. Reoperation of Lake O' the Pines dam releases to restore important aspects of the natural flow regime;
2. Restoring fish assemblages, including repatriation of Paddlefish;
3. Restoring floodplain and riparian forests;
4. Improving water quality in Caddo Lake and its tributaries; and
5. Managing invasive species to maintain recreational opportunities, water quality, and habitat in Caddo Lake.

This chapter summarizes this conservation program, with emphasis on environmental flows management for restoration of native fish communities and their floodplain habitats, Paddlefish reintroduction efforts, and control of exotic invasive species.

Caddo Flows Project

Lake O' the Pines was built for the purposes of flood control, recreation, and water supply. As part of its flood control purpose, the U.S. Army Corps of Engineers (Corps) manages reservoir levels and downstream releases to limit catastrophic flooding in downstream communities and Caddo Lake. This has had typical implications for the flow regime of Big Cypress Bayou, with reduced flow variability, less-frequent flood events, and an overall generalization (Poff et al. 2007; Fitzhugh and Vogel 2011) of the hydrograph (Figure 2). There has also been a reduction in the number of zero flow days and an augmentation of low flows during dry months, due to the need to deliver water to downstream municipal and energy water users.

Many aspects of the Cypress basin ecosystem and biological communities reflect this flow alteration. While the fish community is still healthy and diverse relative to many lower Mississippi River basin streams (Winemiller et al. 2005; Robertson et al.

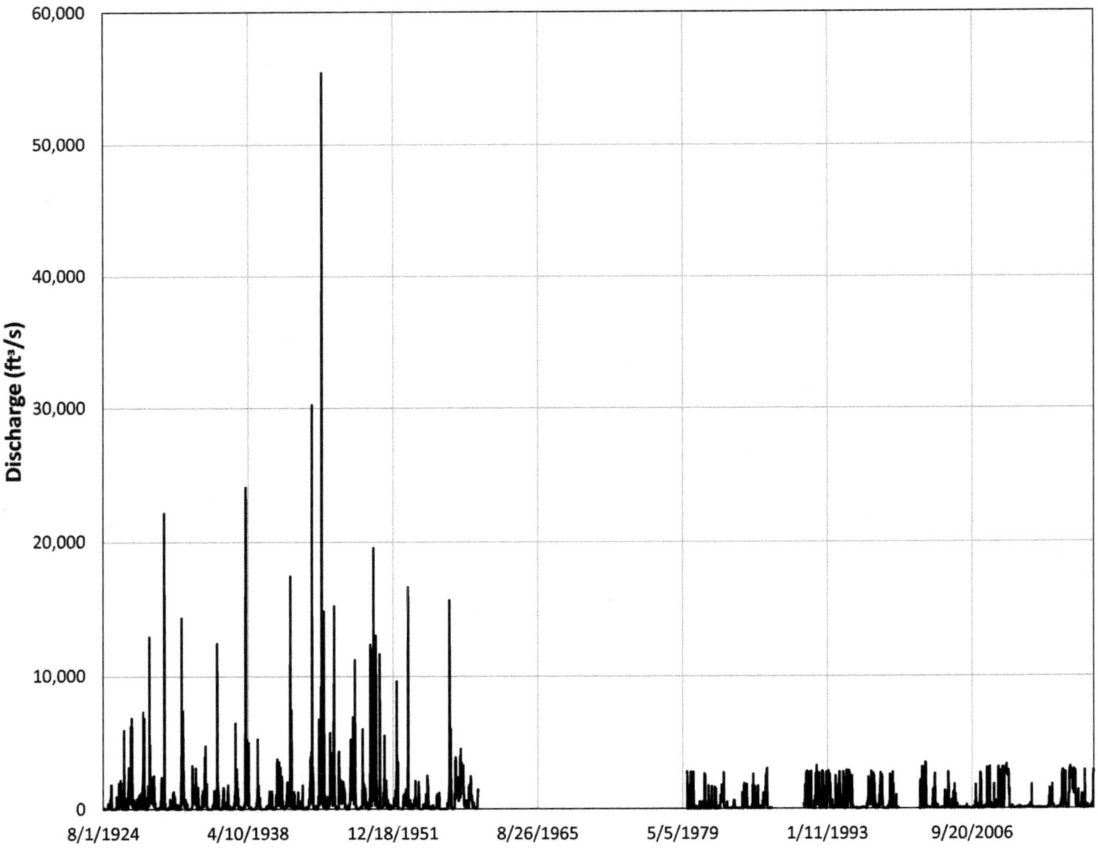

Figure 2. Hydrograph for the U.S. Geological Survey streamflow gauge #07346060 Big Cypress Bayou at Jefferson, Texas showing pre- (1924–1959) and post-Lake O' The Pines (1979–2018) daily peak flows. Lake O' the Pines was completed in 1959, and the streamflow gauge was not in operation from 1959 to 1979.

2016), it has been modified by flow alteration. For example, postdam fish collections show typical patterns resulting from a less variable hydrograph, with a higher proportion of generalist fishes and a decline in fluvial-specialist fishes (Table 1; J. J. Hoover and colleagues, U.S. Army Corps of Engineers, unpublished data). In addition, the loss of Paddlefish from the basin is likely partially attributable to flow regulation subsequent to fragmentation of the watershed by dams. The health of the iconic bald cypress *Taxodium distichum* swamp forest has also been impacted, with loss of as much as a third of its original areal coverage (Keeland and Young 1997; Changxiang et al. 2005).

The Caddo Flows Project is part of a national partnership between the Corps and The Nature Conservancy (TNC), called the Sustainable Rivers Program (SRP; Richter et al. 2006; Warner et al. 2014). This program has the goal of managing dams and other water infrastructure in a manner that balances project purposes while enhancing function of ecosystems. The Corps, TNC, and other partners have engaged deeply at a small number of sites to demonstrate environmental flow management strategies, which are then exportable to other Corps projects. The Cypress basin is one of eight original sites for the program that has now been expanded to 22 nationwide.

The Caddo Flows Project is a multifaceted partnership that includes a diverse array of interests. Primary project partners are the Caddo Lake Institute, the Corps, TNC, and the Northeast Texas Municipal Water District (NETMWD). The Caddo Lake Institute, a local nongovernmental organization whose mission is preservation of Caddo Lake, is the on-the-ground lead on the project and coordinates funding and stakeholder engagement. The Northwest Texas Municipal Water District owns water rights for most of the water in Lake O' the Pines and has been instrumental in providing a path for use of federal funds for the project. Project support is provided by state and federal agencies, local municipalities, industry stakeholders, and others. Significant field research and monitoring has been provided by the U.S. Fish and Wildlife Service (USFWS), the Texas Parks and Wildlife Department (TPWD), the U.S. Geological Survey (USGS), and NETMWD.

Environmental flow management in the Cypress basin has mostly followed an adaptive management process laid out by the Corps and TNC for the SRP (Richter et al. 2006; Warner et al. 2014); the five steps are

1. Literature review synthesizing knowledge of the ecology, hydrology, and other factors in the watershed as they affect flows management;
2. Stakeholder workshops to develop environmental flow recommendations using the literature review, other science, and expert opinion;
3. Flow implementation;
4. Monitoring to evaluate ecosystem response to flow implementation; and
5. Adaptive refinement of the implementation approach following lessons learned via monitoring.

Table 1. Relative abundance of two fish spawning habitat guilds in Big Cypress Bayou during three historical sampling periods.

Reproductive guild	1953–1954	1995	2006
Nonguarders			
Open-substratum pelagophils	22.49	7.25	0.72
Guarders			
Nest-spawner lithophils	7.38	42.58	56.15

This process began in the Cypress basin in 2004 with the orientation meetings and initiation of the literature review (Winemiller et al. 2005) and was followed in 2005 with the first flows workshop, which resulted in initial flow recommendations.

These flow recommendations (Figure 3) follow a building block approach (Tharme and King 1998; Richter et al. 2006), where a flow regime is designed from the ground up to include variability and seasonality in low flows and then adds in episodic high-flow events characteristic of the natural hydrology. Each building block is based on knowledge of ecology, geomorphology, water quality, or other functions of each flow component, including low flows to maintain instream habitats and water quality, high-flow pulses for Paddlefish spawning cues, and bank-full events to maintain geomorphology and habitats. Each flow component also represents a hypothesis of how the ecosystem would be expected to respond to flow management and can be monitored and evaluated in the adaptive management process.

For the Caddo Flows Project, several tools were used to develop flow recommendations, including desktop hydrologic analysis, instream hydraulic habitat modeling, water quality standards, and floodplain inundation modeling. The Indicators of Hydrologic Alteration (IHA; TNC 2007) software was used to define environmental flow components using the predam hydrologic record for the USGS streamflow gauge

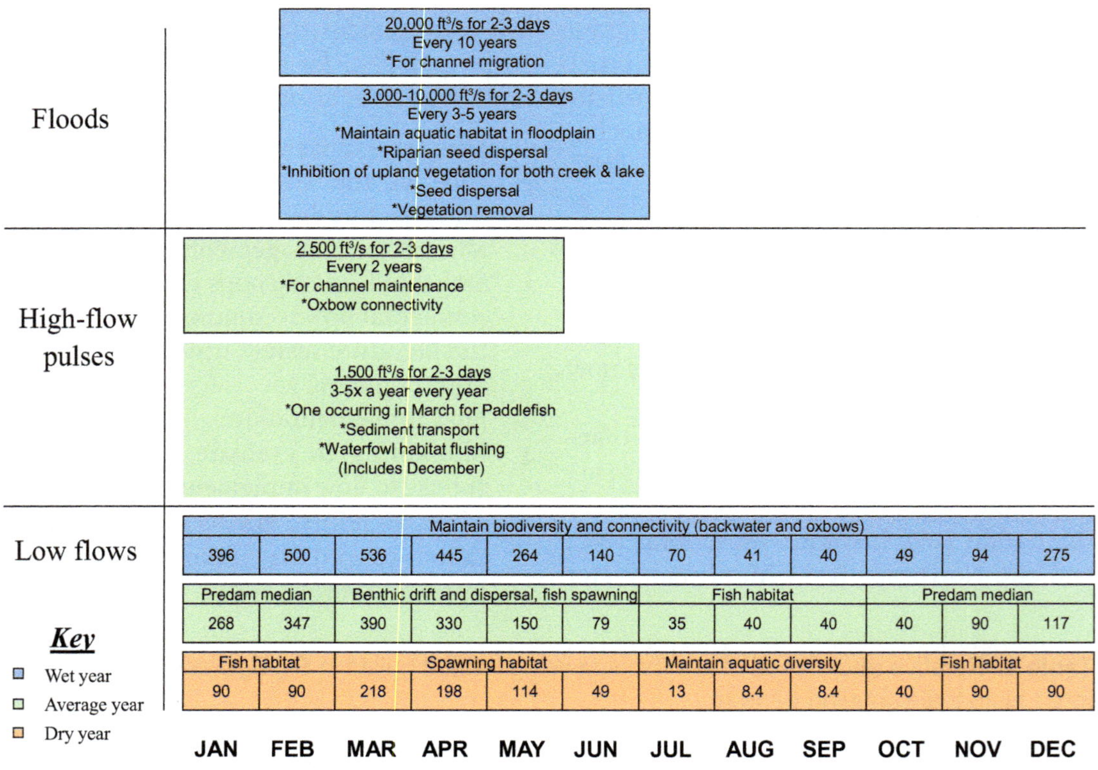

Figure 3. Environmental flow recommendations for the Big Cypress Bayou. Shown are flow magnitudes for monthly low flows and flow magnitude, duration, and frequency for high-flow pulses and floods. Also noted in each box are the ecological or other functions associated with each flow component. Low flows are defined for dry, average, and wet years using 25th, 50th, and 75th percentiles of daily flows by month, respectively.

07346000 Big Cypress Bayou near Jefferson, Texas. All flow components were initially defined using default IHA settings. The flow regime includes low flows defined for three hydrologic conditions (dry, average, and wet), high-flow pulses, and floods. Ecological and other functions were then defined for each flow component, and flow numbers were modified from the IHA results to better match these functions. For example, low flows for some months were changed to optimize instream habitat for fishes based on results of previous site-specific instream flow studies using physical-habitat simulation models (Cloud 1984; Killgore and Hathorn 1987; Killgore and Baker 1996). Resulting flow magnitudes from IHA were increased to 40 or 90 ft^3/s for some late summer and fall months in all three hydrologic conditions to maximize habitats for fluvial-specialist fishes such as Spotted Sucker *Minytrema melanops*, Blackside Darter *Percina maculata*, and Blacktail Shiner *Cyprinella venusta*. Low flows during dry conditions were also increased during August and September to state-defined critical low flows associated with water quality standards (TCEQ 2010).

High-flow pulses from December to June were based upon predam hydrology, but magnitude, duration, and frequency were modified following expert input to ensure sufficient time for fishes to move into or out of secondary channels and oxbow lakes and to allow sufficient time for sediment transport. The 2,500 ft^3/s high-flow pulse recommendation was defined as a bank-full flow event and set at the predam 2-year flood magnitude. A flood flow of 3,000–10,000 ft^3/s is also prescribed to provide additional ecological benefits, including riparian seed dispersal, maintenance of aquatic habitats in the floodplain, and maintenance of riparian vegetation diversity. Even less frequently (10-year return intervals), a flood of 20,000 ft^3/s would be needed to drive channel migration across the floodplain, which is an important mechanism for creating and maintaining habitat for aquatic, riparian, and terrestrial organisms. Flood flows are included in the flow regime for completeness, even though flow releases greater than 3,000 ft^3/s are not currently possible due to physical limitations of the Lake O' the Pines release gates.

Flow recommendations were also developed for two other tributaries, Black Cypress Bayou and Little Cypress Bayou. While these tributaries are not regulated, flow recommendations are instead being used to inform flow protection strategies such as including environmental flow standards in state water planning and permitting and legislative designations as ecologically unique stream segments.

Ecological Monitoring

Ecosystem indicators and a monitoring framework are essential to any environmental flows or other adaptive management or conservation process (Ladson and Argent 2002; Rood et al. 2005; King et al. 2010). The Caddo Flows Project must evaluate success in implementation of flow recommendations, as well as short- and long-term responses of the river channel and the ecosystem. We have adapted a monitoring framework that has been derived from the work at SRP sites and other flow restoration projects (Higgins et al. 2011). The framework sets forth three monitoring phases: implementation, short-term responses, and long-term responses. The implementation phase is an evaluation of the success of implementing the flow recommendations. For example, how often are monthly base flow recommendations met by releases from Lake O' the Pines? Short-term monitoring is an evaluation of how well the system is responding to the flow releases. For example, do base flows result in the expected hydraulic habitats and do habitat types (e.g., riffles and runs) seem to be occupied by the expected fishes? Long-term monitoring is an evaluation of long-term trends in target ecological elements that are key to gauging project success. For example, after 5 years of successful implementation, has fish community structure approached that of the predam period?

The Corps and NETMWD have implemented flow recommendations since 2012, subject to water availability in drought periods and the need to maintain the flood control purpose of the reservoir during wet periods. Low-flow recommendations have been met for 60 of 72 months, with the only gap being in severe drought months of 2012 to 2013, when reservoir levels dropped below trigger levels for drought management (Table 2). High-flow pulses of the 1,500 ft3/s magnitudes have been implemented in all years where they were prescribed but at lower frequencies than recommended (Table 2), and none occurred in 2013. The 2,500 ft^3/s high-flow pulses did not occur during the drought impacted years of 2012 and 2013 but have occurred each year since. Flows of 2,500 and 3,000 ft^3/s (the lower end of the prescribed flood flow) are not currently part of the release agreement but do occur in wet periods such as in 2016, when reservoir storage was evacuated for flood control.

Short- and long-term monitoring have also been initiated in the Cypress basin and track a suite of selected indicators tied to the focal ecological elements and processes in Big Cypress Bayou and Caddo Lake (Table 3). The framework is tied to the ecological, geomorphological, and water quality goals for each flow component and are also informed by the framework used by the Texas

Table 2. Summary of implementation of Big Cypress Bayou flow recommendations from 2012 to 2017. Flows are for the U.S. Geological Survey streamflow gage #07346060 Big Cypress Bayou at Jefferson, Texas. Prescription indicates the target flows, with low flow indicating target magnitudes under three hydrologic conditions ("Dry," "Avg" [for average conditions], or "Wet" years). Implementation columns indicate (1) the number of pulses, if any, implemented each year; and (2) whether low flow targets were met ("No" if not met) and what hydrologic condition flow was released, if implemented. # indicates that for much of 2016 Lake O' the Pines was in flood stage and flow releases were above high-flow pulse levels.

Environmental flow component	Prescription	Years 2012	2013	2014	2015	2016	2017
Floods	20,000 ft^3/s	Not implemented					
	3,000–10,000 ft^3/s	0	0	0	0	1	0
High-flow pulses	2,500 ft^3/s	0	0	0	1	#	1
	1,500 ft^3/s	1	0	2	4	#	1
Low flows	Jan	Dry	No	Avg	Avg	Wet	Dry
	Feb	Dry	No	Avg	Avg	Wet	Avg
	Mar	Dry	No	Avg	Avg	Avg	Dry
	Apr	Dry	No	Avg	Wet	Wet	Dry
	May	Dry	No	Avg	Wet	Wet	Avg
	Jun	Dry	No	Avg	Wet	Wet	Avg
	Jul	Dry	No	Avg	Wet	Wet	Avg
	Aug	Dry	No	Avg	Wet	Wet	Avg
	Sep	Dry	No	Avg	Wet	Wet	Wet
	Oct	Dry	No	Avg	Avg	Avg	Wet
	Nov	No	Dry	Avg	Avg	Avg	Wet
	Dec	No	Dry	Avg	Wet	Avg	Wet

Table 3. Monitoring framework and indicators for short- and long-term monitoring of flow restoration in the Cypress River basin, Texas.

Environmental flow component	Building blocks ecological justification	Predictive models	Short-term monitoring indicators	Long-term monitoring indicators
Floods (high)	Channel migration	Not currently relevant due to limits on flood releases	Not currently relevant due to limits on flood releases	
Floods (low)	Maintain aquatic habitat in floodplain	Floodplain inundation models predicting inundation area of vegetation types and floodplain zones by flow magnitude	Floodplain inundation mapping—areal imagery, other observations Water level monitoring—floodplain groundwater wells, soil moisture	Fish community—fish monitoring in oxbows
	Floodplain and riparian forest health (riparian seed dispersal, inhibition of upland vegetation, vegetation removal)	Floodplain inundation models predicting inundation area of vegetation types by flow magnitude Estimated saturation tolerance limits	Floodplain inundation mapping—area imagery, other observations Water level monitoring—floodplain groundwater wells, soil moisture Floodplain and riparian vegetation monitoring plots—seedling establishment and density	Floodplain and riparian vegetation monitoring plots—sapling, tree establishment and density; native versus invasive sapling, tree species composition
High-flow pulse (high)	Chanel maintenance	Sediment transport models (e.g., Hydrologic Engineering Center's River Analysis System [HEC-RAS], Sediment Impact Analysis Methods [SIAM])	Bedload and suspended load—particle sizes and volume	Sediment composition—particle size distribution in core samples and pebble counts
	Oxbow connectivity	Bank-full predicted based on 1.5-year recurrence interval	Water level monitoring—pressure transducers at oxbow connections	Fish community—fish monitoring in oxbows

Table 3. Continued.

Environmental flow component	Building blocks ecological justification	Predictive models	Short-term monitoring indicators	Long-term monitoring indicators
			Floodplain inundation mapping—areal imagery, other observations	
High-flow pulse (low)	Paddlefish (spawning cue)	No quantitative model; prediction based on known life history	Paddlefish spawning occurrence—Paddlefish present in spawning shoals, spawning observation	Paddlefish recruitment—larval and juvenile presence and density
	Sediment transport	Sediment transport models (HEC-RAS, SIAM)	Bedload and suspended load—particle sizes and volume	Sediment composition—particle size distribution in core samples and pebble counts
Low flows (wet)	Maintain biodiversity and connectivity (backwater and oxbows)	Instream habitat models predicting available habitat at low flows for select species	Instream habitat conditions—mesohabitat composition, hydraulic habitats	Fish community—fish monitoring in all habitats, oxbows
		Inundation models	Fish community—fish monitoring in all habitats, oxbows (short-term changes)	
Low flows (average and dry)	Fish habitat	Instream habitat models predicting available habitat at low flows for select species	Intream habitat conditions—mesohabitat composition, hydraulic habitats Fish community—fish monitoring in all habitats, oxbows (short-term changes)	Fish community—fish monitoring in all habitats

Instream Flow Program (TIFP 2008). An explicit aspect of the framework is also to make hypotheses regarding expected ecological responses to each flow component using predictive models or other tools, then to design a monitoring process to evaluate the predicted responses. Below, we highlight two key elements of this framework and summarize monitoring efforts and results to date.

Fish community monitoring

Native fish communities of Caddo Lake and its tributary streams are a key focus of the flow restoration effort. While Big Cypress Bayou is considered a near-reference stream for the lower Mississippi Valley (K. J. Killgore, U.S. Army Corps of Engineers, personal communication), enhancement of instream habitats to enhance community structure and function is an explicit goal of the current flow recommendations, especially the low-flow components (Figure 3). The fish community of the Cypress basin is diverse, with 90 species having been collected through time, and includes both coastal plain and Mississippi embayment species (Winemiller et al. 2005). Species dependent on floodplain habitats for reproduction (e.g., Paddlefish) and migratory species (e.g., American Eel *Anguilla rostrata*) are present, primarily in the larger streams and rivers, while many smaller streams support largely black water swamp and stream fishes.

To develop fish community restoration targets and monitoring indicators, we compiled available historical fish collection data and evaluated predam community structure and changes over time. While all available collections have been summarized, the most extensive fish collections in the Cypress basin occurred in four phases: (1) sampling by the Texas Game and Fish Commission in the 1950s before construction of Lake O' The Pines, (2) instream flow studies in the 1980s and 1990s during the postdam period (Cloud 1984; Killgore and Hathorn 1987; USACE 1994), and (3) preimplementation monitoring of this study in the late 2000s (also postdam; Braun and Moring 2013). Data are available from these periods for the flow-regulated Big Cypress Bayou, as well as the unregulated Little and Black Cypress bayous. Sampling methods, levels of effort, specific study sites, and climatic and hydrologic conditions vary across collections from the different periods; however, the data are useful for description of predam fish community composition and evaluation of trends in community change postdam.

We use fish family composition to summarize fish community composition and use fish reproductive guilds (Balon 1975, 1981; Simon 1999) as our focal monitoring elements because they more accurately represent the species most likely to be negatively impacted or favored by changes in habitats due to flow regulation. The fish community of all three primary tributaries of the Cypress basin includes members of 19 families and is dominated by cyprinids, centrarchids, poeciliids, fundulids, and percids. While these have been the prominent families across all collections, the proportion of community in each family has changed over time (Figure 4). Before the dam, the fish community was dominated by fluvial specialist cyprinids of primarily the nonguarder open substratum pelagophil and phytolithophil reproductive guilds (Balon 1975, 1981; Simon 1999; Table 1; Figure 5). Collections from the 1990s and 2000s show an increasing abundance of generalist centrarchids in the guarder nest spawner lithophil and polyphil guilds (Table 1; Figure 5). These trends are consistent with expected results of reductions in flow variability and increased low flows that maintain more-prevalent stable pool habitats for sunfishes and less-prevalent riffles and vegetated runs preferred by Cypress basin minnows. Though not to the same degree, this pattern of community structure alteration was also seen in the free-flowing Little and Black Cypress bayous (Braun and Moring 2013).

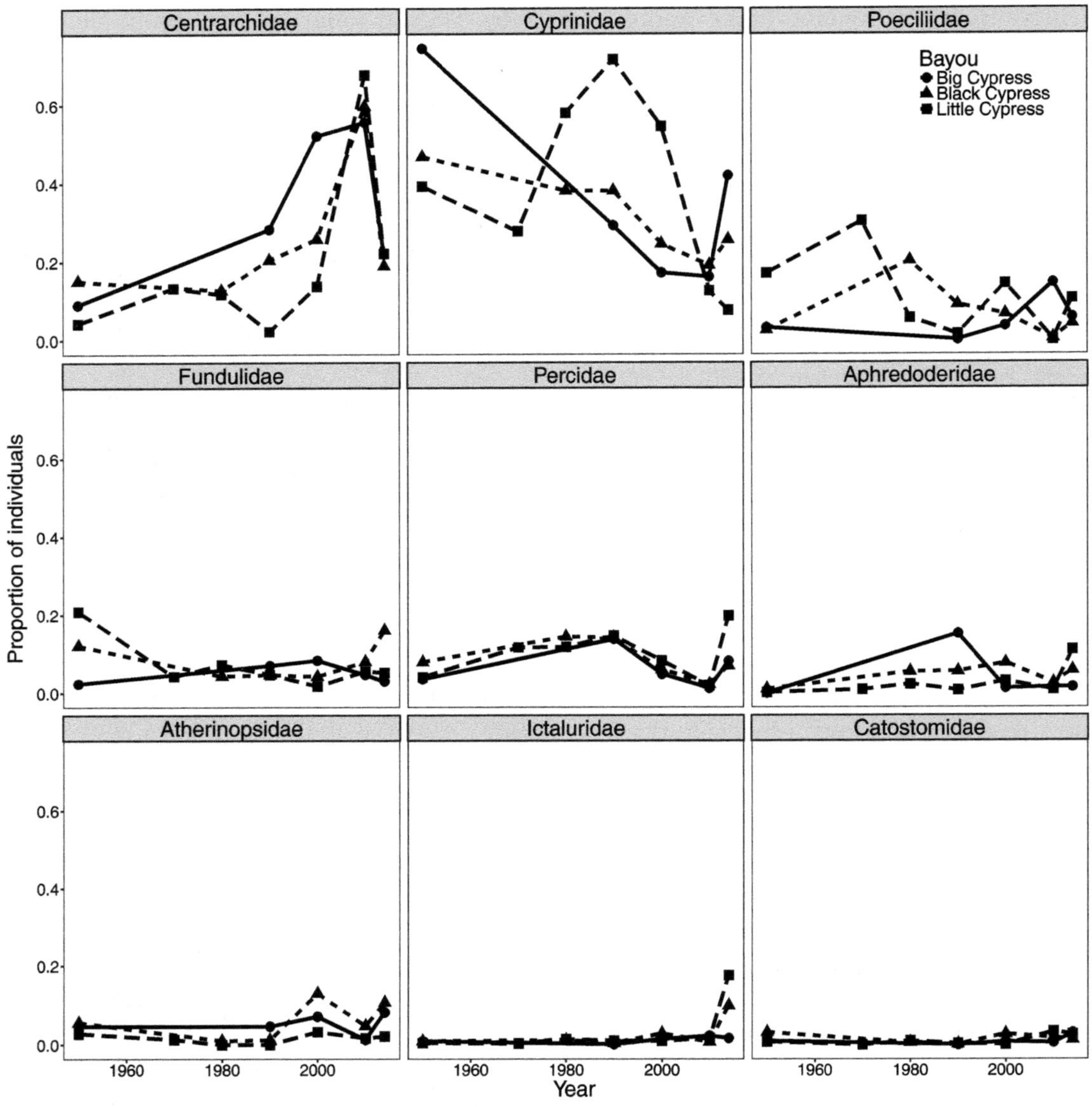

Figure 4. Relative abundance of the nine most abundant fish families in three Cypress River basin tributaries from the 1950s through 2014. Shown are the average relative abundances from all collections by decade and relative abundance from the 2014 postflow implementation monitoring collections.

The goal of fish monitoring following flow implementation is to evaluate any return to predam composition in Big Cypress Bayou and to observe any basinwide trends in fish community in other tributaries. Collections were made by the TPWD in 2014 (Robertson et al. 2016) as the initiation of periodic long-term monitoring of fish community response in Big Cypress Bayou. This sampling was also initiated in Little and Black Cypress bayous, which serve as controls for evaluation of flow restoration response in the Big Cypress fish community. Because there have been declines in some components of the fish community in nonregulated tributaries, it seems likely that other natural factors such as drought and human impacts from land

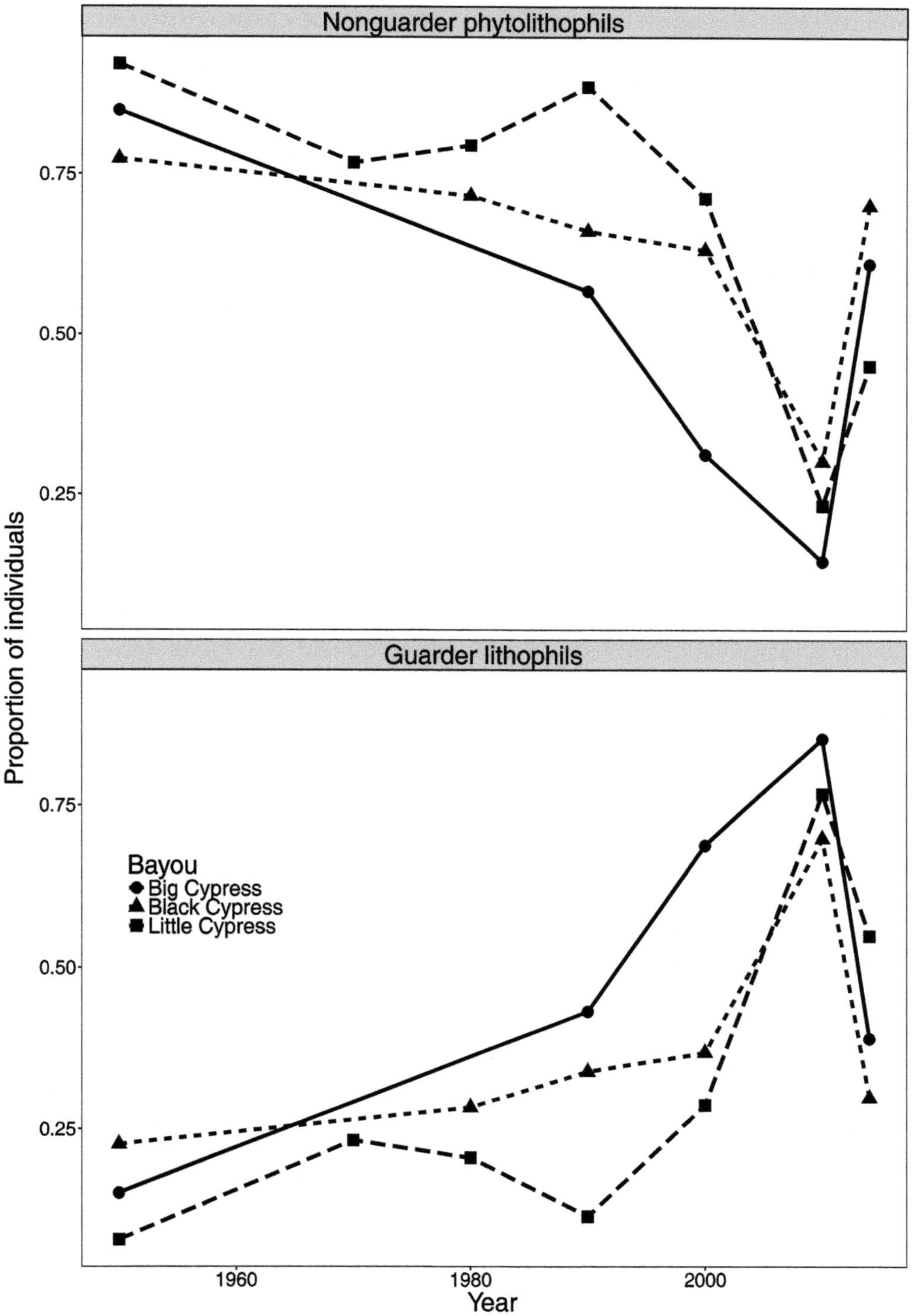

Figure 5. Relative abundance of two fish spawning habitat guilds, nonguarder phytolithophils and guarder lithophils (Balon 1975), in three Cypress River basin tributaries from the 1950s until 2014. Shown are the average relative abundances from all collections by decade from the 1950s until the early 2010s and relative abundance from the 2014 postimplementation monitoring collections.

uses and pollution have also likely affected fish communities.

Fish community monitoring in 2014 included two sites on Big Cypress Bayou and one site each on Little and Black Cypress bayous. These data show some indications of an incremental movement back toward predam community structure in Big Cypress Bayou, but similar changes were also seen in Little and Black Cypress bayous (Figures 4 and 5). Cyprinids were the most abundant predam, particularly in Big Cypress Bayou; centrarchids have increased in all three tributaries since the 1950s, with the highest abundances in the 2000s and 2010s (Figure 4). Community composition in the 2014 collections saw significant changes, with proportions of cyprinids and centrarchids more similar to predam conditions in all three tributaries (Figure 4). Relative abundance of the nonguarder phytolithophils increased from 2010 levels in all three tributaries, most markedly in Little and Big Cypress bayous, and guarder lithophils decreased notably in all three tributaries (Figure 5).

The observation of similar changes in community composition in all three tributaries suggests causation by more than just flow regime releases and may include response to differing hydrologic conditions (e.g., drought recovery) or some aspect of land use and water quality. This may also indicate Big Cypress Bayou functioning as a source for Little and Black Cypress bayous, as much of these streams dried during the 2010 to 2011 drought and recolonization from Big Cypress may have had a significant influence on post drought fish community. Additional and more regular monitoring to be initiated in fall 2018 will further elucidate current fish community structure in the Cypress basin and any flow-mediated biotic response in Big Cypress Bayou.

Floodplain vegetation monitoring

The extensive swamps and bottomland hardwood floodplain forests are some of the most unique characteristics of the Cypress basin and are another focus of the flow restoration work. Healthy floodplain habitats are also vital to the life history of large river fishes, including those of the lower Mississippi River basin and Gulf Coastal Plain (Turner et al. 1994; Bayley 1995; Killgore and Baker 1996; Ward et al. 1999; Winemiller et al. 2000; Winemiller 2005).

Numerous studies have demonstrated that infrequent high flows provide important ecological benefits to maintenance of healthy riparian and wetland forests (Junk et al. 1989; Thoms et al. 2004; Opperman et al. 2010). We evaluated the linkages between flow events and hardwood forests and wetlands based on inundated area relative to flow rates and species composition, as well as the seasonal timing and duration of these flow events. This evaluation included four riparian and floodplain forest-flow guilds (see Figure 6), defined by tree species composition, relative elevation context, and hydroperiod (Huffman and Forsythe 1981; D. D. Diamond, Missouri Resource Assessment Partnership, personal communication). These guilds are (1) lower swamps—dominated by bald cypress, occurring at relatively low surface elevations, and flooding nearly annually; (2) upper swamps—dominated by overcup oak *Quercus lyrata* and requiring flooding every 1 or 2 years, (3) seasonally flooded forests—mixed hardwoods dominated by willow oak *Q. phellos*, and (4) temporarily flooded forests—annual flood probability of 11–50%, mixed hardwoods and pines, and dominated by water oak *Q. nigra* (Figure 6). We used satellite imagery, collected over a full range of flow magnitudes, to produce inundation maps and regression analyses for the three major tributaries (Big Cypress, Little Cypress, and Black Cypress bayous) to Caddo Lake and for Caddo Lake itself. We then estimated flow magnitudes necessary to inundate 75% of the habitat area of each guild in Big Cypress Bayou and used reported data on frequencies and du-

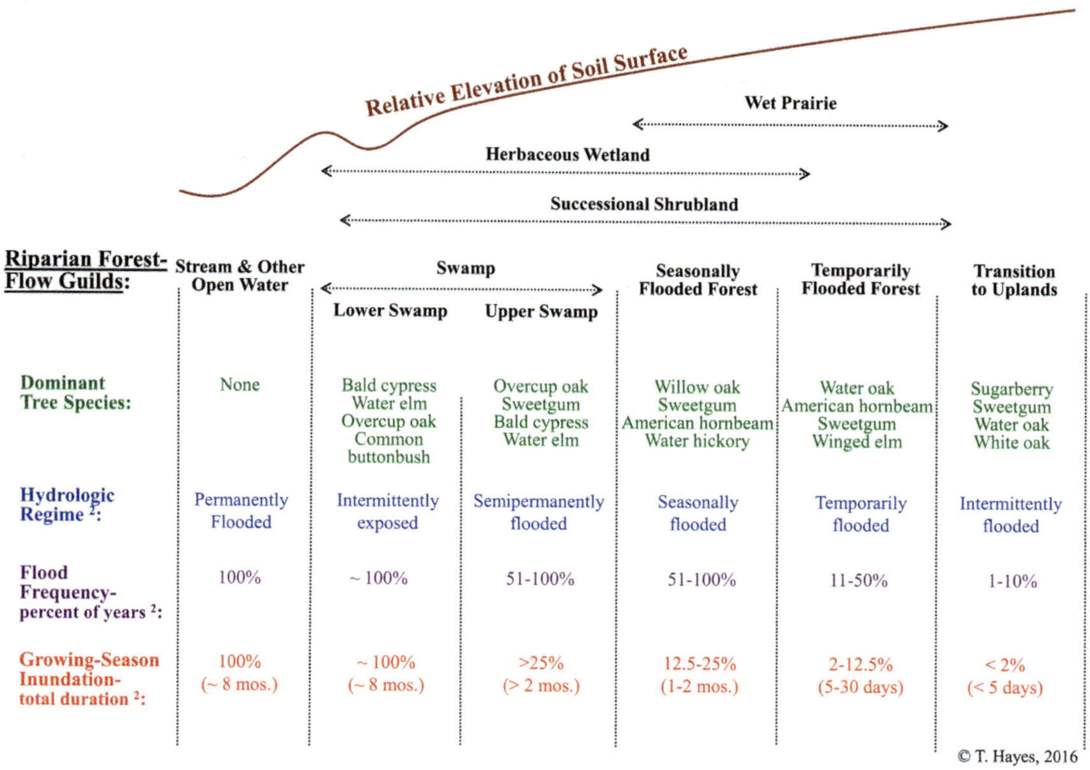

Figure 6. Riparian forest-flow guilds in the Cypress River basin, with notes on landscape context, dominant tree species, hydrologic regime, and typical frequency and duration of inundation. Definition of forest guilds are modified from Huffman and Forsythe (1981) and Diamond, personal communication.

ration of inundation (Trungale and Hayes 2011). This analysis indicates that the prescribed high-flow pulses of 1,500 and 2,000 ft^3/s would provide the flows needed to meet the natural inundation frequencies for lower swamps and occasional inundation of upper swamps and lower elevations within seasonally flooded forests but not the more elevated floodplain, including temporarily flooded forests.

Long-term monitoring of riparian and floodplain forest response to flow implementation is focused on vegetation community composition, survival of tree species, and establishment of seedlings (Table 3). In 2013, we established three permanent study sites on Big Cypress Bayou where vegetation plots are being periodically monitored (Hayes 2016). In 2017, three additional study sites were installed on Little Cypress Bayou as replicate controls for the three sites established in 2013 on Big Cypress Bayou. Comparative analyses of forest-health outcomes between the regulated and unregulated watersheds will enhance adaptive management of ongoing Big Cypress Bayou environmental flows.

At these study sites, we are also monitoring floodplain inundation patterns and soil moisture in order to validate hypotheses of flow magnitudes and duration needed to maintain floodplain habitats and essential ecosystem functions and refine flow releases via adaptive management. Five years of monitoring at these sites has included a period of extreme drought conditions, which resulted in the most severe restriction upon high-flow releases since Lake O' The Pines became operational. From 2012 to 2014, no high-flow pulses occurred (Table 2). Despite this, vegetation plots and soil-moisture monitoring indicate that the lower and upper swamps remain relatively sustainable, even in drought. However, high tree mortality rates and low soil moisture indicate that the higher elevation seasonally and temporarily flooded forest guilds appear unsustainable under extended low-flow periods without any high and flood flows. This provides strong evidence of the importance of the prescribed flow releases for maintaining floodplain forest health.

Long-term resilience of riparian and floodplain forests within the Cypress basin is expected, as higher flow events are scheduled to coincide with riparian tree phenology. For example, annual tree recruitment is mostly determined by spring high flows, which exert a disproportionate influence on seedling establishment and the early stages of succession (Dewey et al. 2006). High flows may also be scheduled to match the timing of seed dispersal and germination of declining tree species identified during forest inventories.

Paddlefish Reintroduction and Habitat Restoration

The USFWS's Texas Fish and Wildlife Conservation Office (TXFWCO) identified the Cypress basin as a priority conservation area. Areas of focus in the Cypress basin include Paddlefish restoration, restoration of aquatic, riparian and floodplain habitats, improvement of water quality and restoration of stream network connectivity.

An intensive Paddlefish restoration initiative was begun by the TXFWCO and TPWD in 2014 (Diaz et al. 2015). In March 2014, 47 Paddlefish from Tishomingo National Fish Hatchery (TNFH) implanted with 1-year radio tags were released at two locations in Big Cypress Bayou. The primary purpose of the reintroduction was to determine the effect, if any, of environmental flow implementation on paddlefish movement in and fidelity to Big Cypress Bayou. In fall of 2014, 2,000 young-of-the-year Paddlefish from TNFH were stocked into the Big Cypress Bayou at Caddo Lake State Park. In April 2016, 21 additional juvenile Paddlefish were radio-tagged and, along with 600 others, were released into Big Cypress Bayou at the Caddo Lake National Wildlife Refuge. To date, radio-tagged Paddlefish have shown a fidelity to Big Cypress Bayou and Caddo Lake. None of the 68 tagged fish were positively documented to have gone over the spillway at the Caddo Lake dam and out of the Cypress watershed. Paddlefish have shown an overall trend of upstream movement during the study period, with the largest number congregating below the Lake O' the Pines spillway (Diaz et al. 2015). High-flow pulses that could serve as Paddlefish spawning cues have occurred in 2016 and 2017 (Table 2), but these fishes are not yet of reproductive age so spawning has not yet been documented.

In the next phase of the reintroduction project, previously released Paddlefish will be recaptured and tagged with long-lasting (7–10 years) acoustic telemetry tags so the fish can be tracked into adulthood. Acoustic telemetry receivers will be placed at known gravel shoals (including those constructed by the Corps to serve as Paddlefish spawning grounds) to detect any Paddlefish using these areas during the presumed spring spawning season. The receivers will also be placed strategically throughout Big Cypress Bayou and at Caddo Lake dam to track movement throughout the system.

In addition to the acoustic-tagged fish, the TXFWCO, TPWD, and Louisiana Department of Wildlife and Fisheries (LDWF) have developed a 10-year Paddlefish stocking plan for the Big Cypress Bayou that will be implemented in 2018 (Table 4). The long-term goal of this project is to re-establish a self-sustaining Paddlefish population in Big Cypress Bayou, downstream of Lake O'

Table 4. Stocking plan for Paddlefish in Cypress River basin, Texas. Shown are planned numbers of individuals to be stocked annually until 2027. Data source: U.S. Fish and Wildlife Service.

Year	Number
2018	14,000
2019	14,000
2020	14,000
2021	11,000
2022	11,000
2023	11,000
2024	11,000
2025	11,000
2026	11,000
2027	11,000

the Pines and into Caddo Lake. The short-term goal of the project is to test the environmental impacts of implementing the experimental environmental flow regime. The TNFH will be producing all Paddlefish for the plan, following the Mississippi Interstate Cooperative Resource Association Paddlefish stocking protocols (MICRA 2012). Wild broodstock will be collected from the Red River, the most proximal source to Big Cypress Bayou. After hatchery production, broodstock will be tagged with coded-wire tags and returned to the Red River where they were captured.

Through the Partners for Fish and Wildlife Program, the USFWS is funding riparian restoration projects that improve connectivity, remove nonnative species, improve water quality, provide management assistance to private landowners, and conduct controlled burns. The TXFWCO and others are currently identifying barriers that prevent or limit fish passage within Cypress basin watersheds. The Partners for Fish and Wildlife Program worked with a private landowner to replace four water crossings with a bridge to help restore access to 11.4 km of Prairie Creek, a Big Cypress Bayou tributary.

Through funding from the National Fish Passage Program, the TXFWCO is currently funding an oxbow reconnection project on Big Cypress Bayou near Jefferson, Texas. The oxbow will serve as nursery habitat for native species as it will connect with Big Cypress Bayou during high-flow pulses, thus allowing entrainment and shelter for juvenile fish and other aquatic species in highly productive habitat. The reconnection will restore functioning floodplain wetlands that have been lost due to historical channelization for navigation and improve water quality. To measure the success of the reconnection project, this oxbow will be stocked with larval or young-of-the-year Paddlefish. The TXFWCO will monitor the fish community in the oxbow periodically to track the growth and survival of the Paddlefish and other aquatic species. If stocked Paddlefish do well in this nursery habitat, additional reconnection projects will be attempted.

Invasive Species Management

Caddo Lake has extensive coverage of numerous nonnative invasive aquatic vegetation species that threaten to block access to many areas of the lake if not managed aggressively. The most problematic species are giant salvinia, water hyacinth, and hydrilla. The TPWD, the LDWF, and local partners, including the Cypress Valley Navigation District and Caddo Biocontrol Alliance, have engaged in vegetation control efforts, with the goal of maintaining recreational opportunities (e.g., fishing, paddling, and touring), especially keeping boat lanes open and improving water quality and wildlife and fisheries habitat.

Giant salvinia was first detected on the Louisiana side of the Caddo Lake in 2006 (Bister and Brice 2010). By the time the infestation was detected, it had expanded to at least 121 ha in the James Bayou arm of the lake (Figure 1). Herbicide treatments were applied immediately in an attempt to reduce the level of giant salvinia in this area. Despite attempts by the TPWD and LDWF, as well as efforts from local stakeholder groups, giant salvinia migrated to the Texas side of the lake during the winter of 2006–2007 and continued to expand in the spring of 2007. A separate infestation was discovered at a private boat ramp on the upper end of the lake on the Texas side in 2007.

Surveys of invasive aquatic vegetation are conducted annually by the TPWD on the Texas side of the lake during late August–early September, at the peak of the growing season, to monitor the coverage of the most problematic species and to assess ongoing management efforts (Table 5). A 2007 vegetation survey (Bister and Brice 2010) of the Texas side of the lake estimated 40 ha of giant salvinia. Giant salvinia continued to expand its coverage to more areas of the lake, and by 2009, coverage had increased to 1,306 ha on the Texas side (Bister and Brice 2010). Extended flood conditions and freezing conditions during winter can reduce the coverage of giant salvinia (Owens et al. 2004). Severe winters in 2009–2010, 2013–2014, and 2017–2018 have resulted in as much as a 95% reduction in aerial coverage relative to previous summer conditions. However, die-offs were not complete, and subsequent mild winters allowed giant salvinia coverage to expand in each of the subsequent years, even with an increase in herbicide application each year.

An integrated approach is used to manage giant salvinia, which includes biological (giant salvinia weevils *Cyrtobagous salviniae*) and chemical herbicide (e.g., glyphosate, diquat, penoxsulam, or flumioxazin) treatments. Weevils are released by the TPWD and Caddo Biocontrol Alliance as a long-term solution for giant salvinia management. Herbicide treatments are used to maintain boater access throughout Caddo Lake, clear shorelines for boat dock use, and reduce large mats of giant salvinia. Nearly 300,000 weevils were released in both 2016 and 2017, and herbicide treatment totaled 1,800 ha in 2016 and 3,700 ha in 2017. With these two methods combined, there have been small-scale control successes, and leveraging these successes to a larger area is a current management goal. For example, Pine Island Pond, a small Caddo Lake embayment, had been virtually inaccessible for several years because of giant salvinia. However, weevil releases reduced giant salvinia coverage to the point that a small herbicide treatment was able to clear most of the remaining plants (Bister 2018).

Water hyacinth was discovered in Caddo Lake in the 1940s, and until recently, some active form of control has been required to keep the population in check. Water hyacinth coverage was estimated at 1,500 ha in 2007, 550 ha in 2008, and 705 ha in 2009. However, the increase in giant salvinia caused water hyacinth coverage to decrease dramatically over the past several years. Coverage of water hyacinth has ranged from 4 to

Table 5. Invasive aquatic vegetation coverage (ha) in the Texas side of Caddo Lake (5,144 ha total) from 2012 to 2017. Surveys were conducted during late August and early September each year. Data source: Texas Parks and Wildlife Department (Wright and Bister 2014; Bister 2018).

Year	Giant salvinia	Water hyacinth	Hydrilla
2012	554	157	1,378
2013	2,428	4	514
2014	973	8	746
2015	1,149	21	913
2016	2,000	66	953
2017	2,150	11	729

157 ha since 2012 (Table 5; Wright and Bister 2014; Bister 2018). Despite the recent decline in coverage, water hyacinth needs to be monitored for a possible increase. Hydrilla was first reported on Caddo Lake in 1993. In 1996, hydrilla had expanded to 235 ha and by 1997 was estimated to cover more than 2,025 ha (Bister 2018). At that time, the infestation was concentrated in the deeper portion of the lake. From available records (Bister 2018), hydrilla declined by 2000 and was reduced to nonproblematic levels by 2001. Records do not reflect any chemical treatments targeting hydrilla on Caddo Lake during this period (1993–2001). Surveys in 2004 documented isolated hydrilla populations on Caddo Lake in shallow, more remote areas less frequented by resource users. The 2005 survey indicated that hydrilla covered 1,015 ha. From 2009 to 2012, hydrilla coverage hovered around 1,620 ha; however, coverage dropped to 514 ha in 2013 due to shading by giant salvinia (Bister 2018). Hydrilla coverage has remained relatively stable (range 729–953 ha) since 2014 (Table 5).

Conservation Challenges and Next Steps

Significant progress has been made in the Cypress basin in understanding and prescribing instream flow needs, in reaching consensus among stakeholders to reoperate releases from Lake O' the Pines, and in initiating monitoring of ecosystem responses to flow implementation. Nevertheless, an active adaptive management process continues to address information gaps and modify flow management efforts. Ongoing research and monitoring priorities to quantify ecosystem responses include (1) long-term monitoring of fish community responses to flows to detect expected movement back to predam community composition; (2) monitoring of survival, movement, and any spawning and recruitment success of Paddlefish; (3) geomorphology and sediment budget research; (4) refinement of instream fish habitat models to improve understanding of fish community responses to flow; and (5) evaluating potential for flow releases to increase drought resiliency of floodplain forest ecosystems.

A focus of future flow-management efforts in the Cypress basin is to strengthen the operational ability to implement flow recommendations at Lake O' the Pines. Approaches to this will include formalizing the ecosystem flow releases in the Lake O' the Pines operation plan and evaluation of strategies to increase the certainty of flow releases through reservoir reallocation and establishment of storage water rights. The project will also explore options to protect flow in other Cypress basin tributaries through water rights and other policy strategies. Project partners also continue to evaluate strategies to address the threat of invasive plants, including the potential modification of the Caddo Lake dam to allow lake level modification to manage invasive plants, enhance water quality, and aid in flood management. This project will also continue to provide lessons learned to a network of other Corps-operated dams through the SRP network.

Acknowledgments

We thank the Caddo Lake Institute and Northeast Texas Municipal Water District for project coordination. We thank Roy Darville, Kevin Mayes, Sarah Robertson, Richard Day, Tom Doyle, Jennifer Wilson, and Bruce Moring for assistance with field studies, data collection, and monitoring. We thank the U.S. Army Corps of Engineers, the U.S Fish and Wildlife Service, the Meadows Foundation, the Caddo Lake Institute, and The Nature Conservancy for project funding.

References

Bagur, J. D. 2001. A history of navigation on Cypress Bayou and the lakes. University of North Texas Press, Denton.

Balon, E. K. 1975. Reproductive guilds of fishes: a proposal and definition. Journal of the Fisheries Research Board of Canada 32:821–864.

Balon, E. K. 1981. Additions and amendments to the classification of reproductive styles in fishes. Environmental Biology of Fishes 6:377–389.

Bayley, P. B. 1995. Understanding large river-floodplain ecosystems. BioScience 45:153–158.

Bister, T. J. 2018. Caddo Lake, 2017 fisheries management survey report. Texas Parks and Wildlife Department, Federal Aid Project Report F-221-M-3, Austin.

Bister, T. J., and M. W. Brice. 2010. Statewide freshwater fisheries monitoring and management program survey report for Caddo Lake, 2009. Texas Parks and Wildlife Department, Federal Aid Report F-30-R-35, Austin.

Braun, C. L., and J. B. Moring. 2013. Baseline assessment of physical characteristics, aquatic biota, and selected water-quality properties at the reach and mesohabitat scale for reaches of Big Cypress, Black Cypress, and Little Cypress bayous, Big Cypress basin, northeastern Texas, 2010–11. U.S. Geological Survey, Scientific Investigations Report 2013-5058, Reston, Virginia.

Changxiang, L., J. A. Neal, C. Scofield, J. Chang, A. K. Ludeke, and C. Frentress. 2005. Classification of land cover and assessment of forested wetlands in the Cypress Creek watershed. Texas Parks and Wildlife Department, Austin.

Cloud, T. 1984. Planning aid report on the aquatic resources of the Cypress Bayou basin, Texas. U.S. Fish and Wildlife Service, Ecological Services, Arlington, Texas.

Dahmer, F. 1995. Caddo was... a short history of Caddo Lake. University of Texas Press, Austin.

Dewey, J. C., S. H. Schoenholtz, J. P. Shepard, and M. G. Messina. 2006. Issues related to wetland delineation of a Texas, USA, bottomland hardwood forest Wetlands 26:410–429.

Diaz, P., M. Montagne, and T. Bister. 2015. Final report for Big Cypress Bayou Paddlefish reintroduction assessment. Final report to Caddo Lake Institute. U.S. Fish and Wildlife Service, San Marcos, Texas.

Fitzhugh, T. W., and R. M. Vogel. 2011. The impact of dams on flood flows in the United States. River Research and Applications 27:1192–1215.

Hayes, T. D. 2016. Big Cypress Bayou monitoring network: forest plot results 2012–2016. Final report, 12/31/16, project # 2016-05, prepared for Caddo Lake Institute. Texas Conservation Science, Austin.

Higgins, J. V., C. P. Konrad, A. Warner, and J. T. Hickey. 2011. A framework for monitoring, reporting and managing dam operations for environmental flows. Version 1.0. The Nature Conservancy, Arlington, Virginia.

Huffman, T., and S. W. Forsythe. 1981. Bottomland hardwood forest communities and their relation to anaerobic soil communities. Pages 187–196 *in* J. R. Clark and J. Benforado, editors. Wetlands of bottomland hardwood forests, Elsevier Scientific Publishing, New York.

Junk, W. J., P. B. Bayley, and R. E. Sparks. 1989. The flood pulse concept in river-floodplain systems. Pages 110–127 *in* D. P. Dodge, editor. Proceedings of the international large river symposium. Canadian Special Publication of Fisheries and Aquatic Sciences 106.

Keeland, B. D., and P. J. Young. 1997. Long-term growth trends of baldcypress (*Taxodium distichum* (L.) Rich.) at Caddo Lake, Texas. Wetlands 17:559–566.

Killgore, K. J., and J. A. Baker. 1996. Patterns of larval fish abundance in a bottomland hardwood wetland. Wetlands 16:288–295.

Killgore, K. J., and P. M. Hathorn. 1987. Application of the habitat evaluation procedures in the Cypress Bayou basin, Texas. U.S. Army Corps of Engineers Waterways Experiment Station, Miscellaneous Paper EL-87-4, Vicksburg, Mississippi.

King, A. J., K. A. Ward, P. O'Connor, D. Green, Z. Tonkin, and J. Mahoney. 2010. Adaptive management of an environmental watering event to enhance native fish spawning and recruitment. Freshwater Biology 55:17–31.

Ladson, A. R., and R. M. Argent. 2002. Adaptive management of e-flows: lessons for the Murray–Darling basin from three large North American rivers. Australian Journal of Water Resources 5:89–101.

MICRA (Mississippi Interstate Cooperative Resource Association). 2012. MICRA Paddlefish stocking protocols. MICRA, Marion, Illinois.

Opperman, J. J., R. Luster, B. A. McKenney, M. Roberts, and A. W. Meadows. 2010. Ecologically functional floodplains: connectivity, flow regime, and scale. Journal of the American Water Resources Association 46:211–226.

Owens, C. S., R. M. Smart, and R. M. Stewart. 2004. Low temperature limits of giant salvinia. Journal of Aquatic Plant Management 42:91–94.

Poff, N. L., J. D. Olden, D. M. Merritt, and D. M. Pepin. 2007. Homogenization of regional river dynamics by dams and global biodiversity implications. Proceedings of the National Academy of Sciences of the United States of America 104:5732–5737.

Ramsar. 1987. Convention on wetlands of international importance especially as waterfowl habitat. Ramsar (Iran), 2 February 1971. UN Treaty Series No. 14583. As amended by the Paris Protocol, 3 December 1982, and Regina Amendments, 28 May 1987.

Richter, B. D., A. T. Warner, J. L. Meyer, and K. Lutz. 2006. A collaborative and adaptive process for developing environmental flow recommendations. River Research and Applications 22:297–318.

Robertson, S., K. Mayes, G. Linam, M. Parker, C. Robertson, A. Grubh, S. Magnelia, and M. Casarez. 2016. Cypress basin bioassessment, Marion and Harrison counties, Texas. Texas Parks and Wildlife Department, River Studies Report No. 24, Austin.

Rood, S. B., G. M. Samuelson, J. H. Braatne, C. R. Gourley, F. M. R. Hughes, and J. M. Mahoney. 2005. Managing river flows to restore floodplain forests. Frontiers in Ecology and Environment 3:193–201.

Simon, T. P. 1999. Assessment of Balon's reproductive guilds with application to Midwestern North American freshwater fishes. Pages 97–121 *in* T. P. Simon, editor. Assessing the sustainability and biological integrity of water resources using fish communities. Lewis Press, Boca Raton, Florida.

TCEQ (Texas Commission on Environmental Quality). 2006. One total maximum daily load for dissolved oxygen in Lake O' the Pines for Segment 0403. TCEQ, Austin.

TCEQ (Texas Commission on Environmental Quality). 2010. Procedures to implement the Texas surface water quality standards. TCEQ, Austin.

TCEQ (Texas Commission on Environmental Quality). 2011. Caddo Lake: watershed protection plan. Fact sheet. TCEQ, Austin.

TCEQ (Texas Commission on Environmental Quality). 2014. 2014 Texas integrated report of surface water quality for the Clean Water Act sections 305(b) and 303(d). TCEQ, Austin.

TIFP (Texas Instream Flow Program). 2008. Texas instream flow studies: technical overview. Prepared by Texas Commission on Environmental Quality, Texas Parks and Wildlife Department, and Texas Water Development Board, Austin.

Tharme, R. E., and J. M. King. 1998. Development of the building block methodology for instream flow assessments, and supporting research on the effects of different magnitude flows on riverine ecosystems. Water Research Commission, Report No. 576/1/98, Pretoria, South Africa.

TNC (The Nature Conservancy). 2007. Indicators of hydrologic alteration (IHA) software. The Nature Conservancy, Arlington, Virginia.

Thoms, M. C., M. Southwell, and H. M. McGinness. 2004. Floodplain-river ecosystems: fragmentation and water resources development. Geomorphology 71:126–138.

Trungale, J., and T. Hayes. 2011. Floodplain inundation analysis interim report. Final report to Caddo Lake Institute, Shreveport, Louisiana.

Turner, T. F., J. C. Trexler, G. L. Miller, and K. E. Toyer. 1994. Temporal and spatial dynamics of larval and juvenile fish abundance in a temperate floodplain river. Copeia 1994:174–183.

USACE (U.S. Army Corps of Engineers). 1994. Red River waterway project, Shreveport, LA, to Daingerfield, TX, reach reevaluation study in-progress review. Appendix 6: aquatic resources. USACE, Vicksburg, Mississippi.

USFWS (U.S. Fish and Wildlife Service). 2015. Final report for Big Cypress Bayou Paddlefish reintroduction assessment. U.S. Fish and Wildlife Service, San Marcos, Texas.

Ward, J. V., K. Tockner, and F. Schiemer. 1999. Biodiversity of floodplain river ecosystems: ecotones and connectivity. Regulated Rivers: Research and Management 15:125–139.

Warner, A. T., L. B. Bach, and J. T. Hickey. 2014. Restoring environmental flows through adaptive reservoir management: planning, science and implementation through the Sustainable Rivers Project. Hydrologic Sciences Journal 59:770–785.

Winemiller, K. O. 2005. Floodplain river food webs: generalizations and implications for fisheries management. Pages 285–309 *in* R. L. Welcomme and T. Petr, editors. Proceedings of the second international symposium on the management of large rivers for fisheries, volume 2. Mekong River Commission, Phnom Penh, Cambodia.

Winemiller, K. O., S. Tarim, D. Shormann, and J. B. Cotner. 2000. Spatial variation in fish assemblages of Brazos River oxbow lakes. Transactions of the American Fisheries Society 129:451–468.

Winemiller, K. O., A. Chin, S. E. Davis, D. L. Roelke, L. M. Romero, and B. P. Wilcox. 2005. Summary report supporting the development of flow recommendations for the stretch of Big Cypress Creek below Lake O' the Pines dam. Report to The Nature Conservancy. Texas A&M University, College Station.

Wright, L. D., and T. J. Bister. 2014. Inland Fisheries Division monitoring and management program survey report for Caddo Lake, 2013. Texas Parks and Wildlife Department, Federal Aid Project Report F-221-M-3, Austin.

American Fisheries Society Symposium 91:367–387, 2019

Lessons from the Driftless Area: The Importance of Human Dimensions Research for Stream Restoration and Fisheries Management

Caroline Gottschalk Druschke*
Department of English, University of Wisconsin–Madison
7187 Helen C. White Hall, 600 North Park Street, Madison, Wisconsin 53706, USA

Eric G. Booth
Department of Civil & Environmental Engineering and Agronomy
University of Wisconsin–Madison
3117 Wisconsin Energy Institute, 1552 University Avenue, Madison, Wisconsin 53726, USA

Abstract.—This chapter shifts a multispecies approach beyond the banks to consider interactions between fish species, stream systems, and the humans that make decisions about riverine landscapes. Building from the idea that successes in stream restoration and fisheries management depend not only on ecological outcomes, but also on manager learning and public support, this chapter uses the Driftless Area region in southwestern Wisconsin, southeastern Minnesota, northeastern Iowa, and northwestern Illinois as a case study. The chapter synthesizes existing research about angler preferences, economic impacts of fisheries, and the human dimensions of stream restoration in the region. It then points to future directions for research, public engagement, and collaboration in the Driftless Area and beyond that could support stream restoration and fisheries management into the future.

Introduction

This chapter builds from our experience working in southwestern Wisconsin's portion of the four-state Driftless Area region, site of the United States' first watershed-based conservation project in Coon Creek in the early 1930s and a central hub of stream restoration activities ever since. While a comprehensive listing of stream projects throughout the region does not currently exist, the Wisconsin Department of Natural Resources (WDNR) owns or holds easements on more than 35,000 acres (14,163 ha) of riparian property along the state's 300-plus Driftless Area streams (WDNR 2013a). In collaboration with local partners and government entities, the WDNR has restored and maintained more than 300 mi (482 km) of Driftless Area streams since 1970 (WDNR 2013a). Regionwide, the Driftless Area Restoration Effort (DARE) has identified 50-plus stream restoration projects they have been involved in to promote habitat for Brook Trout *Salvelinus fontinalis*, Brown Trout *Salmo trutta*, Mottled Sculpin *Cottus bairdii*, and Smallmouth Bass *Micropterus dolomieu*, in addition to supporting aquatic and terrestrial nongame species (DARE, no date). Untold stream projects, small and large, are undertaken by counties, private landowners, and fishing clubs across the Driftless Area portions of Wisconsin, Minnesota, Iowa, and Illinois, and groups like

* Corresponding author: caroline.gottschalk.druschke@wisc.edu

state Departments of Natural Resources and DARE have committed to restoration goals moving forward, including restoration of native riparian vegetation, implementation of direct stream channel restoration, and removal of barriers to fish passage across the region (Mauldin and Hastings 2014). Given the long history of stream restoration across the Driftless Area and sustained energy moving forward, this region offers unique opportunities for exploring and learning not only from the biophysical but also from the human dimensions of stream restoration, including both stakeholder engagement and adaptive management processes and outcomes.

In their landmark paper defining standards for ecologically successful restoration projects, Palmer et al. (2005) distinguished between three axes for the evaluation of river restoration projects: ecological, learning, and stakeholder successes. Ecological success featured five characteristics: basis on a guiding image, measurable ecological improvement, improved resilience, absence of lasting harm, and publicly available pre- and postassessment data. Meanwhile, learning success involved "advances in scientific knowledge and management practices that will benefit future restoration action," and stakeholder success referred to "human satisfaction with restoration outcome" (Palmer et al. 2005:209). The most effective river restoration projects, they argued, meet all three axes of success.

These three axes are central to stream and river restoration, and, in the Driftless Area of southwestern Wisconsin, where we work, increasing effort is going into meeting at least two of these goals: gathering and synthesizing the best available science for restoration work (i.e., ecological success) and sharing that research among managers, researchers, and practitioners in the region (i.e., learning success). But the third axis—stakeholder success—may well be just as critical to restoration project outcomes in the Driftless Area and beyond and is largely understudied across stream restoration literature in the Driftless Area, nationally, and internationally.

Bernhardt et al. (2007) found a positive correlation between community involvement and ecological success in a nationwide study while Druschke and Hychka (2015) found that long-term public engagement played a central role in achieving aquatic restoration project successes in New England—even for projects that were focused primarily on ecological indicators of success. But, as Druschke and Hychka (2015) detailed, "little research explores how to cultivate the sorts of quality public engagement experiences that might contribute to restoration success." And so, while natural resource agencies and organizations (including the WDNR, Trout Unlimited, and the Natural Resources Conservation Service) continue to restore Driftless Area streams for Brook Trout and Brown Trout, and increasingly for nongame species, it is humans who conceive of projects, fund them, enact them, monitor them, and decide whether or not to support them. Likewise, it is humans who have an outsized impact on trout stream quality across the region due to fishing practices, land management practices, and agricultural practices (Figures 1–3). But again, these human impacts, perspectives, and values are largely understudied.

Many restoration projects proceed on generally accepted wisdom about the best ways to design and execute projects, but those views are not always based on research or field-based observation. Meanwhile, the bulk of peer-reviewed research on restoration practices is understandably and necessarily focused on physical and biological attributes. In the Driftless Area, this means there is a focus on science-based restoration practices that might support the restoration of dynamic streams in the face of climate change, including a recent comprehensive science review (Dauwalter 2019) that coincided with the 11th Annual Drift-

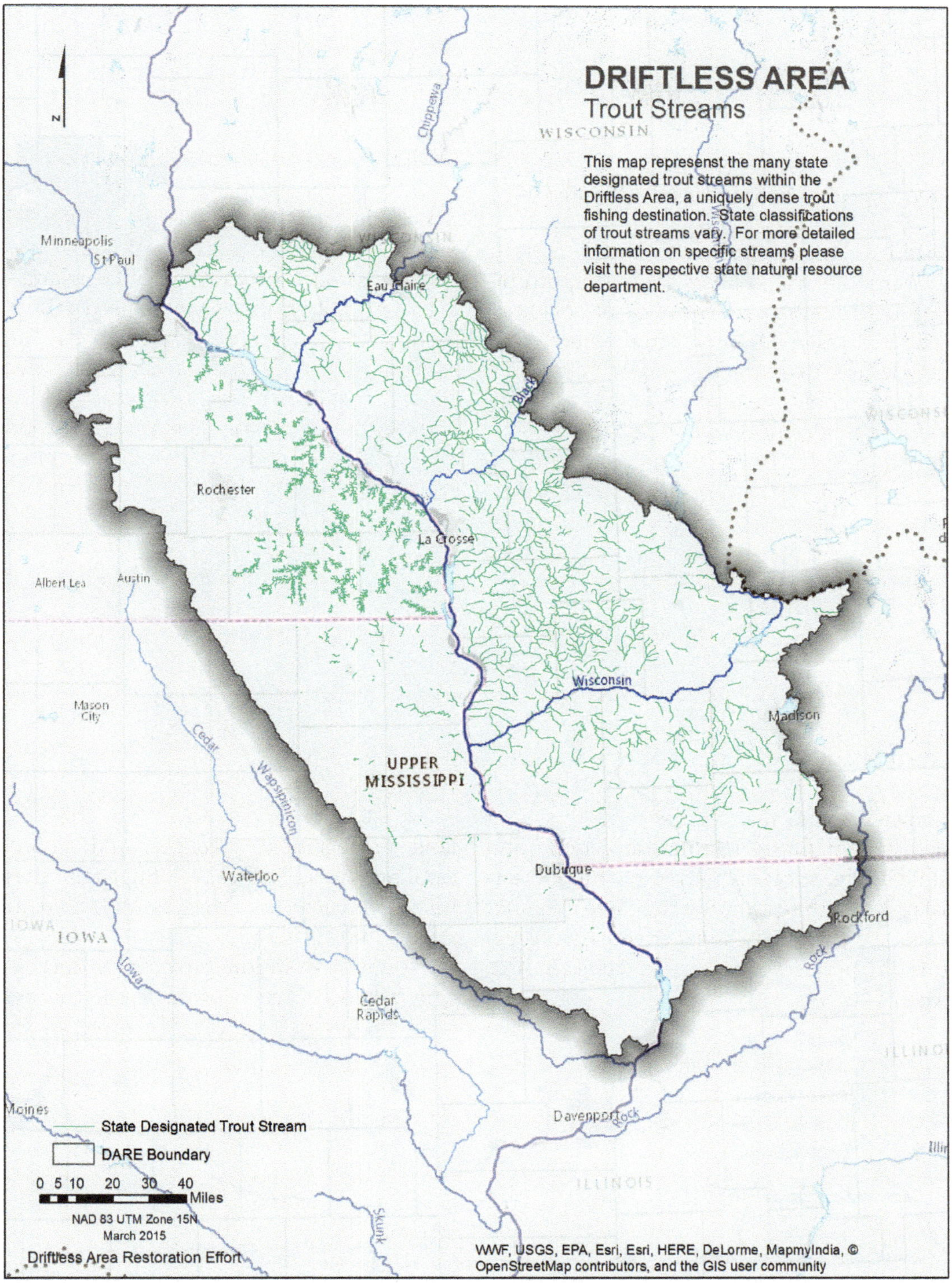

Figure 1. Map of trout streams in the Driftless Area region, encompassing portions of southwestern Wisconsin, southeastern Minnesota, northeastern Iowa, and northwestern Illinois (2015). Credit: Trout Unlimited Driftless Area Restoration Effort.

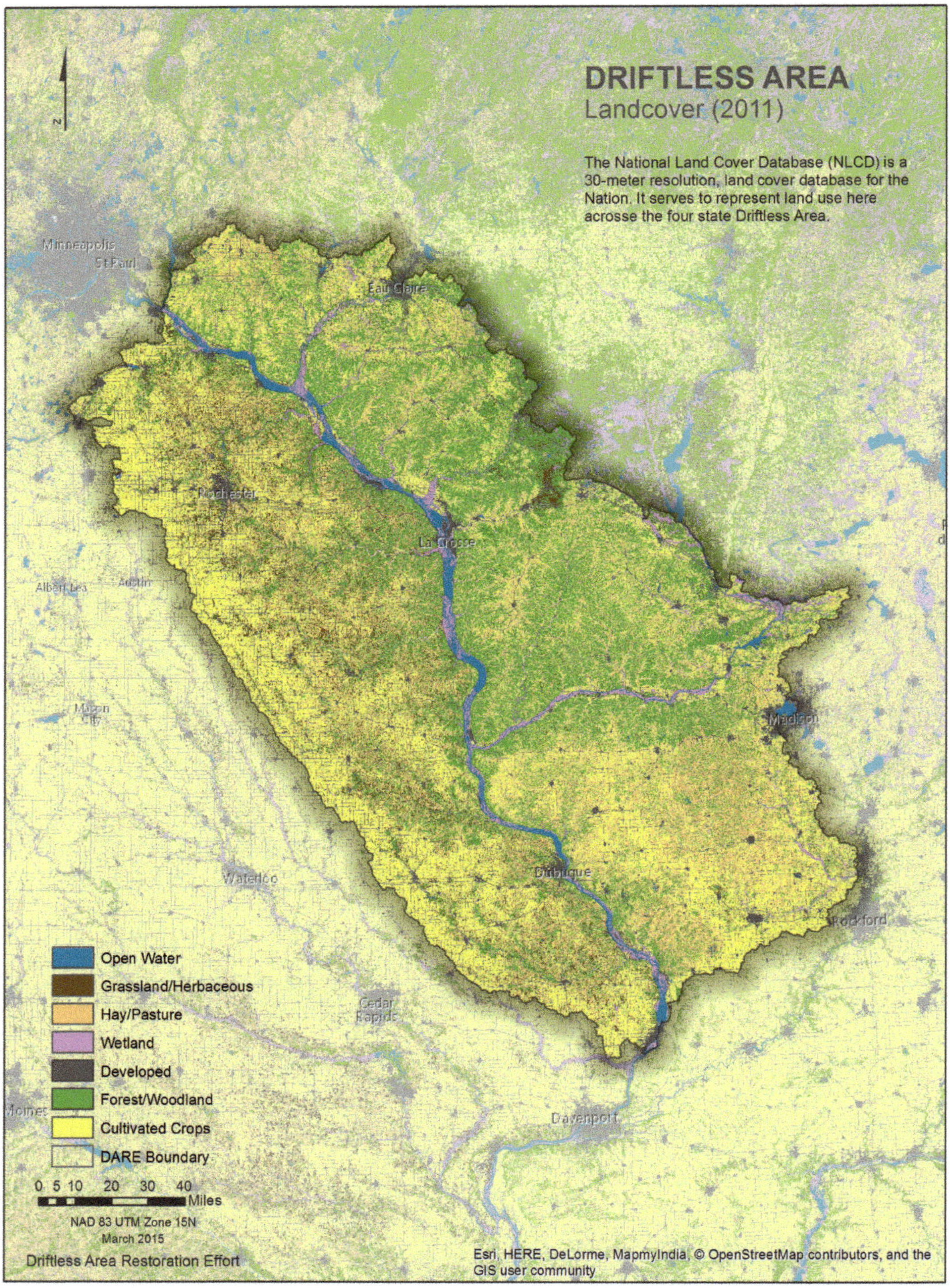

Figure 2. Map of land cover in the Driftless Area region (2011). Credit: Trout Unlimited Driftless Area Restoration Effort.

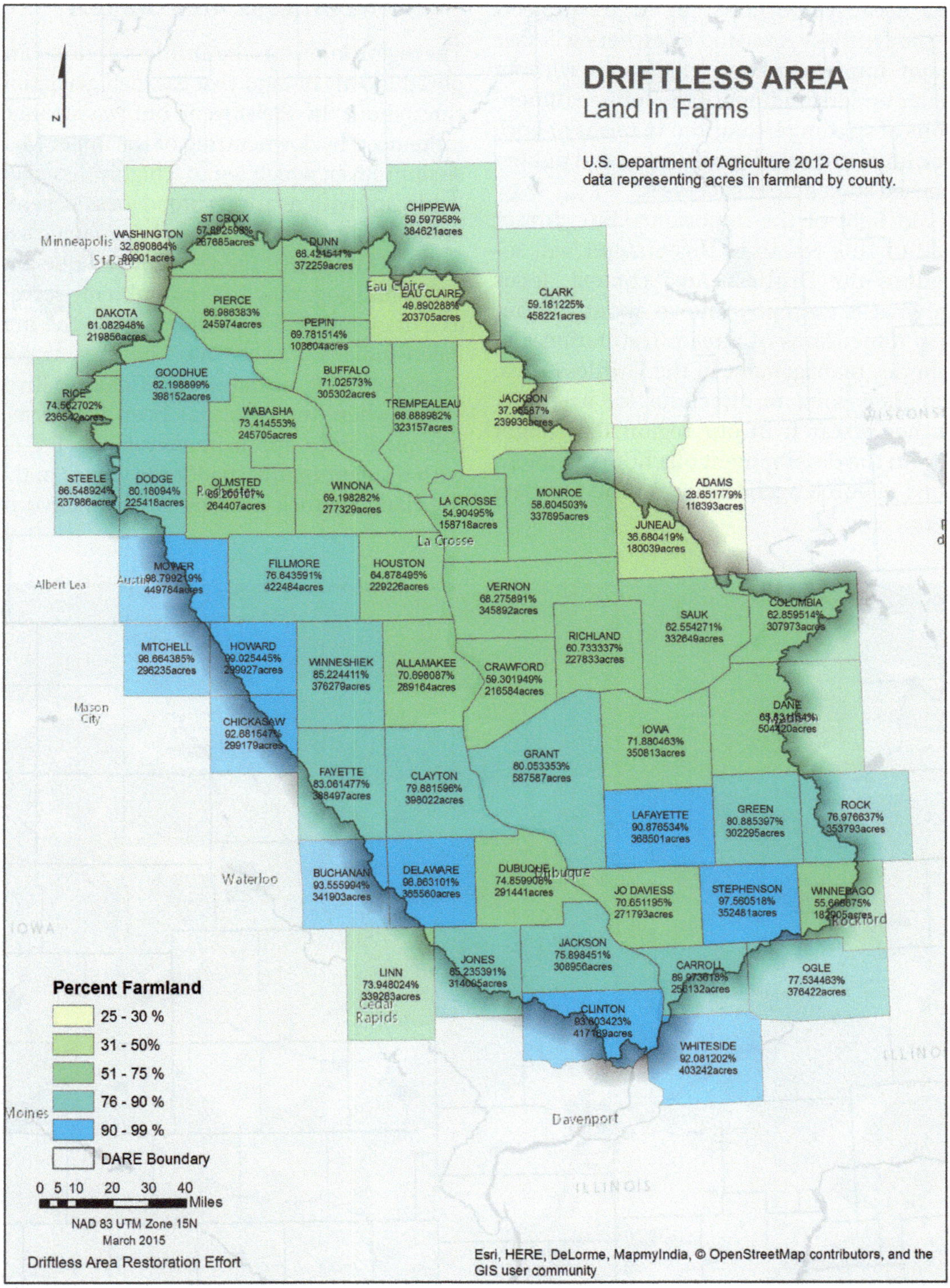

Figure 3. Map of farmland density by county in the Driftless Area region (2015). Credit: Trout Unlimited Driftless Area Restoration Effort.

less Area Symposium. But future projects in the Driftless Area and elsewhere will face major implementation challenges without better understandings of the human dimensions of stream restoration, in terms of both learning successes for managers and public-focused stakeholder successes.

In light of the importance but current lack of this research, this chapter will introduce the Driftless Area context, focus on what is currently known about the human dimensions of stream restoration and fisheries management in the Driftless Area, and then point to directions for necessary further research in our region that might inform the development of a human dimensions research agenda beyond the Driftless Area.

The Driftless Area Context

The region known as the Driftless Area is composed mostly of land that escaped glaciation throughout the Pleistocene but was strongly influenced by downcutting of the upper Mississippi River, which led to a highly dissected landscape with relatively broad loess-covered uplands, steep hillslopes, and deeply incised valleys (Knox 1982; Figure 4). Its unique and beautiful geography has undoubtedly served as a draw for local inhabitants since the first Paleo-Indians moved into the area 12,000 years ago (KVR, no date). The Kickapoo River valley, where we work, has been intensively occupied since approximately 6,000 B.C.E., with residents gathering for the winter in the overhanging rock shelters so distinctive of

Figure 4. Agricultural best management practices, including contour farming and grassed waterways, are visible under cover of snow alongside a soaked floodplain in the Kickapoo River watershed in December 2018.

the topography of the region (KVR, no date). The Driftless Area encompasses the ancestral lands of the Ho-Chunk, Sauk, Dakota, and Potawatomi peoples (Loew 2013; The Ways 2019; Wisconsin First Nations, no date). The majority of people of these sovereign nations were forcibly removed from the region in the early 1800s to make way for Euro-American settlers seeking lead, furs, and agricultural land (Loew 2013). This period of European settlement—specifically, the change in land cover from prairie and oak savanna to cropland and pasture—was accompanied by devastating erosion, flooding, and downstream sedimentation that dramatically altered the geomorphology and habitat of stream-floodplain ecosystems (Happ 1944; Knox 2006). These widespread changes had immediate negative economic consequences, both in terms of agricultural productivity and flood damages, and spawned a new era of land and water conservation highlighted by the creation in 1933 of the nation's first watershed project in the Coon Valley area (Anderson 2002; Leopold 1935).

While land management practices substantially improved as the new conservation ethic spread throughout the region, leading to less erosion (Trimble and Lund 1982) and less flooding (Potter 1991), local streams continued to provide poor habitat for native fish and other wildlife as a consequence of the combination of erosive floods, the burial of riparian wetlands by as much as several meters of sediment, and the subsequent erosion of high vertical banks. Thus, beginning in the 1950s, efforts by local rod and gun clubs were made to enhance stream habitat—primarily for native Brook Trout and introduced Brown Trout—using in-channel structures to provide overhead cover and reduce bank erosion (Vetrano 2017). State and county conservation agencies soon followed in the 1970s and 1980s with more habitat enhancement techniques, including new overhead structures and sloping streambanks to better connect a stream with its floodplain, which led to less erosion and better flood attenuation (Vetrano 2017), though erosion is still a major factor on the region's steep slopes (Figure 5). The focus on instream habitat improvement along with advanced trout stocking programs and continued land management improvements upstream led to a substantial recovery of the fishery by the 1990s and national recognition of the Driftless Area as a major destination for trout anglers (Anderson et al. 2000) (Table 1). Restoration activities continue to be refined and debated today, with added emphases on water quality, management for nongame species, and resilience to climate change; increases in precipitation, coupled with the region's hallmark steep slopes, small streams, and heavy agriculture, are projected to increase flood events in southwestern Wisconsin (WICCI 2011). While these impacts deserve more treatment than this chapter can provide, there is no doubt that increasingly frequent and increasingly severe flooding in the region will affect how restoration projects are approached and prioritized and will impact existing restoration sites in a multitude of ways.

What We Currently Know: Human Dimensions Research in the Driftless Area

Generally speaking, Driftless Area researchers, managers, and practitioners have huge amounts of knowledge about the human dimensions of stream restoration in the region. The Wisconsin and Minnesota Departments of Natural Resources are well known for their deep history of trout stream habitat management, dating back a century. Much work in the region was guided in large part by Ray White and Oscar Brynildson's *Guidelines for Management of Trout Stream Habitat in Wisconsin* (White and Brynildson 1967), a groundbreaking text that contributed to learning successes by offering technical advice based in a philosophy of encouraging a river's natural processes. The history of recreational trout fishing in the

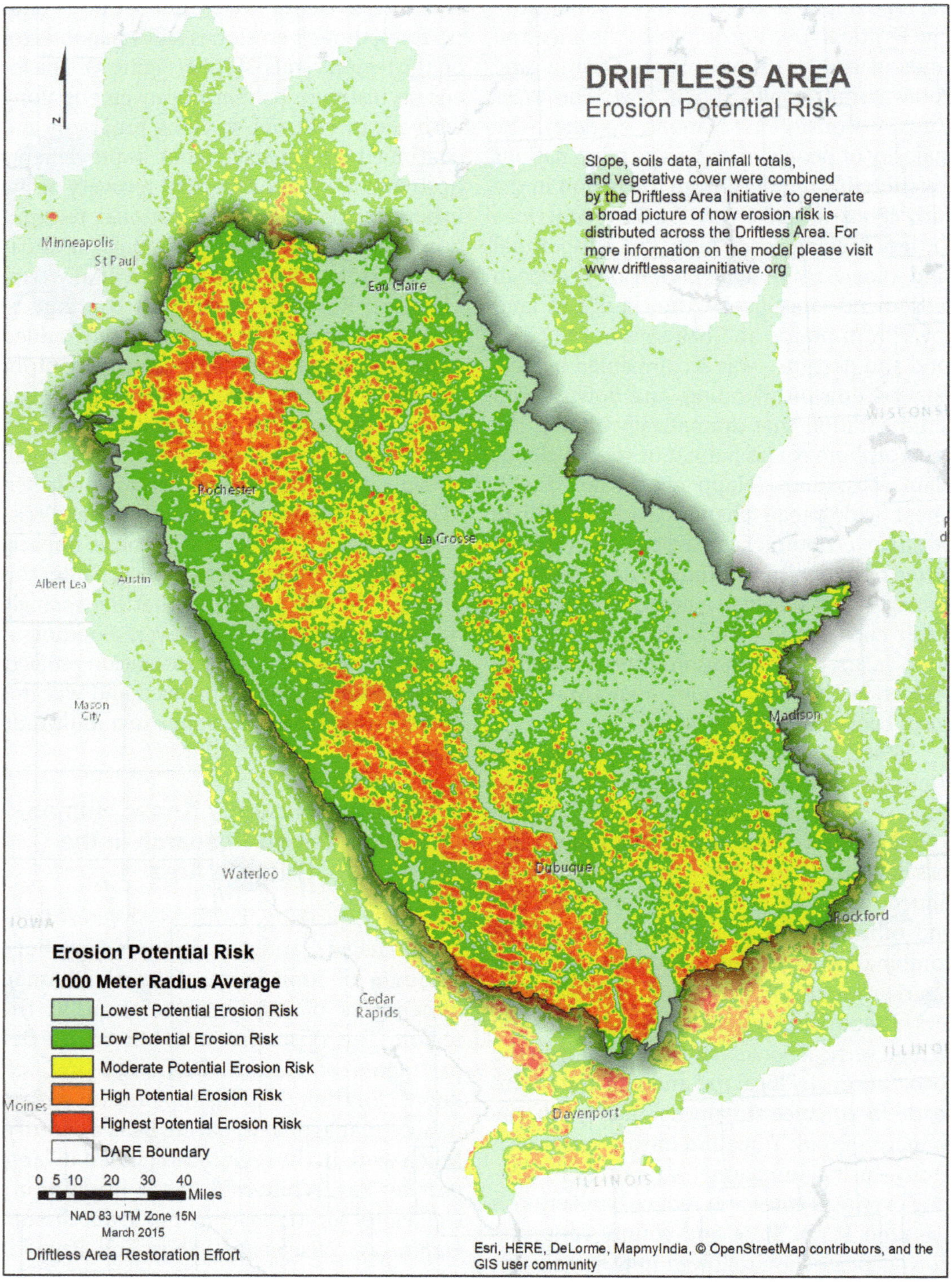

Figure 5. Map of erosion potential risk in the Driftless Area region (2015). Credit: Trout Unlimited Driftless Area Restoration Effort.

Table 1. Change in nonlocal trout angler expenditure patterns in Wisconsin's Kickapoo River valley between 1994 and 1999 (reported in 1997 U.S. dollars using sector specific inflation index). Adapted from Anderson et al. (2000).

	Regional expenditures (nonlocal)	
Spending category	Survey 1 (1994)	Survey 2 (1999)
Lodging	$51,100	$307,300
Groceries	30,300	98,100
Automobile-related	39,100	144,800
Eating/drinking	50,200	263,800
Souvenirs/gifts	3,400	29,900
Entertainment	1,400	18,000
Miscellaneous	15,200	15,500
Fishing supplies	–	99,600
Licenses	–	62,500
Guide services	–	63,500
Total in all like categories	$190,700	$877,500
—percent change	–	360%
Total in all categories	$190,700	$1,103,100
—percent change	–	480%

Driftless Area coupled with an early management orientation in the region mean that managers have been thinking—both explicitly and implicitly—about human aspects of stream restoration for decades.

Referring to work in the Driftless Area's Pecatonica River watershed, Steve Richter, director of conservation programs for The Nature Conservancy in Wisconsin, explained the importance of looking beyond the streambanks to human actors:

> You can't just do stream restoration projects without looking at the practices in the adjacent farm fields. And you can't implement new practices in the field without having strong relationships with farmers and landowners. We took the time to develop strong relationships. That time spent in building relationships leads to bigger outcomes [From Miller 2017.]

As Dieterman and Merten (2016) suggested in their comprehensive history of trout management in southeastern Minnesota,

> Effective and successful fisheries management requires information on the three primary components of a fishery: the biota (primarily fish), their habitat, and the benefits they provide to society. [From Krueger and Decker 1999:16.]

In short, they urge a multispecies approach to fisheries management and stream restoration that includes humans.

And yet, mirroring a national and international trend, there is a lack of published research into the human dimensions of stream restoration in the Driftless Area. As Dieterman and Snook (2015) emphasized, while the Driftless Area region was an early leader in the biological evaluation of stream habitat projects and needs to continue that close biological evaluation with new habitat practices,

> [p]erhaps more importantly, direct tangible benefits of habitat projects for anglers have been less frequently investigated. [From Dieterman and Snook 2015:73.]

They urged more specific, measurable project objectives on both the biological and sociodemographic fronts. This review chapter is a first step towards achieving that goal.

While there is a general lack of peer reviewed literature into the human dimensions of stream restoration in the Driftless Area, there are two clusters of research—angler preferences and economic impacts—that offer a good foundation for building a more robust archive of social science and social-ecological science in this realm. Both clusters are primarily driven by state managers and researchers in the Minnesota, Wisconsin, and Iowa Departments of Natural Resources, with contributions from Trout Unlimited and graduate student researchers in the region.

Trout angler preferences in the Driftless Area

Statewide surveys of angler preferences offer important insight into the possibilities for stream restoration. While the WDNR does not consistently survey anglers in the state, they have conducted several statewide surveys. Schroeder and Fulton (2017) indicated that the Minnesota Department of Natural Resources conducts annual social surveys of angler attitudes; some data from those surveys are included in a variety of reports and manuscripts. Every 5 years, the Iowa Department of Natural Resources surveys 10,000 trout privilege purchasers, though results from those surveys do not seem to be publicly available.

In Minnesota statewide, anglers have consistently placed importance on habitat protection and restoration. A statewide survey of anglers who purchased licenses for the 2003 fishing season found that

> [o]ver three-fourths of respondents felt that improving lake and stream habitat (91.3%) and protecting the land surrounding lakes and streams (83.2%) were important activities. [From Schroeder and Fulton 2005.]

In the Driftless Area portion of southeastern Minnesota specifically, trout anglers linked stream health with agricultural induced erosion, with a majority of anglers indicating that livestock fencing, riparian vegetation, and rip-rap would be at least "very effective" (Vlaming and Fulton 2003). A decade later, a statewide survey of 2014 Minnesota fishing license holders age 18 and older showed similar interest, with respondents rating "protecting the habitat in lakes and streams" as the most important management activity, and "restoring the habitat in lakes and streams" also rating above "important" (4.2 out of 5) (Schroeder 2015:62).

Protection and restoration of trout streams seems to play an important role in supporting and maintaining a strong population of trout anglers. A recent Wisconsin survey of lapsed trout anglers (anglers who did not purchase a trout stamp for 3 years after five consecutive years of purchase) found that quality of the trout fishery was an important factor in the trout angling lapse and recognized that, coupled with external factors, habitat improvement projects can contribute to angler satisfaction (Petchenik 2012). These findings are not divided by region, however, to get a sense of Driftless Area-specific responses. A statewide survey of active Wisconsin trout anglers showed that 56% indicated a preference or requirement for a stream with restored habitat while 74% would prefer not to or would never fish a degraded stream (Petchenik 2014). Petchenik (2014:54) hypothesized that

> the imbalance between these two measures may be one of perception: anglers are more likely able to perceive poor stream habitat but may have more difficulty perceiving stream restoration, particularly if it is an angler's first experience at a stream.

A statewide survey of anglers who purchased Minnesota trout stamps and indicated they fished in southeastern Minnesota found that stream improvement projects most positively affected satisfaction with trout fishing in southeastern Minnesota, with anglers supportive of trout stream easements and fly anglers significantly more supportive of trout stream easements than lure and bait anglers (Schroeder 2014). Anglers were supportive of trout stream easements newly in place, again with fly anglers more supportive (Schroeder 2014). A recent comprehensive study of the economic impact of trout angling in the Driftless Area showed that 88.5% of respondents reported being aware of trout stream preservation and restoration efforts in the region, with almost 80% of that group reporting that past efforts prompted them to be more likely to fish in the region and 72.7% indicating that future trout stream restoration efforts would make them more likely to fish in the region (Anderson 2016).

Minnesota anglers also seem fairly satisfied with state managers' work on habitat protection and restoration. Statewide respondents holding licenses indicated that the Minnesota Department of Natural Resources "performed well at improving lake and stream habitat (68.1%) and protecting the land surrounding lakes and streams (70.1%)" (Schroeder and Fulton 2005). A decade later, Schroeder (2015) reported,

> Respondents felt that the Minnesota DNR was doing well at protecting habitat in lakes and streams, protecting land surrounding lakes and streams, and educating people on how they can help protect lakes and streams.

Schroeder (2015:v) recommended activities for future focus related to habitat management, including

> managing shoreline to protect fish spawning sites, restoring the habitat in lakes and streams, restoring land surrounding lakes and streams that have been damaged/developed, and educating people about lake and stream ecology/habitat.

Schroeder and Fulton's (2017) recent work, based on a survey of Minnesota fishing license holders, reminded readers that management outcomes depend in large part on angler perceptions about those management decisions. Importantly, they found that acceptance of management decisions depended largely on impressions of voice—"the opportunity to express opinions in decision-making processes" (sensu Folger 1977)—and procedural fairness—"i.e., institutional factors like neutrality, respectful treatment, and encouragement of stakeholder voice" (sensu Lind and Tyler 1988; Lawrence et al. 1997; Hamm et al. 2013).

Managing increased fishing pressure—generally and in the wake of habitat restoration projects—may continue to be an issue for state managers in some locations. In Wisconsin's Kickapoo River valley, a two-stage survey (intercept with mail follow-up) and series of focus group interviews conducted with trout anglers in 1994–1995 demonstrated respondent interest in improving fisheries management via management of future fishing pressure and the provision of larger fish, more fish, and greater species diversity on valley streams (Anderson and Marcouiller 1996). A 1999 follow-up to that survey showed that respondents were generally very satisfied with fisheries and river management practices in southwestern Wisconsin, though there continued to be concern about future crowding (Anderson et al. 2000). Creel surveys were conducted during the 2013 season on southeastern Minnesota streams; habitat enhancement projects had occurred on three stream sites within the past 8 years to allow for initial pre- and postproject evaluations, with a fourth site offering a control (Dieterman and Snook 2015). Pre- and postproject comparisons revealed few differences in angler demographics, catch rates, participation, or satisfaction pre- and postproject, with the exception of Trout Run Creek, which saw a 200% increase in angler pressure postproject (Dieterman and Snook 2015). The authors explained the inconclusive results could be due to small preproject sample sizes and recommend future focused creel surveys on a small set of their stream sites for future comparison.

In terms of preferred landscape, statewide in Minnesota, respondents indicated a preference for dense forest adjacent to streams and rivers, natural rocky banks, and rocky streambeds and riverbeds (Schroeder 2015). This result may not hold true for the Driftless Area region, however, whose present-day grassed riparian areas and valley floors with forested hillslopes are somewhat iconic. An earlier survey about trout angling in southeastern Minnesota detailed angler opinions regarding desired stream characteristics. Respondents preferred partial canopy cover and low brush on banks, with views of hills or bluffs, and respondents had a neutral response to the impact of pasture with animals (Vlaming and Fulton 2003). Respondents preferred "medium streams that are 10–25 ft wide, with a mix of both fast and slow water that is usually clear, even in times of high water" (Vlaming and Fulton 2003:87). In Wisconsin, data from the WDNR's Driftless Area Master Plan survey indicated a preference for grass-lined banks over forested or pastured banks (Figure 6), but the survey was targeted only to individuals who signed up to receive updates about the master plan (WDNR 2013b). A 2014 WDNR trout angler survey, meanwhile, indicated a preference for forested banks across the state. Approximately 3 respondents in 10, statewide, indicated they would never or would prefer not to fish a stream that was pastured or mowed (29%) or to fish a stream with an overgrown bank (30%)

Figure 6. Michael Webber takes advantage of the grass-lined banks along a newly restored section of Weister Creek in Wisconsin's Kickapoo Valley Reserve to cast for Brown Trout.

(Petchenik 2014). Statewide, a thin majority of trout anglers needed or preferred forested streambanks (51%), and an equal percentage (51%) preferred not to or would never fish a stream where trees have been removed along the bank (Petchenik 2014). While the Driftless Area-specific sample size was limited, Driftless Area-specific sample responses to the statewide survey, offered directly by Jordan Petchenik (Wisconsin Department of Natural Resources, personal communication), showed more of a preference to fish on pastured or mowed streambanks and more indifference to forested streambanks compared with respondents from other parts of the state.

While not specifically focused on stream management, a recent study of angler preferences for the Minnesota winter fishery showed that fly anglers tended to be specialized on a small group of streams, including Branches of the Whitewater River, and that easy access was one of the common reasons driving angler preference (Snook and Dieterman 2015). This point about angler access might influence future restoration design in terms of designing stream restoration projects that offer ample public access without a decline in fish habitat.

Much of what is known about trout angler preferences in the region comes from reports of state surveys of anglers. A notable methodological exception to that trend is work that emerged from five landowner focus groups conducted in southeastern Minnesota to explore factors influencing riparian and watershed management in the area (Blann 2004). Though the groups varied somewhat based on location and cultural aspects and concerns, emergent themes included strong interests in multigenerational stewardship, coupled with concerns about flooding, erosion, failed agricultural policy, corporatization of agriculture, chemical and livestock pollution, and increasing development (Blann 2004). These interests suggest directions for future research in the region.

Another methodological exception is a pre- and postproject survey-based assessment of a conservation intervention in southeastern Minnesota's Wells Creek watershed (Kelly et al. 2001). While not angling-specific, a 1994 landowner survey gathered baseline data from southeastern Minnesota counties, allowing for comparison between the Wells Creek watershed, other bluffland counties (Goodhue, Wabasha, Olmstead, Winona, Fillmore, and Houston), and other southeastern Minnesota counties (Rice, Steele, Dodge, Freeborn, and Mower). Conservation actions, including social and educational activities related to conservation actions, were introduced in the Wells Creek watershed, and a 1999 survey was used to determine whether any noticeable differences emerged between the Wells Creek watershed, neighboring counties, and other southeastern Minnesota counties. Results demonstrated very few changes in the perceptions and behaviors of landowners over the 5-year span, though it seems possible perceptions and behaviors may change over a longer time period. The study noted that

> Changes that did occur tended to bring the responses of landowners in the bluffland and other southeastern Minnesota counties closer to those in Wells Creek—homogenizing views and actions,

but respondents demonstrated some increasing concerns about increasing development (Kelly et al. 2001:2). Concern with "quality of fish habitat" did not show significant change (Kelly et al. 2001).

A final methodological outlier comes from Epton and Fulton (2002) related to controversial trout management efforts in southeastern Minnesota in the late 1990s. Concerns from the Minnesota Trout Association and Trout Unlimited about results from a 1997 survey related to proposed trout regulation changes in southeastern Minnesota (Anthony 1997) led to the 1998 formation of a stakeholder committee facilitated by the Minnesota Department of Natural Resources and the beginning of a public comment process. Participatory decision making in the process was assessed according to sense of need, agreement on technical boundaries, perceptions of one's own power, and sense of urgency (Epton and Fulton 2002). While all stakeholders agreed on the need for the process and all but one agreed on the technical boundaries, there was a great deal of disagreement about the perception of one's own power and sense of urgency (Epton and Fulton 2002). Participants reported mixed responses about their satisfaction with the process, including satisfaction with outcomes, personal commitment, and willingness to participate again (Epton and Fulton 2002). In terms of procedural justice, respondents mostly agreed that they had a high level of perception of voice and influence but were much more mixed in terms of fairness of outcomes and procedural fairness (Epton and Fulton 2002). Responses about trust in authority, neutrality of authority, respect, pride in participation, and legitimacy of authority were mixed as well (Epton and Fulton 2002). Importantly, as Epton and Fulton (2002:175) explained, "the level of voice or influence given to each of the parties determined their feelings about the fairness of the process used." Despite the general difficulty of finding patterns across the data, Epton and Fulton (2002) recommended the development of future, meaningful opportunities for stakeholders to provide input into decision-making processes in ways that build trust and offer longer-term follow-through.

Economic impacts of trout angling and trout restoration efforts in the Driftless Area

There is a small but relatively thorough body of knowledge about the economic impacts of trout angling in the Driftless Area.

More than 20 years ago, Anderson and Marcouiller (1996) noted the importance of trout angling as a rural economic engine in the Driftless Area region, including both direct and indirect impacts, with research focused specifically on the Kickapoo River valley. Through a two-stage intercept and mail survey and focus groups, mentioned above, the study found that half of trout anglers surveyed were nonlocal and that visiting anglers spent almost US$220,000 during the 1994 season and contributed almost $500,000 to total gross output (Anderson and Marcouiller 1996). The study pointed to past investments in restoration of nearly $330,000 (in 1994 dollars) on the Timber Coulee system and to the impacts of those restoration efforts on supporting increasing spending by out-of-town anglers in the area (Anderson and Marcouiller 1996). A 1999 follow-up to that 1994 survey in the Kickapoo valley demonstrated rapid growth in angling, with double the numbers of trout anglers from 1994 to 1999, including a three-to-one increase in nonlocal anglers and an increase in total expenditures, including a 360% increase in nonlocal angler expenditures (Anderson et al. 2000). Nonlocal anglers spent just over $1 million in the region in 1999, with a total economic impact of $1.5 million.

In Minnesota, a statewide mail survey of Minnesota trout stamp holders focused on the economic and social benefits of coldwater angling. It demonstrated that the southeastern portion of the state accounted for 33.1% of all coldwater angling trips and 75% of stream fishing trips (Gartner et al. 2012). Total direct sales due to stream anglers amounted to over $30 million for the year, with another $18 million in direct income, supporting more than 632 full- and part-time jobs (Gartner et al. 2012).

A comprehensive survey of Driftless Area-wide economic impacts of trout angling was conducted in 2016. Two-thousand surveys were mailed to a representative sample of Wisconsin, Minnesota, and Iowa trout stamp holders who resided in counties not fully within the Driftless Area (1.5% of the total population of estimated trout stamp holders in Wisconsin, Minnesota, and Iowa living outside the Driftless Area), as well as being made available online for mail survey recipients to encourage others to respond online. This yielded useable responses, with the Trout Unlimited Driftless Area Restoration Effort providing expenditure information on restoration projects to complete the analyses. The study estimated the total direct economic impact of restoration and improvement project spending in 2015 across the four-state region at just over $5 million (Table 2) and the total economic impact of fishing to the Driftless Area in 2015 at $703,676,674.50, supporting 6,597 jobs in the region (Anderson 2016). The total economic effect of fishing in the Driftless Area in 2015, including angler spending and restoration work, was $1,627,186,794.79 (Anderson 2016; Table 3).

What We Need to Know: Recommendations for Future Research

While the existing research detailed above focuses on angler perspectives and economic impacts, this work is not nearly as robust as it could be. Presumably, states have a plethora of long-term data from angler surveys that could be analyzed by researchers, and there are a variety of new questions that could be asked about angler perspectives and economic impacts across the region. Further, there are a variety of other questions to be asked of stream restoration projects and a variety of methodologies that could be adopted beyond angler surveys and valuation studies. This chapter closes with suggestions for future human dimensions research in the Driftless Area region that might support stream restoration and fisheries management in the region and offer some guiding principles for efforts beyond the region.

Table 2. Total direct economic impact of restoration and improvement project spending in Driftless Area states. Adapted from Anderson (2016). Data source: Trout Unlimited Driftless Area Restoration Effort.

	Total direct economic impact of restoration and improvement project spending
Iowa	$22,000.00
Minnesota	$1,500,000.00
Wisconsin	$3,500,000.00
Total	$5,022,000.00

Table 3. Total economic effect of angling in the Driftless Area region, including angler spending, secondary effects, and restoration spending. Adapted from Anderson (2016).

Total spending by anglers	$952,146,702.82
Total spending on restoration projects	$5,022,000.00
Total direct spending	$957,168,702.82
Total secondary effects (0.70*total direct spending)	$670,018,091.97
Total effect	$1,627,186,794.79

1. Keep doing what we are doing.
    - Continue research that explores angler perspectives on trout angling in the region.
    - Continue research into economic impacts of angling and stream restoration work.
2. And extend existing work
    - Consider the diverse uses of Driftless Area streams, with special focus on the intersecting needs and impacts of trout angling and livestock grazing.
    - Improve access to existing state data offering important new possibilities for analysis.
    - Existing state surveys and experiential knowledge offer great insights into useful and productive questions that deserve follow-up. Re-

search questions can and should flow from this existing pool of expertise, including state surveys and master plans, public comments, and grounded expertise. Dieterman and Merten (2016), for instance, catalogued historical southeastern Minnesota creel surveys that could be mined for information. A 2013 roving–roving creel survey—in which respondents are intercepted while angling (Pollock et al. 1997)—of 24 southeastern Minnesota trout stream areas identified an opportunity for additional human dimension surveys to identify factors contributing to retention and recruitment of new anglers, young anglers (<16 years old), female anglers, and bait anglers (Dieterman and Snook 2015) while a comprehensive comparison of pre- and posthabitat improvement project creel surveys concluded with a recommendation for future creel surveys focused on a smaller number of stream sites, suggesting that "the compilation of existing data in this report should thus serve to provide more robust data for evaluations of future habitat projects implemented at the other seven stream sites" (Dieter man and Snook 2015). Following Anderson and Marcouiller (1996), managing increasing fishing pressure might also be an important area for follow-up research.
- Increase the amount of research focused on the impacts of specific restoration projects, following the example of Dieterman and Snook (2015), which noted the funding and sample size challenges of assessing sociodemographic and fishery-related benefits on particular streams but offered a comprehensive study design for approaching quantitative research related to angler perceptions pre- and posthabitat project implementation.
- Consider using existing public comments from management plans and meetings as a source of data for management-related research, as well as to guide future research questions. A 1996 survey of southeastern Minnesota trout anglers related to a proposed change in fishing regulations, for instance, includes 15 pages of colorful narrative feedback about trout management that raises issues about access, philosophies of stocking, and elitism, among other issues (Anthony 1997).
- There is great potential for mixed methods and qualitative explorations into the human dimensions of stream restoration in the Driftless Area. Existing research remains in the realm of numbers, featuring basic survey data and economic calculations. Those quantitative data are important—and can be especially useful for supporting arguments (politically and fiscally) for stream restoration projects. But given the deep history and passion of anglers, land managers, and restoration practitioners—and the often-contentious nature of managing this singular and multifunctional landscape—those passions and controversies do not always translate well to quantitative data.
- Likewise, there is a need to build human dimensions explorations of stream restoration in the region in conversation with the vast font of science-based knowledge about Driftless Area hydrology, geomorphology, and biology. This integrative, social-ecological approach will be essential to managing these streams into the future.

3. Focus on adaptive management
   - Consider how human dimensions research and public engagement can support learning successes in adaptive management. There is a need for research that focuses on management expertise and practice, and that need is heightened in the face of the forecasted and yet dynamic and partially uncertain impacts of climate change.
4. Increase the amount of peer-reviewed literature
   - There is a need for an increase in peer-reviewed literature about all aspects of the human dimensions of stream restoration in the Driftless Area. Driftless Area managers have a huge amount of knowledge about the social and managerial aspects of restoration, in addition to their physical and biological knowledge. Extending the peer-review process outside of state agencies would add to the robustness and availability of those data.
5. Build collaborations for richer human dimensions research
   - While much is understood about Trout Unlimited member perspectives on trout angling and stream restoration, there is a need to engage with and study populations outside the Trout Unlimited umbrella, including Driftless Area residents (both anglers and nonanglers) and traveling anglers from outside the region.
   - Continued collaborations with staff from tribal nations, including the Ho-Chunk Nation in our research context, could contribute to a multifaceted understanding of the past, present, and future of stream ecosystems in the Driftless Area.
   - Engaged, participatory research methods can yield important data while also serving to engage broad communities in stream restoration and management. Statewide surveys offer useful insights, but understanding the human dimensions of stream restoration in the Driftless Area poses two paired challenges: (1) Driftless Area-specific data are not always available in state surveys, and (2) state surveys only provide insights in state-specific areas of the Driftless. Coordinating survey efforts across states to provide a multistate understanding of the region would be especially useful.

Moving beyond the Driftless Area

While we have been speaking here to human dimensions research in the context of the Driftless Area region, we think these lessons and recommendations apply much more broadly. Research into the human dimensions of stream restoration and fisheries management lags behind biophysical research and, where it does exist, tends to be similarly focused on angler surveys and estimates of economic impact. This body of work could become more robust by continuing angler surveys and studies of economic impacts; considering the management of trout and other recreational species in wider multispecies assemblages that extend beyond streambanks to include livestock, for instance, and the humans who manage stream systems—both agency resource managers and private landowners; taking mixed-methods approaches to understanding social-ecological dynamics of stream systems and their wider watersheds; and coordinating across state management agencies, tribal nations, and local stakeholders. In the broad context of climate change, urbanization, and nutrient enrichment, among other pressures, this is an urgent moment to learn more about the complex multispecies interactions that shape our stream systems and to gain the knowledge necessary to collaboratively plan for an increasingly dynamic future.

We hope to see more research that builds from existing state and regional archives of data, that focuses on building and studying public engagement with restoration and management, and following the lead of initial innovative research (e.g., Lave et al. 2010; Lave 2012; Druschke 2013; Linton and Budds 2014; Carpenter et al. 2015), that integrates insights from the biophysical and social sciences and the humanities in this context.

Acknowledgments

Thanks to Trout Unlimited's Jeff Hastings for suggesting this paper and Dan Dauwalter for encouraging over many months its development and revision. Bruce Vondracek, Sue Schroeder, and Matt Mitro offered keen insights into a broad range of existing research, but any exclusions or errors are our own. Thank you to the two reviewers who pushed us to refine and extend a number of important points. This work was supported in part by funding from Office of the Vice Chancellor for Research and Graduate Education and the Kickapoo Valley Reforestation Fund (College of Agricultural & Life Sciences), both at the University of Wisconsin–Madison, and an Andrew W. Mellon Foundation New Directions Fellowship.

References

Anderson, A., L. Hewitt, and D. Marcouiller. 2000. Outdoor recreation, community development, and change through time: a replicated study of canoeing and trout angling in southwestern Wisconsin. Center for Community Economic Development, University of Wisconsin–Madison Extension, Madison.

Anderson, A., and D. Marcouiller. 1996. Trout angling and regional development: a case study of southwestern Wisconsin. Center for Community Economic Development, University of Wisconsin–Madison Extension, Madison.

Anderson, D. 2016. Economic impact of recreational trout angling in the Driftless Area. Report to Driftless Area Restoration Effort. Available: www.darestoration.com/documents/Economic%20Impact%20Report%20of%20Trout%20Angling%20in%20the%20Driftless%20Area.pdf. (May 2019).

Anderson, R. 2002. Coon Valley days: a short history of the Coon Creek watershed. Wisconsin Academy Review 48 2:42–48.

Anthony, K. 1997. Survey about proposed changes in trout stream regulations: results and technical report. Minnesota Department of Natural Resources, St. Paul.

Bernhardt, E. S., E. B. Sudduth, M. A. Palmer, J. D. Allan, J. L. Meyer, G. Alexander, J. Follastad-Shah, B. Hassett, R. Jenkinson, R. Lave, J. Rumps, and L. Pagano. 2007. Restoring rivers one reach at a time: results from a survey of U.S. river restoration practitioners. Restoration Ecology 15:482–493.

Blann, K. L. 2004. Landscape-scale analysis of stream fish communities and habitats: lessons from southeastern Minnesota. Doctoral dissertation. University of Minnesota, St. Paul.

Carpenter, S. R., E. G. Booth, S. Gillon, C. J. Kucharik, S. Loheide, A. S. Mase, M. Motew, J. Qiu, A. R. Rissman, J. Seifert, and E. Soylu. 2015. Plausible futures of a social-ecological system: Yahara watershed, Wisconsin, USA. Ecology and Society [online serial] 20(2):10.

DARE (Driftless Area Restoration Effort). No date. Driftless Area data viewer. Available: http://arcg.is/1wsfEI6. (May 2019).

Dauwalter, D. C., editor. 2019. A look back at Driftless Area science to plan for resiliency in an uncertain future: special publication of the 11th annual Driftless Area symposium. Available: www.tu.org/driftless-science-review. (May 2019).

Dieterman, D., and E. Merten. 2016. History of trout management in southeast Minnesota: 1874–2003. Minnesota Department of Natural Resources, Division of Fisheries, St. Paul.

Dieterman, D. J., and V. A. Snook. 2015. The 2013 winter trout fishery on southeast Minnesota streams: historical changes and the influence of angling regulations and instream habitat enhancement projects. Minnesota Department of Natural Resources, Division of Fish and Wildlife, St. Paul.

Druschke, C. G. 2013. Watershed as commonplace: communicating for conservation at the watershed scale. Environmental Communication: A Journal of Nature and Culture 7:80–96.

Druschke, C. G., and K. C. Hychka. 2015. Manager perspectives on communication and public engagement in ecological restoration project success. Ecology and Society [online serial] 20(1):58.

Epton, J. A., and D. C. Fulton. 2002. Evaluating decision processes: summary of case studies of fisheries issues in Minnesota. Minnesota Department of Natural Resources, Division of Fish and Wildlife, St. Paul.

Folger, R. 1977. Distributive and procedural justice: combined impact of voice and improvement on experienced inequity. Journal of Personality and Social Psychology 35:108–119.

Gartner, W. C., L. L. Love, D. Erkkila, and D. C. Fulton. 2012. Economic impact and social benefits study of coldwater angling in Minnesota. Minnesota Department of Natural Resources, St. Paul.

Hamm, J. A., L. M. PytlikZillig, M. N. Herian, A. J. Tomkins, H. Dietrich, and S. Michaels. 2013. Trust and intention to comply with a water allocation decision: the moderating roles of knowledge and consistency. Ecology and Society [online serial] 18(4):49.

Happ, S. C. 1944. Effect of sedimentation on floods in the Kickapoo valley, Wisconsin. Journal of Geology 52:53–68.

Kelly, T., R. Sushak, and P. Jakes. 2001. Changing people's perceptions and behavior through partnerships and education: followup on a case study from Minnesota. U.S. Forest Service, North Central Research Station, St. Paul, Minnesota.

Knox, J. C. 1982. Quaternary history of the Kickapoo and lower Wisconsin River valleys, Wisconsin. Pages 1–65 *in* Quaternary history of the Driftless Area. Geological and Natural History Survey, University of Wisconsin–Madison Extension, Madison.

Knox, J. C. 2006. Floodplain sedimentation in the upper Mississippi valley: natural versus human accelerated. Geomorphology 79:286–310.

Krueger, C. C., and D. J. Decker. 1999. The process of fisheries management. Pages 31–59 *in* C. C. Kohler, and W. A. Hubert, editors. Inland fisheries management in North America, 2nd edition. American Fisheries Society, Bethesda, Maryland.

KVR (Kickapoo Valley Reserve). No date. Archaeology periods. Kickapoo Valley Reserve. Available: http://kvr.state.wi.us/About-Us/History/Archaeology/Archaeological-Periods/. (May 2019).

Lave, R. 2012. Fields and streams: stream restoration, neoliberalism, and the future of environmental science. University of Georgia Press, Athens.

Lave, R., M. Doyle, and M. Robertson. 2010. Privatizing stream restoration in the US. Social studies of science 40:677–703.

Lawrence, R. L., S. E. Daniels, and G. H. Stankey. 1997. Procedural justice and public involvement in natural resources decision making. Society and Natural Resources 10:577–89.

Leopold, A. 1935. Coon Valley: an adventure in cooperative conservation. American Forests 41:205–208.

Lind, E. A., and T. R. Tyler. 1988. The social psychology of procedural justice. Plenum, New York.

Linton, J., and Budds, J. 2014. The hydrosocial cycle: defining and mobilizing a relational-dialectical approach to water. Geoforum 57:170–180.

Loew. P. 2013. Indian nations of Wisconsin: histories of endurance and renewal. Wisconsin Historical Society, Madison.

Mauldin, L., and J. Hastings. 2014. Driftless Area Restoration Effort aquatic conservation plan. U.S. Fish and Wildlife Service, La Crosse, Wisconsin.

Miller, M. L. 2017. More trout, less algae: Wisconsin stream demonstrates benefits of targeted conservation. Cool Green Science, conservation science blog of The Nature Conservancy. Available: https://blog.nature.org/science/2017/09/13/more-trout-less-algae-wisconsin-stream-demonstrates-benefits-targeted-conservation/?utm_source=cgs&utm_medium=archive&utm_campaign=Wildlife. (May 2019).

Palmer, M. A., E. S. Bernhardt, J. D. Allan, P. S. Lake, G. Alexander, S. Brooks, J. Carr, S. Clayton, C. N. Dahm, J. Follstad Shah, D. L. Galat, S. G. Loss, P. Goodwin, D. D. Hart, B. Hassett, R. Jenkinson, G. M. Kondolf, R. Lave, J. L. Meyer, T. K. O'Donnell, L. Pagano, and E. Sudduth. 2005. Standards for ecologically successful river restoration. Journal of Applied Ecology 42:208–217.

Petchenik, J. 2012. Results of the 2011 survey of lapsed Wisconsin inland trout anglers. Wisconsin Department of Natural Resources, Bureau of Fisheries Management, Madison.

Petchenik, J. 2014. Trout fishing in Wisconsin: angler behavior, program assessment and regulation and season preferences. Wisconsin Department of Natural Resources, Bureau of Fisheries Management, Madison. Available: http://dnr.wi.gov/topic/fishing/documents/trout/TroutReportJan2014.pdf. (May 2019).

Pollock, K. H., Hoenig, J. M., Jones, C. M., Robson, D. S., and Greene, C. J. 1997. Catch rate estimation for roving and access point surveys. North American Journal of Fisheries Management 17(1):11–19.

Potter, K. W. 1991. Hydrological impacts of changing land management-practices in a moderate-sized agricultural catchment. Water Resources Research 27:845–855.

Schroeder, S. 2014. A study of trout angler participation and activities in southeastern Minnesota. Minnesota Cooperative Fish and Wildlife Research Unit, Department of Fisheries, Wildlife, and Conservation Biology, University of Minnesota, St. Paul.

Schroeder, S. A. 2015. Fishing and fish habitat in Minnesota. Minnesota Cooperative Fish and Wildlife Research Unit, Department of Fisheries, Wildlife, and Conservation Biology, University of Minnesota, St. Paul.

Schroeder, S. A., and D. C. Fulton. 2005. Fishing in Minnesota: a study of angler participation and activities. Minnesota Cooperative Fish and Wildlife Research Unit, Department of Fisheries, Wildlife, and Conservation Biology, University of Minnesota, St. Paul.

Schroeder, S. A., and D. C. Fulton. 2017. Voice, perceived fairness, agency trust, and acceptance of management decisions among Minnesota anglers. Society and Natural Resources 30:569–584.

Snook, V. A., and D. J. Dieterman. 2015. The 2013 winter trout fishery on southeast Minnesota streams. Minnesota Department of Natural Resources, St. Paul.

Trimble, S.W., and S.W. Lund. 1982. Soil conservation and the reduction of erosion and sedimentation in the Coon Creek basin, Wisconsin. U.S. Geological Survey Professional Paper 1234.

Vetrano, D. M. 2017. Driftless waters: a tale of destruction, renewal and hope for the future, Pages 13–14 *in* B. Carline, editor. Wild Trout Symposium XII: science, politics, and wild trout management. Wild Trout Symposium, Bozeman, Montana.

Vlaming, J., and D. C. Fulton. 2003. Trout angling in southeastern Minnesota: a study of trout anglers. Minnesota Cooperative Fish and Wildlife Research Unit, St. Paul.

The Ways. 2019. Treaty lands map. Wisconsin Public Television Education, Madison. Available: https://theways.org/map. (May 2019).

WDNR (Wisconsin Department of Natural Resources). 2013a. Driftless Area Stream Master Plan Project regional and property analysis, executive summary. WDNR, Publication 071-2013, Madison. Available: https://dnr.wi.gov/files/pdf/pubs/lf/lf0071_ExecSum.pdf. (May 2019).

WDNR (Wisconsin Department of Natural Resources). 2013b. Regional and property analysis for the development of a master plan for Department of Natural Resources' properties along trout and Smallmouth

Bass streams in the Driftless Area. Available: https://dnr.wi.gov/files/pdf/pubs/lf/lf0071_TOC.pdf

White, R. J., and O. M. Brynildson. 1967. Guidelines for management of trout stream habitat in Wisconsin. Wisconsin Department of Natural Resources, Technical Bulletin No. 39, Madison. Available: https://dnr.wi.gov/files/PDF/pubs/ss/SS0039.pdf. (May 2019).

WICCI (Wisconsin Initiative on Climate Change Impacts). 2011. Wisconsin's changing climate: impacts and adaptation. Nelson Institute for Environmental Studies, University of Wisconsin and the Wisconsin Department of Natural Resources, Madison.

Wisconsin First Nations. No date. Ho-Chunk Nation: Hochungra—the people with the big voice. Available: https://wisconsinfirstnations.org/ho-chunk-nation/. (May 2019).

American Fisheries Society Symposium 91:389–399, 2019

Incorporating Predictive Models into a Resilient Fisheries Management Framework: A Brook Trout Perspective

Eric R. Merriam*, J. Todd Petty, and Cory T. Trego
School of Natural Resources, West Virginia University
Morgantown, West Virginia 26506, USA

Abstract.—Increasingly, fisheries managers must make important decisions in complex environments where rapidly changing landscape and climate conditions interact with historical impacts to influence resource sustainability. Successful fisheries management in this setting will require that we adapt traditional management approaches to incorporate information on these complex interacting factors—a process referred to as resilient fisheries management. Large-scale species distribution data and predictive models have the potential to enhance the management of freshwater fishes through improved understanding of how past, present, and future natural and anthropogenic factors combine to determine species vulnerability and resiliency. Here we describe a resilient fisheries management framework that provides guidance on how and when these models can be incorporated into traditional approaches to meet specific goals and objectives for resource sustainability. In addition to elucidating complex drivers of distributional patterns and change, species distribution models can inform the prioritization, application, and implementation of management activities such as restoration (e.g., instream habitat and riparian), protection (e.g., areas where additional land use would result in a change in species distribution), and regulations (e.g., harvest restriction) in a way that informs resiliency to land use and climate change. Although considerable progress has been made with respect to applying species distribution models to the management of Brook Trout *Salvelinus fontinalis* and other aquatic species, there are several areas where a more unified research and management effort could increase the ability of distribution models to inform resilient management. Future efforts should aim to improve (1) data availability, consistency (sampling methodology), and quality (accounting for detection); (2) partnerships among researchers, agencies, and managers; and (3) model accessibility and understanding of limitations and potential benefits to managers (e.g., incorporation into publicly available decision support systems). The information and recommendations provided herein can be used to promote and advance the use of models in resilient fisheries management in the face of continued large-scale land use and climate change.

Introduction

Freshwater fishes have experienced global declines as a result of historical and contemporary anthropogenic activity that far exceed those of other vertebrate groups (Dudgeon et al. 2006; Burkhead 2012). Freshwater systems are highly vulnerable to climate change and associated interactions with

* Corresponding author: emerriam@mix.wvu.edu

other anthropogenic stressors (Heino et al. 2009; Comte et al. 2013). There is a critical need for fisheries managers to adopt strategies that maximize resiliency (i.e., ability to sustain or recover from disturbance; Holling 1973) of at-risk freshwater populations and communities in the face of uncertain future ecological conditions (Biggs et al. 2012).

Several recent resilient fisheries management frameworks (i.e., facilitating ecological resiliency by rapidly adapting management approaches to address changing ecological and social conditions) have been developed that aim to sustain critical ecosystem functions and services (e.g., recreational fisheries) under uncertain future conditions (Pope et al. 2014; Paukert et al. 2016). Resilient management requires utilizing available data to proactively identify and enact management actions with the greatest likelihood of success in the face of uncertainty (e.g., climate change) and then preparing for and adaptively managing through future changes (Wilby et al. 2010; Paukert et al. 2016). Previous frameworks highlight the important role forecasting and prediction play in facilitating resilient fisheries management programs (e.g., aiding in prioritization and selection of protection and/or management priorities; Paukert et al. 2016); however, they tend to focus their discussion on adapting subsequent management activities to maximize resiliency under uncertainty. As a result, little explicit guidance has been provided to fisheries managers regarding how to maximize the utility of models and associated predictions in resilient fisheries management. This represents an important gap in the fisheries literature because models can be a valuable tool for optimizing conservation of many at-risk species, particularly in the face of uncertain future land use and climate change.

Species distribution models have received considerable attention in the ecological and conservation planning literature because of their ability to relate increasingly complex environmental conditions to observed patterns and changes in distributions (Elith and Leathwick 2009) and their capacity to inform fisheries management and conservation (e.g., restoration, protection, prioritization, and regulations). This has been particularly true for the Brook Trout *Salvelinus fontinalis*, which is an important sport fish throughout its native range and an umbrella species for the conservation of coldwater habitats (Kanno and Beazley 2004). Brook Trout populations have undergone significant reductions (reduced or extirpated populations in nearly 70% of historically occupied subwatersheds; Hudy et al. 2008) as a result of large-scale anthropogenic stressors such as land use (Merriam et al. 2018), acid deposition (Hudy et al. 2008), and habitat fragmentation (Letcher et al. 2007), and losses are expected to continue in the face of continued land use (Merriam et al. 2018) and climate change (Flebbe et al. 2006; Snyder et al. 2015; Merriam et al. 2019). Recent efforts to inform the management and conservation of Brook Trout through the use of models highlight the need to better define the role of distribution models in and demonstrate their utility to resilient fisheries management.

Remodeling the Resiliency Framework

Land use and climate change exert considerable influence over all fishery components (i.e., social and ecosystem [habitat and biota]; Figure 1). Social influences (e.g., social pressures that shape policy and management directives) and ecological systems (e.g., positive feedback between warming and release of stored greenhouse gases: Walter et al. 2006) also feedback to drive or alter rates of land use and climate change. Although individual actions of fisheries managers may not be able to directly affect or alter large-scale drivers of land use and climate change from the top down, it is becoming increasingly clear that a watershed-

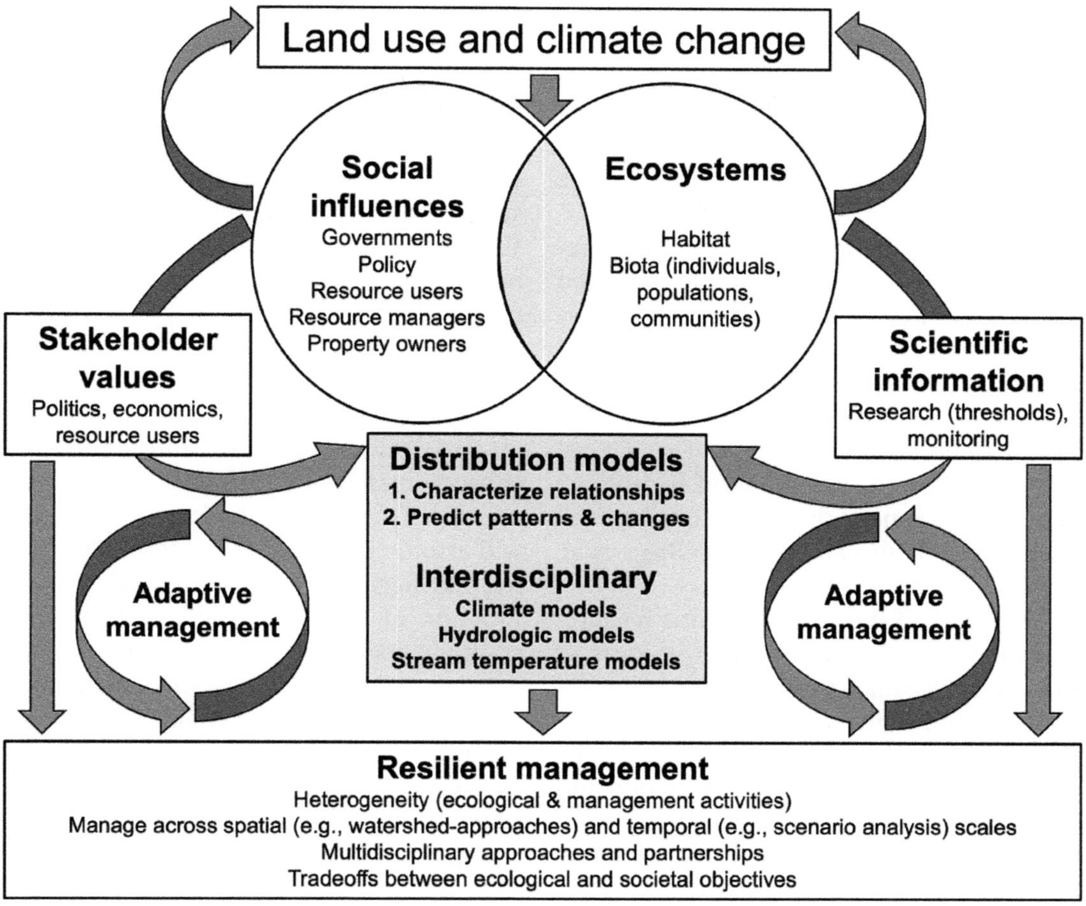

Figure 1. Framework highlighting the critical role models play in informing resilient fisheries management in the face of continued large-scale land use and climate change. This framework is adapted from and builds upon previous efforts by Hansen et al. (2015) and Paukert et al. (2016).

scale approach to fisheries management is needed to increase ecological resilience (Petty and Merriam 2012) and change social perspectives that shape policy decisions and management directives (DeWeber and Wagner 2015b) from the bottom up. As such, resilient fisheries management must consider both social and ecological systems by incorporating stakeholder values and scientific information into a single framework (Hansen et al. 2015; Paukert et al. 2016; Figure 1).

Distribution models can inform the application and implementation of management activities (e.g., restoration, protection, prioritization, and harvest regulations) in a way that informs resiliency to land use and climate change. For example, advances in computational technologies and statistical techniques allow managers to model ecological response to multiple interacting stressors and develop targeted, context-specific management actions that maximize ecological benefit (Petty et al. 2013; Mantyka-Pringle et al. 2016). Large-scale land use and climate change also affect resiliency by altering critical processes occurring across multiple spatial and temporal scales (Pope et al. 2014). Distribution models can be placed within a hierarchical framework (sensu Merovich et al. 2013), enabling managers to incorporate ecological processes occurring across relevant spatial and temporal scales (e.g., dispersal and metapopulation dynamics) into management decisions (Merriam et al. 2019). Distribution models are readily incorporated into scenario analyses, wherein likely future ecological conditions are predicted under various environmental (e.g., land use and climate change; Lynch et al. 2016) and social (e.g., policy and social change; Osterblom et al. 2013) scenarios. Integrating empirical understanding of both the natural (e.g., distribution, climate, and hydrology) and social sciences through modeling will facilitate resilient management (Biggs et al. 2012).

Managing fisheries and aquatic resources will require navigating a great deal of uncertainty (Paukert et al. 2016). Effects of land use and climate change are extremely complex (e.g., interactions between climate change and existing land-use stressors) and constantly changing in response to novel stressors and associated feedback (Heino et al. 2009; Comte et al. 2013). There is also a great deal of uncertainty in model predictions (Wenger et al. 2013; Lourenco et al. 2015). For example, distribution models and predictions often utilize predictions of future climate and ecological condition (e.g., stream temperature) that also have inherent uncertainties (Kwakkel et al. 2015). Thus, it is important to identify and minimize sources of uncertainty to increase the likelihood of successful management (Wenger et al. 2013). Previous resilient management frameworks highlight the need to monitor for and elucidate complex environmental relationships between fisheries and anthropogenic stressors (e.g., thresholds and steady state conditions; Pope et al. 2014) in order to minimize these uncertainties (Paukert et al. 2016). Although such efforts are undoubtedly necessary, managers often cannot afford to wait for results of long-term ecological studies to guide conservation activities that, in many systems, are needed immediately.

Models provide an objective way to identify and initiate management actions that inform resiliency under uncertainty (Wenger et al. 2013). It is important that model results and associated management decisions be placed within a dynamic adaptive management and monitoring framework to minimize potential effects of uncertainty on success of management actions (Figure 1). Managers should further consider numerous candidate strategies (sensu Kwakkel et al. 2015) that incorporate multiple, heterogeneous management activities (Hansen et al. 2015). The potential for long-term success will also be increased if models and resulting management actions are updated as a function of new scientific information on complex response to land use and climate stressors, and evaluations of past management outcomes. Models should not take the place of studies or monitoring programs designed to detect or test hypotheses related to ecological response to land use and climate change. Rather, models can augment and be informed by these efforts to improve management outcomes.

Case study: managing Appalachian Brook Trout

Considerable effort has been put toward modeling current and future patterns and changes in Brook Trout occupancy (Wagner et al. 2013a; DeWeber and Wagner 2015a; Merriam et al. 2018, 2019) and abundance (Kanno et al. 2014, 2015a, 2016) in an effort to facilitate management of this ecologically and socially relevant species (DeWeber and Wagner 2015b). Distribution models have been extremely valuable in characterizing and quantifying environmental drivers of Brook Trout occupancy (e.g., Brown Trout *Salmo trutta* occurrence [Wagner et al. 2013a], land use [Merriam et al. 2018], and climate change [Kanno et al. 2015b]) and abundance (e.g., seasonal weather patterns on adult and juvenile abundance [Kanno et al. 2014, 2015a, 2016; Blum et al. 2018]) across multiple spatial and temporal scales. Abundance data and models have been extended to predict juvenile and adult survival and per-capita recruitment, highlighting the utility of abundance data in identifying demographic drivers of population response to regional environmental factors (Kanno et al. 2015a).

Brook Trout distribution models have been extended to help inform and predict likely outcomes of fisheries management activities. Occupancy models have been used to predict likely distributional response to future land-use development scenarios and identify stream segments and watersheds as protection (e.g., where additional land use results in a predicted loss of Brook Trout occurrence: Merriam et al. 2018) and restoration (e.g., removal of dispersal barriers: Wood et al. 2018, stock exclusion within agricultural watersheds: Sievers et al. 2017, and capturing storm water and sediment runoff and rehabilitating degraded habitats within urbanized watersheds: Yeakley et al. 2014) priorities. Direct incorporation of stream temperature models and future predictions into Brook Trout occupancy models has also enabled researchers and managers to characterize likely distributional change under various climate-change scenarios (DeWeber and Wagner 2015b; Snyder et al. 2015; Merriam et al. 2019). This information can then be used to direct restoration actions designed to mitigate warming temperatures (e.g., riparian planting; Cross et al. 2013) or identify streams as protection priorities (e.g., where groundwater inflows result in stable optimal temperatures: Snyder et al. 2015; Briggs et al. 2018). Abundance models have also been used to identify and direct management of Brook Trout in the face of land use and climate change—albeit to a lesser extent. Blum et al. (2018) modeled and predicted distributional changes in abundance as a function of current and future changes in streamflow and used their model to characterize general management priorities (e.g., conservation of riparian habitat, management of nonnative species, and restoration of habitat connectivity) for stream reaches within the Shenandoah National Park.

DeWeber and Wagner (2015b) demonstrate the utility of linking ecologically based distribution models with social dimensions to improve understanding of climate change impacts by translating predicted losses in Brook Trout occupancy into increases in driving distance to streams with Brook Trout angling opportunities. Their results bolster previous work quantifying and predicting impacts to outdoor recreation (Mendelsohn and Markowski 1999) and associated economic losses as a result of lost Brook Trout habitat under climate change (Ahn et al. 2000).

There has been a recent push to incorporate statistical models within decision support systems to facilitate Brook Trout management (Lynch et al. 2015). Several decision support systems have been developed and made available to aid in the management of Brook Trout habitats and populations. The Spatial Hydro-Ecological Decision System (SHEDS; www.ecosheds.org) is a Web-based application that links data and models, including Brook Trout occupancy, for the northeastern United States (Benjamin Letcher, U.S. Geological Survey, personal communication). A decision support system is available for managing Brook Trout within the Chesapeake Bay watershed. This support system utilizes models published by Merriam et al. (2019) and enables resource managers the ability to visualize predictions and rank streams for restoration or protection priorities under current and future climate scenarios and land-use scenarios (www.fishhabitattool.org). To date, all Brook Trout decision support systems are built around occupancy data and models.

Moving Forward with Models and Brook Trout Management

Recent efforts to model Brook Trout distributions have paved the way for model-informed resilient Brook Trout management in the face of historical, contemporary, and potential future land use and climate change. However, several constraints remain that limit the utility of existing models. There are large portions of the Brook Trout's native range with sparse or absent distribution data (Hudy et al. 2008; DeWeber and Wagner 2015a). This lack of coverage increases model uncertainty and limits our ability to make meaningful inferences and management decisions. Where data do exist, they are often collected by different agencies utilizing different methodologies.

The utility of Brook Trout distribution model outputs has been limited by data quality and structure. Smaller-scale occupancy and abundance modeling efforts have accounted for imperfect detection (e.g., Bayesian hierarchical models; Kanno et al. 2015a, 2016), providing unbiased estimates of occupancy and abundance and enabling inferences regarding changes in distribution over time and in response to land use and climate-change scenarios. Large-scale modeling efforts utilize data from multiple sampling agencies that often use different sampling methodologies with unknown efficiencies. These large-scale models often assume that detection is constant among samples and not confounded with environmental characteristics (DeWeber and Wagner 2015a; Merriam et al. 2018, 2019). Imperfect but constant detection efficiency enables managers to make inferences regarding relative changes over time and accurately characterize patterns of response to environmental characteristics; however, inferences regarding changes in distribution are not appropriate (Guillera-Arroita et al. 2015).

Brook Trout detection varies with sampling gear and environmental characteristics such as stream size, water chemistry, and the presence of nonnative species (Hense et al. 2010; Wagner et al. 2013a). Therefore, it is possible that the assumption of constant detection is not valid or that detection is confounded with environmental variables. Although previous efforts have observed high detection probabilities for Brook Trout (99%, Wagner et al. 2013a; 85%, Kanno et al. 2015b), these estimates reflect data collected by single (or few) agencies across relatively small spatial scales. If detection and occupancy are positively correlated across all sites, models provide accurate ranking of suitability and can be used to prioritize sites for management activities. However, when detection is confounded with environmental conditions, models may not accurately capture the shape of environmental relationships and may provide biased estimates of occurrence and abundance (Denes et al. 2015; Guillera-Arroita et al. 2015). It is vital that managers are aware of the limitations of their data and the associated models. Inappropriate use and/or interpretation can lead to ineffective management decisions and undesired outcomes.

Existing large-scale Brook Trout data sets also lack consistent information on abundance. Abundance data have the potential to provide critical information on population response (e.g., population trends and demographic rates; Wagner et al. 2013b; Kanno et al. 2015a) and likely resilience to land use and climate change (Blum et al. 2018). Improving the availability and consistency of abundance data will enable calculation of population trends over time, allowing managers to better identify at-risk populations and develop and adjust appropriate management activities (e.g., harvest limits; Wagner et al. 2013b). Recent advances in abundance modeling have also enabled simultaneous estimation of Brook Trout population demographic rates (e.g., apparent survival and recruitment; Dail and Madsen 2011; Zipkin et al. 2014; Kanno et al.

2015a); however, these efforts have historically been limited to small-scale mark–recapture studies.

Improving the availability of abundance data will also provide an opportunity to explore the utility of integrated population models to the management of Brook Trout and other fisheries. Integrated population models jointly analyze data on population size (e.g., abundance data) and demographic parameters (e.g., mark–recapture data) (Dail and Madsen 2011; Schaub and Abadi 2011; Zipkin et al. 2014). The application of integrated population models to freshwater systems has been limited. Increasing availability of large-scale abundance data and a large number of mark–recapture studies make Brook Trout a prime candidate to expand the use of integrated population models within freshwater systems.

The need for improved and targeted monitoring has been discussed previously with respect to improving management of fisheries in the face of continued land use and climate change (Lynch et al. 2016; Paukert et al. 2016). However, previous recommendations generally focus on improving our ability to detect and track impacts of large-scale land use and climate change as they occur. We argue that in addition to completing study-specific monitoring and management objectives, sampling could be conducted to increase the potential of the resulting data to inform large-scale models and resilient management. Efforts to obtain estimates of both occurrence and abundance would have the greatest potential to help characterize and manage broad impacts of current and future land use and climate change (e.g., large-scale and long-term trends in populations; Wagner et al. 2013b). When time and resources allow, temporal replication (i.e., seasonally or yearly) can be used to account for imperfect detection, reducing uncertainty in distribution estimates (Zipkin et al. 2014). Finally, sampling through repeated measures (e.g., triple-pass) has additional benefits of enabling managers to estimate demographic rates (e.g., apparent survival and recruitment) (Dail and Madsen 2011; Zipkin et al. 2014; Kanno et al. 2015a, 2016).

Obtaining all aforementioned data will undoubtedly be outside the scope of many individual monitoring and management efforts. However, their potential utility with regard to future management needs is high—particularly as ecosystems continue to be degraded. One way to increase the availability and utility of monitoring data will be through the creation of centralized, open-access databases (Midway et al. 2016). Centralized databases for key at-risk species such as Brook Trout will likely serve several key functions. First, they would enable researchers and managers to identify data gaps, where further monitoring would serve a number of research and management objectives (e.g., better understanding of distribution, decrease uncertainty in distribution estimates). A centralized database would also facilitate collaboration among all participating entities and improve their ability to fulfill specific research and management objectives through interjurisdictional collaboration (Midway et al. 2016).

Strong partnerships and collaborations will be at the center of successful resilient fisheries management efforts (Midway et al. 2016). Improved partnerships and greater communication among researchers and managers will ensure that local knowledge and understanding regarding ecological drivers of Brook Trout distribution are incorporated into models, improving their accuracy, relevance, and likelihood of incorporation into management programs. Improved partnerships among researchers and management agencies will facilitate discussions of how to design sampling efforts to increase benefit to collective (e.g., transboundary management needs; Midway et al. 2016) and shared research and management interests (e.g., quantifying effects of rapidly expanding land use on fishes; Merriam et al. 2018).

Interdisciplinary partnerships will also be critical as fisheries researchers and managers attempt to forecast and manage impacts of climate change on distribution and dynamics of vulnerable fish populations. This could involve collaboration among fisheries managers, climate scientists, and hydrologists. For example, climate change simulations for Brook Trout have been based on mean annual stream temperatures and precipitation rates (Merriam et al. 2019) and not considered changes in frequency and severity of extreme climatic events or predictions of stream flow—both of which likely play an important role in Brook Trout population and habitat response to climate change (Kanno et al. 2015a, 2016; Merriam et al. 2017).

Perhaps most importantly, researchers need to improve partnerships with resource managers in order to make distribution models accessible to those individuals designing and implementing fisheries management programs. Incorporating distribution models into publicly available decision support systems, such as those previously described for Brook Trout (www.fishhabitattool.org; www.ecosheds.org), is one way to facilitate their use by resource managers. It is important to recognize that no one model will be ideal for all management needs. Therefore, decision support systems will be most beneficial if they provide documentation on model strengths and limitations, as well as appropriate management inferences and uses. When developed and used effectively, decision support systems enable managers to predict ecological response to multiple environmental and management scenarios to inform resiliency under uncertainty (Kwakkel et al. 2015; Merriam et al. 2019).

Large-scale species distribution data and models have the potential to enhance the management of freshwater fishes through improved understanding of how past, present, and future natural and anthropogenic factors combine to determine vulnerability and resiliency. Although considerable progress has been made with respect to applying species distribution models to the management of Brook Trout and other aquatic species, there are several areas where a more unified research and management effort could increase the ability of distribution models to inform resilient management. When appropriate, we hope the information and recommendations provided herein will facilitate successful use and incorporation of models into resilient fisheries management, particularly in the face of continued large-scale land use and climate change.

Acknowledgments

We thank Tyler Wagner for providing valuable comments to earlier drafts of this chapter. Support for this work was provided by the Fisheries Program of the U.S. Geological Survey's Ecosystems Mission Area. This work was also partially supported by the National Science Foundation under Award Number OIA-1458952. Use of trade, product, or firm names does not imply endorsement by the U.S. Government.

References

Ahn, S., J. E. de Steiguer, R. B. Palmquist, and T. P. Holmes. 2000. Economic analysis of the potential impact of climate change on recreational trout fishing in the southern Appalachian Mountains: an application of a nested multinomial logit model. Climate Change 45:493–509.

Biggs, R., J. Schluter, D. Biggs, E. L. Bohensky, S. BurnSilver, G. Cundill, V. Dakos, T. M. Daw, L. S. Evans, K. Kotschy, A. M. Leitch, C. Meek, A. Quinlan, C. Raudsepp-Hearne, M. D. Robards, M. L. Schoon, L. Schultz, and P. C. West. 2012. Toward principals for enhancing the resilience of ecosystem services. Annual Review of Environment and Resources 37:421–448.

Blum, A. G., Y. Kanno, and B. H. Letcher. 2018. Seasonal streamflow extremes are key drivers of Brook Trout young-of-the-year abundance. Ecosphere 9:e02356.

Briggs, M. A., Z. C. Johnson, C. D. Snyder, N. P. Hitt, B. L. Kurylyk, L. Lautz, D. J. Irvine, S. T. Hurley, and J. W. Lane. 2018. Inferring watershed hydraulics and cold-water habitat persistence using multi-year air and stream temperature signals. Science of the Total Environment 636:1117–1127.

Burkhead, N. M. 2012. Extinction rates in North American freshwater fishes, 1900–2010. BioScience 62:798–808.

Comte, L., L Buisson, M. Daufresne, and G. Grenouillet. 2013. Climate-induced changes in the distribution of freshwater fish: observed and predicted trends. Freshwater Biology 58:625–639.

Cross, B. K., M. A. Bozek, and M. G. Mitro. 2013. Influences of riparian vegetation on trout stream temperatures in central Wisconsin. North American Journal of Fisheries Management 33:682–692.

Dail, D., and L. Madsen. 2011. Models for estimating abundance from repeated counts of an open population. Biometrics 67:577–587.

Denes, F. V., L. F. Silveira, and S. R. Beissinger. 2015. Estimating abundance of unmarked animal populations: accounting for imperfect detection and other sources of zero inflation. Methods in Ecology and Evolution 6:543–556.

DeWeber, J. T., and T. Wagner. 2015a. Predicting Brook Trout occurrence in stream reaches throughout their native range in the eastern United States. Transactions of the American Fisheries Society 144:11–24.

DeWeber, J. T., and T. Wagner. 2015b. Translating climate change effects into everyday language: an example of more driving and less angling. Fisheries 40:395–398.

Dudgeon, D., A. H. Arthington, M. O. Gessner, Z.-I. Kawabata, D. J. Knowler, C. Leveque, R. J. Naiman, A.-H. Prieur-Richard, D. Soto, M. L. J. Stiassny, and C. A. Sullivan. 2006. Freshwater biodiversity: importance, threats, status, and conservation challenges. Biological Reviews 81:163–182.

Elith, J., and J. Leathwick. 2009. Species distribution models: ecological explanation and prediction across space and time. Annual Review of Ecology, Evolution, and Systematics 40:677–697.

Flebbe, P. A., L. D. Roghair, and J. L. Bruggink. 2006. Spatial modeling to project southern Appalachian trout distribution in a warmer climate. Transactions of the American Fisheries Society 135:1371–1382.

Guillera-Arroita, G., J. J. Lohoz-Monfort, J. Elith, A. Gordon, H. Kujala, P. E. Lentini, M. A. McCarthy, R. Tingley, and B. A. Wintle. 2015. Is my species distribution model fit for purpose? Matching data and models to applications. Global Ecology and Biogeography 24:276–292.

Hansen, G. J. A., J. W. Gaeta, J. F. Hansen, and S. R. Carpenter. 2015. Learning to manage and managing to learn: sustaining freshwater recreational fisheries in a changing environment. Fisheries 40:56–64.

Heino, J., R. Vikkala, and H. Toivonen. 2009. Climate change and freshwater biodiversity: detected patterns, future trends and adaptations in northern regions. Biological Reviews 84:39–54.

Hense, Z., R. W. Martin, and J. T. Petty. 2010. Electrofishing capture efficiencies for common stream fish species to support watershed-scale studies in the central Appalachians. North American Journal of Fisheries Management 4:1041–1050.

Holling, C. S. 1973. Resilience and stability of ecological systems. Annual Review of Ecology and Systematics 4:1–23.

Hudy, M., T. M. Thieling, N. Gillespie, and E. P. Smith. 2008. Distribution, status, and land use characteristics of subwatersheds within the native range of Brook Trout in the eastern United States. North America Journal of Fisheries Management 28:1069–1085.

Kanno, Y., and K. Beazley. 2004. Freshwater fish considerations for aquatic conservation systems planning in Nova Scotia. Proceedings of the Nova Scotian Institute of Science 42:375–391.

Kanno, Y., B. H. Letcher, N. P. Hitt, D. A. Boughton, J. E. B. Wofford, and E. F. Zipkin. 2015a. Seasonal weather patterns drive population vital rates and persistence in a stream fish. Global Change Biology 21:1856–1870.

Kanno, Y., B. H. Letcher, A. L. Rosner, K. P. O'Neil, and K. H. Nislow. 2015b. Environmental factors affecting Brook Trout occurrence in headwater stream segments. Transactions of the American Fisheries Society 144:373–382.

Kanno, Y., B. H. Letcher, J. C. Vokoun, and E. F. Zipkin. 2014. Spatial variability in adult Brook Trout (*Salvelinus fontinalis*) survival within two intensively surveyed headwater stream networks. Canadian Journal of Fisheries and Aquatic Sciences 71:1010–1019.

Kanno, Y., K. C. Pregler, N. P. Hitt, B. H. Letcher, D. J. Hocking, and J. E. B. Wofford. 2016. Seasonal temperature and precipitation regulate Brook Trout young-of-the-year abundance and population dynamics. Freshwater Biology 61:88–99.

Kwakkel, J. H., M. Haasnoot, and W. E. Walker. 2015. Developing dynamic adaptive policy pathways: a computer-assisted approach for developing adaptive strategies for a deeply uncertain world. Climate Change 132:373–386.

Letcher, B. H., K. H. Nislow, J. A. Combs, M. J. O'Donnell, and T. L. Dubreuil. 2007. Population response to habitat fragmentation in a stream-dwelling Brook Trout population. PLOS (Public Library of Science) ONE [online serial] 2(11):e1139.

Lourenco, T. C., A. Rovisco, S. Dessai, R. Moss, and A. Petersen. 2015. Editorial introduction to the special issue on uncertainty and climate change adaptation. Climate Change 132:369–372.

Lynch, A. J., B. J. E. Myers, C. Chu, L. A. Eby, J. A. Falke, R. P. Kobach, T. J. Krabbenhoft, T. J. Kwak, J. Lyons, C. P. Paukert, and J. E. Whitney. 2016. Climate change effects on North American inland fish populations and assemblages. Fisheries 41:346–361.

Lynch, A. J., E. Varela-Acevedo, and W. W. Taylor. 2015. The need for decision-support tools for a changing climate: application to inland fisheries management. Fisheries Management and Ecology 22:14–24.

Mantyka-Pringle, C. S., T. G. Martin, D. B. Moffatt, J. Udy, J. Olley, N. Saxton, F. Sheldon, S. E. Bunn, and J. R. Rhodes. 2016. Prioritizing management actions for the conservation of freshwater biodiversity under changing climate and land-cover. Biological Conservation 197:80–89.

Mendelsohn, R., and M. Markowski. 1999. The impact of climate change on outdoor recreation. Pages 267–288 *in* R. Mendelsohn and J. E. Neumann, editors. The impact of climate change on the United States economy. Cambridge University Press, Cambridge, UK.

Merovich, G. T., J. T. Petty, M. P. Strager, and J. B. Fulton. 2013. Hierarchical classification of stream condition: a house-neighborhood framework for establishing conservation priorities in complex riverscapes. Freshwater Science 32:874–891.

Merriam, E. R., R. Fernandez, J. T. Petty, and N. Zegre. 2017. Can Brook Trout survive climate change in large rivers? If it rains. Science of the Total Environment 607–608:1225–1236.

Merriam, E. R., J. T. Petty, J. Clingerman, and T. Wagner. 2019. Conservation planning at the intersection of landscape and climate change: Brook Trout in the Chesapeake Bay watershed. Ecosphere [online serial] 10:e02585.

Merriam, E. R., J. T. Petty, K. Maloney, J. Young, S. Faulkner, E. Terrence Slonecker, L. E. Milheim, A. Hailegiorgis, and J. Niles. 2018. Brook Trout distributional response to unconventional oil and gas development: landscape context matters. Science of the Total Environment 628–629:338–348.

Midway, S. R., T. Wagner, J. D. Zydlewski, B. J. Irwin, and C. P. Paukert. 2016. Transboundary fisheries science: meeting the challenges of inland fisheries management in the 21st century. Fisheries 41:536–546.

Osterblom, H., A. Merrie, M. Metian, W. J. Boonstra, T. Blenckner, J. R. Watson, R. R. Rykaczewski, Y. Ota, J. L. Sarmiento, V. Christensen, M. Schluter, S. Birnbaum, B. G. Gustafsson, C. Humborg, C.-M. Morth,

B. Muller-Karulis, M. T. Tomczak, and C. Folke. 2013. Modeling social-ecological scenarios in marine systems. BioScience 63:735–744.

Paukert, C. P., B. A. Glazer, G. J. A. Hansen, B. J. Irwin, P. C. Jacobson, J. L. Kershner, B. J. Shuter, J. E. Whitney, and A. J. Lynch. 2016. Adapting inland fisheries management to a changing climate. Fisheries 41:374–384.

Petty, J. T., and E. R. Merriam. 2012. Brook Trout restoration. Nature Education Knowledge 3:17.

Petty, J. T., M. P. Strager, E. R. Merriam, A. E. Maxwell, and P. Ziemkiewicz. 2013. Scenario analysis and the watershed futures planner: predicting future aquatic conditions in an intensively mined Appalachian watershed. Pages 5–19 *in* J. R. Craynon, editor. Environmental considerations in energy production. Society for Mining, Metallurgy, and Exploration 5:5–19.

Pope, K. L., C. R. Allen, and D. G. Angeler. 2014. Fishing for resilience. Transactions of the American Fisheries Society 143:467–478.

Schaub, M., and F. Abadi. 2011. Integrated population models: a novel analysis framework for deeper insights into population dynamics. Journal of Ornithology 152:227–237.

Sievers, M., R. Hale, and J. R. Morrongiello. 2017. Do trout respond to riparian change? A meta-analysis with implications for restoration and management. Freshwater Biology 62:445–457.

Snyder, C. D. N. P. Hitt, and J. A. Young. 2015. Accounting for groundwater in stream fish thermal habitat responses to climate change. Ecological Applications 25:1397–1419.

Wagner, T., J. T. Deweber, J. Detar, and J. A. Sweka. 2013a. Landscape-scale evaluation of asymmetric interactions between Brown Trout and Brook Trout using two-species occupancy models. Transactions of the American Fisheries Society 142:353–361.

Wagner, T., B. J. Irwin, J. R. Bence, and D. B. Hayes. 2013b. Detecting temporal trends in freshwater fisheries surveys: statistical power and important linkages between management questions and monitoring objectives. Fisheries 38:309–319.

Walter, K. M., S. A. Zimov, J. P. Chanton, D. Verbyla, and F. S. Chapin, III. 2006. Methane bubbling from Siberian thaw lakes as a positive feedback to climate warming. Nature (London) 443:71–75.

Wenger, S. J., N. A. Som, D. C. Dauwalter, D. J. Isaak, H. M. Neville, C. H. Luce, J. B. Dunham, M. K. Young, K. D. Fausch, and B. E. Rieman. 2013. Probabilistic accounting of uncertainty in forecasts of species distributions under climate change. Global Change Biology 19:3343–3354.

Wilby, R. L., H. Orr, G. Watts, R. W. Battarbee, P. M. Berry, R. Chadd, S. J. Dugdale, M. J. Dunbar, J. A. Elliott, C. Extence, D. M. Hannah, N. Holmes, A. C. Johnson, B. Knights, N. J. Milner, S. J. Ormerod, D. Solomon, R. Timlett, P. J. Whitehead, and P. J. Wood. 2010. Evidence needed to manage freshwater ecosystems in a changing climate: turning adaptation principles into practice. Science of the Total Environment 408:4150–4164.

Wood, D. M., A. B. Welsh, and J. T. Petty. 2018. Genetic assignment of Brook Trout *Salvelinus fontinalis* reveals rapid success of culvert restoration in headwater streams. North American Journal of Fisheries Management 38:991–1003.

Yeakley J. A., K. G. Maas-Hebner, and R. M. Hughes, editors. 2014. Wild salmonids in the urbanizing Pacific Northwest. Springer, New York.

Zipkin, E. F., J. T. Thorson, K. See, H. J. Lynch, E. H. Campbell Grant, Y. Kanno, R. B. Chandler, B. H. Letcher, and J. A. Royle. 2014. Modeling structured population dynamics using data from unmarked individuals. Ecology 95:22–29.

American Fisheries Society Symposium 91:401–413, 2019

Building a Network of Partners Working to Reconnect Fragmented Streams in the Southeastern United States through Actionable Science

JESSICA GRAHAM* AND KATHLEEN HOENKE
Southeast Aquatic Resources Partnership
1601 Balboa Avenue, Panama City, Florida 32405, USA

Abstract.—There are more than 2.4 million stream km found in the southeastern states that intersect nearly 30 different ecoregions before draining into the Atlantic Ocean or Gulf of Mexico. These streams host one of the greatest diversities of aquatic organisms on earth. In fact, southeastern streams contain more at-risk fish and aquatic invertebrate species than any other region in the United States, containing almost two-thirds of the nation's fish species, more than 90% of the nation's mussel species, and nearly half of the global diversity in crayfish species. Many of these species are endemic to specific drainages and found nowhere else in the world. This level of diversity and endemism increases the sensitivity of these systems to stressors, thus partially explaining the 2–4% annual declines since 1950s. Major threats to aquatic ecosystems in southeastern U.S. streams can be grouped together into interacting categories: land-use changes, incompatible land management practices, nonnative species invasions, habitat fragmentation, flow modifications, and overallocation of water. The Southeast Aquatic Resources Partnership is working across the geography to quantify the degree of fragmentation from both dams and culverts across the landscape and identify opportunities to reconnect rivers and streams through actionable science and effective partnerships.

Aquatic systems are among the most dynamic systems on earth, with interactions between land and water varying over space and time (Jungwirth et al. 2002). These dynamics create a shifting mosaic of habitats with ever-changing location and quality. Natural processes such as flooding and drought work together with woody debris and bedrock to shape river and stream channels (Benda et al. 1998; Bunn and Arthington 2002). The timing and degree of these shifts are driven by a myriad of factors, including precipitation, groundwater, anthropogenic impacts, surrounding land cover, and climatic patterns, that are constantly interacting with each other (Mount 1995). As river channels shift and respond to these interactions, the amount and location of available habitats experience associated changes, creating unique challenges for aquatic organisms to access critical requirements necessary to complete their life cycles (Gagliano et al. 1981). Rivers are not only conduits for fish and other aquatic organisms to pass, but also entire ecosystems that provide a number of ecological, economic, and social services, including our clean water supply for the human population (Postel and Carpenter 1997; Loomis et al. 2000; Bednarek 2001).

There are more than 5.6 million stream kilometers across the conterminous Unit-

* Corresponding author: jessica@southeastaquatics.net

ed States, with 2.4 million km found in the southeastern states (USGS 2013). These streams intersect nearly 30 different ecoregions before draining into the Atlantic Ocean or Gulf of Mexico and host one of the greatest diversities of aquatic organisms on earth. In fact, southeastern streams contain more at-risk fish and aquatic invertebrate species than any other region in the United States (Master et al. 1998), containing almost two-thirds of the nation's fish species, more than 90% of the nation's mussel species, and nearly half of the global diversity in crayfish species. Many of these species are endemic to specific drainages and found nowhere else in the world (Jenkins et al. 2015; Elkins et al. 2016). This level of diversity and endemism increases the sensitivity of these systems to stressors, thus partially explaining the 2–4% annual declines since 1950s (Dudgeon et al. 2006). Extinction rates of freshwater animals in North American streams is estimated to be 4% per decade, a marked increase from terrestrial estimates (Dudgeon et al. 2006).

Major threats to aquatic ecosystems in southeastern U.S. streams can be grouped together into interacting categories: land-use changes, incompatible land management practices, nonnative species invasions, habitat fragmentation, flow modifications, and overallocation of water. Environmental changes occurring at a global scale are superimposed on these categories, thus exacerbating local pressures while creating a higher level of uncertainty in species' responses. Because aquatic systems are highly dynamic and habitat location and availability are constantly changing, unimpeded connectivity of these system is critical to allow movement of organisms to access foraging and spawning areas, shelter, and mates to ensure population persistence. Survival of individual animals, facilitation of reproduction, and maintenance of population continuity are important functions of movement at the population level (Larinier 2001; Limburg and Waldman 2009). Therefore, movement is critical to allow shifts in population responses to extreme conditions brought about by both natural and human-induced perturbations (Freeman 2003). Throughout history, societies have attempted to regulate water resources for use in agricultural, industrial, and domestic purposes, resulting in more than three-quarters of the total discharge of the rivers of the northern third of the world regulated by large dams (Dynesius and Nilsson 1994). The extent of river fragmentation is clear, given that dams today currently retain approximately 10,000 km^3 of water, equivalent to five times the volume of all the world's rivers (Dudgeon et al. 2006). The National Inventory of Dams estimates that just more than 90,000 dams exist in the United States (USACE 2016). This number represents only the larger (>7.6 m) and managed dams in the nation and does not accurately reflect the degree of fragmentation and hydrologic alteration present in aquatic systems, as many smaller structures were installed before any regulations were in place or they fall below the size requirements for regulatory authority (e.g., fishing ponds, amenity lakes, farm ponds, and old mill dams). Many of these dams are no longer used for their intended purpose and may be aging and becoming structurally unsound. The cost of repairing and maintaining obsolete structures can be significant and often more expensive than removing the dams (Loomis 1996; AR et al. 1999). Therefore, identifying these obsolete dams provides an opportunity to examine potential removal that would benefit both the ecosystem and the public. Understanding the level of fragmentation in a given system is important when targeting restoration efforts for native fish and other aquatic biota. However, the number of dams does not tell the complete story of aquatic alteration. Aquatic systems are intersected multiple times by roads, requiring the installation of a structure to allow the passage of water beneath while providing the structural integrity for the safe transportation of vehicles. The southeastern United States has more than 4.8 million km of major roadways intersecting 2.4 million km of streams (U.S. Census Bureau 2017). These intersections require the installation of culverts, bridges, or fords that can move water safely across or below the road alignment. Culverts are likely the most common structures used to move water under a roadway, but often little consideration is given during the design phase to the impacts and disruption of ecosystem processes such as natural hydrology, sediment transport, fish and wildlife passage, or the movement of woody debris (Gibson et al. 2005). Culverts not only fragment the habitat, but also have marked impacts on stream habitat and fauna, with reductions in freshwater mussel populations and fish passage as well as degraded hydrologic conditions (Trombulak and Frissell 2000). The combination of large barriers (such as dams) with small barriers (such as culverts) resulted in more than 170,000 barriers fragmenting the aquatic habitats of the southeastern United States.

Environmental effects of dams and culvert barriers are both immediate as well as subtle. Both dams and culverts can block upstream movement of fish and other organisms. Upstream blockage is not just an issue for diadromous and anadromous fish species and subspecies (e.g., Gulf Sturgeon *Acipenser oxyrinchus desotoi*, American Eel *Anguilla rostrata*, Alabama Shad *Alosa alabamae*, and American Shad *A. sapidissima*), but also for resident native species (e.g., darters and redhorses) that are unable to move freely in the system, thus effectively isolating populations and resulting in a higher likelihood of genetic deterioration and population loss (Haponski et al. 2007; Jelks et al. 2008). Dams and culverts alter hydrologic processes in a more gradual nature, as well with changes in the flood extent, nutrient loads, and sediment transport dynamics that are unique to each river (Johnson 1998). The alteration of hydrologic processes resulting from these anthropogenic barriers is believed to be among the primary causes of declines in aquatic biodiversity (Fullerton et al. 2010).

The Southeast Aquatic Resources Partnership (SARP) is working across the geography to quantify the degree of fragmentation from both dams and culverts across the landscape and identify opportunities to reconnect rivers and streams. The Southeast Aquatic Resources Partnership is a nonprofit organization that focuses on conserving aquatic habitats and the species that depend on them across the 14 southeastern U.S. states and Puerto Rico (Figure 1). In particular, the SARP Connectivity Program brings together partners from all sectors across the geography to reconnect fragmented streams through identification, prioritization, and removal/remediation of barriers to aquatic organisms. This program is composed of three main components that collectively provide collaboration and direction to state-based groups that are working to address connectivity issues. The first component is the Comprehensive Southeast Barrier Inventory (Inventory) that was created by SARP and informed by partners and is the foundation of the program. This Inventory is a large-scale, geographic information system (GIS)-based aquatic barrier inventory consisting of both dams and road-stream crossings. Information on individual documented barriers was gathered from numerous sources across the geography, including state and federal data sets, individual projects that have collected data on small unregulated dams, and personal communication with field biologists. By the end of 2018, the Inventory contained approximately 90,000 data points on dams, much more than the 41,000 documented within the same area through the National Inventory of Dams. (SARP 2017; Figure 2). This Inventory database is housed online

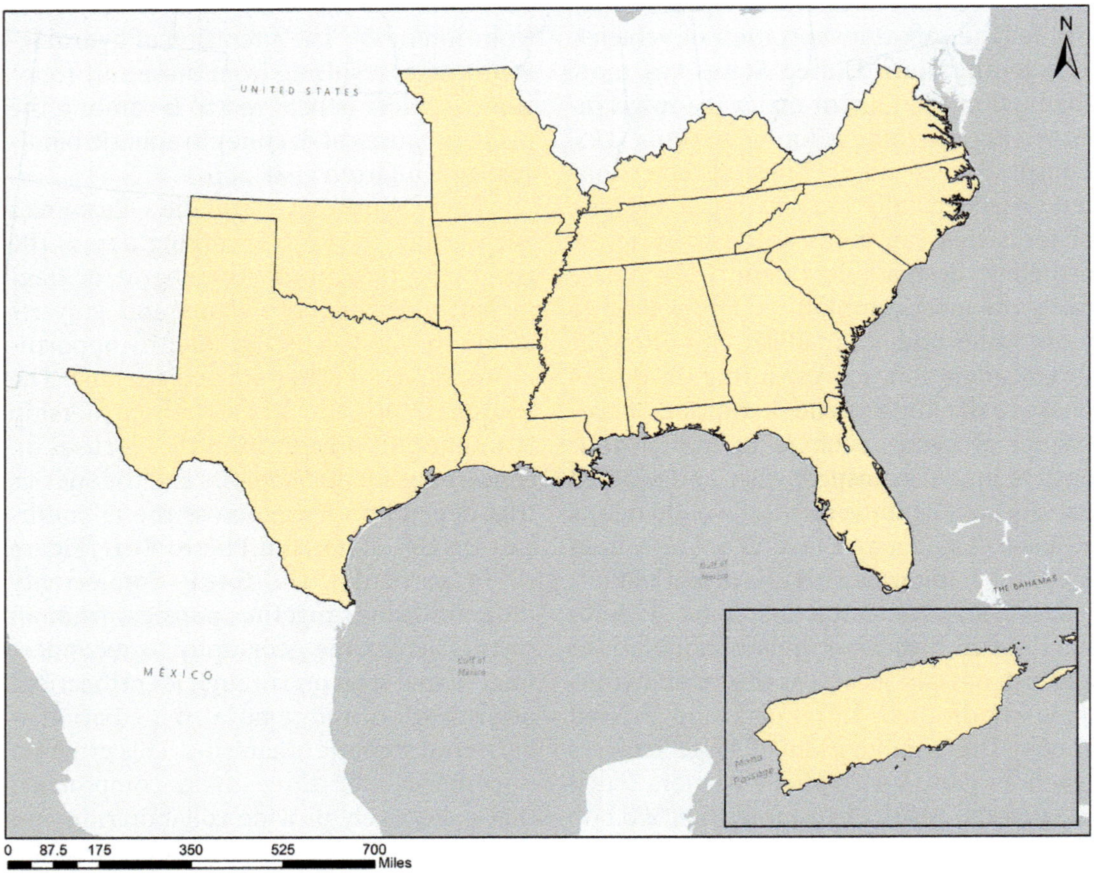

Figure 1. The SARP geography encompasses 14 states across the south-central and southeastern United States, as well as the Commonwealth of Puerto Rico.

and is continually updated as new partners and new data become available. At the present time, more than 100 partners have access to the Inventory, with the ability to add information regarding project feasibility and status. These updates can be viewed instantaneously within numerous state and basin-specific maps that are being used by partners focusing in specific areas.

To further enhance the Inventory beyond just dams and better understand the extent of true aquatic fragmentation, SARP began working with partners to gather additional field data on culverts. By the end of 2018, the Inventory contained more than 1,000,000 data points on road-stream crossings, including more than 86,000 that are known culverts. However, documenting the structure type at a road-stream crossing is only the first step in assessing habitat fragmentation from culverts. Gathering field data on the degree of which a particular culvert creates a barrier is critical when determining whether or not flow and/or passage across the intersection are negatively impacted. Unlike dams that are usually complete barriers to organisms, aquatic organism passage (AOP) at culverts can vary because of the size and species of organisms, and by season and hydrologic conditions during attempted passage (Pess et al. 2005). Field collection is instrumental in determining barrier severity of culverts that lead to more informed decisions when

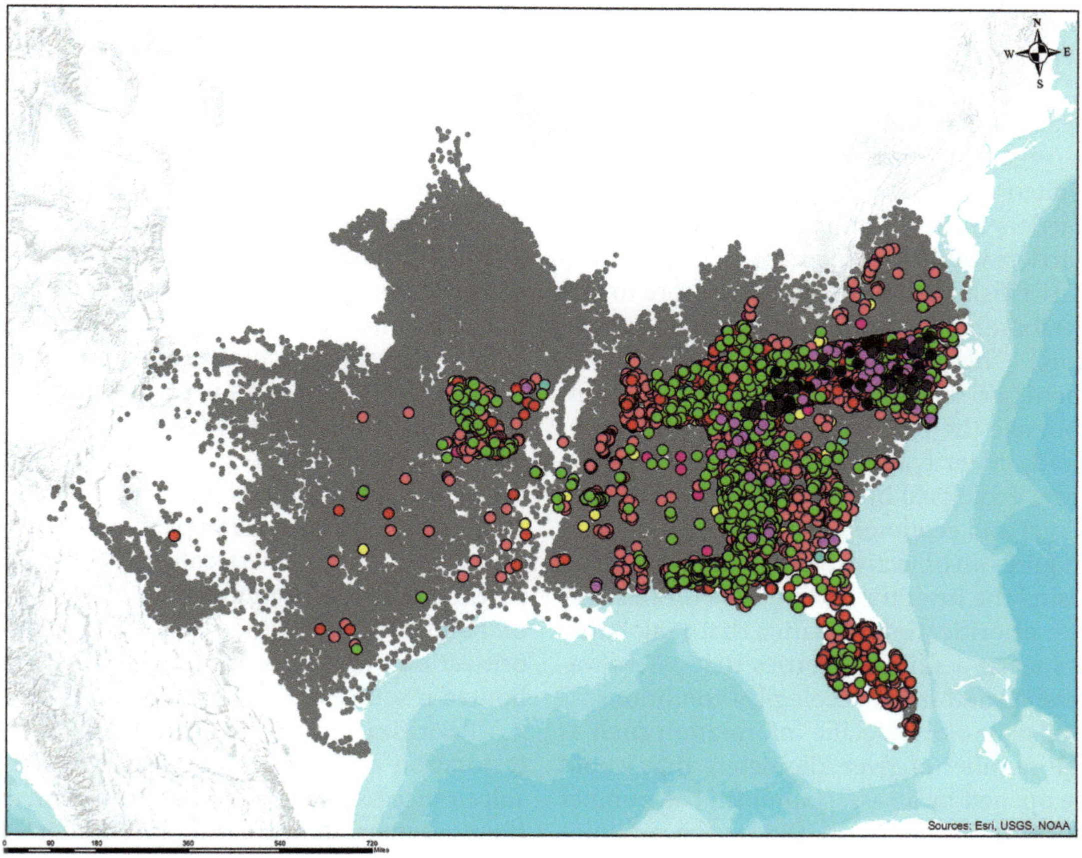

Figure 2. The SARP Inventory hosts information on more than 90,000 dams across the geography. The number of dams per state is illustrated by the color gradient representing the number of documented barriers but does not illustrate the total number of dams that may exist in each state.

targeting restoration projects for benefits of native and at-risk species. Throughout the southeastern United States, numerous partners have assessed road crossings for AOP. However, prior to the development of SARP's Inventory, it was unknown as to where these assessments had occurred throughout the region and were completed using numerous protocols. Over the past 2 years, SARP has worked with partners to gather this information and display it at the regional scale to ensure that assessments completed by partners could be viewable by all partners.

At the end of 2018, the Inventory had approximately 7,000 structures within the 14 states that have been assessed by local, state, and federal partners, using a combination of different protocols. To standardize methods of crossing assessments by willing partners across the Southeast, SARP adopted a protocol originally developed by the North Atlantic Aquatic Connectivity Collaborative that is used to assess road-stream crossing barriers for AOP in the field. This protocol has been shown to decrease the time spent at each crossing while still collecting the necessary information to determine barrier severity (Fair et al. 2018). From 2016 to 2018, SARP has trained more than 90 partners in five states to use the protocol and collect data at road crossings, which has resulted in collection of data at more than 1,000 crossings.

The Inventory indicates that there are nearly 200,000 potential barriers to aquatic organism passage in the southeastern United States alone. The sheer number of barriers makes it difficult to identify which barriers would provide the largest benefit to native fish and be the best candidate to target for removal or remediation (Neeson et al. 2015). Therefore, barrier removal, replacement, or remediation must be prioritized to make the best use of limited restoration funds. Identifying the barriers with both the most ecological benefit and the appropriate social and economic conditions for removal requires an objective prioritization scenario. Using the Inventory, SARP runs various barrier prioritization scenarios that rank barriers using a suite of metrics that assess watershed and stream ecological conditions, as well as project specific information such as the number of river kilometers that would be opened by a passage project, the number of downstream dams, and/or the presence of diadromous or native fish species. These metrics are combined to produce a relative prioritization that can be customized and shared with partners to provide collaborative direction. To date, SARP has completed three regional prioritizations of dams: (1) the SouthEast Aquatic Connectivity Assessment Project, which prioritized dams in the South Atlantic states in partnership with The Nature Conservancy; (2) the Tennessee–Cumberland Aquatic Connectivity Assessment, which spanned the Tennessee and Cumberland River basins in Tennessee; and (3) a prioritization of dams on U.S. Forest Service (USFS) lands across the Southern region (Region 8). Efforts continue to prioritize dams across the SARP geography (Figure 1), funded by a variety of entities, including the Gulf Coastal Plains and Ozarks Landscape Conservation Collaborative, the Texas State Wildlife Grants Program, Florida's Legacy Program, and the U.S. Fish and Wildlife Service's Fish Passage Program. In addition to large-scale regional prioritizations, SARP works to prioritize barriers within smaller watersheds, working closely with local partners. One such example is a recent effort to identify and assess road-stream crossings on national forests in Georgia and South Carolina. Additionally, SARP worked with the USFS Center for Technology Transfer Team to begin conducting surveys in partnership with SARP using this protocol. Through this effort, road-stream crossings have been surveyed in five national forests. These surveys have supported updates and revisions to forest specific plans, including Chattahoochee and Francis Marion–Sumter national forests. The additional data collected by these partners provides a more complete picture of habitat fragmentation for relevant prioritizations, leading to more effective use of limited resources (Jackson and Luken 2010).

Prioritization can provide direction for restoration efforts but does not always result in successful project implementation as there are many local and site-specific details that GIS-based prioritization cannot account for in the analyses. To enhance GIS-based prioritization and minimize the number of infeasible projects rising to the top of the prioritization, partners perform a process called "desktop reconnaissance" on top-ranking dams. This process uses aerial photographs and a standardized coding system to provide additional information regarding surrounding land use for each barrier (e.g., recreational use indicated by docks, boat launches, and swimming areas). This information allows a further refinement of prioritization results based on potential feasibility. Desktop reconnaissance narrows the list of potential projects that can be provided to partners on the ground for further investigation. However, performing reconnaissance on a project often cannot produce usable results without viewing each barrier in the context of the local landscape, such as landowner willingness and community value.

Understanding these dynamics requires a close collaboration with local communities and practitioners to properly characterize feasibility.

Thus, a successful barrier removal or remediation project usually requires multiple partners working together to achieve results. In order to create an environment of success and facilitate successful projects, SARP and partners have been working to build a community of practice surrounding barrier removal through the development of aquatic connectivity teams (ACTs). The ACTs are composed of local, state, and federal partners from many sectors and include on-the-ground project managers addressing aquatic connectivity. These ACTs create a forum that allows resource managers to work together, share resources, disseminate information, and examine regulatory streamlining processes, as well as project management tips and techniques. The Southeast Aquatic Resources Partnership and American Rivers communicate with ACT members on prioritization results, work with team members to perform the additional reconnaissance on top-ranked dams, and continue to provide scientific and technical support to further refine prioritization results. This refined information can be viewed by other team members working together to create a workable list for project managers. Thanks to the strong partnerships within the ACTs, at the end of 2018, SARP and partners have performed desktop reconnaissance on more than 7,500 dams across the southeastern United States (Figure 3). In addition, ACT members have facilitated identification of undocumented dams, as well as conduct field assessments of road-stream crossings using SARP's standardized protocol. Data from all of these activities feed directly back into the Inventory. This cyclical process where ACT members identify undocumented dams and conduct reconnaissance to produce potentially feasible project lists provides momentum to the ACTs. The collaborative forum provided by the ACTs and supported by comprehensive data and prioritization has resulted in more dam and culvert removal/remediation than without this community of practice, where all ACT members are working to reconnect rivers and streams and benefit aquatic species.

Multiscale collaboration among partners working with the SARP Inventory and prioritizations can be specifically viewed through the lens of the ACTs currently operating in North Carolina, Tennessee, Georgia, and Arkansas. The North Carolina Aquatic Connectivity Team (NC ACT) was initiated by American Rivers in 2010, which was prior to the creation of the SARP Inventory and prioritization, resulting in a very partnership-driven ACT. Resource managers at the local and state level within North Carolina formed solid partnerships that resulted in a network of project managers and colleagues. Because these SARP resources were not yet available, this network collaboratively identified and implemented dam removal projects that opportunistically surfaced through word of mouth and partner interactions. Through time, the NC ACT became more efficient at the project implementation process, resulting in a number of projects already planned when the Inventory and subsequent prioritization were released. Therefore, while more than 1,000 dams within the inventory have been further refined through the desktop reconnaissance process in North Carolina, partners are still working on these projects that had been previously initiated. An example of a barrier that was previously identified and prioritized using similar information that is contained in the SARP's Inventory is the Smitherman Mill Dam located in Troy, North Carolina, which was removed in 2013. The removal of this dam opened 8.5 main-stem kilometers and more than 45 km of tributary habitat benefiting dozens of rare and threatened freshwater mussels and the Carolina Redhorse *Moxostoma* sp. cf. *erythrurum*. As

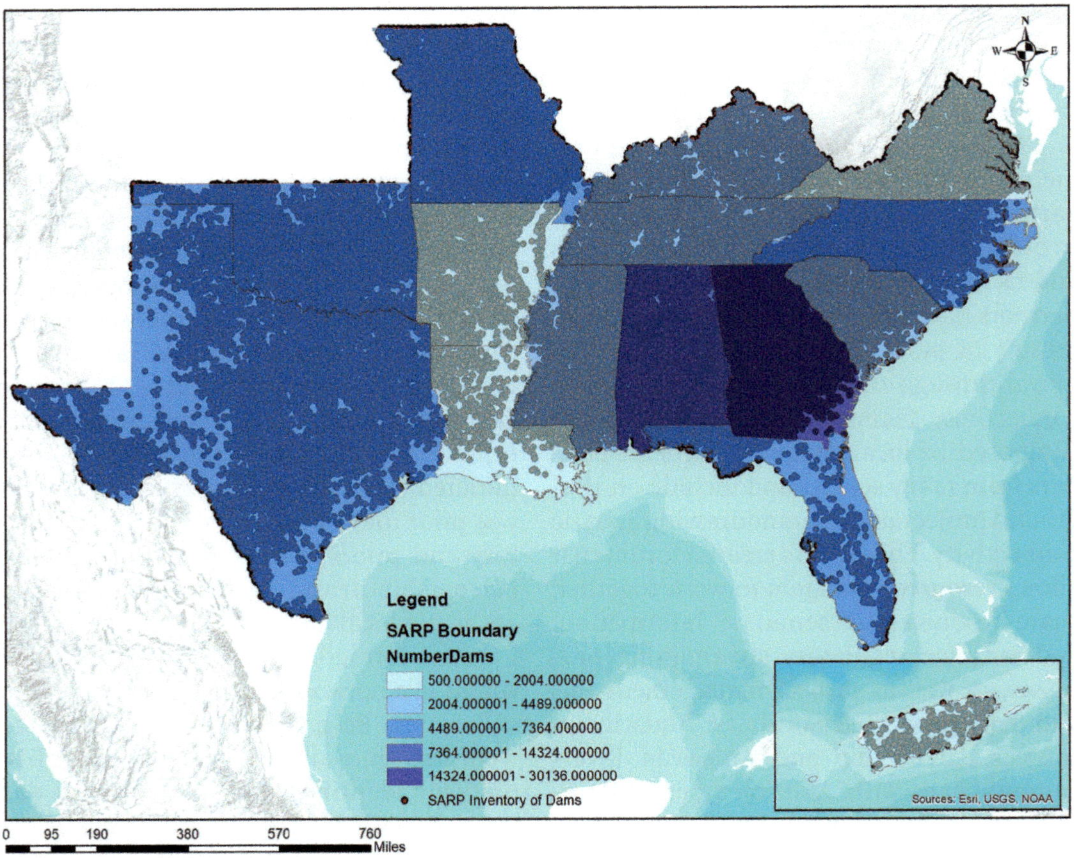

Figure 3. Results from the desktop reconnaissance process that SARP and partners have used to determine potential feasibility relating to individual dams. Colors indicate the degree of feasibility, with warm colors indicating a lack of feasibility and cool colors indicating a greater potential feasibility of a removal project.

these projects are completed, the team will continue to use both a partnership-based and a data-driven approach to implement new projects found by both contacting owners of high-ranking dams within the prioritization and capitalizing on opportunities as they become available. The creation of the SARP Inventory has allowed the NC ACT to facilitate communication and collaboration while tracking project status on the Web map platform, allowing all team members to stay informed while also quantifying the net ecological gains. Prioritization results have also been used to bolster grant proposals to support projects by readily providing metrics on ecological benefits to the habitat and species-specific populations. The utilization of the Inventory by the ACT will likely steer the NC ACT into a more data-driven approach that is readily apparent with ACTs that have been developed following the creation of the Inventory, as described below.

The ACTs in both Tennessee and Georgia began after the creation of the Inventory and are thus currently using a more data-driven approach to identify and implement projects relying heavily on the Inventory and prioritization. Because the Tennessee Aquatic Connectivity Team (TN ACT) has fewer projects in the planning phase, team members sought funding to focus on project identification through the Inventory and prioritization. Using SARP's tools, The Nature Conservancy (TNC) has identified location and feasibility information using the desktop reconnaissance process on more than 1,000 dams. In addition, TNC has contacted landowners of top-ranking dams resulting from this process. Partners continue to work throughout the state to identify high-priority dams and contact landowners to determine feasibility for removal. Although there have not been any projects informed by the SARP Inventory that have been completed by the end of 2018, there are multiple projects currently in the phase of design and permitting. The Sawpauh Mill Dam located in McMinn County, Tennessee is one dam that was identified as a high priority by the Inventory and funded through grants that were secured using metrics provided by the prioritization. The removal of this dam will open 133 km for the threatened Snail Darter *Percina tanasi* and has been approved by the landowners for removal. Because the TN ACT was created more recently than the NC ACT, there are fewer dam removal success stories demonstrating the collaborative power provided by this data-driven process. However, success of collaboration is demonstrated through other efforts using the TN ACT as a forum for discussion, including streamlining regulatory processes such as permit fees associated with the Clean Water Act section 401 certification for dam removal projects.

The Georgia Aquatic Connectivity Team (GA ACT) began in 2015 with a workshop to instruct interested partners on dam removal and is jointly facilitated by TNC and SARP. Members immediately used a data-driven approach to find projects supported by the Inventory. With help from partner contributions of interns, more than 3,000 dams have been manually inspected through the desktop reconnaissance process within the Inventory for potential removal feasibility, and team members are now strategizing how to make landowner contact for top-ranking dams. Following a training on how to use SARP's road-stream crossing protocol within the state, more than 200 road-stream crossings have been assessed in the field within Georgia, and several removal projects are in process, with regular updates by project managers provided to the GA ACT using e-mail list serv. The formation of the GA ACT brought partners together and provided a forum to help facilitate discussions surrounding the removal feasibility of the White Dam, a dam on the University of Georgia's campus. The White Dam was officially removed in the fall of 2018. This project demonstrated the significant coordination and collaboration required for dam removal projects, which were facilitated through the GA ACT. The White Dam was considered to be of historical significance, which required a modification to the complete dam removal process, and a portion of the dam was left in place while allowing the removal of the remaining structure that altered streamflow.

The most recent ACT was formed in Arkansas and has been named, by partners, the Arkansas Stream Heritage Partnership. The impetus for the initiation of this team came from knowledge and experience working with SARP on an individual case-by-case basis. State partners became interested in engaging with SARP's inventory and prioritization work more formally through SARP's Connectivity Program to better align work conducted by multiple agencies. Since the Arkansas Aquatic Connectivity Team's (AR ACT) inception, SARP has been working with American Rivers and the Arkansas Natural Heritage Commission to identify and implement barrier removal and remediation projects through partner engagement. The AR ACT was immediately interested in filling data gaps within the Inventory and identify locations of unregulated dams. Previous to the ACT's formation, SARP partnered with American Rivers to inventory and prioritize dams on USFS land, including those located in Arkansas. U.S. Forest Service biologists and staff were involved in multiple aspects of this previously completed project that prioritized dams on USFS lands. These biologists and staff provided critical knowledge about fragmentation from barriers at the local scale. Through the information provided during the course of this prioritization project, a small dam was discovered on USFS land that if removed would open 9.6 km of Mine Creek, a tributary to the Cossotot River, which is a high-priority river for the Leopard Darter *Percina pantherina*. Through the discovery of the dam and its high ranking through the subsequent prioritization, USFS biologists engaged with partners and attended the inaugural workshop for the AR ACT, which was facilitated by SARP and American Rivers. This workshop brought together partners from various sectors and organizations and created a momentum that had previouslybeen lacking. This momentum has resulted in the immediate engagement from regulators such as the USACE and Arkansas Department of Environmental Quality on the front end of the project, increasing the feasibility for removal. All of these factors resulted in a fast timeline for the removal of the dam, which is slated to occur in 2019.

Despite scientific and technical support provided by SARP and the partnerships built through the creation and facilitation of ACTs, limitations still exist to reconnect rivers and streams. One limitation to project implementation is the lack of capacity and dedicated project managers to successfully coordinate all necessary components of a removal project. To use the inventory to identify a top-ranking barrier and achieve a dam removal or culvert replacement, a project manager must thoroughly research each restoration option and contact landowners to assess landowner willingness. Multiple visits and considerable time are often required to build trust with an owner before a project can move forward to the next phase of implementation. This may be less of an issue in well-established ACTs, where team members from local watershed groups may be well situated to liaise with landowners as they are often familiar with many such individuals and have the necessary trust to move projects forward in a timely manner. However, newer teams may not have fully developed project management capacity, leaving many potential high-ranking dams, identified and ready for further reconnaissance, waiting for a project manager. Even when project managers are found, funding is almost always a major limitation to project implementation. In fact, funding is often unavailable for general project reconnaissance, and liaison with landowners on top-ranking dams is difficult, as it does not guarantee a dam will be removed and metrics will be met. Often, funding is only available for shovel-ready projects, yet these projects cannot exist without prior reconnaissance and landowner collaboration. Thus, opportunistic projects often get implemented before more strategic ones that may provide greater ecological benefits for native fish and aquatic communities.

Sometimes these limitations have been overcome in targeted locations by using a watershed approach where capacity was built through a variety of project types (e.g., barrier removal, upland habitat restoration, and agricultural best management practice implementation). This approach has been successful in the Apalachicola–Chattahoochee–Flint River basin in Alabama, Florida, and Georgia, where various state and federal agencies throughout the basin have been working to conserve listed and at-risk mussel species, as well as the endemic Shoal Bass *Micropterus cataractae*, which has been identified by partners as a keystone species (Birdsong et al. 2015). Partners conducted research to fill research gaps while identifying restoration projects and connectivity issues. However, little to no project implementation had been conducted until funding was secured for a dedicated project manager in the basin. This project manager connected all of the previous work together and created a plan to move previously identified projects towards implementation and hopefully realize benefits to native species. Another watershed approach led by SARP has been ongoing with partners in the Stevens River basin, located in South Carolina. This basin is a targeted basin because of the listed mussels (such as the Carolina heelsplitter *Lasmigona decorata*) and the native Bartram's Redeye Bass *Micropterus* sp. cf. *coosae*. The Southeast Aquatic Resources Partnership collaborated with federal and state partners to strategically identify all of the barriers that had been previously assessed using numerous different protocols. Once collective efforts were consolidated within the inventory, the USFS used SARP's protocol to fill data gaps and assess the road-stream crossings on the Francis Marion National Forest within the basins. Once these surveys were completed, SARP used existing relationships with the U.S. Fish and Wildlife Service, South Carolina Department of Natural Resources, and U.S. Department of Transportation to bring in additional barrier data documented within the Inventory and perform a GIS-based connectivity analysis prioritizing barriers for removal or replacement based on connectivity gains and species access. As a result, the highest priority barriers were targeted for removal by USFS staff.

Research has shown that targeting both dams and culverts when implementing connectivity remediation provides the most ecological gain, particularly when working within the same stream network (Fitzpatrick and Neeson 2018). Thus, a watershed-based approach that incorporates habitat fragmentation resulting from multiple causes, such as dams and culverts, will not only increase ecological gain, but also help build capacity for project management. In an effort to maximize benefits to native fish and other aquatic communities, SARP works with partners to prioritize capacity building in areas designated as SARP's conservation opportunity areas and rely on existing partnerships outside of the ACTs and the collaborative environment the ACTs provide. Only through partnerships can we work together to reconnect our fragmented southeastern U.S. streams.

References

AR (American Rivers), FE (Friends of the Earth), and TU (Trout Unlimited). 1999. Dam removal success stories: restoring rivers through selective removal of dams that don't make sense. AR, FE, and TU, Washington, D.C.

Bednarek, A. T. 2001. Undamming rivers: a review of the ecological impacts of dam removal. Environmental Management 27:803–814.

Benda, L. E., D. J. Miller, T. Dunne, G. H. Reeves, and J. K. Agee. 1998. Dynamic landscape systems. Pages 261–288 *in* S. Kantor, editor. River ecology and management: lessons from the Pacific coastal ecoregion. Springer-Verlag, New York.

Birdsong, T. W., M. S. Allen, J. E. Claussen, G. P. Garrett, T. B. Grabowski, J. Graham, F. Harris, A. Hartzog, D. Hendrickson, R. A. Krause, J. K. Leitner, J. M. Long, C. K. Metcalf, D. P. Philipp, W. F. Porak, S. Robinson, S. M. Sammons, S. Shaw, J. E. Slaughter, IV, and M. D. Tringali. 2015. Native Black Bass Initiative: implementing watershed-scale approaches to conservation of endemic black bass and other native fishes in the southern United States. Pages 363–378 *in* M. D. Tringali, J. M. Long, T. W. Birdsong, and M. S. Allen, editors. Black bass diversity: multidisciplinary science for conservation. American Fisheries Society, Symposium 82, Bethesda, Maryland.

Bunn, S. E., and A. H. Arthington. 2002. Basic principles and ecological consequences of altered flow regimes for aquatic biodiversity. Environmental Management 30:492–507.

Dudgeon, D., A. H. Arthington, M. O. Gessner, Z. Kawabata, D. J. Knowler, C. Leveque, R. J. Naiman, A. Prieur-Richard, D. Soto,

M. L. J Stiassny, and C. A. Sullivan. 2006. Freshwater biodiversity: importance, threats, status, and conservation challenges. Biological Reviews 81:163–182.

Dynesius, M., and C. Nilsson. 1994. Fragmentation and flow regulation of river systems in the northern third of the world. Science 266:753–762.

Elkins, D. C., S. C. Sweat, K. S. Hill, B. R. Kuhajda, A. L. George, and S. J. Wenger. 2016. The southeastern aquatic biodiversity conservation strategy. University of Georgia River Basin Center, Final Report, Athens.

Fair, B., C. Krause, and C. Roghair. 2018. Executive summary of aquatic organism passage surveys on the Long Cane & Andrew Pickens District, South Carolina, and the Chattooga River & Blue Ridge Districts, Georgia 2017. U.S. Forest Service, Southern Research Station, Blacksburg, Virginia.

Fitzpatrick, K. M., and T. M. Neeson. 2018. Aligning dam removals and road culvert upgrades boosts conservation return on investment. Ecological Modelling 368(C):198–204.

Freeman, M. 2003. Ecosystem-level consequences of migratory faunal depletion caused by dams. Pages 255–266 *in* K. E. Limburg and J. R. Waldman, editors. Biodiversity, status, and conservation of the world's shads. American Fisheries Society, Symposium 35, Bethesda, Maryland.

Fullerton, A. H., K. M. Burnett, E. A. Steel, R. L. Flitcroft, G. R. Pess, B. E. Feist, C. E. Torgersen, D. J. Miller, and B. L. Sanderson. 2010. Hydrological connectivity for riverine fishes: measurement challenges and research opportunities. Freshwater Biology 55:2215–2237.

Gagliano, S. M., K. J. Meyer-Arendt, and K. M. Wicker. 1981. Land loss in the Mississippi River deltaic plain. Gulf Coast Association of Geological Societies Transactions 31:295–300.

Gibson, R. J., R. L. Haedrich, and C. M. Wernerheim. 2005. Loss of fish habitat as a consequence of inappropriately constructed stream crossings. Fisheries 30(1):10–17.

Haponski, A. E., T. A. Marth, and C. A. Stepien. 2007. Genetic divergence across a low-head dam: a preliminary analysis using logperch and greenside darters. Journal of Great Lakes Research 33:117–126.

Jackson, S. D., and D. Luken. 2010. Evaluation of a rapid assessment protocol to assess road-stream crossing for aquatic organism passage. Pages 487–500 *in* P. J. Wagner, D. Nelson and E. Murry, editors. Proceedings of the 2009 international conference on ecology and transportation. Center for Transportation and the Environment, North Carolina State University, Raleigh.

Jelks, H. L., S. J. Walsh, N. M. Burkhead, S. Contreras-Balderas, E. Diaz-Pardo, D. A. Hendrickson, J. Lyons, N. E. Mandrak, F. McCormick, J. S. Nelson, S. P. Plantania, B. A. Porter, C. B. Renaud, J. J. Schmitter-Soto, E. B. Taylor, and M. L. Warren, Jr. 2008. Conservation status of imperiled North American freshwater and diadromous fishes. Fisheries 33:372–407.

Jenkins, C., K. S. Van Houtan, S. L. Pimm, and J. O. Sexton. 2015. US protected lands mismatch biodiversity priorities. Proceedings of the National Academy of Sciences of the United States of America 112:5081–5086.

Johnson, W. C. 1998. Adjustment of riparian vegetation to river regulation in the great plains, USA. Wetlands 18:608–618.

Jungwirth, M., S. Muhar, and S. Schmutz. 2002. Re-establishing and assessing ecological integrity in riverine landscapes. Freshwater Biology 47:867–887.

Larinier, M. 2001. Environmental issues, dams and fish migration. FAO (Food and Agriculture Organization of the United Nations) Fisheries Technical Paper 419:45–89.

Limburg, K. E. and J. R. Waldman. 2009. Dramatic declines in North Atlantic diadromous fishes. BioScience 59:955–965.

Loomis, J. B. 1996. Measuring the economic benefits of removing dams and restoring the Elwha River: results of a contingent valuation survey. Water Resources Research 32:441–447.

Loomis, J., P. Kent, L. Strange, K. Fausch, and A. Covich. 2000. Measuring the total economic value of restoring ecosystem services in an impaired river basin: results from a contingent valuation survey. Ecological Economics 33:103–117.

Master, L., S. R. Flack, and B. A. Stein. 1998. Rivers of life: critical watersheds for protecting freshwater biodiversity. The Nature Conservancy, Arlington, Virginia.

Mount, J. F. 1995. California rivers and streams: the conflict between fluvial process and land use. University of California Press, Berkeley.

Neeson, T., M. Ferris, W. Diebal, P. Doran, J. O'Hanley, and P. McIntyre. 2015. Enhancing ecosystem restoration efficiency through spatial and temporal coordination. Proceedings of the National Academy of Sciences of the United States of America 112:6236–6241.

Pess, G., S. Morley, and P. Roni. 2005. Evaluating fish response to culvert replacement and other methods for reconnecting isolated aquatic habitats. Pages 267–276 *in* P. Roni, editor. Monitoring stream and watershed restoration. American Fisheries Society, Bethesda, Maryland.

Postel, S., and S. Carpenter. 1997. Freshwater ecosystem services. Pages 195–214 *in* G. C. Daily, editor. Nature's services: societal dependence on natural ecosystems. Island Press, Washington, D.C.

SARP (Southeast Aquatic Resources Partnership). 2017. Comprehensive southeast aquatic barrier inventory [geodatabase]. Available: www.southeastaquatics.net/sarps-programs/southeast-aquatic-connectivity-assessment-program-seacap. (May 2018).

Trombulak, S. C., and C. A. Frissell. 2000. Review of ecological effects of roads on terrestrial and aquatic communities. Conservation Biology 14:18–30.

U.S. Census Bureau. 2017. 2017 TIGER/Line Shapefiles (machine-readable data files). U.S. Census Bureau, Suitland, Maryland.

USACE (United States Army Corps of Engineers). 2016. National inventory of dams [online database]. Available: http://nid.usace.army.mil/cm_apex/f?p=838:1:0::NO. (May 2018).

USGS (United States Geological Survey). 2013. National hydrography geodatabase: the national map viewer. Available: www.horizon-systems.com/NHDPlus/NHDPlusV2_data.php.

American Fisheries Society Symposium 91:415–430, 2019

Little Tennessee Native Fish Conservation Partnership: Aquatic Conservation on a Landscape Scale

ANDREA LESLIE*
North Carolina Wildlife Resources Commission
645 Fish Hatchery Road, Marion, North Carolina 28752, USA

ERIN MCCOMBS
American Rivers
46 Orchard Street, Asheville, North Carolina, 28801, USA

FRED HARRIS
North Carolina Wildlife Federation
2413 Mary Marvin Trail, Fuquay Varina, North Carolina, 27526, USA

Abstract.—In 2015, the Little Tennessee River basin became the nation's first native fish conservation area. Watersheds designated as native fish conservation areas are managed for the conservation and restoration of native fish and other aquatic species, allowing compatible uses. The Little Tennessee River basin spans three states (Georgia, North Carolina, and Tennessee) and features a diversity of aquatic habitats that include high-elevation coldwater trout streams, warmwater rivers, and large human-made reservoirs. Although this basin is home to a biologically diverse aquatic community, streams have been impacted by a host of stressors, including logging, dams, agriculture, industrial pollutants, piscicides, and development. Some streams impacted in the past now offer restoration opportunities, and numerous efforts are underway to restore native fish and mussels to streams in the Great Smoky Mountains National Park, on U.S. Forest Service land, on the Qualla Boundary of the Eastern Band of Cherokee Indians, and on private lands. More than 20 organizations, including federal and state agencies, industry, and nongovernmental entities, form the Little Tennessee Native Fish Conservation Partnership. The partnership supports work already underway by partners by providing additional funding, technical and educational resources, and a mechanism for collaboration. Perhaps most importantly, the partnership provides a forum to plan and implement watershed conservation on a landscape scale. Partners developed an online conservation mapper, which houses data, maps threats, identifies focal areas for restoration and protection, and ultimately serves as a conservation plan for the watershed. Current efforts to identify habitat restoration and protection projects are underway.

* Corresponding author: andrea.leslie@ncwildlife.org

Introduction

Native fish conservation area concept

In 2015, the Little Tennessee River basin (LTRB) became the nation's first native fish conservation area (NFCA). The LTRB stands out in the southeastern United States as it (1) hosts a rich assemblage of rare and common native aquatic species, (2) is a focus of multiple fish restoration efforts, (3) has a relatively high proportion of public lands compared to the rest of the region, and (4) has a strong set of partners from more than 20 organizations representing three states (including agencies, industry, and nongovernmental organizations from North Carolina, Tennessee, and Georgia) involved in restoration, conservation, and education.

An NFCA is defined as a watershed-scale (fifth-level, 10-digit hydrologic unit code or larger) area where management emphasizes conservation and restoration for long-term persistence of native fishes and other native aquatic species while allowing compatible uses. The concept was developed by Trout Unlimited, the Federation of Fly Fishers, and the Fisheries Conservation Foundation in 2008 (Williams et al. 2011) in response to a significant and steady decline in the diversity of native aquatic organisms throughout the United States, against which traditional conservation practices were having limited effect (Williams et al. 1989). According to the World Wildlife Fund, the average abundance of populations that are monitored in freshwater systems declined by 81% between 1970 and 2012 (World Wildlife Fund 2016).

As identified by Williams et al. (2011) and discussed by Dauwalter et al. (2011), implementation of NFCAs should sustain freshwater habitat complexity, diversity, and connectivity through the conservation of watershed-scale processes; support all life stages for focal fishes and other aquatic species; involve a large-enough watershed to provide for long-term aquatic community persistence; and involve partners that can implement sustainable long-term land and water management. Designation of a watershed as an NFCA should add no new regulations or restrictive-use policies, result in no additional costs or obligations to local tax-payers, and allow existing private land rights. The focus of work for the Little Tennessee River NFCA is to conserve native aquatic resources and the uses they support, strengthen the role of aquatic resources in local economies, and support land managers and local organizations in securing resources for aquatic conservation and management.

Little Tennessee Native Fish Conservation Partnership

The idea of applying the NFCA concept to the LTRB was proposed to resource agencies and watershed conservation nongovernmental organizations in 2013 and 2014 by the North Carolina Wildlife Federation. Interested organizations became part of a steering committee to develop a group structure and basic strategic documents for the effort, which became the Little Tennessee Native Fish Conservation Partnership (hereafter, Partnership). The Partnership formalized the status of the Little Tennessee River basin as an NFCA with a public designation event in 2015, held in the Great Smoky Mountains National Park on Abrams Creek, where partners released federally endangered Citico Darter *Etheostoma sitikuense* and Smoky Madtom *Noturus baileyi* and federally threatened Yellowfin Madtom *N. flavipinnis*.

The Partnership's mission is to conserve and restore habitat in the Little Tennessee River basin for the benefit of native fishes and other aquatic wildlife. Its vision (www.littlet.org/about-the-partnership/) is as follows:

> The aquatic resources of the Little Tennessee River and its tributaries are conserved and restored to the level that native fishes thrive as stable components of diverse ecological communities. Present and future citizens of the basin benefit

> from clean water, outstanding outdoor recreation, and a stable economic base.

Its core values emphasize aquatic habitat conservation and restoration in the basin, education of watershed citizens to foster stewardship, collaboration to achieve common goals, and open communication. Although some partners do engage in advocacy, the Partnership does not advocate for any new regulations or policies.

The Partnership consists of a steering committee, which includes representatives of all partner organizations; an executive committee; and three working groups who carry out the bulk of the Partnership's work—education and outreach, assessment and implementation, and fundraising. Approximately 20 organizations (see Box 1), including state and federal natural resource agencies, the Eastern Band of Cherokee Indians (EBCI), nongovernmental conservation organizations, universities, and industries, are represented on the steering committee. The steering committee meets semiannually to set priorities for the Partnership's efforts, make decisions on funding priorities, and coordinate on various projects.

There is no paid coordinator for the Partnership, and the executive committee carries out coordination activities, such as facilitating meetings, maintaining a Web site, and acquiring funding. The executive committee comprises a chair, two vice-chairs from different states, and two leads for each of the working groups. These roles are voluntary and held for staggered 2-year terms with the option to renew.

Each partner organization carries out its own work in the basin, including monitoring, stream restoration, land protection, and public education. The work of the Partnership adds value to these efforts by providing technical and monetary resources to partner organizations, coordinating collaborative projects, improving internal and external communication, providing a strategic vision, and addressing emerging needs.

Little Tennessee River basin

The LTRB is 6,806 km^2, consisting of three 8-digit hydrologic units and spanning portions of three states in the southern Appalachian Mountains—Georgia, North Carolina, and Tennessee (Figure 1). The Little Tennessee River's headwaters are in northeastern Georgia, and the majority of its length flows through North Carolina and Tennessee, meeting the Tennessee River in Lenoir City

Box 1. Participating organizations in the Little Tennessee Native Fish Conservation Partnership Steering Committee

Little Tennessee Native Fish Conservation Partnership Steering Committee: American Rivers, Conservation Fisheries, Inc., Defenders of Wildlife, Duke Energy, Eastern Band of Cherokee Indians, Fisheries Conservation Foundation, Georgia Department of Natural Resources, Great Smoky Mountains National Park, Mainspring Conservation Trust, The Nature Conservancy, North Carolina Wildlife Federation, North Carolina Wildlife Resources Commission, Tennessee Chapter of the Sierra Club, Tennessee Department of Environmental Conservation, Tennessee Wildlife Resources Agency, Tennessee Valley Authority, Trout Unlimited, University of Tennessee, United States Fish and Wildlife Service, Watershed Association of the Tuckasegee River, and Western Carolina University.

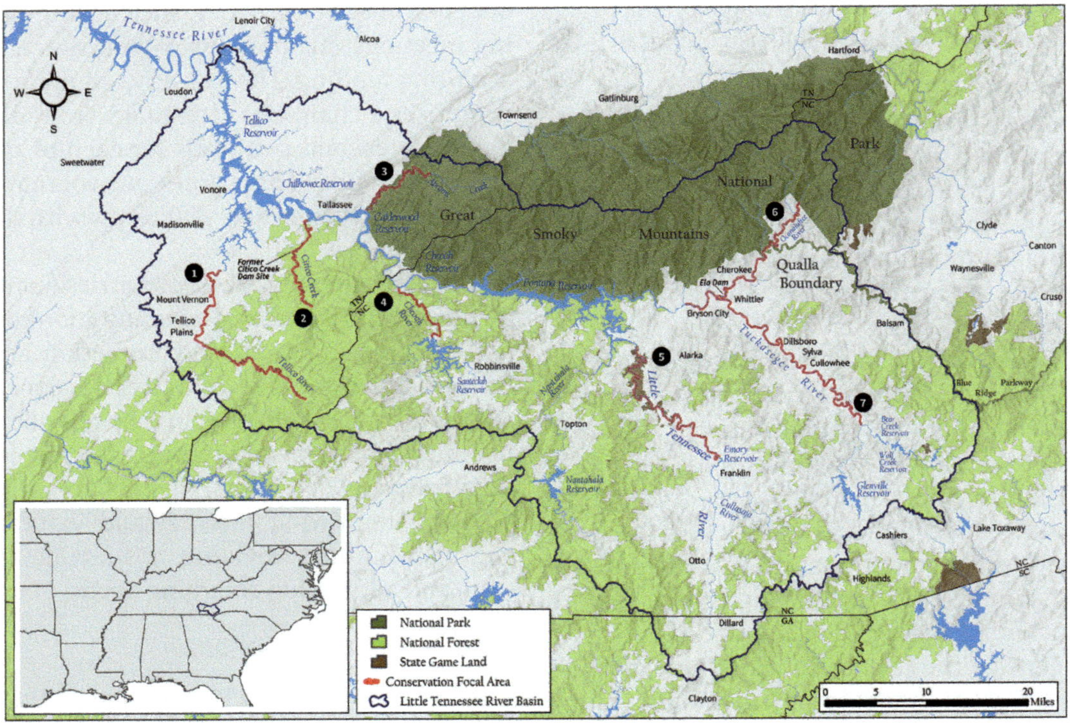

Figure 1. The Little Tennessee River basin and conservation focal areas: (1) Tellico River, (2) Citico Creek, (3) Abrams Creek, (4) Cheoah River, (5) 68-km, free-flowing reach of the Little Tennessee River, (6) Oconaluftee River, and (7) Tuckasegee River.

(64 river kilometers southwest of Knoxville). Approximately 46% of the basin is in federal ownership across the Chattahoochee, Nantahala, and Cherokee national forests, the Great Smoky Mountains National Park, and the Blue Ridge Parkway.

The basin (Figure 1) is home to a biologically rich aquatic community, including 120 species of native fishes, an estimated 30 species of native mussels, 5 species of native snails, and 16 species of native crayfish (Little Tennessee Native Fish Conservation Partnership, unpublished data, www.littlet.org/publications/). Thirty-five of these species have federal and/or state status as endangered, threatened, of special concern, in need of management, or rare. Mollusks have been comparatively undersurveyed, and some species documented in the past 40 years have likely invaded the basin due to impoundments (Isom 1971; Parmalee and Hughes 1993). The number of species of native mussels currently found the basin is less than half of what was present historically (Starnes and Bogan 1988).

Numerous factors drive this biodiversity. Elevations in the basin span a 1,800-m range, from one of the eastern United States' highest peaks (Clingmans Dome, 2,025 m) to 232 m at the Little Tennessee River's confluence with the Tennessee River. The basin features an array of aquatic habitats, including high-elevation seeps, coldwater trout streams, biologically rich warmwater rivers, and large reservoirs. Habitat diversity, redundancy, and connectivity in the river basin help ensure that aquatic life has the needed habitats to support all stages of life.

Geological differences across the basin likely influence species diversity, as well. The

LTRB spans two Omernik Level III ecoregions—the Blue Ridge Mountain and Ridge and Valley ecoregions, the border between them roughly following the North Carolina–Tennessee state lines. Bedrock of the southern Blue Ridge ecoregion is composed mainly of metamorphosed sedimentary and igneous rocks, which are crystalline and can be generalized as interlayered gneisses and schists. Bedrock of the southern Ridge and Valley ecoregion comprises mainly sedimentary rocks, primarily sandstone, siltstone, limestone, and chert. Whereas acidic rocks are more prevalent in the Blue Ridge, the southern Ridge and Valley has both acidic and basic rock types (Rick Wooten, North Carolina Geological Survey, personal communication). These differences in geology drive differences in groundwater and surface water chemistry (e.g., pH and calcium, magnesium, and bicarbonate concentrations) and groundwater recharge and discharge (Swain et al. 2004).

Aquatic communities in the basin have been impacted by numerous stressors since European settlement. Large-scale logging in the late 1800s and early 1900s (Little Tennessee Watershed Association 2011) was followed by the construction of hydropower dams. Between 1919 and 1979, hydropower dams were constructed on the Little Tennessee River main stem and several of its larger tributaries, including the Cheoah River, west and east forks of the Tuckasegee River, and the Nantahala River. The Little Tennessee River itself is 232 km in length and is impounded by six dams, the lower five of which result in an almost contiguous string of reservoirs for the lower 145 km of the river. A free-flowing 68-km reach of the Little Tennessee River in North Carolina supports a great diversity and abundance of native aquatic species (NCWRC 2015), containing all or most of the aquatic species believed to be present prior to European settlement (Figure 1). Industry, development, agriculture, and piscicides have reshaped the aquatic communities of the LTRB, as well (Little Tennessee Watershed Association 2011; Kanno et al. 2016).

Current threats to streams in the LTRB have been documented through a number of watershed planning efforts and include development and associated hydrologic and geomorphic impacts (e.g., sedimentation, storm water runoff, and forest conversion), poor land-use practices (e.g., removal of riparian forest, stream channelization), agricultural pollutants (e.g., nutrients, animal waste, and pesticides), acid deposition, barriers to aquatic passage (existing dams and impassable culverts), other dam and impoundment-related stressors (coldwater releases, peaking flow regulation, and habitat alteration), and climate change (Little Tennessee Nonpoint Source Team et al. 2008; North Carolina Ecosystem Enhancement Program 2011; City of Dillard 2015; NCWRC 2015). Historical impacts have caused local extirpations of native species across the basin, and where habitat and water quality have improved, native species reintroductions are possible. Several long-term fish and mussel community restoration projects have been implemented across the basin, and numerous other restoration efforts have begun more recently.

The LTRB is also an important recreational and economic resource. The economic impact of trout fishing alone can amount to more than US$100 million annually in western North Carolina (NCWRC 2009). Fishing and other recreational activities such as canoeing, whitewater boating, swimming, tubing, and wildlife watching are vital to the region's economy and identity.

Finally, existing partnerships among citizens, local governments, nongovernmental organizations, and agencies have created fertile ground for collaborative progress. Designating the LTRB as an NFCA has bolstered the efforts of these partnerships, supporting a coordinated approach to watershed-level conservation.

Conservation Activities in the Little Tennessee River Basin

While the Little Tennessee Native Fish Conservation Partnership is a young organization, considerable work in aquatic community restoration, land protection, riparian and stream restoration, watershed planning, and environmental education has been underway for decades in the LTRB. This section is not a comprehensive review of all conservation efforts in the basin, but it describes significant aquatic species restoration projects that have been accomplished by partners before the Partnership was established, as well as more recent efforts implemented or supported by the Partnership. It also summarizes a number of habitat restoration and education and outreach efforts accomplished by the Partnership.

Fish and mussel reintroduction efforts

Over the past several decades, partners have been involved in long-term efforts to restore and protect multiple fish species across the basin (Figure 1) through species reintroductions, barrier removals, and riparian restoration projects. This work has demonstrated the effectiveness of captive propagation techniques and the cooperation of multiple partner organizations and agencies across state boundaries. The Partnership supports these efforts by providing funding, increased collaboration, and outreach and education efforts to funders and the general public.

Brook Trout.—Brook Trout *Salvelinus fontinalis* restoration has been a long-term focus in the LTRB. The Great Smoky Mountains National Park's Brook Trout restoration program began in the mid-1980s and has restored Brook Trout populations in 49 km of 13 streams (three of which are in the LTRB) in the park by eliminating Rainbow Trout through angling, piscicides, and/or electrofishing and reintroducing Brook Trout *Oncorhynchus mykiss* (Kanno et al. 2016). In the early 1990s, the Tennessee Wildlife Resources Agency (TWRA) began Brook Trout restoration on streams in the Tellico River watershed; Brook Trout populations in 19 km on five streams have been restored, and additional streams in the Citico Creek watershed are being evaluated for their restoration potential (Bart Carter, TWRA, personal communication). In 2018, the Partnership acquired funding through the Tennessee Valley Authority (TVA) to support the upgrades of a Brook Trout hatchery in eastern Tennessee to support further restoration efforts.

Abrams Creek, Tellico River, and Citico Creek.—In 1986, fish community restoration efforts began in the Great Smoky Mountains National Park; this work occurred in the lower 19-km reach of Abrams Creek, which is effectively isolated between Abrams Falls and the impounded Little Tennessee River. Nearly half of native fish species had been extirpated from this reach (Simbeck 1990), including four federally listed fish species that were the focus of reintroductions—Smoky Madtom (endangered), Yellowfin Madtom (threatened), Citico Darter (endangered), and Spotfin Chub *Erimonax monachus* (threatened). Cooperators included Conservation Fisheries Inc., the North Carolina Wildlife Resources Commission (NCWRC), the TWRA, the U.S. Fish and Wildlife Service (USFWS), the U.S. Forest Service (USFS), the National Park Service, and the University of Tennessee, Knoxville. Citico Creek source populations were used for captive propagation of Smoky Madtom, Yellowfin Madtom, and Citico Darter; all three species were raised in captivity from egg clutches, larval fish, and/or adults (via captive breeding) harvested from Citico Creek. Spotfin Chub were propagated from adults collected from the Little Tennessee River in North Carolina; reproduction, recruitment, and dispersal into suitable habitat were observed for populations of Smoky Madtom, Yellowfin Madtom, and Citico Darter in Abrams Creek, and after 2002, reintroduction efforts were discontinued (Shute et al.

2005; Rakes 2011). Spotfin Chub population reintroductions were also discontinued, as they were not successful, perhaps due to habitat limitations.

In 2003, a similar project was begun to reintroduce the Smoky Madtom, Yellowfin Madtom, Citico Darter, and Spotfin Chub into the Tellico River, also separated from both Citico Creek and Abrams Creek by impoundments on the Little Tennessee River (Rakes 2005). Wild reproduction has been observed by all four species since 2004 in Tellico River (Petty et al. 2011; Rakes 2011). In 2018, the Partnership acquired funding to support efforts to determine recovery potential for Tellico River mussel fauna.

Due to the success of the Abrams Creek and Tellico River reintroductions, there is potential for downlisting the endangered Smoky Madtom and delisting the threatened Yellowfin Madtom, the ultimate goal of these and other recovery efforts. Facilitated genetic exchange among the Citico Creek, Abrams Creek, and Tellico River populations is now a priority (Kulp et al. 2015). In 2017, the Partnership acquired funding through the TVA to support monitoring of Smoky Madtom, Yellowfin Madtom, Citico Darter, and Spotfin Chub in Tellico River and Citico Creek to determine reproduction and dispersal success.

The National Park Service is continuing reintroductions of other extirpated fish species in Abrams Creek in order to establish viable, naturally reproducing populations and reestablish host fish communities for native mussels (Malone et al. 2018), which are dependent on fish hosts for larval development. If host fish reintroductions are successful, extirpated mussels may also be reintroduced. In 2017 and 2018, the Partnership acquired funds through the TVA to support captive propagation and genetic source stock screening for reintroduction of the Blotchside Logperch *Percina burtoni*.

In addition, partners secured funding to reintroduce, tag, and monitor the federally endangered tan riffleshell *Epioblasma aureola* (also *E. florentina walkeri*), as well as the at-risk mussel species Cumberland moccasinshell *Medionidus conradicus* and Tennessee clubshell *Pleurobema oviforme* in Citico Creek. The Virginia Department of Game and Inland Fish mussel hatchery in Marion, Virginia provided juvenile tan riffleshell for reintroduction in 2018, and the remaining two species will be translocated as adults from sustainable source populations in other nearby rivers.

Cheoah River.—In North Carolina, a focus of fish and mussel restoration efforts is a 15-km bypass reach of the Cheoah River between the Santeetlah Dam and the Cheoah Reservoir (Figure 1), where flows were limited to seepage and flood events until improved flow regimes were established in 2005 as part of the license issued by the Federal Energy Regulatory Commission (NCWRC 2000, 2017). Restoration targets include both rare and common species, including the Appalachian elktoe *Alasmidonta raveneliana* (U.S. and North Carolina endangered), slippershell mussel *A. viridis* (North Carolina endangered), wavy-rayed lampmussel *Lampsilis fasciola* (North Carolina species of concern], rainbow *Villosa iris* (North Carolina species of concern), Spotfin Chub, Wounded Darter *Etheostoma vulneratum* (North Carolina species of concern), Olive Darter *Percina squamata* (North Carolina species of concern), and Gilt Darter *P. evides*. In 2018, the Partnership acquired funding to develop propagation protocols for Olive Darter for Cheoah River reintroductions.

Both captive propagation and translocation have been used to reintroduce fish to the Cheoah River, depending on species' abundances and availability of individuals in populations used as sources for reintroductions. However, because most mussel source populations are not robust enough to remove individuals for translocation, mussel community restoration has been largely dependent on captive propagation. Support from the Cheoah Fund and other sources were used to construct the NCWRC Conservation Aquaculture Center at the Marion State Fish Hatchery in Marion, North Carolina in 2008. The center has successfully propagated and cultured all four mussel target species, which have been released to the Cheoah River annually since 2012 (NCWRC 2017). The center serves as a propagation facility for other restoration projects in the LTRB, as well.

Oconaluftee River.—The Oconaluftee River drains 487 km², 60% of which is in the Great Smoky Mountains National Park. The downstream half of the Oconaluftee River runs through the Qualla Boundary of the EBCI (Figure 1). Just upstream of its confluence with the Tuckasegee River, the river is dammed by Ela Dam, which serves as an effective barrier to upstream migration from the Tuckasegee River. Many aquatic species have been extirpated from the lower Oconaluftee River, and the EBCI aims to restore species that are presumed extirpated upstream of the dam (Hickman and LaVoie 2018).

The Sicklefin Redhorse (undescribed *Moxostoma* sp., North Carolina threatened) is the subject of a 2015 candidate conservation agreement (CCA) among the USFWS, the NCWRC, Duke Energy, the TVA, the EBCI, and the Georgia Department of Natural Resources (USFWS et al. 2015). The CCA was developed in order to preclude the need to list the species under the Endangered Species Act by conserving, managing, and improving Sicklefin Redhorse populations. The Sicklefin Redhorse is endemic to the Little Tennessee and Hiwassee River basins in North Carolina and Georgia but has apparently been eliminated from roughly 50% of its former range (adapted from Jenkins 1999). Within the LTRB, Sicklefin Redhorse are found in the Little Tennessee River, the Tuckasegee River, and the 1-km reach of the Oconaluftee River downstream of Ela Dam. Objectives of the CCA include the establishment of at least one additional population in the LTRB.

In 2007, the EBCI, the NCWRC, Conservation Fisheries, Inc., and the USFWS embarked on a project to reintroduce the Sicklefin Redhorse to the Oconaluftee River. Juvenile Sicklefin Redhorse, reared in captivity from eggs collected from Little Tennessee River and Tuckasegee River stock, have been released annually into the Oconaluftee River above Ela Dam. Fish movement has been monitored with radiotelemetry, and fish appear to move downstream over the Ela Dam and into the Tuckasegee River but are unable to move back upstream into the Oconaluftee River (Luke Etchison, NCWRC, personal communication). Additional propagation and reintroduction efforts, population monitoring, and studies of movement patterns, habitat use, and water-quality requirements continue.

The EBCI is evaluating the feasibility of reintroducing four mussel species that have likely been extirpated from the Oconaluftee River—the Appalachian elktoe, slippershell mussel, wavy-rayed lampmussel, and rainbow. Water quality and temperature regime appear suitable for these species, and preferred substrate and fish hosts are present upstream of the Ela Reservoir. All four species occur and are even locally abundant in the main-stem of the Tuckasegee River, to which the Oconaluftee is a major tributary (Figure 1). In 2017, the Partnership acquired funds to support a feasibility study to monitor survival and growth of these four native mussels in the Oconaluftee River using caged individuals.

Habitat restoration

Project prioritization.—Although the Partnership recognized that habitat protection and restoration are critical to native fish conservation, no tools existed to identify and prioritize focal areas or specific projects on a basinwide scale when the Partnership was established. To meet this need, the Assessment and Implementation working group gathered and reviewed existing resources and data relevant to the LTRB. In 2016, the Partnership secured funding from the Duke Water Resources Fund to develop an online conservation mapper, which houses ArcGIS data for the LTRB and provides a common geographic information system analytical platform for all partners through ArcGIS Online. The online conservation mapper was completed in 2017 and is one of the first products produced by and for the Partnership as a whole.

In 2017, American Rivers acquired funding to hire new staff to identify and prioritize restoration and protection projects in the LTRB across North Carolina and Tennessee. This work built upon the conservation mapper and existing watershed plans to create a comprehensive inventory of high priority, feasible restoration and protection projects for project managers to implement. The inventory and prioritization also informs the Partnership where to invest in building capacity for implementation in places lacking project managers or resources.

This project inventory and prioritization is particularly useful in regions of the basin that have existing watershed groups. American Rivers previously prioritized the Tuckasegee River for restoration and protection and established the Tuckasgee River Blue Trail. Blue Trails are rivers adopted by communities to improve access and protect their river and surrounding lands (www.bluetrailsguide.org). The Tuckasegee Visioning Group was established to support the Tuckasgee River Blue Trail. Local and county-level stakeholders and resource managers like Soil and Water Conservation District and Natural Resources Conservation Service staff who would not have engaged in the more geographically expansive Partnership comprised the group along with some members of the Partnership. This watershed-specific group is piloting the implementation of projects identified in the inventory and prioritization effort. Ultimately, the comprehensive inventory of restoration and protection projects across the watershed will allow the Partnership to engage with specific local and county-level partners to implement high-priority projects for the benefit of native aquatic fauna.

Barrier removal.—Barrier removal is recognized by the Partnership as a key means to achieve the conservation objective to connect diverse, high-quality habitat. An early, on-the-ground success was the removal of a dam on Citico Creek in Tennessee's Cherokee National Forest (Figure 1) in 2015. The Citico Creek dam was constructed in 1966 as a "rough fish" barrier to isolate coldwater fish species in Citico Creek from warmwater species in a reservoir downstream. In 2014, the dam was prioritized for removal to address aquatic passage, sediment dynamics, and public safety concerns, particularly for boaters. American Rivers, the USFS, and the USFWS partnered to remove the barrier, using each organization's resources and expertise to implement the project efficiently and at a relatively low cost. American Rivers and the USFS collaborated on the project management, design, and facilitated applications for required project permits while the USFWS provided construction services. This early project "win" among partners increased the momentum in the efforts of the Partnership.

To support the expansion of targeted barrier removal in the LTRB, the Assessment and Implementation working group provided input to a dam removal prioritization scheme in 2016. The Southeast Aquatic Resources Partnership, the USFWS, and American Rivers developed an inventory and prioritization framework for all dams in the LTRB to determine removal potential. The Assessment and Implementation Working Group reviewed and weighted prioritization factors (e.g., occurrence of rare species, number of dams downstream, and number of river miles opened upstream or downstream) that best prioritized projects for implementation. This effort directs partners to work on high-priority dam removals that have the greatest benefit to native fauna. Several dam removal projects are in process as a result of this prioritization effort.

In 2017, Mainspring Conservation Trust, Trout Unlimited, and the NCWRC partnered to remove two barriers on Slatten Branch, a native Brook Trout stream in the Tuckasegee River watershed. The project was funded by Trout Unlimited and matching funds the Partnership acquired through the TVA. Recent monitoring has documented Brook Trout movement into the newly opened stream reach upstream of the former barriers, and project partners have developed relationships with other landowners in the watershed who may participate in other restoration and protection projects.

Shade Your Stream.—In 2012, Mainspring Conservation Trust began a riparian buffer restoration and education program called Shade Your Stream in the Macon County, North Carolina portion of the LTRB. In 2015 and 2016, the Partnership acquired funding through the TVA to expand this program throughout the LTRB. The effort has supported a Web site (http://shadeyourstream.org/), live staking workshops and instructional video, educational signs and brochures, and riparian restoration projects.

Education and Outreach

The Education and Outreach Working Group of the Partnership has developed programs and tools to educate the public about aquatic habitat and fauna of the LTRB and ways the public can engage in environmental stewardship. Three of the major efforts of this working group have been the development of a StoryMap, a North Carolina-based snorkeling program, and educational videos.

In 2017, an ArcGIS StoryMap was developed in conjunction with the development of the online conservation mapper. The StoryMap offers the public an attractive, user-friendly way to learn about the basin and work of the Partnership, with tabs that describe the Little Tennessee River basin, its aquatic species, river access, threats to the river, conservation projects, and how one can get involved (Figure 2; https://arcg.is/0XmDSK). The Partnership updates this StoryMap with additional resources as they become available.

Several partners engage the community with guided snorkeling programs. The USFS developed a snorkeling public education program in the Cherokee National Forest, holding annual snorkeling events in Citico Creek. The USFS also provides a snorkeling curriculum and program toolkit available at FishWatch (www.fs.fed.us/naturewatch/fishwatch.php). In 2016, North Carolina partners began a snorkeling education program modeled after the USFS program. The Partnership acquired funds from the North Carolina Chapter of the American Fisheries Society and the North Carolina Wildlife Federation to purchase snorkeling gear. Using this shared equipment, Mainspring Conservation Trust and the EBCI hold annual snorkeling education programs that help people learn how to identify aquatic animals they observe in the wild and provide basic information on species life history and habitat requirements. The EBCI incorporates snorkeling education into the Cherokee Choices program, which mobilizes the community to confront environmental and biological factors that put Cherokee people at higher risk for diabetes. The cornerstones of this program are listening to the community and providing social support to increase physical activity and promote well-being and healthy choices, which can reduce the risk for obesity and diabetes.

To bring the Little Tennessee River experience to classrooms, living rooms, and smartphones, the Partnership worked with the video production company Freshwaters

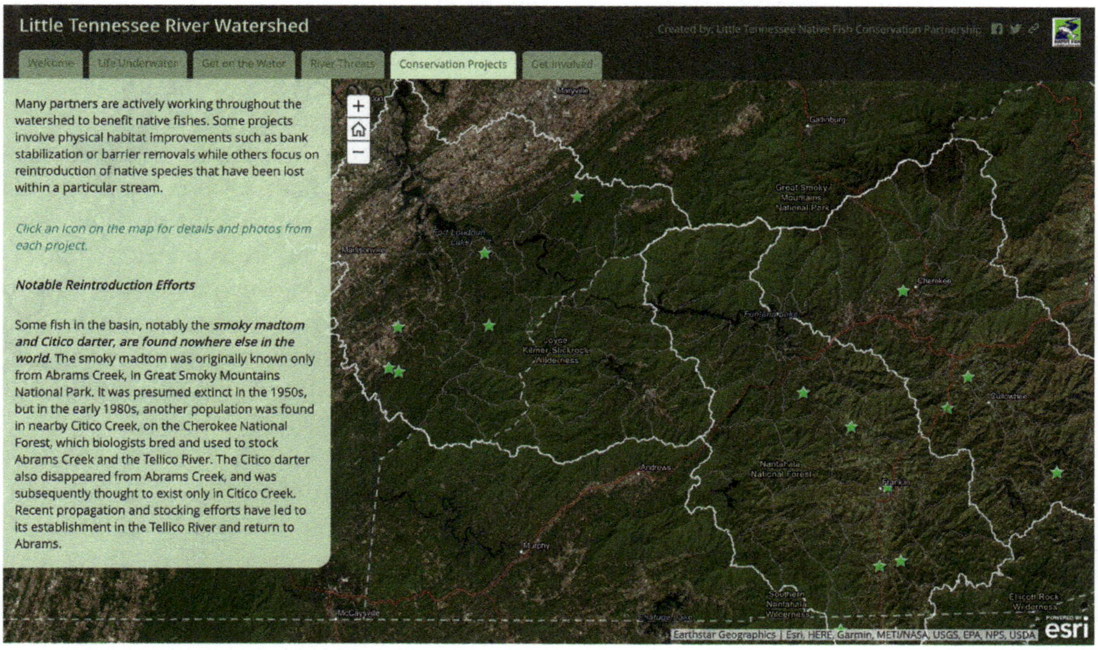

Figure 2. Little Tennessee Native Fish Conservation Partnership StoryMap.

Illustrated to develop a series of high-quality video shorts, 2 to 5 min in length. Films were developed on the following subjects: the Little Tennessee NFCA, nongame fish conservation in the LTRB, Brook Trout recovery on public lands, and riparian stewardship on private lands. The films will be used in classroom settings, showings at local breweries, and other community locations and will be incorporated in the StoryMap and other online communications.

Discussion

Several factors influenced the success of our partnership and perhaps foremost was the wide recognition of the value of aquatic resources inhabiting the Little Tennessee River basin by a variety of institutions (NCWRC 2015; Georgia Department of Natural Resources 2015; Tennessee State Wildlife Action Plan Team 2015). One-hundred-twenty species of fish are present in the basin, and 20 of these species have federal and/or state status as endangered, threatened, of special concern, in need of management, or rare (Little Tennessee Native Fish Conservation Partnership, unpublished data, www.littlet.org/publications/). Several more abundant species are classified as game fish and are highly prized by anglers (e.g., Brook Trout, Smallmouth Bass *Micropterus dolomieu*, and Rock Bass *Ambloplites rupestris*), among others. This mix of prized game and nongame species, common and rare species, results in a broad range of groups and individuals with interest in conserving aquatic resources within the basin.

The use of outside facilitators for the initial meeting was especially helpful in the formation of the Partnership. Representatives of 13 agencies and organizations attended the initial organizational meeting. The facilitators' leadership led to the development of a coherent vision for the Little Tennessee River basin, an agreement to form an NFCA, the establishment of a governing structure, and identification of preliminary working groups. In addition, the facilitators led a discussion of the process of coalition building and the organizational challenges the partnership was likely to face. Overall, facilitators were critical to maintaining the group's focus on desired outcomes and melding their diverse interests into a unified vision for the river basin.

After the formation of a partnership, the diversity of partners' interests became a significant benefit. Sawyer (2007) discussed the creative ability of a group that has diverse perspectives in addressing goals of mutual interest. Due to the river's diverse aquatic resources, a variety of agencies, organizations, and other entities were involved in conservation activities within the basin; many had a history of mutual collaboration and thus were receptive to forming a broader partnership to address common interests. The diverse assemblage of partners enlarged the scope of potential actions, particularly with the emphasis of a nonregulatory approach to achieving conservation objectives. Diverse interests of partners were important in recognizing and addressing potential conflicts likely to arise from a single species management focus, which could adversely affect other species within a biological community.

An atmosphere of trust among the various partners was essential, particularly in the establishment of a vision and goals for the Partnership. The Little Tennessee Native Fish Conservation Partnership was fortunate in that many of our partners have histories of working together, and even from the beginning there were few, if any, issues of mistrust or animosity. As a result, members felt free to speak openly and voice minority opinions without concern of being stigmatized. Trust among partners increased the likelihood that different perspectives on an issue would be discussed and evaluated prior to arriving at a course of action. If mutual trust does not broadly exist within a partnership, a facilitated discussion identifying the obstacles to trust, as well as corrective actions, may be an important first step.

It is difficult to overstate the importance of having nongovernmental partners fully engaged in the partnership. Given that an NFCA is nonregulatory in nature, support of conservation activities by local citizens is critical to success. Members of local, conservation-oriented organizations usually have a knowledge and appreciation of their community's values and culture. Such knowledge can be used to estimate likely community reactions to various conservation objectives and strategies, thus enhancing the efficiency of planning and the efficacy of conservation measures employed.

As the Partnership has grown and proven successful through funding acquisition, technical resource development, and improved coordination, the need for a paid staff person has become apparent. Partners, in addition to their numerous projects not associated with the Partnership, tend to focus on specific partnership activities in which they are directly involved. It became difficult for officers or other partners to prioritize the overall activities of the Partnership and maximize the synergistic potential of the group. One of the partners, American Rivers, received a grant to employ a part-time, term-limited support position to assist in project identification. This position is crucial to helping the Partnership identify and prioritize conservation activities within the basin, but the need for a dedicated general coordinator for the Partnership remains.

Finally, the Partnership demonstrated how collaboration among diverse partners adds value to the individual actions that might have otherwise have been taken. Partners agreed that increasing the level of knowledge and awareness of the basin's aquatic resources among residents was an important objective for the Partnership. The diversity of membership led to a richer set of educational approaches than any one entity was likely to have considered. For example, it is unlikely that the online mapping platform would have been developed by any individual entity. Only through discussions within the Partnership did the need for such a tool become apparent. As previously discussed, the online conservation mapper is used by multiple partners for multiple purposes. Identifying ways to add value to individual efforts is critical to maintenance of interest among partners in the Partnership and commitment to the partnership approach to conservation.

In the future, the Partnership needs to increase participation by organizations in Georgia and Tennessee in order to implement priority activities. Although the Little Tennessee River basin and the Partnership encompasses portions of all three states, to date, the Little Tennessee Native Fish Conservation Partnership activities have been relatively North Carolina-centric. In large part this imbalance is due to much of the free-flowing portion of the river being situated in North Carolina, as are the headquarters of most partners. Still, serious conservation needs exist within the borders of the other states. The Partnership actively solicits additional partners from Tennessee and Georgia and often holds Partnership meetings near or within the borders of those states to encourage more participation. The current effort to inventory and prioritize projects across the basin will hopefully inspire more engagement from partners in the less-represented areas of the basin.

Beyond strengthening geographic scope, success will be determined in large part by attitudes, opinions, and values of the basin's residents. More than half of the land area of the Little Tennessee River basin comprises private properties, and resource conservation is largely a function of the activities conducted on those properties. The videos, online conservation mapper, and Shade Your Stream and other educational programs (e.g., snorkeling) are geared toward changing attitudes and land-use practices in the basin. Such changes, if they occur, are likely to be gradual over an undefined period of time. Building on current educational programs will require additional capacity in terms of financial and human resources.

At this early stage in the Partnership, most partners represent agencies and organizations with paid staff. As the partnership matures and educational efforts pay off, developing an active "Friends of the Little Tennessee" group, made up of basin citizens and businesses, would be an effective way of expanding the partnership and its conservation efforts. After the initial formation of such a group, growth in membership could be an indicator of the efficacy of our education and outreach efforts. Members could be surveyed to identify factors that influenced them to become "Friends," and survey results used as an indicator of the efficacy of individual activities.

Summary

The NFCA concept is a promising approach to conserving the aquatic resources of the Little Tennessee River. This landscape approach likely has utility for other watersheds where emphasis is placed on conserving biological communities. Criteria that are important in selecting target watersheds are outlined above and described by Williams et al. (2011) and include charismatic fish fauna, dedicated partners, and tangible conservation outcomes.

The first selection criterion to determine the suitability of a native fish conservation partnership is whether aquatic biota occur, or potentially occur, that are deemed ecologically, economically, or culturally worth conserving by a segment of society. The Little Tennessee River system supports numerous native species of fishes, mollusks and crayfishes, many of which are endemic, endangered, threatened, or otherwise in need of management attention.

Next, commitment and participation by diverse partners is critical for a successful partnership. Government and tribal agencies are essential in that they have the resources, knowledge, and authority to conduct needed biological investigations and management programs. In our model, the participation of nongovernmental entities such as conservation organizations, academia, utility providers, and businesses is equally essential for the different perspectives and capacity they provide on local issues and conservation planning and implementation. The feature that unifies our partners is the shared mission and vision for conserving the basin's native aquatic resources.

Finally, highly visible projects and outreach efforts are critical to engage the local community in resource conservation. Projects such as fish introductions, habitat improvements, and other actions can often involve citizens with little or no technical expertise in project implementation. Other citizen engagement activities of the Partnership include youth and adult instream experiences, such as wading and snorkeling to observe aquatic fauna in their habitats; showing and producing high-quality videos that illustrate many of the basin's resources; and Web-based materials that allow citizens to better understand and appreciate waters in their communities.

Obviously, there is no single approach to conserving aquatic resources within a river system. The NFCA concept is only one such approach. We hope our experiences described here will stimulate additional thoughts and influence conservation practices in other systems.

Acknowledgments

The authors thank Brian Alford, Bart Carter, Luke Etchison, Andrew Henderson, Jack Henderson, Matt Kulp, Mike LaVoie, Jason Meador, Missy Petty, Pat Rakes, JR Shute, and Rick Wooten for providing information for this chapter. Brian Alford, Laura Craig, Luke Etchison, Jason Meador, Gary Peeples, and Peggy Shute offered thoughtful review. We are thankful for the passion and commitment of the members of the Little Tennessee Native Fish Conservation Partnership.

References

City of Dillard. 2015. Upper Little Tennessee Watershed Management Plan. City of Dillard, Dillard, Georgia.

Dauwalter, D. C., J. S. Sanderson, J. E. Williams, and J. R. Sedell. 2011. Identification and implementation of native fish conservation areas in the upper Colorado River basin. Fisheries 36:278–288.

Georgia Department of Natural Resources. 2015. Georgia State Wildlife Action Plan. Georgia Department of Natural Resources, Social Circle.

Hickman, C. R., and M. J. LaVoie. 2018. Eastern Band Cherokee Indians Wildlife Action Plan: a social-ecological framework to adaptively manage and conserve fish and wildlife. Eastern Band of the Cherokee Indians, North Carolina.

Isom, B. G. 1971. Mussel fauna found in Fort Loudon Reservoir, Tennessee River, Knox County, Tennessee, in December, 1970. Malacological Review 4:127–130.

Jenkins, R. E. 1999. Sicklefin Redhorse (*Moxostoma* sp.), undescribed species of sucker (Pisces, Catostomidae) in the upper Tennessee River drainage, North Carolina and Georgia description, aspects of biology, habitat, distribution, and population status. Unpublished report to the U.S. Fish and Wildlife Service, Asheville Field Office, Asheville, North Carolina and the North Carolina Wildlife Resources Commission, Raleigh.

Kanno, Y., M. A. Kulp, and S. E. Moore. 2016. Recovery of native Brook Trout populations following the eradication of nonnative Rainbow Trout in southern Appalachian mountains streams. North American Journal of Fisheries Management 36:1325–1335.

Kulp, M. A., S. E. Moore, M. Cantrell, S. Chance and G. Moyer. 2015. Threatened and endangered fish translocation to and from Abrams Creek, Citico Creek and Tellico River, Great Smoky Mountains National

Park and Cherokee National Forest: implementation and genetic monitoring plan. National Park Service, Natural Resource Report NPS/GRSM/NRR—2015/930, Fort Collins, Colorado.

Little Tennessee Nonpoint Source Team, World Wildlife Fund, and The Nature Conservancy. 2008. Conservation action plan for the upper Little Tennessee River basin. Tennessee Nonpoint Source Team. Available: www.littlet.org/wpcontent/uploads/2016/05/ULTN-CAP-Final-Report_Oct_2008.pdf. (March 2019).

Little Tennessee Watershed Association. 2011. The state of the streams in the upper Little Tennessee watershed, anniversary edition. Little Tennessee Watershed Association, Franklin, North Carolina.

Malone E. W., J. S. Perkin, B. M. Leckie, M. A. Kulp, C. R. Hurt and D. M. Walker. 2018. Which species, how many, and from where: integrating habitat suitability, population genomics, and abundance estimates into species reintroduction planning. Global Change Biology 24:3729–3748.

NCWRC (North Carolina Wildlife Resources Commission). 2000. Fisheries and Wildlife Management Plan for the Cheoah River basin and the lower Little Tennessee River. NCWRC, Raleigh.

NCWRC (North Carolina Wildlife Resources Commission). 2009. The economic impact of mountain trout fishing in North Carolina. NCWRC, Raleigh.

NCWRC (North Carolina Wildlife Resources Commission). 2015. North Carolina Wildlife Action Plan. NCWRC, Raleigh.

NCWRC (North Carolina Wildlife Resources Commission). 2017. 2017 Cheoah Fund Report. Captive culture and reintroduction of priority fishes and mussels to restore Cheoah River populations. NCWRC, Raleigh.

North Carolina Ecosystem Enhancement Program. 2011. Franklin to Fontana watershed management plan, phase III. Upper Little Tennessee River basin, Swain and Macon counties, NC. North Carolina Department of Environment and Natural Resources, Raleigh.

Parmalee, P. W., and M. H. Hughes. 1993. Freshwater mussels (Mollusca: Pelecypoda: Unionidae) of Tellico Lake: twelve years after impoundment of the Little Tennessee River. Annals of Carnegie Museum 62:81–93.

Petty, M. A., P. L. Rakes, J. R. Shute, and C. L. Ruble. 2011. Captive propagation and population monitoring of rare southeastern fishes: 2010. Final report and Second quarter report for fiscal year 2010, prepared for Tennessee Wildlife Resources Agency, Nashville; the U.S. Fish and Wildlife Service, Washington, D.C; and Cherokee National Forest, Nashville, Tennessee.

Rakes, P. L. 2005. Rare fish reintroductions to the Tellico River: 2004. Progress report to U.S. Fish and Wildlife Service, Asheville, North Carolina.

Rakes, P. L. 2011. Rare fish reintroductions and population monitoring in the Little Tennessee River system: 2010 annual fish translocation report. Prepared for Alcoa Power Generating Inc., Pittsburgh, Pennsylvania.

Sawyer, K. 2007. Group Genius: the creative power of collaboration. Basic Books, New York.

Shute, J. R., P. L. Rakes, and P. W. Shute. 2005. Reintroduction of four imperiled fishes in Abrams Creek, Tennessee. Southeastern Naturalist 4:93–110.

Simbeck, D. J. 1990. Distribution of the fishes of the Great Smoky Mountains National Park. University of Tennessee, Knoxville.

Starnes, L. B., and A. E. Bogan. 1988. The mussels (Mollusca: Bivalvia: Unionidae) of Tennessee. American Malacological Bulletin 6:19–37.

Swain, L. A., T. O. Mesko, and E. F. Hollyday. 2004. Summary of the hydrogeology of the Valley and Ridge, Blue Ridge, and Piedmont physiographic provinces in the eastern United States. U.S. Geological Survey Professional Paper 1422-A.

USFWS (U.S. Fish and Wildlife Service), North Carolina Wildlife Resources Commission, Duke Energy Carolinas, Tennessee Valley Authority, Eastern Band of Cherokee Indians, and Georgia Department of Natural Resources. 2015. Candidate conservation agreement for the Sicklefin Redhorse (*Moxostoma* sp.). USFWS, Asheville, North Carolina.

Tennessee State Wildlife Action Plan Team. 2015. Tennessee State Wildlife Action Plan 2015. Tennessee Wildlife Resources Agency, Nashville.

Williams, J. E., J. E. Johnson, D. A. Hendrickson, S. Contreras-Balderas, J. D. Williams, M. Navarro-Mendoza, D. E. McAllister, and J. E. Deacon. 1989. Fishes of North America endangered, threatened, or of special concern. Fisheries 14:2–20.

Williams, J. E. , R. N. Williams, R. F. Thurow, L. Elwell, D. P. Philipp, F. A. Harris, J. L. Kershner, P. J. Martinez, D. Miller, G. H. Reeves, C. A. Frissell, and J. R. Sedell. 2011. Native fish conservation areas: a vision for large-scale conservation of native fish communities. Fisheries 36:267–277.

World Wildlife Fund. 2016. Living planet report 2016. Risk and resilience in a new era. WWF International, Gland, Switzerland.

American Fisheries Society Symposium 91:431–454, 2019

An Approach for Implementing Large-Scale Watershed Restoration Efforts: Abandoned Mine Drainage Restoration in the West Branch Susquehanna River Watershed, Pennsylvania

SHAWN M. RUMMEL*

Trout Unlimited, Pennsylvania Coldwater Habitat Restoration Program
18 East Main Street, Suite 3, Lock Haven, Pennsylvania 17745, USA

AMY G. WOLFE

Trout Unlimited, Northeast Habitat Program
18 East Main Street, Suite 3, Lock Haven, Pennsylvania 17745, USA

Abstract.—Historic coal mining operations throughout the United States left behind a legacy of waterways polluted by abandoned mine drainage (AMD). Streams polluted by AMD typically have low pH and high concentrations of heavy metals such as aluminum, iron, and manganese, creating toxic environments for aquatic organisms and causing many of these environments to be devoid of aquatic life. This chapter details the large-scale restoration efforts underway since 2004 in the West Branch Susquehanna River watershed in north-central Pennsylvania to remediate AMD. It provides an overview of the historical coal mining throughout the watershed and the types of restoration projects that have been completed. The importance of establishing and maintaining partnerships for large-scale restoration projects is emphasized along with a discussion on how these projects have been monitored and maintained to ensure restoration success. This chapter aims to provide guidance in the development of other large-scale restoration projects through the lessons learned over the past 15 years of restoration in the West Branch Susquehanna River watershed.

Introduction

The West Branch Susquehanna River watershed drains approximately 18,130 km² of mostly forested land in central and north-central Pennsylvania. Although the watershed contains some of the last strongholds of native Brook Trout *Salvelinus fontinalis* within Pennsylvania, coal mining operations that occurred prior to 1977 federal regulations have left approximately 1,900 km of streams within the watershed impaired by abandoned mine drainage (AMD; PA DEP 2018). The most common form of abandoned mine drainage is acid mine drainage, although some discharges from mines may result in alkaline drainage. Since 2004, Trout Unlimited (TU), an organization committed to the conservation, protection, and restoration of North America's coldwater fisheries, has led an effort to restore the watershed in partnership with multiple government agencies, volunteer organizations, municipalities, conservation districts, private industry, and others.

* Corresponding author: shawn.rummel@tu.org

This chapter will focus on the restoration efforts within the West Branch Susquehanna River watershed, with an emphasis on how partnerships were established and maintained, the types of restoration projects that have been completed, and how the success of these projects has been monitored and maintained. The chapter will conclude with a summary of the status of the watershed based on monitoring results, the future direction of the project, the lessons learned over the course of this effort, and implications for other large-scale restoration projects.

Overview of the West Branch Susquehanna River Watershed

The West Branch Susquehanna River is located in central and north-central Pennsylvania (Figure 1), and its watershed drains an area of approximately 18,130 km^2 (7,000 mi^2), with 19,457 km of tributary streams and rivers. It is a major tributary to the Susquehanna River and drains 25.5% of the total Susquehanna River watershed. The watershed is dominated by forested land cover (approximately 83% forested) and other land uses include agriculture (approximately 10%); developed and disturbed lands

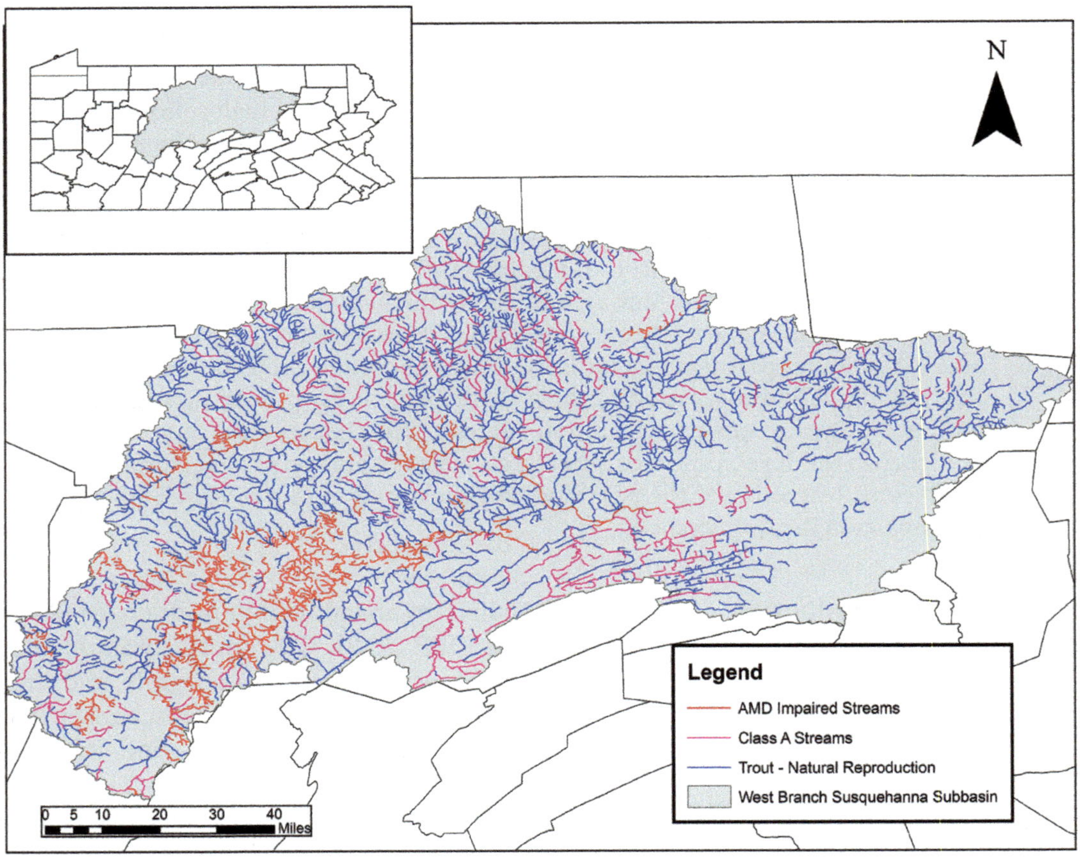

Figure 1. Location of the West Branch Susquehanna River watershed in Pennsylvania. Streams supporting the natural reproduction of trout, Class A-designated trout streams, and abandoned mine drainage-impaired streams are also depicted. Streams not meeting the above criteria are not shown for clarity. Streams are underlain by public property owned by the commonwealth.

(including abandoned mine lands) account for the remaining 7% of the land cover. Approximately 50% of the land within the watershed is publicly accessible through state forest, state park, and state game lands, providing an abundance of outdoor recreation opportunities. The watershed also contains some of the commonwealth's most pristine waterways that are among the best strongholds of Brook Trout in Pennsylvania and the Mid-Atlantic region (Fesenmyer et al. 2017; Rummel et al. 2017). Brook Trout are the only salmonid native to the watershed and have been in decline throughout their native range in the eastern United States (Fesenmyer et al. 2017). The watershed contains 8,160 km of streams that support the natural reproduction of trout, including 1,754 km of Class A trout streams, the highest designation in Pennsylvania (Figure 1). Despite the pristine nature of a large portion of the watershed, the legacy of past mining operations (primarily coal through both underground and surface mines) has left a legacy of more than 1,900 km of waterways impaired by AMD.

Mining History and Abandoned Mine Drainage Legacy in the West Branch Susquehanna River: Status of the Watershed Prior to Restoration Work

According to the Pennsylvania Department of Environmental Protection (PA DEP; PA DEP 2016), the two largest problems affecting the commonwealth's waters are agriculture and abandoned mine drainage. Abandoned mine drainage is the consequence of the historical unregulated coal mining that occurred prior to the establishment of the federal Surface Mining Control and Reclamation Act of 1977. The Surface Mining Control and Reclamation Act established the Office of Surface Mining Reclamation and Enforcement, along with regulations on active coal mines and contained provisions for reclaiming abandoned mine lands and preventing AMD pollution. Current mining operations are now responsible for downstream water quality and land reclamation associated with the mine.

Acid mine drainage is formed when pyrite, a naturally occurring mineral often associated with coal seams, reacts with oxygen and water to produce iron hydroxide and sulfuric acid. The acidic water associated with most mine drainage may also leach metals such as aluminum and manganese from the surrounding bedrock into the water, creating a nonpoint source of pollution typically characterized by low pH, high acidity levels, and elevated concentrations of dissolved aluminum, iron, and/or manganese. These toxic metals can negatively influence the growth rate, development, behavior, and metabolic processes of fishes and other aquatic life (Earle and Callaghan 1998; DeNicola and Stapleton 2016). Abandoned mine drainage pollution can also cause a reduction in the abundance and diversity of benthic macroinvertebrate populations. The metal precipitates can also armor the stream substrate, thereby reducing habitat and diminishing the food supply for aquatic organisms (DeNicola and Stapleton 2016). All but the most pollution tolerant fish and macroinvertebrate species are usually eliminated from AMD-impaired streams.

Coal mining in the West Branch Susquehanna River watershed began in the late 18th century (Sisler 1926). The demand for coal in Pennsylvania peaked during World War I and World War II (PA DEP 2006). During this time, approximately 193 million short tons (175 million metric tons) of coal were being removed from Pennsylvania per year and approximately 200,000 residents were employed in coal mining operations (PA DEP 2006).

Early studies within the watershed document the severity of AMD pollution. A survey of the river in 1931–1932 by the Pennsylvania Fish and Boat Commission found that AMD pollution in the river and several tributaries was severe enough to prohibit the presence of a fishery in Clearfield and Clinton counties (Sorenson 1931, 1932). Further downstream on the river, after the confluence of Bald Eagle Creek (an alkaline tributary), water-quality conditions improved and allowed fish to occupy this section of the river (Sorenson 1931, 1932). Near the city of Williamsport, the pH ranged from 6.2 to 6.8 under normal conditions. However, large fish kills were reported in this area in October 1957, July 1958, September 1960, and October 1962 (Roller 1958; R. Drupieski, 1960 letter to Marlin Wilt, Pennsylvania Department of Health, about fish kill; C. R. Glover, 1957 letter to Raymond Schroll, Jr., Pennsylvania Department of Health, about a fish kill on the West Branch Susquehanna River; M. Wilt, 1958 letter to Walter V. Kohler Pennsylvania Department of Health, about fish killing: West Branch Susquehanna River Williamsport to Jersey Shore; F. Hempt, 1962 letter to Kohler about fish kill) as a result of acid slugs resulting from precipitation events in the western part of the watershed that reduced the pH at Williamsport to 4.5 and elevated concentrations of sulfate and iron (Wilt 1958).

A mine drainage study of the entire Susquehanna River basin from 1964 to 1967 concluded that the headwaters of the West Branch Susquehanna River were acidic until the confluence with Chest Creek in Clearfield County (Federal Water Quality Administration 1968). In 1984, the U.S. Geological Survey (USGS) documented a pH ranging from 3.9 and 4.1 and specific conductance ranging from 330 to 610 µS/cm in the river downstream of Moshannon Creek (Hainly and Barker 1993). The USGS study also determined that Moshannon, Sinnemahoning, Clearfield, and Kettle creeks were the largest tributary sources of acidity and total-recoverable iron to the river (Hainly and Barker 1993).

The historically degraded water quality of the West Branch Susquehanna River had a substantial effect on the fishery in the river. For many years, much of the river was thought to be devoid of any biological life. The 1972 Scarlift report for the West Branch Susquehanna River concluded that the water quality in the river was so degraded that the costs for complete abatement of the pollution could not be economically justified (Pennsylvania Department of Environmental Resources 1972). The report states that "conditions in the study area are such that no more than 30 mi of stream between Barnesboro and Bower could possibly be restored to fishing and recreational use under the most ideal abatement treatment costs for which could easily range from US$20 million to $30 million. This is completely unrealistic in terms of the Federal Water Pollution Control Act benefit values for this reach" (Pennsylvania Department of Environmental Resources 1972). In fact, the Scarlift report identifies that the essential abatement benefit was not for the recovery of the biota in the river itself, but for the protection of the recreational waters of the Curwensville reservoir. In addition, the report indicates that since

> recreation, esthetics, and fishing are the main benefits desired in the West Branch headwater area, these benefits might be made available within the watershed at a justifiable cost by converting one of the tributaries into an improved fishing stream.

A total of 1,893 km of streams within the watershed were listed as impaired by AMD by the PA DEP (2018). This is approximately 66% of the total AMD-impaired mileage in the entire Susquehanna River basin. A 2008 analysis, completed by the Susquehanna River Basin Commission, of historical monitoring stations throughout the watershed described 1,964 AMD discharge monitoring points in the watershed, and a total of 17,022 ha of unreclaimed abandoned mine land features (nearly 23% of the entire commonwealth's share) occur within the watershed (SRBC 2008). Water-quality impairments have been documented as the major limiting factor to biological communities in the watershed, and 90% of the watershed has been characterized as having either excellent or supporting habitat (LeFevre 2003), demonstrating the restoration potential of the watershed if water quality could be improved to support aquatic life.

The Economics of Restoration

The capital costs for remediation projects throughout the West Branch Susquehanna River watershed have more recently been estimated to be at least $110 million; however, the actual amount would vary depending on the type and scale of projects. In addition, another $16 million in annual operation and maintenance costs could be required to maintain the function of treatment systems. These types of costs, like other large-scale restoration projects, seem overwhelming and perhaps prohibitive to restoring watersheds at this scale. However, the economic benefit of restoration projects should also be evaluated to provide policymakers and potential funders guidance on potential returns on investment.

A study by Downstream Strategies, LLC (2008) evaluated the economic benefits for AMD remediation in the West Branch Susquehanna River watershed. This study demonstrated a loss in property values near AMD impaired streams, when compared to unimpaired streams, and the creation of up to 4,120 jobs through restoration work, which would lead to additional tax revenue in the state. In addition, the clean-up of AMD also leads to an increase in drinking water supply options. The study cited that government agencies had already spent $11 million in the watershed to correct problems caused by AMD for drinking water supplies, and cleaner streams would provide more plentiful and cheaper options for public and private drinking water supplies.

The Downstream Strategies, LCC (2008) study concluded that remediating AMD in the West Branch Susquehanna River watershed could generate local economic benefits of up to $616 million for capital expenditures on labor, materials, construction equipment, and so forth and an additional $23 million per year for operation and maintenance of treatment systems. Remediation of AMD in the West Branch Susquehanna River watershed could also provide an additional $22.3 million in sportfishing revenues per year (Downstream Strategies, LCC 2008). The economic benefits to the watershed make a large-scale restoration effort seem much more feasible. However, a watershed scale restoration would not be feasible by a single entity, but would require strong, collaborative partnerships working towards the same goal.

Establishing Partnerships for Restoration

In 1998, TU embraced the significance of AMD problems in the Kettle Creek watershed in Clinton County (one of the major tributaries to the West Branch Susquehanna River) as a component of its nationally renowned Home Rivers Initiative. Since then, TU has been the lead catalyst, working in close partnership with the local Kettle Creek Watershed Association to address severe AMD problems that plague the lower Kettle Creek watershed. Trout Unlimited and its partners have conducted numerous assessments, developed restoration plans, and completed the construction of multiple AMD reclamation and remediation projects.

Trout Unlimited established the West Branch Susquehanna Restoration Initiative in 2004, with the overarching goals of restoring coldwater streams in the watershed and the ultimate recovery of the West Branch Susquehanna River. Through this initiative, TU began working with dozens of organizations and agencies, including multiple government agencies at the state and federal level, volunteer organizations, municipalities, conservation districts, private industry, and others, on a coordinated, strategic, and cost-effective AMD cleanup approach for the entire river basin.

With overwhelming support from the commonwealth, the PA DEP created the West Branch Susquehanna River Task Force in 2004 to bring together all government agency stakeholders and statewide and regional conservation organizations in support of restoring the West Branch Susquehanna River basin. In 2005, TU hosted the first West Branch Susquehanna Restoration Symposium to publicly launch the AMD cleanup initiative. Stemming from facilitated roundtable discussions at the symposium, the West Branch Susquehanna Restoration Coalition was established to represent conservation districts, watershed associations, TU chapters, and other community stakeholders interested in collaborating to restore the watershed. Today, that coalition, whose quarterly meetings continued to be facilitated by TU, represents more than 60 volunteer-led watershed associations, conservation districts, conservation organizations, businesses, and other entities that support the restoration of the West Branch Susquehanna River watershed.

Between 2005 and 2016, TU hosted eight West Branch Susquehanna Restoration Symposium events. On average, more than 100 people participated in each event representing government agencies, watershed and sportsmen associations, conservation organizations, academic institutions, private industry, and local elected officials. The symposium serves as a forum for the exchange of ideas regarding AMD abatement in the region and provides an excellent opportunity for networking among volunteers, policymakers, technical experts, students, and others interested in restoring land and water impacted by AMD. The symposium also offers tours to local AMD project sites and/or workshops to help build the capacity of the groups working to restore AMD polluted streams in the watershed.

Abandoned Mine Drainage Restoration Projects in the West Branch Susquehanna River Watershed

Overview of project types completed to date

Each discharge of AMD can vary in its chemical composition, and therefore, there is no single prescription for remediation. Each site must be evaluated individually, and variables such as water chemistry, flow regimes, landscape variables such as slope and the amount of land available, land ownership, and access to the project site all dictate the type and design of treatment systems. Typical AMD restoration projects may be divided into three main categories; passive treatment, active treatment, and surface reclamation of abandoned mine lands. The underlying water-quality goal of each of these categories is to elevate pH to a level such that the dissolved metals precipitate out of solution (generally a minimum of 4.5 to reduce dissolved aluminum and iron), leading to a discharge of water with a near neutral pH, low dissolved metal concentrations, decreased acidity, and increased alkalinity.

Passive treatment abates AMD via naturally occurring chemical and biological processes that require minimal operation and maintenance and is the most common type of treatment employed in Pennsylvania. Passive treatment generally uses a combination of wetlands, limestone-filled channels or ponds, and ponds containing both limestone and organic material to treat AMD polluted water (Figure 2). The design of these systems varies from site to site as previously discussed and routine monitoring (quarterly or at a minimum annually) is required to ensure that the system is effectively treating the raw AMD discharge water and to determine when maintenance of the system may be required.

Active treatment is generally used for

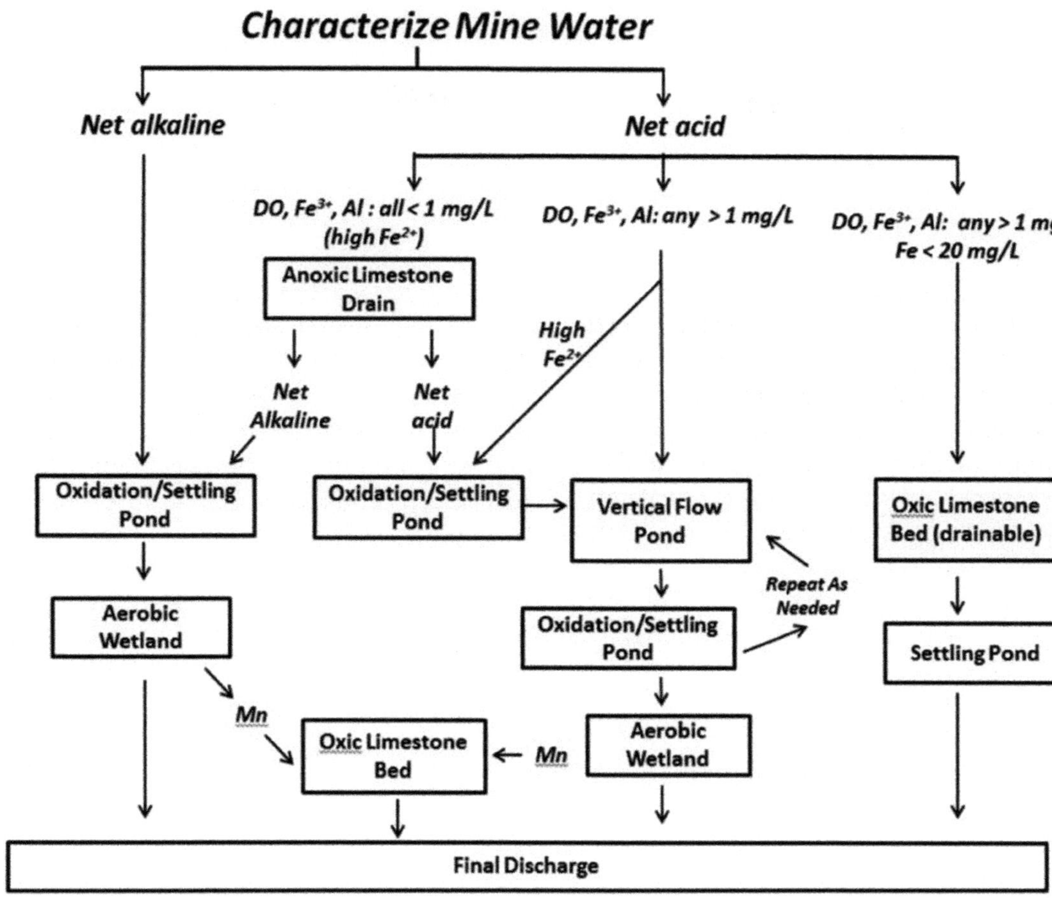

Figure 2. Schematic of decision tree for the design of passive treatment systems (adapted from R. Hedin, T. Weaver, N. Wolfe, and G. Watzlaf, Hedin Environmental, paper presented at the 2013 Annual National Association of Abandoned Mine Land Programs Conference).

AMD pollution with high flows and limited space available for a treatment system. Typically, active treatment refers to the continuous application of alkaline material to increase the pH of the AMD polluted water, accelerate the rate of chemical oxidation of ferrous iron, and precipitate dissolved metals (Johnson and Hallberg 2005) (Figure 3). The materials used to accomplish this chemical reaction may vary, but limestone, hydrated lime, pebble quicklime, soda ash caustic soda, or ammonia are commonly used. Active treatment requires constant maintenance and has higher operating costs than passive treatment but can be effective in certain situations.

Reclaiming abandoned mine lands increases precipitation infiltration rates on sites and reduces runoff and sedimentation to stream channels on these landscapes by enhancing vegetative cover (Figure 4). Alkaline material is often added to the soil during the reclamation process, promoting vegetation growth and enhancing the buffering capacity of the soils. Land reclamation projects may also improve wildlife populations by providing increased habitat availability and forage opportunities as discussed later in the chapter.

Between 2004 and 2005, more than $113 million was spent through various state and federal grant programs and other funding

Figure 3. Lime dossers, a form of active treatment, installed in the Dents Run watershed.

sources to address AMD pollution in the watershed. To date, more than 80 restoration projects, a combination of the broad treatment categories previously described, have been implemented throughout the West Branch Susquehanna River watershed (Figure 5), primarily on coldwater tributaries to the river.

Monitoring Progress toward Recovery at Multiple Scales

Monitoring the effectiveness is a critical component to any restoration project. It is valuable for adaptive management of projects and often required to meet the deliverables of a project for funders. For most large-scale efforts with multiple individual restoration projects being implemented, monitoring may occur at several scales: the individual project, stream reach, and watershed scale. As previously mentioned, for AMD restoration, monitoring individual treatment projects is vital to ensuring their continued effectiveness. However, to document the impacts on water quality and biological communities downstream of treatment and to understand the cumulative effects that these projects are having on the larger ecosystem, a broader monitoring plan is needed.

Small-sale abandoned mine drainage remediation case study: Twomile Run watershed

Although not all AMD treatment systems in the West Branch Susquehanna River watershed and their associated watersheds have been monitored in detail with respect to

Figure 4. **(A)** Twomile Run surface reclamation site prior to restoration; **(B)** Twomile Run surface reclamation area following restoration.

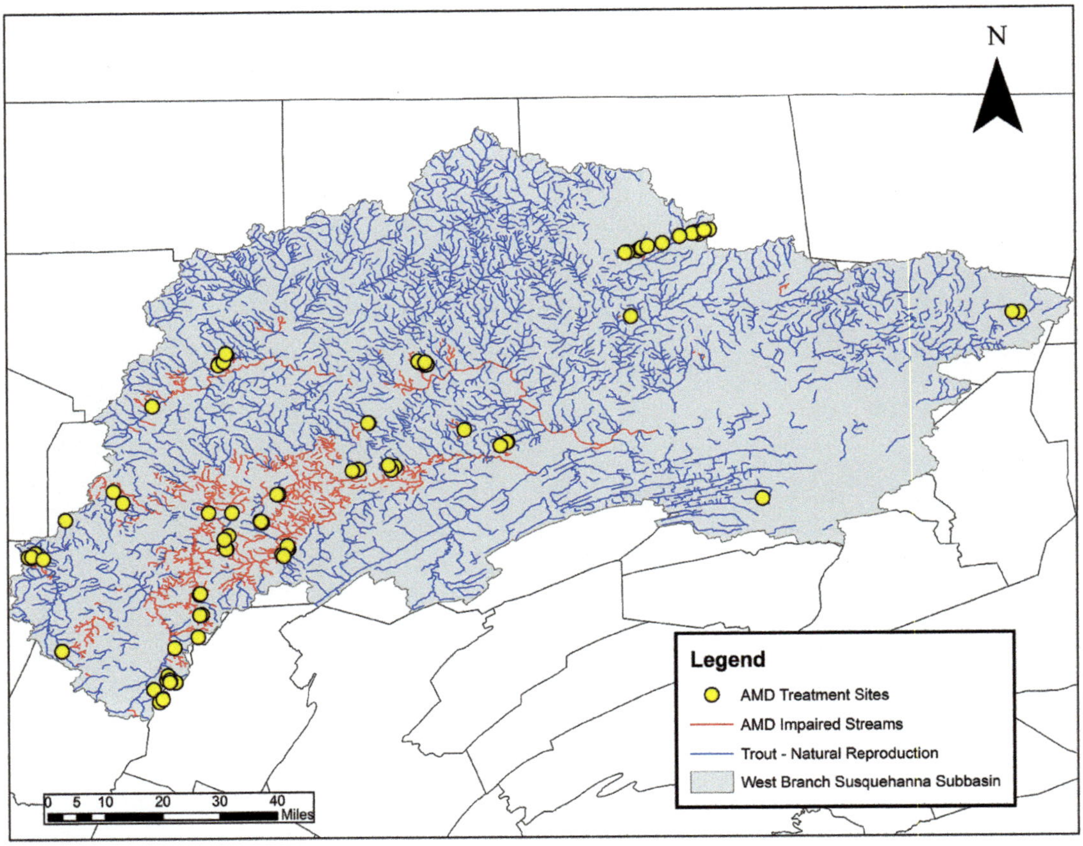

Figure 5. Location of abandoned mine drainage treatment within and surrounding the West Branch Susquehanna River watershed. Abandoned mine drainage-impaired streams are shown in red and streams that support the natural reproduction of trout are shown in blue.

biological communities or instream water quality, the AMD restoration work in the Twomile Run watershed (Figure 6) provides a case study of small scale restoration efforts within a coldwater system. Twomile Run flows for 7.28 km and is a tributary to Kettle Creek (Figure 7), which drains to the West Branch Susquehanna River. It drains an area of approximately 23 km^2 and AMD produced by historical mining polluted 3.7 km of the main stem and an additional 4.8 km of tributaries. Twomile Run has three main tributaries (Huling Branch, Middle Branch, and Robbins Hollow). Huling Branch, Middle Branch, and the main stem of Twomile Run each contain substantial Brook Trout populations upstream of their AMD impairments, which serve as source populations of Brook Trout following remediation of the AMD-impaired reaches (Figure 6A).

The planning phase of the project included an aerial infrared survey to determine the location of AMD sources in the watershed, detailed water-quality sampling, and the development of a restoration plan for the area. In 2000, a passive treatment system was installed along Middle Branch to treat two highly acidic discharges with high aluminum and iron concentrations. Trout Unlimited, in partnership with the Kettle Creek Watershed Association, established a long-term water-quality monitoring protocol to evaluate the system's efficiency. Within 1 year of sampling, it was determined

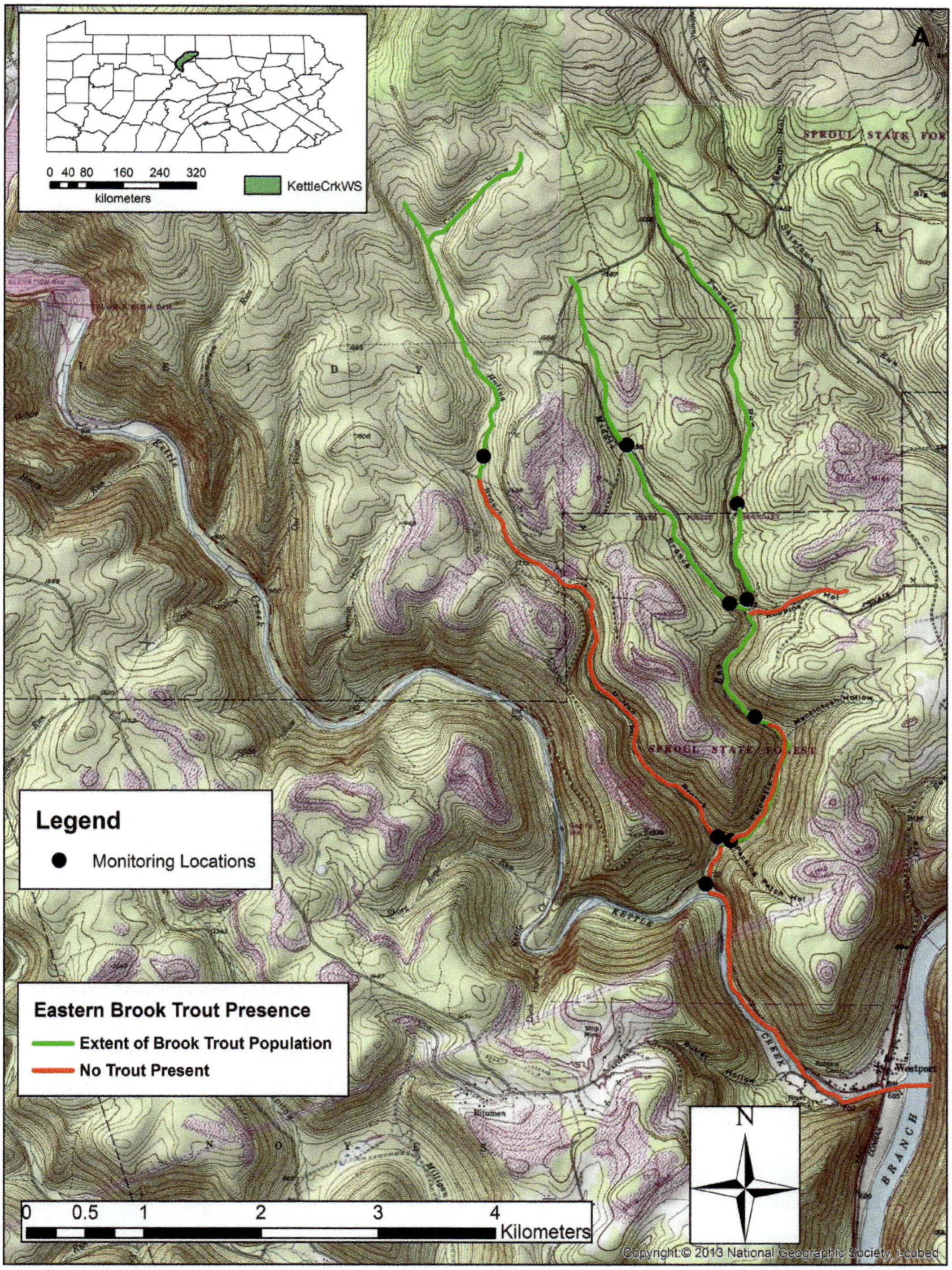

Figure 6. **(A)** Detail of Brook Trout distribution (shown as green lines) in the Twomile Run watershed prior to restoration projects. **(B)** Current distribution of Brook Trout populations in the Twomile Run watershed (shown as green lines). Project sites are depicted as yellow triangles and black dots represent monitoring locations for water-quality sampling, benthic macroinvertebrate-community surveys, and fish-population surveys.

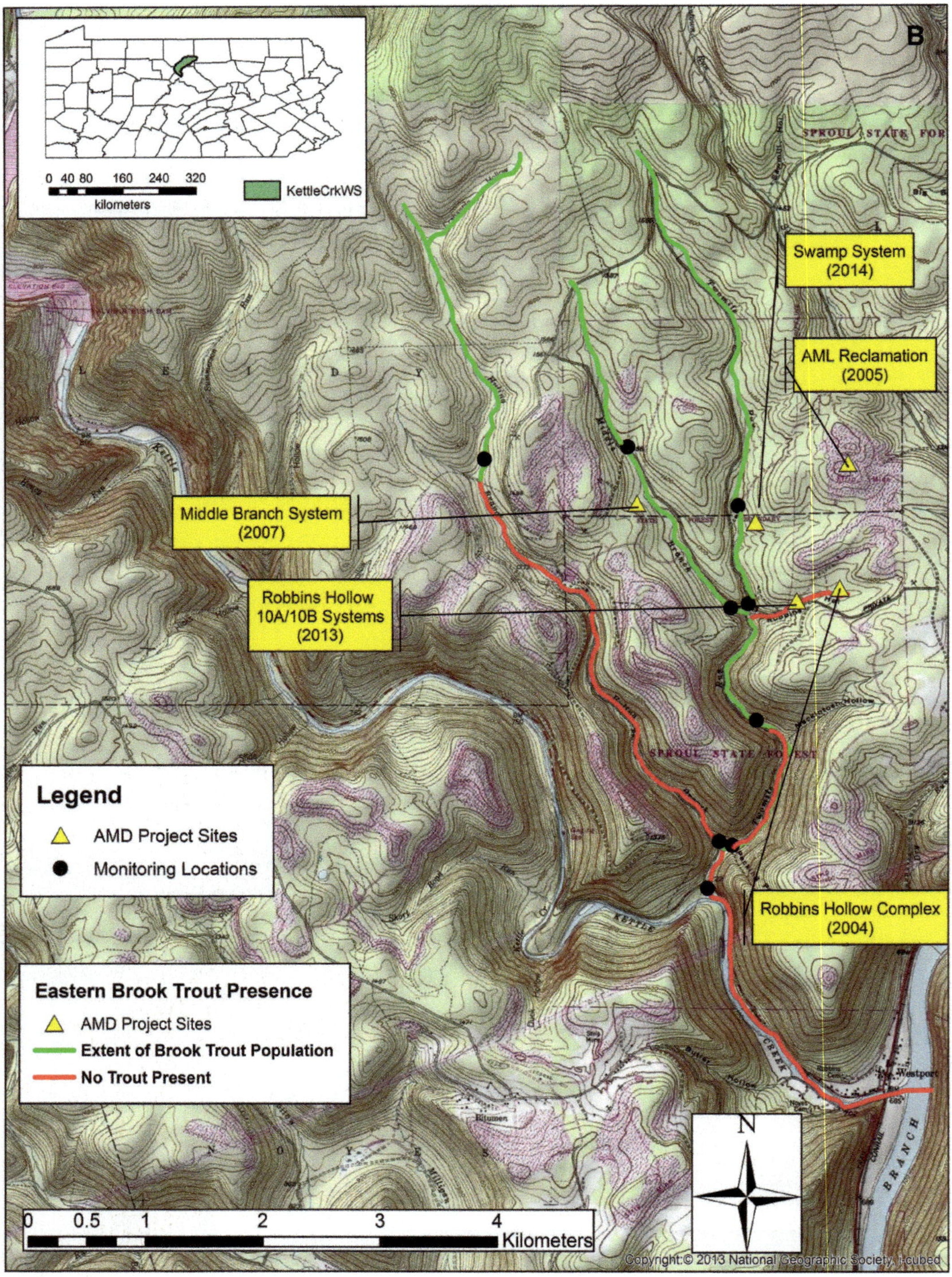

Figure 6. Continued.

Figure 7. The confluence of Twomile Run (from left) with Kettle Creek. Note the iron (orange) and aluminum (white) precipitate formed as the abandoned mine drainage-impaired Twomile Run mixes with the clean water of Kettle Creek.

that the system was being overloaded during peak flows, resulting in a decline in its performance (Hedin Environmental 2007), further demonstrating the value and need for monitoring plans for treatment system evaluation. Based upon the monitoring results, the system was rehabilitated in 2007 and, upon completion, was treating AMD with an initial pH of 3.2 and high metal concentrations and discharging water with a pH of 7.1 and reduced metal concentrations into Middle Branch, significantly improving the water quality in Middle Branch downstream of the treatment system (Table 1).

The improvements in water quality in Middle Branch downstream of the treatment system also provided an opportunity to document the recovery of Brook Trout populations and benthic macroinvertebrate communities to this once lifeless reach. A single Brook Trout was documented in this reach in 2010. Since 2010, the density of Brook Trout within the reach downstream of the treatment system has increased and is currently comparable to the unimpaired reference reach upstream of the treatment system (Figure 8). Similarly, benthic macroinvertebrates have also begun to recover throughout the previously impaired reach. The Pennsylvania index of biological integrity was used to assess the overall health of the stream (Figure 9). The Pennsylvania index of biological integrity includes the total abundance, dominant taxa abundance, taxa richness, percent Ephemeroptera, Plecoptera, Tricoptera taxa, Shannon diversity index, Hilsenhoff biological index, and percent sensitive individuals (PA DEP 2012) In total, an additional 1.9 km of coldwater habitat that was previously impaired was made available to Brook Trout and other aquatic species as a result of the Middle Branch passive treatment system.

At the Hydrologic Unit Code 10 watershed scale of Twomile Run, AMD remediation work continued in an effort to restore the watershed as a Brook Trout fishery and lessen the water-quality impacts to Kettle Creek and subsequently the West Branch Susquehanna River. Treatment systems were completed in the Robbins Hollow drainage in 2004 and included a complex of five passive treatment systems. The goal of these systems was to mitigate the acidity from Robbins Hollow entering Twomile Run, rather than to recover Brook Trout to Robbins Hollow.

A land reclamation project was completed in 2005 on an abandoned surface mine in the headwaters of Twomile Run, which included recontouring 23 ha and the addition of an alkaline byproduct. The project and associated new vegetation growth successfully reduced flow, acidity, and metal loading to Twomile Run by 30–50% (Trout Unlimited 2010).

In 2012, the Swamp passive treatment system was installed to mitigate AMD on the upper portion of Twomile Run (Figure 6). This treatment system effectively improved the water chemistry of Twomile Run downstream of the treatment system (Table 2). Additional treatment on Robbins Hol-

Table 1. Mean water-quality parameters downstream of treatment system outlet for Middle Branch over time.

	1995–2000	2001–2006	2007–2013
pH	4.21	5.29	6.00
Alkalinity (mg/L)	1	3	10
Acidity (mg/L)	42	16	3
Iron (mg/L)	0.2	0.1	0.1
Manganese (mg/L)	1.7	0.6	0.1
Aluminum (mg/L)	5.0	0.7	0.2

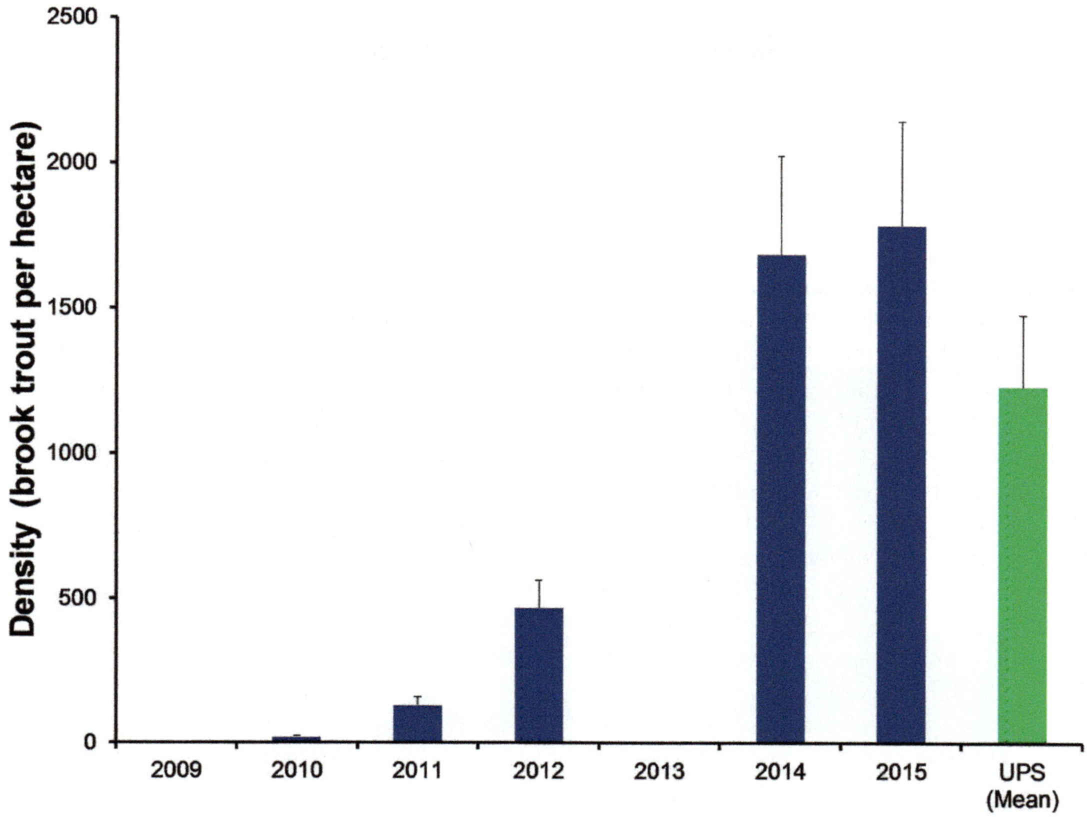

Figure 8. Density of Brook Trout (fish per hectare) in Middle Branch, downstream of the AMD treatment system over time. Mean density (fish per hectare) for the reference site located upstream of the abandoned mine drainage impairment on Middle Branch is shown by the green bar. No data were collected in 2013 and 2016 due to site access and high flows, respectively.

low, completed in 2013, further improved the water chemistry in Robbins Hollow entering Twomile Run (Table 3). In the first 3 years following water-quality improvements, the cumulative effect of these treatment systems within the Twomile Run watershed has been an additional 4.8 km of Brook Trout occupied coldwater habitat and the reconnection of two Brook Trout populations that were previously isolated upstream of the AMD in Middle Branch and Twomile Run (Figure 6).

Monitoring cumulative effects of small-scale remediation projects at the watershed scale

Despite the efforts to restore the West Branch Susquehanna watershed, there had never been a concerted effort to measure the improvements at a watershed scale. In 2009, TU completed a watershed-wide assessment of water chemistry and fishery communities to document changes in water quality and biological conditions, as well as to establish a benchmark of current conditions, including benthic macroinvertebrate communities, so that future remediation efforts may be evaluated.

Water-quality samples were obtained from 55 AMD impaired tributaries to the West Branch Susquehanna River and 12 main-stem river sites at high base flow and low base flow conditions in 2009 and compared to results from a U.S. Geological Survey study completed in 1984 (Hainley

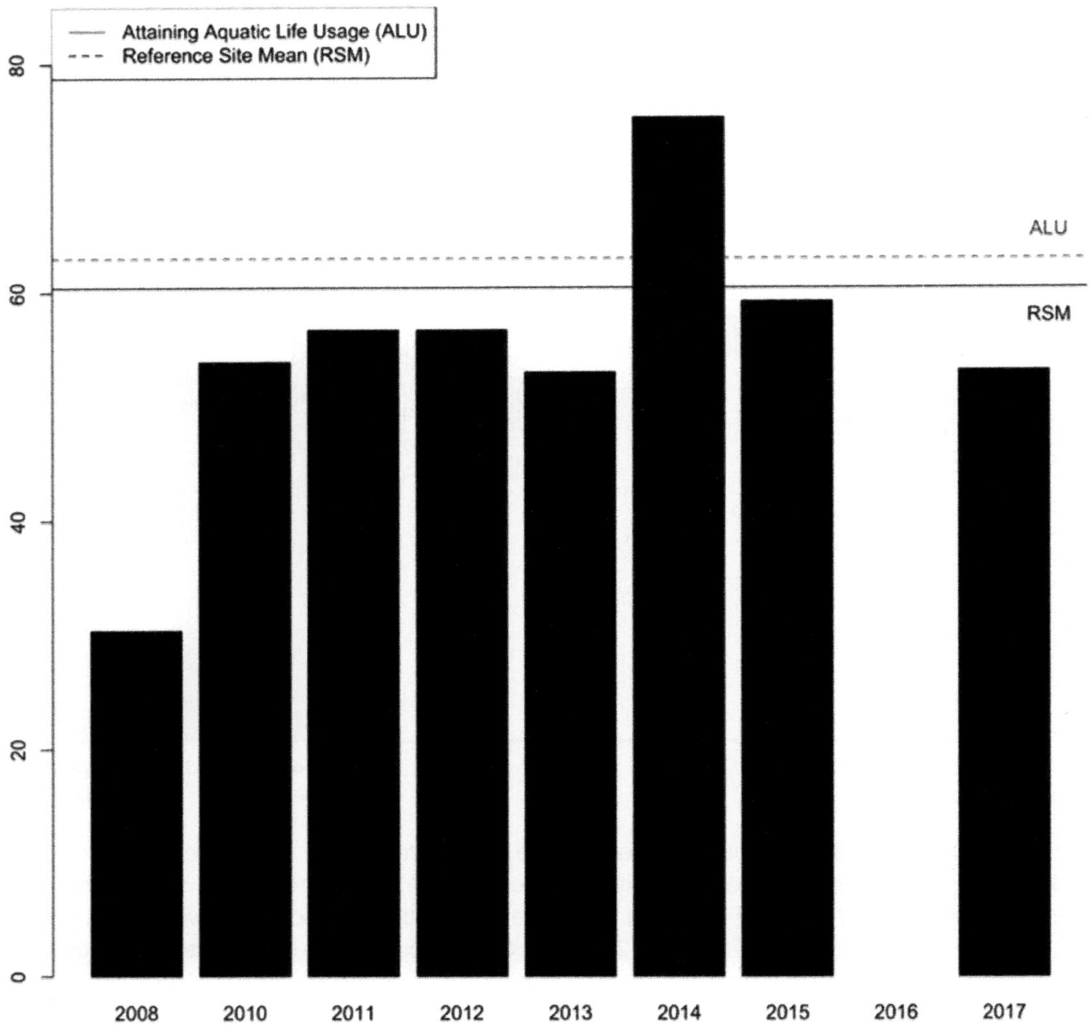

Figure 9. Benthic macroinvertebrate Pennsylvania index of biological integrity (IBI) scores for Middle Branch downstream of the abandoned mine drainage (AMD) treatment system over time. Solid horizontal line depicts the mean IBI score for the reference site located on Middle Branch upstream of the treatment system and AMD impairment. The dashed green horizontal line depicts the IBI score threshold for a site to meet the attaining life use criteria established by the Pennsylvania Department of Environmental Protection. No data were collected in 2016 due to high flow conditions.

and Barker 1993). Results of this study indicated significantly improved water quality and biological conditions compared to historical conditions. Eighty-five percent of the 55 tributaries sampled had a higher pH than previous sampling in 1984, and 79%, 68%, and 92% of the tributaries were lower in acidity, iron, and aluminum concentrations, respectively, compared to the 1984 data. As previously mentioned, the 1972 Scarlift report observed acidic conditions along the entire length of main stem of the West Branch Susquehanna River. According to the 2009 sampling results, the river was in a net alkaline state along its entire main stem, and concentrations of acidity, iron and

Table 2. Mean water-quality parameters for Twomile Run upstream of confluence with Middle Branch and downstream of the "Swamp" passive treatment system over time.

	1999–2004	2005–2012	2013–2016
pH	4.0	3.9	6.3
Alkalinity (mg/L)	1	0	15
Acidity (mg/L)	64	46	−2
Iron (mg/L)	0.9	1.2	0.3
Manganese (mg/L)	5.5	2.4	1.6
Aluminum (mg/L)	7.7	4.9	1.1

aluminum had all decreased over the 25-year period.

The fishery of the West Branch Susquehanna River also responded to the improvements in water-quality conditions. Fishery surveys completed by the Pennsylvania Fish and Boat Commission at 10 main-stem river sites throughout the section of the river that previously did not support a fishery showed a two- to a fivefold increase in fish species diversity in 2009 compared to 1999 surveys (Detar and Hollender 2005; Table 4). Total catch of individuals at each of the 10 sites also increased (Table 4). Thirty-five fish species were collected in 2009 among the 10 sites, including five species that were not observed during previous surveys, including Mimic Shiner *Notropis volucellus*, Central Stoneroller *Campostoma anomalum*, Shorthead Redhorse *Moxostoma macrolepidotum*, Green Sunfish *Lepomis cyanellus*, and Greenside Darter *Etheostoma blennioides*. In addition to the increased diversity of the fishery, multiple age-classes, including juveniles, were observed for most species, indicating that reproduction is occurring within the river.

The distribution of fishes also changed between 1999 and 2009. Bluntnose Minnow *Pimephales notatus*, a pollution tolerant species, occurred at 3 of the 10 sample sites in 1999 but was collected at 9 sites in 2009. Pollution-intolerant species, such as Northern Hog Sucker *Hypentelium nigricans*, River Chub *Nocomis micropogon*, and Longnose Dace *Rhinichthys cataractae*, also occurred with greater abundance at 9 of the 10 sample sites in 2009 compared to 1999.

The improvements documented in the 2009 study were attributed to a combination of factors that included reclamation activities, better permitting for mining projects, passive and active treatment projects, and the gradually diminishing amount of pyrite available for oxidation over time (natural attenuation). Some abandoned mine lands that had usable coal remaining have been remined. Remining operations have

Table 3. Mean water–quality parameters for Robbins Hollow downstream of treatment system complex over time.

	1995–2004	2005–2010	2013
pH	3.8	4.5	5.3
Alkalinity (mg/L)	0	1	6
Acidity (mg/L)	79	58	18
Iron (mg/L)	2.3	1.3	0.4
Manganese (mg/L)	7.5	7.0	4.9
Aluminum (mg/L)	8.8	6.5	2.9

Table 4. Total catch and total abundance at fishery survey sites on the main stem of the West Branch Susquehanna River. Sites are listed from most upstream to downstream.

	Total catch		Species richness	
Site	1998	2009	1998	2009
Emeigh Run	57	150	6	10
Shyrock Run	206	417	17	19
McGees Mills	113	143	14	14
Bower	141	167	11	13
Hogback	234	504	19	20
Clearfield	40	113	6	10
Egypt	8	115	5	11
Deer Creek	12	135	6	14
Burns Run	9	45	5	12
Hyner	13	420	3	16

improved water quality from historical discharges at the mined areas since Surface Mining Control and Reclamation Act regulations hold mining companies accountable for the water quality and land reclamation at current mine sites. The improvements between the historical data sets used for comparisons to the 2009 data are remarkable considering the 1972 Scarlift report suggested that improvements at this scale would be nearly impossible. However, the results also indicate that the full potential of restoring the watershed has not yet been realized, and it may only be in the beginning stages of recovery.

During this 10-year period, only 10 of the 68 tributaries samples met water-quality standards established in Pennsylvania's Chapter 93 issued under sections 5 and 402 of The Clean Streams Law water-quality criteria concentrations. In addition, although the fishery survey results were encouraging, fish species diversity and total abundance in the West Branch Susquehanna River remained relatively low compared to other non-AMD impaired sections of the river.

Since the completion of the 2009 TU study of the watershed, efforts to remediate AMD throughout the West Branch Susquehanna River watershed have continued. Since 2010, a minimum of $30 million on AMD remediation projects has been used throughout the watershed, including several large-scale active treatment systems, large abandoned mine land reclamation projects, and active and passive AMD treatment systems on tributaries. Therefore, it is important to evaluate and continue to document the environmental benefits of those investments and to also reevaluate where future remediation efforts are most needed.

In 2017, with funding from the PA DEP's Growing Greener Program, TU began to replicate the 2009 study to further document water quality and biological recovery of the system. Although much of the study remains to be completed, preliminary water-quality results suggest that improvements have continued throughout the watershed. For example, mean low base flow pH of all replicated samples has increased from 5.18 in 2009 to 5.46 in 2017.

In addition, this latest project will also provide the next steps to ensure that AMD restoration projects continue to enhance the overall water quality and biological communities in the watershed, including the development of a long-term monitoring strategy (with a citizen science component) at the watershed scale and a prioritization of subwatersheds for future implementation of

AMD restoration projects. This latest evaluation of the watershed will also provide an assessment of restoration progress, to date, through the addition of reference sites with no known impairments and strong Brook Trout populations. The project is slated to be completed in the winter of 2019.

Role of the citizen scientist in monitoring

A component to monitoring the effectiveness of AMD treatment systems and recovery of the watershed that has not been fully implemented is the use of a citizen-based monitoring program. Although citizens of the watershed have been actively involved with their local watershed groups, TU chapters, and other organizations that have completed monitoring and restoration projects, a concerted effort has not been organized to maximize the benefits of citizen science. Over the past decade, public participation has increased drastically as state and federal agencies have developed protocols for citizen science monitoring of stream condition and water quality (Nerbonne and Vondracek 2003; Newman et al. 2012). As a member-based organization with approximately 300,000 members and supporters, TU has the ability to successfully organize and lead such efforts.

In Pennsylvania, TU volunteers and other concerned citizens have been monitoring water quality in response to the expansion of shale gas development in the commonwealth (Williams et al. 2016). The goals of this program were to (1) identify and report pollution events resulting from shale gas development and (2) collect baseline water quality data in areas projected for future development. These goals are being met by conducting outreach to TU chapters and other outdoor groups, training volunteers in water quality monitoring protocols, providing the necessary equipment, providing technical support to volunteers collecting data, and maintaining an online results database (Williams et al. 2016). This program, as well as other citizen-based monitoring programs, has increased the monitoring capabilities for state agencies while providing a means for the local community to study their home watersheds (Williams et al. 2016). A similar monitoring program could be developed in relation to AMD in the West Branch Susquehanna River watershed and other areas. Tulloch et al. (2013) stated that citizen science is particularly effective when data are collected over longer spatial and temporal scales. Given the large scale of the project area, citizen-based monitoring may be the most cost-effective method to collect the data needed to ensure restoration success and protect the recovery that has occurred to date.

Multispecies benefits of restoration in the West Branch Susquehanna River watershed

Historical mining operations have also had a detrimental effect on numerous other wildlife species beyond aquatic ecosystems. Negative impacts associated with habitat loss due to mining activity, particularly the loss of forested habitat, in birds, mammals, amphibians, and reptiles, have been well-documented in the scientific literature. Wildlife species that rely on the health of aquatic ecosystems are also adversely impacted by AMD pollution. For example, Louisiana waterthrush *Seiurus motacilla* inhabiting riparian habitat along acidified streams exhibited reduced breeding densities, increased territory size, and a delayed clutch initiation when compared to unimpacted streams (Mulvihill et al. 2008).

Abandoned mine drainage remediation and abandoned mine land reclamation has also been documented to have beneficial effects on several wildlife species. Buehler and Percy (2015) provides a literature review on the wildlife response to both mining and postmining reclamation in the eastern United States.

In avian communities, there is generally a shift from forest bird communities to those associated with early successional habitats due to mining activities. Land reclamation has increased numbers of Henslow's *sparrows* Ammodramus henslowii and/or grasshopper sparrows *Ammodramus savannarum*, two species of conservation concern, as well as north harriers *Circus cyaneus* and short-eared owls *Asio flammeus* (Vukovich 2004; Stauffer et al. 2011). Game bird species such as eastern wild turkey *Meleagris gallopavo* have also been shown to inhabit reclaimed mine lands, and their densities observed on reclaimed sites has been higher than densities on reference areas (Rice 1986).

Elk *Cervus canadensis* have been reintroduced into Pennsylvania, and the West Branch Susquehanna River watershed contains the highest densities of elk within the state. Numerous studies have been completed on elk on reclaimed mine lands in Kentucky (see Buehler and Percy 2015). Reclaimed sites serve as important forage sites for elk (Schneider et al. 2006; Lupardus et al. 2011). Land reclamation sites in the West Branch Susquehanna River watershed (such as the project in the Twomile Run watershed previously described) are commonly planted with elk forage species to promote their use by elk. Each fall, visitors travel to the region to have the opportunity to view the elk, providing tourism dollars to the local economy. The relatively open areas of reclaimed mine lands compared to the surrounding forests may provide an increased opportunity for wildlife viewing in accessible areas. In addition, the Pennsylvania elk population has increased to a level sustainable for a limited hunting season, providing jobs for local guides, increasing food and lodging revenues in the area, and increasing revenue to the Pennsylvania Game Commission through license sales and entry into the lottery system to draw an elk hunting permit.

Studies have also shown several species of small mammals, bats, and reptiles to utilize these sites (see Buehler and Percy 2015 for review). In addition, wetlands associated with mine sites and passive treatment remediation of AMD are inhabited by amphibians that would not otherwise have been present at the site (Timm and Meretsky 2004; Lannoo et al. 2009). Wetlands created for remediation also provide additional habitat for wetland bird species and wetland plants.

Lessons Learned and Implications for Other Large-Scale Restoration Projects

The past 14 years of restoration projects within the West Branch Susquehanna River watershed have provided the opportunity to realize individual project success as well as setbacks. Once considered an impossible goal, both economically and practically, the recovery of the AMD-impaired West Branch Susquehanna River and its tributaries now seems to be attainable. The lessons learned while working towards this recovery may be applied to most, if not all, large-scale restoration projects. In a broad context, two major lessons of this work are prominent when considering other projects of this scale: establishing and fostering partnerships, and the importance of long-term monitoring of projects at various scales.

Lesson #1: partnerships

Probably the most important key to the success of the project, to date, has been the establishment and fostering of partnerships. A restoration project of this scale would not have been possible through the actions of a single entity. Both the individual efforts and the collaboration among state and federal agencies, county conservation districts, environmental consulting firms, colleges and universities, private foundations, nongovernmental organizations, community member-based watershed associations, and concerned local citizens have been key in the success of the restoration process. The diversity of partners has allowed for greater funding opportunities ranging from small planning grants to larger grants for treatment system design and implementation. Partnerships also provide a diversity in expertise ranging across several disciplines, including engineering, environmental science, ecology, wildlife and fisheries biology, forestry, mining operations, construction, hydrology, geology, policy, and fundraising, among others, allowing for a greater exchange of ideas and more thorough planning.

Establishing collaborative partnerships for large-scale restoration projects alone is not enough. Relationships among partners must be fostered and stakeholders in restoration must be kept informed and up to date on progress. Regular meetings of the West Branch Susquehanna Restoration Coalition, as well as the symposiums hosted within the watershed, have provided regular venues, which are open to the public, where stakeholders can get information on individual project updates, share research findings and needs, and discuss issues that may have arisen over the course of a project.

Collaborative partnerships are also vital to the planning phases of large-scale restoration initiatives. Haphazardly implementing small-scale restoration projects on the ground with the hope of having long-term and/or large-scale cumulative effects is unlikely to succeed. Rather, restoration goals should be clearly stated and projects should be prioritized to maximize their potential impact. This is especially critical when dealing with limited funding opportunities in landscapes where the need for on the ground project implementation seems to be almost infinite. In the case study of the West Branch Susquehanna River, numerous restoration plans have been developed at a variety of geographic scales. Applying the principals of adaptive management to the process, these plans have been consistently updated and reevaluated as new data become available and restoration techniques improved.

The restoration work in the West Branch Susquehanna River watershed provides a good example of how strategic planning and implementation of individual restoration projects can lead to large-scale changes in water quality and biological recovery in polluted watersheds. The vast majority of projects that have been completed within the West Branch Susquehanna River watershed have occurred on first- to third-order coldwater streams. The benefit of these treatment systems has not only been the nearly immediate improvement of water quality within those streams, but also a decrease in the overall loadings of acidity and heavy metal concentrations to the larger river system. Hence, the water-quality benefits that have led to improvement in the coldwater fisheries of tributaries have also yielded benefits that helped to improve the warmwater fisheries and aquatic communities inhabiting the larger streams and downstream reaches of the river itself.

Lesson #2: long-term monitoring

Effective monitoring of large-scale restoration projects is also critical to ensure their success and must occur at several geographic scales. However, the monitoring component of projects can often be the most difficult component to fund since most grants and funding programs operate on a 2- or 3-year basis from project inception to project completion. Routine water-quality monitoring plans for individual AMD treatment systems are essential to provide regular data on the effectiveness of the treatment, the detection of any issues within the system, and feedback on when maintenance of the system is needed. Monitoring at a variety of scales has also been important in documenting restoration success and understanding how aquatic systems respond to treatment. Monitoring efforts have included studies from the reach scale to the more recent West Branch Susquehanna River watershed scale to help guide restoration goals and evaluate restoration effectiveness.

The indicators used in monitoring programs should be thoroughly considered to determine those most appropriate for a given project. Monitoring metrics will vary depending on the project type, the overall goals of restoration, the timeline needed to demonstrate results following project implementation, requests from funders, and funding availability for monitoring. For AMD remediation projects, the most commonly selected indicators include water-quality parameters such as flow, pH, specific conductance, acidity, and total or dissolved concentrations of aluminum, iron, and manganese. Benthic macroinvertebrate community composition and fish population metrics such as density, biomass, and distribution are common indicators to monitor biological recovery. Water-chemistry parameters offer a relatively short time scale to document improvements post-treatment while biological metrics may take several years to see signs of improvement and may require more long-term monitoring to demonstrate significant trends and therefore additional funding. As biological indicators integrate water quality over time, they often better represent the long-term improvement to water quality, whereas routine water-chemistry sampling represents conditions at a single point in time and requires numerous samples before trends can be detected.

A possible method to increase the monitoring capacity of large-scale restoration projects is the use of citizen science. Although some protocols used in monitoring plans will always require the use of professionally trained staff, citizen-based monitoring may be a useful tool for routine monitoring and/or the maintenance of data loggers in the field, thereby reducing the costs of implementing a monitoring plan.

Although the West Branch Susquehanna River and its tributaries appear to be on a trajectory toward recovery, the efforts in the watershed are not yet finished. There are currently several projects underway, but numerous AMD polluted tributaries to the West Branch Susquehanna River still exist where restoration efforts have not yet begun. Safeguarding and increasing the funding that is available to plan and construct new AMD remediation projects, as well as to monitor and maintain the existing treatment systems, is vital to ensuring that the improvements made to date are sustained and that the trajectory of recovery continues for future generations.

Acknowledgments

We would like to acknowledge Kathleen Lavelle and Allison Lutz for their contributions on data organization and assistance with data analysis. Funding for this project was provided in part by the R.K. Mellon Foundation.

References

Buehler, D. A., and K. L. Percy. 2015. Coal mining and wildlife in the eastern United States: a literature review. Available: www.researchgate.net/publication/267246672_Coal_Mining_and_Wildlife_in_the_Eastern_United_States_A_Literature_Review. (March 2019).

DeNicola, D. M., and M. G. Stapleton. 2016. Using macroinvertebrates to assess ecological integrity of streams remediated for acid mine drainage. Restoration Ecology 24:656–667.

Detar, J., and B. Hollender. 2005. Reinventory of the Babb Creek watershed (309A) tributary to Pine Creek, Tioga County. Pennsylvania Fish and Boat Commission, Bellefonte, Pennsylvania.

Downstream Strategies, LLC. 2008. An economic benefit analysis for abandoned mine drainage remediation in the West Branch Susquehanna River watershed, Pennsylvania. Submitted by Downstream Strategies, LLC to Trout Unlimited, Mill Hall, Pennsylvania.

Earle, J., and T. Callaghan. 1998. Effects of mine drainage on aquatic life, water uses, and man-made structures. Pages 4.1–4.10 *in* K.

B. C. Brady, M. W. Smith, and J. Schueck, editors. Coal mine drainage prediction and pollution prevention in Pennsylvania. Pennsylvania Department of Environmental Protection, 5600-BK-DEP2256, Harrisburg.

Federal Water Quality Administration. 1968. Mine drainage study of the Susquehanna River basin. U.S. Environmental Protection Agency, Washington, D.C.

Fesenmyer, K. A., A. L. Hawk, S. M. Rummel, M. Mayfield, S. L. McFall, and J. E. Williams. 2017. Eastern Brook Trout conservation portfolio, range-wide habitat integrity and future security assessment, and focal area risk and opportunity analysis. Trout Unlimited, Arlington, Virginia.

Hainley, R. A., and J. L. Barker. 1993. Water quality of the upper West Branch Susquehanna River and tributary streams between Curwensville and Renovo, Pennsylvania, May and July 1984. U.S. Geological Survey, Water Resources Investigations Report 90-4011, Reston, Virginia.

Hedin Environmental. 2007. Middle Branch Passive Treatment System Rehabilitation Project. Prepared by Hedin Environmental for Trout Unlimited, Arlington, Virginia and the Kettle Creek Watershed Association, Cross Fork, Pennsylvania.

Johnson, D. B., and K. B. Hallberg. 2005. Acid mine drainage remediation options: a review. Science of the Total Environment 338:3–14.

Lannoo, M. J., V. C. Kinney, J. L. Heemeyer, N. J. Engbrecht, A. L. Gallant, and R. W. Klaver. 2009. Mine spoil prairies expand critical habitat for endangered and threatened amphibian and reptile species. Diversity 1:118–132.

LeFevre, S. 2003. West Branch Susquehanna subbasin survey: a water quality and biological assessment. Susquehanna River Basin Commission, Harrisburg, Pennsylvania.

Lupardus, J. L., L. I. Muller, and J. L. Kindall. 2011. Seasonal forage availability and diet for reintroduced elk in the Cumberland Mountains, Tennessee Southeastern Naturalist 10:53–74.

Mulvihill, R. S., F. L. Newell, and S. C. Latta. 2008. Effects of acidification on the breeding ecology of a stream-dependent songbird, the Louisiana waterthrush (*Seiurus motacilla*). Freshwater Biology 53:2158–2169.

Nerbonne, J. F., and B. Vondracek. 2003. Volunteer macroinvertebrate monitoring: assessing training needs through examining error and bias in untrained volunteers. Journal of North American Benthological Society 22:152–163.

Newman, G., A. Wiggins, A. Crall, E. Graham, S. Newman, and K. Crowston. 2012. The future of citizen science: emerging technologies and shifting paradigms. Frontiers in Ecology and the Environment 10:298–304.

PA DEP (Pennsylvania Department of Environmental Protection). 2006. Pennsylvania annual reports on mining activities. Available: http://coalmininghistorypa.org/annualreport/2006/bituminous.htm. (April 2019).

PA DEP (Pennsylvania Department of Environmental Protection). 2012. A benthic macroinvertebrate index of biotic integrity for wadeable freestone riffle-run streams in Pennsylvania. PA DEP, Division of Water Quality Standards, Harrisburg.

PA DEP (Pennsylvania Department of Environmental Protection). 2016. 2016 final Pennsylvania Integrated Water Quality Monitoring and Assessment report: Clean Water Act Section 305(b) and 303(d) List. PA DEP, Harrisburg.

PA DEP (Pennsylvania Department of Environmental Protection). 2018. Integrated list non attaining. Pennsylvania Spatial Data Access. Available: www.pasda.psu.edu/uci/DataSummary.aspx?dataset=888. (March 2019).

Pennsylvania Department of Environmental Resources. 1972. West Branch Susquehanna River Mine Drainage Pollution Abatement Project Operation Scarlift. Pennsylvania Department of Environmental Resources, Harrisburg.

Rice, F. H. 1986. Response of wild turkeys to surface coal mining in north central Pennsylvania. Master's thesis. Pennsylvania State University, University Park.

Roller, M. 1958. Memorandum report: Fish killing West Branch Susquehanna River Williamsport to Lewisburg. Available from the Pennsylvania Department of Environmental Protection, Northcentral Regional Office, Williamsport.

Rummel, S. M., K. Fesenmyer, A. Haak, M. Mayfield, M. Hudy, M., and J. Williams. 2017. Identifying high-value "marginal" Brook Trout populations using a conservation portfolio approach. Pages 185–194 *in* B. Carline, editor. Wild Trout XII: science, politics, and wild trout management: who's driving and where are we going? Wild Trout Symposium, Bozeman, Montana.

Schneider, J., D. S. Maehr, K. J. Alexy, J. J. Cox, J. L. Larkin, and B. C. Reeder. 2006. Food habits of reintroduced elk in southeastern Kentucky. Southeastern Naturalist 5:535–547.

Sisler, J. D. 1926. Bituminous coal fields of Pennsylvania part II. Detailed description of coal fields. Mineral Resource Report 6. Department of Environmental Resources. Bureau of Topographic and Geologic Surveys. Pennsylvania Geological Survey, Middletown.

Sorenson, D. 1931. Commonwealth of Pennsylvania Board of Fish Commissioners stream survey report: West Branch Susquehanna River, Clearfield County, Pennsylvania. Pennsylvania Fish and Boat Commission, Harrisburg.

Sorenson, D. 1932. Commonwealth of Pennsylvania Board of Fish Commissioners stream survey report: West Branch Susquehanna River, Clinton County, Pennsylvania. Pennsylvania Fish and Boat Commission, Harrisburg.

Stauffer, G. E., D. R. Diefenbach, M. R. Marshall, and D. W. Brauning. 2011. Nest success of grassland sparrows on reclaimed surface mines. Journal of Wildlife Management 75:548–557.

Timm, A., and V. Meretsky. 2004. Anuran habitat use on abandoned and reclaimed mining areas of southwestern Indiana. Proceeding of the Indiana Academy of Science 113:140–146.

Trout Unlimited. 2013. Kettle Creek watershed, Clinton County, PA abandoned mine drainage projects 1998–2013. Available: www.kettlecreek.org/uploads/2/5/6/0/25607137/kettle_amd_project_summaries_2.pdf. (April 2019).

Tulloch, A. I. T., H. P. Possingham, L. N. Joseph, J. Szabo, and T. G. Martin. 2013. Realizing the full potential of citizen science monitoring programs. Biological Conservation 165:128–138.

Vukovich, M. 2004. Use of reclaimed surface mine grasslands by northern harriers, short-eared owls, and other raptors: abundance, behavior, and nesting success. Master's thesis. Eastern Kentucky University, Richmond.

Williams, J. E., S. Rummel, J. Lemon, M. Barney, K. Smith, K. Fesenmyer, and J. Shoen. 2016. Engaging a community of interest in water quality protection: anglers monitoring wadeable streams. Journal of Soil and Water Conservation 71:114A–119A.

American Fisheries Society Symposium 91:455–478, 2019

Restoration of Diadromous Fish Populations in Tributaries of Waquoit Bay, Cape Cod, Massachusetts

STEPHEN T. HURLEY*
Massachusetts Division of Fisheries and Wildlife
195 Bournedale Road, Buzzards Bay, Massachusetts 02532, USA

Abstract.—Waquoit Bay is a coastal estuary located on the south side of Cape Cod. The primary rivers feeding the bay, the Quashnet and Childs rivers, are small, coldwater, groundwater-fed streams. Most of the watersheds of both rivers were originally set aside in the 1600s as a plantation for the Native American Mashpee Wampanoag tribe. The rivers were heavily modified in the late 1700s by the building of mill dams and later in the 1800s by cranberry agriculture. The anadromous Brook Trout *Salvelinus fontinalis* fisheries in both rivers were acclaimed in the early 1800s. Anadromous river herring *Alosa* spp. runs were created on both streams by connecting the streams to Johns Pond, a natural kettle hole pond. After anadromous Brook Trout populations declined due primarily to habitat loss, efforts were initiated in the 1950s to restore anadromy to Brook Trout in Cape Cod rivers by overstocking with hatchery Brook Trout. After this project, land protection along the river started with the purchase of abandoned cranberry bogs. Both rivers were heavily stocked with Brown Trout *Salmo trutta* in the 1970s and 1980s to create a sea-run Brown Trout fishery. In 1976, Trout Unlimited began an ongoing habitat improvement project in the Quashnet River. In the 1970s and 1980s, the rapid development of Cape Cod threatened the watershed. In 1988, the Waquoit Bay National Estuarine Research Reserve was formed and the Commonwealth of Massachusetts purchased land in the watershed to preserve it as open space. As part of the purchase agreement, a potential well site was reserved, which led to studies by the U.S. Geological Survey on the hydrology of the Quashnet River and the impact of potential wells. In the early 1990s, fisheries management shifted away from the stocking of Brown Trout to focus on the native Brook Trout fishery. The Mashpee National Wildlife Refuge, a consortium of landowners centered on Waquoit Bay, was formed in 1995. In 1997, the contaminant ethylene dibromide from the former Otis Air Force Base Superfund site was found to be entering the upper Quashnet River. This led to the creation of a system of berms and groundwater extraction systems. The failure of part of the berm system led to concerns about fisheries impacts, and a restoration plan was developed. A Brook Trout passive integrated transponder tagging project was initiated on the Quashnet River in 2007, and the Brook Trout population has been annually sampled since 2000. In 2008–2010, adult wild Brook Trout from the Quashnet River were transplanted to the Childs River and a wild Brook Trout population was reestablished. Nitrogen loading from the watershed has become a major issue for the Waquoit Bay estuary, causing algae blooms and water-quality impacts. The fisheries of the

* Corresponding author: steve.hurley@mass.gov

Waquoit Bay tributaries have been protected and enhanced by an ongoing combination of land protection, fisheries management and research activities, and habitat improvements involving a wide variety of partners. Watershed development and potential climate change continue to threaten both the estuarine resources of Waquoit Bay and the native freshwater and diadromous fisheries of its tributaries.

Introduction

Massachusetts is one of the original 13 U.S. colonies, and fisheries have been an important part of its history. Cape Cod was named for the abundance of the Atlantic Cod *Gadus morhua* off its shores, and the Pilgrims that arrived in Plymouth in 1620 were taught by the Native American Tisquantum (Squanto) to fertilize their crops with the abundant anadromous river herring *Alosa* spp., which were easily caught during their spring spawning runs. The native diadromous fish of Massachusetts included the catadromous American Eel *Anguilla rostrata* and anadromous Alewife *Alosa pseudoharengus*, American Shad *A. sapidissima*, Atlantic Salmon *Salmo salar*, Atlantic Tomcod *Microgadus tomcod*, Blueback Herring *A. aestivalis*, Brook Trout *Salvelinus fontinalis*, Rainbow Smelt *Osmerus mordax*, Sea Lamprey *Petromyzon marinus*, Striped Bass *Morone saxatilis*, and White Perch *M. americana* (Table 1).

The early abundance of these migratory fishes was soon impacted by the building of dams and other obstructions on coastal rivers, land clearing for agriculture and industry, and pollution caused by the industrial revolution. The Massachusetts Commissioners on Fisheries was formed in 1866 following the loss of the salmon and shad runs in the Connecticut and Merrimack rivers caused by the construction of the Turners Falls, Holyoke, and Lawrence dams (Cardoza 2015). The descendants of that agency, the Massachusetts Division of Fisheries and Wildlife (MDFW) and the Massachusetts Division of Marine Fisheries (MDMF) continue, with numerous partners, to restore and protect the fish populations of the Commonwealth of Massachusetts.

The history of these fisheries and the challenges in protecting and restoring them in the face of almost 400 years of development are well illustrated by the story of Waquoit Bay, a shallow coastal estuary located on the south side of Cape Cod, Massachusetts in the towns of Falmouth and Mashpee. The major freshwater input to the bay is from groundwater and groundwater-fed streams, including the Quashnet River, Childs River, and four small freshwater tributaries: Red Brook, Bournes Pond Outlet, Abigails Brook, and Dutchmans Creek (Louis Berger Group 2010). The history of the diadromous fisheries of Waquoit Bay demonstrates some of the factors that have impacted freshwater and migratory fish populations in the northeastern United States and offers some hope of success in halting and reversing the decline of fisheries in the region.

Geographic Area

Cape Cod

Cape Cod is located in the Atlantic Coastal Pine Barrens ecoregion and was formed about 10,000–25,000 years ago after the retreat of the Laurentide glaciers, which once covered New England (Portnoy et al. 2001). The upper Cape is dominated by two terminal moraines, upon which the major highways Route 6 and Route 28 are routed, and a large outwash plain known as the Mashpee Pitted Plain. Cape Cod has close to 1000 kettle hole ponds, clear rounded ponds formed by melting blocks of ice (Eichner et al. 2003). Linear valleys, known as spring sapping valleys (Schwarzman 2002), were carved by erosion of meltwater streams into the approximately 100-m-deep outwash plains. These valleys

Table 1. Freshwater and diadromous fish of the Waquoit Bay watershed.

Family	Common name	Scientific name
Anguillidae	American Eel	*Anguilla rostrata*
Clupeidae	Blueback Herring	*Alosa aestivalis*
	Alewife	*A. pseudoharengus*
Salmonidae	Rainbow Trout	*Oncorhynchus mykiss*
	Brown Trout	*Salmo trutta*
	Brook Trout	*Salvelinus fontinalis*
Osmeridae	Rainbow Smelt	*Osmerus mordax*
Esocidae	Chain Pickerel	*Esox niger*
Cyprinidae	Golden Shiner	*Notemigonus crysoleucas*
Catostomidae	White Sucker	*Catostomus commersonii*
Ictaluridae	Brown Bullhead	*Ameiurus nebulosus*
Gadidae	Atlantic Tomcod	*Microgadus tomcod*
Cyprinodontidae	Sheepshead Minnow	*Cyprinodon variegatus*
	Banded Killifish	*Fundulus diaphanus*
	Mummichog	*F. heteroclitus*
	Striped Killifish	*F. majalis*
Atherinidae	Atlantic Silverside	*Menidia menidia*
Gasterosteidae	Fourspine Stickleback	*Apeltes quadracus*
	Threespine Stickleback	*Gasterosteus aculeatus*
	Ninespine Stickleback	*Pungitius pungitius*
Moronidae	White Perch	*Morone americana*
	Striped Bass	*M. saxatilis*
Centrarchidae	Pumpkinseed	*Lepomis gibbosus*
	Bluegill	*L. macrochirus*
	Smallmouth Bass	*Micropterus dolomieu*
	Swamp Darter	*Etheostoma fusiforme*
Percidae	Tessellated Darter	*E. olmstedi*
	Yellow Perch	*Perca flavescens*

were later partially flooded by sea level rise, which occurred about 5,000 years ago. Small groundwater-fed streams are formed when groundwater levels intersect the valley floors.

The glaciation of Massachusetts resulted in a greatly limited freshwater fish fauna (Hartel et al. 2002), which is even more limited on Cape Cod, a peninsula surrounded by the sea. The tributary streams were likely colonized by freshwater fish from the Georges Bank Refugium, by anadromous and euryhaline fish species and the catadromous American Eel (Hartel et al. 2002). Due to the outwash sands, the fresh waters of Cape Cod are infertile and slightly acidic, with a conductivity of 40–150 μS/cm.

The natural landscape of Cape Cod was dominated by pitch pine *Pinus rigida* and oak *Quercus* spp. forests with other hardwoods, including maple *Acer* spp. and beech *Fagus grandifolia*, occurring in the river valleys (Schwarzman 2002). These woodlands were subject to periodic fires, caused by lightning strikes as well as those that were purposely set by Native Americans to increase game abundance.

Waquoit Bay

Waquoit Bay is a shallow (mean depth at low tide = 0.3 m, mean tide range = 0.6 m) coastal estuary located on the south side of Cape Cod in the towns of Falmouth and Mashpee

(Figure 1). The substrate is primarily sandy silt and detritus with patches of gravel and shell fragments. Summer water temperatures range from 19°C to 26°C and salinity is 27–34 parts per thousand (Hughes et al. 2002). The bay is separated from the marine environment of Nantucket Sound by a barrier beach with historically variable openings (Geist 1996). Dedicated riprapped outlets are now in place at Eel Pond and the middle of Waquoit Bay. It is hydrologically connected with the Eel Pond estuary on the west and tidal ponds, including Hamblin Pond, Jehu Pond, and Sage Lot Pond, on the east.

Theodore Lyman III had a colorful description of Waquoit Bay in the Massachusetts Commissioners of Fisheries (1870) report:

> Waquoit Bay... is one of the most singular of the land-locked fiords on the south shore of Cape Cod—an irregular, many-lobed body of salt water, which is rendered in its upper part brackish by fresh streams, and which empties in the sea only through a single very narrow cut in the low sand barrier which shuts out the ocean. The influent brooks, dignified by the names of Quoshnet and Child's Rivers, are of the clear cold water peculiar to that sandy country, which everywhere pours out its springs near the surface. It is these spring brooks and these extensive shallows of salt or brackish water, protected from the ocean and teeming with small crustacea, that render this region so renowned for its trout, which there get a size, fatness and flavor rarely found anywhere else. For similar reasons the alewives here find a proper breeding ground, by passing through the sheltered bay and ascending the brooks to the ponds at their sources. The scup, too, and the Striped Bass, eminently estuary fishes, delight in the grateful cover and the abundant food.

The watershed of Waquoit Bay is distinguished less by its surface water watershed than by its groundwater watershed, which extends through Mashpee and Falmouth into the lower parts of Sandwich and incorporates part of the former Otis Air Force Base and Ashumet and Johns ponds (Figure 2). The bay receives most of its freshwater flow from the Quashnet and Childs rivers and several smaller tributaries, including Red Brook, Abigails Brook, and Bournes Pond outlet.

Quashnet River

The Quashnet River is a low-gradient (1.4 m/km), 8-km, groundwater-fed stream that flows from Johns Pond in Mashpee through abandoned cranberry bogs to the tidal Moonakiss River and the Waquoit Bay estuary in Falmouth. The substrate is primarily sand with some patches of gravel and detritus. Median flow at the Martin Road U.S. Geological Survey (USGS) gage is 0.6 m^3/s, with a minimum flow of 0.3 m^3/s and a maximum flow of 1 m^3/s in 2010. Drainage area is 11 km^2 and specific conductance ranges from 77 to 175 μS/cm, mean 110 μS/cm (Max Holmes, Cape Cod Rivers Observatory, personal communication). The resident fish population in the Quashnet River is primarily composed of American Eel, Brook Trout, Tessellated Darter, and White Sucker, with occasional freshwater- and estuarine-derived species from Johns Pond and the Waquoit Bay estuary.

Childs River

The Childs River flows for 3.7 km at a gradient of 3.1 m/km from Johns Pond in Mashpee to Eel Pond in Falmouth through abandoned cranberry bogs and is impounded by a small dam just below Carriage Shop Road. The Childs River is much smaller than the Quashnet River, with flows ranging from 0.05 to 0.15 m^3/s. The resident fish population in the Childs River is American Eel and Brook Trout, with occasional freshwater- and estuarine-derived species from the Falmouth Rod and Gun Club ponds and the Waquoit Bay estuary.

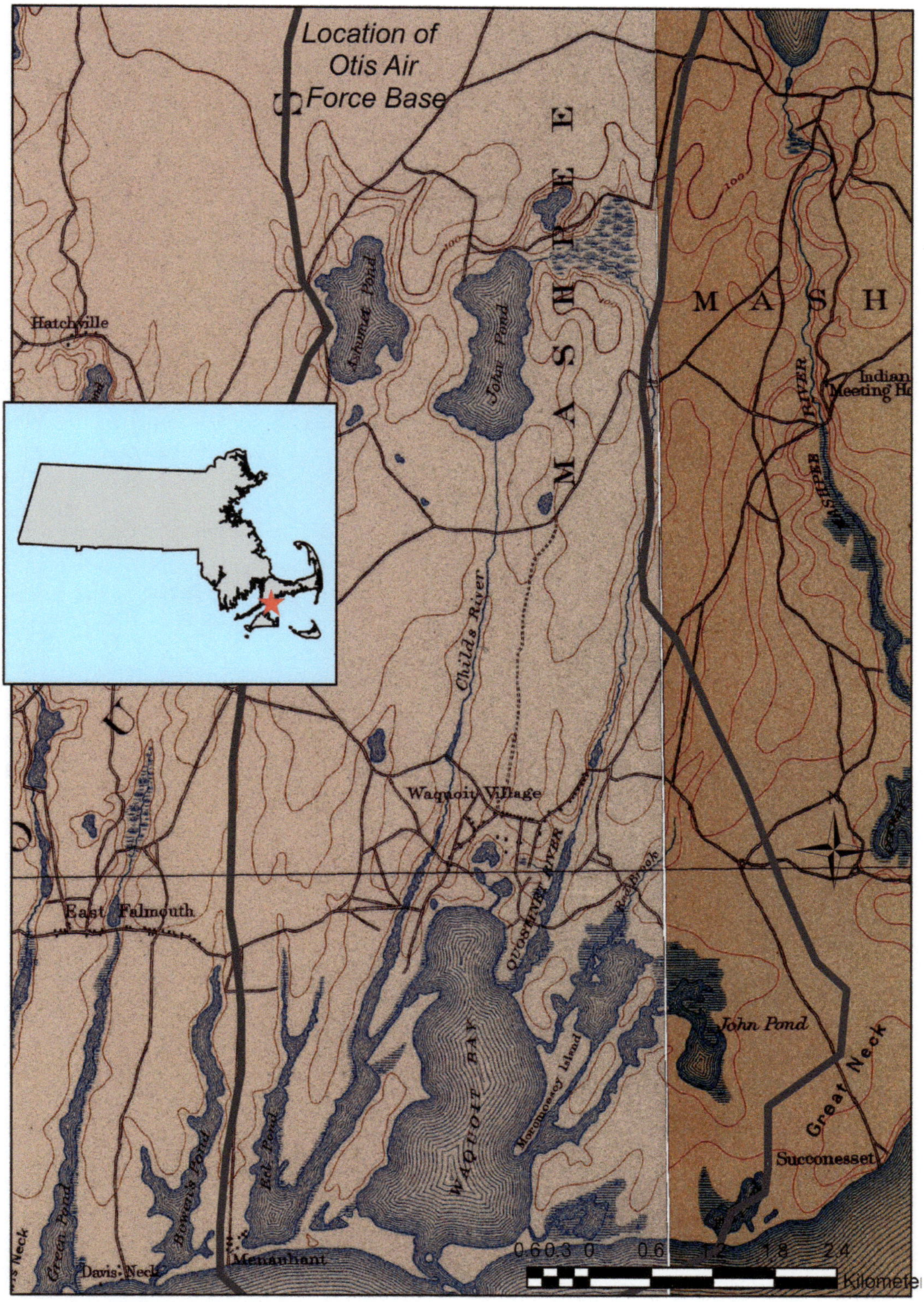

Figure 1. Section of 1893 U.S. Geological Survey topographic maps with Waquoit Bay, Cape Cod, Massachusetts overlaid with watershed (groundwater) outline.

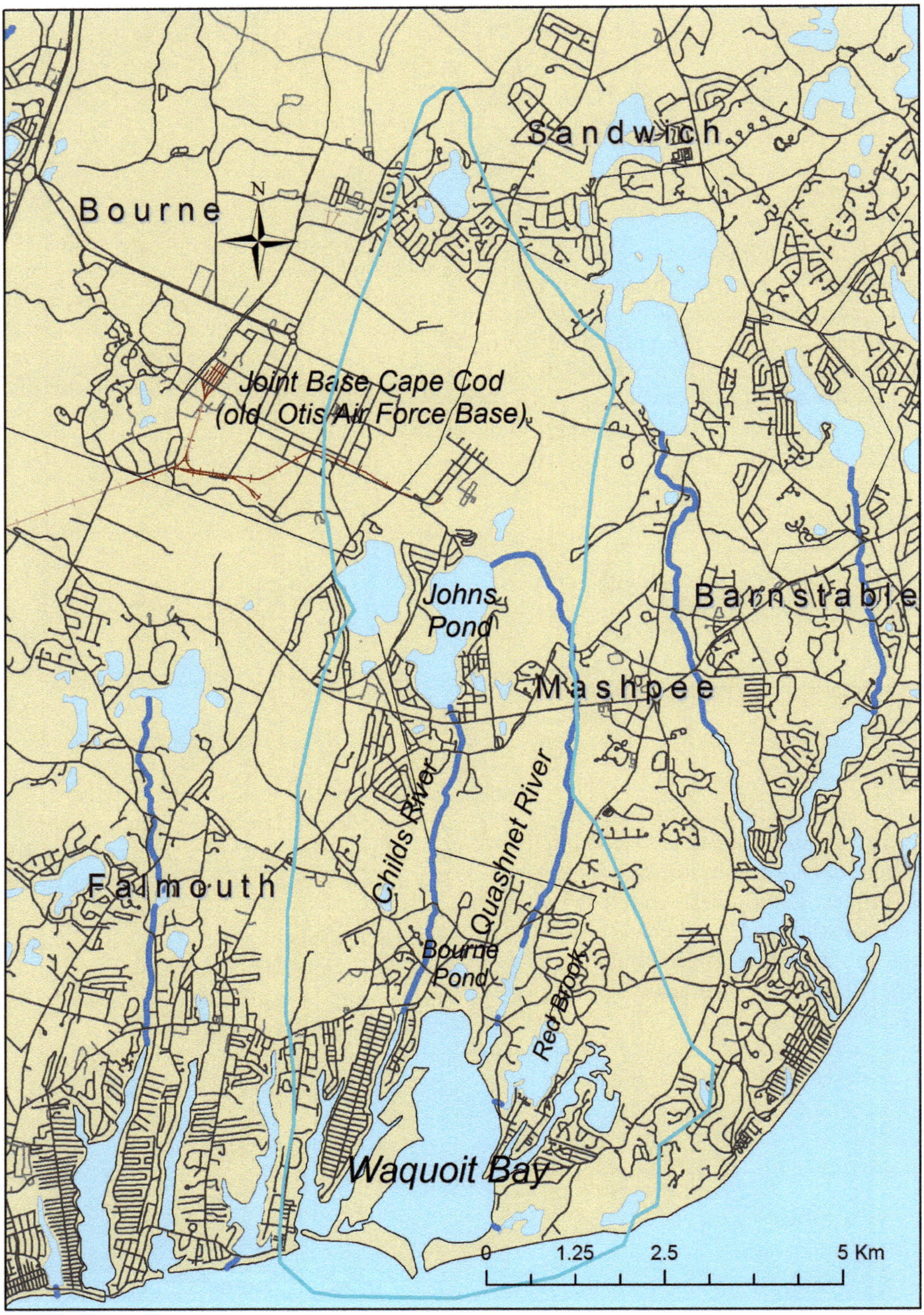

Figure 2. The watershed (groundwater) outline of Waquoit Bay and the present road network.

Johns Pond

Johns Pond is a 128.2-ha natural kettle hole pond with an average depth of 7 m and a maximum depth of 19.8 m. Transparency is generally good, extending to 4–6 m in the summer, and aquatic vegetation is scarce, limited primarily to the cove areas at the pond's southern end. The bottom is composed primarily of sand, although there are some areas of rubble and gravel. The pond has 6 km of shoreline, which is heavily developed with residential homes and beaches. The pond drains into the Quashnet River at the north end through a controlled outlet and fish ladder, and once flowed through a gated outlet to the Childs River in the south end.

The fish community consists of typical warmwater species for the area, including Largemouth Bass *Micropterus salmoides*, Smallmouth Bass, Yellow Perch, White Perch, Banded Killifish, Tessellated Darter, Brown Bullhead, Golden Shiner, White Sucker, Pumpkinseed, and Bluegill (Table 1). Migratory fish include Alewife and American Eel.

Smaller tributaries

Red Brook flows 2.2 km at a gradient of 1.9 m/km from its headwaters to Hamblin Pond and connects to Waquoit Bay by the tidal Little River. Red Brook is composed of a series of abandoned cranberry bogs that are now flooded. Flow ranges from 0.02 to 0.06 m^3/s. Mullan (1957) described Red Brook as a short stream originating in a spring-fed Atlantic white cedar *Chamaecyparis thyoides* swamp that flowed through run-down semi-abandoned bogs. Trout fishing was good until 1947, and Mullan (1957) noted that a few Brook Trout may persist in spring holes in the swamp.

Abigails Brook flows 1.4 km through abandoned cranberry bogs at a gradient of 1.9 m/km from Witch and Jim ponds to Great River and Jehu Pond. A small freshwater outlet from Bourne Pond (4 ha) flows into Bog Pond before entering the tidal Caleb Pond. The other freshwater inputs to the Waquoit Bay are primarily shoreline and subsurface seeps and intermittent freshwater tributaries.

History and Threats to the Watershed

Early history and European colonization

Native Americans of the Wampanoag tribe of the Algonquian grouping colonized the area about 5,000 years BCE and established year-round and seasonal settlements in the Waquoit Bay area (Church 2013). The primary food resources were abundant shellfish supplemented by wild game, saltwater and freshwater fish, and agriculture (corn *Zea mays* and other crops). A Wampanoag legend attributed the formation of the Santuit River, a similar system to the Waquoit Bay tributaries, to "a great trout from the sea" that formed the channel by pushing up from the sea to hear the songs of a Wampanoag princess (Works Progress Administration 1937).

In the late 1500s and early 1600s, the Cape Cod area was visited by occasional fishermen and early explorers who brought diseases that decimated whole villages and greatly reduced the Native American population. Major landscape changes to the area started in 1620 with the arrival of the Pilgrims to Plymouth, about 32 km to the north. Their arrival was closely followed by the arrival of the Puritans to the Boston area in 1628, and European colonization spread throughout Massachusetts as the population grew and more immigrants arrived. As the settlers need for land increased, the remaining Native Americans were reduced to isolated villages and the district of Mashpee was formed by the Plymouth court in 1665. This district encompassed most of the Waquoit Bay watershed and the tribe was then known as the South Sea Indians.

Construction of mills and dams

While poor soils limited agricultural development of the watershed, the demand for grain mills and sawmills resulted in the first dams being built in the late 1700s. Potential mill sites on the Quashnet and Childs rivers led to the lower portions of these rivers and nearby lands being annexed to the town of Falmouth in April 1725 (Lynch et al. 1986). A dam was constructed on the Quashnet River near the head of tide at the present day Martin Road for a grist mill around 1795. After the mill burned down, the dam was rebuilt for a shingle mill in 1832 (Oldenbourg 2010). On the Childs River, a dam was built before 1825 for a carding mill and rebuilt in 1855 to power a carriage shop (Deyo 1890). Massachusetts is now estimated to have more than 2,000 dams on its rivers, not including small barriers formed by cranberry bog sluices, road culverts, and other obstructions.

Cranberry culture

Cranberry *Vaccinium macrocarpon* is a native berry used by the Wampanogs. Cranberry culture first began on Cape Cod in 1816. After the mill on the Quashnet River burned on July 8, 1894, the former impoundment was converted to cranberry bogs. An Atlantic white cedar swamp known as Whitcombs cedar swamp was cleared and converted into cranberry bogs. By the 1920s and 1930s, almost the entire length of the river had been converted into cranberry bogs.

Cranberry bog culture requires clearing riparian trees that would shade the berries, building small dams (dikes) and constructing main, perimeter, and lateral ditches to control water levels. Sand, fertilizer, and pesticides are applied to encourage vine growth and control pests. Sand was usually obtained from the valley walls, which caused a widening of the river valleys. Rotenone was commonly used as a cranberry bog pesticide in the 1950s and caused documented fish kills in the Mashpee River in 1952. In 2000, a fish and frog kill due to Diazinon, a cranberry bog pesticide, was reported in the upper Childs River. The removal of the riparian canopy, stream-channel alterations, and water-level manipulations led to warming water and declines in trout abundance in these coastal streams. Cranberry culture also impacted most of the Childs River, Abigails Brook, and Red Brook.

Nitrogen loading

The estuarine resources of Waquoit Bay have suffered from algae blooms and hypoxia events that are believed to be linked to increased nitrogen loading (U.S. Environmental Protection Agency 2002). The primary controllable nitrogen source (60–85%) is believed to be from onsite septic systems (Howes et al. 2013), which offer little control of nitrogen given Cape Cod's sandy soils. Over the past 40 years, the Mashpee and Falmouth area have experienced some of the highest rates of development in Massachusetts with the installation of individual septic systems at most of these sites. Portions of the watershed had a total maximum daily load for nitrogen established in 2006 by the Massachusetts Department of Environmental Protection and the U.S. Environmental Protection Agency. The total maximum daily load for the Quashnet River was established as 27 kg/d, 25 kg/d for Great River/Jehu Pond and 2 kg/d for Little River/Hamblin Pond (MDEP 2006). Background concentrations of nitrogen in the watershed are about 0.05 mg/L while the Childs River average concentration is 0.21 mg/L and the Quashnet River is 0.20 mg/L. Recent reports indicate that the watershed needs a 53% reduction in nitrogen loading to achieve water quality targets (Cape Cod Commission 2017).

Nitrogen input to estuaries has been implicated in the decline of eelgrass *Zostera* spp. due to increased algae growth. Eelgrass provides aquatic habitat to many species, including bay scallops *Argopecten irradians irradians*. Sea-run Brook Trout may have utilized eelgrass beds during their fall estuarine migration. The integrity of estuarine fish communities has degraded and many nekton species, such as bay scallops, have declined in abundance as eelgrass beds have declined in Waquoit and Buttermilk bays (Hughes et al. 2002).

Howes et al. (2013) noted that restoration of the Waquoit Bay estuary will require a reduction in the current levels of nitrogen loading and management to reduce future inputs. Due to the high cost of known control measures (i.e., sewers and wastewater treatment plants), alternative actions have been discussed. One idea is to use the denitrifying actions of instream ponds to reduce nitrogen loads; however, this may have a negative impact on coldwater habitat. Another nonstructural method would be increased shellfish cultivation of species like quahogs (northern quahog *Mercenaria mercenaria*) or oysters (eastern oyster *Crassostrea virginica*) that could provide filtration services to reduce nitrogen levels. Presently, the only practical solution is installation of sewers in parts of the watershed that experience the highest levels of nitrogen loading.

Otis Air Force Base impacts and ethylene dibromide

Johns Pond and the Quashnet River have been impacted by groundwater contamination plumes from the former Otis Air Force Base Superfund site (now known as Joint Base Cape Cod) and a high incidence of Brown Bullhead in Johns and Ashumet ponds with tumors (papillomas) from an unknown cause were noted during surveys in 2002 (Baumann et al. 2008). In 1997, the fuel additive ethylene dibromide (EDB) was discovered in the surface water of the upper Quashnet River in cranberries and in fish tissue of White Sucker and Brook Trout. This contamination from Otis Air Force Base was associated with the testing of fuel dump valves on Constellation reconnaissance aircraft that provided early warning of potential Soviet Union bomber attacks during the Cold War. Due to concerns about possible EDB in cranberries, which could be used for baby food, the cranberry crop was incinerated and plans were developed to control the pollution.

In 1999, the Air Force Center for Environmental Excellence (now Air Force Civil Engineer Center) constructed earthen berms to separate cranberry bog operations from contaminated surface water and built wells and a granular activated carbon treatment plant to clean the groundwater and discharge treated water. During final construction of the berms, a failure and subsidence of a berm occurred in June 1999. The subsidence of a portion of the berm occurred in an area of the bog built over a deep peat hole (a kettle hole that later became an Atlantic white cedar swamp). The berm has continued to subside and degrade fisheries habitats. Numerous studies to correct the damage to the river by the berm subsidence led to a plan to rechannel the river nearer to mineral soils at the edge of the now abandoned bogs (Woods Hole Group 2007), which would improve degraded Brook Trout and river herring habitat.

Although trace amounts of EDB were detected in the river for years, no ecological or human health risks were determined to exist at the time. Areas where groundwater containing EDB was upwelling into the bogs were used by Brook Trout as spawning areas and provided coldwater habitat. Considerable review and technical assistance was provided by MDFW staff to prevent habitat degradation during cleanup activities and during review of remediation proposals, one of which included a 30-acre (12 ha) pond, which would have degraded existing Brook Trout habitat.

Trout Research and Management

History of sea-run Brook Trout

The large size and abundance of sea-run Brook Trout on Cape Cod attracted some of the earliest sportfishing anglers to the area's streams. A medical doctor named Dr.

Jerome V. C. Smith in 1833 wrote one of the first books by an American about fishes and angling in the United States, entitled *A Natural History of the Fishes of Massachusetts*. As a scientific work, it was later panned by more knowledgeable ichthyologists such as D. H. Storer and Theodore Gill (Hartel et al. 2002), but it remains a valuable historical reference on the abundance and location of Cape Cod's Brook Trout fisheries. Smith (1833) reported for the Childs River,

> It was on the eighteenth of May 1829, that two persons, one of whom was an English gentleman .... took seventy fine sea trout, weighing thirty eight pounds, a number of the largest weighed about a pound and a half.

Although Smith (1833) thought highly of the Childs and Mashpee rivers, he also noted,

> There is another river which flows into Waquoit Bay, known by the Indian name of Quashnut. In its tide waters, the sea-trout are sometimes collected in considerable numbers. It is at best however an uncertain place, and never can amount to much as a trout stream on account of the mill which obstructs their progress to its upper waters, to say nothing of the stealthy practice of seining, before mentioned.

Sea trout was an early name for the anadromous form of Brook Trout, now known as salters or sea-run Brook Trout.

In 1840, Samuel Trescott Tisdale, noted angler and introducer of black bass to Massachusetts, reported on a 3-week fishing trip to southeastern Massachusetts and Cape Cod in the *Spirit of the Times*, focusing on the Mashpee River but noting other trout streams, including the Childs River:

> May 28th, started for Agawam, and on our way crossed with several promising streams, into all of which we expected 'to drop.' The first of these is known as Childs River-a place which in former times disputed claims with Marshpee – and nearing this the order was given to holt...We rigged with reluctance and moved from place to place with little luck, and as all hopes of success seemed failing, we were on the point of "giving up," when a jerk was made by the admiral (for he was now promoted), followed which came fluttering to the shore a large sized trout, confirming beyond all doubt, the opinion which he had so strenuously maintained...With buoyant hearts we began afresh, and as the young blood rolled in, the old trout rolled out. Until we piled up 37 old salters, drawn from the deepest run of the river. [From Tisdale 1840.]

Another early fishing report in the 1843 *Spirit of the Times* reporting catching 100 trout from the Quashnet River:

> The 27th of the month brought our last week to a close, and ever anxious to explore untrodden ground, we took a look at a river, said to be full of Brook Trout, known as the Quoshnet River...Most of the entire lengths may be forded but the track is through a tangled thicket and difficult to make rapid headway. At very sandy place we could see hundreds of trout, some apparently of large size darting back and forth and appearing to be as plenty as herrings. We fished for an hour or two and caught upwards of a hundred of them, just for the sport, and then departed. [From Keay 1998.]

Early Brook Trout management and regulations

The concern for the decline in sea-run Brook Trout fisheries was in evidence by laws passed to protect trout as early as 1849 for the Mashpee River and in 1864 for the Quashnet River (Massachusetts Secretary of the Commonwealth 1887). The 1849 act established an open season for trout fishing from April 1 to September 15, a prohibition against taking of trout except by hook and line, and established the requirement for a permit from the town of Mashpee, with a fine of 50 cents per trout taken in violation of the law. The 1864 Quashnet River act established a similar open season from March 20 to September 15 and increased the fine to US$20 per violation. Many of the better trout streams in southeastern Massachusetts and Cape Cod in the mid- to late 1800s to the early 1900s were preserved or reserved for fishing by payments of fees to landowners. The fishing rights to the Quashnet and other rivers were leased by the town of Mashpee for an annual fee of up to $125 in the early 1900s (Golder 1994).

Decline of sea-run Brook Trout

A quote from the Massachusetts Commissioners of Fisheries (1869) report illustrates the importance of Cape Cod streams and their sea-run Brook Trout.

> Cape Cod is especially adapted to the culture of the finest trout for it is not the celebrated Marshpee River alone which is capable of producing them. Every tiny creek or rill that empties into salt water is available for their culture. The trout of the Marshpee after spawning in the river go down stream and by December 1st are all collected at its mouth where numerous springs gushing from the high banks furnish a plentiful and equible water here they lie during the winter sometimes even buried in the mud all but the head. They are then dark and in poor condition. In sunny warm days they venture forth and are sometimes killed from being caught in shoal water by a cold snap when the ice makes round them before they are aware of their danger. In the middle or end of March they push out into the brackish water of the bay going back and forth with the tide and feeding voraciously on the small shrimps which there abound. About the middle of May they begin to run up the river and they are then in beautiful condition with dark green backs and silvery sides and the roof of the mouth light colored. The great run of prime fish is in June. During the summer heats they lie in the deep holes under the banks and gradually lose their sea coat taking on the spots and mottles of a Brook Trout from which they cannot readily be distinguished when kept for a year in confinement. So far as is known all so called Brook Trout will go to the sea if they get a chance but will not therefore resemble each other. When a trout is highly fed he becomes pig bodied and his head seems diminutive (Marshpee variety) but if his food must be worked for his body becomes long and graceful and the head is conspicuous (Red Brook variety) finally if he be starved for part of the year the head becomes disproportionate and the body is small. The growth of sea going trout is rapid and a marked fish from Quashnett Brook which was put in Marshpee River weighing 4 oz was taken in two years and had gained one pound in that time.

When Theodore Lyman III conducted his survey of weir-net fisheries off Waquoit Bay in the 1870s, the former catches of Brook Trout in weir fisheries in Vineyard Sound was a memory, as no trout were captured during his survey and a comment was made by Abishai Phinney:

> My personal observation has been mostly on the trout and halibut. These are fishes, the scarcity of which has never been attributed to the weirs, and the trout now are very scarce with us indeed. [From Lyman 1872.]

In the 1950s, small populations of Brook Trout still existed in some Cape Cod streams but at a fraction of their historical abundance. A noted Woods Hole Oceanographic Institution scientist named Dr. John H. Ryther did some of the earliest scientific studies on Cape Cod's salter trout, which was summarized by Mullan (1958) in a publication titled *The Sea Run Brook Trout Fishery of Cape Cod*. Dr. Ryther experimented with overstocking these streams with hatchery Brook Trout from the Sandwich State Fish Hatchery with the hypothesis, based on some Pacific salmon studies, that overstocking would force more trout out to sea to become salters. No great increase in salter numbers were noted, but some straying between streams was documented. In 1956, the MDFW purchased 12.5

ha of former cranberry bog that bordered about 1.6 km on both sides of the Quashnet River to the north of Martin Road in order to protect the Brook Trout. This was one of the first land purchases by the state to protect a trout stream.

Hatchery Brook Trout and stocking

Due to the decline of Brook Trout in Massachusetts waters caused by mill dam development and overfishing, some of the first experiments in trout culture were carried out in nearby Sandwich at Shawme Ponds to the north by Captain Nathaniel Atwood in 1856. Although the eggs were successfully fertilized, they died of fungus before hatching (Massachusetts Commissioners 1857). Trout culture is still an important component of supplying the demand for trout fishing in Massachusetts. The Sandwich State Trout Hatchery on the north shore of Cape Cod celebrated its 100th anniversary of supplying trout to Massachusetts anglers in 2012. Annually supplying about 27,000 kg of trout, it is the primary supply for the 27 ponds and two streams currently stocked on Cape Cod. Brook Trout were identified as a species of greatest conservation need in the 2015 Massachusetts State Wildlife Action Plan (MDFW 2015).

The Quashnet River was stocked as early as 1901 (and probably before) with Brook Trout and on multiple occasions in later years. It was stocked annually between 1958 and 1981 with Brook Trout, Brown Trout *Salmo trutta*, or Rainbow Trout *Oncorhynchus mykiss* and was also stocked with landlocked Atlantic Salmon *Salmo salar* in 1972–1973. The Quashnet River was also stocked with Rainbow Smelt in 1918 and White Perch in 1922. Between 1907 and 1931, the Childs River was stocked with Brook Trout, White Perch, and Rainbow Smelt (Mullan 1957). The Childs River was annually stocked by the Falmouth Rod and Gun Club in the 1950s with Brook Trout, Brown Trout, and Rainbow Trout. Johns Pond is annually stocked with Rainbow Trout, Brook Trout, and tiger trout (Brown Trout × Brook Trout hybrid) and was stocked in the past with Brown Trout.

Brown Trout stocking and management

By the 1960s, fisheries managers and anglers had essentially given up on salter Brook Trout, and management focused on the introduced European Brown Trout, which grows to a larger size and is able to tolerate higher water temperatures. The Quashnet River was annually stocked with Brook Trout and/or Brown Trout until 1975, when the MDFW began monitoring sea-run Brown Trout. Based on successful stocking of Connecticut coastal streams using Scottish, English, and Danish strains of anadromous Brown Trout, the MDFW "looked to sea-run Brown Trout as a vehicle for better utilization of our estuarine resources" (Bergin 1984). Cape Cod salter streams were heavily stocked (15,000–25,000 fish per year) with hatchery Brown Trout from 1975 to 1988. Several 2.7–3.6-kg sea-run Brown Trout were caught by anglers throughout this time period. In the late 1970s and early 1980s, a program was developed to introduce new strains of Brown Trout and recapture returning sea-run Brown Trout in the fall for broodstock collection and spawning at the Sandwich State Fish Hatchery (Bergin 1990). Brown Trout ranging from 15 to 23 cm were heavily stocked and the breeding program was initially successful with an increase in sea-run Brown Trout returns from 0.5% to 1.5%. However, it was officially terminated in the late 1980s following annual declines in Brown Trout returns. The Childs and Quashnet rivers continued to be stocked with Brown Trout. Bergin (1984) believed that the sea-run Brown Trout fishery kept fishing pressure off anadromous Brook Trout. The potential for competition was recognized, but an abundant forage base in the form of young-of-year Alewife and abundant habitat other than Brook Trout habitat were thought to be neutralizing factors.

In the early 1990s, the MDFW decided to stop the stocking of Brown Trout in some streams to focus on management of native Brook Trout. Brown Trout stocking was terminated in the nearby Mashpee River after 1992 and in the Santuit River after 1993. The Quashnet River was last stocked with Brown Trout in 1994, and the Childs River Brown Trout stocking was reduced to a token amount and then terminated after 2007. Brown Trout populations rapidly declined (Figure 3), even though Brown Trout reproduction and a few tiger trout hybrids were noted in previous surveys.

An increasing focus on restoring sea-run Brook Trout led to a revision of trout harvest regulations in 2000. Prior to this, in an effort to protect sea-run Brown Trout broodstock, Brown Trout harvest was set at a two fish daily bag limit while eight native Brook Trout could be harvested. Catch-and-release regulations were expanded to the entire length of the Quashnet River, and the sea-run Brown Trout streams (including the Childs River) were changed to the regulatory category of major rivers, which reduced the trout bag limit to three fish per day.

Brook Trout genetics

In the mid-1990s, interest increased in restocking the Quashnet River with different strains of sea-run Brook Trout. It was de-

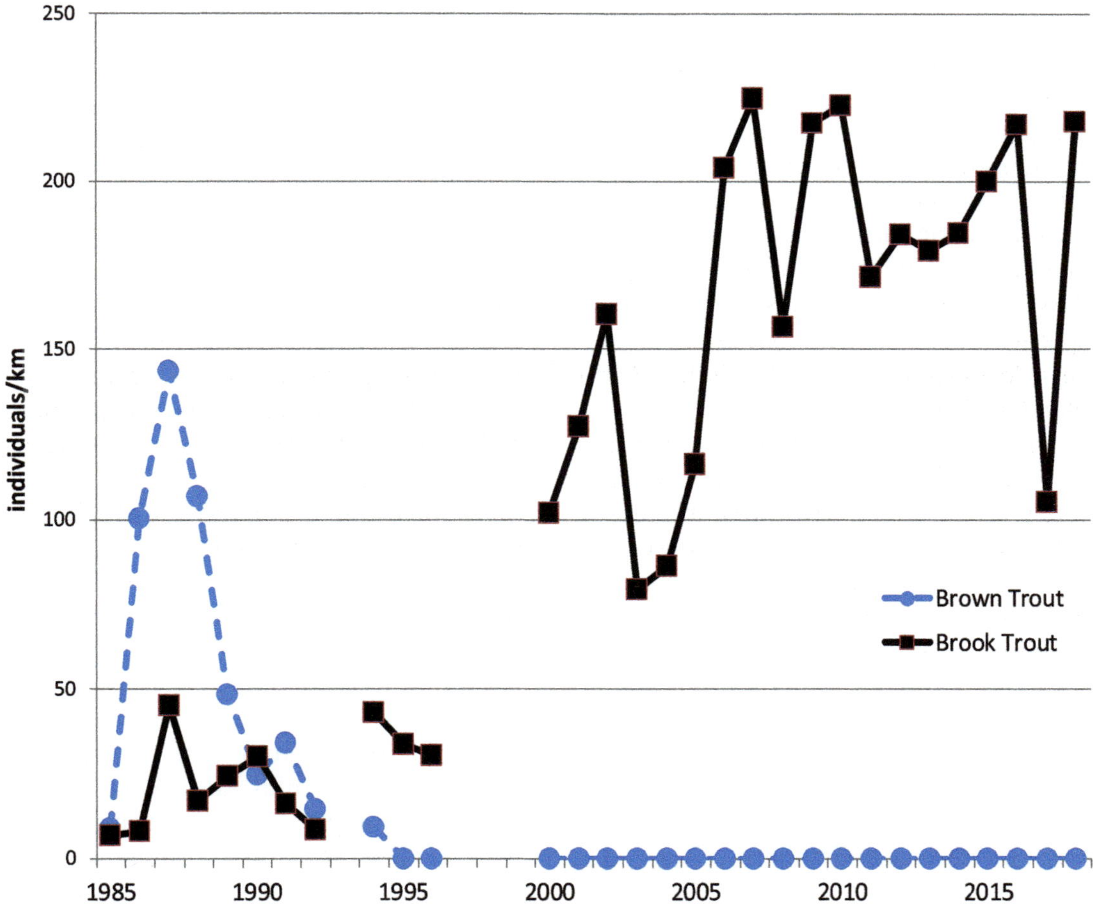

Figure 3. Fall electrofishing catch per effort (individuals per kilometer) of Brook Trout and Brown Trout in the Quashnet River, 1985–2018.

cided the best way to spend Trout Unlimited (TU) funds prior to stocking was to commission a review study of anadromous Brook Trout throughout their range. Ryther (1997) found that anadromous Brook Trout had declined throughout their range due to habitat loss, overfishing, and competition with introduced species.

The status and origin of the strains of wild Brook Trout in Cape Cod coastal streams after years of historical stocking was an open question that led to desires for stocking strains of nonnative Brook Trout that were known to be anadromous. The genetics of Brook Trout collected from the Quahsnet River in 2003 were compared to Brook Trout collected from the Mashpee River, Santuit River, Red Brook (Wareham/Plymouth), and Connetquot River on Long Island (Annett et al. 2012). Brook Trout from each stream were genetically different from the Sandwich State Fish Hatchery Brook Trout, indicating that there was little evidence that heavy Brook Trout stocking had caused genetic introgression with hatchery fish. Each stream possessed a distinct Brook Trout population, which suggests that locally adapted populations likely persisted in the Quashnet River and other streams (Annett et al. 2012).

Brook Trout tagging

In 2007, the Waquoit Bay National Estuarine Research Reserve (WBNERR), MDFW, and Cape Cod Chapter of TU started a passive integrated transponder (PIT) tagging study on Brook Trout in the Quashnet River, with design and construction assistance from the USGS Conte Anadromous Fish Research Center. This study has expanded to a total of seven antennas on three rivers: Quashnet River (n = 3), Childs River (n = 1), and Red Brook in Wareham/Plymouth (n = 3). As of January 2019, in the Quashnet River, 3,873 trout have been tagged, 741 trout have been recaptured on one to five occasions, and 93,094 antenna detections of 2,394 fish have been made in the Quashnet River. In the Childs River, 885 trout have been tagged, 167 have been recaptured on one to four occasions, and 31,403 antenna detections of 328 fish have been made. Future publications are planned on the movements and growth of PIT-tagged Brook Trout in these rivers.

Brook Trout transplantation

The success of Brook Trout restoration efforts on the Quashnet River from habitat improvement and management changes led to interest in restoring a Brook Trout population to the Childs River by the Falmouth Rod and Gun Club. In 2006, an extensive electrofishing survey of the Childs River was conducted and a single adult Brook Trout was captured (probable migrant from the Quashnet River). This survey, along with previous surveys, confirmed that a reproducing population of Brook Trout no longer existed in the Childs River. Stream temperature monitoring using Onset Hobo loggers had identified the Childs River as one of the colder streams on Cape Cod.

Increased Brook Trout populations in the Quashnet River (Figure 3) and the PIT tagging study offered an opportunity to restore nearby extirpated Brook Trout populations through the transplant of wild Brook Trout broodstock. In June 2008, the Quashnet River was electrofished and 19 adult Brook Trout (146–249 mm) were PIT-tagged and transplanted into the lower reaches of the Childs River. Further introductions of 29 Brook Trout were made in May 2009 and 37 trout in May 2010. Excellent summer growth (53–96 mm) of recaptured trout was noted from the initial transplantation. Successful Brook Trout year-classes were produced in the Childs River from 2009 to 2018, as documented by annual spring and fall electrofishing surveys. Limited movement of PIT-tagged Brook Trout has been detected between the Childs River and the Quashnet River through the Waquoit Bay estuary.

Other Diadromous Fishes

Herring runs

In 1850, Mashpee authorized the digging of a channel to Johns Pond to improve the herring run in the Quashnet River. In 1867, shares were issued for the Waquoit Herring River Company and shareholders dug a canal from the south end of Johns Pond to the Childs River to produce a herring run. This fishery peaked in 1872 when 180 barrels were captured and the average catch was 80–100 barrels a year. A barrel of river herring in Maine was calculated at about 400 herring for historical reference purposes (Hall et al. 2012). Reports were made by this company to the Massachusetts Tax Commissioners Department until 1918. Belding (1921) estimated that the 1912 production of the Childs River was 90 barrels while the Quashnet River production was estimated at 25 barrels. Belding (1921) commented for the Childs River that "[a] larger fishery could have been maintained if it had not been for the cranberry bogs" and for the Quashnet River that "[o]wing to the fact that the stream is entirely in the hands of the cranberry bog owners, and the fishery is artificial, further development is impossible."

Some of the first scientific studies to protect diadromous fish in Massachusetts occurred near Waquoit Bay. A petition with more than 5,000 signatures to prohibit weir fisheries, which were thought to impact the river herring runs, was presented to the Massachusetts Legislature (Lyman 1871), and Lyman (1876) reported,

> The following experiment carried out for a single locality by the Commissioners in 1871 will illustrate the method. Waquoit Bay is a shallow irregular land locked many lobed fiord having at its upper end influent brooks and communicating at its lower end with the sea by a very narrow passage through a sand barrier. Five hundred yards to the west of this passage a weir projects 1,400 ft into the sea and apparently stretches across the line of approach of the alewives which are said to come in spring from the westward. Complaint was made that this weir caught the alewives that were seeking to run into Waquoit Bay in order to enter the brooks at its head and thus pass into ponds where they spawned. The complaint seemed so reasonable and consistent as scarcely to need proof and partly because the case promised to be simple the Commissioners hired the weir for one season and put a man in charge who kept a daily record of fish taken and of temperature of air and water. The result was this first that about four times as many alewives passed into the brooks at the bay's head as were captured in the weir outside the mouth of the bay secondly that of the alewives that sought the brooks none were taken by the weir on their passage from the sea into the bay. In other words the proved result was diametrically opposed to the untried theory.

River herring (Alewife and Blueback Herring) utilize Waquoit Bay and associated tributaries for their annual spawning runs. Recent run-size estimates, based on volunteer counts using MDMF methodology (Nelson 2006) for the Quashnet River, have ranged from 40,854 (±12,447) in 2014 to 8,252 (±4,470) in 2017 (J. Muramoto, Association for the Preservation of Cape Cod, personal communication). Reback et al. (2004) listed seven obstructions on the Quashnet River and stated that cranberry bog water use sometimes resulted in lower than adequate pond levels for herring out-migration, which has likely contributed to herring-run declines. Cranberry growing has now been abandoned on both the Quashnet and Childs rivers, and the Childs River outlet is not maintained for herring passage. Reback et al. (2004) suggested that the Childs River herring run suffered from access problems due to a subsurface outlet at Johns Pond and a deteriorated fish ladder at the Carriage Shop ponds. Future resources to support the conservation and management of a herring run in Waquoit Bay should be directed to the Quashnet River, with continued efforts to maintain connectivity through the outlet from Johns Pond to Waquoit Bay (Reback et al. 2004).

The river herring run into Johns Pond from the Quashnet River is managed by the town of Mashpee with the assistance of the MDMF, and the river is monitored for difficulties in fish passage. One common management activity on Massachusetts river herring runs by town herring wardens is the annual clearing of brush and debris from river channels before the herring migration. Concerns about the legality of this process under wetlands regulations and sometimes overzealous clearing of woody habitat for trout and other species led to the development of Stream Channel Maintenance Protocols for Diadromous Fish Passage (MDMF 2016). These protocols are used to guide restoration efforts to balance the needs for river herring passage with aquatic habitat for native Brook Trout.

American Eel

One of the most abundant year-round diadromous fish in the Waquoit Bay tributaries is the catadromous American Eel. American Eels are common in Massachusetts rivers, streams, and ponds connected to the ocean (Hartel et al. 2002), but their abundance has declined rangewide (Haro et al. 2000). The juvenile form of American Eel, called glass eels, are commonly observed migrating up the Waquoit Bay tributaries in the spring. American Eels continue to dominate the fish fauna in most of the Waquoit Bay tributaries, but their historical abundance was much higher. An indication of the former abundance of this resource is given by Goode (1887):

> The eel fishery gives employment every winter to about one hundred men belonging in Waquoit and East Falmouth. A large proportion of eels taken are caught in Waquoit Bay. About 300 barrels are shipped to New York annually .... No considerable fisheries are carried on at East Falmouth except in winter when about thirty or forty men engage in spearing eels.

American Eel would commonly winter in the mud in Waquoit Bay and other estuaries where local residents would drill holes in the saltwater ice to probe the mud with multi-tined, long-handled spears until a concentration of eels were located. As recently as 1967, more than 455 kg of eels were harvested from the bay itself (Curley et al. 1971), but by 2009, the overall Massachusetts landings had declined to historic lows of about 909 kg, statewide (Shepard 2015). This practice is still alive only in the memories of current residents and local tribe members due to the decline of American Eels and the lack of safe saltwater ice in recent decades.

White Perch

White Perch are still found throughout the Waquoit Bay watershed, but the anadromous populations are believed to have declined in abundance. As recently as the 1980s, the early-spring White Perch runs were heavily fished, but interest in the fishery has declined as the runs have diminished. Anadromous populations of White Perch were trapped and transferred in large numbers from the coastal ponds in Falmouth and Martha's Vineyard to inland waters in the early 1900s. These anadromous populations became landlocked in many Massachusetts ponds (Hartel et al. 2002), resulting in smaller individuals compared to anadromous White Perch.

Striped Bass

Striped Bass are occasionally observed in the Quashnet River and are common in Waquoit Bay during the spring, summer, and fall where they support an active sport fishery. Spawning has not been documented north of the Hudson River. However, juvenile Striped Bass are often observed in limited numbers in the freshwater sections of coastal streams in Massachusetts during the spring, summer, and early fall.

Rainbow Smelt

Rainbow Smelt once ranged as far south as Virginia and may have entered almost every coastal stream in Massachusetts (Hartel et al. 2002). Rainbow Smelt were captured in small numbers in 1967–1968 in Waquoit Bay with shrimp trawls and beach seines (Curley et al. 1971). Rainbow Smelt were stocked in the Quashnet River in the early 20th century, but spawning runs have not been observed in the Waquoit Bay tributaries or other Cape Cod streams (Enterline et al. 2012 and Brad Chase, MDMF, personal communication). Landlocked populations, introduced by stocking of eggs or fry, still occur in Higgins Pond in Brewster and were once found in other Cape Cod ponds. They can still be found in limited numbers in the Weweantic River in Wareham in Buzzards Bay to the southwest but have greatly declined throughout Massachusetts and their southern range (Chase 2009; Enterline et al. 2012). Their potential for restoration to the Waquoit Bay tributaries is low due to declines in the southern populations of Rainbow Smelt and limited spawning habitat.

Conservation and Restoration

Land protection

The rapid development of Cape Cod, and particularly the Mashpee and Falmouth area, led to the designation of an area of critical environmental concern in 1979, which increased environmental oversight by state permitting agencies. Proposed development of Washburn Island led to the formation of Citizens for the Protection of Waquoit Bay, an environmental advocacy group, in 1981. In 1984, with the efforts of the Citizens for the Protection of Waquoit Bay and others, the WBNERR was formed. This is one of 29 reserves designated by the National Oceanic and Atmospheric Administration to protect and study coastal estuarine systems in partnership with state and local groups.

On July 28, 1988, with the support of TU and the Quashnet Coalition, the MDFW, and the Massachusetts Department of Environmental Management (now the Massachusetts Department of Conservation and Recreation) purchased 146 ha of land surrounding the Quashnet River from the Forward Development Corporation for $10 million. The land is comanaged by the Massachusetts Department of Conservation and Recreation and the MDFW. The commonwealth's purchase was subject to an easement for a 0.007 m^3/s well site reserved by the Forward Development Corporation to supply a nearby planned 180-unit condominium development and to allow the Mashpee Water District to extract water in the volume permitted by the commonwealth. These potential water withdrawals raised direct concerns by the MDFW about the impacts of future wells on riverine resources. There were also concerns about the supplied water being used to fuel development of up to 1,214 ha of open space in the town of Mashpee, which provided fisheries and wildlife habitats.

Concerns about Waquoit Bay and water-quality impacts to its fisheries have led to many land purchases for open space by a variety of partners. As of 2019, about 15% of the 6,079-ha Waquoit Bay watershed (Figure 4) has been protected as open space by the Commonwealth of Massachusetts, the Falmouth Rod and Gun Club, the town of Mashpee, the U.S. Fish and Wildlife Service (USFWS), and other partners of the Mashpee National Wildlife Refuge (formed in 1995).

Groundwater studies

A USGS gaging station was established on the Quashnet River in response to concerns of the MDFW and TU about the impact of future water extraction on the river and its trout fisheries. Baevsky (1991) reported a link

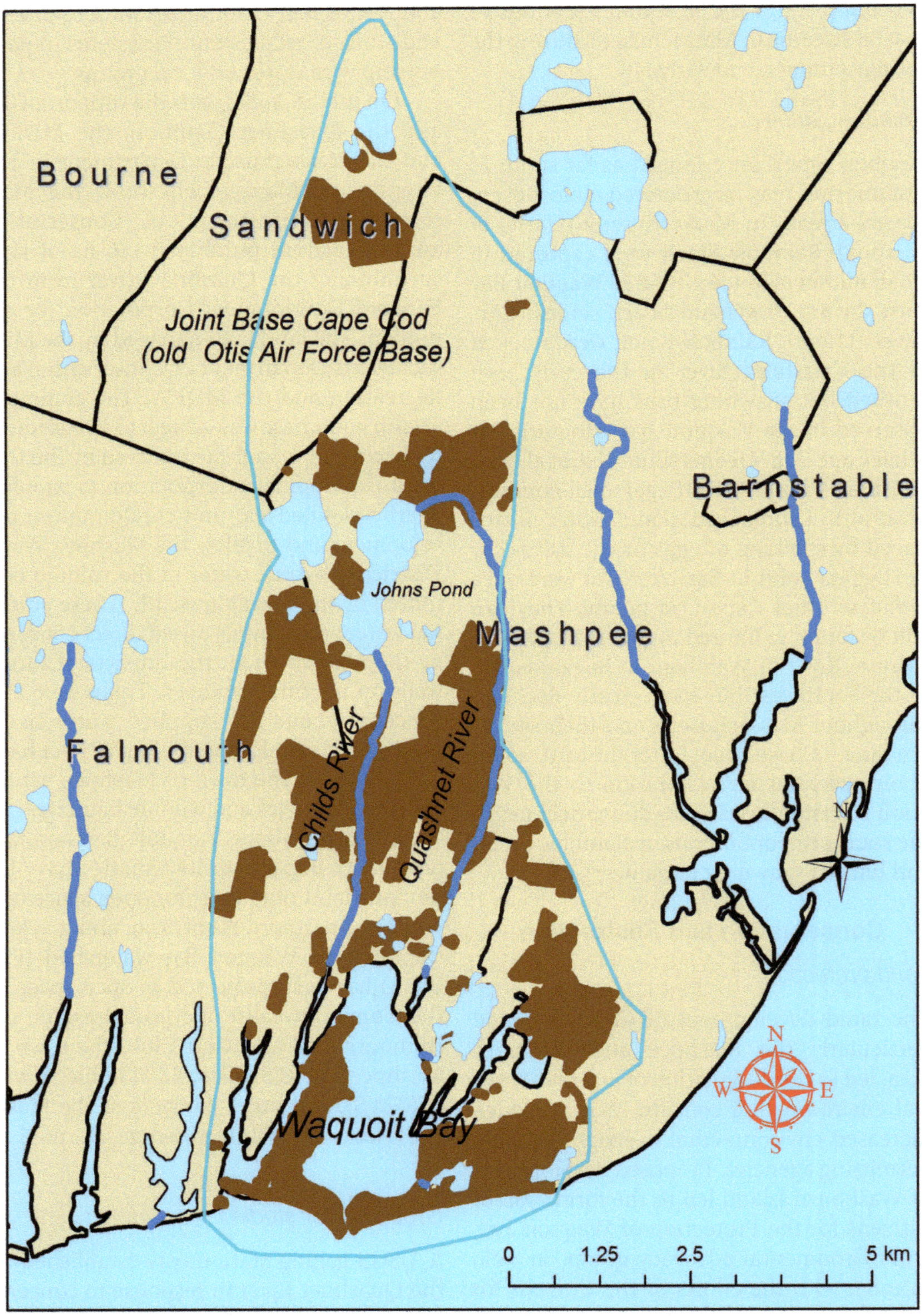

Figure 4. Protected areas (darker) as of 2018 within the Waquoit Bay watershed (groundwater) overlaid on a regional map.

between increased cranberry bog releases and Brown Trout spawning in the Quashnet River and recommended more detailed analysis of groundwater and streamflow to assess potential impacts of groundwater withdrawals. Barlow and Hess (1993) found that a well drawing 0.04 m^3/s would reduce the river flow by intercepting groundwater that would have been supplied to the river. In 1994, the MDFW determined that the well would have a negative impact on fisheries resources by reducing groundwater flow in areas used for Brook Trout spawning and approval for the well was denied. The Mashpee Water District later developed two municipal wells along Turner Road to the west between the Quashnet and Childs rivers. In 2013, there was continued interest by the Mashpee Water District in the potential well sites despite having developed the two new wells and taking over another smaller well site along the Quashnet River near Route 151.

Trout Unlimited improvements

Planning of a habitat restoration project for the Quashnet River was initiated by TU in 1974–1975, and the first instream habitat work began in 1976. By this time, the abandoned cranberry bogs purchased by the MDFW in 1954 were overgrown with brush and lacked a defined stream channel in many areas. One of the first projects involved stabilizing a failed culvert bank at Martin Road. Still underway, this project has devoted more than 24,000 man hours to cutting brush, placing the brush on the banks to build up the streambanks, and building bank hides and deflectors (Bergin and Smith 1992). The Cape Cod Chapter of TU has been the primary driving force behind the restoration and has received numerous national awards (Golder 1994). The MDFW and TU also helped stop the filling of wetlands as part of an expansion of the Quashnet Valley Golf Course along the upper reaches of the river in 1983. To help enhance Brook Trout recovery efforts, the river between Route 151 and Route 28 was designated a catch-and-release area by the MDFW in February 1986. A survey completed in 2007 by the WBNERR documented more than 300 habitat improvement structures in the lower river. Presently, improvement efforts have been focused on planting trees and building more structures of cedar logs and boulders to increase habitat availability.

Stream restoration activities

In addition to the restoration activities of TU, the water level in the former impoundment above the MDFW-owned bogs was gradually lowered from 2000 to 2014 by removing boards in the dam to reduce summer water temperatures and improve fish passage. The remaining boards were removed with the assistance of the Student Conservation Association in 2014. While restoration continued throughout the watershed, state agencies and local nongovernmental organizations desired a more comprehensive look at habitat restoration projects and future opportunities in the Waquoit Bay tributaries. The Waquoit Bay stream restoration plan was completed in 2010 (Louis Berger Group 2010) and prioritized ecological restoration opportunities across the watershed. This plan facilitated the removal of a culvert on Abigails Brook and removal of two failing concrete bog flumes on the Quashnet River by the USFWS. Future priorities in the report include breaching or removal of the Carriage Shop dam on the Childs River and the rechanneling and wetland restoration on the upper Quashnet River. Both of these projects continue to move forward to the permitting stage.

Bournes Pond outlet

A small freshwater tributary of Waquoit Bay originates in Bournes Pond (a small kettle hole pond) and flows through a shallow impoundment (Bog Pond) before entering the tidal Caleb Pond. Alewife were historically in this small stream with records from

Lyman (1872), indicating that 25,000 Alewives had been taken in the mid-1800s. By 2009, only a few river herring were noted. The town of Falmouth Department of Natural Resources and the nearby landowner expressed interest in reviving the fishery. The MDMF undertook a study to assess the river herring run and developed a plan for the removal of defective culverts and flumes and the construction of a fish ladder, which was designed by USFWS engineers. Restoration projects were completed in 2017 (Brad Chase, MDMF, personal communication), and these measures should allow increased passage by river herring and American Eels.

Childs River restoration

Restoration efforts in the Childs River have been largely led by the Falmouth Rod and Gun Club. This organization was instrumental in the initial success of the Brook Trout transplantation project and has worked to pursue ownership/management control of the cranberry bogs formerly owned by the town of Falmouth. Restoration plans were developed in 2017–2018 that included previously discussed options (Louis Berger Group 2010) to mitigate surface-water warming (Figure 5) caused by the ponds owned by the Rod and Gun Club on Carriage Shop Road. Plans are currently being developed to re-

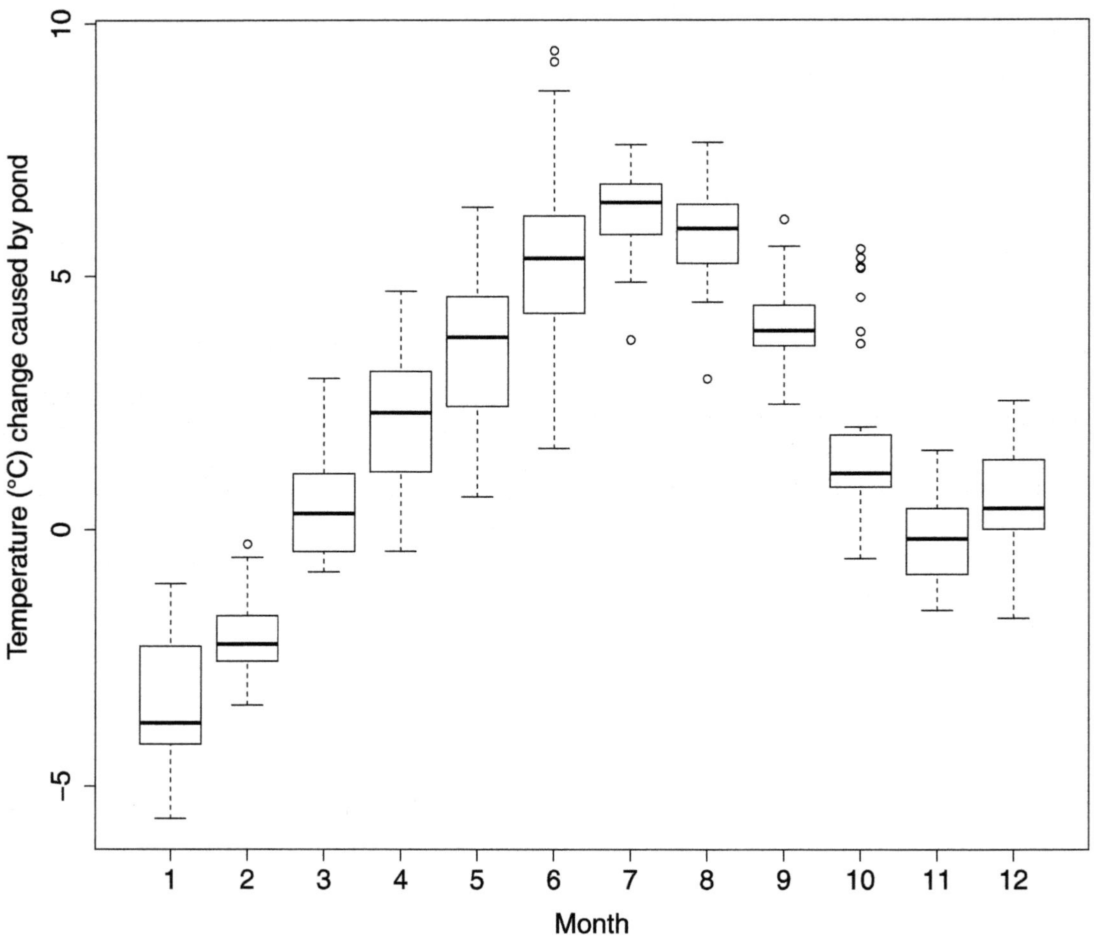

Figure 5. Increase in temperature (°C) caused by small ponds on the Childs River, Falmouth, detected by recording electronic thermographs placed above and below the ponds.

move the dam and ponds, relocate the channel from the center of the bogs to the side, and restore riparian and wetland habitats.

Restoration and climate change

Climate change, in the form of changes in precipitation patterns, storm frequency, and water temperatures, is embedded in the history of Waquoit Bay tributaries and Cape Cod. Climate change will ultimately affect the diadromous fisheries of Cape Cod, with many of these effects already apparent. Southern species are now found north of Cape Cod, and growing seasons are lengthening. The man-made connection to headwater ponds may impact the coldwater fisheries of Waquoit Bay. These ponds, a reflection of groundwater levels, have profound effects on downstream temperatures. High groundwater levels coupled with higher summer temperatures and constrained pond-front infrastructure may cause increased summer spillover flows and result in the reduction or elimination of coldwater habitats in the tributaries. The apparent loss of the Brook Trout population in the nearby Santuit River, discovered in 2015, may be a harbinger of the future for other sea-run Brook Trout fisheries of Cape Cod. Past and ongoing efforts to protect the watershed of Waquoit Bay and implement instream habitat improvements may delay or prevent these impacts.

Summary

Waquoit Bay has suffered from large-scale landscape changes and other human impacts that have caused declines in the abundance of diadromous fishes. Landscape preservation, management changes, and habitat improvements have supported population increases in wild Brook Trout, maintenance of Alewife and Blueback Herring runs, and aquatic passage for American Eels. Data are lacking on other diadromous species of Cape Cod and Waquoit Bay, such as Rainbow Smelt, Sea Lamprey, White Perch, and Atlantic Tomcod. These species have all suffered range wide declines likely caused by large-scale environmental change and have limited potential for restoration to the Waquoit Bay tributaries.

Recognizing the linkage between water quality in Waquoit Bay and its watershed, federal, state and local agencies; private conservation groups; and the general public have worked together to preserve a large portion of the Waquoit Bay watershed. Land protection may offer the best hope for preservation of diadromous fisheries by protecting water quality and adequate streamflow. Many challenges remain to the continued presence of the diadromous and freshwater fisheries of the Waquoit Bay tributaries in the form of climate change, nutrient and contaminant loading, and other man-made changes in hydrology. At the local and regional scale, a large amount of resources has been directed to protecting this coastal watershed. Only our descendants will be able to decide if the past, present, and future efforts to preserve and protect the environment of Waquoit Bay and its natural resources were sufficient.

Acknowledgments

Thanks are due to the many people that have worked to protect Waquoit Bay and its rivers over many years, including Fran Smith and many other members of Trout Unlimited, Brendan Annett and Jim Rassman and the staff of the Waquoit Bay National Estuarine Research Reserve, and Steve Baker and other members of the Falmouth Rod and Gun Club. This paper has benefited from the input of two anonymous reviewers.

References

Annett, B., G. Gerlach, T. L. King, and A. R. Whiteley. 2012. Conservation genetics of remnant coastal Brook Trout populations at the southern limit of their distribution: population structure and effects of stocking. Transactions of the American Fisheries Society 141:1399–1410.

Baevsky, Y. H. 1991. Physical and water quality characteristics affecting trout spawning habitat in the Quashnet River, Cape Cod, Massachusetts. U.S. Geological Survey Water-Resources Investigations Report 91-4045, Reston, Virginia.

Barlow, P. M., and K. M. Hess. 1993. Simulated hydrologic responses of the Quashnet River stream-aquifer system to proposed ground-water withdrawals, Cape Cod, Massachusetts. U.S. Geological Survey, Water-Resources Investigations Report 93-064, Reston, Virginia.

Baumann, P. C., D. R. LeBlanc, V. S. Blazer, J. R. Meier, S. T. Hurley, and Y. Kiryu. 2008. Prevalence of tumors in Brown Bullhead from three lakes in southeastern Massachusetts, 2002. U.S. Geological Survey, Scientific Investigations Report 2008-5198, Reston, Virginia.

Belding, D. L. 1921 (reprint 1964). A report upon the Alewife fisheries of Massachusetts. Reprint 1964, Massachusetts Division of Fish and Game, Boston.

Bergin, J. D. 1984. Massachusetts coastal trout management. Pages 137–142 *in* F. Richardson and R. H. Hamre, editors. Wild Trout III: proceedings of the symposium, Yellowstone National Park, September 24–25, 1984. Wild Trout Symposium, Bozeman, Montana.

Bergin, J. D. 1990. An evaluation of five strains of Brown Trout. Pages 66–71 *in* J. C. Borawa, editor. Brown Trout workshop biology and management. American Fisheries Society, Southern Division, Trout Committee, Bethesda, Maryland.

Bergin, J., and F. Smith. 1992. A simple, inexpensive technique to stabilize stream banks on low gradient, sandy streams. Pages 19–20 *in* M. Marcinko and D. Graaff, editors. Proceedings of the East Coast trout culture and management workshop. Pennsylvania State University, State College.

Cape Cod Commission. 2017. Watershed report: Upper Cape, Waquoit Bay, Falmouth, Mashpee and Sandwich. Cape Cod Commission, Barnstable, Massachusetts.

Cardoza, J. 2015. The Massachusetts Division of Fisheries and Wildlife, 1866–2012. Massachusetts Division of Fisheries and Wildlife, Westboro, Massachusetts.

Chase, B. C. 2009. The spawning habitat of anadromous Rainbow Smelt: trouble at the tidal interface. Pages 859–862 *in* A. J. Haro, K. L. Smith, R. A. Rulifson, C. M. Moffitt, R. J. Klauda, M. J. Dadswell, R. A. Cunjak, J. E. Cooper, K. L. Beal, and T. S. Avery, editors. Challenges for diadromous fishes in a dynamic global environment. American Fisheries Society, Symposium 69, Bethesda, Maryland.

Church, N. 2013. Waquoit Bay: prehistory, history and natural history. Spritsail 27(2):20–27, Woods Hole Museum, Falmouth, Massachusetts.

Curley, J. R., R. P. Lawton. J. M. Hickey, and J. D. Fiske. 1971. A study of the marine resources of the Waquoit Bay–Eel Pond estuary. Massachusetts Division of Marine Fisheries, Monograph Series No. 9, Boston.

Deyo, S. l. 1890. History of Barnstable County, 1620–1890. H.W. Blake and Company, New York.

Eichner, E. M., T. C. Cambareri., G. Belfit., D. McCaffery, S. Michaud, B. Smith, and M. Fenn. 2003. Cape Cod pond and lake atlas. Cape Cod Commission Water Resources Office, Barnstable, Massachusetts.

Enterline, C. L., B. C. Chase, J. M. Carloni, and K. E. Mills. 2012. A regional conservation plan for anadromous Rainbow Smelt in the U.S. Gulf of Maine. Maine Department of Marine Resources, Augusta.

Geist, M. A. 1996. The ecology of the Waquoit Bay National Estuarine Research Reserve. Waquoit Bay National Estuarine Research Reserve, Falmouth, Massachusetts.

Goode, G. B. 1887. The fisheries and fishery industries of the United States. Section II: A geographical review of the fisheries industries and fishing communities for the year 1880. U.S. Commission on Fish and Fisheries, Washington, D.C.

Golder, R. J. 1994. Profiles in conservation: Francis Smith, reshaper of rivers. Trout, (summer):51–54.

Hall, C. J., A. Jordaan, and M. G. Frisk. 2012. Centuries of anadromous forage fish loss:

consequences for ecosystem connectivity and productivity. BioScience 62:723–731.

Haro, A., W. Richkus, K. Whalen, A. Hoar, W. Busch, S. Lary, T. Brush, and D. Dixon. 2000. Population decline of the American Eel: implications for research and management. Fisheries 25(9):7–16.

Hartel, K. E., D. B. Halliwell, and A. E. Launer. 2002. Inland fishes of Massachusetts. Massachusetts Audubon Society, Lincoln.

Howes, B., R. Samimy, D. Schlezinger, E. Eichner, S. Kelley, J. Ramsey, and P. Detjens. 2013. Linked watershed-embayment approach to determine critical nitrogen loading thresholds for the Waquoit Bay and Eel Pond embayment systems, towns of Falmouth and Mashpee Massachusetts. Massachusetts Estuaries Project, School of Marine Science and Technology, University of Massachusetts, Dartmouth.

Hughes, J. E., L. A. Deegan, J. C. Wyda, M. J. Weaver, and A. Wright. 2002. The effects of eelgrass habitat loss on estuarine fish communities of southern New England. Estuaries 25:235–249.

Keay, D. L. 1998. The history of the Quashnet River valley. Waquoit Bay Estuarine Research Reserve, Falmouth, Massachusetts.

Louis Berger Group. 2010. Waquoit Bay restoration master plan, Falmouth and Mashpee, Massachusetts. Report prepared for the Waquoit Bay National Estuarine Research Reserve. Louis Berger Group, Needham, Massachusetts.

Lynch, G. E., J. F. Lynch, and J. Beatty. 1986. Waquoit. Pages 402–419 *in* M. L. Smith, editor. 1986. The book of Falmouth: a tricentennial celebration: 1686–1986. Falmouth Historical Commission, Falmouth, Massachusetts.

Lyman, T. 1871. On the possible exhaustion of the sea fisheries. Pages 60–74 *in* Massachusetts Commissioners of Inland Fisheries. Sixth annual report. Wright and Potter, Boston.

Lyman, T. 1876. Sketch of the progress of fish culture. Pages 49–69 *in* Massachusetts Commissioners on Inland Fisheries. Tenth annual report, year ending Jan. 1. 1876. Wright and Potter, Boston.

Massachusetts Commissioners. 1857. Report of the Massachusetts Commissioners appointed under resolve of 1856, concerning the artificial propagation of fish with other documents. William White, Printer to the state, Boston.

Massachusetts Commissioners of Fisheries. 1869. Report of the Commissioners of Fisheries for the year ending January 1, 1869. Wright and Potter, Boston.

Massachusetts Commissioners of Fisheries. 1870. Report of the Commissioners of Fisheries for the year ending January 1, 1870. Wright and Potter, Boston.

MDEP (Massachusetts Department of Environmental Protection). 2006. Final Quashnet River, Hamblin Pond, Little River, Jehu Pond, and Great River in the Waquoit Bay system total maximum daily loads for total nitrogen. MDEP, Boston.

MDFW (Massachusetts Division of Fisheries and Wildlife). 2015. Statewide wildlife action plan. MDFW, Westboro.

Massachusetts Secretary of the Commonwealth. 1887. Laws relating to inland fisheries in Massachusetts, 1623–1886. Wright and Potter, Boston.

Mullan, J. W. 1957. A fisheries investigation of Cape Cod coastal streams and the Swift River below Quabbin Reservoir. Massachusetts Division of Fisheries and Game, Final Report FA Project F-1-R-5, Westboro.

Mullan, J. W. 1958. The sea run or "salter" Brook Trout (*Salvelinus fontinalis*) fishery of the coastal streams of Cape Cod, Massachusetts. Massachusetts Division of Fish and Game, Bulletin 17, Boston.

Nelson, G. 2006. A guide to statistical sampling for the estimation of river herring run size using visual counts. Massachusetts Division of Marine Fisheries, Technical Report 25, Boston.

Oldenbourg, N. D. 2010. Living off the land—a photo and timeline essay. Spritsail 24(2):10–26, Woods Hole Museum, Falmouth, Massachusetts.

Portnoy, J. W., M. G. Winkler, P. R. Sanford, and C. N. Harris. 2001. Kettle pond data atlas for Cape Cod National Seashore: paleoecology and modern water chemistry. U.S. Department of Interior, National Park Service, Cape Cod National Seashore, Wellfleet, Massachusetts.

Reback, K. E., P. D. Brady, K. D. McLaughlin, and C. G. Milliken. 2004. A survey of anadromous fish passage in coastal Massachusetts. Part 2: Cape Cod and the islands. Massachusetts Division of Marine Fisheries, Technical Report TR-16, Boston.

Ryther, J. H. 1997. Anadromous Brook Trout: biology, status and enhancement. Trout Unlimited, Arlington, Virginia.

Schwarzman, B. 2002. The nature of Cape Cod. University Press of New England, Lebanon, New Hampshire.

Shepard, S. L. 2015. American Eel biological species report. U.S. Fish and Wildlife Service, Hadley, Massachusetts.

Smith, J. V. C. 1833 (1970 reprint). Natural history of the fishes of Massachusetts, embracing a practical essay on angling. Freshet Press, Rockville Center, New York.

Tisdale, S. T. 1840. Trouting expedition to "the Cape." Spirit of the Times (August 1):258–259.

Woods Hole Group. 2007. Quashnet River channel relocation and restoration study. Woods Hole Group, Falmouth, Massachusetts.

Works Progress Administration. 1937. Massachusetts: a guide to its places and people. Riverside Press, Cambridge, Massachusetts.

U.S. Environmental Protection Agency. 2002. Waquoit Bay watershed ecological risk assessment: the effect of land-derived nitrogen loads on estuarine eutrophication. U.S. Environmental Protection Agency, Office of Research and Development, National Center for Environmental Assessment, Washington, D.C.

American Fisheries Society Symposium 91:479–514, 2019

Challenges and Opportunities to Apply Process-Based Restoration at Scales Appropriate to Anadromous Fishes

Kris M. Homel*
Oregon Department of Fish and Wildlife, West Region Office
17330 Evelyn Street, Clackamas, Oregon 97015, USA

Chris M. Lorion
Oregon Department of Fish and Wildlife
4034 Fairview Industrial Drive SE, Salem, Oregon 97302, USA

Benjamin J. Clemens
Oregon Department of Fish and Wildlife, Corvallis Fish Research
28655 Highway 34, Corvallis, Oregon 97333, USA

Abstract—Habitat restoration in streams is critical to recovery of imperiled fishes. In the Pacific Northwest of North America, stream habitat restoration has focused primarily on anadromous salmonids *Oncorhynchus* spp. listed under the Endangered Species Act. Substantial restoration efforts have focused on the Oregon Coast evolutionary significant unit of Coho Salmon *O. kisutch*. However, many of these efforts have been local in scale (i.e., particular reaches within relatively small streams, high up in watersheds) and do not address habitat-forming processes at the watershed scale. Other anadromous species such as Chum Salmon *O. keta* and Pacific Lamprey *Entosphenus tridentatus* that co-occur in these watersheds are rarely targeted in habitat restoration yet are thought to benefit from restoration efforts for Coho Salmon. However, Chum Salmon and Pacific Lamprey tend to occur lower in watersheds than Coho Salmon, inhabit freshwater for different periods of time, and are therefore challenged by different limiting factors. As a result, benefits from restoration efforts for Coho Salmon may not be fully realized by Chum Salmon and Pacific Lamprey. Mounting evidence suggests that process-based restoration may yield substantial biological dividends for anadromous fishes. Although process-based restoration is still an area of active research, it does possess a stronger and more holistic conceptual foundation than restoration strategies aimed at creating particular channel forms. Process-based restoration addresses the geological, physical, chemical, climatic, and ecological processes that interact to form habitat mosaics to which anadromous fishes have adapted. Understanding and working within these processes throughout entire watersheds and downstream into the estuary has the potential to benefit multiple anadromous fishes and substantially improve the ecological functions of watersheds. We explore the range of habitat relationships expressed among three imperiled anadromous fishes—Coho Salmon, Chum Salmon, and Pacific Lamprey—and present case studies to illustrate the importance of implementing process-based restoration to address limiting factors in their freshwater habitats.

* Corresponding author: kristen.m.homel@state.or.us

We conclude by suggesting three main strategies to address restoration challenges and limitations: forming multidisciplinary collaborations of restoration practitioners, investing in education and outreach to build larger and more diverse constituencies, and developing spatial and organizational tools to expand restoration efforts to the watershed scale.

Introduction

Habitat restoration in streams is a critical component of management and recovery of imperiled fishes, particularly salmonids (Bernhardt et al. 2005). Freshwater habitat is considered a limiting factor for all anadromous salmonids (*Oncorhynchus* spp.) in the Pacific Northwest of North America, and efforts to restore habitat have been ongoing for more than two decades (LCFRB 2010; ODFW 2010, Appendix A; NMFS 2013, 2016). Substantial advancements have been made in restoration practices, from early work involving small-scale, engineered projects (e.g., anchored large wood structures, bank revetments) to the current science prioritizing restoration of habitat-forming processes at the watershed scale (Ebersole et al. 1997; Bohn and Kershner 2002; Bond and Lake 2003; Beechie et al. 2010; Perring et al. 2015). Despite these advancements, no anadromous fishes have been removed from listing under the Endangered Species Act (ESA; ESA 1973). One conclusion from this is that habitat restoration alone is insufficient to recover ESA-listed fishes (Bisson et al. 2009). Another conclusion is that there is a disconnect between the types of restoration projects that have been completed and those that are required to promote recovery of anadromous fishes.

Effective habitat restoration requires an understanding of how habitat is created and how fishes respond to that habitat at various spatial scales. A few physical and biological processes create the range of habitats within streams (Knighton 1998). This physical habitat template (hereafter, "physical template") can be described as the range of physical characteristics positioned throughout a watershed (Southwood 1977, 1988). In this physical template, large-scale features (e.g., underlying geology, topography of the stream network, and climate) control stream discharge and sediment load (Knighton 1998; Benda et al. 2004; Buffington 2012). These, in turn, influence local channel forms (Knighton 1998), such as pool-riffle sequences (Frissell et al. 1986; Montgomery and Buffington 1997; Montgomery 1999). Biological processes also shape channel form (Beechie et al. 2010). For example, large wood in the riparian zone can stabilize banks and eventually recruit to the stream, changing sediment transport locally and creating new habitat features in the channel (Gurnell et al. 1995; Montgomery et al. 2003; Collins et al. 2012). Local habitat can be influenced by features that occur both upstream (e.g., tributary junctions; Frissell et al. 1986; Benda et al. 2004) and downstream (e.g., dams; Frissell et al. 1986; Pringle 1997). Collectively, the hierarchical, longitudinal, and spatial organization of the physical template control the range of habitat features that may exist in particular locations and over space and time (Warren and Liss 1980; Poff and Ward 1990).

In response to variation in the physical template, aquatic species and life stages of species partition use of habitat types across scales (Schlosser and Angermeier 1995; Fausch et al. 2002). For example, stream fishes can complete their life cycles at various spatial and temporal scales (Schlosser and Angermeier 1995; Fausch et al. 2002; Wolter et al. 2016). Freshwater, resident fishes may inhabit a single tributary (Bjornn and Reiser 1991), whereas anadromous fishes spend different portions of their life cycles in freshwater, estuarine, and marine habitats (Quinn and Myers 2004). Further variation exists among anadromous fishes, with some expressing an ocean-type life history and migrating as fry while others exhibit a stream-type life history and have extended rearing as juveniles or prespawn adults. Because the physical template is comprised of diverse habitats, many species and life stages within each species contribute to biodiversity within the ecosystem (Gorman and Karr 1978; Ebersole et al. 2003).

A major challenge underlying stream restoration is determining how to translate complex relationships between habitat-forming processes at multiple scales and species adaptations to those processes into salient restoration actions. In some cases, restoration may focus on replicating historical conditions that led to species colonization (Ebersole et al. 1997; Cluer and Thorne 2014). In other cases, restoration may focus on creating a new range of conditions—if restoring historical conditions is precluded by anthropogenic activities such as dams or other development (Bernhardt and Palmer 2007). Selection of a set of conditions (i.e., the form of the channel being restored) implies selection of a stage in the evolution of stream channels (Schumm et al. 1984; Simon and Hupp 1986). This selection may be influenced by the shifting baseline associated with how the restoration practitioner defines "natural" (Humphries and Winemiller 2009). It may also be informed by considering how channel morphology evolves over time (Walter and Merritts 2008) and which channel form might be considered a starting point for that evolution (Cluer and Thorne 2014).

Restoration of physical habitat may depend on restoring historical biodiversity. For example, the American beaver *Castor canadensis* acts as an ecosystem engineer by altering stream environments through creation and maintenance of beaver dams and ponds (Rosemond and Anderson 2003). These ponds may function as rearing habitat for juvenile Coho Salmon *Oncorhynchus kisutch* (Pollock et al. 2004) and other species. Unlike human-engineered projects, these beaver ponds function dynamically in the landscape. Floods may disrupt the ponds, but beaver can recreate these again or build dams in new locations. Other studies have shown the effect of marine-derived nitrogen input from decaying salmon carcasses on the growth of riparian vegetation (Helfield and Naiman 2001). In Southeast Alaska, the mean annual basal area growth of Sitka spruce *Picea sitchensis* near spawning streams was three times greater than Sitka spruce in sites without salmon carcasses (Helfield and Naiman 2001). This creates a feedback between the abundance of salmon in a stream, the growth of trees in the associated riparian zone, their eventual recruitment into the stream, and the resulting habitat created in the stream, which supports more salmon (Helfield and Naiman 2001). These types of physical and biological feedbacks are an important consideration in restoration planning.

The diversity of the physical template and the biodiversity of ecosystems within watersheds can result in increased resilience through diverse responses to perturbation among life histories, populations, and species (Den Boer 1968; Hilborn et al. 2003; Greene et al. 2010; Schroeder et al. 2016). Resilience may be promoted through access to diverse, interconnected habitats that retain their naturally dynamic function (Bisson et al. 2009; Waldman et al. 2016). Where ecosystems are currently intact, creating protected, natural preserves may be sufficient to retain biodiversity (Williams et al. 2011). However, where ecosystems are degraded, resilience may be achieved by restoring the capacity of the physical template to create diverse habitats (Poff and Ward 1990; Ebersole et al. 1997). This is the essence of process-based restoration.

Process-based restoration is organized around two principles: (1) ecosystems are dynamic, and (2) restoring variability and hydrologic connectivity of stream networks creates a range of conditions to which species have adapted (Poff et al. 1997; Kondolf et al. 2006; Bisson et al. 2009; Beechie et al. 2010). Depending on the magnitude of degradation in a particular watershed and the scales at which physical and biological processes function, process-based restoration may take years or decades to be fully implemented (Wiens 1989). The restored processes are not targeted to the biological needs of particular species. By restoring the capacity of the physical template to create and maintain habitats and ecological processes, process-based restoration indirectly addresses biological requirements of multiple life stages, life histories, and species.

Process-based restoration at watershed scales is increasingly recognized as being more effective (Beechie et al. 2010) than restoration of specific channel forms at small spatial scales (e.g., Rosgen 1997). However for several reasons, application of restoration is often constrained to local, small-scale projects (Christian-Smith and Merenlender 2010) that focus on the biological requirements of a single species (Roni et al. 2002). Restoration practitioners may be constrained by geopolitical boundaries and the necessity of working at local scales (Bond and Lake 2003) although watershed processes may function at larger spatiotemporal scales. Alternatively, underlying causes of disturbance may be out of the purview of the practitioner to address (e.g., human development in a floodplain). These geopolitical and purview challenges affect where projects can be implemented within a watershed. Restoration in some locations may not contribute to improving watershed processes, optimize benefits to multiple species, or effectively address limiting factors (Barnas et al. 2015). These restoration constraints are, and will increasingly be, challenging as threats from climate change (Mote et al. 2003; Bisson et al. 2009; Beechie et al. 2012), development (Hudy et al. 2008), and human population growth (Postel 2000) exacerbate existing limiting factors pertaining to habitat (Williams et al. 2011).

In this chapter, we explore the range of habitat relationships expressed among imperiled anadromous fishes and the importance of implementing process-based restoration to address limiting factors in freshwater habitat. We present case studies on three anadromous fishes that were historically abundant in the Pacific Northwest (USA): Coho Salmon, Chum Salmon *Oncorhynchus keta*, and Pacific Lamprey *Entosphenus tridentatus*. These species co-occur and coevolved within the same physical templates, yet each differs with respect to biology, the portion of their life cycle spent in freshwater, life history expression (including the extent of habitat use in the upper, middle, and lower portions of watersheds), limiting factors, and the status of conservation planning and implementation. The case studies illustrate a range of restoration results and considerations, from successful small-scale projects improving headwater habitat of Coho Salmon to restoration challenges and critical uncertainties for Chum Salmon and Pacific Lamprey. Through examination of these case studies, we identify physical, spatial, and social factors that challenge restoration implementation and success, where success is defined as measurable increases in the biological metrics relating to population viability (McElhany et al. 2000) observed after versus before restoration.

We conclude by suggesting three main strategies to address restoration challenges and limitations. The first suggestion is to build and foster multidisciplinary collaborations among professionals in social, geological, biological, and political sciences and use this diverse expertise to address complex questions and challenges (e.g., Bottom et al. 2009). These collaborations also allow for the scaling up of discrete restoration projects completed at local scales (which alone would be ineffective) into coordinated, process-based restoration and monitoring at larger spatial scales. The second suggestion is to invest in education and outreach to build and diversify the constituency supporting watershed restoration. The third suggestion is to develop new spatial and organizational tools to facilitate restoration planning at watershed scales. The steps involved are achievable and work within the constraints of local organizations to realize results beyond their purview or boundaries.

Case Studies

The following three species-specific case studies cover the biology, abundance trends, limiting factors, management practices, and the results of these practices. Each case study varies in content based upon the available information, geographical extent of the populations, and status of recovery planning and implementation. We suggest that the available information is a general proxy for the level of societal interest and subsequent scientific and management interests. For example, Oregon coast Coho Salmon (OCCS) have been the focus of harvest, and they have been intensively studied, monitored, and managed (Figure 1). As such, the information in that case study briefly discusses biology and abundance and focuses in more detail on management that has been, and is being, implemented. By comparison, Columbia River Chum Salmon have received much less interest in harvest after an early period of heavy exploitation and therefore less interest in management (Figure 1). As a result, the case study on Columbia River Chum Salmon focuses on biology and new conservation and recovery goals. Finally, Pacific Lamprey are currently harvested primarily by Native American tribes and the stock structure is thought to be broad and not constrained to a general region or particular basin, unlike OCCS or Columbia River Chum Salmon. Although changing, Pacific Lamprey are not of great interest to many nontribal members (Figure 1). The case study on Pacific Lamprey therefore provides broad, but not deep, coverage on all topics.

Oregon coast Coho Salmon

The OCCS evolutionary significant unit (ESU) consists of 21 independent populations and 36 dependent populations distributed in coastal watersheds of Oregon, from the Necanicum River in the north to the Sixes River in the south (Lawson et al. 2007; Figure 2; Table 1). Independent populations generally correspond to the larger basins or subbasins in the ESU that have sufficient potential habitat for long-term population persistence. Dependent populations occur in smaller basins and likely depend on immigrants from nearby independent populations for long-term persistence (Nicholas et al. 2005). Geographic population structure is weak within the ESU due to relatively high migration rates among basins (Ford et al. 2004). Genetic studies suggest that the percentage of immigrants in OCCS populations varies from 2% to more than 30%, depending on location (Johnson and Banks 2008).

Coho Salmon along the Oregon coast predominantly exhibit a 3-year life cycle in which adult fish spawn in small tributaries during November–January, juveniles rear in streams for 1 year, and smolts out-migrate to the ocean the following spring. Coho Salmon then spend around 18 months in the ocean before returning to freshwater to spawn (Table 1; Nickelson and Lawson 1998). Three variants to this life history exist. They are juvenile Coho Salmon (Figure 3) that rear in lakes (Lawson et al. 2007) or estuary environments (Miller and Sadro 2003; Jones et al. 2014a) before out-migrating to the ocean and precocious males (jacks) that reside for 6 months in the ocean before returning to spawn. Juvenile Coho Salmon rearing in streams exhibit seasonal shifts in preferred habitat and are primarily found in pool habitats during the summer (Nickelson et al. 1992). By contrast, juvenile Coho Salmon

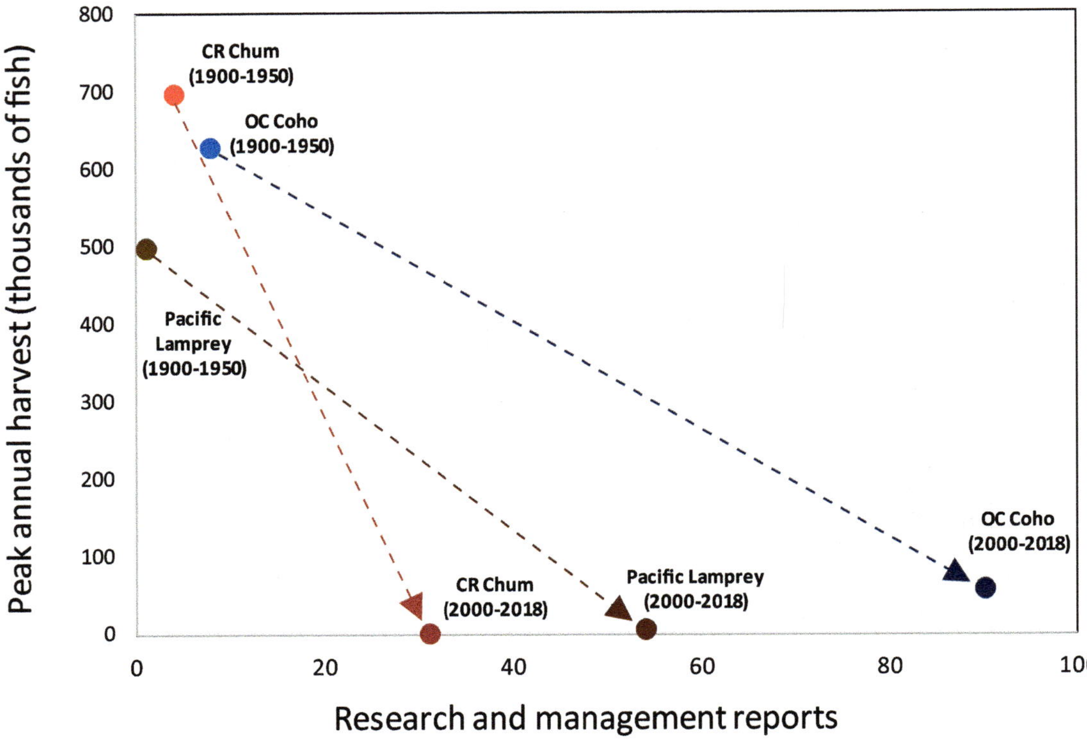

Figure 1. Comparison of peak annual harvest and number of research and management reports for Oregon coast (OC) Coho Salmon, Columbia River (CR) Chum Salmon, and Pacific Lamprey during two periods, 1900–1950 and 2000–2018. Peak annual harvest is for naturally produced fish (restricted to catch at Willamette Falls for Pacific Lamprey). Research and management report total is a minimum count that includes peer-reviewed journal articles, state agency reports, state and federal conservation plans, and university theses related in whole or part to the target species and evolutionary significant unit (for Coho Salmon and Chum Salmon) or the state of Oregon (for Pacific Lamprey). Reports were found using Google Scholar based on search terms relevant to each species, using the Oregon State University library database, and through the Oregon Department of Fish and Wildlife, Washington Department of Fish and Wildlife, and NOAA Fisheries Web sites.

express a strong preference for deep, low-velocity habitats with abundant cover (Bustard and Narver 1975), including beaver ponds and off-channel pools (Nickelson et al. 1992) during the winter. During the marine phase of their life history, OCCS are found off the coasts of Oregon, northern California, and Washington (Weitkamp and Neely 2002). Cycles of marine productivity in this area strongly influence smolt survival and adult recruitment (Nickelson 1986; Lawson 1993), and indices of marine conditions can explain more than 80% of the variability in the abundance of naturally produced Coho Salmon from Oregon (Koslow et al. 2002). This is due in part to a correlation between climate factors affecting marine survival and those affecting freshwater survival (Lawson et al. 2004).

Historical estimates of abundance for naturally produced OCCS may have been as high as 1–2 million fish per year when ocean conditions were favorable (ODFW 2007). These large runs, along with major hatchery programs, supported economically important ocean and river fisheries for decades.

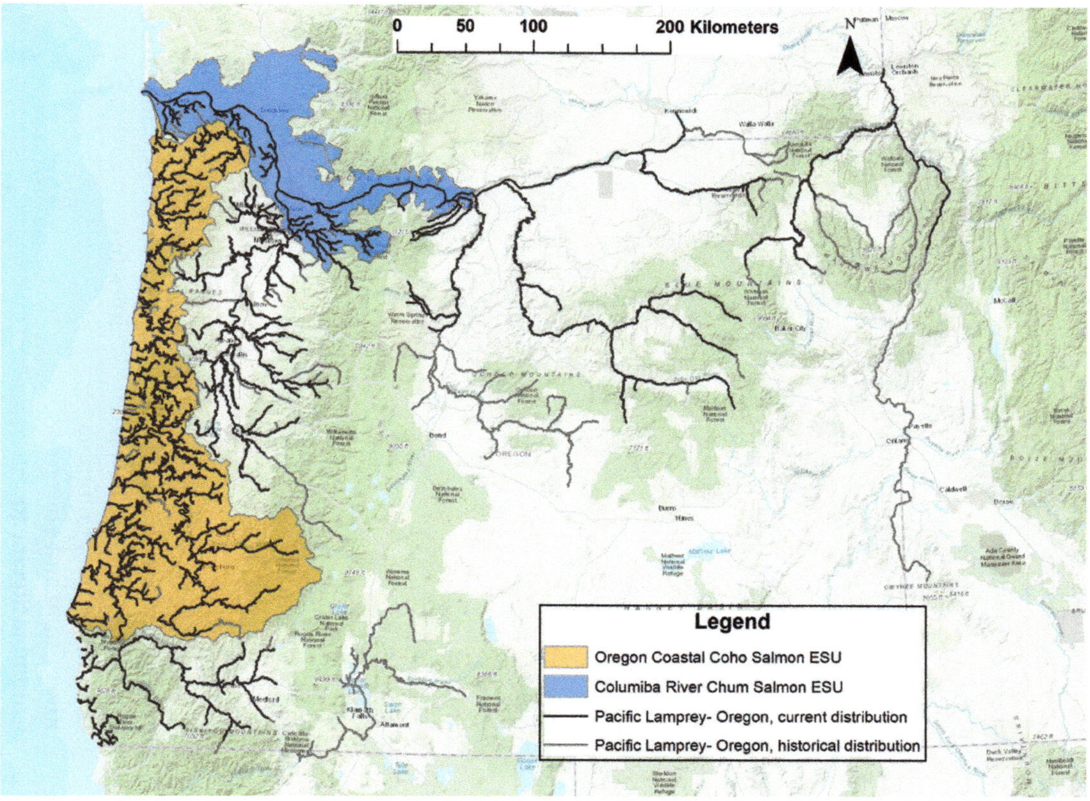

Figure 2. Evolutionary significant units (ESUs) for Oregon coast Coho Salmon and Columbia River Chum Salmon and current and historical distribution of Pacific Lamprey within Oregon (the distribution of Pacific Lamprey in Washington and Idaho is not shown on the map).

Ocean harvest of Coho Salmon in the Oregon production index area from California to southern Washington ranged from 1.0 to 2.5 million fish annually in the 1960s and 1970s (Nickelson 1986). Subsequently, wild OCCS populations declined to 15,000–25,000 fish during periods of poor marine survival in the 1980s and 1990s. This resulted in an initial threatened listing for OCCS under the ESA in 1998. The original listing identified several major factors contributing to the threatened status, including habitat degradation, harvest impacts, and hatchery releases (NMFS 1998). Habitat degradation is due to a variety of historical and current land-use management practices that have reduced stream complexity (Nicholas et al. 2005). Legal challenges regarding the status of the hatchery and wild components of OCCS led to delisting in 2001. Status reviews by the Oregon Department of Fish and Wildlife (ODFW) and the National Marine Fisheries Service (NMFS) determined that OCCS were viable and not likely to become endangered in the foreseeable future (Nicholas et al. 2005; NMFS 2006). However, a legal challenge to this finding was upheld in federal court and led to OCCS again being listed as threatened under the ESA in 2008. This designation was supported by subsequent NMFS reviews and represents that current status of the ESU (NMFS 2016).

In response to the decline of OCCS, federal, state, regional, and local partners have undertaken major efforts to conserve and recover this ESU. Significant management

Table 1. Comparison of biology, historical factors affecting status, management, and habitat restoration foci for Oregon coast Coho Salmon, lower Columbia River Chum Salmon, and Pacific Lamprey. N/A = not applicable.

	Oregon coast Coho Salmon	Lower Columbia River Chum Salmon	Pacific Lamprey
Biology and life cycle			
Evolutionary significant unit structure	21 independent + 35 dependent populations	16–17 independent populations; 90% extirpated	Poorly understood, may be panmictic across range
Reproductive strategy	Semelparous	Semelparous	Semelparous
Home to natal streams?	Yes	Yes	No
Peak-spawn timing	November–January	October–December	March–June
Fecundity	1,500–3,500	2,000–4,000	98,000–238,400
Freshwater residence period	Generally 1 year	2–3 weeks	<1 year and up to 9 years
Estuary residence period	<1 week to several months	1–4 weeks	Unknown
Ocean residence period	18 months (6 months for jacks)	3–5 years	≤3.5 years
Historical factors affecting status			
Overfishing	Important	Important	Contributing
Habitat degradation	Critical	Critical	Critical
Fish passage barriers	Contributing	Contributing	Critical
Hatchery impacts	Important	Contributing	N/A
Exotic species introductions	Contributing	Contributing	Contributing
Management			
Conservation status:			
Federal	Federally listed as threatened	Federally listed as threatened	N/A
Hatchery releases	Yes	Yes	No
Directed fishery	Limited harvest in some locations	No	Limited harvest in some locations
Tribal recovery plans	n/a	N/A	CRITFC 2011
Federal recovery plan	National Marine Fisheries Service Oregon Coast Coho Recovery Plan	Recovery Plan for Lower Columbia Chinook, Lower Columbia Coho, Columbia River Chum and Lower Columbia Steelhead	Pacific Lamprey Conservation Initiative

Table 1. Continued.

	Oregon coast Coho Salmon	Lower Columbia River Chum Salmon	Pacific Lamprey
Management (continued)			
State conservation plan	Oregon Coast Coho Conservation Plan	Lower Colmbia Conservation and Recovery Plan	Draft conservation in external review
Habitat restoration			
Focus	Stream habitat complexity, riparian restoration, fish passage	Artificial spawning channels, reconnect tidal, vegetated habitats	Species-specific: passage; value-added with salmonid restoration
Location	Primarily small to medium streams	Downstream portion of rivers, estuary	Variable
Citations	Sandercock 1991; Lawson et al. 2007; Nickelson and Lawson 1998; Nicholas et al. 2005; NMFS 2016	ODFW 2010; LCFRB 2010	Kan 1975; Beamish 1980; Clemens 2011; Clemens et al. 2013; Dawson et al. 2015

Figure 3. Juvenile Coho Salmon. Image credit: Derek Wiley.

changes included reducing hatchery smolt releases from a peak of 35 million fish in the early 1980s down to less than 1 million fish by 2005 (Nicholas et al. 2005) and sharply reducing fishery impacts on wild populations. Harvest impact rates on wild OCCS, which peaked at more than 80% in the past, have generally been less than 20% since the 1990s (NMFS 2016). The state of Oregon also increased efforts to protect and restore habitat for salmon statewide through the Oregon Plan for Salmon and Watersheds (OPSW, Nicholas et al. 2005).

Habitat restoration in coastal watersheds has been a key part of the OPSW, and millions of U.S. dollars have been spent annually since 1997 on instream habitat projects (Figure 4), riparian restoration, and removal of fish passage barriers in basins that support OCCS (OWEB 2017). The OPSW also established a robust habitat monitoring program for wadeable streams of Oregon throughout the ESU to track status and trends in stream habitat (Anlauf et al. 2011). Much of the restoration work in coastal watersheds has been coordinated by local watershed councils that prioritize projects based on limiting factors for specific populations. Based on considerable research (Nickelson et al. 1992; Solazzi et al. 2000; Rodgers et al. 2005), loss of stream habitat complexity (e.g., beaver ponds, secondary channels, and pools with large wood accumulations) has been identified as the primary limiting factor for most OCCS populations (Nicholas et al. 2005). Water quality, particularly water temperature, is a secondary limiting factor for most populations (Nicholas et al. 2005; NMFS 2016).

Changes in fisheries management have alleviated threats posed by overfishing and hatchery production to the viability of OCCS (NMFS 2016) and improved the productivity of wild populations (Buhle et al. 2009; Jones et al. 2018). Restoration projects aimed at increasing habitat complexity have significantly improved habitat for OCCS in stream reaches (Jones et al. 2014b) and small watersheds (e.g., Solazzi et al. 2000; Johnson et al. 2005) throughout the ESU. Restoration work has also opened up hundreds of kilometers of stream habitat to OCCS through barrier removal and provided access to estuarine rearing areas (e.g., Jones et al. 2014a; Brown et al. 2016; OWEB 2017). Since the original listing in 1998, OCCS populations have displayed cycles of high and low abundance with wild spawner escapement estimates ranging from as low as 50,000–60,000 fish to a high of greater than 350,000 fish.

Although restoration efforts in various streams have improved habitat in targeted

Figure 4. Large wood placement sites in Oregon coastal streams. Wood addition is a common restoration technique to improve stream habitat for Coho Salmon. Image credits: Derek Wilson, Erik Suring.

areas, little evidence exists for an overall improving trend in stream habitat conditions for the OCCS ESU (Stout et al. 2012; NMFS 2016; Strickland et al. 2018). Likewise, an analysis of life-cycle data indicated that the relatively high spawner abundance of OCCS observed in recent years appeared to be driven largely by increased ocean survival and not by improved freshwater production (Falcy and Suring 2018). Given the current status of stream habitat in the ESU, uncertainty remains about whether sufficient high-quality rearing habitat exists to sustain OCCS populations through an extended period of poor ocean conditions (NMFS 2016). Anticipated effects of global climate change may also negatively impact the quantity and quality of rearing habitat for juvenile OCCS in the future (Wainwright and Weitkamp 2013). The most recent ESA recovery plan for OCCS emphasizes the importance of restoring the ecosystem processes that sustain complex rearing habitat for juvenile OCCS, including the importance of restoring connections between streams and floodplains (NMFS 2016). Much of this work will have to be accomplished on private lands, which surround 80% of the stream habitat with high intrinsic potential for OCCS production (Burnett et al. 2007).

Columbia River Chum Salmon

The Columbia River Chum Salmon ESU consists of 17 populations (Myers et al. 2003) distributed from the mouth of the Columbia River upstream to The Dalles Dam at river kilometer (RKM) 309 (Nehlsen et al. 1991; Figure 2; Table 1). Populations are grouped into three ecological regions (also referred to as "strata" in recovery plans; ODFW 2010, Appendix A; NMFS 2013): the Coast (associated with the Coast Range), the Cascades (associated with the Cascade Range), and the Gorge (associated with the Columbia River Gorge). Genetic structure occurs among ecological regions, but not typically among populations within a region (Small et al. 2006), following isolation-by-distance (Wright 1943). This has been described for Chum Salmon at various spatial scales in other watersheds (Olsen et al. 2008; Petrou et al. 2014; Small et al. 2014). The genetic basis for population delineations is uncertain because 90% of populations are extirpated (Myers et al. 2003). Hence population boundaries defined for fall-run Chinook Salmon *Oncorhynchus tshawytscha* were applied to the Chum Salmon ESU (Myers et al. 2003).

Columbia River Chum Salmon predominantly exhibit a fall-run life cycle. Adults return to spawn at ages 3 to 5 (Salo 1991; Figure 5) and enter the lower Columbia River in October (WDF et al. 1993, cited by Johnson et al. 1997; Table 1). An estimated 90.0–97.8% of Chum Salmon home to their natal streams (Sakano 1960; McQuarrie and Bailey 1980; Table 1), although straying may be higher between streams in close proximity (e.g., 2 km; Tallman and Healey 1994) or as a consequence of incorrectly identifying population boundaries. Spawning peaks around mid- to late November (WDF et al. 1993, cited by Johnson et al. 1997). Chum Salmon spawn in low gradient habitats (Salo 1991; Montgomery et al. 1996), at the downstream ends of tributaries (Fulton 1970) and in the mainstem Columbia River (Fulton 1970), within large gravel and cobble beds (Duker 1977; Morrow 1980; Hale et al. 1985; Garland et al. 2003). In some cases, spawning is associated with upwelling groundwater (Sano 1966; in Schroder 1974; Tautz and Groot 1975; Hale et al. 1985). Fry out-migrate in early spring, shortly after emerging from the gravel (Bakkala 1970; Scott and Crossman 1973; Hale et al. 1985). Chum Salmon fry occupy the estuary anywhere from a few weeks (Mason 1974; Beacham and Starr 1982; Hale et al. 1985) to months (Herrmann 1971; Forsberg et al. 1977; Pearcy et al. 1989; Table 1). Growth during this estuary or early ocean phase is critical to their marine survival (Healey 1982a; Johnson et al. 1997; Kaeriyama et al. 2007). Juvenile Chum Salmon inhabit the near-

Figure 5. Chum Salmon actively spawning in a river on the Oregon coast. Image credit: Derek Wiley.

shore environment (Johnson et al. 1997) in the ocean as they migrate north to feeding grounds in the Gulf of Alaska (Salo 1991; Urawa et al. 2009). The majority of the life cycle occurs in the ocean (Table 1).

During historical periods of high abundance, Columbia River Chum Salmon spawning may have extended upstream as far as the Walla Walla River (RKM ~470; Nehlsen et al. 1991; Johnson et al. 1997). Monitoring data at lower Snake River dams (FPC 2018) indicates that some Chum Salmon migrated above Little Goose Dam, 638 RKM from the mouth of the Columbia River. During the 1920s, spawning populations were estimated to exceed 1 million adults (NPPC 1986, Appendix D) based on commercial harvest of more than 3.85 million kg of Chum Salmon (Johnson et al. 1997). Beginning in the 1930s and extending into the 1940s, Chum Salmon experienced precipitous declines in abundance and distribution. Causes for decline included overharvest, loss of spawning habitat and access to spawning habitat, altered hydrology, changes to the estuary (e.g., through diking, dredging, and alterations to the food web), and predation (ODFW 2010, Appendix A). Currently, several hundred to several thousand Chum Salmon return each year to spawn. Most spawning now occurs downstream of Bonneville Dam (RKM 235; McElhany et al. 2007) and in the Grays River basin, Washington—the only two populations not considered extirpated (Kostow 1995). In response to these declines, Chum Salmon were listed as threatened under the ESA in 1999 (NMFS 1999).

Following listing, both the ODFW and the Washington Department of Fish and Wildlife initiated recovery efforts for the Columbia River Chum Salmon ESU (WDFW 2004; LCRFRB 2010; ODFW 2010, Appendix A; NMFS 2013). Recovery planning in Oregon has been hampered by limited historical and contemporary data on distribution and abundance. Thus, it has not been possible to conduct a thorough assessment of survival risks using the viable salmon population approach (McElhany et al. 2000) that was applied to other salmon species in the Columbia River (ODFW 2010, Appendix A). Also, it was not possible to generate specific recovery goals (ODFW 2010, Appendix A; NMFS 2013). Despite the lack of data or recovery goals, both state agencies have independently developed a recovery strategy, entailing five objectives: (1) harvest restrictions, (2) habitat restoration and creation of artificial spawning channels, (3) development of a conservation broodstock, (4) supplementation and reintroduction, and (5) researching and addressing limiting factors (WDFW 2004; LCRFRB 2010; ODFW 2010, Appendix A; NMFS 2013). These steps were designed to collectively re-establish self-sustaining, naturally reproducing populations of Chum Salmon. Each of these objectives is explained below.

Harvest of Chum Salmon in the Columbia River was prohibited in 1992 in Oregon and 1995 in Washington. Since 1995, annual incidental harvest has been restricted to 1–5% of the Chum Salmon return to the Columbia basin. Depending on where harvest occurs, the effective harvest rate on a particular population may exceed 5%. The ocean harvest rate for Chum Salmon off the Oregon Coast is unknown but assumed to be zero (NMFS 2008). Harvest is no longer considered to be a limiting factor.

Extensive habitat surveys have been completed to identify the distribution, quality, and quantity of suitable habitats and where restoration is required. Because Chum Salmon spawn in the lower portions of watersheds (Fulton 1970), restoration of those habitats will require restoring upstream processes that create spawning habitat and addressing other limiting factors like erosion or water quality issues in freshwater through large-scale, integrated projects. Within potential spawning areas for Chum Salmon, large-scale restoration has been limited to a few rivers where other focal species (like Coho Salmon) occur and where a substantial portion of the land is managed by state and federal agencies. No large-scale restoration projects have been implemented to specifically improve spawning habitat for Columbia River Chum Salmon in Oregon. In the majority of Chum Salmon spawning streams, potential restoration sites occur on private property. Substantial effort will be required to build stewardship and conservation agreements prior to implementing restoration in these areas. In contrast, efforts to recover estuary-rearing habitat are ongoing. Numerous projects have been completed to breach dikes and levees to inundate former tidal habitats and many more are planned (LCEP 1999). Although these projects have focused on creating habitat to benefit juvenile Chinook Salmon, the created habitat may also benefit Chum Salmon by improving the estuary food web and the complexity of available habitat.

The restoration component of recovery planning for Chum Salmon also includes creating artificial spawning channels to increase abundance and buffer against catastrophic loss of this ESU in the short term. Several spawning channels have been built in Washington tributaries to the Columbia River, and a hydrological evaluation of a new site in Oregon is underway (Figure 6). These channels are designed to contain a suitable composition of gravel and small cobble and sufficient depth, flow, and temperature to maximize the survival of Chum Salmon eggs. Estimates of egg-to-fry survival in these channels vary but typically exceed 40% (Schroder 1974; Schroder 1977; Cowan 1991), making them high-production areas with the potential to serve as sources for recolonizing adjacent, restored habitats.

With current abundance of Columbia River Chum Salmon at such low levels, natural recolonization into restored habitat is likely insufficient to re-establish functioning populations. The spatial extent where adults could spawn is large relative to the number of adult Chum Salmon. Thus, a high probability exists that recolonizing spawners may not find each other. This low spawner density problem, termed the "Allee effect," can result in small populations disappearing in the wild when they drop below the critical density threshold associated with higher encounter probabilities (Courchamp et al. 1999; Deredec and Courchamp 2007). Therefore, the third objective of the recovery strategy has been to develop a conservation broodstock to serve as a source for supplementation and reintroduction efforts. To this end, a conservation broodstock was initiated in 2010 using eggs collected from wild-origin adults from the Grays River, a large,

Figure 6. Spawning channel evaluation site on Sundial Island at the confluence of the Sandy River and the Columbia River, Oregon. The white line traces the path of a relic channel that could be converted into an artificial spawning channel for Chum Salmon. Arrows highlight locations where a spawning channel could connect to the Sandy River.

genetically similar population in Washington (Small et al. 2011). Annual releases have ranged from 100,000 to 200,000 fed fry.

Broodstock returns to Big Creek Hatchery in Oregon have been sufficient to initiate supplementation into extant, low-abundance populations and reintroduction into extirpated populations. Supplementation and reintroduction efforts have included releasing adult Chum Salmon to spawn naturally and rearing eyed eggs in remote site incubators. Remote incubation of eyed eggs was an effective technique for producing large, acclimated fry releases, whereas adult outplanting was useful for identifying (1) the condition of the habitat, (2) freshwater survival rates, and (3) where habitat restoration may be needed.

Over the past 15 years, marine survival rates of Columbia River Chum Salmon (which include survival in the Columbia River and estuary) have ranged from 0.01% to 1.63% and averaged 0.34% (K. M. Homel, unpublished data; Hillson and Grobelny 2017). This is substantially lower than rates reported for other fall-run Chum Salmon populations in Washington, British Columbia, and Alaska (range = 0.1–3%; Hunter 1959; Fukuwaka and Suzuki 2000; average = 1.8%; Salo 1991). During favorable ocean conditions, adult Chum Salmon may be abundant and broadly distributed in freshwater. However, much of that increased distribution is into areas with poor habitat. Therefore production of juvenile Chum Salmon is limited by the availability and quality of spawning habitat. During unfavorable ocean conditions, returns of adult Chum Salmon are typically very low and spawners concentrate in areas of high-quality spawning habitat. In these years, abundance is limited by the marine environment, suggesting restoration of freshwater habitats, alone, would be insufficient to recover populations. In 9 of the past 15 years, marine survival rates were less than 0.2% and were considered poor because production was below replacement. Consequently, the ODFW is combining efforts to restore spawning habitat (to address freshwater survival limitations) with research into factors contributing to the regionally poor marine survival observed in Columbia River Chum.

The last component of the ODFW's recovery strategy focuses on identifying and addressing limiting factors (ODFW 2010, Appendix A). Spawning habitat availability and quality were quantified for populations in the coastal region, and this information is being incorporated into restoration plans. Current efforts focus on researching how pathogens, habitat quality, and food web alterations might limit survival of Chum Salmon in the lower Columbia River or estuary. Over time, different factors may limit survival or reproduction in response to variable abiotic and biotic conditions. Therefore, limiting-factors research must be approached in an adaptive management framework. This allows iterative identification of to understand which factors are currently most limiting and which could become limiting as factors are addressed or if the climate regime shifts. This concept of recovery extends beyond restoring the physical locations occupied by Chum Salmon; it requires thinking about the entire dimension of habitat, from water quality to food availability, community interactions, and beyond.

The future direction of ODFW's Chum Salmon recovery efforts will include significant investment in education and outreach, particularly among private landowners. This outreach will be critical for implementing restoration of spawning habitat at a scale sufficient to improve freshwater survival rates for Chum Salmon. Outreach will also contribute to efforts to address potential limiting factors in the estuary environment. Many of the values associated with recovery of Chum Salmon relate to how they function as a keystone species, so increasing their abundance could improve ecosystem function and resilience. Through education and outreach, the public can better understand these values and adopt behaviors that facilitate recovery actions (e.g., landowners participating in large-scale restoration). Ultimately, an increase in public interest in Chum Salmon is required to develop management, policy, and funding decisions that more fully support recovery.

Pacific Lamprey

Pacific Lamprey is an anadromous, jawless fish from an ancient lineage (Petromyzontiformes; Renaud 2011; Docker et al. 2015; Potter et al. 2015; Figure 7). This species is native to the North Pacific Ocean and surrounding freshwater habitats and has co-occurred with Coho Salmon and Chum Salmon for millennia (Wydoski and Whitney 1979; Moyle 2002; Figure 2). In the Pacific Northwest, Pacific Lamprey were historically abundant. Dam counts at the lowermost dam on the Columbia River, Bonneville Dam (RKM 235), indicated that up to nearly half-a-million adult Pacific Lamprey occurred during the late 1950s and early 1960s. Similarly, harvest at Willamette Falls in the Willamette River (a major tributary to the Columbia River) was greater than 500,000 fish during the late 1940s (Kostow 2002). At both locations, the number of Pacific Lamprey plummeted sharply in subsequent decades. In the early 1990s, concerns over the biological status of Pacific Lamprey led Native American tribes to champion their importance and urge actions to stop their decline.

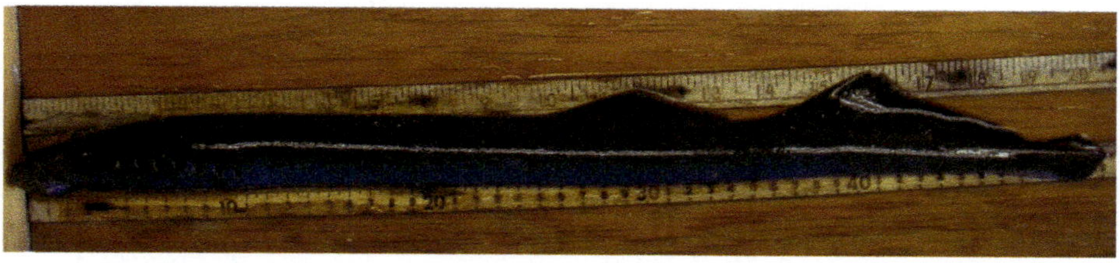

Figure 7. Adult Pacific Lamprey. Image credit: Benjamin Clemens.

In 1993, the ODFW listed Pacific Lamprey as a sensitive species; by 1996, the Oregon Fish and Wildlife Commission approved restrictions of harvest and harvest methods, and in 2002, the commission approved a ban of commercial harvest (Kostow 2002; Clemens et al. 2017). In 2003, interest in the plight of Pacific Lamprey culminated in a petition for an ESA (1973) listing. However, the 90-d finding by the U.S. Fish and Wildlife Service (USFWS) concluded that listing was not justified, due to a paucity of information on their biology, ecology, and threats (USFWS 2004).

The life span of Pacific Lamprey is estimated at 4–13 years, with most of it spent in freshwater (~3–9 years), including 3–7 years as eyeless larvae and 0–2 years as return migrants from the ocean (Clemens et al. 2010, 2017; Dawson et al. 2015; Table 1). As larvae, Pacific Lamprey reside in burrows within soft sediments where they feed on diatoms and other micro-organisms before undergoing a significant morphological and physiological transformation into eyed juveniles with sharp teeth and a rasping tongue conducive to a parasitic feeding in the ocean (McGree et al. 2008; Dawson et al. 2015; Manzon et al. 2015). Up to 3.5 years of life is spent as a juvenile in the ocean where Pacific Lamprey parasitize fishes and whales prior to ceasing feeding and migrating back into freshwater in various states of reproductive maturation to spend 0–2 years migrating to the spawning grounds (Beamish 1980; Clemens et al. 2013; Table 1). Pacific Lamprey do not home to their natal streams or show stock structure in the classical sense of the definition exemplified by salmonids (Groot and Margolis 1991; Quinn and Myers 2004; Quinn 2005; Goodman et al. 2008; Spice et al. 2012) that can lead to strong genetic differentiation (Hendry et al. 2004; Quinn and Myers 2004). Rather, they home to the bile acids produced by larval lamprey, which may signal that a particular basin or habitat is suitable for procreation (Robinson et al. 2009; Clemens et al. 2010; Yun et al. 2011). However, growing evidence suggests that the presence of larval Pacific Lamprey is not necessary to draw adult Pacific Lamprey into basins, but rather that Pacific Lamprey can recolonize habitats that have been blocked by dams (Clemens et al. 2017; Moser and Paradis 2017; Jolley et al. 2018) or the aftermath of volcanic eruptions (Lin et al. 2008). Pacific Lamprey are highly fecund compared with Coho and Chum salmons (Table 1) and, like salmon, die after spawning (Johnson et al. 2015; Clemens 2019).

Pacific Lamprey are found primarily in larger, low-elevation streams (Gunckel et al. 2009) but are otherwise widely distributed in river basins along the West Coast of North America from Baja, Mexico north to the Bering and Chuchki seas off Alaska (Renaud 2008, 2011). In Oregon, Pacific Lamprey are found in tributaries to the Snake and Columbia rivers and in coastal systems (Figure 2). They can occur more than 1,200 km inland from the ocean into Idaho (IDFG 2011; McIlraith et al. 2015). However, up to eight dams in the Columbia and Snake rivers (and other limiting factors) limit the distribution and abundance of Pacific Lamprey in the upper portions of these basins.

Pacific Lamprey were highly abundant in Oregon during the late 1800s and early 1900s, followed by a rapid decline during the 1950s and 1960s and then again in the 1970s (CRITFC 2011; Sheoships 2014). Among the many limiting factors and threats identified for Pacific Lamprey in freshwater, the estuary, and the ocean, the following are consistently noted: (1) passage barriers, (2) degradation of stream and floodplain habitats, (3) dredging and excavation of river habitats, (4) dewatering, (5) perturbations to water quality and quantity, and (6) lack of awareness (Kostow 2002; Moyle et al. 2009, 2013; Luzier et al. 2011; Clemens et al. 2017). Of these, passage barriers to up- and downstream passage and screening are the most discrete limiting factor, and in some areas, they may be the most important limiting factor (Kostow 2002; Moyle et al. 2009; Luzier et al. 2011; Clemens et al. 2017). In addition, Pacific Lamprey were aggressively harvested for various uses in the early to mid-1900s (hatchery fish meal, biological supply companies, bait for trapping mink, eradication in some streams through poisoning) and, by these means, were likely overharvested in some areas (Close et al. 1995, 2002).

Interest in conservation of Pacific Lamprey has increased substantially over the past two decades. Until recently, lamprey have not been a focus for research and monitoring (Close et al. 2002; Kostow 2002; Mesa and Copeland 2009; Clemens et al. 2017). An increasing number of conservation plans, information reviews, highlights of particular information gaps, and scientific and management actions have recently occurred (reviewed in Clemens et al. 2017). Of particular regional importance to the Snake and Columbia River basins (and therefore large parts of the Pacific Northwest), are the Pacific Lamprey Tribal Restoration Plan set forth by the Columbia River Inter-Tribal Fish Commission (CRITFC 2011) and the federally coordinated Pacific Lamprey Conservation Initiative (Luzier et al. 2009 and 2011; USFWS 2012). Another draft conservation plan for lamprey by the ODFW is in the external review process.

The Tribal Pacific Lamprey Restoration Plan for the Columbia River basin was written by CRITFC (2011). This restoration plan focuses on actions to address key uncertainties and threats for Pacific Lamprey. Six objectives were identified in the tribal plan as important to restoring Pacific Lamprey: (1) improvements in passage (Figure 8), (2) habitat improvement, (3) supplementation through releases of hatchery-reared larval lamprey and translocation of adult Pacific Lamprey from below dams on the Columbia River upstream to previously inhabited spawning areas in the upper basin, (5) researching the effects of contaminants and taking actions to reduce contaminants, (6) public education and outreach, (7) research, and (8) collaboration and coordination.

The USFWS' Pacific Lamprey Conservation Initiative is a regional, multistate strategy driven by partnerships to improve the status of Pacific Lamprey throughout its range in North America. Through this initiative, the USFWS coordinates and implements research, conservation, and restoration actions. This initiative is composed of an assessment and template for conservation measures (Luzier et al. 2011), a conservation agreement (USFWS 2012), and regional implementation plans with collaborations among multiple partners (USFWS 2012, 2015). Interest in lamprey across many sectors has grown, and 26 entities are signatories to the Pacific Lamprey Conservation Agreement. The conservation agreement is a cooperative effort among entities to improve habitats, reduce threats, and improve population status for Pacific Lamprey (USFWS 2012).

It is too early to tell how adherence to conservation plans and approaches have worked. Abundance data are lacking for large regions of the Pacific Northwest. Therefore, inferences are made from adult counts at the few dams where this data are

Figure 8. A lamprey passage structure at Winchester Dam on the North Umpqua, Oregon (RKM 190). Winchester Dam is approximately 5 m in height from the river bottom to the precipice. This dam forms a lowhead barrier to upstream movements of fishes. The images are different vantage points of the same passage structure and dam. The black arrows indicate the passage route for adult Pacific Lamprey attempting to ascend the dam en route to spawning areas upstream. Lamprey enter the structure from the bottom, adhering to the vertical metal face with their sucker mouths and moving upward to the first resting box before ascending the enclosed, inclined ramp to the collection box at the top of the dam precipice. Adult Pacific Lamprey are not able to readily pass most adult fishways constructed for passage of adult Pacific Salmon (PLTWG 2017); therefore, passage structures similar in concept to this one have been designed, constructed, and installed to promote upstream passage of these fish in various locations in the Pacific Northwest (e.g., Moser et al. 2011). Image credits: Julie Reed, Oregon Department of Fish and Wildlife.

recorded, and then the relative abundance is inferred across other regions that do not have dams and/or counting mechanisms by which to collect these data. In these ways, the relative abundance of Pacific Lamprey is difficult to gauge. Finally, existing data from adult counts at dams appear to indicate a periodicity to abundance estimates, making it difficult to assess whether relative abundance is increasing, decreasing, or stable into the foreseeable future. Whatever the case, adult counts at dams are but a small fraction of the historical abundance of the 1940s and 1950s.

Synthesis

Habitat restoration approaches: what has been done?

During the past 20 years stream restoration efforts have increased substantially throughout the Pacific Northwest (Barnas et al. 2015; PNSHP 2019). The primary motivation for this has been the listing of many stocks of anadromous salmonids under the ESA and the recognition that habitat degradation is often a primary cause of population declines (Katz et al. 2007; Barnas et al. 2015). Restoration projects vary widely, from riparian planting on short stream reaches to dam removal that may affect access to hundreds of kilometers of stream habitat (Katz et al. 2007). Most restoration projects are expected to benefit a variety of resident and anadromous fishes, yet these often target a specific limiting factor for a focal fish species. For OCCS, the emphasis has been on projects that address habitat complexity, water quality, and fish passage in small- to -medium-sized streams (Figure 4). For Columbia River Chum Salmon, the emphasis has been on restoring estuaries and creating suitable spawning habitat in low-gradient streams and rivers (including the construction of artificial spawning channels; Figure 6), although few projects have been planned and implemented. For Pacific Lamprey, restoration has primarily focused on improving passage at artificial barriers (Table 1). Like Chum Salmon, there has been limited restoration specifically targeting Pacific Lamprey. Restoration targeting biological requirements of salmonids may benefit Pacific Lamprey (Streif 2009), although specific restoration projects for Pacific Lamprey probably involve increasing complexity of large watersheds (Clemens et al. 2017).

As our three case studies illustrate, habitat restoration requirements and the potential impact of restoration on recovery vary considerably among anadromous fishes depending on the portion of their life cycle spent in freshwater, the location in the watershed (i.e., upper, middle, or lower), and the specific freshwater habitats used. For OCCS, most spawning and rearing currently occurs in small- to -medium sized (wadeable) streams (Figure 9), and mortality during the freshwater period can account for more than half of the total life-cycle mortality for Coho Salmon (Bradford 1995). As a result, restoration focusing on tributary streams is a critical component of conservation and recovery for this ESU (Nicholas et al. 2005). By contrast, tributary restoration efforts are likely to have limited direct benefits for Chum Salmon, which spend more than 90% of their life cycle in estuarine and marine environments (Table 1). During that limited freshwater residency, Chum Salmon spawn and temporarily rear in the downstream portion of rivers in low-gradient habitats. These locations in the watershed may be degraded by many discrete perturbations upstream (such as landslides), and restoration requires coordinated effort to restore habitat-forming processes throughout the watershed. In addition, because of the larger estuary and marine component of the life cycle and associated poor marine survival, restoration of freshwater habitat alone is insufficient to recover Chum Salmon (Figure 9).

Identification of restoration needs for Pacific Lamprey is more difficult for at least

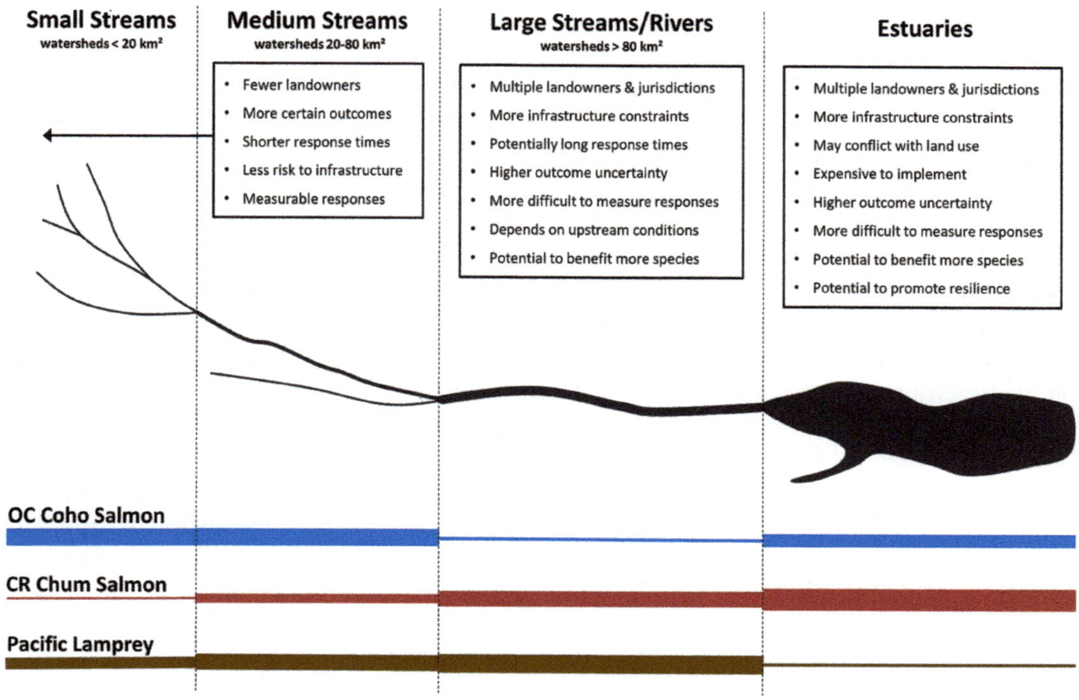

Figure 9. Generalized conceptual diagram of watersheds, with notes on the advantages and constraints for restoration in different areas of the watershed and the relative importance of these areas for Oregon coast (OC) Coho Salmon, Columbia River (CR) Chum Salmon, and Pacific Lamprey. The thickness of the colored horizontal bars indicates the relative importance of each area for spawning or rearing. For Pacific Lamprey, there is uncertainty about the relative importance of different habitat types, particularly estuaries.

three reasons, including (1) a lack of understanding of how changes to habitats correspond with early-life-stage survival and production, (2) a lack of information about the relative contributions of various habitats to overall production of juveniles out-migrating to the ocean, and (3) uncertainty about how an absence of homing to natal streams and subsequent population structuring in watersheds affects the response of Pacific Lamprey to habitat restoration in time and space. Life-stage survival rates for Pacific Lamprey in the wild are not understood (Clemens et al. 2017). However, based on their high fecundity, long freshwater rearing period (Table 1), and limited information on juvenile production, a high proportion of total life-cycle mortality for Pacific Lamprey may occur in freshwater. An unknown amount of mortality of Pacific Lamprey occurs in the ocean. Knowledge of habitat relationships for Pacific Lamprey has increased considerably in recent years, yet models that indicate how changes in habitat influence larval lamprey survival and abundance are lacking at all levels, from river reaches to the watershed level. Uncertainty also exists about the relative contributions of habitats in wadeable streams versus larger rivers to overall larval abundance. However, the distribution of spawning adult Pacific Lamprey suggests that main-stem river habitats are important and that restoration in tributaries alone is unlikely to achieve conservation goals.

The scale of watershed restoration needed to achieve broad population uplift for Pacific Lamprey is unknown. Straying is widespread among salmon populations (Quinn 1993; Keefer and Caudill 2014), yet homing still produces a feedback between freshwater habitat quality and spawner abundance in these fishes. This may not be the case for Pacific Lamprey, except at very large scales. To achieve broad population uplift, we hypothesize that Pacific Lamprey may benefit more from a larger geographical scale of restoration of a mosaic of habitats within and across many watersheds. This hypothesis will be difficult to assess.

Restoring watershed processes: what is the goal?

Building on the premise we described in the Introduction to this chapter, the preceding synthesis, and Figure 9, we suggest that habitat restoration to support recovery of all anadromous fishes in watersheds in the Pacific Northwest will require a process-based approach implemented at a watershed scale. This approach directly restores the physical, chemical, and biological processes that create and maintain the freshwater and estuarine habitats utilized by anadromous fishes (Beechie et al. 2010). Restoring watershed processes in large watersheds necessarily starts with smaller tributaries, as these provide organic and inorganic materials that structure stream habitats and support food webs downstream (e.g., Vannote et al. 1980; Kiffney et al. 2006). Concerted efforts have been made to apply process-based restoration in headwaters within public lands through addressing sources of sediment and large wood (e.g., road decommissioning, riparian planting; Reeves et al. 2006). However, further implementation of process-based restoration throughout target watersheds and on privately owned lands will be critical to achieving sufficient restoration to support recovery of anadromous fishes (e.g., Burnett et al. 2007).

A strong emphasis has been placed on restoration and monitoring in small- to medium-sized streams on the Oregon coast and in the lower Columbia River due to the relevance to OCCS recovery. Headwater projects are particularly relevant to Coho Salmon because the targeted processes overlap with seasonal habitats occupied by Coho Salmon. Achieving restoration goals in relatively small watersheds is more feasible than larger watersheds due to the smaller number of landowners, greater certainty about the stability and functionality of instream habitat manipulations, and fewer constraints from infrastructure (Figure 9). The outcomes of restoration work in smaller watersheds can also be quantified more readily in established habitat models for Coho Salmon (e.g., Jones et al. 2014b). Restoration at small- to medium-sized scale alone is unlikely to achieve desired outcomes for larger streams and rivers without careful consideration of the spatial arrangement of restoration sites and their collective effects (Palmer 2009).

Watershed restoration that restores the ecological functions of main-stem river habitats for anadromous fishes would require a much broader coalition of partners, greater societal engagement, and much longer time periods (i.e., decades to centuries) than achieved to date (Figure 9). The importance of coordinated planning for multiple anadromous fishes has been recognized in conservation plans prepared for Oregon coastal basins (ODFW 2014) and the lower Columbia River (NMFS 2013). These plans specifically highlight the need to restore watershed processes at large spatial scales to protect and enhance large tributary and main-stem habitats for Chum Salmon and Chinook Salmon. For Pacific Lamprey, an even larger perspective is probably required (see above). Large, functioning watersheds may still be the building block for conservation of Pacific Lamprey.

In addition to implementing watershed restoration at both ESU and continental scales, a need exists to integrate estuaries into watershed restoration (Figure 9). All anadromous fishes must pass through this transitional habitat, and the condition of estuaries has a strong influence on survival and life history diversity (Magnusson and Hilborn 2003; Bottom et al. 2005; Flitcroft et al. 2019). Estuarine rearing is important for species like Chum Salmon and Chinook Salmon that out-migrate as subyearlings (Healey 1982a; Pearcy et al. 1989; Bottom et al. 2005). Estuaries also play a role in Coho Salmon recovery and resilience (Koski 2009; Jones et al. 2014a). Even species that spend very little time in the estuary may experience significant mortality while migrating through this habitat (Schreck et al. 2006; Clements et al. 2012; Romer et al. 2012). Estuary restoration is challenging due to high costs and other societal constraints (Figure 9) and uncertainty about the efficacy of alternative restoration strategies (Bottom et al. 2005). Restoration efforts in freshwater are unlikely to have desired outcomes without commensurate improvements in estuary habitat, especially for Chum Salmon.

Future directions: strategies to achieve process-based restoration of watersheds

Successful implementation of large-scale restoration projects depends on application of best science practices and a shift in public perception of process-based restoration at watershed to continental scales. We suggest three strategies to achieve this goal. The first strategy is to create interdisciplinary coalitions of restoration practitioners to both address the complexity of restoration and integrate local efforts into process restoration. The second strategy is to invest in education and outreach in order to build constituencies based on ecological values of restoration. The final strategy we propose is to develop new spatial and organizational tools in an effort to address critical uncertainties in the application (as opposed to the concept) of process-based restoration.

Scaling up local restoration efforts through strategic planning and implementation will be necessary to achieve process-based restoration in large watersheds. Individual organizations are constrained by their purview (e.g., water-quality focus, habitat restoration, and disease assessments) and organizational boundaries (e.g., watersheds or county line). By building multidisciplinary collaborations among organizations at the watershed scale, discrete restoration projects at local scales can be coordinated into process-based restoration at watershed scales. Moreover, the resulting collaboration contains the diverse expertise required to address complex questions and challenges (e.g., Bottom et al. 2009). In Oregon, strategic planning of this kind has begun through the Coast Coho Business Plan and Focused Investment Partnerships (OWEB 2017). Much more work is needed in this area with regards to developing spatially explicit analyses to prioritize restoration throughout watersheds. There is also a need to develop coalitions of restoration practitioners to coordinate restoration across multiple scales and land ownerships. Given the uncertainties about process-based restoration at these scales, it would be useful to conduct restoration efforts in an adaptive management framework, informed by long-term monitoring data.

Effective restoration at a scale relevant to all anadromous fishes in the Pacific Northwest will require stronger connections between social and ecological systems that recognize the importance of biodiversity and range of ecosystem services provided by functioning watersheds (Postel and Thompson 2005; Bottom et al. 2009). Societal values and priorities surrounding the conservation of anadromous fishes have shifted greatly over the past 100 years. Broad support exists to recover OCCS to the point where populations provide ecological, cultural, and economic benefits (ODFW 2007). Columbia River Chum Salmon, in contrast, lack a strong constituency due to their low abundance and the presence of other fishes more favored for harvest in the Columbia

River. Tribal interests in the conservation of Pacific Lamprey have been a driving force behind their management (Close et al. 2002; CRITFC 2011; Clemens et al. 2017). Education and outreach about conservation benefits beyond harvest opportunities, like improved water quality and the conservation of biodiversity, could greatly expand the constituencies supporting watershed- and continental-scale efforts in restoration. Developing these constituencies may also provide improved opportunity to implement restoration projects on private property—a critical step for restoring processes at the watershed scale (Burnett et al. 2007).

Achieving the restoration approach outlined above involves considerable technical challenges in restoration. Although a process-based, watershed-scale approach is increasingly recognized as the most effective way to achieve restoration goals (Roni et al. 2002; Beechie et al. 2010), a major gap still exists between theory and practice (Palmer 2009). A full discussion of research needs is beyond the scope of this chapter, but some important topics include improving species distribution models (Herlihy et al. 2006; Elith and Leathwick 2009), mapping habitats more comprehensively (e.g., Benda et al. 2011), and developing process-based models to predict restoration outcomes (and then monitoring those outcomes to compare with model results; Lake et al. 2007). Other factors such as the prevalence of pathogens or predators also need to be evaluated as part of a holistic approach towards planning watershed restoration, as these factors can affect distribution, abundance, and how fishes interact with physical habitat (Bond and Lake 2003; Lawrence et al. 2014). These products would contribute to watershed prioritization and provide more realistic expectations regarding the time scales and potential outcomes of watershed restoration. They could also provide more effective ways of incorporating climate change projections into restoration planning. The topics presented above are areas of active research and many resources exist to aid in prioritizing restoration projects (Lake et al. 2007). We suggest that investment into creating multidisciplinary partnerships of restoration practitioners who access the newest prioritization and organizational tools would improve implementation and effectiveness of habitat restoration at scales large enough to benefit anadromous fishes.

Acknowledgments

The Oregon Department of Fish and Wildlife supported the preparation of this chapter. Additional financial support was provided by the Pacific Coastal Salmon Recovery Fund through the Oregon Watershed Enhancement Board. We thank Erik Suring, Tom Friesen, Renee Coxen, and Jim Brick (all with the Oregon Department of Fish and Wildlife) for providing thorough reviews of this chapter prior to publication. Two anonymous reviewers also provided detailed suggestions. These reviews strengthened and clarified the message in this chapter.

References

Anlauf, K. J., W. Gaeuman, and K. K. Jones. 2011. Detection of regional trends in salmonid habitat in coastal streams, Oregon. Transactions of the American Fisheries Society 140:52–66.

Bakkala, R. G. 1970. Synopsis of biological data on the Chum Salmon, *Oncorhynchus keta*, (Walbaum 1792) U.S. Fish and Wildlife Service Circular 315.

Barnas, K. A., S. L. Katz, D. E. Hamm, M. C. Diaz, and C. E. Jordan. 2015. Is habitat restoration targeting relevant ecological needs for endangered species? Using Pacific salmon as a case study. Ecosphere [online serial] 6(7):110.

Beacham, T. D. and P. Starr. 1982. Population biology of Chum Salmon, *Oncorhynchus keta*, from the Fraser River, British Columbia. U.S. National Marine Fisheries Service Fishery Bulletin 80:813–826.

Beamish, R. J. 1980. Adult biology of the River Lamprey (*Lampetra ayresi*) and the Pacific Lamprey (*Lampetra tridentata*) from the Pacific coast of Canada. Canadian Journal of Fisheries and Aquatic Sciences 37:1906–1923.

Beechie, T., H. Imaki, J. Greene, A. Wade, H. Wu, G. Pess, P. Roni J. Kiimball, J. Stanford, P. Kiffney, and N. Mantua. 2012. Restoring salmon habitat for a changing climate. River Research and Applications 29:939–960.

Beechie, T. J., D. A. Sear, J. D. Olden, G. R. Pess, J. M. Buffington, H. Moir, P. Roni, and M. M. Pollock. 2010. Process-based principles for restoring river ecosystems. Bioscience 60:209–222.

Benda, L., N. L. Poff, D. Miller, T. Dunne, G. Reeves, G. Pess, and M. Pollock. 2004. The network dynamics hypothesis: how channel networks structure riverine habitats. BioScience 54:413–427.

Benda, L., D. Miller, and J. Barquín. 2011. Creating a catchment scale perspective for river restoration. Hydrology and Earth System Sciences 15:2995–3015.

Bernhardt, E. S., M. A. Palmer, J. D. Allan, G. Alexander, K. Barnas, S. Brooks, J. Carr, S. Clayton, C. Dahm, J. Follstad-Shah, D. Galat, S. Gloss, P. Goodwin, D. Hart, B. Hassett, R. Jenkinson, S. Katz, G. M. Kondolf, P. S. Lake, R. Lave, J. L. Meyer, T. K. O'Donnell, L. Pagano, B. Powell, E. Sudduth. 2005. Synthesizing U.S. river restoration efforts. Science 308:636–637.

Bernhardt, E. S., and M. A. Palmer. 2007. Restoring streams in an urbanizing world. Freshwater Biology 52:738–751.

Bisson, P. A., J. B. Dunham, and G. H. Reeves. 2009. Freshwater ecosystems and resilience of Pacific salmon: habitat management based on natural variability. Ecology and Society [online serial] 14(1):45.

Bjornn, T. C. and D. W. Reiser. 1991. Habitat requirements of salmonids in streams. Pages 83–138 *in* W. R. Meehan, editor. Influences of forest and rangeland management on salmonid fishes and their habitats. American Fisheries Society, Special Publication 19, Bethesda, Maryland.

Bohn, B. A., and J. L. Kershner. 2002. Establishing aquatic restoration priorities using a watershed approach. Journal of Environmental Management 64:355–363.

Bond, N. R., and P. S. Lake. 2003. Local habitat restoration in streams: constraints on the effectiveness of restoration for stream biota. Ecological Management and Restoration 4:193–198.

Bottom, D. L., C. A. Simenstad, J. Burke, A. M. Baptista, D. A. Jay, K. K. Jones, E. Casillas, and M. H. Schiewe. 2005. Salmon at river's end: the role of the estuary in the decline and recovery of Columbia River salmon. NOAA Technical Memorandum NMFS-NWFSC-68.

Bottom, D. L., K. K. Jones, C. A. Simenstad, and C. L. Smith. 2009. Reconnecting social and ecological resilience in salmon ecosystems. Ecology and Society [online serial] 14(1):5.

Bradford, M. J. 1995. Comparative review of Pacific salmon survival rates. Canadian Journal of Fisheries and Aquatic Sciences 52:1327–1338.

Brown, L. A., M. J. Ewald, L. S. Brophy, and S. van de Wetering. 2016. Southern Flow Corridor baseline effectiveness monitoring: 2014. Prepared for Tillamook County by Oregon Estuary Technical Group, Institute for Applied Ecology, Corvallis.

Buffington, J. M. 2012. Changes in channel morphology over human time scales. Pages 435–463 *in* M. Church, P. M. Biron, and A. G. Roy, editors. Gravel-bed rivers: processes, tools, environments. Wiley, Chichester, UK.

Buhle, E. R., K. K. Holsman, M. D. Scheuerell, et al. 2009. Using an unplanned experiment to evaluate the effects of hatcheries and environmental variation on threatened populations of wild salmon. Biological Conservation 142:2449–2455.

Burnett, K. M., G. H. Reeves, D. J. Miller, S. Clarke, K. Vance-Borland, and K. R. Christiansen. 2007. Distribution of salmon-habitat potential relative to landscape characteristics and implications for conservation. Ecological Applications 17:66–80.

Bustard, D. R., and D. W. Narver. 1975. Aspects of the winter ecology of juvenile Coho Salmon (*Oncorhynchus kisutch*) and steelhead trout (*Salmo gairdneri*). Journal of the Fisheries Research Board of Canada 32:667–680.

Christian-Smith, J., and A. M. Merenlender. 2010. The disconnect between restoration goals and practices: a case study of watershed restoration in the Russian River basin, California. Restoration Ecology 18:95–102.

Clemens, B. J. 2011. The physiological ecology and run diversity of adult Pacific Lamprey, *Entosphenus tridentatus*, during their freshwater spawning migration. Doctoral dissertation. Oregon State University, Corvallis.

Clemens, B. J. 2019. A call for standard terminology for lamprey life stages. Fisheries 44:243–245.

Clemens, B. J., R. J. Beamish, K. C. Coates, M. F. Docker, J. B. Dunham, A. E. Gray, J. E. Hess, J. C. Jolley, R. T. Lampman, B. J. McIlraith, M. L. Moser, J. G. Murauskas, D. L. G. Noakes, H. A. Schaller, C. B. Schreck, S. J. Starcevich, B. Streif, S. J. van de Wetering, J. Wade, L. A. Weitkamp, and L. A. Wyss. 2017. Conservation challenges and research needs for Pacific Lamprey in the Columbia River basin. Fisheries. 42:268–280.

Clemens, B. J., T. R. Binder, M. F. Docker, M. L. Moser, and S. A. Sower. 2010. Similarities, differences, and unknowns in biology and management of three parasitic lampreys of North America. Fisheries 35:580–594.

Clemens, B. J., S. J. van de Wetering, S. A. Sower, and C. B. Schreck. 2013. Maturation characteristics and life history strategies of the Pacific Lamprey, *Entosphenus tridentatus*. Canadian Journal of Zoology 91:775–788.

Clements S., T. Stahl, and C. B. Schreck. 2012. A comparison of the behavior and survival of juvenile Coho Salmon (*Oncorhynchus kisutch*) and steelhead trout (*O. mykiss*) in a small estuary system. Aquaculture 362–363:148–157.

Close, D. A., M. S. Fitzpatrick, and H. W. Li. 2002. The ecological and cultural importance of a species at risk of extinction, Pacific Lamprey. Fisheries 27:19–25.

Close, D. A., M. Fitzpatrick, H. Li, B. Parker, D. Hatch, and G. James. 1995. Status report of the Pacific Lamprey (*Lampetra tridentata*) in the Columbia River basin. Annual Report of the Oregon Cooperative Fish and Wildlife Research Unit to the Bonneville Power Administration, Portland, Oregon.

Cluer, B., and C. Thorne. 2014. A stream evolution model integrating habitat and ecosystem benefits. River Research and Applications 30:135–154.

Collins, B. D., D. R. Montgomery, K. L. Fetherston, and T. B. Abbe. 2012. The floodplain large-wood cycle hypothesis: a mechanism for the physical and biotic structuring of temperate forested alluvial valleys in the North Pacific coastal ecoregion. Geomorphology 139–140:460–470.

Courchamp, F., T. Clutton-Brock, and B. Grenfell. 1999. Inverse density dependence and the Allee effect. Trends in Ecology and Evolution 14:405–410.

Cowan, L. 1991. Physical characteristics and intragravel survival of Chum Salmon in developed and natural groundwater channels in Washington. Pages 109–124 *in* J. Colt and R. J. White, editors. Fisheries bioengineering symposium. American Fisheries Society, Symposium 10, Bethesda, Maryland.

CRITFC (Columbia River Inter-Tribal Fish Commission). 2011. Tribal Pacific Lamprey restoration plan for the Columbia River basin. CRITFC, Portland, Oregon. Available: www.critfc.org/wp-content/uploads/2012/12/lamprey_plan.pdf. (August 2016).

Dawson, H. A., B. R. Quintella, P. R. Almeida, A. J. Treble, and J. C. Jolley. 2015. The ecology of larval and metamorphosing lampreys. Pages 75–137 *in* M. F. Docker, editor. Lampreys: biology, conservation and control. Springer, Fish and Fisheries Monograph Series 31, New York.

Den Boer, P. 1968. Spreading of risk and stabilization of animal numbers. Acta Biotheoretica 18:165–194.

Deredec, A. and F. Courchamp. 2007. Importance of the Allee effect for reintroductions. Écoscience 14:440–451.

Docker, M. F., J. Hume, and B. J. Clemens. 2015. A surfeit of lampreys: preface. *In* M. F. Docker, editor. Lampreys: biology, conservation and control. Chapman Hall, Fish and Fisheries Monograph Series 31, New York.

Duker, G. J. 1977. Nest site selection by Chum Salmon (*Oncorhynchus keta*) in a spawning channel. Master's thesis. University of Washington, Seattle.

Ebersole, J. L., W. J. Liss, and C. A. Frissell. 1997. FORUM: Restoration of stream habitats in the western United States: restoration as reexpression of habitat capacity. Environmental Management 21:1–14.

Ebersole, J. L., W. J. Liss, and C. A. Frissell. 2003. Thermal heterogeneity, stream channel morphology, and salmonid abundance in northeastern Oregon streams. Canadian Journal of Fisheries and Aquatic Sciences 60:1266–1280.

Elith, J., and J. R. Leathwick. 2009. Species distribution models: ecological explanation and prediction across space and time. Annual Review of Ecology, Evolution, and Systematics 40:677–697.

ESA (Endangered Species Act). 1973. U.S. Code, volume 16, sections 1531–1544.

Falcy, M. R., and E. Suring. 2018. Detecting the effects of management regime shifts to dynamic environments using multi-population state-space models. Biological Conservation 221:34–43.

Fausch, K. D., C. E. Torgersen, C. V. Baxter, and H. W. Li. 2002. Landscapes to riverscapes: bridging the gap between research and conservation of stream fishes: a continuous view of the river is needed to understand how processes interacting among scales set the context for stream fishes and their habitat. BioScience 52:483–498.

Flitcroft, R. L., I. Arismendi, and M. V. Santelmann. 2019. A review of habitat connectivity research for Pacific salmon in marine, estuary, and freshwater environments. Journal of the American Water Resources Association 55:430–441.

Ford, M. J., D. Teel, D. M. Van Doornik, D. Kuligowski, and P. W. Lawson. 2004. Genetic population structure of central Oregon coast Coho Salmon (*Oncorhynchus kisutch*). Conservation Genetics 5:797–812.

Forsberg, B. O., J. A. Johnson, and S. M. Klug. 1977. Identification, distribution, and notes on food habits of fish and shellfish in Tillamook Bay, Oregon. Oregon Department of Fish and Wildlife, Resource Section, Federal Aid progress report, Salem.

FPC (Fish Passage Center). 2018. Adult salmon dam count - annual totals data query [online database]. http://fpc.org/web/apps/adultsalmon/Q_adultcounts_annualtotals-query.php. (August 2018).

Frissell, C. A., W. J. Liss, C. E. Warren, and M. D. Hurley. 1986. A hierarchical framework for stream habitat classification: viewing streams in a watershed context. Environmental Management 10:199–214.

Fukuwaka, M., and T. Suzuki. 2000. Density-dependence of Chum Salmon in coastal waters of the Japan Sea. North Pacific Anadromous Fish Commission Bulletin 2:75–81.

Fulton, L. A. 1970. Spawning areas and abundance of steelhead trout, Coho, Sockeye, and Chum salmon in the Columbia River basin—past and present. NOAA Technical Report NMFS SSRF 618.

Garland R. D., K. F. Tiffan, D. W. Rondorf, J. Skalicky, and D. R. Anglin. 2003. Assessment of chum and fall Chinook Salmon spawning habitat near Ives and Pierce islands in the Columbia River. Report to the Bonneville Power Administration, Contract 00004701, Portland, Oregon.

Goodman, D. H., S. B. Reid, M. F. Docker, G. R. Haas, and A. P. Kinziger. 2008. Mitochondrial DNA evidence for high levels of gene flow among populations of a widely distributed anadromous lamprey *Entosphenus tridentatus* (Petromyzontidae). Journal of Fish Biology 72:400–417.

Gorman, O. T., and J. R. Karr. 1978. Habitat structure and stream fish communities. Ecology 59:507–515.

Greene, C. M., J. E. Hall, K. R. Guilbault, and T. P. Quinn. 2010. Improved viability of populations with diverse life-history portfolios. Biology Letters 6:382–386.

Groot, C. and L. Margolis. 1991. Pacific salmon life histories. University of British Columbia Press, Vancouver.

Gunckel, S. L., K. K. Jones, and S. E. Jacobs. 2009. Spawning distribution and habitat use of adult Pacific and Western Brook lampreys in Smith River, Oregon. Pages 173–190 *in* L. R. Brown, S. D. Chase, M. G. Mesa, R. J. Beamish, and P. B. Moyle, editors. Biology, management, and conservation of lampreys in North America. American Fisheries Society, Symposium 72, Bethesda, Maryland.

Gurnell, A. M., K. J. Gregory, and G. E. Petts. 1995. The role of coarse woody debris in forest aquatic habitats: implications for management.

Hale, S. S., T. E. McMahon, and P. C. Nelson. 1985. Habitat suitability index models and instream flow suitability curves: Chum Salmon. U.S. Fish and Wildlife Service, Biological Report 82(10.108).

Healey, M. C. 1982a. Juvenile Pacific salmon in estuaries: the life support system. Pages 315–341 *in* V. S. Kennedy, editor. Estuarine comparisons. Academic Press, New York.

Healey, M. C. 1982b. Timing and relative intensity of size-selective mortality of juvenile Chum Salmon (*Oncorhynchus keta*) during early sea life. Canadian Journal of Fisheries and Aquatic Sciences 39:952–957.

Helfield, J. M., and R. J. Naiman. 2001. Effects of salmon-derived nitrogen on riparian forest growth and implications for stream productivity. Ecology 82:2403–2409.

Hendry, A. P., V. Castric, M. T. Kinnison and T. P. Quinn, 2004. The evolution of philopatry and dispersal: homing versus straying in salmonids. Pages 52–91 *in* A. P. Hendry and S. C. Stearns, editors. Evolution illuminated: salmon and their relatives. Oxford University Press, Oxford, UK.

Herlihy, A. T., R. M. Hughes, and J. C. Sifneos. 2006. Landscape clusters based on fish assemblages in the conterminous USA and their relationship to existing landscape classifications. Pages 87–112 *in* R. M. Hughes, L. Wang, and P. W. Seelbach, editors. American Fisheries Society, Symposium 48, Bethesda, Maryland.

Herrmann, R. B. 1971. Food of juvenile Chinook and Chum Salmon in the lower Chehalis River and upper Grays Harbor. Pages 59–82 *in* Grays Harbor cooperative water quality study 1964–1966. Washington State Department of Fisheries, Technical Report 7, Olympia.

Hilborn, R., T. P. Quinn, D. E. Schindler, and D. E. Rogers. 2003. Biocomplexity and fisheries sustainability. Proceedings of the National Academy of Sciences of the United States of America 100:6564–6568.

Hillson, T. D. and J. A. Grobelny. 2017. Chum Salmon Enhancement in the lower Columbia River: development of an integrated strategy to implement habitat restoration, reintroduction and hatchery supplementation in the tributaries below Bonneville Dam. Report to Bonneville Power Administration, contract numbers 67595 and 73502, Portland Oregon.

Hudy, M., T. M. Thieling, N. Gillespie, and E. P. Smith. 2008. Distribution, status, and land use characteristics of subwatersheds within the native range of Brook Trout in the eastern United States. North American Journal of Fisheries Management 28:1069–1085.

Humphries, P., and K. O. Winemiller. 2009. Historical impacts on river fauna, shifting baselines, and challenges for restoration. BioScience 59:673–684.

Hunter, J. G. 1959. Survival and production of Pink and Chum salmon in a coastal stream. Journal of the Fisheries Research Board of Canada 16:835–886.

IDFG (Idaho Department of Fish and Game). 2011. The status of Pacific Lamprey (*Entosphenus tridentatus*) in Idaho. IDFG, Boise, Idaho.

Johnson, M. A., and M. A. Banks. 2008. Genetic structure, migration, and patterns of allelic richness among Coho Salmon (*Oncorhynchus kisutch*) populations of the Oregon coast. Canadian Journal of Fisheries and Aquatic Sciences 65:1274–1285.

Johnson, N. S., T. J. Buchinger, and W. Li. 2015. Reproductive ecology of lampreys. Pages

265–303 *in* M. F. Docker, editors. Lampreys: biology, conservation and control. Springer, Fish and Fisheries Monograph Series 31, New York.

Johnson, O. W., W. S. Grant, R. G. Kope, K. Neely, F. W. Waknitz, and R. S. Waples. 1997. Status review of Chum Salmon from Washington, Oregon, and California. NOAA Technical Memorandum NMFS-NWFSC-32.

Johnson, S. L., J. D. Rodgers, M. F. Solazzi, and T. E. Nickelson. 2005. Effects of an increase in large wood on the abundance and survival of juvenile salmonids (*Oncorhynchus* spp.) in an Oregon coastal stream. Canadian Journal of Fisheries and Aquatic Sciences 62:412–424.

Jolley, J. C., G. S. Silver, J. E. Harris, and T. A. Whitesel. 2018. Pacific Lamprey recolonization of a Pacific Northwest river following dam removal. River Research and Applications 34:44–51.

Jones, K. K., T. J. Cornwell, D. L. Bottom, L. A. Campbell, and S. Stein. 2014a. The contribution of estuary-resident life histories to the return of adult *Oncorhynchus kisutch*. Journal of Fish Biology 85:52–80.

Jones, K. K., K. Anlauf-Dunn, P. S. Jacobsen, M. Strickland, L. Tennant, and S. E. Tippery. 2014b. Effectiveness of instream wood treatments to restore stream complexity and winter rearing habitat for juvenile Coho Salmon. Transactions of the American Fisheries Society 143:334–345.

Jones, K. K., T. J. Cornwell, D. L. Bottom, S. Stein, K. J. Anlauf-Dunn. 2018. Population viability improves following termination of Coho Salmon hatchery releases. North American Journal of Fisheries Management 38:29–55.

Kaeriyama, M., A. Yatsu, H. Kudo, and S. Saitoh. 2007. Where, when, and how does mortality occur for juvenile Chum Salmon *Oncorhynchus keta* in their first ocean year? North Pacific Anadromous Fish Commission Technical Report 7:52–55.

Kan, T. T. 1975. Systematics, variation, distribution and biology of lampreys of the genus *Lampetra* in Oregon. Doctoral dissertation. Oregon State University, Corvallis.

Katz, S. L., K. Barnas, R. Hicks, J. Cowen, and R. Jenkinson. 2007. Freshwater habitat restoration actions in the Pacific Northwest: a decade's investment in habitat improvement. Restoration Ecology 15:494–505.

Keefer, M. L., and C. C. Caudill. 2014. Homing and straying by anadromous salmonids: a review of mechanisms and rates. Reviews in Fish Biology and Fisheries 24:333–368.

Kiffney, P. M., C. M. Greene, J. E. Hall, and J. R. Davies. 2006. Tributary streams create spatial discontinuities in habitat, biological productivity, and diversity in mainstem rivers. Canadian Journal of Fisheries and Aquatic Sciences 63:2518–2530.

Knighton, D. 1998. Fluvial forms and processes. Oxford University Press, New York.

Kondolf, G. M., A. J. Boulton, S. O'Daniel, G. C. Poole, F. J. Rahel, E. H. Stanley, E. Wohl, A. Bång, J. Carlstrom, C. Cristoni, H. Huber, S. Koljonen, P. Louhi, and K. Nakamura. 2006. Process-based ecological river restoration: visualizing three-dimensional connectivity and dynamic vectors to recover lost linkages. Ecology and Society [online serial] 11(2):5.

Koski, K. V. 2009. The fate of Coho Salmon nomads: the story of an estuarine-rearing strategy promoting resilience. Ecology and Society [online serial] 14(1):4.

Koslow, J., A. Hobday, and G. Boehlert. 2002. Climate variability and marine survival of Coho Salmon (*Oncorhynchus kisutch*) in the Oregon production area. Fisheries Oceanography 11:65–77.

Kostow, K., editor. 1995. Biennial report on the status of wild fish in Oregon. Oregon Department of Fish and Wildlife, Salem.

Kostow, K. 2002. Oregon lampreys: natural history status and problem analysis. Oregon Department of Fish and Wildlife, Salem.

Lake, P. S., N. Bond, and P. Reich. 2007. Linking ecological theory with stream restoration. Freshwater Biology 52:597–615.

Lawrence, D. J., B. Stewart-Koster, J. D. Olden, A. S. Ruesch, C. E. Torgersen, J. J. Lawler, D. P. Butcher, and J. K. Crown. 2014. The interactive effects of climate change, riparian management, and a nonnative predator on stream-rearing salmon. Ecological Applications 24:895–912.

Lawson, P. W. 1993. Cycles in ocean productivity, trends in habitat quality, and the restoration of salmon runs in Oregon. Fisheries 18:6–10.

Lawson, P. W., E. A. Logerwell, N. J. Mantua, R. C. Francis, and V. N. Agostini. 2004. Environmental factors influencing freshwater survival and smolt production in Pacific Northwest Coho Salmon (*Oncorhynchus kisutch*). Canadian Journal of Fisheries and Aquatic Sciences 61:360–373.

Lawson, P. W., E. P. Bjorkstedt, M. W. Chilcote, C. W. Huntington, J. S. Mills, K. M. S. Moore, T. E. Nickelson, G. H. Reeves, H. A. Stout, T. C. Wainwright, and L. A. Weitkamp. 2007. Identification of historical populations of Coho Salmon (*Oncorhynchus kisutch*) in the Oregon coast evolutionarily significant unit. NOAA Technical Memorandum NMFS-NWFSC-79.

LCEP (Lower Columbia Estuary Partnership). 1999. Lower Columbia River Estuary Program comprehensive conservation and management plan. LCEP, Portland, Oregon. Available: www.estuarypartnership.org/1999-management-plan. (April 2019).

LCRFB (Lower Columbia River Fish Recovery Board). 2010. Washington lower Columbia salmon recovery and fish and wildlife subbasin plan. LCRFB, Vancouver, Washington. Available: https://docs.wixstatic.com/ugd/810197_dec0786f0c474a1192fa14b45bde9be3.pdf. (April 2019).

Lin, B., Z. Zhang, Y. Wang, K. P. Currens, A. Spidle, Y. Yamazaki, and D. A. Close. 2008. Amplified fragment length polymorphism assessment of genetic diversity in Pacific Lampreys. North American Journal of Fisheries Management 28:1182–1193.

Luzier, C. W., J. Brostrom, C. Cook-Tabor, D. H. Goodman, R. D. Nelle, K. Ostrand, H. Schaller, and B. Streif. 2009. Proceedings of the Pacific Lamprey Conservation Initiative work session—October 28–29, 2008. U.S. Fish and Wildlife Service, Portland, Oregon. Available: www.fws.gov/columbiariver/publications/Lamprey_Conservation_Proceedings_Final_09.pdf. (August 2016).

Luzier, C. W., H. A. Schaller, J. K. Brostrom, C. Cook-Tabor, D. H. Goodman, R. D. Nelle, K. Ostrand, and B. Streif. 2011. Pacific Lamprey (*Entosphenus tridentatus*) assessment and template for conservation measures. U.S. Fish and Wildlife Service, Portland, Oregon.

Magnusson, A., and R. Hilborn. 2003. Estuarine influence on survival rates of Coho (*Oncorhynchus kisutch*) and Chinook Salmon (*Oncorhynchus tshawytscha*) released from hatcheries on the U.S. Pacific coast. Estuaries and Coasts 26:1094–1103.

Manzon, R. G., J. H. Youson, and J. A. Holmes. 2015. Lamprey metamorphosis. Pages 139–214 *in* M. F. Docker, editor. Lampreys: biology, conservation and control. Springer, Fish and Fisheries Monograph Series 31, New York.

Mason, J. C. 1974. Behavioral ecology of Chum Salmon fry (*Oncorhynchus keta*) in a small estuary. Journal of the Fisheries Research Board of Canada 31:83–92.

McElhany, P., M. H. Rucklelshaus, M. J. Ford, T. C. Wainwright, and E. P. Bjorkstedt. 2000. Viable salmonid populations and the recovery of evolutionarily significant units. NOAA Technical Memorandum NMFS-NWFSC-42.

McElhany, P., M. Chilcote, J. Myers, and R. Beamesderfer. 2007. Viability status of Oregon salmon and steelhead populations in the Willamette and lower Columbia basins. Prepared for the Oregon Department of Fish and Wildlife, Salem and National Marine Fisheries Service, Silver Spring, Maryland.

McIlraith, B. J., C. C. Caudill, B. P. Kennedy, C. A. Peery, and M. L. Keefer. 2015. Seasonal migration behaviors and distribution of adult Pacific Lampreys in unimpounded reaches of the Snake River basin. North American Journal of Fisheries Management 35:123–134.

McGree, M., J. Stone, and T. A. Whitesel. 2008. Metamorphosis, growth, and survival of larval Pacific Lamprey reared in captivity.

Transactions of the American Fisheries Society 137:1866–1878.

McQuarrie, D. W. and D. D. Bailey. 1980. Adult returns to Thornton Creek. Pages 36–44 *in* Proceedings of the 31st annual northwest fish culture conference. Available from the Washington Department of Fish and Wildlife, Olympia.

Mesa, M. G. and E. S. Copeland. 2009. Critical uncertainties and research needs for the restoration and conservation of native lampreys in North America. Pages 311–321 *in* L. R. Brown, S. D. Chase, M. G. Mesa, R. J. Beamish, and P.B. Moyle, editors. Biology, management, and conservation of lampreys in North America. American Fisheries Society, Symposium 72, Bethesda, Maryland.

Miller, B. A., and S. Sadro. 2003. Residence time and seasonal movements of juvenile Coho Salmon in the ecotone and lower estuary of Winchester Creek, South Slough, Oregon. Transactions of the American Fisheries Society 132:546–559.

Montgomery, D. R., J. M. Buffington, N. P. Peterson, D. Schuett-Hames, and T. P. Quinn. 1996. Stream-bed scour, egg burial depths, and the influence of salmonid spawning on bed surface mobility and embryo survival. Canadian Journal of Fisheries and Aquatic Sciences 53:1061–1070.

Montgomery, D. R. and J. M. Buffington. 1997. Channel-reach morphology in mountain drainage basins. Geological Society of America Bulletin 109:596–611.

Montgomery, D. R. 1999. Process domains and the river continuum. Journal of the American Water Resources Association 35:397–410.

Montgomery, D. R., B. D. Collins, J. M. Buffington, and T. B. Abbe. 2003. Geomorphic effects of wood in rivers. Pages 21–48 *in* S. V. Gregory, K. L. Boyer, and A. M. Gurnell, editors. The ecology and management of wood in world rivers. American Fisheries Society, Symposium 37, Bethesda, Maryland.

Morrow, J. E. 1980. The freshwater fishes of Alaska. Alaska Northwest Publishing Company, Anchorage.

Moser, M. L., M. L. Keefer, H. T. Pennington, D. A. Ogden, and J. E. Simonson. 2011. Development of Pacific Lamprey fishways at a hydropower dam. Fisheries Management and Ecology 18:190–200.

Moser, M. L., and R. L. Paradis. 2017. Pacific Lamprey restoration in the Elwha River drainage following dam removals. American Currents 42:3–8.

Mote, P. W., E. A. Parson, A. F. Hamlet, K. N. Ideker, W. S. Keeton, D. P. Lettenmaier, N. J. Mantua, E. L. Miles, D. W. Peterson, D. L. Peterson, R. Slaughter, and A. K. Snover. 2003. Preparing for climatic change: the water, salmon, and forests of the Pacific Northwest. Climatic Change 61:45–88.

Moyle, P. B. 2002. Inland fishes of California. University of California Press, Berkeley.

Moyle, P. B., L. R. Brown, S. D. Chase, and R. M. Quiñones. 2009. Status and conservation of lampreys in California. Pages 279–292 *in* L. R. Brown, S. D. Chase, M. G. Mesa, R. J. Beamish, and P. B. Moyle, editors. Biology, management, and conservation of lampreys in North America. American Fisheries Society, Symposium 72, Bethesda, Maryland

Moyle, P. B., J. D. Kiernan, P. K. Crain, and R. M. Quiñones. 2013. Climate change vulnerability of native and alien freshwater fishes of California: a systematic assessment approach. PLOS (Public Library of Science) ONE [online serial] 8(5):e63883.

Myers, J., C. Busack, D. Rawding, and A. Marshall. 2003. Historical population structure of Willamette and lower Columbia River basin Pacific salmonids. NOAA Technical Memorandum NMFS-NWFSC-73.

Nehlsen, W., J. E. Williams, and J. A. Lichatowich. 1991. Pacific salmon at the crossroads: stocks at risk from California, Oregon, Idaho, and Washington. Fisheries 16:4–21.

Nicholas J., B. McIntosh, and E. Bowles. 2005. Oregon coastal Coho assessment part 1: synthesis. Oregon Department of Fish and Wildlife, Salem.

Nickelson, T. E. 1986. Influences of upwelling, ocean temperature, and smolt abundance on marine survival of Coho Salmon (*Oncorhynchus kisutch*) in the Oregon production area. Canadian Journal of Fisheries and Aquatic Sciences 43:527–535.

Nickelson, T. E., and P. W. Lawson. 1998. Population viability of Coho Salmon, *Oncorhynchus kisutch*, in Oregon coastal basins: application of a habitat-based life cycle model. Canadian Journal of Fisheries and Aquatic Sciences 55:2383–2392.

Nickelson, T. E., J. D. Rodgers, S. L. Johnson, and M. F. Solazzi. 1992. Seasonal changes in habitat use by juvenile Coho Salmon (*Oncorhynchus kisutch*) in Oregon coastal streams. Canadian Journal of Fisheries and Aquatic Sciences 49:783–789.

NMFS (National Marine Fisheries Service). 1998. Endangered and threatened species: threatened status for the Oregon coast evolutionarily significant unit of Coho Salmon. Federal Register 63:153(10 August 1998):42587–42591.

NMFS (National Marine Fisheries Service). 1999. Endangered and threatened species: threatened status for two ESUs of Chum Salmon in Washington and Oregon. Federal Register 64:57(25 March 1999):14508–14517.

NMFS (National Marine Fisheries Service). 2006. Endangered and threatened species: withdrawal of proposals to list and designate critical habitat for the Oregon coast evolutionarily significant unit of Coho Salmon. Federal Register 71:12(19 January 2006):3033–3048.

NMFS (National Marine Fisheries Service). 2008. NOAA Fisheries 2008–2017 United States v. Oregon management agreement. May 5, 2008. NMFS, Silver Spring, Maryland.

NMFS (National Marine Fisheries Service). 2013. Lower Columbia River plan for salmon and steelhead. NMFS, West Coast Region, Portland, Oregon. Available: www.westcoast.fisheries.noaa.gov/protected_species/salmon_steelhead/recovery_planning_and_implementation/lower_columbia_river/lower_columbia_river_recovery_plan_for_salmon_steelhead.html. (May 2014).

NMFS (National Marine Fisheries Service). 2016. Recovery plan for Oregon coast Coho Salmon evolutionarily significant unit. NMFS, West Coast Region, Portland, Oregon.

NPPC (Northwest Power Planning Council). 1986. Compilation of information on salmon and steelhead losses in the Columbia River basin, March 1986. NPPC, Portland, Oregon.

ODFW (Oregon Department of Fish and Wildlife). 2007. Oregon coast Coho conservation plan for the state of Oregon. ODFW, Salem.

ODFW (Oregon Department of Fish and Wildlife). 2010. Lower Columbia River conservation and recovery plan for Oregon populations of salmon and steelhead. ODFW, Salem.

ODFW (Oregon Department of Fish and Wildlife). 2014. Coastal multi-species conservation and management plan. ODFW, Salem.

Olsen, J. B., B. G. Flannery, T. D. Beacham, J. F. Bromaghin, P. A. Crane, C. F. Lean, K. M. Dunmall, and J. K. Wenburg. 2008. The influence of hydrographic structure and seasonal run timing on genetic diversity and isolation-by-distance in Chum Salmon (*Oncorhynchus keta*). Canadian Journal of Fisheries and Aquatic Sciences 65:2026–2042.

OWEB (Oregon Watershed Enhancement Board). 2017. 2015–17 biennial report for the Oregon plan for salmon and watersheds. OWEB, Salem.

Palmer, M. E. 2009. Reforming watershed restoration: science in need of application and applications in need of science. Estuaries and Coasts 32:1–17.

Pearcy, W. J., C. D. Wilson, A. W. Chung, and J. W. Chapman. 1989. Residence times, distribution, and production of juvenile Chum Salmon, *Oncorhynchus keta*, in Netarts Bay, Oregon. U.S. National Marine Fisheries Service Fishery Bulletin 87: 553–568.

Perring, M. P., R. J. Standish, J. N. Price, M. D. Craig, T. E. Erickson, K. X. Ruthrof, A. S. Whiteley, L. E. Valentine, R. J. Hobbs. 2015. Advances in restoration ecology: rising to the challenges of the coming decades. Ecosphere [online serial] 6(8):1–25.

Petrou, E. L., J. E. Seeb, L. Hauser, M. J. Witteveen, W. D. Templin, and L. W. Seeb. 2014. Fine-scale sampling reveals distinct isolation by distance patterns in Chum Salmon (*Oncorhynchus keta*) populations occupying a glacially dynamic environment. Conservation Genetics 15:229–243.

PLTWG (Pacific Lamprey Technical Workgroup). 2017. Practical guidelines for incorporating adult Pacific Lamprey passage at fishways. June 2017. U.S. Fish and Wildlife Service, Washington, D.C. Available: www.fws.gov/pacificlamprey/mainpage.cfm. (April 2019).

Northwest Fisheries Science Center. 2019. Pacific Northwest Salmon Habitat Project database [online database]. Available: www.webapps.nwfsc.noaa.gov/apex/f?p=409:13::::::. (February 2019.)

Poff, N. L., J. D. Allan, M. B. Bain, J. R. Karr, K. L. Prestegaard, B. D. Richter, R. E. Sparks, and J. C. Stromberg. 1997. The natural flow regime. BioScience 47:769–784.

Poff, N. L., and J. V. Ward. 1990. Physical habitat template of lotic systems: recovery in the context of historical pattern of spatiotemporal heterogeneity. Environmental Management 14:629–645.

Pollock, M. M., G. R. Pess, T. J. Beechie, and D. R. Montgomery. 2004. The importance of beaver ponds to Coho Salmon production in the Stillaguamish River basin, Washington, USA. North American Journal of Fisheries Management 24:749–760.

Postel, S. L. 2000. Entering an era of water scarcity: the challenges ahead. Ecological Applications 10:941–948.

Postel, S. L. and B. H. Thompson Jr. 2005. Watershed protection: capturing the benefits of nature's water supply services. Natural Resources Forum: A United Nations Sustainable Development Journal 29: 98–108.

Potter, I. C., H. S. Gill, C. B. Renaud, and D. Haoucher. 2015. The taxonomy, phylogeny, and distribution of lampreys. Pages 35–74 *in* M. F. Docker, editor. Lampreys: biology, conservation and control. Springer, Fish and Fisheries Monograph Series 31, New York.

Pringle, C. M. 1997. Exploring how disturbance is transmitted upstream: going against the flow. Freshwater Science 16:425–438.

Quinn, T. P. 1993. A review of homing and straying of wild and hatchery-produced salmon. Fisheries Research 18:29–44.

Quinn, T. P. and K. W. Myers. 2004. Reviews in Fish Biology and Fisheries 14:421–442.

Quinn, T. P. 2005. The behavior and ecology of Pacific salmon and trout. University of Washington Press, Seattle.

Quinn, T. P. and K. W. Myers. 2004. Anadromy and the marine migrations of Pacific salmon and trout: Rounsefell revisited. Reviews in Fish Biology and Fisheries 14:421–442.

Reeves, G. H., J. E. Williams, K. M. Burnett, and K. Gallo. 2006. The aquatic conservation strategy of the Northwest Forest Plan. Conservation Biology 20:319–329.

Renaud, C. B. 2008. Petromyzontidae, *Entosphenus tridentatus*: southern distribution record, Isla Clarión, Revillagigedo Archipelago, Mexico. Check List 4:82–85.

Renaud, C. B. 2011. Lampreys of the world: An annotated and illustrated catalogue of lamprey species known to date. Food and Agriculture Organization of the United Nations, Species Catalogue for Fishery Purposes No. 5, Rome.

Robinson, T. C., P. W. Sorensen, J. M. Bayer, and J. G. Seelye. 2009. Olfactory sensitivity of Pacific Lampreys to lamprey bile acids. Transactions of the American Fisheries Society 138:144–152.

Rodgers, J. D., K. K. Jones, A. G. Talabere, C. H. Stein, and E. H. Gilbert. 2005. Oregon coast Coho habitat assessment, 1998–2003. Oregon Department of Fish and Wildlife, OPSW-ODFW-2005-5, Salem.

Romer, J. D., C. A. Leblanc, S. Clements, J. A. Ferguson, M. L. Kent, D. Noakes, and C. B. Schreck. 2012. Survival and behavior of juvenile steelhead trout (*Oncorhynchus mykiss*) in two estuaries in Oregon, USA. Environmental Biology of Fishes.

Roni, P., T. J. Beechie, R. E. Bilby, F. E. Leonetti, M. M. Pollock, and G. R. Pess. 2002. A review of stream restoration techniques and a hierarchical strategy for prioritizing restoration in Pacific Northwest watersheds.

North American Journal of Fisheries Management 22:1–20.

Rosemond, A., and C. Anderson. 2003. Engineering role models: do non-human species have the answers? Ecological Engineering 20:379–387.

Rosgen, D. L. 1997. A geomorphological approach to restoration of incised rivers. *In* S. S. Y. Wang, E. J. Langendoen, and F. D. Shields, Jr., editors. Management of landscapes disturbed by channel incision: stabilization, rehabilitation, restoration. University of Mississippi, The Center for Computational Hydroscience and Engineering, Oxford.

Sandercock, F. K. 1991. Life history of Coho Salmon (*Oncorhynchus kisutch*). Pages 394–445 *in* C. Groot and L. Margolis, editors. Pacific salmon life histories. UBC Press, Vancouver.

Sakano, E. 1960. Results from marking experiments on young Chum Salmon in Hokkaido, 1951–1959. Scientific Reports of the Hokkaido Salmon Hatchery 915:17–38. (In Japanese, English summary.)

Salo, E. O. 1991. Life history of Chum Salmon, *Oncorhynchus keta*. Pages 231–309 *in* C. Groot and L. Margolis, editors. Pacific salmon life histories. UBC Press, Vancouver.

Sano, S. 1966. Salmon of the North Pacific Ocean—part III: a review of the life history of North Pacific salmon. 3. Chum Salmon in the Far East. International North Pacific Fish Commission Bulletin 18:41–57.

Schlosser, I. J., and P. L. Angermeier. 1995. Spatial variation in demographic processes in lotic fishes: conceptual models, empirical evidence, and implications for conservation. Pages 360–370 *in* J. L. Nielsen, editor. Evolution and the aquatic ecosystem: defining unique units in population conservation. American Fisheries Society, Symposium 17, Bethesda, Maryland.

Schreck, C. B., T. P. Stahl, L. E. Davis, D. D. Roby, and B. J. Clemens. 2006. Mortality estimates of juvenile spring–summer Chinook Salmon in the lower Columbia River and estuary, 1992–1998: evidence for delayed mortality? Transactions of the American Fisheries Society 135:457–475.

Schroder, S. L. 1974. Assessment of production of Chum Salmon fry from the Big Beef Creek spawning channel. Annual Report Anadromous Fish Project. University of Washington, Fisheries Research Institute, FRI-UW-7411, Seattle.

Schroder, S. L. 1977. Assessment of production of Chum Salmon fry from the Big Beef Creek spawning channel. Completion Report Anadromous Fish Project. University of Washington, Fisheries Research Institute, FRI-U W-7718, Seattle.

Schroeder, R. K., L. D. Whitman, B. Cannon, and P. Olmsted. 2016. Juvenile life-history diversity and population stability of spring Chinook Salmon in the Willamette River basin, Oregon. Canadian Journal of Fisheries and Aquatic Sciences 73:921–934.

Schumm, S. A., M. D. Harvey, and C. C. Watson. 1984. Incised channels: morphology, dynamics, and control. Water Resources Publications, Littleton, Colorado.

Scott, W. B. and E. J. Crossman. 1973. Freshwater fishes of Canada. Fisheries Research Board of Canada Bulletin 184.

Sheoships, G. 2014. Pacific Lamprey *Entosphenus tridentatus*: Integrating ecological knowledge and contemporary values into conservation planning, and stream substrate associations with larval abundance in the Willamette River basin, Oregon, U.S.A. Master's thesis. Oregon State University, Corvallis.

Simon, A., and C. R. Hupp. 1986. Channel widening characteristics and bank slope development along a reach of Cane Creek, West Tennessee. Pages 113–126 *in* S. Y. Subitzky, editor. Selected papers in the hydrologic sciences. U.S. Geological Survey Water-Supply Paper 2290.

Small, M. P., A. E. Frye, J. F. Von Bargen, and S. F. Young. 2006. Genetic structure of Chum Salmon (*Oncorhynchus keta*) populations in the lower Columbia River: are Chum Salmon in Cascade tributaries remnant populations? Conservation Genetics 7:65–78.

Small, M. P., B. Glaser, T. Hillson, and C. Bowman. 2011. Population genetic structure and recovery of Chum Salmon in the lower Columbia River. Project report by Washington Department of Fish and Wildlife, Molecular Genetics Lab, Olympia.

Small, M. P., T. H. Johnson, C. Bowman, and E. Martinez. 2014. Genetic assessment of a summer Chum Salmon metapopulation in recovery. Evolutionary Applications 7:266–285.

Solazzi, M. F., T. E. Nickelson, S. L. Johnson, and J. D. Rodgers. 2000. Effects of increasing winter rearing habitat on abundance of salmonids in two coastal Oregon streams. Canadian Journal of Fisheries and Aquatic Sciences 57:906–914.

Southwood, T. R. E. 1977. Habitat, the templet for ecological strategies? Journal of Animal Ecology 46:337–365.

Southwood, T. R. E. 1988. Tactics, strategies and templets. Oikos 52:3–18.

Spice, E. K., D. H. Goodman, S. B. Reid, and M. F. Docker. 2012. Neither philopatric nor panmictic: microsatellite and mtDNA evidence suggest lack of natal homing but limits to dispersal in Pacific Lamprey. Molecular Ecology 21:2916–2930.

Stout, H. A., P. W. Lawson, D. L. Bottom, T. D. Cooney, M. J. Ford, C. E. Jordan, R. G. Kope, L. M. Kruzic, G. R. Pess, G. H. Reeves, M. D. Scheuerell, T. C. Wainwright, R. S. Waples, E. Ward, L. A. Weitkamp, J. G. Williams, and T. H. Williams. 2012. Scientific conclusions of the status review for Oregon coast Coho Salmon (*Oncorhynchus kisutch*). NOAA Technical Memorandum NMFS-NWFSC-118.

Streif, B. 2009. Considering Pacific Lamprey when implementing instream activities. Pages 255–268 *in* L. R. Brown, S. D. Chase, M. G. Mesa, R. J. Beamish, and P. B. Moyle, editors. Biology, management, and conservation of lampreys in North America. American Fisheries Society, Symposium 72, Bethesda, Maryland.

Strickland, M. J., K. Anlauf-Dunn, K. Jones, and C. Stein. 2018. Winter habitat condition of Oregon coast Coho Salmon populations, 2007–2014. Oregon Department of Fish and Wildlife, Information Report 2018-01, Salem.

Tallman, R. F., and M. C. Healey. 1994. Homing, straying, and gene flow among seasonally separated populations of Chum Salmon (*Oncorhynchus keta*). Canadian Journal of Fisheries and Aquatic Sciences 51:577–588.

Tautz, A. F., and C. Groot. 1975. Spawning behavior of Chum Salmon (*Oncorhynchus keta*) and Rainbow Trout (*Salmo gairdneri*). Journal of the Fish Research Board of Canada 32:633–642.

Urawa, S., S. Sato, P. A. Crane, B. Agler, R. Josephson, and T. Azumaya. 2009. Stock-specific ocean distribution and migration of Chum Salmon in the Bering Sea and North Pacific Ocean. North Pacific Anadromous Fish Commission Bulletin 5:131–146.

USFWS (U.S. Fish and Wildlife Service). 2004. Endangered and threatened wildlife and plants; 90-day finding on a petition to list three species of lampreys as threatened or endangered. Federal Register 69:257(27 December 2004):77158–77167.

USFWS (U.S. Fish and Wildlife Service). 2012. Conservation agreement for Pacific Lamprey (Entosphenus tridentatus) in the states of Alaska, Washington, Oregon, Idaho, and California. USFWS, Washington, D.C. Available: www.fws.gov/pacificlamprey/Documents/Pacific%20Lamprey%20Conservation%20Agreement.pdf. (August 2016).

USFWS (U.S. Fish and Wildlife Service). 2015. Implementation of the conservation agreement for Pacific Lamprey in the states of Alaska, Washington, Oregon, Idaho, and California: interim operating guidelines, May 7, 2015. USFWS, Washington, D.C.

Vannote, R. L., G. W. Minshall, K. W. Cummins, J. R. Sedell, and C. E. Cushing. 1980. The river continuum concept. Canadian Journal of Fisheries and Aquatic Sciences 37:130–137.

Wainwright, T. C., and L. A. Weitkamp. 2013. Effects of climate change on Oregon coast Coho Salmon: habitat and life-cycle interactions. Northwest Science 87:219–242.

Waldman, J., K. A. Wilson, M. Mather, and N. P. Snyder. 2016. A resilience approach can improve anadromous fish restoration. Fisheries 41:116–126.

Walter, R. C., and D. J. Merritts. 2008. Natural streams and the legacy of water-powered mills. Science 319:299–304.

Warren, C. E., and W. J. Liss. 1980. Adaptation to aquatic environments. Pages 15–40 *in* R. T. Lackey and L. A. Nielsen, editors. Fisheries management. Wiley, New York.

WDF (Washington Department of Fisheries), WDW (Washington Department of Wildlife), and WWTIT (Western Washington Treaty Indian Tribes). 1993. 1992 Washington State salmon and steelhead stock inventory (SASSI). Washington Department of Fish and Wildlife, Olympia.

WDFW (Washington Department of Fish and Wildlife). 2004. Hatchery and genetic management plan for the lower Columbia River Chum Salmon ESU—Grays River, WA. WDFW, Seattle. Submitted to the National Oceanic and Atmospheric Administration, Silver Spring, Maryland.

Weitkamp, L., and K. Neely. 2002. Coho Salmon (*Oncorhynchus kisutch*) ocean migration patterns: insight from marine coded-wire tag recoveries. Canadian Journal of Fisheries and Aquatic Sciences 59:1100–1115.

Wiens, J. A. 1989. Spatial scaling in ecology. Functional Ecology 3:385–397.

Williams, J. E., R. N. Williams, R. F. Thurow, L. Elwell, D. P. Philipp, F. A. Harris, J. L. Kershner, P. J. Martinez, D. Miller, G. H. Reeves, C. A. Frissell, and J. R. Sedell. 2011. Native fish conservation areas: a vision for large-scale conservation of native fish communities. Fisheries 36:267–277.

Wolter, C., A. D. Juijse, and P. Parasiewicz. 2016. Temporal and spatial patterns of fish response to hydromorphological processes. River Research and Applications 32:190–201.

Wright, S. 1943. Isolation by distance. Genetics 28:114–138.

Wydoski, R. S. and R. R. Whitney. 1979. Inland fishes of Washington. University of Washington Press, Seattle.

Yun, S., A. J. Wildbill, M. J. Siefkes, M. L. Moser, A. H. Dittman, S. C. Corbett, W. Li, and D. A. Close. 2011. Identification of putative migratory pheromones from Pacific Lamprey. Canadian Journal of Fisheries and Aquatic Sciences 68:2194–2203.

American Fisheries Society Symposium 91:515–564, 2019

New Approaches to Providing Streamflow for Fisheries in the American West: Embracing Prior Appropriation and the Marketplace

Rob Van Kirk* and Brandon Hoffner
Henry's Fork Foundation
Post Office Box 550, Ashton, Idaho 83420, USA

Amy Verbeten
Friends of the Teton River
Post Office Box 768, Driggs, Idaho 83422, USA

Scott Yates
Trout Unlimited
321 East Main Street, Suite 411, Bozeman, Montana 59715, USA

Abstract.—Most rivers in the American West are highly managed for irrigation and other uses, but they also support popular trout fisheries and other ecologically valuable resources. Traditionally, streamflow to support these resources has been conceptually based on minimum streamflow and natural hydrologic regimes while the prior appropriation doctrine is generally considered an impediment to providing these flows. Successes often occur in situations with substantial federal nexus. However, protective federal laws and policies usually do not apply to popular nonnative trout fisheries or to nonlisted native species. Furthermore, many western trout fisheries and important native fish populations occur in highly altered dam tailwaters and in agricultural landscapes where hydrologic regimes have been more subtly altered by a century of irrigation diversion, seepage, and groundwater return flow. In these cases, traditional ecological and administrative viewpoints can fail to provide appropriate instream flows and may be detrimental to existing fish and wildlife resources. We propose a new paradigm based on the viewpoints that fisheries and related resources can be maintained in regulated hydrologic regimes and in watersheds dominated by private-land agriculture and that prior appropriation can be a pragmatic tool for providing instream flow at times and places where it addresses ecologically limiting factors. Application of this paradigm requires collaboration among irrigators and other water users, agricultural producers, government agencies, and conservation organizations. New strategies for providing ecological streamflow include capitalizing on the difference between physical and "paper" water, providing high-resolution water-supply information to water users and managers, actively managing groundwater and surface water together, and developing market mechanisms to change irrigation practices. We provide examples of these strategies from the Snake River basin and discuss transferability of our approach to other watersheds.

* Corresponding author: rob@henrysfork.org

Introduction

Most rivers in the American West are managed for irrigation, hydroelectric generation, and other uses, but they also support popular trout fisheries and ecologically valuable resources. Fisheries advocates have worked for decades to provide adequate streamflow to support these resources. However, the prior appropriation doctrine of water allocation is usually an impediment, exceptions being cases in which federal laws and policies supersede state water resource management. Because many important trout fisheries in the West are found in fully appropriated rivers and do not enjoy substantial federal protection, agencies and nongovernmental organizations (NGOs) with missions to maintain coldwater fisheries must develop new strategies to provide environmental flows, especially in the face of a changing climate. In this chapter, we present four such strategies in the context of a new paradigm of science and policy for providing instream flow.

We begin by reviewing scientific, legal, and administrative aspects of providing environmental flows, identifying those components central to our approach. Next, we describe two relatively new concepts in the conservation of aquatic ecosystems in the West: (1) that in many instances irrigation efficiency does not create water savings, and (2) that in many locations throughout the West, traditional irrigation practices have created hybrid human-natural hydro-ecosystems of high value for fish, wildlife, and related recreation. We conclude the review by describing uncertainties in predicting effects of climate change in regulated hydrologic systems. We then use two case studies from the upper Snake River basin in Idaho and Wyoming to illustrate our paradigm and specific strategies, which are based around

- administrative flexibility in prior appropriation,
- science-based improvements in water management,
- conjunctive management of groundwater and surface water, and
- incentive-based approaches to reducing irrigation demand.

The chapter ends with a comparison and transferability of our case studies to other water conservation efforts in the West, highlighting the central role of locally based collaboration among diverse stakeholders as a key to success.

Providing Streamflow for Fisheries in the West: A Critical Review

The science of environmental flows

Since the 1970s, fisheries biologists have recognized the need to provide adequate streamflow for fish and other aquatic resources in regulated rivers. Early methods for determining what are now termed ecological or environmental flows focused on the minimum flow necessary to maintain aquatic life or on some other fixed flow that provided a given amount of habitat (e.g., Tennant 1976; Petts 2009). Through the 1990s, the instream flow incremental methodology, based on the physical habitat simulation system (PHABSIM), was the most widely used tool for determining instream flow needs (Gillilan and Brown 1997). Within a regulatory process, the determination of streamflow requirements based on PHABSIM is frequently a negotiated settlement that pits flow-based habitat availability against maximum use of the water resource (Bovee et al. 1998). The resulting flow requirements are usually a prescription for static flows that benefit a fixed set of species and life-history stages (Railsback 2016).

Meanwhile, river ecologists recognized that streams and their floodplains are dynamic ecosystems driven by spatial and temporal variability in streamflow (Poff and Ward 1989; Poff and Ward 1990; Richter et al. 1997a). This recognition led to formal definition of a river's hydrologic regime as magnitude, frequency, duration, timing, and rate of change of streamflow and to the recommendation that restoration of natural-flow regimes should be the fundamental basis of managing river ecosystems (Poff et al. 1997). The natural-flow paradigm is based on empirical evidence that anthropogenic alteration of natural hydrologic regimes affects all aspects of aquatic and riparian ecosystem function (Lloyd et al. 2004; Petts 2009; Poff and Zimmerman 2010). For example, hydrologic alteration has resulted in widespread replacement of native fish species by nonnatives (Bain et al. 1988; Hughes and Noss 1992; Scheidegger and Bain 1995; Richter et al. 1997b; Freyrer and Healey 2003).

Often conducted with the indicators of hydrologic alteration (Richter et al. 1996), assessments of hydrologic alteration and recommendations for restoration of natural flow regimes have been applied to rivers regulated by dams (Hesse and Mestl 1993; Richter et al. 1998; Benjamin and Van Kirk 1999; Galat and Lipkin 2000; Wildhaber et al. 2000; Magilligan and Nislow 2001; Singer 2007). River regulation by dams is well known to result in simplification of the stream channel and floodplain (Ligon et al. 1995; Collier et al. 1996; Magilligan et al. 2003) and loss or decadence of native riparian vegetation (Merigliano 1996; Patten 1998; Nilsson and Berggren 2000; Stromberg 2001; Nislow et al. 2002). However, aquatic habitat can also be substantially compromised by seasonal diversion of water from streams that are not regulated by dams (Deitch et al. 2009; Bauer et al. 2015). Restoration of natural hydrologic regimes has resulted in documented recovery of native fish assemblages (Travnichek et al. 1995) and riparian areas (Molles et al. 1998; Rood et al. 2003) in individual cases. However, ecological response to hydrologic alteration is variable and inconsistent across systems, hampering efforts to predict whether restoration of the natural-flow regime will accomplish ecological objectives (Lloyd et al. 2004; Petts 2009; Poff and Zimmerman 2010). More recent analysis at the ecoregion-scale found that designer flows that provide key hydrologic components can produce greater ecological response with the available water than mimicking the natural-flow regime (Chen and Olden 2017). The designer flow concept is an important component of our approach.

Prior appropriation systems of water allocation

In the 18 states (including Alaska) that lie wholly or partially west of the 100th meridian, water is a public resource managed by the state, which issues rights to individuals or to entities such as irrigation districts and cities to use water when it is available (McCormick 1994). Allocation of surface water in these states is governed at least in part by the prior appropriation doctrine widely known as "first in time is first in right" (Fiege 1999; Behnampour 2011; Toll 2011). Under this doctrine, which originated among miners during the California Gold Rush, the first user to divert water from a stream and put it to beneficial use has priority over all subsequent users (Reisner and Bates 1990; Gillilan and Brown 1997). Irrigation, mining, stock water, municipal, and instream flow are now commonly recognized beneficial uses. Historically, the priority date of a water right was the date on which the water was first diverted and put to beneficial use. In more recent decades, priority dates for diversion of natural streamflow are determined by date of application to the state water resources agency or by an adjudication decree. In this chapter, we use the common definition of "natural streamflow" or "natural flow" as the flow that would be present in the river without impoundments, diversions, and injections of out-of-basin water (Gillilan and Brown 1997; Fiege 1999; Olenichak 2015). Prior to construction of reservoirs and other water infrastructure, natural flow was the only source of water available to claim and divert. In addition to "first in time" and "beneficial use," the third fundamental tenet of prior appropriation is "use it or lose it," which specifies that all or part of a water right can be subject to forfeiture if it not diverted and applied to a beneficial use (Reisner and Bates 1990; Fereday and Creamer 2010; Behnampour 2011). Forfeiture is intended to maximize use of the water resource and prevent speculation (Anderson and Kraft 2012).

Typical components of a water right include priority date, source of water, diversion rate or annual volume, place of use, beneficial use(s), and season of use. In a typical situation, a stream's natural flow is diverted, most commonly for irrigation, by a number of users on that stream. As streamflow recedes from peaks produced by rain events or snowmelt and is no longer sufficient to meet all rights, priority dates determine the order in which users must curtail diversion. The prior appropriation system also applies to water stored in reservoirs. Generally, the priority date of storage water is authorization date of reservoir construction or incorporation date of an irrigation entity formed to administer the storage water. Water accrues to a storage account when that right is in priority. Once stored under the applicable right, this water can be released and used at a later time when needed. Most western states also apply prior appropriation to allocation of groundwater (Blomquist et al. 2001), but the degree to which states administer surface water and groundwater conjunctively (as a single entity) varies widely. Because the effect of groundwater use on other water users is attenuated in space and time, conjunctive administration cannot be done in real time and requires use of computer models of groundwater flow (Cosgrove and Johnson 2004; Johnson et al. 2008).

The prior appropriation doctrine has long been considered an impediment to providing instream flow (Gillilan and Brown 1997). Historically, instream flow was not considered a beneficial use, in part because the concept of beneficial use is based on diversion of water from the stream (Reisner and Bates 1990). Most states now recognize instream flow, but because this recognition occurred only in recent decades, instream flow rights are junior in priority to almost all other rights and are therefore ineffective at keeping water in the stream when it is needed most—when supplies are low (Gillilan and Brown 1997). In Idaho, for example, there are 291 water rights for minimum streamflow, all with priority dates of 1976 or later. By comparison, 1,131 surface water rights in the upper Snake River basin alone, including all of the large storage rights, are senior to all of the minimum-flow rights in the whole state.

As a result, some advocates for instream flow protection have proposed that modification or even elimination of prior appropriation is necessary to attain equitable distribution of water resources in the West, essentially to move water from perceived economically inefficient uses (e.g., irrigation of low-value crops such as hay) and toward uses with higher socioeconomic benefit, including environmental flows (Keenan et al. 1999; Howe and Goemans 2003; Toll 2011). Reallocation of "conserved water" is a fundamental concept among proponents of reforming western water institutions (Reisner and Bates 1990; Gleick 2010; Sabo et al. 2010; Schaible et al. 2010). Because irrigated agriculture is by far the largest use of water in the West (Maddock and Hines 1995; Frisvold and Konyar 2012; Averyt et al. 2013), changing irrigation practices, in theory, provides the largest opportunity for water savings (Reisner and Bates 1990; Howe and Goemans 2003; Behnampour 2011). This theory is based on the assumption that increasing irrigation efficiency will reduce consumptive use of water, and the difference between the amount of water withdrawn and that required for crop needs is "new water" that can be reallocated. Unfortunately, at the river-basin scale, hydrologic and economic analyses show this assumption to be false

(Grafton et al. 2018). The majority of water diverted in excess of crop needs returns to the hydrologic system via surface or groundwater pathways, where it becomes all or part of the water supply for other users (Jensen 2007; Ward and Pulido-Velázquez 2008; Qureshi et al. 2010). In other words, the "excess" water of one user is not wasted because it has already been allocated to another user and may in fact contribute to streamflow as it travels from the point of return downstream to the next user. Furthermore, increased irrigation efficiency rarely decreases consumptive use of water across a river basin because more efficient application increases crop production per unit of land area at the field scale (Huffaker 2008; Ward and Pulido-Velázquez 2008), an individual water user will always gain economic benefit by applying some fraction of water saved through efficiency toward increased crop yield (Contor and Taylor 2013), and water not diverted from a stream by one user becomes available to another, who can use it to expand irrigation in space and/or time, thereby increasing crop production across the irrigated region (Huffaker 2008).

However, when accompanied by appropriate economic incentives and administrative mechanisms, improved irrigation efficiency can increase instream flow at the scale of individual tributaries and stream reaches. Trout Unlimited (TU) has assisted landowners in converting flood irrigation to gated pipe or sprinkler systems along small western trout streams, often identifying or providing economic incentives for landowners to leave the saved water in the stream rather than use it to increase production. Once water is saved at the field scale and left in the stream, the challenge then becomes ensuring that the water is not simply diverted by the next user. In most cases, a water user farther downstream in the basin will eventually divert the saved water, but fishery objectives can be met if the water flows through a target reach before being diverted. We refer to this as "shepherding" the saved water through the target reach, which requires creative administrative mechanisms that depend on local hydrogeology, water-rights accounting procedures, and spatial distribution of water rights across the basin. For example, water saved in the upper Green River basin in Wyoming is shepherded to the large reservoirs downstream on the Colorado River, where it is diverted and used. However, the saved water flows through headwater streams in the upper basin and benefits trout fisheries there.

Despite the challenges to providing instream flow under prior appropriation, the doctrine is so firmly embedded in western state constitutions and laws that pragmatic approaches to providing instream flow must work within the prior appropriation system. By far, the most promising mechanisms to accomplish this are administrative transfers from one use to another, temporary leases or transfers through water-supply banks, and market-based transactions (Keenan et al. 1999; Howe and Goemans 2003; Knapp et al. 2003; Wildman and Forde 2012; Payne et al. 2014). Although these concepts have been around for decades (Palmer 1991; Gillilan and Brown 1997), their full potential to provide instream flow has not been realized due to a variety of legal, administrative, sociological, and economic factors (McCormick 1994; Keenan et al. 1999; Contor 2010; Wildman and Forde 2012). Our approach relies heavily on creative ways to use transfers, leases, and markets to provide instream flow.

Advantages and limitations of federal policies

While the state institution of prior appropriation can hinder efforts at providing instream flow, substantial federal nexus can provide mechanisms for success. The Federal Power Act (FPA 1920) provides the best example of a federal mechanism that can provide environmental flows, particularly on large and mid-sized rivers. Privately owned and operated hydroelectric power facilities generally fall under the jurisdiction of the Federal Power Act (FPA) and therefore under licensing authority of the Federal Energy Regulatory Commission (FERC). The provisions of the FPA give leverage to federal agencies with a diverse array of authorities related to protection of fish and wildlife species, including those listed under the Endangered Species Act (ESA 1973) and to management of lands along river corridors that are impacted by hydropower activities (FPA 1920). State agencies that issue and oversee water rights and manage fish and wildlife also have certain conditioning authority in FERC licensing (FPA 1920). For example, a state environmental quality department can require that a FERC licensee maintain dissolved oxygen no lower than the state standard. Finally, conservation groups can also participate and, at times, use the FPA to enforce federal and state conditions. Over the past 30 years, the Pacific Northwest and parts of California have been ground zero for FERC proceedings and fishery protection, and conservation gains—including the quantity and timing of streamflows—have been the product of complex multiparty settlement agreements, dam decommissioning and removal efforts, and litigation (Becker 2006; Allen 2010; Blumm and Erickson 2012). Because hydropower licenses usually range from 30 to 50 years and the nature of river operations is often experimental in order to balance power production and fishery protection, adaptive management provisions are critical (Williams and Brown 2012; Chaffin and Gosnell 2017).

Outside of the FERC context, federal authorities and laws provide limited ability to change river operations and benefit fish and wildlife. A large swath of western ranch and farm fields are irrigated with water stored and delivered by the U.S. Bureau of Reclamation (USBR). While many of these facilities also produce power, they are not regulated by FERC. And while there are applicable federal laws such as the National Environmental Policy Act (1970) and Fish and Wildlife Coordination Act (1934) that provide a nexus to fish and wildlife protection, such laws are process-oriented and, standing alone, have generally been unsuccessful in significantly modifying historic river operations or drastically improving conditions (Houck 1981; Veiluva 1981). Each USBR facility is authorized for a specific purpose, and throughout the West, the primary purpose is usually irrigation, with a variety of recognized secondary uses such as hydropower, flood control, municipal and industrial water supply, fish and wildlife, and recreation (Benson 1997). While USBR facilities are also required to comply with the Endangered Species Act (ESA), many high-quality nonnative fisheries and other coldwater resources in the West—particularly those in tailwaters—fall outside of the high-stakes regulatory environment of the ESA. In these places, where most legal options to protect and enhance trout fisheries are of limited value both in form and substance, collaboration and innovation are critical to success. Rivers joining the Henrys Fork (Figure 1) on this list of venerate western fishing destinations list include the North Platte, San Juan, Green, Provo, Owyhee, Arkansas, and Bighorn (Gunn and Gunn 2013).

Moving beyond a litigation-based or legal paradigm is a reality in many parts of the West when it comes to water. Even a large proportion of FERC-required flow conditions and regimes are based on collaborative, multistakeholder negotiated agreements, which provide certainty for all parties and avoid prolonged permitting processes, delayed conservation gain, and loss of power generation opportunity (Hydropower Reform Coalition 2017, 2018). Further, the current political climate limits the ability of federal and state agencies to address climate change. Significant changes to federal and state law, especially those that relate to water storage and delivery, are, at best, a pipe

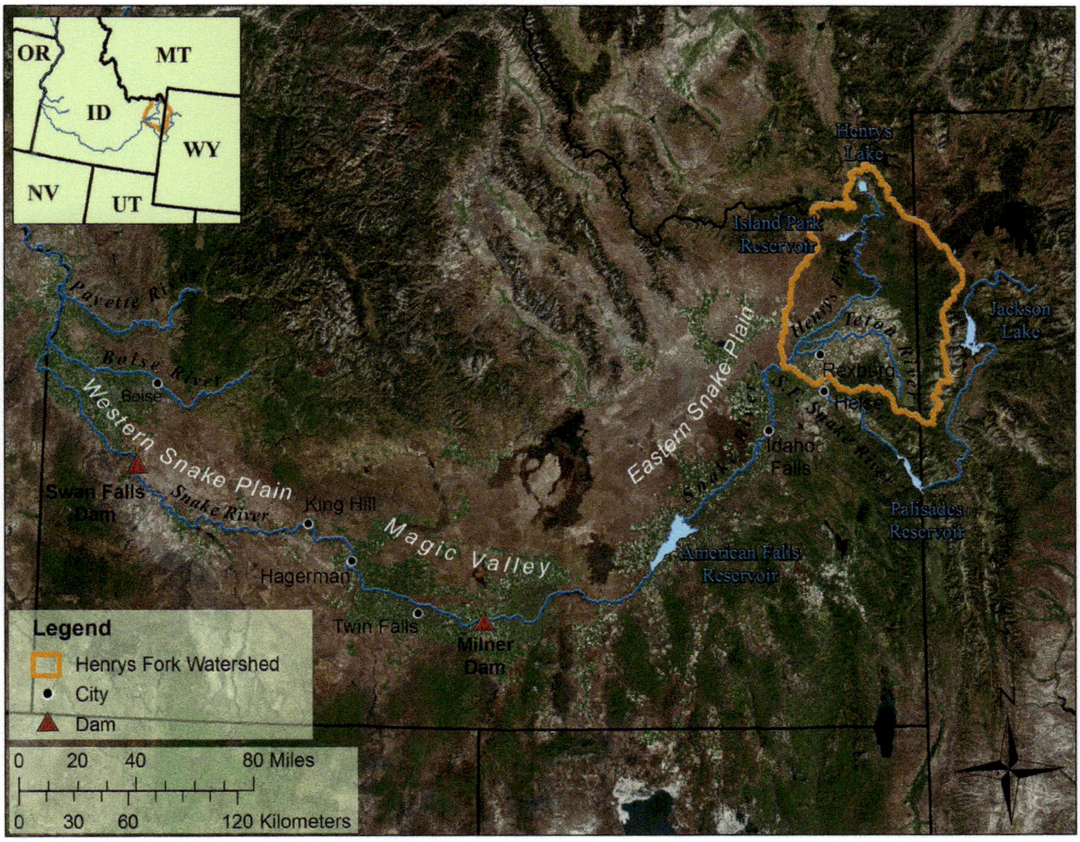

Figure 1. Map of upper Snake River basin.

dream and, at worst, likely to detract from the time and effort needed to spur collaboration and innovation necessary for functional change. Our approach moves away from overreliance on federal legal authority and toward development of collaborative solutions that include multisector benefits.

Ecohydrology of western irrigation systems

As reviewed above, the most obvious and widely understood hydrological effects of irrigation in the West are low streamflow downstream of diversions and large-scale alteration of flow regimes by dams. However, many popular trout fisheries and ecologically important native fish populations are found in stream reaches that are affected primarily by irrigation withdrawal and return flow. These can be main-stem river reaches far enough downstream of dams that their impact is relatively small, or they can be smaller tributaries that are unregulated by dams but provide irrigation water to adjacent private land. In these settings, there is little or no federal nexus, hydrologic alteration is subtle and not easily quantifiable, and strategies such as managed reservoir flow releases are not available. Irrigation in these areas often uses 19th-century infrastructure and practices, most notably conveyance in unlined canals and ditches and application via flooding (Figure 2). Although details vary with climate, crop types, and geology, the primary hydrologic characteristic of these traditional irrigation systems, particularly in alluvial river valleys, is substantial interaction of groundwater and

Figure 2. Traditional flood irrigation in the Henrys Fork watershed.

surface water (Fernald and Guldan 2006; Kendy and Bredehoeft 2006; Fernald et al. 2007; Jensen 2007; Ochoa et al. 2007). Canal and irrigation-application seepage in these systems can range from 20% to 75% of total water withdrawn (Thompson 1988; Battikhi and Abu-Hammad 1994; Zaladis et al. 1997; Lankford 2006), and recharge incidental to irrigation can be a major if not the dominant source of aquifer recharge (Whitehead 1994, 1996; Planert and Williams 1995; Robson and Banta 1995; Johnson et al. 1999; Peterson 2011). In these cases, irrigation return flows take subsurface pathways back to streams, creating a natural storage mechanism without the ecological effects of dams. Total consumptive use of water in these systems can be as little as 10–20% of the total water supply, especially in geologic settings where canal seepage is very high (Peterson 2011; USBR 2012a).

Stream hydrographs altered by diversion and subsequent return through shallow aquifers show reductions in peak flow and increases in late-summer and winter base flow relative to natural conditions, with the greatest base flow increases occurring where flood irrigation dominates (Figure 3). Irrigation seepage increases local groundwater supply and substantially increases the influence of groundwater emergence on surface hydrologic regimes (Strange et al. 1999; Venn et al. 2004). On the one hand, this subtle hydrologic alteration could negatively affect native fish species that evolved in flashy snowmelt-dominated natural flow regimes, especially in the presence of nonnative species (Fausch et al. 2001). On the other hand, irrigation-enhanced groundwater returns can provide ecologically important wetland habi-

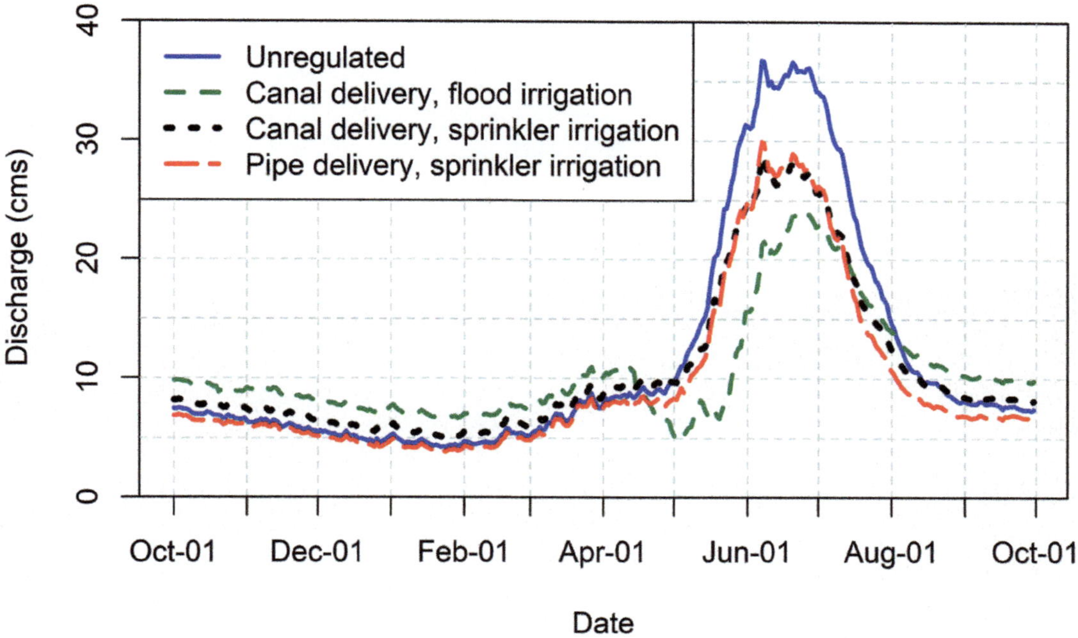

Figure 3. Modeled mean annual hydrographs of streamflow in the upper Teton River, Idaho, under different irrigation scenarios. Model applies different irrigation scenarios to hydrologic conditions observed over 1979–2008 using a one-dimensional, analytical model of groundwater flow similar to that described in Boggs et al. (2010). Unregulated scenario assumed that all diverted water remained in stream channels. Flood-irrigation scenario assumed that all available streamflow up to capacity of canal system was diverted throughout the summer and that diverted water in excess of evapotranspiration recharged the local aquifer. Canal-sprinkler scenario is actual condition over the period of record. Pipeline-sprinkler scenario assumed diversion of the amount of water needed to meet crop evapotranspiration or all available streamflow, whichever was smaller, and no conveyance losses. See Peterson (2011) for details.

tats (Peck and Lovvorn 2001) and popular spring-creek trout-fishing opportunities.

Taken together, irrigation-enhanced wetlands, groundwater-fed streams, riparian habitats along the canals themselves, and traditional low-intensity agricultural land uses have created hybrid human-natural ecosystems in many places that have high ecological and sociological value (Figure 4; Fiege 1999; Crifasi 2005; Fernald et al. 2006, 2007; Wiener et al. 2008). Increases in irrigation efficiency in these ecosystems have had unintended negative hydrological and ecological consequences (Johnson et al. 1999; Kendy and Bredehoeft 2006), prompting the Natural Resources Conservation Service, a primary funder and facilitator of irrigation efficiency projects since the 1970s, to rethink its own policies. In 2017, the agency held a webinar, targeted at its employees in the West, that discussed unintended but potential negative effects of irrigation efficiency improvements on fish and wildlife habitat, water supply, and local economies (NRCS 2017). In the context of rapidly growing cities and large-scale industrial agriculture elsewhere in the West, these unique landscapes are becoming more important to fish and wildlife conservation but are also becoming increasingly threatened by exurban development (Figure 5), which can negatively affect local and regional hydrology as well as so-

Figure 4. Human-natural irrigated landscape in Henrys Fork watershed. Note riparian vegetation along canals, functional riparian areas along the river, and high groundwater filling ponds and old river channels.

ciological institutions (Franklin and Narayanan 1988; Fereday and Creamer 2010; Liegel 2011; Baker et al. 2014). In these landscapes, water management and land conservation are highly linked (Riebsame et al. 1997; Baker et al. 2014), and "natural" ecological and hydrological conditions are difficult to define (Shallat 2000). Our water-management paradigm acknowledges that these hybrid landscapes can be ecologically important, depending on location, degree of functionality, and types of organisms supported.

Effects of climate change

Climate change was not a consideration when early methods for assessing instream flow needs were developed 40 years ago. Now, its effects are intrinsically linked to water management and planning in the West (Gleick 2010; Sabo et al. 2010; Miller and Piechota 2011). The changing climate of the West is characterized by more rain and less snow, declining snowpack retention, earlier snowmelt, increased year-to-year variability in temperature and precipitation, and increased intensity and frequency of droughts (Hamlet et al. 2005; Stewart et al. 2005; Mote 2006; Luce and Holden 2009; Fyfe et al. 2017; Mote et al. 2018). In unregulated stream systems, this combination of factors has moved timing of peak runoff earlier in the spring, decreased summertime flows, and increased the duration of the summer low-flow period (Regonda et al. 2005; Moore et al. 2007; Rood

Figure 5. Development on formerly irrigated land in Teton Valley, Idaho and Wyoming.

et al. 2007; Kunkel and Pierce 2010; Mayer and Naman 2011; Leppi et al. 2012). Effects of climate change on trout depend on life history, but the most consistently observed negative effects are related to low summertime streamflow, but not to temperature per se (Kovach et al. 2016). In addition to direct effects on fish populations themselves, decreased summertime streamflow will increase conflicts among agricultural use, aquatic conservation, and the recreational fishing industry in the West (Al-Chokhachy et al. 2017).

In regulated river systems, effects of climate change are difficult to distinguish from those of water use and management (Huntington 2006). For example, reductions in late-summer streamflow that might be attributable to climate change in unregulated rivers may actually be due in part to increased irrigation efficiency and increases in groundwater pumping in regulated systems (Van Kirk and Naman 2008). In many cases, trends in regional management of water may have much larger effects on natural hydrologic regimes than global change (Lattenmaier et al. 1999; Marcarelli et al. 2010; Naik and Jay 2011; Moore et al. 2012). Furthermore, climate change can interact with regulated hydrology in dynamic and unexpected ways. In the Snake River basin, early-season diversion has increased since the early 1970s in association with increased temperature and decreased precipitation, whereas late-summer diversion of surface water has decreased because of lower available supply (Hoekema and Sridhar 2011). Shortages in surface water supply during the summer increase reliance on groundwater and incentivize efficiency in use of surface water, ultimately reducing groundwater supply and creating a negative feedback loop in which surface water shortages and groundwater pumping continue to increase (Hoekema and Sridhar 2013).

Water rights appropriated in the 19th century, infrastructure built in the first half of the 20th century, and system management that evolved during the mid-20th century were based on quantity, timing, and predictability of water availability in the climatic regime of that time. As water availability shifts in response to climate change, existing physical and administrative systems may be insufficient to deliver reliable water supplies to existing users, let alone provide streamflow for fish. To what degree this possibility is realized depends on the type of water use and on buffering capacity of groundwater resources and existing reservoir systems (Lettenmaier et al. 1999). Our paradigm incorporates adaptation to the effects of climate change on both natural and regulated hydrology.

A new paradigm, new strategies

Our review suggests that providing instream flow has had mixed success in prior appropriation states and cannot rely on federal laws in many cases. Restoration of natural hydrologic regimes may not provide the greatest ecological response per unit of available water and can even be counterproductive in human-natural systems that have evolved since the late 19th century around groundwater recharge incidental to traditional irrigation. Although trends in local and regional water use and management can outweigh the effects of climate change on hydrologic regimes, natural hydrologic regimes are shifting as a result of climate change, and climate and water management can interact in unexpected ways.

We propose a paradigm for providing instream flow that is based on five fundamental concepts:

1. Designer flows that address ecologically limiting flow components can produce better outcomes for specific species and resources than restoration of the natural flow regime.
2. Increases in irrigation efficiency must be accompanied by appropriate economic incentives and administrative mechanisms to ensure that water is actually saved at the farm/ranch scale, that the saved water increases streamflow at times and places beneficial to fish, and that loss of incidental groundwater recharge does not have negative ecological consequences somewhere else in the river basin.
3. Hybrid human-natural hydroecosystems have evolved around traditional irrigation practices in many locations, and some of these have ecological value not despite alteration of natural flow regimes, but rather because of alteration.
4. Changes in water management can have greater effects than climate change in regulated systems, but effects of climate change on regulated systems can be greater and more complex than effects on natural hydrology.
5. Collaborative solutions with benefits to multiple stakeholders have greater potential to provide instream flows than litigious or regulatory approaches.

Within this paradigm, our case studies illustrate effectiveness of four specific strategies:

- Capitalizing on administrative flexibility within prior appropriation systems,
- Using science and technology to improve management precision,
- Actively managing groundwater and surface water together, and
- Using incentive-based mechanisms to reduce irrigation demand.

Study Area

We provide case studies from two watersheds in the upper Snake River basin: Henrys Fork of the Snake River and the Teton

River. These two rivers are managed and administered as part of a large system of conjunctively administered surface and groundwater rights that stretches from the Snake River headwaters in Yellowstone National Park downstream to Swan Falls Dam south of Boise (Figure 1). Thus, we provide an introduction to water management in the upper Snake River basin before focusing on the case-study watersheds.

Upper Snake River basin

The upper Snake River basin is the 93,000 km^2 (36,000 mi^2) drainage area of the Snake River upstream of King Hill, Idaho (Figure 1). Annual water supply in the upper Snake River basin averages 15,000 Mm^3 (11 M acre-feet), 46% of which originates upstream of Heise, Idaho in the Snake River headwaters and 23% of which is supplied by the Henrys Fork. The Eastern Snake Plain Aquifer (ESPA), which underlies the eastern Snake River plain from King Hill upstream to the southwestern extent of the Henrys Fork watershed, is a key hydrogeologic feature of the basin. Surface water—including streams and the surface water irrigation system—is highly connected with groundwater on the ESPA, which is the sole source of drinking water for all cities and residences on the plain and is a large source of irrigation water for agriculture (Johnson et al. 1999; IWRB 2009; TNC 2014). Water generally flows through the aquifer from northeast to southwest and discharges to the Snake River in the vicinity of American Falls Reservoir and in the Thousand Springs area near Hagerman, Idaho. Water levels in and discharge from the ESPA have been declining for 60 years due to a combination of decreased recharge incidental to surface water irrigation and increased groundwater pumping (Johnson et al. 1999; IWRB 2009; Boggs et al. 2010).

Agriculture is by far the single largest economic sector in the upper Snake River basin, producing an estimated US$10 × 10^9 in goods and services (IWRB 2009). Of 12,000 km^2 (2.9 M acres) of crop lands in the basin, more than 10,000 km^2 (2.1 M acres) are irrigated. By area, the most widely grown crops are hay (32%), wheat (23%), barley (16%), potatoes (10%), and corn (9%). Most of the hay and corn is feed for dairy cows in the Magic Valley, which is now a major international producer of dairy products. Global companies that source raw agricultural products from the upper Snake River basin or process food in the basin include General Mills, Anheuser-Busch, Miller-Coors, Chobani, and Clif Bar. On an annual average, irrigation withdrawals total 9,700 Mm^3 (7.3 M acre-feet) of surface water from the upper Snake River and its major tributaries and 1,600 Mm^3 (1.2 M acre-feet) of groundwater from the ESPA. An additional 2,000 Mm^3 (1.6 M acre-feet) of surface water and groundwater are withdrawn in smaller tributary basins (IWRB 2009; TNC 2014). The Thousand Springs area near Hagerman supports a major aquaculture industry. Together, agriculture and aquaculture account for 98% of water withdrawal in the basin (TNC 2014).

Most of the surface water in the basin is administered by Idaho Water District 01 (WD01), exceptions being smaller tributaries. The WD01 system consists of seven USBR storage reservoirs, one private reservoir (Henrys Lake), more than 350 points of diversion, and thousands of kilometers of canal. Irrigators upstream of Idaho Falls hold most of the senior natural-flow rights, with priorities in the 1870s through 1890s, including those on the lower Henrys Fork and Teton River (Figure 1). Irrigators in the Magic Valley hold the largest junior natural-flow rights, with priorities around 1900, and divert from the main-stem Snake River. The storage reservoirs were added to the irrigation system after private entities had already claimed most of the available natural flow (Fiege 1999; Slaughter and Wiener 2007). The Magic Valley irrigators hold the most senior storage rights, in Jackson Lake (1910) and American Falls Reservoir (1921), and the upper-valley irrigators hold the junior storage rights, in Henrys Lake (1923), Island Park Reservoir (1935), and Palisades Reservoir (1939). Groundwater rights are the most junior in the basin, with priorities in the 1950s through early 1990s.

The modern era of reservoir system management in the upper Snake River basin began in 1956, with the completion of Palisades Reservoir. A large dam on the Teton River was completed in 1976 but failed catastrophically as the reservoir was being filled for the first time and has not been rebuilt (Palmer 1991; Reisner 1993). Palisades Reservoir contains nearly 30% of the total reservoir system capacity of 5,100 Mm^3 (4.1 M acre-feet) and provides a large amount of flexibility in managing the reservoir system because it sits between Jackson Lake and American Falls Reservoir, was authorized for hydroelectric generation and flood control in addition to irrigation, and contains some unallocated space that stores water for a basinwide rental pool (WD01 2018). Irrigation-rights holders anywhere in the basin can rent water from what is termed the "common" pool for a nominal administrative fee, in years when the pool has filled. The rental pool can also serve as a quasi-market for storage water, into which irrigation entities can lease storage water they will not need in a given season and from which water-short users can rent water. Private irrigation entities can enter into lease-rental agreements for any negotiated price, as long as the administrative fee is paid and the water-right transaction is recorded by WD01. Because price is negotiable in these private transactions, the rental pool has some aspects of a market. However, it is not a true market because only entities that already hold water rights can participate. In particular, conservation organizations and agencies that do not hold water rights cannot rent water for any purpose.

Administrative irrigation season runs from April 1 through October 31 in most of the basin and storage season for the remainder of the year. Although most reservoir storage occurs during the winter, maximum reservoir storage is generally reached after capture of snowmelt in the spring and early summer. Storage releases meet a large fraction of total diversion after mid-July (Figure 6). Milner Dam is the accounting point for surface water in WD01, which means that as long as all stored, delivered, and diverted water upstream of Milner can be accounted to an appropriate water right, actual water in the river at any given location other than Milner Dam does not need to match the mix of storage and natural flow that should theoretically be present at that location (Olenichak 2015). This divorce of "physical" and "paper" water allows flexibility to optimize use of storage water for irrigation, storing as much as possible during the winter and delivering as little as possible during the summer. The only formal environmental flow provision that affects winter storage operations is delivery of at least 525 ft^3/s (14.9 m^3/s) from Minidoka Dam for protection of the ESA-listed Snake River physa snail *Haitia natricina* (USBR 2014). This minimum flow applies only to the 18.5 km (11.5 mi) of river immediately upstream of Milner Reservoir (Figure 1).

The management objective for the entire system is zero flow at Milner (IWRB 2012; Tuthill et al. 2013). Water flows past Milner Dam only when supply (natural flow) exceeds reservoir capacity and irrigation demand or when storage water is deliberately released for downstream uses (Figure 6), including flow to benefit downstream migration of anadromous salmonids in the lower Snake River (USBR 2017). Generally, flow is zero at Milner only during irrigation season, since all flow delivered to Milner Reservoir during the summer is diverted for irrigation, and the reservoir is too small to store reach gains and the Minidoka minimum flow over the winter. Flow at Milner was zero for extended periods during the irrigation

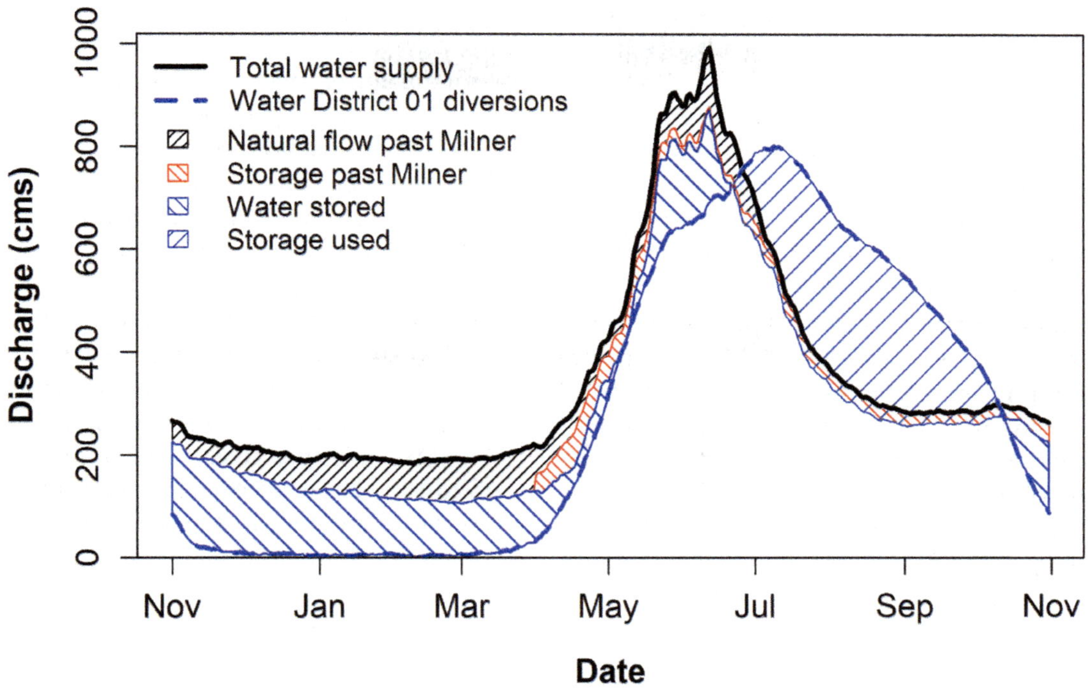

Figure 6. Mean annual hydrograph of water supply, storage and delivery in Idaho Water District 01. Period of record is 1988–2014.

season in 15 of the 35 years between 1980 and 2014. From a water-rights perspective, rights downstream of Milner are usually met by tributary inflow and groundwater discharge from the ESPA. From an ecological perspective, the zero-flow objective results in degraded water quality and limited fishery value for many miles downstream of Milner. Native anadromous fish were extirpated from the Snake River from Milner Dam downstream to Hells Canyon (the mid-Snake) decades ago by construction of hydropower dams. Current fisheries in that reach are primarily supported by nonnative warm- and coolwater species. Because of the low fishery value of the mid-Snake and improbability of meaningful restoration of native coldwater species upstream of Hells Canyon, fisheries conservation interests in Idaho generally accept the reality of the zero-flow objective and consider it as one of the key administrative peculiarities of the upper Snake River basin that allows high-quality coldwater fisheries to persist throughout most of the river upstream of Milner.

Three other management peculiarities relevant to fisheries conservation result from the difference between physical and paper water. The first is storage of physical water late in the irrigation season when storage rights are not in priority. This is usually done in the upstream-most reservoirs in the system. As long as physical water is not needed to meet diversion needs downstream, storing water does not interfere with delivery and can be done because water stored in an upstream reservoir can always be delivered downstream at a later time if needed. The stored water is part of total system carryover, which is accounted to storage rights in priority at the end of irrigation season. A second example is diversion of storage water at a point to which the storage cannot physically be delivered. For example, a user in the Henrys Fork can divert available physical water from the river that is accounted on paper to the Palisades rental pool. The physical water is delivered from Palisades Reservoir to the Henrys Fork confluence, at which point storage delivery matches diversion both physically and on paper. The third is a consequence of the high connectivity between surface and groundwater. The system is operated so that several river reaches, including the lower Teton River and lower Henrys Fork, convey less water than is required on paper to meet downstream rights and are sometimes even intentionally dried up. Groundwater returns to the river, called "reach gains," add sufficient water to the river between these dry reaches and downstream users.

Ten major administrative and legal actions have changed water management and use in the upper Snake River basin over the past 35 years.

- **1984: Swan Falls Settlement Agreement.** To resolve several legal disputes, the state of Idaho reached a settlement with Idaho Power that guaranteed a minimum flow in the Snake River at the Swan Falls power plant south of Boise (Jones 2016). At the time of the settlement, discharge from the ESPA at Thousand Springs and other reach gains downstream of Milner were sufficient to meet the minimum flow. Excess water was allocated to new irrigation. In 2015, due to decreased ESPA discharge, the state, for the first time since the settlement, was required to deliver rented storage water downstream of Milner Dam to meet the minimum flow.
- **1987–2014: Snake River Basin Adjudication.** To meet one of the outcomes of the Swan Falls negotiations, the state adjudicated every water right in the Snake River basin, more than 158,000 in all (Vonde et al. 2016). Rights were modified, confirmed, combined, and eliminated through the adjudication process.
- **1992: Lower Snake River salmon listed under the ESA.** To aid downstream migration of smolts, the USBR delivers up to 647 Mm^3 (487,000 acre-feet) of water to the lower river each spring (USBR 2017). Around 270 Mm^3 (200,000 acre-feet) of this is delivered from the upper Snake River storage system, rented by the USBR from the Palisades rental pool.
- **1993: ESPA pumping moratorium.** In response to declining aquifer levels, the state issued a moratorium on new rights to pump groundwater from the ESPA, with the exception of wells for individual-dwelling domestic use.
- **1994: Conjunctive Management Rules.** The conjunctive management rules mandated and provided guidance for administration of surface water and groundwater in the upper Snake River basin as a single supply of water under a common priority system (Tuthill et al. 2013). Adoption of the rules moved groundwater rights from minimal regulation within their own priority system to full administrative regulation within the basinwide system of surface water and groundwater rights.
- **1990s–present: Water calls against groundwater users.** The conjunctive management rules opened the door for senior surface-water users to issue water delivery calls against junior groundwater users on the ESPA.
- **2009: Eastern Snake Plain Aquifer Comprehensive Management Plan.** The long-term goal to stabilize the ESPA is a 800 Mm^3/year (600,000 acre-feet/year) change in the water budget through a combination of managed aquifer recharge and decreased groundwater pumping (IWRB 2009). The state operates a public managed recharge program with an annual objective of 330 Mm^3 (250,000 acre-feet)
- **2012: State Water Plan update.** This revision of the plan emphasized management and recovery of aquifers statewide (IWRB 2012).

- **2015: Settlement of water call.** A decade-long water call by a group of Magic Valley surface water users against groundwater users was settled through negotiation. The primary provision required groundwater users to cut annual pumping by 320 Mm^3 (240,000 acre-feet) or mitigate continued pumping with delivery of rented storage water or with an equal amount of managed aquifer recharge beyond that conducted by the state.
- **2016: ESPA Groundwater Management Area** was designated by the state to facilitate coordinated recovery of the aquifer.

Listing of Snake River steelhead (anadromous Rainbow Trout *Oncorhynchus mykiss*) and salmon is the only of these actions that involved other states, in that case Oregon and Washington. An interstate compact between Idaho and Wyoming, signed in 1949, appropriated all but 4% of Snake River water originating in Wyoming to the state of Idaho, effectively giving Idaho full authority for allocating and managing water in the upper Snake River basin.

Henrys Fork and Teton River watersheds

The 8,300-km^2 (3,200 mi^2) Henrys Fork watershed lies at the transition between the eastern Snake River plain and high-elevation mountain ranges and plateaus of the Continental Divide and Teton Range (Figure 7). Although the Teton River is a tributary to the Henrys Fork, the fisheries, ecological characteristics, and conservation needs of the two are very different. The two watersheds are hydrologically linked by the Crosscut Canal, which delivers Island Park Reservoir storage water to the Teton River. The Henrys Fork supports economically important nonnative Rainbow Trout and Brown Trout *Salmo trutta* fisheries (Loomis 2006; Grunder et al. 2008) in a river system that is altered by two large storage reservoirs and by major hydroelectric power plants at Island Park, Ashton, and Chester dams (Figure 7). Large irrigation canals divert about 1,500 Mm^3 (1.1 M acre-feet) annually out of annual water supply of 3,300 Mm^3 (2.5 M acre-feet; USBR 2012a). Senior natural-flow rights, more than 270 Mm^3 (200,000 acre-feet) of storage, and a long growing season allow consistent production of high-value crops. Relatively little of the agricultural land along the Henrys Fork, Fall River, and lower Teton River has been lost to development (Baker et al. 2014). By contrast, the upper Teton River watershed is a stronghold of native Yellowstone Cutthroat Trout *Oncorhynchus clarkii bouvieri* and is free of storage reservoirs and large hydroelectric plants. Teton Valley is about 300 m (1,000 ft) higher in elevation than the growing regions of the Henrys Fork, limiting production of high-value crops. Most water rights in the upper Teton River are junior to those downstream, and annual diversion, mostly into small canals and pipe systems, totals 120 Mm^3 (90,000 acre-feet) out of an annual supply of 400 Mm^3 (300,000 acre-feet). Development has fragmented and replaced a substantial amount of formerly irrigated agricultural land in Teton Valley (Figure 5; Baker et al. 2014). Consumptive use of irrigation water in the entire Henrys Fork watershed is less than 25% of total diversion and about 12% of total water supply (USBR 2012a, 2015).

Case Studies: Henrys Fork and Upper Teton River

The case studies are presented from the perspective of two local watershed conservation organizations, the Henry's Fork Foundation (HFF) and Friends of the Teton River (FTR).

Henrys Fork

The Henrys Fork is really four rivers according to most anglers who ply its waters, plus a fifth reach that provides sparse angling opportunity. Scientists and fisheries managers also think about the Henrys Fork

Figure 7. Map of Henrys Fork watershed, including Teton subwatershed.

in terms of these different river segments, which are defined by distinctive differences in geology, hydrology, and ecology (Figure 7). Upstream of Island Park Reservoir, Big Springs feeds the Henrys Fork, and its nearly constant flow is cold and clear most of the year. Aquatic productivity is limited by low phosphorus concentrations (R. Van Kirk, unpublished data), and use of the river by large numbers of nonangling recreational floaters often conflicts with the tranquil flatwater fishing experience sought by anglers. Fishing clubs established in the early 20th century line the riverbanks, and the historic buildings of these encampments hold old stories of large fish, prereservoir conditions, and rides on the train from Salt Lake City to Island Park (Van Kirk and Gamblin 2000). From Island Park Dam to Mesa Falls, Harriman State Park is a dry-fly-fishing nirvana bounded on either end by very productive canyon stretches, as measured by trout size and abundance (Angradi and Contor 1989). Rainbow Trout grow rapidly in this aquatic system because of nutrient enrichment from Island Park Reservoir (Van Kirk, unpublished). Brown Trout are not found upstream of Mesa Falls.

Below Mesa Falls, the Henrys Fork continues down another canyon, captures additional flow from the spring-fed Warm River, and then flows out onto the eastern Snake plain. Although management of Island Park Reservoir upstream reduces flow in this reach during storage season and increases it during periods of storage delivery, unregulated tributary inflow minimizes the effect of reservoir management in this reach, leading to relatively consistent flows across years and stable populations of wild Rainbow and Brown trouts. This attracts heavy fishing pressure throughout the year between Warm River and Chester, Idaho. Irrigation diversions, groundwater return flows, and braided channels characterize much of the lower river from St. Anthony to the North Fork Teton River. Brown Trout dominate the fishery there, and low summertime flows limit fish populations. Complex channels and beautiful cottonwood riparian forest are constant reminders that providing consistent water to this section would provide great returns when it comes to trout and other aquatic and riparian resources. The fifth Henrys Fork stretches from the North Fork Teton River downstream to the South Fork Snake River confluence, a reach severely degraded by deposition of sediment from the Teton Dam collapse. Angling pressure is light in this section, and while the opportunity for restoration is high, the cost would be astronomical to bring this section back to its full potential.

For advocates of high-quality wild-trout fisheries on the Henrys Fork, the complexity and diversity of the river provide constant management challenges but also opportunities. A balancing act is required to keep the reach immediately below Island Park Reservoir and the segment near St. Anthony simultaneously productive as wild-trout fisheries. Issues occur elsewhere on the river, but it is not often that management decisions can have opposite effects for one river reach versus another as it does for these two. Complexity weaves through all aspects of management of the Henrys Fork, as differential nutrient availability, effect of Island Park Reservoir on water quality, climate change, short-term flow disruptions at hydropower facilities, canal entrainment, a growing human population, and new basin-wide water management scenarios interact within and across river reaches. While the concepts have been modified over the years and the model refined under changing external conditions, conservation groups and agencies have found success on the Henrys Fork by sticking to two core principles: use science and technology as the basis for conservation and water management strategies, and effectively collaborate with other watershed stakeholders, a mode of operation we refer to as “science-based collaboration.”

The fishery.—Although the Henrys Fork has been popular among anglers since the early 1900s, its current status among fly fishermen as a world-class wild-trout fishery dates back to the early 1970s (Van Kirk and Griffin 1997; Van Kirk and Gamblin 2000). The reach of the river most revered among fly fishermen is the 24 km (15 mi) immediately downstream of Island Park Reservoir, including the Harriman State Park reach of the river, which is known for prolific aquatic insect hatches (Lawson 2012; McDaniel 2012). The Idaho Department of Fish and Game conducts an annual population estimate in the Box Canyon reach, immediately downstream of Island Park Dam, which serves as an index of the trout population from the dam to Mesa Falls. From the early 1900s to 1977, nonnative Rainbow Trout were stocked throughout the Henrys Fork, and Box Canyon Rainbow Trout density at the end of this period was 3,643 fish/km (Figure 8). In 1978, stocking ceased in the Island Park-to-Mesa Falls reach, and it has been managed since then as a wild-trout fishery (Van Kirk and Gamblin 2000). From 1978 to 1994, the population continued to receive supplementation from wild and hatchery Rainbow Trout migrating downstream from Island Park, and Rainbow Trout density averaged 2,528 fish/km over that time period. In 1994, a hydroelectric facility with

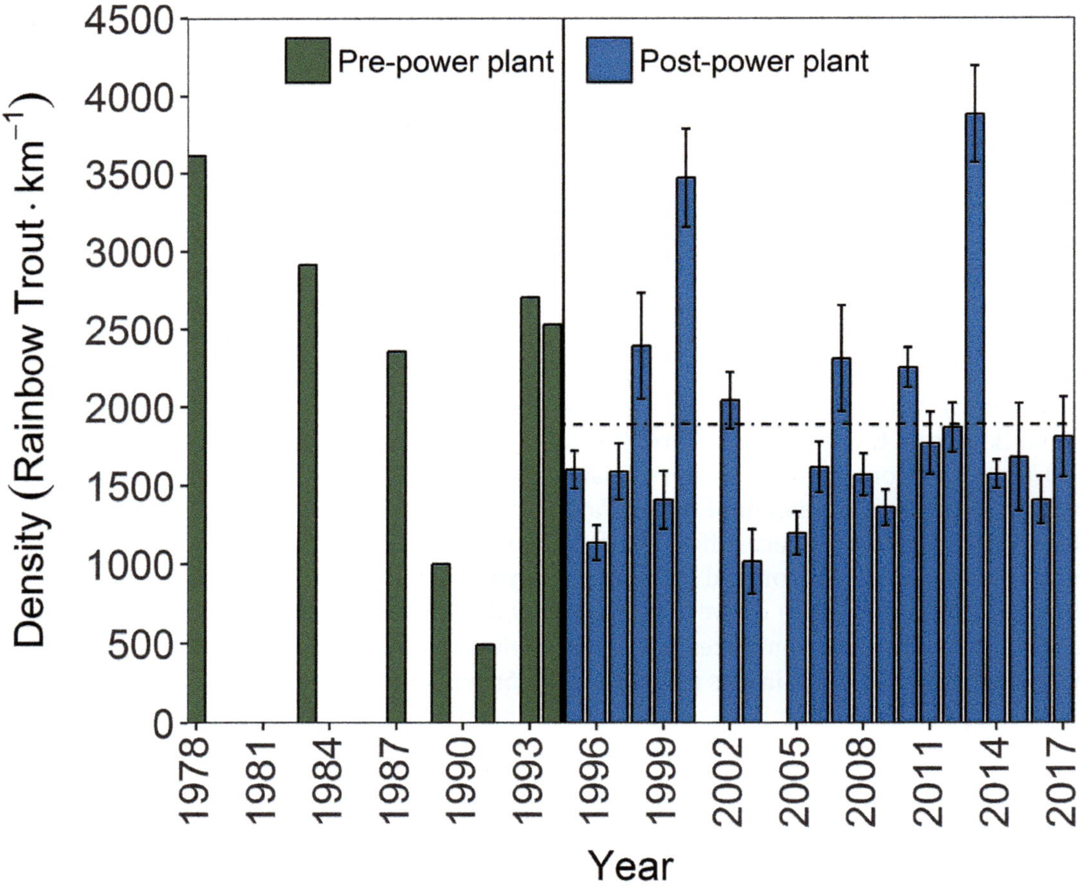

Figure 8. Density of Rainbow Trout in the Box Canyon reach of the Henrys Fork before and after construction of the Island Park hydroelectric facility. The dashed line is mean density of Rainbow Trout since construction of the power plant.

fish screens was installed on the dam, greatly reducing opportunities for fish passage to the river downstream. Since then, Rainbow Trout density has averaged 1,857 fish/km (Figure 8). In the 1990s, HFF-sponsored research showed that habitat availability for juvenile Rainbow Trout during winter was the factor most likely limiting juvenile survival and recruitment (Smith and Griffith 1994; Griffith and Smith 1995; Meyer 1995; Meyer and Griffith 1997). Subsequent research showed that availability of winter habitat was positively dependent on flow downstream of Island Park Reservoir during the coldest part of the winter, December through February (Gregory 2000; Mitro et al. 2003; Garren et al. 2006).

Island Park Reservoir was authorized and built in the 1930s for the sole purpose of storing and delivering irrigation water for Fremont-Madison Irrigation District (FMID). Over the 80 water years since the reservoir first stored water, an average of 100 Mm^3 (80,000 acre-feet) has been stored and delivered each year out of a capacity of 167 Mm^3 (135,000 acre-feet), leaving 41% of the reservoir's capacity remaining as carryover at the end of irrigation season. Reservoir carryover can vary highly from year to year (e.g., 15% of capacity in 2016 followed by 81% of capacity in 2017). Winter outflow is primarily dependent on reservoir volume at the end of irrigation season (Benjamin and Van Kirk 1999), so trout abundance is lower following years with high storage delivery. For decades, HFF, TU and others have advocated for winter flows as high as possible under the constraints of physically filling Island Park Reservoir. Over the past 40 years, winter flows—and hence trout populations—have ebbed and flowed with cycles of water availability. Prior to 2015, anglers generally accepted these supply-driven cycles, became accustomed to reduced trout populations during years of low water supply and increased populations during wet years, and were satisfied with long-term conditions. Indeed, more than half of anglers interviewed in a 2014 survey conducted on the Harriman State Park reach rated their fishing experience above average as compared with their previous experience on that reach of river. Although water supply in 2014 was well below average (Figure 9), the population of 3-year-old trout that year was high, following very strong recruitment of 2-year-olds in 2013 as a result of high winter flows in 2011–2012 (Figures 8 and 9), potentially explaining the high angler satisfaction in 2014. However, the same survey showed that anglers valued noncatch aspects of the fishery, such as habitat quality, number of rising fish, and quality of insect hatches, much more than they valued the number of fish they actually caught (Laatsch et al. 2017), providing a possible explanation why cycles in trout abundance did not lead to widespread dissatisfaction with the fishery over the long term. As long as the noncatch aspects of the fishery were maintained, anglers were satisfied.

Collaboration: Henry's Fork Watershed Council.—In September of 1992, complete drawdown of Island Park Reservoir mobilized 45,000–90,000 metric tons (50,000–100,000 tons) of sediment that had accumulated on the reservoir bottom and delivered it downstream into the Henrys Fork (Van Kirk and Griffin 1997; Van Kirk and Gamblin 2000), abruptly halting trout fishing season, degrading aquatic habitat, and heightening concerns about water management. Although the reservoir was drawn down to less than 10% of capacity through delivery of irrigation water, the Idaho Department of Fish and Game elected to take advantage of the low reservoir to undertake a nongame fish eradication treatment and requested full drawdown to facilitate that treatment. Angler and community frustration arising from this event led to a search for alternative management approaches, culminating in organization of the Henry's Fork Watershed Council (hereafter, Council) in 1993

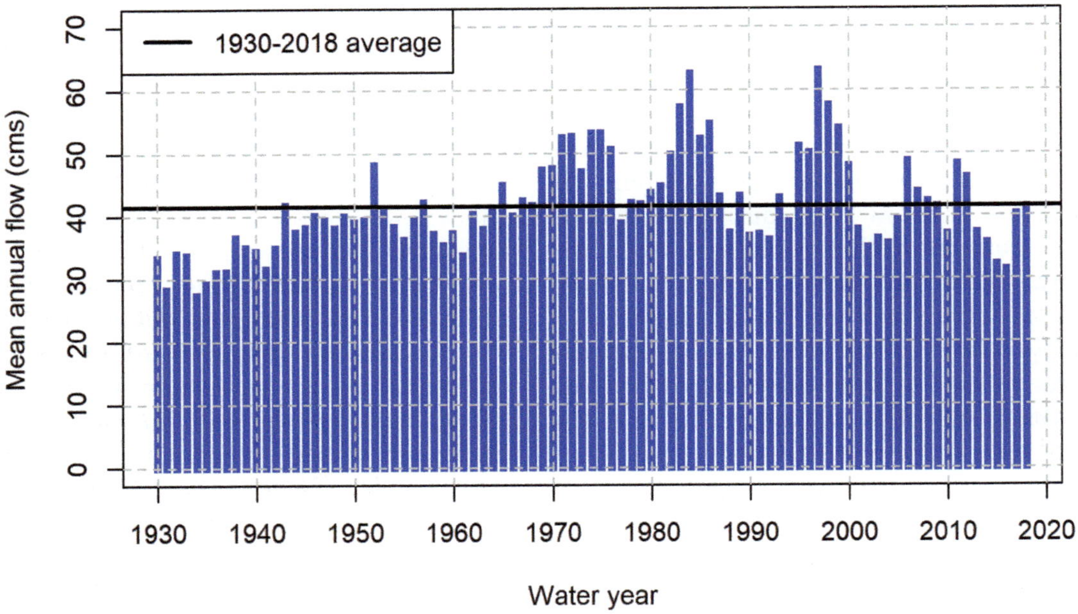

Figure 9. Mean annual natural streamflow in the upper Henrys Fork watershed (Henrys Lake to Ashton).

(Van Kirk and Griffin 1997; Weber 2000). Chartered by the Idaho Legislature, the Council provides a forum and a process for building consensus among diverse stakeholders, including local, state, and federal agencies, farmers and ranchers, environmentalists, and recreationists. The Council is jointly facilitated by HFF, representing 2,500 members, and FMID, representing 1,700 area farmers. The Council has been meeting regularly 6–8 times each year since its formation, and meetings are attended by an average of 40 individuals.

Water quality and water supply have long been primary issues for the Council, which recently served as the stakeholder work group for the Henry's Fork Basin Study, a multiyear planning process conducted by the USBR in conjunction with the Idaho Water Resource Board. The basin study produced a suite of potential water-management alternatives, including expansion of surface-water storage in the watershed (USBR 2015). The Council also reviewed and endorsed a long-term water quality monitoring network operated by HFF, provided input to development of the Eastern Snake Plain Comprehensive Aquifer Management Plan, and continues to discuss and endorse projects related to groundwater management and aquifer recharge. Most importantly, the Council has been able to facilitate effective water management, addressing the needs of local farmers and ranchers while protecting fisheries and other aquatic resources, even during recent drought. This has been achieved primarily through increased communication among agencies, science-based collaboration involving NGOs and agencies, and trust and relationships built slowly over many years. Building trust, initially through community building exercises and lengthy lunch and refreshment breaks, was the Council's most important goal during its first few years, a process goal it shared with other collaborative watershed groups throughout the West (Griffin 1999; Kenney 1999; Weber 2000). Later, Council discussions led to more formal outcomes.

As part of a program to transfer small federal facilities to local irrigation entities, the federal government transferred title to Chester Dam, the Crosscut Canal, and several wells and well permits from the USBR to FMID (Fremont-Madison Conveyance Act of 2003). Earlier versions of the title transfer included Island Park Dam, but Council discussions revealed lack of broad stakeholder support for transfer of the dam. Also as a result of Council discussions, the act opened the door for creative and collaborative water management. Buried in the language of the bill is the following sentence:

> Within 60 days of enactment of this Act, in collaboration with stakeholders in the Henry's Fork watershed, the Secretary shall initiate a drought management planning process to address all water uses, including irrigation and the wild-trout fishery, in the Henry's Fork watershed.

The resulting drought management plan, signed in 2005, does not specify any particular flow targets but guarantees a permanent seat at the flow management table for HFF and other NGOs (Joint Committee 2005). In addition to HFF and TU, the other signatories to the plan were The Nature Conservancy (TNC), FMID, the USBR, and the North Fork Reservoir Company, owner of Henrys Lake. However, meetings are open to the public, and most meetings are also attended by representatives of the Idaho Department of Fish and Game, Water District 01, and the Fall River Rural Electric Cooperative, operator of the hydroelectric plant at Island Park Dam.

Fishery enhancement.—Since the 1990s, HFF has devoted most of its fisheries restoration and enhancement efforts to increasing the amount of winter habitat available to juvenile trout immediately downstream of Island Park Dam. Of many concepts investigated and tested (Gregory 2000), two had the most promise. One was to increase winter outflow from Island Park Dam through science-based collaboration and administrative flexibility in the WD01 water-rights accounting system. The second was to restore the ability of juvenile trout to access high-quality winter habitat in the Buffalo River upstream of a small hydroelectric power dam near its confluence with the Henrys Fork (Figure 7). Neither of these two concepts would have even gotten off the ground without a flow objective based on science, federal actions, and collaboration. With these ingredients, they became HFF's signature projects from 2005 through 2016.

In the winter of 2005–2006, the Drought Management Planning Committee first implemented a new strategy to maximize outflow from Island Park Reservoir over the December–February period. The strategy is to begin storing water as early as mid-September, once local irrigation demand drops enough that it can be met by reach gains downstream of the reservoir. This physical storage in Island Park Reservoir occurs when the Island Park storage right is not in priority. Rather, the stored water usually accrues to the American Falls 1921 right once carryover is calculated at the end of irrigation season, providing a good example of how streamflow can be provided by exploiting flexibility in the prior-appropriation system. Research has shown that low outflow from Island Park Dam has no negative effect on trout during the fall, when aquatic vegetation provides habitat and increases water depth per unit of streamflow (Griffith and Smith 1995; Gregory et al. 2011; Kuzniar et al. 2017). Autumn outflow from Island Park Dam was as low as 0.06 m^3/s (2 ft^3/s) during some of these studies, leaving the ~5.7 m^3/s (~200 ft^3/s) inflow from the Buffalo River as the primary source of streamflow in the river channel downstream. Thus, outflow is reduced in the early fall to store as much water as possible by the end of November. This allows higher December–February outflow than would be possible without the early storage. We emphasize again that the drought management plan does not mandate any minimum flow level or course of action; rather, it provides a legally required process in which to discuss options based on basinwide conditions and on data that HFF and others bring to the table. The only firm requirement is that the parties meet at least four times each year, coinciding with annual hydrologic and agricultural cycles. Any results beyond the meetings themselves are due to relationship-building, trust, applied science and technology brought to the table primarily by the conservation NGOs, and administrative flexibility. Much of the trust has been built within the construct of the Council.

In the early 2000s, the small Buffalo River hydroelectric plant, originally built to generate power for construction of Island Park Dam, was due for relicensing by FERC. A fish ladder already existed on the power dam, but it was designed to pass only large spawning-sized fish. State and federal agencies, HFF, and Fall River Electric negotiated license conditions that required construction and monitoring of a new fish ladder that could allow juvenile trout to migrate from the Henrys Fork upstream to access winter habitat in the unregulated, spring-fed Buffalo River. The new ladder was completed in the fall of 2005, coincident with the first winter of the formal drought management planning process. Operation and monitoring of the ladder has been shared by HFF and Fall River Electric, with HFF conducting most of the monitoring and related research.

Given the large amount of resources HFF invested in both drought management planning and the Buffalo River fish ladder between 2005 and 2016, HFF conducted a rigorous quantitative analysis of the effectiveness of these two programs (Morrisett 2016; Oldemeyer et al. 2017). This analysis showed that these two projects were successful at increasing the Rainbow Trout population over what it would have been without the higher winter flows and Buffalo River fish passage. Unfortunately, the combined total effect of the two management actions was only a 9% increase in the population—3% due to the fish ladder and 6% due to flow management—in a population with an interannual coefficient of variation of 38% (Oldemeyer et al. 2017). Hundreds of thousands of dollars in resources resulted in a response that could not be distinguished statistically from background variability.

Climate change and decline in angling experience.—Beginning in 2015, anglers became much less satisfied with their Harriman State Park fishing experience and much more vocal in their dissatisfaction with the cyclical nature of the fishery below Island Park Reservoir. Although reduced trout abundance has certainly contributed to this increase in dissatisfaction, reduced water clarity and earlier and higher irrigation season delivery from Island Park Reservoir appear to have contributed as much or more (Laatsch et al. 2017). High summertime flow and increased turbidity, the latter caused by export of sediment and organic matter from the reservoir during these high-flow periods, reduce opportunities for anglers to fish to rising trout during June and July—the fishing experience that made the Henrys Fork famous. Whereas in the past, irrigation storage and delivery were of concern to HFF and its constituency primarily through their effects on winter flow, the effects of storage and delivery are now equally relevant to water quality and the early-season fishing experience, which, in turn, is important to small businesses in Island Park that derive a substantial portion of their annual income from fishing during June and July. Anglers who previously had faith that fishing conditions would improve during the next wet cycle have now become convinced that the fishing experience in Harriman State Park and adjacent reaches will continue to deteriorate.

The sentiment among anglers is generally corroborated by HFF's scientific analysis. Through its hydrologic modeling and water quality monitoring programs, the HFF science team has quantified increases in turbidity, earlier irrigation delivery, and lower reservoir carryover and linked them to a variety of physical, chemical, and biological factors, including earlier snowmelt; longer, hotter growing seasons; warmer water temperatures; cyanobacteria blooms in the reservoir; and cumulative effects of sediment and nutrient deposition in the reservoir over its 80-year life (Van Kirk, unpublished). When natural flow watershed-wide, including that from the Fall and Teton rivers, is sufficient to meet irrigation demand, storage delivery is minimal, water quality remains good, and winter flow downstream of Island Park Dam can provide good survival of juvenile trout. However, in this recruitment-limited fishery, where maximum trout lifespan is 6 years (Angradi and Contor 1989), a consistently high trout population can be maintained only when winter flow provides good juvenile survival in at least 3 years out of 5. Over the 50-year history of the modern Henry's Fork fishery, sequences of two or more consecutive years of high water supply have become shorter, and intervening drought has become more severe (Figure 9). Furthermore, mean April–June temperature at the eight snow survey sites in the watershed with long periods of record has increased at a watershed-averaged rate of 0.67°C/decade (1.2°F/decade) since 1989, a statistically significant trend (Van Kirk, unpublished). Mean April–June temperature is a strong negative predictor of timing of snowmelt runoff, and earlier runoff misaligns water supply with irrigation demand, requiring greater and earlier delivery of storage water from Island Park Reservoir, contributing to water-quality degradation and decreased angling experience. Whether these trends will continue into the future is unknown, but the climate-change literature cited in our review indicates that our region can expect higher interannual variability in water supply, increasing frequency and severity of droughts, and earlier snowmelt, all of which are already apparent in the Henrys Fork watershed. The low-hanging fruit of slight increases in winter flow, fish passage, and tributary restoration are not enough to meet the objectives of fisheries managers or expectations of anglers in a changing climate.

At the same time as angling conditions have declined in the reach of the Henrys Fork immediately downstream of Island Park Dam, angling pressure and the demand for high-quality wild-trout fishing have increased substantially in the lower watershed, particularly on Fall River and the Henrys Fork near St. Anthony, downstream of the largest irrigation diversions. Currently, the only way to limit delivery of Island Park storage water is to maintain lower flows in the river at St. Anthony, ultimately pitting the quality of the fishery between Island Park Dam and Mesa Falls against that in the lower watershed. Conversely, increasing flows to benefit fisheries in the St. Anthony reach during the warmest period of the summer requires high delivery from Island Park Reservoir, which negatively impacts the fish population and the fishing experience in Box Canyon and Harriman State Park. Under current system administration, increased delivery from Island Park Reservoir occurs only if water is needed downstream of the Henrys Fork and therefore passes through the St. Anthony reach without being diverted. This is most likely to occur when some of the physical water in Island Park Reservoir belongs to the senior American Falls Reservoir storage right, illustrating why management of fisheries in the Henrys Fork depends on water rights, irrigation need, and reservoir management 100 mi (161 km) downstream.

Against the backdrop of climate change, the chemistry and physics of an aging reservoir, and increased demand for quality fishing in the lower watershed, HFF is challenged to maintain quality wild-trout fisheries in the Henrys Fork within physical and administrative systems that are much different than those present when HFF was founded in 1984. Power plants on the main river at Island Park (constructed in 1994) and Chester (constructed in 2011) have added infrastructural and operational complexity, and the most important administrative and legal actions affecting water management in the Snake River basin have occurred in the years since the founding of HFF.

Drought of 2016: a turning point.—The summer of 2016 provided an excellent example of how increased system complexity and climate change can interact in unexpected ways, with nearly disastrous consequences for all stakeholders. It also provides an example of how science-based collaboration can be effective even in extreme conditions. In the spring of 2016, snowpack was near average across much of the upper Snake River, and water supply looked adequate to meet demand. Irrigation entities that projected having surplus storage negotiated 80 Mm^3 (60,000 acre-feet) of private rental-pool leases with groundwater users eager to meet the terms of the recent settlement with the surface-water users. The rental water, reallocated to groundwater recharge, was stored in Palisades Reservoir on paper, but 27 Mm^3 (20,000 acre-feet) were diverted from the lower Henrys Fork during the middle of irrigation season, requiring physical delivery of water from Island Park Reservoir over and above what was required to meet irrigation demand. This demand was higher than originally anticipated because snow melted much earlier than average, and less than 2.5 cm (1 in) of rain fell during the 3-month heart of the growing season. Had managers adhered to the 28-m^3/s (1,000 ft^3/s) flow target at St. Anthony recommended by the Drought Management Planning Committee earlier in the summer, only 13 Mm^3 (10,000 acre-feet; 7.5% of capacity) would have remained in the reservoir at the end of irrigation season. To avoid what could have resulted in a sedimentation event similar to the one that occurred when the reservoir was drawn down in 1992, the Drought Management Planning Committee held an emergency meeting in early August. Based on projected irrigation demand, groundwater return flows, and water quality data, the committee concluded that the optimal strategy for all stakeholders was to temporarily reduce the flow target to 24 m^3/s (850 ft^3/s) at St. Anthony. This action allowed full delivery of the managed recharge water but saved 13 Mm^3 (10,000 acre-feet) in Island Park Reservoir, increasing carryover to 15%. This prevented a large release of sediment from the reservoir, kept turbidity low enough to maintain tolerable fishing conditions for the rest of the summer, and increased winter flow by 1.4 m^3/s (50 ft^3/s).

In retrospect, water users and managers throughout the basin regretted the unintended negative consequences of using rented storage water for in-season managed recharge. Irrigation entities that leased their projected storage surplus to the groundwater users in the spring ended up running out of water for their own irrigation needs later in the summer. Fremont-Madison Irrigation District was a willing partner in the managed recharge effort, agreeing to deliver the Henrys Fork portion of the leased storage water to a managed recharge facility within its boundaries for a nominal fee. However, this delivery ended up jeopardizing subsequent refill of its own storage in Island Park Reservoir. Water managers in the USBR did not realize until late in the summer that high demand for storage water was due in part to delivery of rental water for aquifer recharge and were challenged to avoid degradation of water quality and recreational experience at nearly every reservoir in the basin. Some water managers and policymakers called the use of surface storage water for managed aquifer recharge a case of "robbing Peter to pay Paul," which had not been the intent of the settlement between surface and groundwater users.

Meanwhile, despite the positive short-term outcome of the emergency drought management planning meeting on Island Park Reservoir management, conservation organizations could only stand on the sidelines and watch the basinwide situation unfold, since they were not parties to the settlement agreement, did not hold water rights that would have allowed them to participate in the market for rental water, and had no legal or administrative mechanisms at their disposal to influence private water lease agreements. However, this feeling of helplessness, along with intense displeasure on the part of anglers and public criticism of HFF for not preventing the poor fishing conditions and high reservoir drawdown of 2016, forced HFF and its partners to critically review their conservation strategies and ask some tough questions. In the case of HFF, the questions boiled down to this:

> How do you positively impact a complex river when management decisions can have both a negative and positive effect on different sections of the river simultaneously, the challenges facing the river are escalating, and the effort put forth thus far has resulted in a modest 9% increase in wild Rainbow Trout populations?

New strategies.—It is clear that management of Island Park Reservoir is key to maintaining quality fisheries in the Henrys Fork, but the reservoir is—plain and simple—an irrigation storage and delivery facility. Its mandate is different from many other federal reservoirs around the West that are authorized for additional uses such as flood control and hydropower, often with an allowance for fish and wildlife needs. Any management scheme for Island Park Reservoir must first and foremost store and deliver water for Idaho farmers within the prior appropriation framework provided by the Idaho Constitution. The Henry's Fork Basin Study (USBR 2015) provided a few alternatives, including expanded storage in existing reservoirs, construction of a new off-stream storage reservoir, and even construction of a small version of Teton Dam. However, the honest assessment among all stakeholders is that any type of additional surface water storage in the Henrys Fork basin is not economically and politically feasible and may never be again. Instead, HFF, TU, and FTR are pursuing three new strategies, all implemented through science-based collaboration: (1) improvements in science and technology to increase water-management precision, (2) conjunctive management, and (3) reduction in consumptive use of water through emerging water markets and alternative crop management.

Enhanced communication of science and technology is part of the answer. In 2017, the HFF science team developed a dynamic, stochastic computer simulation model of natural streamflow, reservoir storage and delivery, river reach gains and losses, and irrigation diversion that predicts daily stream and reservoir conditions across the entire watershed for the entire irrigation season, using only inputs known on April 1 (Van Kirk, unpublished). The primary goal of the model is to predict content of Island Park Reservoir at the end of the water year. In 2017, the model predicted this variable to within 1%. In 2018, the model overpredicted end-of-season reservoir content by 7%. Working with the science team, HFF's communications team disseminates model output to diverse stakeholders through varied and appropriate media. More than 180 individuals, including water managers in the USBR, WD01, and FMID, as well as fishing outfitters and guides, receive up-to-date daily water-supply reports from HFF every workday morning year-round. In addition to helping outfitters, guides, and anglers plan fishing trips, this information helps optimize management of the Henrys Fork water system, with the ultimate goal of minimizing summertime delivery of water from Island Park Reservoir while providing at least subsistence flows in the lower watershed. Currently, subsistence flow values are based on the Tennant (1976) method and set at 35% of the mean annual unregulated flow in the Henry's Fork at St. Anthony and in Fall River. These targets are around 50% of the mean annual regulated flow. Future research will refine these lower-watershed flow targets based on direct measurements of habitat availability and water temperatures. In 2017, when water supply was above average, model output and daily updates saved almost 5,000 acre-feet of Island Park Reservoir storage water through precise management, while meeting flow objectives in the lower watershed on all but 5 d during irrigation season. During 2018, when water supply was average over the whole water year and only 85% of average during the summer, the reservoir ended the water year 73% full, compared with 43% full in previous water years with similar supply. Flow objectives in the lower watershed were met on 80 out of the 90 d during which storage water was being delivered.

Given the state's managed aquifer recharge objectives and the need to mitigate groundwater pumping under the settlement, conjunctive management has great potential for improving streamflow and water quality, as long as summertime recharge does not tax the reservoir system as it did in 2016. The state's public recharge program avoids using storage water by diverting only excess natural flow that would have otherwise spilled over Milner Dam (CH2M and HFF 2016), essentially using the aquifer instead of surface reservoirs as a storage mechanism. One of the primary aquifer recharge sites is located in the Henrys Fork watershed, and about 33% of the discharge response from recharge at that site is realized as increased reach gains to the lower Henrys Fork. Thus, managed recharge provides a strategy by which both streamflow and water temperature in the lower Henrys Fork during late summer can be improved, using water diverted during years and seasons when supply is good. In the long run, this will not only reduce the amount of storage delivery needed from Island Park Reservoir but could also help mitigate the effects of climate change, particularly low flows and high water temperatures during late summer. Despite potential pitfalls, such as delivery of physical storage water for recharge under private lease agreements, privately managed recharge provides opportunities to reduce diversion of surface water during the middle of the summer. If some minor modification were made to the current rental-pool rules (WD01 2018), storage water saved during the summer by reduction in crop use could be rented and diverted for managed recharge during the fall, winter, and spring, when reach gains would be sufficient to support the diversion. Thus, physical delivery of water reservoirs would not be required, only paper withdrawal of the rented water. Recharge from this rented water would be used to mitigate groundwater use, essentially moving crop production from surface-irrigated land along key stream reaches to groundwater-irrigated land on the ESPA.

More generally, HFF and its partners are working with local agricultural producers to reduce irrigation demand through incentive-based mechanisms that are voluntary, that compensate producers for revenues potentially lost because of reduced irrigation, and that recognize the socioeconomic benefits of both agriculture and functioning aquatic systems to local communities. Consistent with its science-based approach, HFF is developing a geographic information system tool that will quantify the potential for demand reduction on a given parcel of land to meet specific hydrologic objectives, primarily reducing delivery of storage water from Island Park Reservoir and increasing summertime streamflow in the lower watershed. To get the program started, only irrigators who have a relationship with HFF were contacted to participate, but as the program has expanded, irrigation entities and individuals are being contacted based on the prioritization results. Despite continual discovery of administrative and economic hurdles, HFF, TU, FTR, and TNC are committed to pioneering this program. Out of the gates, irrigation withdrawal was reduced by 1.3 Mm^3 (970 acre-feet) in 2017. The most important administrative element in the Henrys Fork watershed is that FMID is the sole owner of storage water rights in Island Park Reservoir, so the success of the program depends on finding win-win-win solutions that benefit FMID, its irrigators, and the fishery. Beyond water leases and forbearance, HFF and partners are exploring on-the-farm changes in crop production, soil-health enhancement, and alternative crop markets as components of an overall suite to benefit instream flow. To move the program forward, HFF and FTR jointly hired a landowner outreach manager to plan, implement, and promote innovative on-farm soil and water management techniques. The person hired for the position grew up in the local area, is a former farmer himself, and has earned the trust of both agriculture and conservation interests through years of work in water management. As the Council experience has taught us, no program of this type would be successful without trust and firsthand knowledge of the local area. In his first 6 months on the job, the manager has contacted over a dozen individual landowners and producers, selected through a combination of word-of-mouth recommendations from other supportive producers and the parcel prioritization exercise.

All water conservation efforts and agreements must be carefully designed so that saved water either physically stays in the reservoir or flows through specific stream reaches at times when it addresses flow limitations. Fortunately, flexibility in water rights accounting, the need for groundwater mitigation on the ESPA, and the large senior storage right in American Falls Reservoir make it possible that the inevitable use of the saved water by other users can occur downstream of the targeted stream reaches. Comprehensive and coordinated management of surface water and groundwater at the farm, canal company, irrigation district, and state levels is necessary to provide durable sustainability for our wild-trout populations and agricultural economy at a meaningful scale. As our watershed-level program grows and more coordination is required across the whole basin, the current single staff position may not be sufficient to both work one on one with individual producers and develop administrative water-transaction mechanisms at the basin and state levels. Funding the program—both to support staff and to fund specific leases and forbearance—will be a major challenge over the next few years. Initial efforts will target philanthropic and corporate funding that can be invested at the individual farm level. However, long-term efficacy of the program will require self-sustaining funding mechanisms such as irrigator-to-irrigator leases and increased farm revenue from development of new markets for agricultural products. In our watershed, quinoa and specialty malting barley have potential to increase value per acre over traditional crops while using less water. By funding the program through a combination of philanthropic donations, investment by corporations seeking to meet sustainability objectives, direct water leases, and markets for new agricultural products, the long-term cost is shared by conservation and recreation interests willing to pay directly for improved streamflow and its benefits, consumers loyal to corporations with strong sustainability commitments, groundwater users in need of mitigation water to maintain their production, and consumers of specialty products.

Upper Teton River

Originating on the west slope of the Teton Range, the Teton River flows more than 100 km (62 mi) to its confluence with the Henrys Fork. The Teton River is one of very few remaining within the entire Greater Yellowstone Ecosystem that supports widespread and abundant populations of native trout (Gresswell 2011; Meyer et al. 2014). Though less renowned than the neighboring Henrys Fork and South Fork, the Teton River recreational fishery is regionally acclaimed and is known to both local and destination anglers as an iconic setting in which to catch Yellowstone Cutthroat Trout (YCT) on a dry fly. What is perhaps most unique about this native fishery is its persistence within a landscape that is primarily privately owned and has been significantly altered from its natural state. The locally based conservation organization FTR is working with a diverse collaborative to employ an innovative approach to protecting and restoring a thriving native YCT fishery in the Teton River watershed.

Conservation issues.—Similar to many watersheds in the West, a variety of interrelated land use, water use, and fisheries issues are playing out in the Teton River Watershed. At the start of the 20th century, tributaries to the upper Teton River flowed continuously from high-elevation headwaters to the Teton River in all but the driest years (Van Kirk, unpublished) and supported both fluvial and resident life history forms of YCT. As agriculture was established in Teton Valley, irrigation canals and ditches were built to divert and distribute streamflow according to agricultural needs. This changed the Teton River hydrograph from one dominated by snowmelt to one dominated by groundwater, with lower peak and higher base flow (Figure 3). Cumulative impacts on water quality from fertilizer application and livestock have resulted in elevated nitrogen levels in ground and surface water (IDEQ 2003; Cosgrove and Taylor 2007) while more recent development pressures have resulted in bank stabilization and channelization of tributary streams, destruction of riparian vegetation, and loss of connectivity between stream channels and their floodplains. Brook Trout *Salvelinus fontinalis* and Rainbow Trout were introduced to the Teton watershed beginning in the early 20th century (Dunham et al. 2002), and the altered hydrology, combined with degraded habitat conditions, favored the introduced species, which have invaded all but the most isolated headwater tributaries. The catastrophic failure of Teton Dam and rapid emptying of the reservoir irreversibly altered the geomorphology of the Teton River Canyon and the native fishery that inhabited it. Climate change adds a new layer of complexity to maintaining and restoring native trout in the Teton River.

Unsurprisingly, these events led to significant declines in the native YCT population. Between 1999 and 2003, Idaho Department of Fish and Game surveys found that YCT populations declined in all portions of the main-stem Teton River above the Teton Dam, despite increasingly restrictive regulations enacted in 1990 designed to protect the native fish (Meyer et al. 2006, 2014). In response to declines in water quality, habitat, and the fishery, a group of concerned local residents founded FTR in 2001. A significant focus of the group's effort has been researching the native YCT population, with the goal of understanding limiting factors for the population and developing strategies for reversing its decline. This work was further catalyzed in 2009 by designation as a model watershed by the Bonneville Environmental Foundation, and funding support to engage in a long-term research and monitoring effort.

The natural hydrologic regime paradigm suggests that YCT benefit when streamflow most closely follows the natural hydrograph (Gresswell et al. 1994; Gresswell et al. 1997; Fausch et al. 2001; Hauer and Lorang 2004; Gresswell 2011), and FTR initially identified tributary dewatering, due to agricultural diversions, as the biggest factor contributing to YCT decline in the watershed. Restoration of the natural hydrologic regime was a primary conservation focus for FTR. However, recent scientific analysis by FTR and our partners, including HFF, has indicated that this factor, alone, is not as significant as once thought (Van Kirk, unpublished). Other factors including nonnative species invasion, climate change, increased stream temperature, and drought may, cumulatively, have a greater impact on the success of native YCT populations than hydrologic alteration alone. This has been accompanied by an evolving understanding of the value of working agricultural lands and the human and ecological systems that have developed around a groundwater-dominated hydrologic regime. The organization has also formally acknowledged the practical reality that restoration to presettlement hydrologic conditions may simply not be logistically feasible in light of private land ownership, changing climatic conditions, and increasing human population growth and development pressure. As a result, FTR has intentionally made a significant paradigm shift in its approach to water management and YCT conservation. This approach is premised on the reality that the region is facing ever increasingly significant water challenges and issues that traditional water management approaches and philosophies are unable to address. Friends of the Teton River's overall approach has shifted to one that recognizes the dynamic ecological, social, and political drivers affecting water use in the region and that relies on creative, local solutions to reduce conflicts over water through the implementation of collaborative conservation efforts.

Concurrent with this intentional strategic shift by FTR, leaders within the Teton Valley agricultural water-user community have been faced with the reality that basin-wide water management and administration, including stabilization of the ESPA, will impact this region, likely resulting in mitigation and water supply requirements for municipalities, changes to the timing of surface water delivery calls, and impacts associated with the formation of the ESPA Groundwater Management Area. Failure to address these factors will likely result in continued declines in Idaho's aquifer and river levels, continued pressure to convert farmland to subdivisions, and ultimate loss of the economic, cultural, and ecological benefits associated with large expanses of working open spaces stewarded by generational family farmers.

Collaboration: Teton Water Users Association.—In response to these issues, leaders from FTR, the Teton Soil Conservation District, and the Teton County Farm Bureau joined in 2015 to form a diverse working group, called the Teton Water Users Association (TWUA). The mission of the TWUA is to bring together individuals who can, collectively, identify water management solutions that satisfy the needs of all constituents within the community—farmers who depend on water for crop and livestock production, municipalities that require clean and adequate water for residents, and conservation interests seeking water for fish and wildlife.

To date, the TWUA has outlined a Teton Watershed Restoration Plan designed to optimize outcomes for all three interest groups (agricultural, municipal/county, and conservation). The primary concerns identified by TWUA members are local water-level reductions of up to 17 m (55 ft) in Teton Valley's local aquifer (Nicklin Earth and Water 2003), reduction in base flow in the Teton River, and decreased wetland and instream habitat and migratory corridors for fish and wildlife. Cumulatively, these trends raise concerns about long-term water availability for municipal and residential use, tributary and river stream flows for fish and other wildlife, and water availability for agricultural production. Furthermore, the incremental conversion of land from agriculture to exburban use has negatively impacted Teton Valley's prime economies, including agriculture and outdoor recreation. Stabilizing irrigation water supply will help keep agriculture financially viable, ultimately maintaining existing agricultural land and the open space, wildlife habitat, and ecological functionality it provides. To address these concerns, the specifically stated goals of the Teton Watershed Restoration Plan are to

1. Stabilize the Teton Valley aquifer, thereby protecting municipal and residential water supply;
2. Insulate farmers against changes in water availability and increase water-supply reliability, particularly during times of drought;
3. Maintain valuable wetland habitat and create tributary streamflow conditions beneficial for trout; and
4. Quantifiably increase base flow in the Teton River.

New strategies.—The primary implementation strategy within this plan is to restore traditional flood-irrigation practices by repairing and recommissioning canals and other infrastructure that have gone unused since conversion to pipelines and sprinklers, thereby increasing groundwater recharge incidental to irrigation in Teton Valley. Flood irrigation will occur in the spring, when snowmelt provides an abundant water supply, and be applied to pasture and hay fields. Later in the summer, when water supply becomes limited, irrigators will use their more efficient pipe and sprinkler systems to maximize utility of available water on higher-value crops such as potatoes and grain. The strategy of increasing groundwater recharge by restoring traditional flood irrigation—versus other options such as development of a formal managed recharge program—was selected for three reasons: it is based on traditional agricultural practices in Teton Valley, it is a cost-effective way to store water in the local aquifer for use later in the summer, and it can be implemented with existing water rights. Historically, irrigators in Teton Valley diverted as much water as possible during runoff and spread it across as much land as possible, subirrigating the lower parts of the valley and maintaining higher base flow, which met the needs of downstream irrigators later in the summer and allows upstream irrigators to continue to divert remaining streamflow as it recedes (Kilburn 1964). Because of the high elevation of Teton Valley, runoff occurs well before crop needs exceed available supply, so the diversion of water during runoff essentially stores it in the aquifer to provide a supply later in the summer when crop demand is highest. This practice was gradually abandoned starting around 1970 because of labor cost savings and increased crop yields associated with pipe and sprinkler systems. Financial and technical assistance provided by government agencies helped facilitate the conversion from flood to sprinkler irrigation. Over time, traditional canal and ditch irrigation systems have become less functional due to fragmentation of agricultural lands and socioeconomic changes within canal companies (Figure 5; Liegel 2011; Baker et al. 2014). However, the old canal and ditch systems are still present over most of the valley and can be recommissioned at much lower financial and environmental cost than developing new managed aquifer recharge sites and infrastructure or building surface storage reservoirs. Current pipe and sprinkler systems can still be used during late summer to retain the benefits of irrigation efficiency to agricultural producers.

Equally important, this strategy can be implemented by creatively using existing water rights and administrative rules within prior appropriation and WD01 accounting. The water rights that can be used are irrigation rights with priorities in the 1890s to early 1900s. These allow diversion of natural flow starting on April 15, coincident with onset of runoff (Figure 3), and application of that water to land associated with those rights, regardless of irrigation method. Development of a formal managed recharge program would require new water rights, with groundwater recharge as a stated beneficial use. Obtaining such rights can require a difficult legal process since it often requires filing new water rights applications or changing the intended use of existing rights, each of which can be slowed by protests from other water users and lengthy administrative review processes. Furthermore, the new rights would have a priority date junior to all existing rights in the upper Snake River basin and therefore have limited utility. However, incidental recharge, which is utilized in this project, depends only on creative management of existing irrigation water rights. In years when the upper Snake Reservoir system fills early in the spring, diversion of water in Teton Valley, at the headwaters of the system, can prevent spill of water out of the basin at Milner Dam. Temporary storage in the Teton Valley aquifer allows in-basin use of this water to benefit multiple resources, subject of course to the well-accepted reality of greatly degraded ecological function downstream of Milner. In years when the upper Snake River system does not fill, the priority dates of Teton Valley irrigation rights are senior to all storage rights in the system and therefore can legally be diverted when in priority regardless of storage needs downstream. Although the TWUA is currently pursuing incidental recharge for the reasons listed here, it has discussed the future possibility of obtaining new water rights for managed aquifer recharge because of additional systemwide market opportunities associated with managed recharge, in years when those rights are in priority.

Development of this strategy was strongly informed by science and technology. The HFF science staff provides consulting and technical services to the TWUA, including projection of hydrologic response to incidental recharge in Teton Valley using a previously developed groundwater-surface water model (Peterson 2011; USBR 2012b). This groundwater-surface water modeling work indicates that, by restoring traditional water diversion and flood irrigation practices in the Teton Valley to the greatest extent possible, the TWUA can take tangible steps toward stabilizing the local aquifer and increasing base flow in the Teton River and, in so doing, address multiple levels of water need and begin to proactively plan for Teton Valley's water future. This model, as well as an earlier version, confirms that the hydrogeologic properties of the Teton Valley alluvial aquifer make this possible—water recharged to the aquifer during early spring runoff increases aquifer discharge on time scales of months rather than decades, increasing flow in the Teton River during late summer months critical for both farmers and fish (Van Kirk and Jenkins 2005). Restoring the full extent of flood irrigation that existed in the 1960s will not be possible due to exurban development on formerly irrigated lands, but modeling suggests that enough flood irrigation is possible under current conditions to increase stream base flow. Science and technology will continue to be used, both to identify particular canals and fields that can feasibly be used for flood irrigation and to assess effectiveness of the program.

This project recognizes the direct connection between groundwater and surface water, actively manages these resources together and makes additional water available to support regional conjunctive management goals, including stabilization of the ESPA. Water users in the upper Teton Valley have relatively junior water rights when compared to those downstream, near Rexburg. Therefore, the upper Teton Valley water users are subject to water calls by the senior users each year. Because bolstering the aquifer in the spring will improve late-season streamflow in the Teton River, water calls will be delayed, allowing Teton Valley water users to continue diverting later into the summer, as occurred historically (Kilburn 1964). Additionally, by quantifiably increasing base flow in the Teton River and providing additional natural flow to lower Teton River irrigators during mid- to late-summer, the project will decrease need for delivery of Island Park Reservoir storage water through the Crosscut Canal. This predicted outcome intimately links TWUA goals with those of HFF, one reason that FTR and HFF have jointly funded the landowner outreach staff position.

The project utilizes economic incentives, paid directly to agricultural producers and irrigation entities, to promote water management practices that increase incidental recharge. For example, during the pilot phase of the program, resources will be required to renovate canal and ditch systems. Costs are yet unknown, since this renovation will have to occur within the overlay of roads, utilities, and structures built over the past few decades. As in the Henrys Fork case, the economic incentives will initially be provided by external grants and philanthropic donations. Over the long term, project partners plan to actively develop a self-sustaining market. When the market is fully developed, incentive payments will come from a combination of conservation buyers and downstream farms and municipalities that benefit from the decreased spring runoff and increased late-summer flows. As described above, private managed aquifer recharge funded by groundwater users in need of mitigation is being conducted in the upper Snake River basin, but at the time of this writing, we are unaware of any efforts to actively create a market for incidental recharge.

It is important to note that this project has been designed specifically to address factors that limit trout populations in the Teton River watershed, as identified by FTR's intensive research and monitoring efforts. The streamflow increases predicted by groundwater–surface water modeling are expected to benefit trout populations, including native YCT, in the Teton River by increasing streamflow and decreasing water temperature during July, August, and September. In addition, the project is expected to support wetland habitat along the Teton River, benefiting a host of species. During development of the project, state and federal partners, including the Idaho Department of Environmental Quality, Idaho Department of Fish and Game, U.S. Forest Service, and U.S. Fish and Wildlife Service, collaborated to develop a document that provides a concise picture of the desired conservation outcomes in the Teton River watershed, as applied to YCT (FTR 2016: Appendix B). The document clearly describes the applicable threats, necessary conservation measures, and stream priorities necessary for the protection and restoration of YCT in the Teton River watershed. Especially notable is the identification of specific tributaries in which diversion for recharge may be used to intentionally maintain low-flow barriers to upstream movement of nonnative species, protecting distinct populations of genetically pure YCT in isolated headwaters. In other tributaries, where invasion by Cutthroat Trout × Rainbow Trout hybrids or other nonnative trout has already occurred, restoration of streamflow connectivity, or a combination of flow restoration and recharge, is prioritized. The document functions to coordinate the interests of these various entities and to prioritize strategies appropriate for specific tributaries and can be used to ensure that aquifer recharge efforts promote the maintenance and recovery of YCT in the Teton River watershed in the future.

Despite development and vetting of these stream-specific strategies, healthy skepticism remains in the scientific community as to the relative impact and associated consequences for native trout populations resulting from aquifer recharge. Specifically, there is concern that further manipulation of the hydrograph may negatively impact YCT populations. Friends of the Teton River and our partners are currently working with leading scientists in the field of YCT biology and management to collect and analyze data relevant to this question. In 2010, FTR formally empaneled many of the region's most notable fisheries biologists, fisheries resource managers, and climatologists to serve as a formal technical team to review, analyze, and interpret research, as well as to design and analyze monitoring efforts, in order to advise FTR and the TWUA on whether specific recharge activities, in specific tributaries and during specific times and quantities, are likely to, or are actually measured to, impact YCT. This ad hoc technical team meets at least annually, and more often as needed. This effort is designed to specifically encourage adaptive management of the project and to continue to maintain critical scientific review.

In a region where new-West conservation values are often pitted against old-West lifestyles, in a place where intensive farming and ranching comes into conflict with modern-day recreational pursuits, this project draws our community together to generate positive conservation outcomes. It has forged new and unlikely partnerships, opened constructive dialog between traditionally feuding interests, and provided a vehicle by which to develop a water resource plan that will join, not divide, the community. It has afforded Teton Valley's water users—irrigators cities, county, and conservation interests—a means by which to plan for the future and generate collaborative solutions that address the goals of a diverse community. Scientific modeling demonstrates that family farmers and ranchers who actively manage water in a manner consistent with their existing water rights and the principles of conjunctive management can stabilize municipal water supplies and create much-needed refugia for native fish and wildlife provided by enhanced wetlands, higher base flow, and lower summertime water temperatures. By using the incentive-based approach, the community as a whole will be able to recognize and economically support the quantifiable contributions that ranching and farming make to our community, reversing the typical narrative. Additionally, this project helps promote ancillary land-use benefits. By encouraging incidental recharge, farmers and ranchers are also encouraged to keep large swaths of land undeveloped so that incidental recharge efforts can be conducted. Those large land tracts are invaluable for big-game winter range, migratory corridors, and the integrity of stream channels and floodplains. In these ways, the project is not only a win-win-win for ecological, farming, and municipal interests, it is a means to help heal the divides in the community and sustain the economies and community values upon which this region was settled.

Comparison of Case Studies to Other River Basins

The efforts in the Henrys Fork basin are part of a larger web of collaborative approaches emerging from traditional and protracted natural resource conflict throughout the West. Interestingly, some of the most visible and effective collaborations seem to be emerging from areas where the smoke has recently cleared after high-profile and divisive water scarcity issues. That does not mean conflict is over in the examples cited below, but from the ashes opportunity has risen, and real discussions emerged to address a multiplicity of long-term water supply and management issues accounting for a variety of historic consumptive and nonconsumptive water use needs. This is certainly true for places like the Yakima and Colorado River basins, where stakeholders have been working on large-scale and long-term approaches to avoid future disagreements and provide much needed certainty for both traditional water users and river systems. Like the Henrys Fork, stakeholders in these and other drainages around the West are identifying a range of alternatives such as nonregulatory, voluntary, and compensated demand reduction; infrastructure modernization; groundwater recharge and storage; expansion of existing storage; water market creation; and innovative water transfers to satisfy a large range of consumptive and conservation needs.

Yakima

One high-profile western example of large-scale collaboration, the Yakima Basin Integrated Plan (YBIP) is a balanced package of potential actions that will return thousands of salmon and steelhead to the basin annually, improve water quality and quantity, and support a healthy agricultural and recreation economy. The plan was endorsed by a diverse coalition of conservation groups, irrigators; farmers; sportsmen and women; local, state, and federal governments; and the Confederated Tribes and Bands of the Yakama Nation. While the ESA issues and compliance formed a backdrop and incentive for the Yakima negotiations, the solutions package in the YBIP was not formed from litigation or federally required actions. Instead, stakeholders sat at the table for several years and melded a variety of new storage, expansion of existing storage, tributary habitat restoration including flow, and land-acquisition components into what equates to a phased long-term watershed health program.

The Yakima package itself is unique in that it pushes potential new storage projects into the future based on climate expectations and water need while identifying early action items such as expanding existing storage and restoring tributary flows (USBR and WDE 2012). The financial model is also decidedly less federal than historic approaches. Washington State is committed to 50% of the YBIP implementation costs, and an innovative finance approach also challenges the historic USBR model of federal-borne construction costs and amortized low-interest irrigation district repayment (WDE 2018). Irrigation districts are proposing to finance, build, and operate one of the major water-supply projects in the YBIP. Finally, early action items have focused resources on acute challenges in the basin, such as low tributary flows. During the 2015 drought, partnerships built through the YBIP resulted in rapid action through the Kittitas Reclamation District to provide flows in streams that would have otherwise run dry, securing important habitat for salmon and steelhead (WDE 2016). Further, the Kittitas Reclamation District was recently awarded a federal WaterSmart grant and is working with TU to develop innovative water market strategies as part of that effort. Although the USBR basin study in the Henrys Fork watershed has not generated a coordinated, large-scale package of options beyond those in the basin study document itself, FTR received a WaterSmart grant to support TWUA activities, and the success of stakeholder-driven processes in the Yakima basin mirrors that in the Henrys Fork and Teton watersheds.

Colorado River

Like the Yakima, the Colorado River basin has faced years of drought. The stakes are especially high on the Colorado because the basin is so large and serves such a diverse group of water user communities, including the large metro areas of Denver, Las Vegas, Phoenix, and San Diego. A host of challenges face the basin, including increased water demands from growing cities, a changing climate of prolonged drought and decreased snowpack, and declining water levels in Lakes Powell and Mead that, if not checked, could trigger future water calls and conflict (USBR 2012c). Water use in the basin is also divided according to a complex interstate compact that essentially allocated water based on negotiations that occurred between Upper Basin (Colorado, New Mexico, Utah, and Wyoming) and Lower Basin (Arizona, California, and Nevada) states in the 1920s. All this led Eric Kuhn, longtime general manager of the Colorado River Water Conservancy District, to note,

> [t]he Colorado River is already fully used. In the very near future, the demand for the river's resources will far exceed the available supply. In order to meet the needs of people and aquatic dependent species and habitats, new ways of thinking and doing business will be essential. [From Alliance for Water Efficiency 2012.]

A massive basin study was completed by reclamation in 2012 and identified that under certain climate change scenarios, the large storage reservoirs so critical to reliability could run very low, putting system operations at risk (USBR 2012c). In addition to municipal water conservation and targeted new and expanded existing storage, demand management was identified as an additional tool to maintain reservoir levels in the face of drought and climate change. Shortly thereafter, in 2014, USBR and the four largest municipal water providers in the Colorado River basin announced the System Conservation Pilot Program (SCPP), to begin developing tools for responding to long-term drought conditions. The purposes of the SCPP for the Upper Basin included funding and testing voluntary, demand-management measures that could ultimately be used to insulate a variety of system operations, including maintaining water in Lake Powell above the minimum levels needed to ensure compliance with Colorado River compacts and to maintain hydropower generation at the reservoir. Funding for SCPP during its first 5 years has varied depending on the availability of municipal and federal funding. In 2015–2016, the SCPP conserved approximately 14.1 Mm^3 (11,400 acre-feet) of water with about 32 projects. Roughly 75% of the water was conserved through temporary, split-, or late-season fallowing; ranchers and farmers irrigated for part but not all of the potential irrigation/production season and were compensated for production lost during the remainder of the season. In the 2016–2017 round, water users submitted 47 applications for SCPP projects, with a potential 25 Mm^3 (20,000 acre-feet) of water savings in Wyoming alone. That year, the fourth of the pilot, there were $7.5 million in applications and only $1.8 million available for funding

The SCPP has also provided an opportunity for conservation groups like TU and TNC to work with agricultural producers to help create the market necessary for demand reduction and management to work. Both groups had been working in the headwater reaches of the basin for years. In the Upper Basin, TU has worked closely with producers, state agencies, and the Upper Colorado River Commission to successfully implement the SCPP. Much of this work has involved setting the table for water-market discussions by working with ranchers and farmers to repair aging irrigation infrastructure and improve water delivery for agricultural purposes. Such work is successful because it is pragmatic, voluntary, and non-regulatory—designed to benefit both people and fish. Designing creative ways to reduce demand via market creation dovetails nicely with broader on-ranch habitat restoration and cattle and farm operations.

However, demand management is a classic chicken-and-egg scenario in many northern Rocky Mountain states, including those in the upper part of the Colorado River basin. The states of Colorado, Utah, and Wyoming lack explicit statutory provisions to protect water not only downstream of nonparticipating head gates, but also between states to ensure that when lands are enrolled in the SCPP or any other type of demand reduction program and taken temporarily out of production, the saved water can be shepherded and accounted for in Lake Powell. However, the beauty of any type of pilot program is that it involves risk but can also fuel innovation. In the case of the SCPP, significant numbers of landowners have chosen to participate and some have become advocates for either continuation of the pilot or the institutionalization of a long-term demand-management program. While hardly unanimous, some water-management officials and program funders agree that significant market demand exists. Some large water districts like Grand Valley Water Users in Colorado and Carbon Canal Company in Utah are participating, and individual ranchers along upper Green River tributaries in Wyoming have signed up in remarkable numbers.

It is also important to note that despite the limitations of existing law and the traditionally narrow confines of the prior appropriation doctrine, market demand continues unabated. If all the successful SCPP applications are funded and implement during the 2018 field season, 40 Mm^3 (30,000 acre-feet) of water would be available in the Upper Basin for delivery to Lake Powell. Trout Unlimited has also been partnering with producers in the upper Green River to develop a tributary model of demand-management participation. Pursuant to this model, either all or a large majority of the water users on a tributary participate, thereby eliminating the risk under current law that flows will not be shepherded past nonparticipating head gates. This is not dissimilar to the rotational water sharing provisions available under some western state water laws. The approach has created a competitive advantage for water users willing to work together to maintain or increase market share in the pilot program. It is also worth noting that the late-season flows restored on these small tributaries are not based on agency-mandated minimum streamflow or planned natural flow regime, but they are significantly higher than recent averages and occur during the late summer when fish need it the most. Trout Unlimited hired an engineering firm to quantify flow improvements on one upper Green River tributary during the 2018 season, but results were not available at the time of this writing. As the program moves from pilot to permanence, streamflow monitoring and accounting will be critical. Water agencies and participants will be developing protocols to verify not only that participating ranches and farms reduce water use, but that such water is accounted for in multiple locations and shepherded to Lakes Powell and Mead. The pilot program was focused primarily on market establishment, but any type of extended pilot, future demonstration, or permanent program will have to deal more surgically with monitoring and shepherding water both inter- and intrastate.

The demand-reduction concept shows great promise in the West because it is a voluntary, incentive-based tool that not only addresses ongoing water shortages, but can also offset economic and environmental impacts of those shortages. Water leased under such a program remains tied to the land and keeps producers whole, which has great benefits for both agriculture and coldwater fisheries. Landowners in parts of Wyoming, Utah, and Colorado have participated in and benefitted from a program that attaches a value to the nondiversion of their water rights during low-flow conditions, improving flows in Colorado River tributaries and sending water downstream for overall system reliability. Demand-management tools have the potential to create drought resilience across agricultural, municipal, and fishery interests in chronically water-short basins. While demand reduction is still in its infancy in the Henrys Fork watershed, the Colorado River basin provides examples of successful approaches and promise for similar success on the Henrys Fork and Teton River.

Transferability of the Upper Snake River Experience to other Watersheds

The concepts on which our paradigm and strategies are based have evolved through decades of experience in the upper Snake River basin. Our literature review, as well as comparison of our experience with those in the Yakima and Colorado River basins, suggests that at least some components of the upper Snake River experience are applicable to other river basins in the West. In fact, the designer-flow concept was developed and tested outside of the Snake River basin, and effects of climate change on water supply are central to water resources planning and management throughout the West. However, three specific components of the upper Snake River experience critical to implementation of our strategies may not be present in any given watershed around the West. The first is trust and collaboration facilitated by a stakeholder group that has been meeting regularly for 25 years. Trust is obviously necessary for producers to be open to voluntary irrigation demand reduction (even if compensated) and for stakeholders with different perspectives to collaborate, but it is also necessary for irrigation entities and water-management agencies to use science and technology produced by NGOs as the basis for management decisions and strategies. Although the Council was created from a citizen-led effort, the Council would probably not have been as effective and durable without state and federal presence in water and natural resource management in the watershed and without voluntary participation of locally based personnel from these agencies. However, we emphasize that the Council is not led by agency personnel and that federal facilities constitute only a small part of the irrigation infrastructure in the watershed.

The second key ingredient in our approach is substantial, observable, and well-understood interaction between surface water and groundwater. Managers and policymakers in the upper Snake River basin recognized the importance of interactions between groundwater and surface water in the 1970s, commissioning a line of scientific investigation that continues to this day. Research in the upper Snake River basin has contributed greatly to the larger scientific literature on hydrologic aspects of irrigation seepage, the paradox of irrigation efficiency, and managed aquifer recharge. Although our literature review cites many cases from around the West where interactions between groundwater and surface water are highly connected and are major components of hydroecological systems, the upper Snake River basin is arguably the single largest irrigation system in the United States where interactions between groundwater and surface water are central to system management and administration. Moreover, routine water-management discussions held in local coffee shops and in the statehouse reflect widespread recognition of groundwater and surface water as the same resource. Full adoption of conjunctive water management and water-rights administration in the upper Snake River provides a unique, flexible, and perhaps enviable framework for water conservation that does not exist in most other river basins in the West. However, practitioners in other basins could develop a framework for managing groundwater and surface water together by starting with a comprehensive hydrogeologic assessment.

Third, water management in the upper Snake River basin is more flexible than in many other western irrigation systems because of Idaho's particular interpretation of prior appropriation as it has evolved through court cases, adjudication proceedings, and administrative practice. Storage of water in headwater reservoirs out of priority is a good example of flexibility that has been exploited to benefit fish and wildlife. In other western states, water in a given reservoir can be stored only when the right to fill that reservoir is in priority (CDWR 2016). If this policy were implemented in the upper Snake River basin, streamflow immediately downstream of most reservoirs in the system would be zero when the storage rights in a particular reservoir were in priority. This practice would also leave the reservoir system unfilled in many years. Although the same flexibility we enjoy in the upper Snake River basin may not be available in other basins or states, our overall strategy of seeking solutions within prior appropriation, as it is practiced locally, may still prove useful in other locations.

Conclusions

In the upper Snake River basin, regulatory approaches to providing instream flow have not led to much success because they are adversarial by nature and because legal and administrative tools are not available in most locations. This has led the three watershed conservation organizations represented in this chapter to develop innovative, collaborative approaches that fully engage water users, water managers, and other river stakeholders as equal partners. In the Henrys Fork watershed, this engagement was facilitated largely by willingness of stakeholders to embrace the consensus-based approach of the Henry's Fork Watershed Council, but also by acceptance of science and technology brought to the table by the NGOs. That science—particularly in the area of groundwater–surface water interactions—has pushed us away from the concepts of minimum streamflow and natural hydrologic regime and toward pragmatic, site- and resource-specific hydrologic objectives. Collaborative development and implementation of conservation strategies based on science has been and will likely continue to be successful in the areas of water operations and native trout restoration. It remains to be seen whether new science and technology will be successful in implementing managed aquifer recharge to meet local rather than basinwide objectives and in developing incentive-based demand reduction on a scale large enough to provide meaningful streamflow improvements. Current administrative constraints on water exchanges and availability of water for managed recharge are likely to provide the greatest impediments to success in these areas. While doubling down on science-based collaboration was natural for HFF and FTR, both of these organizations underwent some serious and soul-searching discussions internally and with each other to reach the point of fully embracing incentive-based approaches and development of sustainable water markets as tools for improving streamflow for fish and other aquatic organisms. Some, including the lead author of this chapter, argued strongly against these approaches on the grounds that a fully developed water market would simply make it easier for water to be transferred out of the Snake River basin to the highest bidders, most likely urban centers to our south and west. This argument concluded that neither small agricultural producers nor conservation groups could compete in that market. In the end, based on careful analysis of water-rights law and administration, economic and political reality, and the threats to trout fisheries posed by climate change, HFF and FTR joined their larger partners, TU and TNC, in embracing market concepts as the best opportunity for maintaining coldwater ecosystems in the western United States. The question before us at this point is whether angling constituencies and conservation philanthropists will bring enough monetary resources—and patience—to the table as we develop true markets for water that will sustain both agriculture and aquatic systems over the long term.

Acknowledgments

Lisa Pelly and Cory Toye from Trout Unlimited contributed material on the Yakima and upper Colorado River basins. Brian Hodge from Trout Unlimited contributed information regarding the site-specific nature of determining localized ecological value of water-use efficiency projects. Bryce Oldemeyer and Christina Morrisett of Henry's Fork Foundation prepared maps and some of the graphics. The lead author prepared of most of the literature review as part of work funded by U.S. Department of Agriculture award 2008-51130-19555 to the lead author and by National Science Foundation award 1038973 to Douglas Jackson-Smith. Comments of two anonymous reviewers greatly improved the manuscript.

References

Al-Chokhachy, R., A. J. Sepulveda, A. M. Ray, D. P. Thoma, and M. T. Tercek. 2017. Evaluating species-specific changes in hydrologic regimes: an iterative approach for salmonids in the Greater Yellowstone Area (USA). Reviews in Fish Biology and Fisheries 27:425–441.

Allen, D. N. 2010. The Klamath Hydroelectric Settlement Agreement: federal law, local compromise, and the largest dam removal project in history. Northwest Journal of Environmental Law and Policy 16:427–468.

Alliance for Water Efficiency. 2012. Supply shortages looming: Colorado River Basin Water Supply & Demand Study. Available: www.allianceforwaterefficiency.org/colorado-basin-study-released-2012.aspx. (March 2019).

Anderson, P. R., and A. J. Kraft. 2012. Why does Idaho's water law regime provide for forfeiture of water rights? Idaho Law Review 48:419–446.

Angradi, T., and C. Contor. 1989. Henrys Fork fisheries investigations. Idaho Department of Fish and Game, Federal Aid in Sport Fish Restoration, Project No. F-71-R-12, Subproject III, Jobs 7a and 7b, Job Completion Report for 1986–1987, Boise.

Averyt, K., J. Meldrum, P. Caldwel, G. Sun, S. McNulty, A. Huber-Lee, and N. Madden. 2013. Sectoral contributions to surface water stress in the coterminous United States. Environmental Research Letters [online serial] 8:035046.

Bain, M. B., J. T. Finn, and H. E. Brooke. Stream flow regulation and fish community structure: Ecology 69:382–392, 1988.

Baker, J. M., Y. Everett, L. Liegel, and R. Van Kirk. 2014. Patterns of irrigated agricultural land conversion in a western USA watershed: implications for landscape-level water management and land-use planning. Society and Natural Resources: An International Journal 27:1145–1160.

Battikhi, A. M., and A. H. Abu-Hammad. 1994. Comparison between the efficiencies of surface and pressurized irrigation systems in Jordan. Irrigation and Drainage Systems 8:109–121.

Bauer S., J. Olson, A. Cockrill, M. van Hattem, L. Miller, M. Tauzer, and G. Leppig. 2015. Impacts of surface water diversions for marijuana cultivation on aquatic habitat in four northwestern California watersheds. PLOS (Public Library of Science) ONE [online serial] 10(9):e0120016.

Becker, D. H. 2006. The challenges of dam removal: the history and lessons of the Condit Dam and potential threats from the 2005 Federal Power Act amendments. Environmental Law 36:811–868.

Behnampour, L. C. 2011. Reforming a western institution: How expanding the productivity of water rights could lessen our water woes. Environmental Law 41:201–233.

Benjamin, L., and R. W. Van Kirk. 1999. Assessing instream flow and reservoir operations on an eastern Idaho river. Journal of the American Water Resources Association 35:899–909.

Benson, R. D. 1997. Whose water is it? Private rights and public authority over Reclamation project water. Virginia Environmental Law Journal 16:363–427.

Blomquist, W., T. Heikkila, and E. Schlager. 2001. Institutions and conjunctive water management among three western states. Natural Resources Journal 41:653–683.

Blumm, M. C., and A. B. Erickson. 2012. Dam removal in the Pacific Northwest: lessons for the nation. Environmental Law 42:1043–1100.

Boggs, K. G., R. W. Van Kirk, G. S. Johnson, J. P. Fairley, and P. S. Porter. 2010. Analytical solutions to the linearized Boussinesq equation for assessing the effects of recharge on aquifer discharge. Journal of the American Water Resources Association 46:1116–1132.

Bovee, K. D., B. L. Lamb, J. M. Barthalow, C. B. Stalnaker, J. Taylor, and J. Henriksen.

1998. Stream habitat analysis using the instream flow incremental methodology. U.S. Geological Survey, Biological Resources Division Information and Technology Report USGS/BRD-1998-0004, Reston, Virginia.

CDWR (Colorado Division of Water Resources). 2016. General administration guidelines for reservoirs. CDWR, Denver.

CH2M and HFF (Henry's Fork Foundation). 2016. Eastern Snake Plain Aquifer (ESPA) review of comprehensive managed aquifer recharge program. Final Report to Idaho Water Resource Board, Boise, Idaho.

Chaffin, B. C., and H. Gosnell. 2017. Beyond mandatory fishways: federal hydropower relicensing as a window of opportunity for dam removal and adaptive governance of riverine landscapes in the United States. Water Alternatives 10:819–839.

Chen, W., and J. D. Olden. 2017. Designing flows to resolve human and environmental water needs in a dam-regulated river. Nature Communications [online serial] 8:2158.

Collier, M., R. H. Webb, and J. C. Schmidt. 1996. Dams and rivers: a primer on the downstream effects of dams. U.S. Geological Survey Circular 1126.

Contor, B. A. 2010. Status of ground water banking in Idaho. Journal of Contemporary Water Research and Education 144:29–36.

Contor, B. A., and R. G. Taylor. 2013. Why improving irrigation efficiency increases total volume of consumptive use. Irrigation and Drainage 62:273–280.

Cosgrove, D. M., and G. S. Johnson. 2004. Transient response functions for conjunctive water management in the Snake River plain, Idaho. Journal of the American Water Resources Association 40:1469–1482.

Cosgrove, D. M., and J. Taylor. 2007. Preliminary assessment of hydrogeology and water quality in ground water in Teton, County, Idaho. Idaho Water Resources Research Institute Technical Report 07-002. Boise.

Crifasi, R. R. 2005. Reflections in a stock pond: are anthropogenically derived freshwater ecosystems natural, artificial, or something else? Environmental Management 36:625–639.

Deitch, M. J., G. M. Kondolf, and A. M. Merenlender. 2009. Surface water balance to evaluate the hydrological impacts of small instream diversions and application to the Russian River basin, California, USA. Aquatic Conservation: Marine and Freshwater Ecosystems 19:274–284.

Dunham, J. B., S. B. Adams, R. E. Schroeter, and D. C. Novinger. 2002. Alien invasions in aquatic ecosystems: toward an understanding of Brook Trout invasions and their potential impacts on inland Cutthroat Trout. Reviews in Fish Biology and Fisheries 12:373–391.

ESA (Endangered Species Act). 1973. U.S. code, volume 16, sections 1531–1544.

Fausch, K., Y. Taniguchi, S. Nakano, G. D. Grossman, and C. R. Townsend. 2001. Flood disturbance regimes influence rainbow trout invasion success among five holarctic regions. Ecological Applications 11:1438–1455.

FPA (Federal Power Act). 1920. U.S. code, volume 16, sections 791–825.

Fish and Wildlife Coordination Act. 1934. U.S. code, volume 16, sections 667–667e.

Fereday, J. C., and M. C. Creamer. 2010. The maximum use doctrine and its relevance to water rights administration in Idaho's lower Boise River basin. Idaho Law Review 47:67–111.

Fernald, A. G., T. T. Baker, and S. J. Guldan. 2007. Hydrologic riparian, and agroecosystem functions of traditional acequia irrigation systems. Journal of Sustainable Agriculture 30:147–171.

Fernald, A. G., and S. J. Guldan. 2006. Surface water–groundwater interactions between irrigation ditches, alluvial aquifers, and streams. Reviews in Fishery Science and Aquaculture 14:79–89.

Fiege, M. 1999. Irrigated Eden: the making of an irrigated agricultural landscape in the American West. University of Washington Press, Seattle.

Franklin, D. R., and R. Narayanan. 1988. Trends in western United States agriculture: irrigation organizations. Water Resource Bulletin 24:1289–1294.

Fremont-Madison Conveyance Act of 2003. Public Law 85, 108th Congress (30 September 2003).

Freyrer, F., and M. P. Healey. 2003. Fish community structure and environmental correlates in the highly altered southern Sacramento–San Joaquin Delta. Environmental Biology of Fishes 66:123–132.

FTR (Friends of the Teton River). 2016. Expansion of and existing watershed group: improving ecological resilience, conserving water, and reducing conflicts through formation of the Teton River Advisory Council. Final Report to U.S. Bureau of Reclamation, WaterSmart Grant #R13AP80029, Denver, Colorado.

Frisvold, G. B., and K. Konyar. 2012. Less water: how will agriculture in southern mountain states adapt? Water Resources Research [online serial] 48:W05534.

Fyfe, J. C., C. Derksen, L. Mudryk, G. M. Flato, B. D. Santer, N. C. Swart, N. P. Molotch, X. Zhang, H. Wan, V. K. Arora, J. Scinocca, and Y. Jiao. 2017. Large near-term projected snowpack loss over the western United States. Nature Communications [online serial] 8:14996.

Galat, D. L., and R. Lipkin. 2000. Restoring ecological integrity of great rivers: historical hydrographs aid in defining reference conditions for the Missouri River. Hydrobiologia 422–423:29–48.

Garren, D., W. C. Schrader, D. Keen, and J. Fredericks. 2006. Regional fisheries management investigations, upper Snake region. Idaho Department of Fish and Game, Federal Aid in Fish Restoration, Program F-71-R-28, 2003 Annual Performance Report, Boise.

Gillilan, D. M., and T. C. Brown. 1997. Instream flow protection. Island Press, Washington, D.C.

Gleick, P. H. 2010. Roadmap for sustainable water resources in southwestern North America. Proceedings of the National Academy of Sciences of the United States of America 107:21300–21305.

Grafton, R. Q., J. Williams, C. J. Perry, F. Molle, C. Ringler, P. Steduto, B. Udall, S. A. Wheeler, Y. Wang, D. Garrick, and R. G. Allen. 2018. The paradox of irrigation efficiency: higher efficiency rarely reduces water consumption. Science 361:748–750.

Gregory, J. S. 2000. Winter fisheries research and habitat improvements on the Henrys Fork of the Snake River. Intermountain Journal of Sciences 6:232–248.

Gregory, J. S., A. M. Emery, and J. N. DeRito. 2011. Survival and movement of adult rainbow trout during winter and spring in the Henrys Fork of the Snake River. Intermountain Journal of Sciences 17:1–13.

Gresswell, R. E. 2011. Biology, status and management of the Yellowstone Cutthroat Trout. North American Journal of Fisheries Management 31:782–812.

Gresswell, R. E., W. L. Liss, and G. L. Larson. 1994. Life-history organization of Yellowstone Cutthroat Trout (*Oncorhynchus clarki bouvieri*) in Yellowstone Lake. Canadian Journal of Fisheries and Aquatic Sciences 51(Supplement 1):298–309.

Gresswell, R. E., W. J. Liss, G. L. Larson, and P. J. Bartlein. 1997. Influence of basin-scale physical variables on life history characteristics of Cutthroat Trout in Yellowstone Lake. North American Journal of Fisheries Management 17:1046–1064.

Griffin, C. B. 1999. Watershed councils: An emerging form of public participation in natural resource management. Journal of the American Water Resources Association 35:505–518.

Griffith, J. S., and R. W. Smith. 1995. Failure of submersed macrophytes to provide cover for Rainbow Trout throughout their first winter in the Henrys Fork of the Snake River. North American Journal of Fisheries Management 15:42–48.

Grunder, S. A., T. J. McArthur, S. Clark, and V. K. Moore. 2008. 2003 Economic survey report. Idaho Department of Fish and Game, Report 08-129, Boise.

Gunn, T., and W. Gunn. 2013. 50 Best tailwaters to fly fish. Stonefly Press, West Palm Beach, Florida.

Hamlet, A. F., P. W. Mote, M. P. Clark, and D. P. Lettenmaier. 2005. Effects of temperature and precipitation variability on snowpack trends in the western United States. Journal of Climate 18:4545–4561.

Hauer, F. R., and M. S. Lorang. 2004. River regulation, decline of ecological resources, and potential for restoration in a semi-arid lands river in the western USA. Aquatic Sciences 66:388–401.

Hesse, L. W., and G. E. Mestl. 1993. An alternative hydrograph for the Missouri River based on the precontrol condition. North American Journal of Fisheries Management 13:360–366.

Hoekema, D. J., and V. Sridhar. 2011. Relating climatic attributes and water resources allocation: a study using surface water supply and soil moisture indices in the Snake River basin, Idaho. Water Resources Research [online serial] 47:W07536.

Hoekema, D. J., and V. Sridhar. 2013. A system dynamics model for conjunctive management of water resources in the Snake River basin. Journal of the American Water Resources Association 49:1327–1350.

Houck, O. 1981. Judicial review under the Fish and Wildlife Coordination Act: a plaintiff's guide to litigation. Environmental Law Reporter 11:50043–50050.

Howe, C. W., and C. Goemans. 2003. Water transfers and their impacts: lessons from three Colorado water markets. Journal of the American Water Resources Association 39:1055–1065.

Huffaker, R. 2008. Conservation potential of agricultural water conservation subsidies. Water Resources Research [online serial] 44(7):W00E01.

Hughes, R. M., and R. F. Noss. 1992. Biological diversity and biological integrity: current concerns for lakes and streams. Fisheries 17(3):11–19.

Huntington, T. G. 2006. Evidence for intensification of the global water cycle: review and synthesis. Journal of Hydrology 319:83–95.

Hydropower Reform Coalition. 2017. Federal Power Act. Available: www.hydroreform.org/policy/fpa. (March 2019).

Hydropower Reform Coalition. 2018. Citizen toolkit for effective participation, appendix C: settlement agreements. Available: www.hydroreform.org/hydroguide/hydropower-licensing/citizen-toolkit-for-effective-participation/Appendix_C. (March 2019).

IDEQ (Idaho Department of Environmental Quality). 2003. Teton River subbasin assessment and total maximum daily load. IDEQ, Idaho Falls Regional Office, Idaho Falls.

IWRB (Idaho Water Resource Board). 2009. Eastern Snake Plain Aquifer Comprehensive Aquifer Management Plan. IWRB, Boise.

IWRB (Idaho Water Resource Board). 2012. Idaho State Water Plan. IWRB, Boise.

Jensen, M. E. 2007. Beyond irrigation efficiency. Irrigation Science 25:233–245.

Johnson, G. S., B. A. Contor, and D. M. Cosgrove. 2008. Efficient and practical approaches to groundwater right transfers under the prior appropriation doctrine and the Snake River example. Journal of the American Water Resources Association 44:27–36.

Johnson, G. S., W. H. Sullivan, D. M. Cosgrove, and R. D. Schmidt. 1999. Recharge of the Snake River Plain Aquifer: transitioning from incidental to managed. Journal of the American Water Resources Association 35:123–131.

Joint Committee. 2005. Henry's Fork drought management plan. Fremont-Madison Irrigation District, St. Anthony, Idaho.

Jones, J. 2016. A little dam problem. Caxton Press, Caldwell, Idaho.

Keenan, S. P., R. S. Krannich, and M. S. Walker. 1999. Public perceptions of water transfers and markets: describing differences in water use. Communities Society and Natural Resources 12:279–292.

Kendy, E., and J. D. Bredehoeft. 2006. Transient effects of groundwater pumping and surface-water-irrigation returns on streamflow. Water Resources Research [online serial] 42:W08415.

Kenney, D. S. 1999. Historical and sociopolitical context of the western watersheds movement. Journal of the American Water Resources Association 35:493–503.

Kilburn, C. 1964. Ground water in the upper

part of Teton Valley Teton counties, Idaho and Wyoming. U.S. Geological Survey Water-Supply Paper 1789.

Knapp, K. C., M. Weinberg, R. Howitt, and J. F. Posnikoff. 2003. Water transfers, agriculture, and groundwater management: a dynamic economic analysis. Journal of Environmental Management 67:291–301.

Kovach, R. P., C. C. Muhlfeld, R. Al-Chokhacy, J. B. Dunham, B. H. Letcher, and J. L. Kershner. 2016. Impacts of climatic variation on trout: a global synthesis and path forward. Reviews in Fish Biology and Fisheries 26:135–151.

Kunkel, M. L., and J. L. Pierce. 2010. Reconstructing snowmelt in Idaho's watershed using historic streamflow records. Climatic Change 98:155–176.

Kuzniar, Z. J., R. W. Van Kirk, and E. B. Snyder. 2017. Seasonal effects of macrophyte growth on Rainbow Trout habitat availability and selection in a low-gradient, groundwater-dominated river. Ecology of Freshwater Fish 26:653–665.

Laatsch, J., R. Van Kirk, C. Morrisett, K. Manishin, and J. DeRito. 2017. Angler perception of fishing experience in a highly technical catch-and-release fishery: how closely does perception align with biological reality? Pages 47–53 *in* R. F. Carline, and C. LoSapio, editors. Wild Trout XII: Science, politics, and wild trout management: who's driving and where are we going? Wild Trout Symposium, Bozeman, Montana.

Lankford, B. 2006. Localising irrigation efficiency. Irrigation and Drainage 55:345–362.

Lattenmaier, D. P., A. W. Wood, R. N. Palmer, E. F. Wood, and E. Z. Stakhiv. 1999. Water resources implications of global warming: a U.S. regional perspective. Climate Change 43:537–579.

Lawson, M. 2012. Fly-fishing guide to the Henry's Fork. Stackpole Books, Mechanicsburg, Pennsylvania.

Leppi, J. S., T. H. DeLuca, S. W. Harrar, and S. W. Running. 2012. Impact of climate change on August stream discharge in the central-Rocky Mountains. Climatic Change 112:997–1014.

Lettenmaier, D. P., A. W. Wood, R. N. Palmer, E. F. Wood, and E. Z. Stakhiv. 1999. Water resources implications of global warming: a U.S. regional perspective. Global Change 43:537–579.

Liegel, L. H. 2011. Landscapes in transition: Exploring the intersections between land use planning and water management in Henry's Fork watershed, Idaho. Master's thesis. Humboldt State University, Arcata, California.

Ligon, F. K., W. E. Dietrich, and W. J. Trush. 1995. Downstream ecological effects of dams: a geomorphic perspective. BioScience 45:183–192.

Lloyd, N., G. Quinn, M. Thoms, A. Arthington, B. Gawne, P. Humphries, and K. Walker. 2004. Does flow modification cause geomorphological and ecological response in a rivers? A literature review from an Australian perspective. Cooperative Research Center for Freshwater Ecology, Technical Report 1/2004, Canberra, Australia.

Loomis, J. 2006. Use of survey data to estimate economic value and regional economic effects of fishery improvements. North American Journal of Fisheries Management 26:301–307.

Luce, C. H., and Z. A. Holden. 2009. Declining annual streamflow distributions in the Pacific Northwest United States, 1948–2006. Geophysical Research Letters [online serial] 36:L16401.

Maddock, T. S. and W. G. Hines. 1995. Meeting future public water supply needs: a southwest perspective. Water Resources Bulletin 31:317–329.

Magilligan, F. J., and K. H. Nislow. 2001. Long-term changes in regional hydrologic regime following impoundment in a humid-climate watershed. Journal of the American Water Resources Association 37:1551–1569.

Magilligan, F. J., K. Nislow, and B. E. Graber. 2003. Scale-independent assessment of discharge reduction and riparian disconnectivity following flow regulation by dams. Geology 31:569–572.

Marcarelli, A. M., R. W. Van Kirk, and C. V. Baxter. 2010. Predicting effects of hydrologic alteration and climate change on ecosystem metabolism in a western U.S. river. Ecological Applications 20:2081–2088.

Mayer, T. D., and S. W. Naman. 2011. Streamflow response to climate as influenced by geology and elevation. Journal of the American Water Resources Association 47:724–738.

McCormick, Z. 1994. Institutional barriers to water marketing in the West. Water Resources Bulletin 30:953–961.

McDaniel, J. 2012. Fly fishing the Harriman Ranch of the Henry's Fork of the Snake River: lessons learned and friends made sight fishing to selective trout. Whitefish Press, Cincinnati, Ohio.

Merigliano, M. F. 1996. Ecology and management of the South Fork Snake River cottonwood forest. U.S. Bureau of Land Management, Idaho State Office Technical Bulletin 96-9, Boise.

Meyer, K. A. 1995. Experimental evaluation of habitat use and survival of Rainbow Trout during their first winter in the Henry's Fork of the Snake River, Idaho. Master's thesis. Idaho State University, Pocatello.

Meyer, K. A., and J. S. Griffith. 1997. First-winter survival of Rainbow Trout and Brook Trout in the Henrys Fork of the Snake River, Idaho. Canadian Journal of Zoology 75:59–63.

Meyer, K. A., E. I. Larson, C. L. Sullivan, and B. High. 2014. Trends in the distribution and abundance of Yellowstone Cutthroat Trout and nonnative trout in Idaho. Journal of Fish and Wildlife Management 5:227–242.

Meyer, K. A., D. J. Schill, J. A. Lamansky Jr. M. R. Campbell, and C. C. Kozfkay. 2006. Status of Yellowstone Cutthroat Trout in Idaho. Transactions of the American Fisheries Society 135:1329–1347.

Miller, W. P., and T. C. Piechota. 2011. Trends in western U.S. snowpack and related upper Colorado River basin streamflow. Journal of the American Water Resources Association 47:1197–1210.

Mitro, M. G., A. V. Zale, and B. A. Rich. 2003. The relation between age-0 Rainbow Trout (*Oncorhynchus mykiss*) abundance and winter discharge in a regulated river. Canadian Journal of Fisheries and Aquatic Sciences 60:135–139.

Molles, M. C., Jr., C. S. Crawford, L. M. Ellis, H. M. Valett, and C. N. Dahm. 1998. Managed flooding for riparian ecosystem restoration. BioScience 48:749–756.

Moore, J. N., A. S. Arrigoni, and A. C. Wilcox. 2012. Impacts of dams on flow regimes in three headwater subbasins of the Columbia River basin, United States. Journal of the American Water Resources Association 48:925–938.

Moore, J. N., J. T. Harper, and M. C. Greenwood. 2007. Significance of trends toward earlier snowmelt runoff, Columbia and Missouri basin headwaters, western United States. Geophysical Research Letters 34:L16402.

Morrisett, C. 2016. Buffalo River fish ladder 2006–2016 comprehensive report. Henry's Fork Foundation, Ashton, Idaho.

Mote, P. W. 2006. Climate-driven variability and trends in mountain snowpack in western North America. Journal of Climate 19:6209–6220.

Mote, P. W., S. Li, D. P. Lettenmaier, M. Xiao, and R. Engel. 2018. Dramatic declines in snowpack in the western U.S. npj Climate and Atmospheric Science [online serial] 1:2.

Naik, P. K., and D. A. Jay. 2011. Human and climate impacts on Columbia River hydrology and salmonids. River Research and Applications 27:1270–1276.

National Environmental Policy Act. 1970. U.S. code, volume 42, sections 4321–4347.

NRCS (Natural Resources Conservation Service). 2017. Effects of irrigation efficiency improvements at multiple scales [Webinar training]. Available: https://usdanrcs.adobeconnect.com/penetilowb16/.

Nicklin Earth and Water. 2003. Ground-water model for the upper Teton watershed. Project completion report to Cascade Earth Sciences, Inc., Pocatello, Idaho.

Nilsson, C., and K. Berggren. 2000. Alterations of riparian ecosystems caused by river regulation. BioScience 50:783–792.

Nislow, K. H., F. J. Magilligan, H. Fassnacht, D. Bechtel, and A. Ruesink. 2002. Effects of dam impoundments on the flood regime of natural floodplain communities in the upper Connecticut River. Journal of the American Water Resources Association 36:1533–1548.

Ochoa, C. G., A. G. Fernald, S. J. Guldan, and M. K. Shukla. 2007. Deep percolation and its effects on shallow groundwater level rise following flood irrigation. Transactions of the American Society of Agricultural and Biological Engineers 50:73–81.

Oldemeyer, B. J. Flinders, C. Morrisett, and R. Van Kirk. 2017. Long-term effectiveness of flow management and fish passage on the Henry's Fork Rainbow Trout population. Pages 125–131 *in* R. F. Carline, and C. LoSapio, editors. Wild Trout XII: Science, politics, and wild trout management: who's driving and where are we going? Wild Trout Symposium, Bozeman, Montana.

Olenichak, T. 2015. Concepts, practices, and procedures used to distribute water within Water District #1 upper Snake River basin Idaho. Idaho Department of Water Resources, Water District 01, Idaho Falls.

Palmer, T. 1991. The Snake River: window to the West. Island Press, Washington, D.C.

Patten, D. T. 1998. Riparian ecosystems of semi-arid North America: diversity and human impacts. Wetlands 18:498–512.

Payne, M. T., M. G. Smith, and C. J. Landry. 2014. Price determination and efficiency in the market for South Platte basin ditch company shares. Journal of the American Water Resources Association 50:1488–1500.

Peck, D. E., and J. R. Lovvorn. 2001. The importance of flood irrigation in water supply to wetlands in the Laramie basin, Wyoming, USA. Wetlands 21:370–378.

Peterson, K. 2011. An analytical model of surface water/groundwater interactions in a western watershed experiencing changes to water and land use. Master's thesis. Humboldt State University, Arcata. California.

Petts, G. E. 2009. Instream flow science for sustainable river management. Journal of the American Water Resources Association 45:1071–1086.

Planert, M., and J.S. Williams. 1995. Ground water atlas of the United States: California, Nevada. U.S. Geological Survey, Hydrologic Atlas HA 7301-B, Reston, Virginia.

Poff, N. L., J. D. Allen, M. B. Bain, J. M. Karr, K. L. Prestegaard, B. D. Richter, R. E. Sparks, and J. C. Stromberg. The natural flow regime a paradigm for conservation and restoration. BioScience 47:769–784, 1997.

Poff, L. N., and J. V. Ward. 1989. Implications of streamflow variability and predictability for lotic community structure: a regional analysis of stream flow patterns. Canadian Journal of Fisheries and Aquatic Sciences 46:1805–1817.

Poff, L. N., and J. V. Ward. 1990. Physical habitat template of lotic systems: recovery in the context of historical pattern of spatiotemporal heterogeneity. Environmental Management 14:629–645.

Poff, N. L., and J. K. Zimmerman. 2010. Ecological responses to altered flow regimes: a literature review to inform the science and management of environmental flows. Freshwater Biology 55:194–205.

Qureshi, M. E., K. Schwabe, J. Connor, and M. Kirby. 2010. Environmental water incentive policy and return flows. Water Resources Research 46:W04517.

Railsback, S. F. 2016. Why it is time to put PHABSIM out to pasture. Fisheries 41:721–725.

Regonda, S. K., B. Rajagopalan, M. Clark, and J. Pitlick. 2005. Seasonal cycle shifts in hydroclimatology over the western United States. Journal of Climate 18:372–384.

Reisner, M. 1993. Cadillac Desert: the American West and its disappearing water. Penguin Books, New York.

Reisner, M., and S. Bates. 1990. Overtapped oasis: reform or revolution for western water. Island Press, Washington, D.C.

Richter, B. D., J. B. Baumgartner, D. P. Braun, and J. Powell. 1998. A spatial assessment of hydrologic alteration within a river network. Regulated Rivers Research and Management 14:329–340.

Richter, B. D., J. B. Baumgartner, J. Powell, and D. P. Braun. 1996. A method for assessing hydrologic alteration within ecosystems. Conservation Biology 10:1163–1174.

Richter, B. D., J. B. Baumgartner, R. Wigington, and D. P. Braun. 1997a. How much water does a river need? Freshwater Biology 37:231–249.

Richter, B. D., D. P. Braun, M. A. Mendelson, and L. L. Master. 1997b. Threats to imperiled freshwater fauna. Conservation Biology 11:1081–1093.

Riebsame, W. E., J. Wescoat, and P. Morrisette. 1997. Western land use trends and policy: implications for water resources. Report to the Western Water Policy Review Advisory Commission, National Technical Information Service, Springfield, Virginia.

Robson, S. G., and E. R. Banta. 1995. Ground water atlas of the United States: Arizona, Colorado, New Mexico, Utah. U.S. Geological Survey, Hydrologic Atlas HA 7301-C, Reston, Virginia.

Rood, S. B., C. R. Gourley, E. A. Ammon, L. G. Heki, J. R. Klotz, M. L. Morrision, D. Mosley, G. G. Scoppettone, S. Swanson, and P. L. Wagner. 2003. Flows for floodplain forests: a successful riparian restoration. BioScience 53:647–656.

Rood, S. B., J. Pan, K. M. Gill, C. G. Franks, G. M. Samuelson, and A. Shepherd. 2007. Declining summer flows of Rocky Mountain rivers: changing seasonal hydrology and probable impacts on floodplain forests. Journal of Hydrology 349:397–410.

Sabo, J. L., T. Sinha, L. C. Bowling, G. W. Schoups, W. W. Wallender, M. E. Campana, K. A. Cherkauer, P. L. Fuller, W. L. Graf, J. W. Hopmans, J. S. Kominoski, C. Taylor, S. W. Trimble, R. H. Webb, and E. E. Wohl. 2010. Reclaiming freshwater sustainability in the Cadillac Desert. Proceedings of the National Academy of Sciences of the United States of America 107:21263–21270.

Schaible, G. D., C. S. Kim, and M. P. Aillery. 2010. Dynamic adjustment of irrigation technology/water management in western U.S. agriculture: toward a more sustainable future. Canadian Journal of Agricultural Economics 58:433–461.

Scheidegger, K. J., and M. B. Bain. 1995. Larval fish distribution and microhabitat use in free-flowing and regulated rivers. Copeia 1995:125–135.

Shallat, T. 2000. Ecology in policymaking: water and the restoration of America's Snake River plain. Water Policy 2:327–341.

Singer, M. B. 2007. The influence of major dams on hydrology through the drainage network of the Sacramento River basin, California. River Research and Applications 23:55–72.

Slaughter, R. A., and J. D. Wiener. 2007. Water, adaptation, and property rights on the Snake and Klamath rivers. Journal of the American Water Resources Association 43:308–321.

Smith, R. W., and J. S. Griffith. 1994. Survival of Rainbow Trout during their first winter in the Henrys fork of the Snake River, Idaho. Transactions of the American Fisheries Society 123:747–756.

Stewart, I. T., D. R. Cayan, and M. D. Dettinger. 2005. Changes toward earlier streamflow timing across western North America. Journal of Climate 18:1136–1155.

Strange, E. M., K. D. Fausch, and A. P. Covich. 1999. Sustaining ecosystem services and processes in human-dominated watersheds: biohydrology and ecosystem processes in the South Platte River basin. Environmental Management 24:39–54.

Stromberg, J. C. 2001. Restoration of riparian vegetation in the south-western United States: importance of flow regimes and fluvial dynamism. Journal of Arid Environments 49:17–34.

Tennant, D. L. 1976. Instream flow regimens for fish, wildlife, recreation and related environmental resources. Fisheries 1(4):6–10.

TNC (The Nature Conservancy). 2014. Upper Snake River basin water stewardship assessment. TNC, Arlington, Virginia

Thompson, S. A. 1988. Patterns and trends in irrigation efficiency. Water Resource Bulletin 24:57–63.

Toll, M. 2011. Reimagining western water law: time-limited permits based on a comprehensive beneficial use doctrine. Colorado Law Review 82:595–637.

Travnichek, V. H., M. B. Bain, and M. J. Maceina. 1995. Recovery of a warmwater fish assemblage after the initiation of a minimum-flow release downstream from a hydroelectric dam. Transactions of the American Fisheries Society 124:836–844.

Tuthill, D. R., P. J. Rassier, and H. N. Anderson. 2013. Conjunctive management in Idaho: public-private partnerships & conjunctive management of surface and ground water. The Water Report 108:1–11, Envirotech Publications, Eugene, Oregon.

USBR (U.S. Bureau of Reclamation). 2012a. Henrys Fork Basin Study water needs assessment. USBR, Pacific Northwest Region, Technical Report PN-HFS-001, Boise, Idaho.

USBR (U.S. Bureau of Reclamation). 2012b. Henrys Fork Basin Study conservation alternatives. USBR, Pacific Northwest Region, Technical Report PN-HFS-006, Boise, Idaho.

USBR (U.S. Bureau of Reclamation). 2012c. Colorado River Basin Water Supply and Demand Study. USBR, Lower Colorado Regional Office, Boulder City, Nevada.

USBR (U.S. Bureau of Reclamation). 2014. Biological assessment for Bureau of Reclamation operations and maintenance in the Snake River basin above Brownlee Reservoir on the Snake River physa snail (*Haitia* [*Physa*] *natricina*). USBR, Snake River Area Office, Boise, Idaho.

USBR (U.S. Bureau of Reclamation). 2015. Henrys Fork Basin Study final report. USBR, Snake River Area Office, Boise, Idaho.

USBR (U.S. Bureau of Reclamation). 2017. 2017 salmon flow augmentation and program and other activities associated with the NOAA Fisheries Service 2008 Biological Opinion and Incidental Take Statement for operations and maintenance of Bureau of Reclamation projects in the Snake River basin above Brownlee Reservoir. USBR, Pacific Northwest Region, annual progress report, Boise, Idaho.

USBR (U.S. Bureau of Reclamation) and WDE (Washington Department of Ecology). 2012. Final Programmatic Environmental Impact Statement Yakima River Basin Integrated Water Resource Management Plan Benton, Kittitas, Klickitat, and Yakima counties, Washington. USBR, Columbia-Cascades Area Office and WDE, Central Region Office, Yakima, Washington.

Van Kirk, R., and A. Jenkins. 2005. Hydrologic alteration in the upper Teton watershed and its implications for Cutthroat Trout restoration. Project completion report for Friends of the Teton River, Driggs, Idaho.

Van Kirk, R. W., and S. W. Naman. 2008. Relative effects of climate and water use on base-flow trends in the lower Klamath basin. Journal of the American Water Resources Association 44:1035–1052.

Van Kirk, R. W., and M. Gamblin, M. 2000. History of fisheries management in the upper Henry's Fork watershed. Intermountain Journal of Sciences 6:263–284.

Van Kirk, R. W., and C. B. Griffin. 1997. Building a collaborative process for restoration: Henrys Fork of Idaho and Wyoming. Pages 253–276 *in* J. E. Williams, M. P. Dombeck, W. Elmore, and C. A. Wood, editors. Watershed restoration: principles and practices. American Fisheries Society, Bethesda, Maryland.

Veiluva, M. 1981. The Fish and Wildlife Coordination Act in environmental litigation. Ecology Law Quarterly 9:489–517.

Venn, B. J., D. W. Johnson, and L. O. Pochop. 2004. Hydrologic impacts due to changes in conveyance and conversion from flood to sprinkler irrigation practices. Journal of Irrigation and Drainage Engineering-ASCE 130:192–200.

Vonde, A. Y., C. M. Bromley, M. M. Carter, A. L. Courtney, S. E. Hamlin-Nygard, H. A. Hensley, S. J. Kilminster-Hadley, M. C. Orr, D. I. Stanish, and C. J. Strong. 2016. Understanding the Snake River basin adjudication. Idaho Law Review 52:53–222.

Ward, F. A., and M. Pulido-Velázquez. 2008. Water conservation in irrigation can increase water use. Proceedings of the National Academy of Sciences of the United States of America 105:18215–18220.

WDE (Washington Department of Ecology. 2016. 2015 Drought response – summary report. WDE, Olympia.

WDE (Washington Department of Ecology) 2018. Cost Estimate and Financing Plan Yakima River Basin Integrated Water

Resource Management Plan. WDE, Olympia.

WD01 (Water District 01). 2018. Water District 1 rental pool rules. Idaho Department of Water Resources, WD01, Idaho Falls, Idaho.

Weber, E. P. 2000. Cooperative watershed management and research: the case of the Henry's Fork Watershed Council. Intermountain Journal of Sciences 6:293–311.

Whitehead, R. L. 1994. Ground water atlas of the United States: Idaho, Oregon, Washington. U.S. Geological Survey, Hydrologic Atlas HA 7301-H, Reston, Virginia.

Whitehead, R. L. 1996. Ground water atlas of the United States: Montana, North Dakota, South Dakota, Wyoming. U.S. Geological Survey, Hydrologic Atlas HA 7301-I, Reston, Virginia.

Wiener, J. D., K. A. Dwire, S. K. Skagen, R. R. Crifasi, and D. Yates. 2008. Riparian ecosystem consequences of water redistribution along the Colorado Front Range. Water Resources Impact 10(3):18–21.

Wildhaber, M. L., V. M. Tabor, J. E. Whitaker, A. L. Allert, D. W. Mulhern, P. J. Lamberson, and P. L. Powell. 2000. Ictalurid populations in relation to the presence of a main-stem reservoir in a Midwestern warmwater stream with emphasis on the threatened Neosho Madtom. Transactions of the American Fisheries Society 129:1264–1280.

Wildman, R. A., Jr., and N. A. Forde. 2012. Management of water shortage in the Colorado River basin: evaluating current policy and the viability of interstate water trading. Journal of the American Water Resources Association 48:411–422.

Williams, B. K., and E. D. Brown. 2012. Adaptive management: the U.S. Department of the Interior applications guide. U.S. Department of the Interior, Washington, D.C.

Zaladis, G., X. Dimitriadis, A. Antonopolous, and A. Gerakis. 1997. Estimation of a network irrigation efficiency to cope with reduced water supply. Irrigation and Drainage Systems 11:337–345.

American Fisheries Society Symposium 91:565–588, 2019

The Weber River Partnership: How Fish Gained Relevance through a Recently Formed Watershed Group

Paul D. Thompson*
Utah Department of Natural Resources
1594 West North Temple, Suite 3310, Salt Lake City, Utah 84114, USA

Paul C. Burnett
Trout Unlimited
5279 South 150 East, Ogden, Utah 84405, USA

Abstract—The Weber River is primarily known as a blue-ribbon Brown Trout *Salmo trutta* fishery; however, this river also supports populations of two jeopardized fishes, Bonneville Cutthroat Trout *Oncorhynchus clarkii utah* and Bluehead Sucker *Catostomus discobolus*. At least one population of Bonneville Cutthroat Trout in the Weber River provides an important and popular local fishery and expresses a fluvial life history where main-stem individuals grow large (300–500 mm total length) and migrate into small tributaries for spawning. Bluehead Suckers currently occur in the main stem of the Weber River, where they travel distances of 20 km between spawning and overwintering habitats. The habitat for both species has been fragmented by more than 300 barriers composed of irrigation diversions, road crossings, and utility stream crossings. Beginning in 2010, the Utah Division of Wildlife Resources and Trout Unlimited began undertaking barrier removal for native fish as a priority conservation action. Initially, the effort to reconnect habitat was slow and the lack of relationships with stakeholders such as water users, government agencies, private landowners, and utility companies was hampering progress with habitat reconnection. New barriers were being built at a faster rate than barriers were being removed. To build these relationships, a steering committee was formed to secure a small grant, hire a consulting firm, organize stakeholder meetings to identify broad stakeholder priorities, and write a watershed plan that ultimately identified Bonneville Cutthroat Trout and Bluehead Sucker as priority conservation targets. The watershed plan and subsequent stakeholder meetings developed a framework for the Weber River Partnership. The partnership holds an annual symposium where larger watershed issues are discussed. The symposium also provides a platform where all stakeholders can understand the activities occurring throughout the watershed and where there are opportunities to collaborate. The Weber River Partnership has provided a forum where fisheries managers have told the story of Bonneville Cutthroat Trout and Bluehead Sucker and the importance of habitat connectivity. Through collaborative relationships with nontraditional partners, the relevance of fisheries in the Weber River has been realized. Further relevance in the watershed is evidenced by the development of a wide range of on-the-ground actions. Fish passage has been re-established at three main-stem and four tributary

* Corresponding author: paulthompson@utah.gov

barriers. Additional projects are in various stages of development, including a large fish ladder that will be built as part of a Federal Energy Regulatory Commission relicensing project at a small hydroelectric dam, and we continue to be contacted by water users with interest in developing irrigation diversion reconstruction projects that incorporate fish passage.

Introduction

Many native fishes have steadily declined in the 20th century across western North America in both distribution and abundance (Williams et al. 1989; Moyle and Leidy 1992). Most declines have been attributed to the construction of dams and diversions, degradation of habitat, and the introduction of nonnative fish species. Dams have changed environmental conditions by altering the natural hydrograph and stream water temperatures, preventing the movement of fishes to different and necessary habitats, and homogenizing instream habitat (Vanicek et al. 1970; Martinez et al. 1994; Gido et al. 2013). Small irrigation diversion dams are numerous throughout many western drainages and not only remove water, but also fragment fish populations and strand fish in canals and agriculture fields (McAda 1977; Mueller and Marsh 2002; Carlson 2006). In addition, barriers can block fish from accessing preferred environments in other reaches and critical spawning areas (Compton et al. 2008). Many native fish populations cannot persist in these fragmented habitats, and the ones that do may go through genetic bottlenecks over time, which further decreases fitness (Bessert and Orti 2008).

Weber River

The Weber River, a tributary to Great Salt Lake, occurs in a 6,413-km^2 watershed in north-central Utah (Figure 1). It originates near Reids Peak (3,569 m above mean sea level) and flows 201 km northwest to Great Salt Lake (1,280 m above mean sea level). The riparian community is typical of an interior stream in the western United States and is composed primarily of willows *Salix* spp., narrowleaf cottonwood *Populus angustifolia*, and Fremont's cottonwood *P. fremontii*. The Weber River drainage is composed primarily of private lands that are utilized for agricultural production, including livestock grazing and farming; the headwater tributaries are U.S. Forest Service lands. The natural hydrograph is dominated by snowmelt runoff, with the highest annual discharge occurring during April–late June and periods of base flow during July–March. Water resources in the Weber River are heavily regulated by the management of seven large irrigation or flood-control reservoirs that have a dam spillway height of ≥19.5 m. The seven reservoirs dictate base flow within segments of the main-stem Weber River, and base flow can fluctuate greatly depending on the location within the watershed and the time of year (Webber et al. 2012).

Fish habitat in the Weber River has been heavily degraded and highly fragmented. Land management actions and direct stream channel modifications resulting from flood control, irrigation diversions, and transportation infrastructure have simplified almost all of the 201 km of a historically complex river system (Portugal et al. 2016). The construction of the interstate highway system in the 1960s left a significant mark on the Weber River by straightening and reducing channel length in Henefer Valley from 15.6 to 14.1 km, a permanent loss of 1,423 m of habitat (Barton and Winger 1973). Also lost with stream length was habitat complexity; formerly complex channel planforms were replaced by a straight, ditch-like channel. A recent basinwide habitat assessment indicated that approximately 52% of the length of all streams in the Weber River watershed were in moderate to poor condition, and

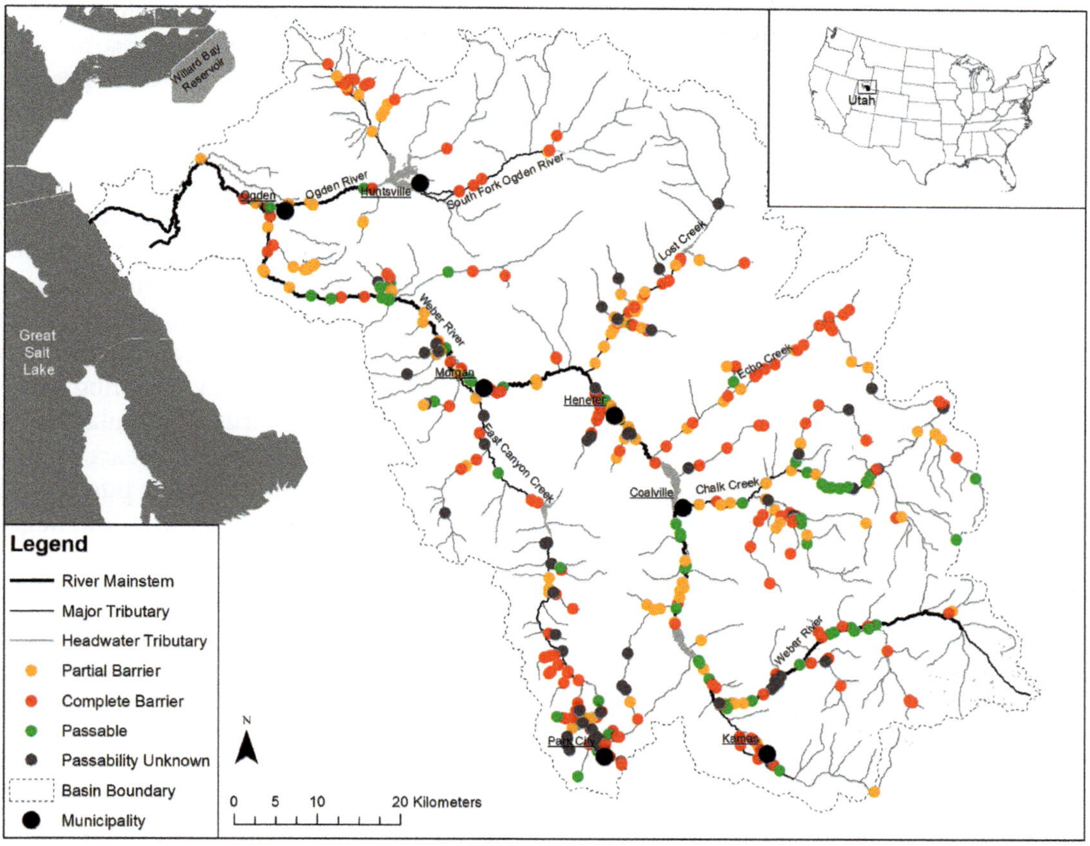

Figure 1. Location of the Weber River, Utah, showing barriers to fish movement within the Weber River watershed.

the entire length of the Weber River and its major tributaries were rated moderate to poor (Portugal et al. 2016). The concurrent construction of water diversions, dams, small impoundments, and road crossings have resulted in 316 fish migration barriers throughout the watershed (Figure 1), fragmenting the habitat that remains. Consequently, two native fishes, the Bonneville Cutthroat Trout (BCT) *Oncorhynchus clarkii Utah*, the only native trout to the Weber River, and the Bluehead Sucker *Catostomus discobolus*, continue to be seriously imperiled by these habitat impacts because they cannot move between habitats and the habitat they occupy is degraded. Nevertheless, the resilience of these native species continues to elevate them in priority and indicates a continued need to develop aquatic conservation projects in the Weber River.

In addition to physical impacts to the Weber River and its tributaries, the future of water is a continuing and increasing concern, particularly when viewed through the lens of climate change and increasing human demands. Climate models continue to predict changes in the snow-to-rain ratio in watersheds like the Weber River throughout the western United States with an earlier runoff season and more precipitation falling as rain during the spring (Christensen et al. 2004). But more importantly, additional water demands in the spring and fall and hotter summer conditions may lead to dramatic increases in water use (Udall and Overpeck 2017). Coupled with predicted population growth in northern Utah, the proposition of providing water for a broadening diversity of needs in the Weber River watershed, including agricultural, municipal, and hydropower uses, as well as values supported by system water, such as recreational fisheries, native aquatic species conservation and water recreation, is becoming progressively more challenging. For example, the Ogden–Clearfield metropolitan area was the 16th fastest growing area from 2000 to 2010 according to U.S. census figures, and the watershed is projected to grow in population from 500,000 to 1.8 million by 2020. The Provo–Orem metropolitan area was identified as the fastest growing area in the same census, growing at a 43.5% rate with no change in future trajectory. Water from the upper Weber River is diverted out of the watershed into the Provo River to support additional projected growth along the Wasatch Front. Population growth in Utah is confounded by the fact that all of the available water rights in the Weber River watershed are currently appropriated.

Despite the extensive degradation in the past and threats to the future, the Weber River and its tributaries provide habitats, opportunities, and resources that are essential to the local communities of northern Utah and form the backbone of our shared priorities and values in this watershed. The Weber River supports one of the most popular Brown Trout *Salmo trutta* stream fisheries in the state (Lilieholm et al. 2017), and combined with important native fish populations like the Bluehead Sucker, many partners have prioritized the restoration actions in the Weber River for decades.

Bluehead Sucker

Bluehead Suckers have a relatively wide distribution and occur in the Colorado and Green, upper Snake, Bear, and Weber rivers (Sublette et al. 1990; Sigler and Sigler 1996). Like most native fishes in the western United States, the Bluehead Sucker has experienced a decrease in abundance and is now restricted to approximately 50% of its historical range (Bezzerides and Bestgen 2002). Where Bluehead Suckers still remain, little is known regarding population size or basic vital rates, especially outside the Colorado and Green rivers, likely because of its cryptic nature and lack of sportfish status. Bluehead Suckers still persist in the upper Snake River in Idaho, although their range appears restricted (Meyer et al. 2012) and in the Bear and Snake rivers in Wyoming (Carlson 2006), but the status of populations is unknown and distribution appears spotty. Andreasen (1973) reported that portions of the Bear, Ogden, and Weber River drainages in the Bonneville Basin were occupied by Bluehead Suckers. Extensive surveys from 2004 to 2017 in Utah discovered small, remnant populations in the Snake River drainage and the Weber River; however, no populations have been observed in Utah's portion of the Bear River.

Early taxonomists separated Bluehead Suckers that occurred in the Colorado and Green rivers from those occurring in the Snake, Bear, and Weber rivers. The latter group was considered a distinct species, the Green Sucker *Pantosteus virescens* (Snyder 1924; Sigler and Miller 1963). In contemporary times, these fish have been considered Bluehead Suckers (Sigler and Sigler 1996). Recent genetic analyses of the Bluehead Sucker complex has demonstrated differences between populations occurring in the Colorado and Green rivers and those occurring in the Snake, Bear, and Weber rivers, once again bringing to question whether these fish are unique and should be considered a distinct species (Hopken et al. 2013; Unmack et al. 2014).

Webber et al. (2012) reported on two recently discovered populations of Bluehead Sucker in the Weber River, and at that time, population declines had already been noted in the furthest upstream population, likely due to a variety of factors, some irreversible. Coldwater dam releases from Rockport Reservoir have restricted this population of Bluehead Sucker to the lowest 1 km of a 16-km reach between Rockport and Echo reservoirs. Emerging fry drift a short distance into the downstream reservoir, which is dominated by nonnative fish. As this population was monitored more recently, additional declines were observed and recruitment had not been documented for 10 years. These declines triggered the Utah Division of Wildlife Resources (UDWR) and the Utah Three Species Team to move this population further downstream to a reach of the Weber River where it may persist and contribute to a viable spawning population. Webber et al. (2012) described the population of Bluehead Sucker found lower in the Weber River as small, but stable because recent recruitment had been observed. More recent monitoring within the lower population has demonstrated the same downward trend in population abundance (Figure 2) with minimal recruitment (Figure 3). The population of Bluehead Sucker in the Weber River as a whole has experienced a recruitment bottleneck due to habitat homogenization, largely a result from past channelization impacts throughout the watershed (Maloney 2017). If these downward population trends are to be reversed, increasing juvenile rearing habitat must be a large component of the conservation effort.

Bonneville Cutthroat Trout

During the past century, Cutthroat Trout *Oncorhynchus clarkii* have experienced large, rangewide reductions in distribution and abundance due to a combination of competition and hybridization with nonnative trout, habitat loss, and fragmentation and, in some cases, disease and overharvesting (Behnke 1992; Duncan and Lockwood 2001; Fausch 2008). At one time, genetically pure populations of BCT were thought to be extinct (Duff 1988) and reports from state agencies and universities in the mid- to late 1970s reported that few genetically pure populations of BCT remained (Sigler and Miller 1963; Holden et al. 1974; Behnke 1976; Hickman 1978). In 1979, the American Fisheries Society listed BCT as threatened and then as endangered in 1989 (Duff 1996). In 1995, Utah formed an interagency conservation team for BCT, which developed a conservation agreement and strategy for Utah

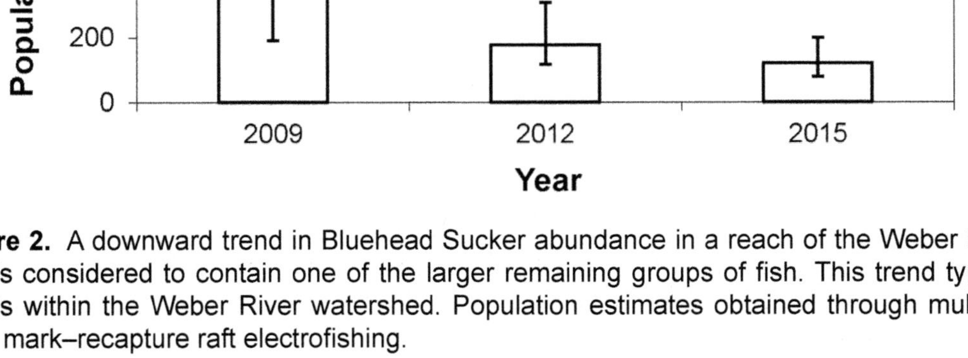

Figure 2. A downward trend in Bluehead Sucker abundance in a reach of the Weber River that is considered to contain one of the larger remaining groups of fish. This trend typifies trends within the Weber River watershed. Population estimates obtained through multiple-pass mark–recapture raft electrofishing.

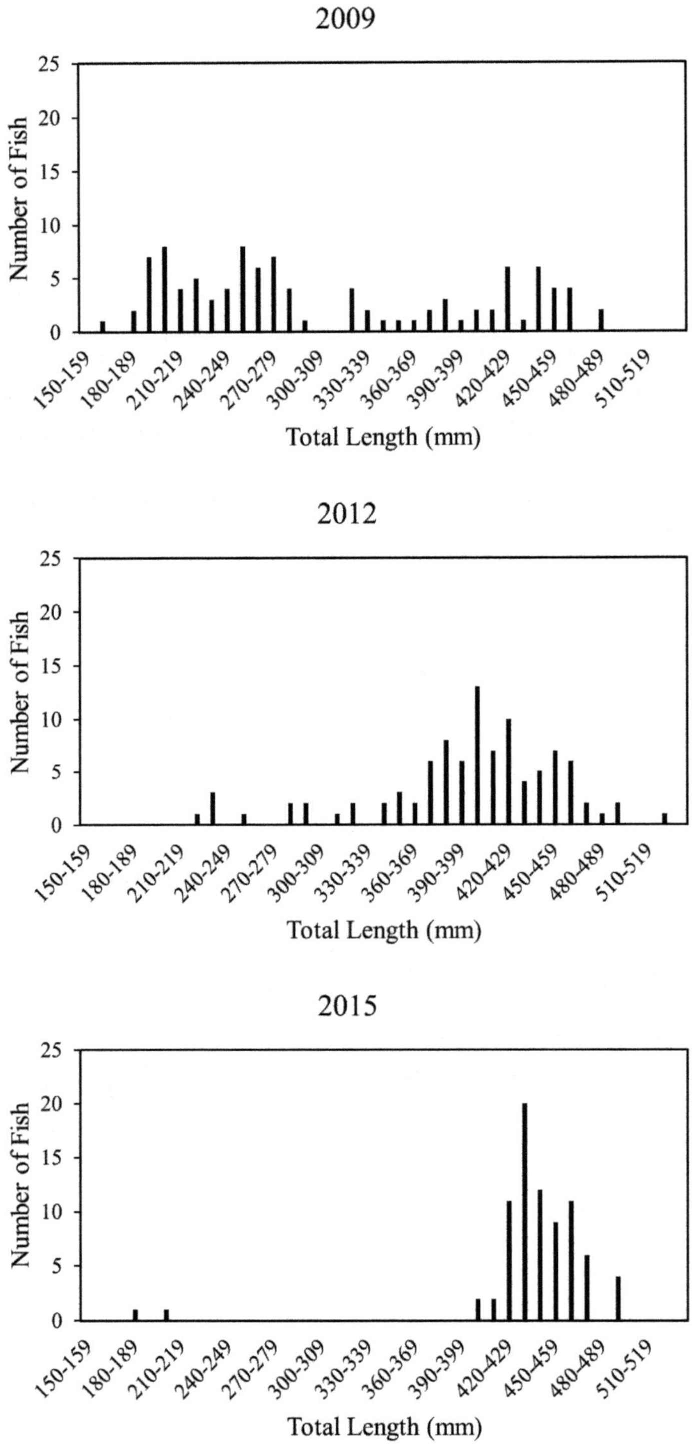

Figure 3. Length-frequency distribution of Bluehead Sucker in a reach of the Weber River that is considered to contain one of the larger remaining groups of fish. Most populations in the Weber River are characterized by larger, older fish with little recruitment. Data were collected while obtaining multiple-pass population estimates.

(Lentsch et al. 1997). The conservation team soon grew to encompass the entire range of BCT, leading to the development of a range-wide conservation agreement and strategy (Lentsch et al. 2000). The formation of these conservation teams stimulated additional surveys for the species, reintroductions into historical habitat, and habitat restoration projects, which improved the status of BCT (May and Albeke 2005). These efforts contributed to a U.S. Fish and Wildlife Service not warranted finding in 2001 from a 1998 petition to list BCT as a threatened species (USFWS 2001).

Currently, in the Weber River watershed, BCT occupy approximately 55% of their historic range, and 43 conservation populations occur throughout the watershed (BCT Status Assessment Database, Utah Division of Wildlife Resources). While BCT metapopulations still persist in the Weber River (e.g., Thompson 2000), no populations exhibiting fluvial life history characteristics were known as of the mid- to late 2000s. Most populations remaining are considered resident, meaning the BCT can fulfill their life history requirements in a single stream reach. Fluvial fish require interconnected habitats that allow the population to travel large distances to find suitable spawning habitat, often in tributaries. Only a handful of fluvial BCT populations remain throughout their range, including populations in the Logan River, Utah (Budy et al. 2012) and the Bear River, Idaho–Wyoming (Colyer et al. 2005), so protecting this life history strategy in the few areas where it persists is critical. In 2010, the UDWR observed a concentration of larger BCT (e.g., >300 mm total length [TL]) in the lower reaches of the Weber River in Morgan Valley. While BCT populations were known to occupy several tributary streams in this valley, biologists did not believe that downstream movement of BCT from these tributaries could explain the concentration of fish in the main-stem Weber River. In 2011, the UDWR, Trout Unlimited (TU), and Utah State University partnered to study this population of BCT. Between 2011 and 2018, more than 2,800 BCT were tagged with passive integrated transponder (PIT) tags, and between five and nine antennas were placed in the eight tributary streams each year between April and June, which is the BCT spawning period. This effort documented (1) a relatively large population of BCT in the main-stem Weber River (densities ranged from 200 to 300 fish/stream kilometer), (2) BCT movements within the Weber River and into tributaries during the spawning period, and (3) multiple barriers to fish movement within the main-stem Weber River and the spawning tributaries (Figure 4).

Need for a Partnership

Despite its value to local communities and economy and despite the popularity and importance of its fisheries, the Weber River remains one of Utah's most degraded and at-risk watersheds due to the widespread watershed degradation in both the main-stem and tributary systems described previously (Portugal et al. 2016). The consequence of lacking a comprehensive partnership was further highlighted as biologists were discovering the first documented fluvial BCT population in the Weber River. Habitat reconnection efforts were initiated quickly (where willing partners were identified) to begin modifying irrigation diversions and road crossings to provide fish passage. One main-stem and two tributary barriers were removed between 2012 and 2014, which initially encouraged biologists with how quickly on-the-groundfish passage restoration efforts could proceed. Unfortunately, the opportunity to improve fish passage at important irrigation structures had been missed during this same time frame. These projects had been completed without the fisheries managers' knowledge, demonstrating the lack of communication between biologists and water managers. With these new and rebuilt structures, habitat fragmentation within the BCT population had

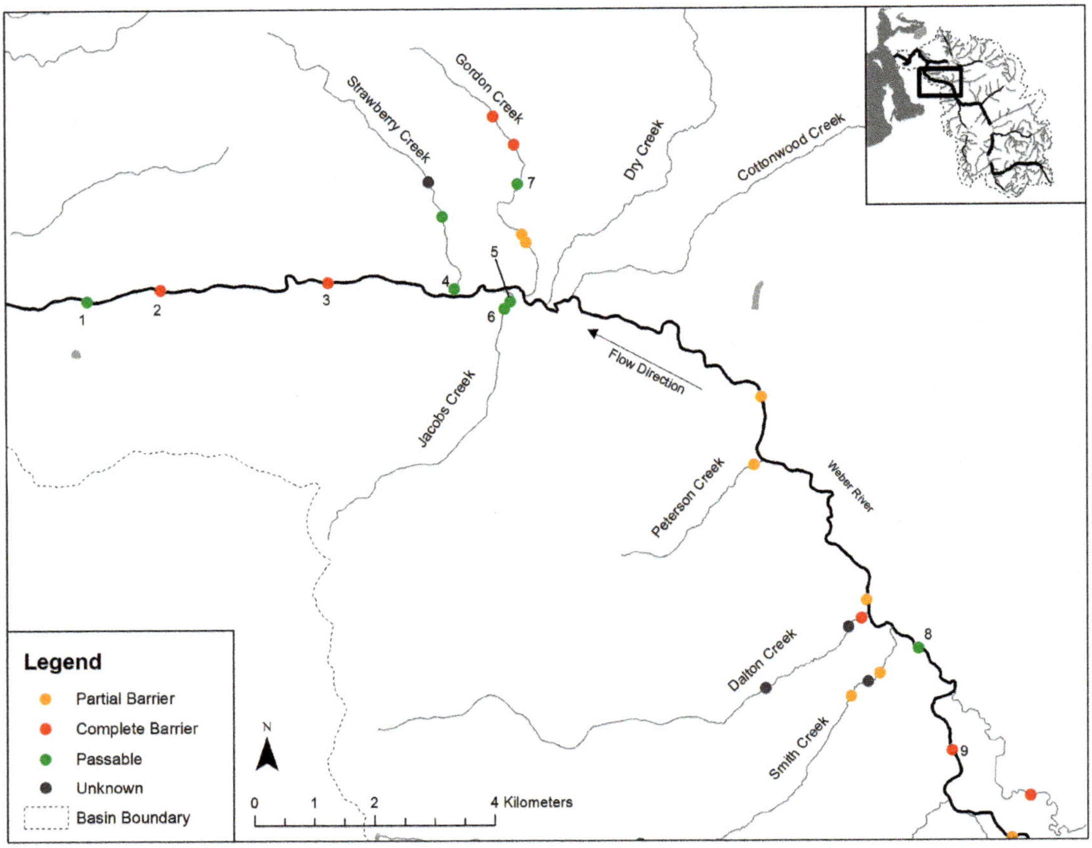

Figure 4. A map showing barriers in the fluvial Bonneville Cutthroat Trout reach. This map of the priority reach of the Lower Weber River highlights the important features described in this chapter: (1) Lower Weber River Diversion project completion 2011; (2) Davis and Weber Diversion still a complete barrier to fish movements; (3) PacifiCorp Weber Hydro Diversion Dam, which is currently under relicensing; (4) Strawberry Creek Culvert, where a fish ladder was constructed in 2016; (5) Lower Jacobs Creek culvert, which was replaced in 2013; (6) Upper Jacobs Creek culvert was modified in 2017; (7) Gordon Creek bridge location, which was modified in 2014; (8) Morgan Diversion, which was removed during the spring of 2017; and (9) the Stoddard Diversion, which is currently the upstream end of the Bonneville Cutthroat Trout population within this reach of the Weber River.

actually increased instead of been reduced. Many stakeholders (including water managers) did not understand what the fisheries managers were trying to accomplish, and without communication, significant reconnection and restoration efforts could not be accomplished. Restoration practitioners also had a knowledge that the watershed had been degraded and fragmented in the past, but there was no framework through which comprehensive assessments and prioritization could be made. Our project successes would likely be limited without a comprehensive planning and assessment tool.

Additionally, the 2011 water year was an extreme flood year, with 25–50-year return-interval events occurring throughout the watershed. The community responses to the flood event were both uncoordinated and extremely harmful to important fish habitats throughout the entire watershed, further reversing any positive restoration

Table 1. A list of contributors to the Weber River Partnership and the specific projects highlighting the diversity of partners that have contributed to each project and the partnership as a whole. NGO = nongovernmental organization.

Organization	Organization type	Weber River Partnership	Strawberry Creek Fish Ladder	Lower Jacobs Creek Culvert	Upper Jacobs Creek Culvert	Lower Weber Diversion Modernization	Gordon Creek Fish Passage	Morgan Diversion Removal	Weber River Hydro Relicensing	Bluehad Sucker Life History Study	Bluehead Sucker Brood Development
Destination Sports	Commercial	X									
Dominion Energy	Commercial			X	X						
Orvis	Commercial		X								
Pacificorp	Commercial	X							X	X	
Patagonia	Commercial		X								
RiverRestoration.org	Commercial	X				X					
U.S. Bureau of Reclamation	Federal	X	X			X			X	X	X
Natural Resources Conservation Service	Federal	X						X			
U.S. Fish and Wildlife Service	Federal	X	X		X	X	X		X		
U.S. Forest Service	Federal	X							X		
Utah Reclamation Mitigation and Conservation Commission	Federal										X
U.S. Geological Survey	Federal	X								X	
Morgan County	Local	X									
Ogden City	Local	X								X	
Desert Fish Habitat Partnership	NGO					X					
Lawrence T. and Janet T. Dee Foundation	NGO	X	X		X	X		X			
National Fish and Wildlife Foundation	NGO	X	X		X		X	X	X		
Summit Land Conservancy	NGO	X									
Trout Unlimited	NGO	X	X	X	X	X	X	X	X	X	
Western Native Trout Initiative	NGO		X		X		X				
Private Citizens	Private	X		X	X		X	X			
Utah Department of Agriculture and Food	State	X						X			

Table 1. Continued.

Organization	Organization type	Weber River Partnership	Strawberry Creek Fish Ladder	Lower Jacobs Creek Culvert	Upper Jacobs Creek Culvert	Lower Weber Diversion Modernization	Gordon Creek Fish Passage	Morgan Diversion Removal	Weber River Hydro Relicensing	Bluehad Sucker Life History Study	Bluehead Sucker Brood Development
Utah Division of Water Quality	State	X							X		
Utah Department of Transportation	State		X								
Utah Division of Water Rights	State	X						X			
Utah Division of Wildlife Resources	State	X	X	X	X	X	X	X	X	X	X
Utah Department of Natural Resources	State	X								X	X
Utah State University	University	X								X	
Brigham City Hyrdo	Water supplier/ utility		X			X					
Davis and Weber Counties Canal Company	Water supplier/ utility	X								X	
Provo River Water Users Association	Water supplier/ utility	X								X	
Snyderville Basin Water Reclamation District	Water supplier/ utility	X									
South Weber Irrigation Company	Water supplier/ utility	X				X					
Uintah Central Irrigation Company	Water supplier/ utility	X				X					
Weber Basin Water Conservancy District	Water supplier/ utility	X	X		X	X			X	X	
Weber River Water Users Association	Water supplier/ utility	X	X								

trends from the previous decades. The slow pace of habitat restoration and reversal of the trend due to actions from the 2011 flood event, increasing demands for water, and climate change have highlighted the need to improve collaboration throughout the watershed. A group of staff from several natural resources organizations recognized this challenge and sought to formalize a watershed partnership and develop a watershed plan in order to sustain the Weber River watersheds diverse values. Without a watershed plan with broad stakeholder acceptance, we were concerned that a continued imbalance in how water resources were being allocated and impacted would benefit some values (e.g., economic development) at the cost of other values (e.g., ecological and recreational value of fisheries).

Development of the Partnership

To help address the various challenges described above in the Weber River watershed, agencies, nongovernmental organizations, and individuals with diverse backgrounds and interests worked together to develop the Weber River Partnership. Partners have each contributed to the development of the Weber River Partnership and the Weber River Watershed Plan by funding representatives' participation and committing time and resources. Many partners also have provided funding to implement specific projects that provide benefits to the diverse interests in each project (Table 1).

The Weber River Partnership draws it strength from the diverse composition of individuals representing broad interests and backgrounds, ranging from representatives of state agencies such as the UDWR, Utah Department of Agriculture and Food, Utah Division of Water Quality, U.S. Department of Agriculture, Natural Resources Conservation Service, water users such as the Weber River Water Users Association and Weber Basin Water Conservancy District, municipalities such as Ogden City, and nonprofits such as Summit Land Conservancy and TU. The Weber River Partnership has been in existence as a voluntary, informal partnership since 2013, when a steering committee wrote the Weber River Watershed Plan, which was completed in 2014. The Bonneville Environmental Foundation was a critical partner in the facilitation and development of the plan. The Weber River Watershed Plan serves as the guiding framework to help prioritize our conservation and restoration efforts to improve the watershed's long-term health, measured by the status of specific conservation targets, which are categorized into three main elements: water resources, native aquatic species, and ecological systems (Weber River Partnership 2014). The Weber River Watershed Plan is scientifically informed and includes input from a broad range of resource experts and community stakeholders. The plan identifies 14 threats to the conservation targets that are pervasive throughout the Weber River watershed. The most threatening challenges are those that negatively affect values across the entire watershed and require a comprehensive effort in order to develop lasting and effective solutions. For example, stream habitat fragmentation caused by infrastructure cannot be solved by fisheries biologists alone. Modernizing infrastructure requires a cooperative partnership among the stakeholders that depend on that infrastructure. Although many physical threats were rated as "very high," a key and overarching threat was the lack of durable watershed leadership and communication structure. While developing the watershed plan, it became clear that many of the physical threats to the watershed were a result of these social deficiencies. As an informal partnership, progress has been made in addressing these deficiencies, and we aspire to have independent, but collaborative leadership to improve the effectiveness and credibility of the partnership as it moves forward. The 2014 watershed plan highlight that many of the other difficult challenges can be more effectively addressed with effective watershed leadership.

Through the development of the Weber River Watershed Plan, the steering committee identified four broad strategies to help address the challenges within the watershed. These strategies include communications, policy, land and water use, and habitat restoration. The four strategies should not be viewed as separate silos but should be considered in a hierarchical manner. For example, communications are the basis upon which all other strategies rely. If communications are not effective among the watershed stakeholders, then effective watershed policy cannot be developed or efforts to change policy are more likely to be biased towards particular interests. Further, land and water use cannot be developed in a consistent and sustainable manner if effective policy and communications do not exist. Finally, habitat restoration is ineffective if no or insufficient land-use strategies are in place to protect restoration measures. Currently, individual project partners rely upon small-scale funding and coordinate on a voluntary basis. As a watershed partnership, the hope is to catalyze a more durable partnership by investing in partner capacity to develop an official, chartered leadership structure and establish recognized nonprofit status. This would allow the partnership to solicit funds and continue to engage the broad base of stakeholders throughout the watershed. This leadership is essential to continuing collaborative work in the watershed and provides a durable framework for the future. Although the initial group of partners involved in the Weber River Partnership were committed to collaboration, continuing the effort remains the most significant challenge, particularly as individuals advance in their careers and to new positions and thereby are no longer able to participate in the partnership.

It is readily understood that most of the challenges to successfully developing a cohesive watershed group in the Weber River are based on social values. Conflict has occurred for generations among the stakeholders, ranging from water user conflicts with water conservancy districts to conflicts flaring over out-of-basin water deliveries. Members of the Weber River Partnership have already experienced events when agricultural producers in the watershed have expressed their skepticism of people and organizations with an interest other than traditional agricultural values. In terms of restoration and management, the status quo, characterized by small-scale individual projects and activities, is both easy and inexpensive to implement. Most small water users, landowners, and even agencies act exclusively in their own interests without considering their operations in the context of the larger watershed. This becomes challenging when considering larger-scale restoration, which is potentially more expensive upfront and involves complex long-term planning with results that may not be immediately visible. Certain actions may be perceived as hostile to traditional management strategies. Ideas such as diversion consolidation, efficiency improvements, and fish passage have met resistance in the past, but through this partnership, many of the more complex challenges are becoming more tractable to address.

Partnership Successes

Fish have gained relevance through the development of the Weber River Partnership. Ten years ago, fisheries managers largely were working independently from other stakeholders in the watershed, especially with regard to habitat restoration projects. Past projects primarily consisted of working with a few willing landowners to stabilize eroding streambanks and add instream structures to concentrate fish. These efforts were completed in return for public access for anglers to a small stream reach. While fish passage was a restoration need, positive efforts along this front were primarily with these individual landowners who had failing irrigation structures, and oftentimes, these projects materialized out of opportunity and not priority. At this time, fish managers did not have a relationship with the local water districts and neither side really understood what the other was trying to accomplish. The Weber River Partnership provided a mechanism for better communication among stakeholders in the watershed, including conversations between these two groups. Initially, there was some hesitation to work together, as water users felt the fish biologists just wanted to find a way to take their water and fish biologists felt that irrigators did not care about the fish. The more the two groups talked, however, the more they began to understand what the other was trying to accomplish and more importantly why they were doing what they were doing. Simple communication produced immediate results.

Through the partnership we were able to leverage our broad expertise to develop some needed data products. We completed a watershed-wide assessment of fish migration barriers in 2013, using subjective methods based on high-resolution aerial imagery and identified more than 226 irrigation diversions and small dams (≤5 m in height) throughout the Weber River watershed. These structures, coupled with road and utility crossings, and other impoundments have resulted in 316 barriers to fish movement across the watershed (Table 2; Figure 1). Some of these barriers may be more permeable than others, depending on the time of the year and stream flow. One hundred and ninety-six of these structures are thought to be complete barriers to upstream fish movement (red symbols in Figure 1). The Weber River Partnership also coordinated and reviewed the 2016 basinwide habitat assessment and restoration-potential study that has provided guidance to project prioritization (Portugal et al. 2016). The Weber River partnership also has catalyzed several efforts, which are described below.

Echo Dam fish salvage

Ten days following a Bluehead Sucker and BCT presentation at the first Confluence Symposium in 2014, the Weber River Water Users Association (hereafter, "the Association") called the UDWR. They were going to be diverting water around the spillway pool downstream from one of the major impoundments on the Weber River, Echo Res-

Table 2. The current numbers of barriers still present within the Weber River watershed. Structures are classified as three main types: culverts or road crossings, diversion structures, and impoundments. The extent of the blockage is characterized as partial, complete, passable, and needs further assessment.

	Culvert	Diversion	Impoundment	Total
Partial barrier[a]	33	77	10	120
Complete barrier[b]	43	72	81	196
Passable[c]	6	57	1	64
Needs further assessment[d]	40	20	6	66
Total	122	226	98	446

[a] Structure is sometimes passable depending upon water conditions.
[b] Structure is never passable.
[c] Structure does not appear to impede fish movements
[d] The extent of the blockage was indeterminate based on the initial study and needs to be evaluated in detail.

ervoir to complete a maintenance project on the concrete basin. Before the water was entirely drained from the basin, the Association wanted to know if the UDWR had any interest in sampling the spillway pool and salvaging Bluehead Sucker and BCT. The courtesy phone call was greatly appreciated as little communication had transpired between these two groups prior to this point. A UDWR sampling crew met the Association a few days later and both groups sampled the spillway pool. Because Bluehead Sucker and BCT were found while sampling, this effort quickly turned into a show-and-tell teaching opportunity, further strengthening the relationship between the two groups.

During the salvage effort, 10 previously untagged Bluehead Sucker were sampled and moved downstream a few hundred meters. Nine of these fish ranged from 200 to 244 mm TL. This is the size range that has largely been absent during Bluehead Sucker sampling efforts on the Weber River (see Figure 3). In fact, only six juvenile Bluehead Suckers had been captured earlier that year directly downstream of this spillway pool while completing a three-pass mark–recapture population estimate on a 14.1-km reach of the Weber River. The multiple-pass effort captured and PIT-tagged 96 Bluehead Suckers, so this salvage effort equated to salvaging close to 10% of the population in this river reach. Similarly, only 34 BCT had been captured and PIT-tagged during this sampling effort, so the movement of seven untagged BCT downstream from the spillway pool equated to approximately 15–20% of this population. The salvage effort was a great opportunity to tell the Weber River fishery story to the Association, but the effort also ended up salvaging a significant number of these two fish species that were not abundant in this reach of the Weber River.

Improving Fish Passage

In 2010, prior to the formation of the Weber River Partnership and during the time when biologists with TU and the UDWR were initiating studies to better understand the fluvial BCT population, we were approached by water companies to collaborate on irrigation diversion reconstruction projects. The fisheries interests provided funding and expertise to the projects, which helped build a relationship between the two groups. The fisheries group was encouraged to see reconnection progress happening so quickly, even while we were still learning about the movement requirements of the fluvial BCT population. We drew on the research and observations of fluvial BCT in the Bear River watershed to guide our initial habitat reconnection projects; however, a lack of a comprehensive assessment of barriers limited our ability to be strategic in these initial projects.

Even though we were beginning to build credible relationships with water users, a general lack of communication between the water and fisheries communities hampered reconnection. While a few reconnection projects were being completed, several other water projects were being proposed and constructed or reconstructed, which introduced new and solidified some older obstructions to fish movement. A frustrating development occurred when a large irrigation diversion on the Weber River main stem was rebuilt during this time frame. This project had been in the works for more than a year between the design and reconstruction stages, but fisheries managers were unaware of it until the diversion had been rebuilt. This lack of communication was one of the driving factors that motivated fisheries managers to join the small, core group that helped form the Weber River Partnership. Once the partnership had developed and fisheries biologists had the opportunity to share their message, water companies began making phone calls to the fish biologists when they were considering diversion modifications. Today, some water districts are collaborating with fisheries interests, asking if fish barrier removal projects can be incorporated into water infrastructure grants, which provides more comprehensive and competitive proposals for the water users and facilitates fish passage projects.

Main-stem diversions and Pacificorp relicensing.—PacifiCorp's Weber River Hydroelectric Project is located in Weber Canyon. This small, 5-m-high dam holds water to create a 3.2-ha reservoir forebay (Figure 5), which provides the water to fill the penstock flow line for downstream power generation. The dam serves as a complete fish barrier separating two important Bluehead Sucker populations and the recently discovered fluvial BCT population. The hydroelectric facility is scheduled to complete Federal Energy Regulatory Commission (FERC) relicensing in 2020, and PacifiCorp began soliciting public input and organizing working groups several years prior to the final relicensing date. The Weber River Partnership was just gaining traction as PacifiCorp was beginning the scoping process, but communication and collaboration among stakeholders in the watershed had already improved. Partly due to the new cooperative climate within the watershed, PacifiCorp petitioned FERC to relicense using the alternative relicensing process (ALP). The ALP is infrequently used as this process requires consensus among stakeholders during the entire relicensing process. Because consensus can be difficult to achieve, the integrated relicensing process, which accounts for disagreement among stakeholders, is FERC's default relicensing process and is typically utilized. In 2017, the Weber relicensing process was the only license utilizing the ALP anywhere in the United States.

After a couple of scoping meetings, fish passage at the hydroelectric facility (and

Figure 5. Pacificorp Weber River Hydroelectric Dam in Weber Canyon.

maintaining the existing instream flow requirements) and improved recreation facilities emerged as the primary mitigation measures championed by both stakeholders and PacifiCorp. The cooperative study on the BCT population in this reach of the Weber River had demonstrated that a fluvial life history form persists, as these fish were traveling as much as 15 km per year to spawn. The PacifiCorp facility (along with another full-spanning irrigation dam unrelated to and just downstream of the powerhouse) fragmented the BCT population with limited movement downstream and no movement upstream except under nontypical, infrequent operating conditions. A small number of BCT (n = 30) were documented to move upstream past the facility during the winter when the dam was operated in a specific way (with the low-level gate open) and when streamflow was less than 3 m^3/s. During the BCT study, PacifiCorp had drained the forebay to remove materials that had made the low-level gate inoperable; consequently, it had not been used for several years. This gate allows water to flow through the base of the dam, whether the forebay is full or empty, and can be utilized when winter streamflow is too low to generate power and also to maintain the instream flow requirements. The fact that BCT did move upstream through the dam after many years of being isolated from the upstream population further demonstrated the importance of fish passage at the PacifiCorp facility, especially since no spawning tributaries exist downstream of the PacifiCorp Dam.

As part of FERC relicensing, PacifiCorp organized a Fisheries Working Group (FWG), which met consistently for the better part of 2 years. The group had participants from the UDWR, Utah Department of Environment Quality Division of Water Quality, TU, USFWS, U.S. Forest Service, and PacifiCorp. The FWG examined existing fisheries data, determined that a fish ladder was needed to allow upstream fish movement, and helped to select and then worked with a consultant to determine the conceptual fish ladder design that would best meet existing and future fish and site location needs. Target velocities for the fish ladder were velocities that would allow fish passage of juvenile Bluehead Sucker (<150 mm TL). A vertical-slot fish ladder style was chosen by the FWG, and a conceptual design drawing for this preferred alternative was developed in consultation with the FWG. The FWG also cooperatively designed and PacifiCorp funded a three-phase study to look at fish entrainment in the PacifiCorp hydroelectric intake and penstock (Pacificorp–Hydro Resources 2017). Overall, the collaborative efforts of the various stakeholders to the Weber Hydroelectric Project relicensing process will result in a fish ladder that will ensure restoration of the movement of at least two fish species of concern in the Weber River, and improve the health and stability of this fishery.

Working with Pacificorp to provide fish passage through their relicensing of the Weber Hydroelectric facility has been part of a larger effort to reconnect main-stem habitat. Fisheries managers also have been in contact with all the other main-stem diversion owners/operators within the lower Weber River reach containing the fluvial BCT population and important Bluehead Sucker population. While many diversions remain fish barriers in this reach, fish passage has been provided at two main-stem diversions since 2010 (Figure 4). One main-stem diversion that provided irrigation water to a small ranch was removed entirely. Multiple partners came together and funded a new pivot to service the agricultural land and the removal and revegetation of the diversion site. The water right was moved downstream where water was pumped directly from the river to service the new pivot.

A second, 2-m-high main-stem diversion located near Ogden was reconstructed, which included a fish passage channel around the diversion and screening on both irrigation canals that were serviced by the diversion. This diversion was originally constructed in the 1930s and its failure in the near future was a real possibility. The two water companies serviced by the diversion were hopeful of getting financial assistance through the U.S. Bureau of Reclamation's (USBOR) WaterSMART Grant to assist with the expensive rebuild. The UDWR and TU were contacted and fish passage became the conservation tie needed to make this grant competitive. Ten partners/funders came together for this project and the result was a US$500,000 fish-friendly irrigation diversion rebuild. A fish trap was opportunistically operated in the side channel during a 2-year period and fluvial BCT and juvenile Bluehead Suckers (along with all species in the Weber River) were documented to move upstream past this diversion, representing the first fish movements upstream at this location in more than 70 years.

Tributary reconnection.—Ten diversions or road crossings were identified as partial or complete barriers to fluvial BCT in the lower Weber River during the BCT research and the barrier assessment. Similar to the main-stem Weber River, barriers to BCT still persist in spawning tributaries, but reconnection progress has been significant. Four barriers have been removed in priority tributaries while funding to remove two additional barriers has been secured (Figure 4). The first tributary work was completed on Gordon Creek. A bridge crossing had resulted in a 1.5-m drop that was a barrier to resident BCT and likely impeded fluvial fish at lower streamflows. The landowners on both sides of the stream agreed to the project. To improve fish passage at this location, the stream was reconstructed with a step pool complex downstream from the barrier. The smaller drops with holding pools allowed salmonids of all sizes to access spawning habitat upstream of the bridge (Figure 6).

Jacobs Creek is one of the most important spawning tributaries, based on BCT detections on antennas during the PIT tag study. Two road crossings spaced approximately 150 m apart, the first one 213 m upstream of the confluence with the Weber River, were each partial barriers to migrations of spawning BCT, and each blocked the upstream movements of spawning BCT. In tandem, these partial barriers acted as a complete impediment to spawning fish attempting to access 2 km of spawning habitat upstream of the road crossings. The lower barrier was a road crossing that consisted of two undersized, perched culverts. The fish passage barrier upstream was another problem road culvert that consisted of a perched culvert with steep grade and no holding pools downstream.

In 2012, Questar Gas (now Dominion Energy) submitted a stream alteration permit to reconstruct their natural gas pipeline crossing on Jacobs Creek directly downstream from the lower road crossing. Questar Gas was contacted to discuss the project, and they agreed to improve passage at the road crossing while they had equipment in the area. This barrier removal was funded completely by Questar Gas, and the result was the placement of an appropriate-sized culvert with baffles for the road crossing and two hardened structures downstream to help control the grade and prevent future erosion. The barrier at the upper culvert was addressed in 2017 through a collaborative project including the UDWR, TU, Dominion Energy, and the Weber Basin Water Conservancy District. Monitoring of PIT-tagged fluvial BCT indicated that under the best spring runoff flow conditions, approximately half of the BCT that attempted to move upstream past this road crossing were successful. Rock, equipment, and operator were provided in-kind by the Weber Basin Water Conservancy District who owned the service road with the problem culvert. The result from this project was a restructuring of the stream downstream from the culvert with hardened structures to reduce grade, create holding pools, and reduce the jump

Figure 6. Gordon Creek after channel reconstruction to remove a fish barrier.

required into the culvert. This project was completed in 2017, and future monitoring of the fluvial BCT population will shed light on its success.

Fish passage at Strawberry Creek proved to be the most challenging project to date. Upstream fish passage in this stream had been severed since the early 1960s when Interstate 84 was built. For the past 50+ years, a 115-m cement box culvert with a 5% slope had prevented any fish from moving upstream. This culvert was located 150 m upstream of the confluence with the Weber River, so without access further upstream, spawning habitat for BCT was extremely limited. This project was planned and discussed for several years with the Utah Department of Transportation, who had concerns about reducing the capacity of their culvert with a fish ladder. Another concern was compromising the concrete in the culvert by drilling and securing a structure. The final design agreed upon was a vertical slot fish ladder that would be fabricated in 3-m sections, bolted together and glued to one side of the cement box culvert (Figure 7). The Strawberry Creek fish ladder was installed and evaluated during the 2017 BCT spawning window. Antennas were placed upstream and downstream of the fish ladder, and BCT were able to move upstream through the culvert for the first time in more than 50 years. One smaller 200-mm-TL BCT even moved up during the 6 weeks that antennas were in operation. This fish ladder was funded by a diverse array of partners (Table 1).

Bluehead Sucker Life History Study

Bluehead Suckers had not been observed in the Weber River while sampling from

Figure 7. Strawberry Creek's vertical-slot fish ladder.

the 1980–1990s, and at that time, they were thought to have been extirpated from this watershed. Targeted Bluehead Sucker surveys were completed in the 2000s, rediscovering small populations of this species. Population monitoring within the five reaches where this species persisted has indicated that small populations remain, but declining trends have been observed in many reaches (Figure 2), and limited to no recruitment within these population segments (Figure 3) was as much of a concern as the declining populations. Fisheries managers were unsure how to reverse the declining trends within the populations because little to no information was known about the demographics or life history requirements of this species in the Weber River. To move forward with the management of this species, a life history study was needed, but obtaining funding for this study was the challenge. Because Bluehead Sucker is not a sport fish, many funding sources were not available and many of the needs of the remaining 140 species of greatest conservation need identified in the Utah Wildlife Action Plan (UDWR 2015) were prioritized higher than this study. Fisheries managers in the Weber River realized that a nontraditional funding approach was needed to complete this work.

Some seed money was secured from the Endangered Species Mitigation Fund (ESMF), which is a fund comprised largely from a Utah water tax and administered by the Utah Department of Natural Resources. These monies were enough to entice a researcher to bring on a graduate student to study Bluehead Sucker in the Weber River, even though the project was only half funded. The seed money gave fisheries managers a year to secure an additional $70,000 for this study. During this time, the Bluehead Sucker was gaining relevance with stakeholders in the Weber River watershed. The development of the Weber River Restoration Plan (Weber River Partnership 2014) identified Bluehead Sucker as a conservation focal species. In addition, a Bluehead Sucker presentation was given at the first annual Weber River Watershed Symposium. Fisheries managers were talking to all stakeholders in the watershed about Bluehead Suckers, and the bottom line message in all of these efforts was the need to work together now to stabilize and/or increase Bluehead Sucker populations in the Weber River so we can keep this species from needing protection under the Endangered Species Act. This message resonated well, especially within the water community. The money needed to round out the graduate project was secured from the following groups: TU, Ogden City, Pacificorp, USBOR, Weber Basin Water Conservancy District, Davis and Weber Counties Canal Company, Weber River Water Users Association, and the Provo River Water Users Association. Before the development of the Weber River Partnership, many of these nontraditional funders were not aware that Bluehead Sucker was a species of fish, let alone that they occurred in the Weber River. They have begun to see the value of helping this species in their local river and especially keeping this fish from needing protection under the Endangered Species Act.

The graduate study provided much needed information, including locations where Bluehead Sucker spawn in the Weber River and the substrate required for spawning. Since the spawning bars have been identified, fisheries managers are now able to help protect them from future modifications such as new utility line crossings. The study also identified where juvenile Bluehead Sucker rearing habitats remained, and it quantified the habitat characteristics at these slow-velocity habitats (Maloney 2017). This has provided a template for habitat-restoration biologists to incorporate more of these habitats into their stream-restoration projects.

Bluehead Sucker Brood Development

The graduate study provided a path forward to improve Bluehead Sucker populations in the Weber River. This new information and monitoring efforts indicated that these fish needed larger, interconnected mainstem habitats. While monitoring, Bluehead Sucker were PIT-tagged, and some longer movements of greater than 10 km between spawning and overwintering habitats further demonstrated the need for connected habitat. While the Weber River is highly fragmented (Figure 1), TU and the UDWR were having success reconnecting large portions of the main-stem Weber River for BCT, as well as Bluehead Sucker. Despite these successes, fisheries managers feared that the populations of Bluehead Sucker in the Weber River might be too depressed to respond positively to the recent habitat improvements. This evidence supported the development of a Weber River broodstock, which managers thought would be needed to increase population densities and allow Bluehead Sucker to realize the habitat benefits and persist into the future.

Conservation of Bluehead Sucker has gained momentum within the Weber River Partnership, and the recent genetic information indicating that the Bluehead Suckers in the Weber River are genetically distinct from their conspecific populations in the Colorado and Green rivers prompted the UDWR to prioritize the development of a brood for the Weber River. Bluehead Sucker had not been cultured within the state of Utah prior to this effort; consequently, the first step was to consult with adjacent states where limited culture efforts for this species had begun and then develop an informal culture plan for the Weber River. A portion of a sport fish hatchery was recommissioned to accommodate this new brood, and some funding from the ESMF and UDWR general fund had been secured, but additional monies were needed to move forward with the brood for any length of time.

To begin development of the brood, raft electrofishing over spawning locations identified in the graduate study was completed to collect ripe fish. These fish were transported downstream to hatchery personnel who spawned the fish maximizing crosses and genetic material. In general, ripe female Bluehead Suckers were the limiting factor on a given spawning day, so multiple males were spawned with a single female after male sperm motility was confirmed under a microscope. Fish were then transported back upstream, floated down to the spawning areas where they were captured, and released. This process was repeated over the weeks of the spawn in order to maximize genetic material within the brood for each population segment. The water hardened eggs were transported to the recommissioned hatchery where the different spawning crosses containing distinct genetic material have been held and grown separately. Funding was needed to assist with the wild egg take, as well as hatchery operations for this new brood. In 2016, the USBOR, who had not provided funding for fish projects in the Weber River until the development of the Weber River Partnership, did provide funding to assist with the brood development (as well as other fish passage projects), allowing for the first egg take in 2017. Heavy runoff during 2017 restricted the egg take to one Bluehead Sucker population, but more than 45,000 eggs were collected over five different spawning days. Four brood groups from this population remain in the hatchery close to 2 years later; these fish represent genetic contributions from 32 individual Bluehead Suckers. Spawning success was similar in 2018, and these brood lots and other lots developed from the remaining Weber River Bluehead Sucker population segments will be used to produce larger numbers of fish for stocking back into the Weber River. Most recently, funding from the Utah Reclamation Mitigation and Conservation Commission has been earmarked to assist with the Weber River brood development. This is the first fish specific funding provided to the Weber River from the Utah Reclamation Mitigation and Conservation Commission. These multiple partners to the brood development will allow the continuation of this brood into the foreseeable future.

Conclusions

The Weber River Partnership has provided an extremely valuable framework for communication, education, and collaboration within the Weber River watershed. The project successes highlighted in this chapter, and many that have not been discussed, were made possible by bringing groups of people with seemingly different sets of values together to find the areas where those value sets overlapped. Finding that space where the shared values exist has been the key to the success of the Weber River Partnership, but it has taken effort and much work remains. Most of the participants and participating organizations in the Weber River Partnership do so voluntarily. Voluntary participation is considered a keystone feature of developing successful paths forward when Utah's communities are faced with complicated challenges. That volunteerism also extends to the leaders of the Weber River Partnership, who also must carve out capacity to ensure that this effort continues. With that in mind, the most significant challenge moving forward is that of sustaining the partnership and collaboration over time as individuals advance in their careers and new people replace them. Our current effort is focused on developing capacity through grants to secure a building block of funding to hire professional staff, specifically a coordinator to help keep the partnership moving forward.

Acknowledgments

We would like to thank the original Weber River Partnership planning group who saw the need and had the vision for the partnership. This group consisted of Kent Sorenson, Ben Nadolski, and Paul Thompson (UDWR), Paul Burnett (TU), Kari Lundeen (Utah Division of Water Quality), and Jake Powell (Kamas Valley Conservation District). We would like to thank Robert Warren (Bonneville Environmental Foundation), who provided guidance and expertise on forming watershed groups and facilitating initial stakeholder meetings within the Weber River. We would like to thank all participants listed in this chapter and any that we may have missed for participating in the partnership. Last, we would like to thank all funders of fishery-related projects in the Weber River watershed. The fisheries work described in this paper was completed by the UDWR Northern Regional Office with help from many partners, including the Weber Basin Anglers TU Chapter.

References

Andreasen, J. K. 1973. Reproductive life history of *Catostomus ardens* and *Catostomus discobolus* in the Weber River, Utah. Master's thesis. Brigham Young University, Provo, Utah.

Barton, J. R., and P. V. Winger. 1973. A Study of the channelization of the Weber River, Summit County, Utah. Report to the Utah Division of Wildlife Resources. Brigham Young University, Provo, Utah.

Behnke, R. J. 1976. Summary of information on the status of the Utah or Bonneville Cutthroat Trout, *Salmo clarki utah*. Report to the U.S. Forest Service, Wasatch-Cache National Forest, Salt Lake City, Utah.

Behnke, R. 1992. Native trout of western North America. American Fisheries Society, Monograph 6, Bethesda, Maryland.

Bessert, M. L., and G. Orti. 2008. Genetic effects of habitat fragmentation on Blue Sucker populations in the upper Missouri River. Conservation Genetics 9:821–832.

Bezzerides, N., and K. R. Bestgen. 2002. Status review of Roundtail Chub *Gila robusta*, Flannelmouth Sucker *Catostomus latipinnis*, and Bluehead Sucker *Catostomus discobolus* in the Colorado River basin. Colorado State University, Larval Fish Laboratory, Technical Report 118, Fort Collins.

Budy, P., S. Wood, and B. Roper. 2012. A study of the spawning ecology and early life history survival of Bonneville Cutthroat Trout. North American Journal of Fisheries Management 32:436–449.

Carlson, A. J. 2006. Watershed scale habitat use and canal entrainment by Bonneville Cutthroat Trout in the Smiths Fork–Bear River drainage. Doctoral dissertation. University of Wyoming, Laramie.

Christensen, N. S., A. W. Wood, N. Voisin, D. P. Lettenmaier, and R. N. Palmer. 2004. The effects of climate change on the hydrology and water resources of the Colorado River basin. Climatic Change 62:337–363.

Colyer, W. T., J. L. Kershner, and R. H. Hilderbrand. 2005. Movements of fluvial Bonneville Cutthroat Trout in the Thomas Fork of the Bear River, Idaho–Wyoming. North American Journal of Fisheries Management 25:954–963.

Compton, R. I., W. A. Hubert, F. J. Rahel, M. C. Quist, and M. R. Bower. 2008. Influences of fragmentation on three species of native warmwater fishes in a Colorado River basin headwater stream system, Wyoming. North American Journal of Fisheries Management 28:1733–1743.

Duff, D. A. 1988. Bonneville Cutthroat Trout: current status and management. Pages 121–127 *in* R. E. Gresswell, editor. Status and management of interior stocks of Cutthroat Trout. American Fisheries Society, Symposium 4, Bethesda, Maryland.

Duff, D. A. 1996. Conservation assessment for inland Cutthroat Trout status and distribution. U.S. Forest Service, Intermountain Region, Ogden, Utah.

Duncan, J. R., and J. L. Lockwood. 2001. Extinction in a field of bullets: a search for causes in the decline of the world's freshwater fishes. Biological Conservation 102:97–105.

Fausch, K. D. 2008. A paradox of trout invasions in North America. Biological Invasions 10:685–701.

Gido, K. B., D. L. Propst, J. D. Olen, K. R. Bestgen. 2013. Multidecadal responses of native and introduced fishes to natural and altered flow regimes in the American Southwest. Canadian Journal of Fisheries and Aquatic Sciences 70:554–564.

Hickman, T. J. 1978. Systematic study of the

native trout of the Bonneville Basin. Master's thesis. Colorado State University, Fort Collins.

Holden, P. B., W. White, G. Sommerville, D. Duff, R. Gervais, and S. Gloss. 1974. Threatened fishes of Utah. Proceedings of the Utah Academy of Sciences Arts and Letters 51:46–55.

Hopken, M. W., M. R. Douglas, and M. E. Douglas. 2013. Stream hierarchy defines riverscape genetics of a North American desert fish. Molecular Ecology 22:956–971.

Lentsch, L., Y. Converse, and J. Perkins. 1997. Conservation agreement and strategy for Bonneville Cutthroat Trout (*Oncorhynchus clarki utah*) in the state of Utah. Utah Department of Natural Resources, Publication Number 97-19, Salt Lake City.

Lentsch, L. D., C., A. Toline, J. Kershner, J. M. Hudson, and J. Mizzi. 2000. Range-wide conservation agreement and strategy for Bonneville Cutthroat Trout (*Oncorhynchus clarki utah*). Utah Department of Natural Resources, Publication 00-19, Salt Lake City.

Lilieholm, R. J., J. M. Keating, and R. S. Krannich. 2017. 2016 Utah angler periodic survey. Report to the Utah Division of Wildlife Resources, Utah State University, Logan.

Maloney, B. C. 2017. Evaluating habitat-based niche requirements and potential recruitment bottlenecks for imperiled Bluehead Sucker (*Catostomus discobolus*). Master's thesis. Utah State University, Logan.

Martinez, P. J., T. E. Chart, T. A. Trammel, J. G. Wullschleger, and E. P. Bergersen. 1994. Fish species composition before and after construction of a main stem reservoir on the White River, Colorado Environmental Biology of Fishes 40:227–239.

May, B. E. and S. Albeke. 2005. Range-wide status of Bonneville Cutthroat Trout (*Oncorhynchus clarki utah*) 2004. Utah Department of Natural Resources, Publication Number 05-02, Salt Lake City.

McAda, C. W. 1977. Aspects of the life history of three catostomids native to the upper Colorado River basin. Master's thesis. Utah State University, Logan.

Meyer, K. A., J. A. Lamansky, Jr., D. J. Schill, and D. W. Zaroban. 2012. Non-game fish species distribution and habitat associations in the upper Snake River basin of Idaho. Idaho Game and Fish Department, Report Number 12-01, Boise.

Moyle, P. B. and R. A. Leidy. 1992. Loss of biodiversity in aquatic ecosystems: evidence from fish faunas. Pages 127–169 *in* P. G. Fielder and S. K. Jain, editors. Conservation biology: the theory and practice of nature conservation, prevention, and management. Chapman and Hall, New York.

Mueller, G. A., and P. C. Marsh. 2002. Lost, a desert river and its native fishes: a historical perspective of the lower Colorado River. U.S. Geological Survey, USGS/BRD/ITR-2002-0010, Denver, Colorado.

Pacificorp–Hydro Resources. 2017. Final technical reports Weber Hydroelectric Project relicensing FERC Project No. 1744. Submitted by Pacificorp–Hydro Resources, Salt Lake City, Utah to Federal Energy Regulatory Commission, Washington, D.C.

Portugal, E. W., W. Macfarlane, J. Gilbert, J. Gilbert, S. M. Shahverdian, and J. M. Wheaton. 2016. Geomorphic assessment of the Weber River watershed following a River Styles framework. Prepared for the Utah Division of Wildlife Resources, Utah State University, Logan.

Sigler, W. F. and R. R. Miller. 1963. Fishes of Utah. Utah Department of Fish and Game, Salt Lake City.

Sigler, W. F. and J. W. Sigler. 1996. Fishes of Utah: a natural history. University of Utah Press, Salt Lake City.

Snyder, J. O. 1924. Notes on certain catostomids of the Bonneville system, including the type of *Pantosteus virescens* Cope. Proceedings of the U.S. Natural Museum 64:1–6.

Sublette, J. E., M. D. Hatch, and M. S. Sublette. 1990. The fishes of New Mexico. University of New Mexico Press, Albuquerque.

Thompson, P. 2000. Bonneville Cutthroat Trout (*Oncorhynchus clarki utah*) surveys in the Chalk Creek (sections 02-03) drainage, 1998–1999. Utah Department of Natural Resources, Publication 00-11, Salt Lake City.

Udall, B., and J. Overpeck. 2017. The twenty-first century Colorado River hot drought and implications for the future. Water Resources Research 53:2404–2418.

UDWR (Utah Division of Wildlife Resources). 2015. Utah Wildlife Action Plan 2015–2025. Utah Department of Natural Resources, Publication Number 15-14, Salt Lake City.

Unmack, P. J., T. E. Dowling, N. J. Laitinen, C. L. Secor, R. L. Mayden, D. K. Shiozawa, and G. R. Smith. 2014. Influence of introgression and geological processes on phylogenetic relationships of western North America Mountain Suckers (*Pantosteus*, Catostomidae). PLOS (Public Library of Science) ONE [online serial] 9(3):e90061.

USFWS (U.S. Fish and Wildlife Service). 2001. Endangered and threatened wildlife and plants: 12-month finding for a petition to list the Bonneville Cutthroat Trout as threatened throughout its range. Federal Register 66:195(9 October 2001):51362–51366.

Vanicek, C. D., R. H. Kramer, and D. R. Franklin. 1970. Distribution of Green River fishes in Utah and Colorado following closure of Flaming Gorge Dam. Southwestern Naturalist 14:297–315.

Webber, P. A., P. D. Thompson, and P. Budy. 2012. Status and structure of two populations of the Bluehead Sucker (*Catostomus discobolus*) in the Weber River, Utah. The Southwestern Naturalist 57:267–276.

Weber River Partnership. 2014. Weber River Restoration Plan. Weber River Partnership, Ogden, Utah.

Williams, J. E., J. E. Johnson, D. A. Hendrickson, S. Contreras-Balderas, J. D. Williams, M. Navarro-Mendoza, D. E. McAllister, and J. E. Deacon. 1989. Fishes of North America: endangered, threatened, or of special concern. Fisheries 14:2–20.

American Fisheries Society Symposium 91:589–647, 2019

Collaboration, Commitment, and Adaptive Learning Enable Eradication of Nonnative Trout and Establishment of Native Westslope Cutthroat Trout into One-Hundred Kilometers of Cherry Creek, a Tributary to the Madison River, Montana

PATRICK T. CLANCEY*[1]
21 South Alta Street, Helena, Montana 59601, USA

BRADLEY B. SHEPARD[1]
B.B. Shepard and Associates
65 9th Street Island Drive, Livingston, Montana 59047, USA
and
Department of Ecology, Montana State University
Post Office Box 173460, Bozeman, Montana 59717, USA

CARTER G. KRUSE
Turner Enterprises, Inc.
901 Technology Boulevard, Bozeman, Montana 59718, USA

SCOTT A. BARNDT
Custer Gallatin National Forest
Post Office Box 130, Bozeman, Montana 59771, USA

LEE NELSON
Montana Fish, Wildlife & Parks
1420 East Sixth Avenue, Helena, Montana 59624, USA

BRUCE C. ROBERTS[2]
4355 Vesta Drive, Helena, Montana 59602, USA

R. BEAUREGARD TURNER
Turner Enterprises, Inc.
901 Technology Boulevard, Bozeman, Montana 59718, USA

Abstract.—A collaborative group of fisheries managers and researchers (Cherry Creek Working Group) took advantage of an 8-m waterfall, 100 km of upstream habitat, and a 3 ha-lake to eradicate nonnative trout and introduce native Westslope Cutthroat Trout (WCT) *Oncorhynchus clarkii lewisi* in Cherry Creek, a tributary to the Madison River. This project was part of a larger, broad-scale effort to restore WCT within the Madison River basin. The project was logistically and politically complex and required long-term commitments by

* Corresponding author: patclancey5@gmail.com
[1] Retired, Montana Fish, Wildlife and Parks
[2] Retired, U.S. Forest Service

state and federal agencies, a private landowner, nongovernmental partners and university researchers. We describe and discuss the social, legal, and logistical challenges that arose during this project and provide our perspective on why this project succeeded in spite of these challenges. Administrative and legal challenges delayed implementation of the project for several years, but all challenges were resolved in favor of the project's collaborators. Over a 12-year period, nonnative trout occupying the area were eradicated using piscicides, and more than 39,000 WCT eyed eggs and fry were introduced into four geographic areas (phases) that were separated by natural or temporary fish barriers. Population recovery, measured by comparing pre- and post-treatment trout densities and mean sizes, appeared to occur in 3–4 years. We summarize research that documents the effects of piscicides on nontarget species and the expansion of introduced WCT and their progeny to fill all available habitats, along with lessons learned that are helpful to others designing species conservation efforts of similar scale and complexity.

Introduction

At the time of initial European exploration into western North America, Westslope Cutthroat Trout (WCT) *Oncorhynchus clarkii lewisi* were distributed across most accessible waters draining the northern portion of the Rocky Mountains in Idaho, Montana, and Wyoming, USA (Figure 1), and southern Alberta and British Columbia, Canada, including a few disjunct areas to the west (Behnke 1992). Subsequently, humans have affected the quality, availability, and connectivity of their habitat; introduced numerous nonnative species into waters they historically occupied; and occasionally overharvested them. By the early 1990s, WCT had been lost from an estimated 40% of their historical habitats, and populations with little to no evidence or suspicion of hybridization occupied less than 25% of their historical habitats in the United States (Shepard et al. 2005). The decline in WCT throughout their native range led to several petitions for listing them in the United States under the Endangered Species Act (ESA; e.g., USFWS 2003), being listed as threatened in Canada (COSEWIC 2006) and designated as a species of special conservation concern by many fish and habitat management agencies. The Montana Natural Heritage Program lists WCT as "apparently secure and/or suspected to be declining" globally (G4) but a "species of concern" (S2) within the state.

In Montana, WCT are native in most waters west of the Continental Divide, and east of the Divide they are native in the Missouri River drainage upstream of the Marias River and in some disjunct island mountain ranges in central Montana (Figure 1). They historically occupied about 28,000 km of habitat within the Missouri River basin, but by the 1990s, they occupied only about 1,700 km (6% of historical stream kilometers) and were primarily restricted to relatively small headwater tributaries (Shepard et al. 1997, 2003). In the Madison River drainage, WCT historically occupied slightly less than 2,000 km, but by the 1990s they only occupied about 80 km (4% of historical stream kilometers; Shepard et al. 2003). The level of introgression of many of these populations was unknown. We know with near certainty that by 2003, genetically pure aboriginal Madison WCT occupied only about 4 km of stream in two locations, each with base flows of less than 0.03 m^3/s. Habitat alteration and competition and introgression with nonnative species were primary agents of these declines.

Montana Fish, Wildlife & Parks (MFWP) is the agency responsible for conservation, stewardship, and management of all fish species within Montana. To address the

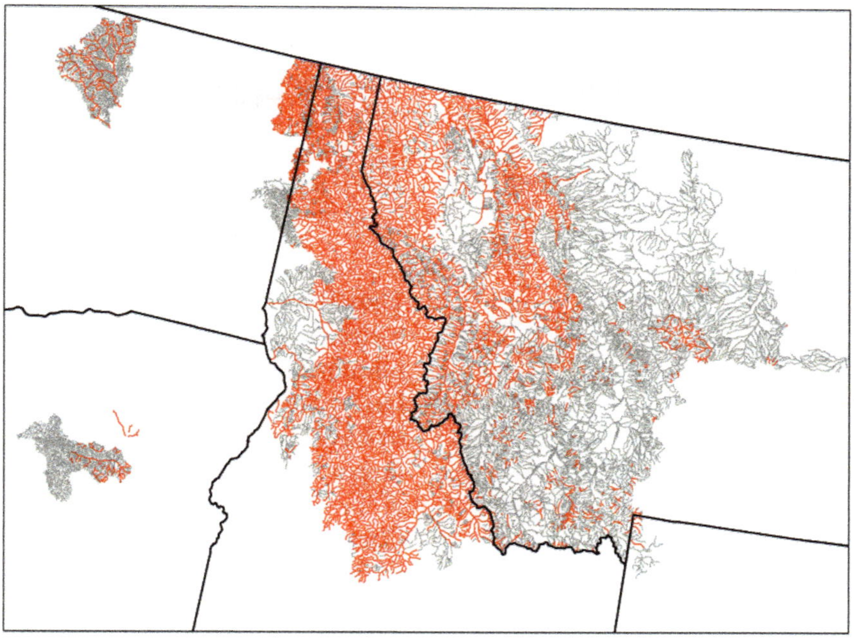

Figure 1. Westslope Cutthroat Trout historical (gray) and current (2005) distribution, from Shepard et al. 2005.

declining status of WCT, MFWP held a series of open forum meetings beginning in 1996 when the state's governor convened a conservation summit and invited representatives from state and federal agencies, Indian tribes, agricultural producers, resource extraction industries, conservation organizations, and private landowners to attend and become involved in WCT conservation. These meetings led to the formation of a Westslope Cutthroat Trout Steering Committee, composed of representatives of many of these varied interests, and a Technical Committee, composed of fisheries professionals who served as technical advisors to the Steering Committee. The Steering Committee used scientific information provided by the Technical Committee and considered social and economic issues to make policy recommendations to MFWP.

One such policy was a formal *Memorandum of Understanding and Conservation Agreement for Westslope Cutthroat Trout in Montana* (WCTMOU; MFWP 1999). The WCTMOU identifies conservation and restoration goals and objectives for WCT in Montana, and states,

> The management goal for Westslope Cutthroat Trout in Montana is to ensure the long-term, self-sustaining persistence of the subspecies within each of the five major river drainages they historically inhabited in Montana (Clark Fork, Kootenai, Flathead, upper Missouri, and Saskatchewan), and to maintain the genetic diversity and life history strategies represented by the remaining populations.

Specific objectives detailed in the WCTMOU are

1. To protect all genetically pure WCT populations,
2. To protect introgressed (<10% introgressed) populations,
3. To ensure the long-term persistence of WCT within their native range,
4. To provide technical information, administrative assistance, and financial

resources to assure compliance with listed objectives and encourage conservation of WCT, and

5. To design and implement an effective monitoring program by the year 2002 to document persistence and demonstrate progress towards goal.

Objective 3 further states,

> The long-term persistence of Westslope Cutthroat Trout within their native range will be ensured by maintaining at least ten population aggregates throughout the five major river drainages in which they occur, each occupying at least 50 mi of connected habitat.

The Cherry Creek project offered an opportunity to restore WCT to nearly 100 km (60 mi) of connected habitats and help meet this WCTMOU objective for long-term persistence of the subspecies.

The WCTMOU was updated in 2007 and coupled with a similar document for Yellowstone Cutthroat Trout (YCT) *Oncorhynchus clarkii bouvieri* called the Memorandum of Understanding and Conservation Agreement for Cutthroat Trout in Montana (CTMOU; MFWP 2007). Additional conservation, tribal, agency and resource extraction interest groups joined with the original signatories to support the updated conservation agreement.

In this chapter we describe and summarize the genesis of the Cherry Creek project, the challenges faced and how they were resolved, the methodology of our piscicide treatments and fish and embryo translocations, why we believe the project was successful, how we incorporated research and monitoring efforts into the project, and the monitoring and research results to date. Because the Cherry Creek project is one of the largest native trout conservation projects successfully completed to date, it was instrumental in paving the way for additional large-scale native trout restoration projects in Montana and across the range of interior Cutthroat Trout. One important goal of the project's collaborators (Cherry Creek Working Group [CCWG]) is sharing our experiences and lessons we learned while implementing this project. We summarize these lessons in each section of the project for ease of reference.

Project Area

The Cherry Creek drainage is a fifth-level hydrologic unit code (HUC) waterway (HUC number 1002000714; USGS 2018) and is the largest tributary volumetrically in the lower 65 km of the Madison River. The project area ranges in elevation from approximately 2,652 m mean sea level (msl) at the headwaters of Cherry Lake to 1,539 m msl at the barrier waterfall, and approximately 1,350 m msl at its confluence with the Madison River. There are an estimated 145 km of stream above an 8-m barrier waterfall (Cherry Creek Falls; Figure 2), mostly in first- and second-order drainages (National Hydrography Dataset; USGS 2013). The waterfall is approximately 13 km upstream of the Cherry Creek confluence with the Madison River.

Historically fishless above the barrier waterfall, the only known fish introduction in the project area occurred in 1924, when fry of an unknown species were planted in Cherry Lake by a forest ranger along with the Flying D Ranch manager and his 10-year-old son (as related to MFWP personnel by the son when he was 90 years old). The following year, the boy caught YCT in Cherry Lake. We suspect that Rainbow Trout *Oncorhynchus mykiss* and Brook Trout *Salvelinus fontinalis* were introduced into the drainage by ranch hands or sportsmen over time. Based on pretreatment distributions determined by electrofishing, we estimate that about 100 of the 145 km of streams in the project area were occupied by nonnative Rainbow Trout, YCT, and/or Brook Trout prior to eradication (Figure 3). No other fish species occupied the project area.

Land ownership within the project area is both public and private. Phase 1 (Figure

Figure 2. Cherry Creek waterfall at stream kilometer 13. This waterfall is the downstream extent of the project area.

3) of the project area is primarily managed by the Gallatin National Forest (GNF), while Turner Enterprises, Inc.'s (TEI) Flying D Ranch encompasses most of Phases 2–4. State of Montana lands are also present. Turner Enterprises, Inc. manages the Flying D Ranch for bison production, timber extraction, and commercial hunting while promoting conservation and restoration of native species and their habtats. Conservation work has included riparian and range restoration, forest management, wolf and native fish restoration, and other similar activities. About 44,500 of the 47,400 ha of the ranch are under conservation easement with The Nature Conservancy. The GNF lands are managed primarily for recreation and livestock grazing. A small portion of the project area (Cherry Lake, its inlet streams, and a short section of its outlet stream) lies within the Lee Metcalf Wilderness Area.

Due to the large size of the project area, we broke the area into smaller sections, or phases, for treatment. Natural or manmade barriers were used to separate the phases and prevent fish in untreated downstream phases from migrating upstream into treated phases between years (Figure 3). Phase 1 was isolated from the remainder of the watershed by two natural waterfalls. We converted two irrigation weir sites into barriers between Phases 2 and 3 (hereafter, the Phase-2 barrier) and Phases 3 and 4 (hereafter, the Phase-3 barrier).

The Phase-2 barrier site was a removable pin-and-plank irrigation diversion structure with concrete sidewalls and apron. With planks removed, this weir did not impede upstream movement of aquatic organisms prior to modification in November 2004 for this project. To make this site a fish passage barrier, six 20 cm × 20 cm × 6 m treated beams were bolted together, placed with a backhoe against the upstream side of the concrete sidewalls, and covered with fine-mesh coconut fiber mat to capture

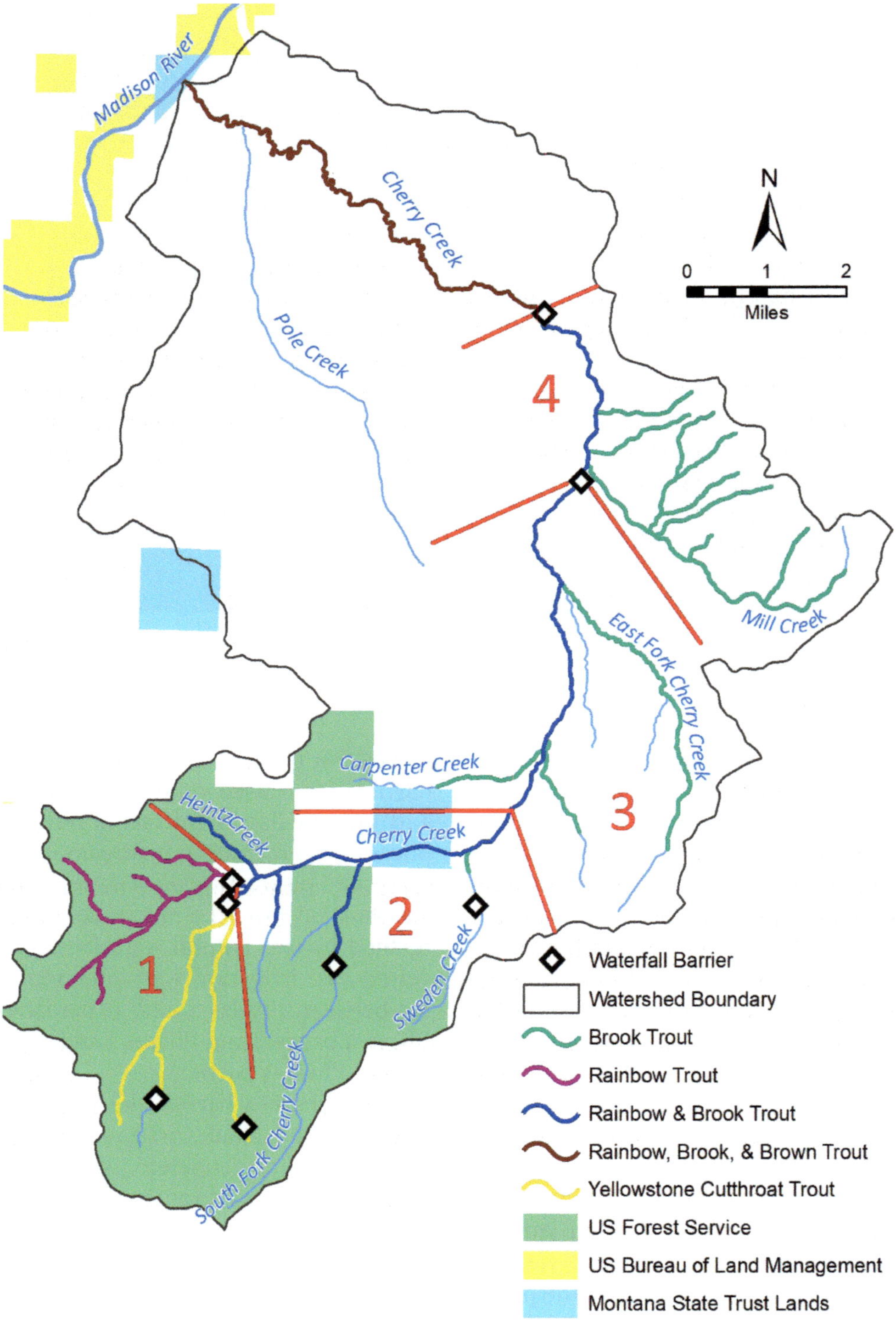

Figure 3. Map of the Cherry Creek drainage showing 2002 pretreatment fish distributions and phases. The downstream extent of the project area is a waterfall 13 stream kilometers upstream of the Cherry Creek mouth. The only fish species in the project area were Brook Trout, Rainbow Trout, and Yellowstone Cutthroat Trout.

and hold fine sediment to prevent seepage between and around the beams. Height of the constructed barrier was ~1.2 m, the maximum site topography would allow but not sufficient to prevent fish passage. Thus, a screen constructed of angle iron and aluminum electrical conduit was bolted at an angle to the downstream face of the barrier in July 2005 to block fish attempting to leap over the barrier. The angled screen allowed most debris to wash off. Although of modest height, we never saw or documented a fish passing over this temporary barrier with the screen attached during the project. Just upstream from this barrier site, a small side channel that flowed around the barrier site was disconnected from the main stem by placing concrete highway jersey barriers at the upstream end of the side channel and covering with fine-mesh coconut fiber mat secured along their base with stream gravel. This effectively dewatered the side channel.

The Phase-3 barrier site included a permanent 1.2-m concrete irrigation weir spanning the entire Cherry Creek stream channel with a 5-m-long downstream concrete apron, as well as a large off-channel spring that created a side channel of about 0.14 m^3/s adjacent to Cherry Creek starting above the concrete weir site and entering Cherry Creek immediately below the weir. Although the spring channel is isolated from the main stream channel above the weir during low flow periods, we were concerned that overbank flows could connect the spring channel to Cherry Creek and fish passage around the weir barrier; thus, a small, wooden temporary barrier was constructed near the mouth of the spring channel in summer 2006, about 1 year prior to the initial chemical treatment of Phase 3. This irrigation diversion site is a partial impediment to upstream aquatic organism passage, and the structure has caused streambed deposition that, over time, resulted in significant aggradation of the stream channel bed upstream of the weir. In September 2008, a 30-cm beam was attached to the top of this weir, raising its height to 1.5 m. Modeling based on water depth on the apron and fish jumping ability indicated that the barrier height alone was adequate to prevent upstream fish passage; thus, a screen was not attached to the downstream face of the Phase-3 barrier at that time.

The geology around this Phase-3 barrier site creates a unique localized flow and temperature dynamic in Phase 4 of Cherry Creek. During July and August, much of the flow in Cherry Creek above the concrete weir goes subsurface and the channel can be completely dry in August during normal or below average flow years. Subsurface water loss begins several km above the barrier site but increases substantially as it approaches the barrier. When present, the dewatered reach is typically 1–1.5 km in length just above the barrier site. Much of the water returns to the surface in a number of in- and off-channel springs within the immediate vicinity of the weir, including the spring we constructed the barrier on in 2006, contributing approximately 0.45 m^3/s to stream discharge within a 180-m reach upstream of and adjacent to the weir. These springs moderate water temperature, maintaining the stream temperature downstream of this site at about 10–13°C during the summer months.

Once we felt sure that we had achieved nonnative trout eradication, these two temporary barrier sites were restored to their original pretreatment condition in 2011. The Phase-2 main- and side-channel barriers were removed, the 30-cm beam was removed from the top of the Phase-3 barrier, and the wooden barrier on the adjacent spring channel was removed and all disturbed areas reclaimed. Consideration was given to completely removing the concrete weir at the Phase-3 barrier, but the 1.2-m differential in streambed elevation would have led to significant sediment transport, turbidity, and channel change, even if weir removal was done incrementally. Removal of the weir could also have affected the springs that maintain streamflow and cooler water temperatures year-round through much of Phase 4. Alternatively, in 2016, two large rock step pools were constructed downstream of the concrete weir (Figure 4). These pools improved aquatic organism passage by reducing the jump height over the weir from about 1.2 m to about 46 cm.

Mistakes, learning experiences, and innovations

We initially thought the existing 1.2-m irrigation weir between Phases 3 and 4 would serve as a fish movement barrier, but we discovered adult fish, primarily Rainbow Trout, in the lower portion of Phase 3 in spring 2008 and 2009 after Phase-3 chemical treatments in 2007 and 2008. The 30-cm beam was secured to the top of the concrete weir in September 2008. Nevertheless, during spring runoff in 2009, fish were observed jumping from the standing wave at the toe of the barrier and swimming up the laminar water flow and over the Phase-3 barrier, necessitating a third piscicide treatment of a portion of Phase 3 in 2009. An inclined screen was subsequently added to the barrier in July 2009, and after this screen was added, no additional fish were documented passing the barrier. We discovered that even though modeling may conclude that a barrier will effectively prevent fish passage, it does not guarantee that fish will not be able to pass the structure. Monitoring and direct observation were critical to identifying the ability of fish to pass this partial barrier and for modifying it into a complete barrier. Having the resources in place to be able to quickly address the problem was also critical.

With the experience gained on the Cherry Creek project, we feel that we can now chemically treat larger areas more expeditiously than we did during this project, potentially reducing or eliminating the need for barriers while reclaiming large subdrainages, drainages, or watersheds for native fish conservation.

Project Genesis and Environmental Analyses

An important catalyst for the Cherry Creek project occurred about 10 years before the project was conceived. In the late 1980s, whirling disease *Myxobolus cerebralis* was introduced into the Madison River, likely through an illegal Rainbow Trout introduction. Routine monitoring of trout populations in the Madison River detected a widespread and significant decline of the

Figure 4. Pre (left) and post (right) photos of Phase-3 barrier showing the results of step pool construction.

Rainbow Trout population upstream of Ennis, Montana by 1991, and by 1996, the number of Rainbow Trout larger than 300 mm decreased from a long-term average of 620/km to 125/km (Clancey 1997). Angler success, commercial outfitting, and other businesses in Ennis and surrounding communities suffered significantly due to the diminished Rainbow Trout population in the Madison River. While MFWP and others were making substantial efforts to address the effects of whirling disease, MFWP considered native trout conservation as part of the solution to restore the trout population and, subsequently, angling in the Madison River. After careful consideration, MFWP decided to restore native WCT in selected tributaries to the Madison River. This effort was designed to accomplish two objectives: (1) to initiate a long-term WCT conservation and restoration program that contributed toward meeting the goals outlined in the developing WCTMOU, and (2) to provide source populations of trout potentially less affected by whirling disease that could provide fish to the Madison River via downstream fish movement to diversify and improve recreational angling opportunities. Montana Fish, Wildlife & Parks expected that chemical removal of nonnative fish would be necessary in some locations to achieve the probable goals of the developing WCTMOU and the WCT conservation and restoration program.

In 1997, MFWP began a widespread investigation of Madison River tributaries upstream of Ennis Reservoir to assess their fish populations, species compositions, and habitat suitability and identify streams that could be included in WCT restoration efforts (Clancey 1998a; Sloat et al. 2000). After learning of MFWP's conservation effort through a local newspaper article, Ted Turner, founder and chairman of TEI, directed his staff to contact MFWP through Dr. Calvin Kaya at Montana State University to inquire about the feasibility of including the Cherry Creek drainage in the program. Dr. Kaya had worked with TEI's Flying D Ranch and MFWP in 1994 to conduct an experimental Arctic Grayling *Thymallus arcticus* introduction into a section of Cherry Creek to assess how they would interact with the existing sympatric nonnative Rainbow Trout and Brook Trout populations. Monitoring the following year showed that there were no Arctic Grayling remaining in Cherry Creek. On June 25, 1997, representatives of MFWP and TEI, along with Dr. Kaya, met in Ennis to discuss the possibility of exploring the Cherry Creek drainage to determine its feasibility for supporting populations of WCT, and possibly Arctic Grayling and other native nongame fish species. These discussions led to an agreement between MFWP and TEI that allowed MFWP to lead a short-term, intensive effort to survey the upper Cherry Creek basin to assess its suitability for WCT.

Cherry Creek is one of only five tributaries to the lower 65 km of the Madison River (below Ennis Reservoir) and is one of the largest tributaries in the Madison drainage in terms of drainage area, stream discharge, and stream miles. Cherry Creek Falls (Figure 2) is a critical feature that made Cherry Creek particularly attractive for a large-scale WCT restoration effort. Above the waterfall are approximately 145 km of stream (National Hydrography Dataset; USGS 2013) and a 3-ha headwater lake (average depth 4.5 m; maximum depth 12 m). Brook Trout, Rainbow Trout, and YCT were the only fish species found above the waterfall—none of which are native to the Madison River drainage.

Nine people surveyed a significant portion of the upper Cherry Creek drainage (Figure 3) from August 25 to 29, 1997. During this time, fish species composition and distribution surveys were conducted, stream discharges were measured, habitat characteristics were observed and noted as to whether they were suitable for WCT, and initial consideration was given to the logistical effort necessary to manage a crew conducting a large-scale fish eradication project in this remote area.

Upon completion of the initial surveys, two fisheries professionals who had experience conducting chemical eradication projects visited the site to offer their assessments. Dr. Bruce Rosenlund of the U.S. Fish and Wildlife Service (USFWS) in Colorado had conducted many similar, but smaller, projects in Colorado, including in Rocky Mountain National Park. Dr. Rosenlund spent three days with members of the CCWG assessing the drainage. Dr. Robert Gresswell of the USFWS, who had conducted a similar project in Arnica Creek in Yellowstone National Park (Gresswell 1991), spent one day in the area with project leaders. They both concluded that a physical effort (using netting and electrofishing) to eradicate nonnative fish would fail due to the size and complexity of the watershed, but that a chemical eradication effort could be successful if a dedicated effort could be sustained. Among the difficulties they cited were the ability to sustain a large-scale, multiyear effort and how to manage water and North American beavers *Castor canadensis* in several large beaver pond complexes. Based upon the feasibility work, the assessments of these experienced biologists, and additional discussion between MFWP and TEI about expanding the effort to potentially include other native aquatic species such as Arctic Grayling, Rocky Mountain Sculpin *Cottus bondi* (formerly Mottled Sculpin *C. bairdii*) and western pearlshell *Margaritifera falcata*, a decision was made to move forward with an environmental assessment (EA) analyzing the Cherry Creek project.

The state of Montana requires an EA be done for all state actions that can potentially impact public health and welfare or the environment (Montana Environmental Policy Act). For the Cherry Creek project, this process began in December 1997 when public scoping meetings were held in Bozeman, Ennis, and Three Forks. A draft EA was developed and released in April 1998 (Bramblett 1998), with a 32-d public review and comment period. The state's EA was developed and written to also meet National Environmental Policy Act (NEPA) standards, allowing the GNF to use the same analysis as the basis for its decision making regarding application of piscicides and related use of mechanized equipment within the Lee Metcalf Wilderness portion of the project area. The EA considered several alternatives for restoration of native fish in Cherry Creek, ranging from physical removal of nonnative fish to the introduction of WCT and Arctic Grayling without the removal of nonnative fish, but ultimately concluded that chemical eradication (i.e., application of piscicides) of nonnative fish had the highest likelihood of success. The public was notified through press releases and legal notices in several southwestern Montana newspapers, through live and recorded radio programs, and by mailing or delivering 120 copies of the EA to sporting and conservation organizations, government agencies, and interested parties. During this comment period, public meetings were held in Bozeman, Three Forks, and Butte to take comment on the EA. Copies of the EA were available at several MFWP offices in southwest Montana and at the public meetings. After considering public comments, a final EA and formal decision notice were finalized in July 1998 (Clancey 1998b; USFS 1998).

The U.S. Environmental Protection Agency (EPA) authorizes two compounds to be used as piscicides—Antimycin A and rotenone. Both occur naturally in the environment but must be combined with other carrier agents that enhance mixing and dispersal in water in order to effectively function as a piscicide. Both piscicides were analyzed in the EA and authorized for use on the project. The product names used during the Cherry Creek project are Fintrol (Antimycin A) and CFT Legumine (rotenone). Rotenone powder was also used in some locations such as springs and seeps and was administered following the procedures described by Finlayson et al. (2000).

As required by state statute, MFWP applied for and received a permit from the Montana Department of Environmental Quality (DEQ) that allowed a short-term exemption to state water-quality standards for application of piscicides for the removal of "undesirable and nuisance aquatic species" in the project area. Additionally, MFWP applied for and received a short-term exemption from the DEQ for turbidity exceedance that might occur during installation of temporary fish movement barriers used to divide the project area into phases to facilitate piscicide application across years. Finally, MFWP received two authorizations from the GNF for use of mechanized equipment and application of piscicides in the Lee Metcalf Wilderness. All these permits were procured by August of 1998.

During the planning and development for this project, we recognized a unique opportunity to incorporate research within the project. Too often practitioners of conservation projects fail to collect information that allows objective evaluation of success or failure (Pullin and Knight 2001; Sheller et al. 2006; Anderson et al. 2014). We agreed to conduct project activities in a way that maintained enough scientific rigor and allowed collection of data to compare fish population characteristics (e.g., density and size structure) before and after restoration; to assess project impacts on nontarget organisms; to evaluate how WCT donor source (e.g., nearest neighbor single population versus genetically variable broodstock) influenced population recovery and project success, and to measure survival, condition, and movement of translocated individuals. This information was intended to inform future restoration and translocation efforts and provide information on donor-source selection criteria.

Parallel to the beginning of the Cherry Creek project, the Sun Ranch, a privately owned property in the upper Madison Valley, entered into a formal agreement in 1999 to assist MFWP with WCT conservation and recovery by building and operating a small conservation hatchery, dubbed the Sun Ranch Hatchery (SRH). This facility incubates eggs taken from wild donor populations and has the capacity to rear a limited number of the fry during their first summer to provide both eyed eggs and young-of-the-year WCT for introduction into restoration sites. The Sun Ranch also constructed a brood rearing pond to develop a Madison River/upper Missouri River WCT broodstock. Some of the fry reared in the facility's fry-rearing troughs are released into the Sun Ranch Pond (SRP) each year to maintain the brood and incorporate as much genetic diversity into the brood pond as feasible. The Sun Ranch facility was an important component of the Cherry Creek project because its isolation allowed for wild donors to be used with a minimal risk of transferring disease into state-owned hatchery facilities, and its operators could focus solely on meeting conservation needs in Cherry Creek. This facility is also playing an important role in WCT restoration in Yellowstone National Park, Wyoming.

Mistakes, learning experiences, and innovations

Project opportunities such as Cherry Creek are rare at this scope and scale. Private entities can be creative and instrumental in helping public resource agencies be more nimble or flexible in finding solutions (e.g., TEI and the SRH). Common ground and mutually desired outcomes arising from different places (e.g., ethical beliefs, legal obligations, and recreational opportunities) can create collaborators from diverse interests such as state, federal, and private entities.

Interestingly, delays in implementation of the project (next section) provided opportunities for us to become more educated and gain experience with piscicide application. During the several years the project was delayed, we were able to participate in similar but much smaller piscicide projects in Idaho and Crater Lake National Park. The field experience gained while assisting with these projects removed some of the mystique of handling, applying, and detoxifying piscicides and managing large field crews.

Issues, Concerns, and Legal Challenges

During the public scoping process in 1997, many conservation-minded individuals and groups supported the project, but it was not without controversy. There were several reasons for opposition, but the primary issues raised included a general disdain for Ted Turner and the perceived cozy collaboration between the state of Montana and TEI, concerns regarding the effects of piscicides on the environment and human health and livelihoods, opposition to the removal of an existing wild trout population, and perceived impacts to the designated wilderness area that encompassed a small portion of the Cherry Creek headwaters. Although piscicide use for restoration work was not new to fisheries management in the late 1990s, piscicides had not been commonly used for native trout conservation in many Rocky Mountain states at that time (with a few notable exceptions such as Stuber et al. 1988 and Gresswell 1991). Montana was no exception, and in large part due to an unfamiliarity with piscicides in general, some members of the public were skeptical of their proposed use in Cherry Creek.

As the environmental review process proceeded and the EA was released for public comment, a small number of vocal individuals began insistently objecting to the project, often arguing unrelated issues involving Mr. Turner's environmental and personal beliefs, lack of public access to the privately held TEI properties, and a 1996 land trade between TEI and the state of Montana that consolidated blocks of public and private lands. As project opponents threatened administrative and legal challenges via letters to local newspapers and by word of mouth, some regional MFWP administrators became uneasy with the project. Consideration was given to abandoning the proposal, but key MFWP administrators at the state level, including the agency director, supported the project and were committed to seeing the public environmental review process run its course.

Project opponents petitioned the USFWS to list the "Cherry Creek cutthroat" as a distinct population segment of YCT, stating that it had evolved in a harsh headwater environment, was resistant to whirling disease, successfully competed with nonnative species; and was genetically pure. However, the petition did not provide any scientific data to substantiate these claims and the USFWS could find no evidence to support the petition. The USFWS did not consider this population of YCT unique because fish stocked from this same strain are found in numerous other mountain lakes with similar conditions. They stated that the source of the YCT initially stocked in Cherry Lake and its outflow originated from MFWP's Yellowstone River Trout Hatchery in Big Timber. Fish from this hatchery were released in mountain lakes in the Missouri River drainage and for many years were routinely stocked in fishless mountain lakes throughout Montana. The individual who stocked these YCT in Cherry Lake in 1924, on his 10th birthday, stated that he did not know the source of the fish. Yellowstone Cutthroat Trout are not considered native to Cherry Creek and Cherry Lake since the area is well outside the fish's historical range of the Yellowstone and Snake River drainages. Additionally, the Cherry Creek basin is within the known historical habitat (i.e., the Madison River) of the Westslope subspecies of Cutthroat Trout.

Written response to the EA included 21 individuals or groups supporting the proposal, 40 opposed, and 222 individuals that signed an opposing form letter. One state agency submitted a letter of comment without taking a stance on the proposal. Comments and questions were wide-ranging and often not pertinent to the proposed project. Some comments were legitimately concerned with issues such as the effect of piscicides on nontarget organisms, the use of a gas-powered outboard motor on Cherry Lake within the Lee Metcalf Wilderness, or concern over the loss of public access and angling opportunity on the national forest. On the other hand, similar to the public scoping process, many comments were tangential extensions of ongoing land management and wildlife-related issues with MFWP, TEI's Flying D Ranch, or Mr. Turner himself. Others questioned the environmental review process and the responsibility of federal agencies in that process while a few tried to connect the project with their belief that the United Nations was creating a one-world government and Mr. Turner was complicit because he had pledged to make a large donation to the United Nations.

In the July 6, 1998 decision notice approving the proposed project, the CCWG provided a response to the comments submitted by the public regarding the EA (Clancey 1998b). Comments were sorted into 17 separate categories, including concerns about the effects of the piscicide on water quality and nontarget organisms, project impacts on federally designated wilderness, genetic integrity of the donor fish and disease concerns, privatization of water and wildlife, public awareness of the project, impacts of introduced WCT on land management activities and recreational opportunities, and impacts on hunting and fishing, including to an existing hunting outfitter business. Some commenters felt it was a conflict for MFWP to enter into a formal agreement with TEI for project funding or questioned the authority of the regional forester to direct GNF involvement in the project. Other issues raised were not pertinent to the project, such as questioning how a fish named the Westslope Cutthroat Trout could be native east of the Continental Divide.

Chemical treatment was scheduled to begin in August 1998 but was halted the morning of the first day due to last minute logistical and scheduling concerns raised by TEI. These concerns were resolved relatively quickly but not in time to salvage fieldwork in 1998, so the project was rescheduled for August 1999. This delay led to a cascade of legal filings and decisions that prevented fieldwork until 2003. It also provided our first lesson on the importance of clear and consistent communication among CCWG members.

Prior to the 1999 start date, a few of the most vehement project opponents contacted both of Montana's U.S. senators and Montana's lone U.S. representative, asking them to intervene and stop the project by requiring the GNF to rescind or not authorize necessary permits or decisions for project related actions on forest lands, and to intervene with Montana state agencies that had permitting authority over various aspects of the project. These requests were not simultaneous but were made in a serial manner, apparently to maximize the time necessary for the CCWG to address them and was planned by opponents to delay project implementation for as long as possible. All three Congressional offices contacted MFWP and the GNF to gather information about the project, and one U.S. senator sent a staff member from Washington D.C. for a 2-d field visit. That senator sent a letter to project leaders acknowledging their openness about the project and thanking them. Ultimately, all three offices responded that the project was legal, that proper federal procedures had been followed, and that the project was within the state's authority. Simultaneous with the congressional inquiries, project opponents teamed with an attorney who represented cyanide heap leach gold mines in Montana to legally challenge the Cherry Creek project. Their premise was that if mining companies cannot release pollutants into a Montana stream without a Montana Pollution Discharge Elimination System (MPDES) permit, then MFWP should not be allowed to either. They threatened a lawsuit against the DEQ that triggered the DEQ to conduct a separate EA assessing the legality and public safety of using piscicides. During this EA process, the DEQ suspended their previously issued permits to apply the chemicals in the project area. Montana Fish, Wildlife & Parks did not object to the permit suspension while the DEQ conducted its analysis.

The DEQ EA was released on July 29, 1999. Public comment was accepted through September 14, 1999. On October 14, 1999, the DEQ issued a decision authorizing and reissuing the permits to conduct the chemical treatments with no modifications. In their decision document, the DEQ identified and responded to 64 comments, many of which were redundant to those addressed in MFWP's 1998 EA. Montana Fish, Wildlife & Parks had stated that if the DEQ's EA resulted in reissuance of the permits, chemical treatment of Cherry Lake may be conducted as late as October or November 1999; however, the opponent's attorney appealed the DEQ decision within minutes of it being issued. He challenged the DEQ EA on five points: (1) the DEQ erred in not requiring MFWP to apply for and receive an MPDES permit, as is required for industrial effluent; (2) the DEQ erred by failing to apply the state's nondegradation statute, a violation of the state's constitution, which guarantees its citizens the right to a "clean and healthful environment"; (3) the DEQ erred in granting a series of short-term exemptions to certain state statues for a multiyear project; (4) the DEQ failed to determine if the project was necessary; (5) and the DEQ failed to determine if the nonnative trout to be eradicated were "undesirable and nonnative species."

The Montana Board of Environmental Review (BER) considered the appeal. The BER appointed a hearings examiner who gathered information about the project and reviewed pertinent state statutes and administrative rules. The hearings examiner issued his opinion on July 11, 2000, recommending summary judgment to the DEQ, essentially dismissing all five points of contention. On September 28, 2000, the BER issued summary judgment to the DEQ but stipulated a 33-d stay of execution to allow the appellant the opportunity to file a petition for judicial review in the Montana District Court. On October 31, 2000, the appellant filed the petition against the BER, DEQ, and MFWP. The elements of the lawsuit were that (1) since the permit from the DEQ to apply piscicide was not subject to nondegradation statue, it is a violation of the (Montana) constitutional clean and healthy environment provision; (2) application of the piscicide to outstanding resource waters (within wilderness area) violates statutory and constitutional provisions (federal regulations); (3) since the project was to be completed over a period of up to 10 years, it violated the short-term nature of the permit that allows short-term exceedance of certain water-quality parameters; and (4) the DEQ did not independently determine the necessity of the project.

Montana Fish, Wildlife & Parks submitted a response to the petition on November 29, 2000. After a series of fact finding, the FWP filed its argument for summary judgment on August 6, 2001, arguing, among other things, that the project supports both the Clean Water Act and Montana citizens' right to a clean and healthful environment by restoring a native fish to state waters. On March 28, 2002, the Montana District Court ruled for the agencies (Montana District Court 2002), finding (on the four elements listed above) (1) that the existing and beneficial uses of the water in question would be preserved upon completion of the activity, which included application of EPA-registered pesticides in accordance with the labels; (2) the same determination as item 1; (3) that though the project is anticipated to be underway for up to 10 years, the actual application of the piscicides would occur for no more than a few hours each day for 3 weeks or less in any year. (The permits are valid for only 1 year and MFWP must apply for a new permit each year, which provides the DEQ the opportunity to review and approve, modify, or deny the permit on an annual basis.); and (4) that it is not the DEQ's responsibility to determine whether the proposed activity itself is necessary, but instead to determine if an exemption from water-quality standards is necessary to conduct the proposed activity.

With the favorable ruling from the Montana District Court, MFWP planned to implement the project in August 2002. In early June, project opponents filed a 60-d notice of intent to sue MFWP in the Federal District Court of Montana. On July 3, in face of the threatened lawsuit, MFWP made the decision to again postpone the project to avoid the cost of preparing for and initiating the project only to have it suspended by court order, should a court decide that was appropriate. Project opponents were notified of MFWP's decision to postpone but then subsequently did not file the threatened lawsuit.

During Montana's 2001 legislative session, a state senator from the Bozeman area publicized his intent to introduce a bill into the legislature that would specifically prohibit expenditure of state money for piscicide applications on the Cherry Creek project. Montana Fish, Wildlife & Parks administrators feared the bill could be expanded by other legislators to prohibit other native fish conservation or restoration actions. The newly appointed MFWP director proposed to the senator that MFWP would not spend any state money on piscicide application for the project if he would not introduce his bill, and the legislator agreed. Consequently, TEI committed additional funding to the project to support MFWP's field staff time and expenses for piscicide applications. Other MFWP expenditures related to the project, such as planning, administrative and monitoring costs, were not affected by this agreement.

In 2003, MFWP notified project opponents that the project was scheduled to begin August 1, and opponents filed a 60-d notice of intent to sue on June 1. However, this time, MFWP initiated the project as scheduled on August 1 and the lawsuit was filed. The federal district court judge dismissed the suit due to a technical error by the litigants and their bad-faith tactics. Montana Fish, Wildlife & Parks negotiated with the litigants to allow completion of first-year treatments by August 20, 2003, and the litigants agreed not to seek an injunction against the project in 2003. Litigants re-filed the federal lawsuit on October 31, 2003. Elements of the lawsuit were (1) as used during the project, piscicides meet the federal Clean Water Act (CWA) definition of "pollutant" and are being applied from a "point source," therefore subject to an NPDES/MPDES review. (Parties agreed on the point-source issue, so the remaining issue was whether the piscicide or its residue were pollutants); and (2) the CWA and the Federal Insecticide, Fungicide, and Rodenticide Act (FIFRA) are in conflict over exemption of FIFRA-approved pesticides from certain provisions of the CWA because discharge of pesticides, even for their intended purpose, constitutes a waste because not all the pesticide applied is absorbed by the target organisms, but also by nontarget organisms.

The Federal District Court of Montana ruled in March 2004 (U.S. District Court for the District of Montana 2004) that (1) the piscicide was applied in full accordance with its label in a manner that does not constitute pollution, it performed as intended and dissipated rapidly, leaving no trace of residue. Because the piscicide functioned as intended, it is not a pollutant or a chemical waste, therefore not subject to NPDES/MPDES review; (2) there is no conflict between the CWA and FIFRA based on the EPA's valid reasoning to coordinate the CWA and FIFRA, and on the EPAs long-standing administration of the CWA and FIFRA (EPA 2003).

The judge's ruling went on to state that one of the provisions of the CWA is to restore biological integrity to the nations waters and that the Cherry Creek Project

> comports four-square with the stated objective of the CWA to restore biological integrity of the drainage by removal of invasive pest species of trout and reintroduction of native species of trout.

On June 14, 2004, project opponents appealed the federal district court's ruling to the Ninth Circuit Court of Appeals. A three-judge panel of the Ninth Circuit considered the appeal and on September 8, 2005 ruled (1) since the piscicide (Antimycin A) was applied to the waters of Cherry Creek in accordance with its FIFRA label, with no residual or unintended effect, it is not a waste and therefore not a pollutant for CWA purposes; and (2) therefore, an NPDES permit was not necessary (U.S. Court of Appeals for the Ninth Circuit 2005).

After initial chemical treatments in 2003, the project was conducted as scheduled through the remaining years of the legal challenges.

It should be noted that a case in the U.S Sixth Circuit Court of Appeals in 2009 tested a similar situation (National Cotton Council et al. v. EPA) in which that appellate court vacated the EPA's 2006 Final Rule on Aquatic Pesticides and found that point-source discharges of biological pesticides and chemical pesticides that leave a residue into waters of the United States were pollutants under the CWA. As a result of that decision, NPDES permits are now generally required for these types of discharges as of October 31, 2011. Montana developed an NPDES permit system that has been used since 2012 to permit and conduct piscicide projects and is compliant with the EPA's requirements.

Mistakes, learning experiences, and innovations

Throughout the years of administrative and legal challenges, strong leadership among all the partners was critical to maintain focus, continuity, and commitment to completing project goals. This required consistent communication between members of the CCWG and within each agency. Montana Fish, Wildlife & Park's legal staff coordinated closely with project leaders to ensure the technical accuracy of the legal filings, which allowed the courts to clearly see the legitimacy of the project. Project leaders and the collaborating parties did not try to gloss over or hide any potential drawbacks or uncertainties related to the project or to chemical treatments. Doing so would have been damaging to the credibility of the CCWG and possibly led to termination of the project.

Collaboration and Delegation

An important element of the success of the Cherry Creek project was the collaboration between the three primary partners of the CCWG. The success of this collaboration over the duration of the project can be attributed in large part to consistent staff (e.g., little turnover in permanent staff over the course of the project); a shared vision important to all partners; clearly defined fiscal, logistical, and operational roles and responsibilities for each partner; and a common work ethic. Perhaps most importantly, the process of defining roles and the tools used to do that resulted in clear communication and trust among the partners and individual biologists.

The partners used a variety of tools to coordinate their roles at different management levels. These tools were essential to translating both broad formal agreements (e.g., both MFWP and the U.S. Forest Service (USFS) were signatories to the WCTMOU) and mutual interests in native fish conservation (declared as part of the missions of TEI, the GNF, and MFWP) to actual activities on the ground. Both the tools and the relationships that were formed in using them served to maintain partner engagement and coordination throughout the course of the project, including project challenges and setbacks, and postproject monitoring.

The first tool used by the partners was a formal memorandum of understanding (MOU) laying out the framework for the Cherry Creek project. Specifically, the MOU described project objectives; responsibilities of each entity, including planning, permitting, and respective authorities; and established an Incident Command System (ICS) structure to manage project implementation. Although not legally required, this MOU was requested by leadership of all partners, in no small part because this was the first restoration project in Montana using piscicides for fish removal on such a large scale that involved multiple partners across a mixed ownership landscape. The USFS suggested use of the ICS, given the agency's experience using this structure to safely manage logistically challenging and interjurisdictional work, and other partners concurred.

In general, the MOU defined the partners' roles by statutory responsibility. For example, by State of Montana statutes, MFWP holds authority for fish and wildlife management within Montana on lands of most ownerships. Federal lands can be an exception, but the USFS defers fish and wildlife management to states in cases where those actions are consistent with federal law, where those actions are not specifically delegated by federal law to the USFS, or by agreement with the states. Thus, MFWP took the lead in authorizing the overall Cherry Creek project as a fisheries management action. However, because of multijurisdictional boundaries (private, state, federal, and wilderness) within the project area, all three parties had the discretionary authority to approve separate aspects of the overall project.

Montana Fish, Wildlife & Parks was responsible for approving and overseeing the removal of nonnative trout, stocking genetically pure WCT, and applying piscicides and associated detoxifying agent outside the wilderness boundary. Under the direction of MFWP, TEI hired a consultant to write the 1998 draft EA analyzing and disclosing the effects of the overall project as required by the Montana Environmental Policy Act and NEPA, along with compiling a literature review on the effectiveness and potential effects of rotenone and Antimycin A.

The USFS retained decision authority for pesticide application and mechanized equipment use within the Lee Metcalf Wilderness, as delegated to the USFS by the Wilderness Act of 1964 (U.S. Code, volume 16, sections 1131–1136). The Association of Fish and Wildlife Agencies agreement (USFS et al. 2006) provides guidance to state fish and wildlife management agencies, the USFS, and Bureau of Land Management personnel for the management of fish and wildlife populations in wilderness areas in accordance with the Wilderness Act. Under the Wilderness Act, designated land-management officials can approve the use of otherwise prohibited tools or activities (such as pesticides or mechanized equipment) if traditional wilderness management tools or activities cannot achieve the desired objectives or if otherwise prohibited tools could be used with less impact to wilderness character than traditional tools. To approve prohibited tool use, the GNF was required to analyze the effects of the use of those tools on wilderness character against more traditional tools and activities through the Minimum Requirements Decision Guide, separately from NEPA. For this project, application of pesticides, nonlanding helicopter flights for restocking, outboard boat motors, electrical pumps, and scuba diving equipment were analyzed and approved. A

NEPA analysis, decision notice, and finding of no significant impact were required for the USFS to authorize a pesticide use permit for the application of piscicides within wilderness.

Turner Enterprises, Inc. owners and managers of the Flying D Ranch retained authority for those activities that occurred within the boundaries of the ranch beyond fisheries management actions. Turner Enterprises, Inc. handled logistical concerns such as authorizing access for project activities; permitting, constructing, and maintaining temporary fish movement barriers; implementing beaver and beaver dam management; coordinating food, transportation, and lodging; purchasing and maintaining project equipment; and hiring temporary field staff to assist with the project. Turner Enterprises, Inc. has also conducted much of the pre- and posttreatment population monitoring under permit from MFWP.

The MOU provided the framework within which a second set of tools nested: partnership agreements (PA) and interagency agreements allowing the partners to transmit funds from one entity to the next and to further define responsibilities beyond statutory roles. One such PA was between MFWP and TEI, allowing the transfer of funds to cover many MFWP expenses for the project. The GNF entered into a second PA with MFWP and TEI, defining the roles of the partners on all lands within the project area, including financial allocations. This PA used the relatively newly enacted Wyden Amendment authority (Public Law 109-54, Section 434, passed in 1998), which authorized the USFS to use USFS resources on private lands when there is a partnership action with a clear benefit to public resources, such as fish and wildlife and their habitats managed by the USFS. This was the first time that the GNF used Wyden authority, and it allowed the GNF to maintain the same roles in support of the project throughout project implementation regardless of landownership where activities were occurring. Turner Enterprises, Inc. and GNF also consummated an access PA that allowed GNF biologists access, for purposes related to project activity, across private land to the forest via previously closed routes. Finally, the GNF entered into an interagency agreement with the Bureau of Land Management so that boxes of hand-held radios could be ordered from the National Interagency Fire Center. This allowed all project personnel to carry a radio, greatly facilitating field operations, including piscicide application and detoxification.

Project partners held annual coordination meetings to review roles and responsibilities as spelled out by the MOU and agreements, as well as to plan operations for the upcoming field season. At these meetings, adjustments were made as needed, including identifying whether formal adjustments needed to be captured in the MOU or agreements.

Finally, the ICS was a tool also defined under the MOU, but its use was tailored to project implementation by a written set of ICS documents covering all aspects of project roles and responsibilities. The ICS is commonly used by state, federal, and local agencies to manage natural disaster and other complex incident responses (Figure 5), as the ICS structure can be molded to achieve very specific project needs. As such, some of these roles and responsibilities aligned directly with the formal authorities of different partner representatives, but others corresponded with skill sets of project staff or were eliminated for this project because they were unnecessary.

The ICS is a standardized approach to the command, control, and coordination of incident response, providing a common hierarchy within which personnel from multiple agencies can be effective. The ICS was initially developed to address problems of interagency responses to wildfires but slowly evolved as the preferred coordination model to address similar issues related

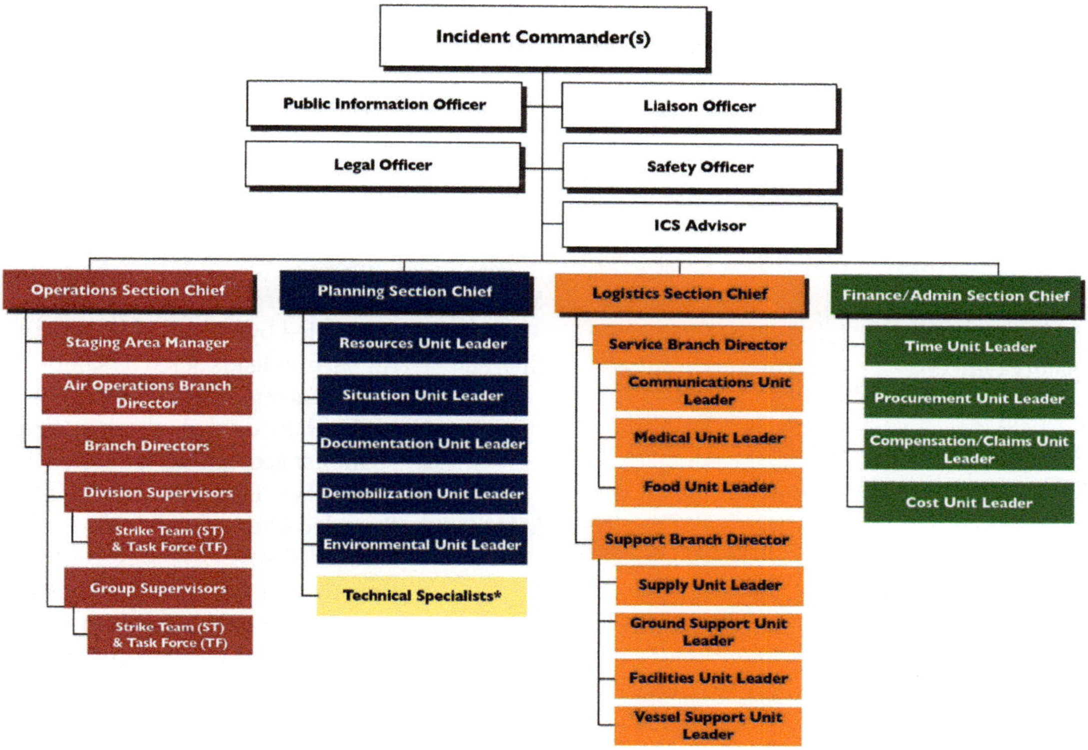

Figure 5. Structure of the Incident Command System and command, command staff, general staff, and substaff positions.

to the responses of other disasters ranging from active shootings to hazardous materials scenes throughout the United States. During a large-scale and rapid response to a natural disaster, an individual (or a unified group) typically occupies one of the command, general, and substaff positions without collateral duties to minimize distractions. Because this project was planned and coordinated over a relatively long period of time, lead partners from MFWP, TEI, and the GNF could be involved in multiple collateral duties without compromising the safety of crew members or the effectiveness of the treatment.

For the most part, finance, procurement, and administration section duties were shared among all partner leads prior to implementation, as previously spelled out in the project-specific MOU and PAs. Several units or substaff positions, such as air operations, were not required to implement this project. Table 1 describes the assigned duties for the Cherry Creek project organized using the ICS organizational structure.

In addition to ICS roles, standard ICS procedures were adapted for this project, some of which are discussed here. First, the equivalent of preseason incident management team meetings were held to reaffirm and review roles, discuss lessons learned, and incorporate any changes in ICS structure or operations. Second, during the actual project, daily briefings for project personnel, with reports and input from various ICS leads, were critical for detailing a day's treatment objectives, communication procedures, personnel assignments, and safety concerns. Third, project leads from each agency held evening planning meetings to review daily operations, adjust tactics, address personnel or safety issues, and prepare

Table 1. Various partnership responsibilities as related to the levels of the Incident Command System. MFWP = Montana Fish, Wildlife & Parks; GNF = Gallatin National Forest; TEI = Turner Enterprises, Inc.

Command and general staff	Substaff positions	Lead	Responsibilities
Incident Commander		MFWP	Management oversight of the implementation of the overall project.
Public Affairs Officer		MFWP	Public information releases were handled by MFWP Region 3 staff.
Safety Officers		Shared	All partner leads acted as safety officers regardless of affiliation. Maintained a daily check in/check out list accounting for crew members during and end of shift.
		MFWP	Provided licensed pesticide coordinator
			Provided all implementation and safety training as related to the various chemicals and associated personal protective equipment (PPE).
		GNF	Established an emergency communication and evacuation plan.
			Provided all other safety training using the agencies job hazard analysis protocol, including "Bear Aware" and bear spray training.
Operations Section	Divisions/ Strike Teams	Shared	All partnership entities provided personnel and volunteers to implement treatment operations. Personnel were assigned to individual drip stations or backpack spraying teams.
		MFWP	Provided oversight and operation of the detoxification station.
Planning Section	Resources	TEI/GNF	Assigned personnel to daily tasks. Provided morning safety and operation briefings. Maintained daily inventories of personnel and equipment.
	Situation	MFWP	Monitored stream flows and calculated pesticide application rates.
Logistics Section	Communications	GNF	Provided radios for all personnel, radio use training, and a general communication plan.

Table 1. Continued.

Command and general staff	Substaff positions	Lead	Responsibilities
	Medical	GNF	Provided medical supplies.
	Food	TEI	Provided meals for crew members.
	Supply	TEI	Provided and maintained supplies such as drip stations, hip boots, bear spray, PPE, and backpack sprayers.
	Facilities	TEI	Provided lodging facilities.
	Ground Support	Shared	Each partner provided transportation for their own employees.

the next day's plan. Fourth, when personnel were divided into separate groups to treat different portions of stream during a day, leaders were assigned to each group; these leaders checked in with the incident commander during the day. The incident commander delegated this role to another project lead in his absence so that operations could be maintained. Fifth, radio communication was critical to effective choreography of operations, from logistically coordinating crews of up to 35 people hiking to setting up along various stream reaches during a day, choreography of chemical application (coordination of drip station timing, notifying backpack sprayers of sites that needed sprayed, etc.), resolving safety issues (e.g., lightning storms, bear sightings, and accidental pepper spray discharges), and managing crew egress from the field. Sixth, project leaders ensured that at least one first aid kit and qualified first aid person was with each group of chemical applicators. Finally, project leads conducted a review of a year's operations so that lessons learned could be incorporated to adjust the next year's operations.

Another type of agreement that can be useful when management agencies and private landowners consider collaborative conservation of species under ESA consideration, such as WCT was at the time of this project, is a Candidate Conservation Agreement with Assurances (CCAA). It was clear at the beginning of the Cherry Creek project that regardless of the outcome, WCT could potentially be listed under the ESA. Endangered Species Act listing can bring federal oversight to activities on private land that would not otherwise be considered if the species had not been restored on that property prior to its listing. While the "cloud" of federal oversight does not concern TEI, it can and does intimidate many private landowners. The CCAA is designed to address conservation needs of a species that potentially may become listed by asking landowners to voluntarily conduct conservation actions that will protect the population of that species on their private lands and could ultimately preclude the need for listing. In return, the landowner is provided assurances that they will not be subject to future regulatory obligations beyond what was agreed to at the time they entered into the CCAA, even if the species is listed. In an effort to expand WCT restoration on private lands in Montana, MFWP developed and formally entered into a WCT CCAA with the USFWS in 2007. Turner Enterprises, Inc. signed onto MFWP WCT CCAA through a certificate of inclusion (COI) in June 2009. Although the Cherry Creek project was already well underway, the COI affirmed that if TEI allowed MFWP to establish WCT in the Cherry Creek project area, TEI would not be held to additional regulatory obligations if WCT were listed under the ESA in the future. Further, the COI preemptively permitted any incidental take of WCT that might occur during regular ranching or recreational activities if the species was listed. While a CCAA is for species that have not yet been listed, the Safe Harbor Agreement process can provide similar assurances to private landowners working with already listed species.

Mistakes, learning experiences, and innovations

A well-thought-out MOU, PA, and ICS structure led to the continuity between partners and years and was the basis for clear communication. The need for lengthy planning and coordination became less necessary as the project progressed. With adequate time, training, and oversight, a planned proactive project such as the Cherry Creek project can be successfully completed and accomplished with no major injuries to personnel or property. Individuals can be assigned collateral duties within the ICS structure, not compromising the safety of crew members and effectiveness of the treatment. It is imperative that the treatment personnel have direct communication with each other and with those attending the detoxification facilities. Therefore, it is recommended that temporary radio repeaters be placed, instead of using the person-to-person relay system to communicate up and down the project area. Though we did not have the need for them during the Cherry Creek project, we recommend that trained backcountry emergency management technicians be within communication and able to be responsive to incidents in a timely manner.

Piscicide Treatments

Each year's preparation for conducting chemical treatments included using backpack electrofishers to confirm the upstream distribution of nonnative trout in the phase to be treated, measuring stream discharges to estimate the quantity of piscicide necessary for treatment, installing staff gauges to monitor changes in stream discharge, using EPA-compliant tracer dyes to determine mixing and travel time of stream flow, and capturing and posting sentinel fish at specific points based on bioassay results. Some activities could be conducted weeks before chemical application (e.g., fish distributions), others no more than 24 h prior to treatment (e.g., measuring or monitoring stream discharge). Additionally, training on command structure, logistics, equipment, communications, backcountry safety, food and waste management, and piscicide use and safety was provided to crews each year prior to the initial treatment and as new crew members arrived on the project site.

The volume of Cherry Lake was estimated from a hand-drawn map made by GNF personnel a decade or more prior to the project. We used a hand-held depth finder transported by a raft prior to treatment of the lake to re-estimate the volume of the lake, which confirmed the earlier GNF measurements.

Prior to application of piscicides, streamflow travel time was measured by applying enough tracer dye to create a visible plume at the most upstream point chemical would be applied. The leading edge of the dye plume was tracked and renewed when necessary to maintain an easily visible plume. Sequentially numbered, orange-painted wooden stakes and flagging were used to mark each 15-min interval for Fintrol and 30-min interval for CFT Legumine along the stream. The upstream starting point was labeled as station 0; then, subsequent stations were sequentially numbered as 1, 2, 3, and so forth. Thus, if stations were marked every 30 min of stream travel, station 1 would be 0.5 h downstream of the starting point, station 2 would be 1 h downstream, and so on. Determining travel time and marking the stations was important to plan drip station spacing but also allowed pretreatment detection of side-channel water, springs, and small tributaries that needed to be sprayed, as personnel conducting the dye tests necessarily had to walk each stream section that was to be treated.

Stream discharge measurements were conducted prior to applying the tracer dye. A graduated 30- or 60-cm metal staff gauge was installed at each measuring site. If the gauge reading changed significantly between dye testing and piscicide application, a new stream discharge measurement was taken and used to recalculate the piscicide quantity necessary for treating that section. Stream discharge was measured using either a USGS model 622 type AA current meter or a USGS model 625 pygmy current meter, since renamed model 6200 and 6205, respectively. Stream discharges measured throughout the project area ranged from less than 0.001 m^3/s to more than 0.85 m^3/s.

Pretreatment bioassays were conducted in a representative section of stream to determine the persistence of the piscicides, minimum lethal concentrations, and our ability to detoxify them with potassium permanganate. Bioassays were conducted for EPA-registered piscicides Fintrol (active ingredient [a.i.] Antimycin A, 10%) and CFT Legumine (a.i. rotenone, 5%). The Antimycin A bioassay was conducted August 1, 2003; the rotenone bioassay was conducted on July 29, 2007. We did not feel it was necessary to conduct bioassays every year or in every phase as preliminary water chemistry analysis suggested that pertinent parameters (temperature, alkalinity, and pH) were relatively consistent throughout most of the project area. We stated in the EA that in areas where we suspected antimycin may have reduced effectiveness, we may need to use rotenone to achieve an effective treatment. In our discussions of piscicide concentration throughout this chapter, we will be describing the concentration of Antimycin A or rotenone a.i., not the quantity of the Fintrol or CFT Legumine formula. When discussing the product in more general terms, we will refer to them by their product names of Fintrol and CFT Legumine.

To determine the persistence of Antimycin A in a stream, four sentinel fish were placed in net holding bags or plastic buckets with holes drilled in them (i.e., flow-through buckets) at 15-min streamflow travel time intervals downstream for a total distance of 4 h below a single piscicide application point. Antimycin A was applied at a test concentration of 12 parts per billion (ppb) for 6 h 40 min, and the sentinel fish were monitored for up to 48 h to determine the exposure time necessary to achieve 100% mortality.

An Antimycin A serial dilution bioassay was conducted simultaneously with the travel time bioassay. Six plastic buckets with sentinel fish were placed 10-min streamflow travel time downstream of the piscicide application point. All six buckets were placed in the stream so the stream water would regulate the water temperature in the buckets, but stream water did not mix with water in the buckets (nonflow through buckets). About an hour after initiation of the persistence bioassay, untreated water in each bucket was replaced with the appropriate amount of treated stream water to achieve a range of Antimycin A concentrations in the buckets (12, 10, 8, 4, 2, and 0 ppb). Sentinel fish in each bucket were monitored, and their condition was noted hourly. Water in the buckets was aerated with portable aquarium pumps, and 3 gal of water in each bucket was replaced every hour with the appropriate ratio of treated and untreated stream water to maintain the test concentration.

In the Antimycin A persistence bioassay, 100% mortality was achieved as far as 90 min downstream but required almost 45 h to occur at that station (Table 2). Sentinel fish in the serial dilution bioassay showed complete mortality of all fish at all test concentrations except the control (i.e., 0 ppb; Table 3). In general, the lower the test concentration, the more time required to achieve complete

Table 2. Rainbow Trout mortality downstream from a station applying 12 ppb Antimycin A (4.18 ml Fintrol) to a 0.0014 m^3/s stream during bioassays. The station operated on August 1, 2003, from 1250 to 1930 hours (6 h 40 min).

Travel time (minutes below station)	Time post initial exposure (h : min)	Fish condition (dead/unstable/upright)	Percent mortality
15	6:25	2/1/1	50
	21:00	4/0/0	100
30	6:14	0/1/3	0
	20:44	4/0/0	100
60	5:48	0/0/4	0
	20:12	0/1/3	0
	24:06	0/3/1	0
	30:19	1/2/1	25
	46:04	4/0/0	100
90	5:31	0/0/4	0
	19:59	0/0/4	0
	25:36	0/0/4	0
	29:51	0/0/4	0
	45:52	4/0/0	100
120	4:52	0/0/4	0
	19:07	0/0/4	0
	23:36	0/0/4	0
	29:30	0/0/4	0
	45:29	0/0/4	0
180	4:08	0/0/4	0
	18:00	0/0/4	0
	22:06	0/0/4	0
	28:25	0/0/4	0
	43:53	0/0/4	0
240	3:11	0/0/4	0
	5:30	0/0/4	0
	21:06	0/0/4	0
	27:25	0/0/4	0
	42:54	0/0/4	0

mortality. After consideration of the bioassay results, we decided to apply 10 ppb Antimycin A every 30 min of streamflow time for an 8-h duration. Bioassays indicated that this would result in complete mortality of fish within an acceptable time and reduced the risk of fish surviving a lower Antimycin A concentration or a less-frequent application interval. When determining treatment dosages and intervals based on bioassay results, it is important to consider the additional stress the sentinel fish may be under from capture and holding activities and the potential impact on bioassay results. One might expect the sentinel fish to be more susceptible to the effects of a piscicide than free-swimming fish, thus being reasonable but conservative in selection of treatment

Table 3. Fish condition (dead/unstable/upright) of sentinel Rainbow Trout exposed to various bioassay concentrations of Antimycin A for 6 h 40 min.

Antimycin A concentration (ppb)	Time post initial exposure (h : min)										
	1:00	2:00	3:00	4:00	5:00	6:00	6:40	21:18	25:20	29:23	47:52
0	0/0/4	0/0/4	0/0/4	0/0/4	0/0/4	0/0/4	0/0/4	0/0/4	0/0/4	0/0/4	0/0/4
2	0/0/4	0/0/4	0/0/4	0/0/4	0/0/4	0/0/4	0/0/4	2/1/1	3/0/1	3/0/1	4/0/0
4	0/0/4	0/0/4	0/0/4	0/0/4	0/0/4	0/0/4	0/0/4	4/0/0			
8	0/0/3	0/0/3	0/0/3	0/1/2	1/0/2	3/0/0					
10	0/0/4	0/0/4	0/0/4	1/3/0	2/2/0	4/0/0					
12	0/0/4	0/0/4	0/0/4	1/3/0	4/0/0						

dosage and interval based on bioassay results is warranted.

A detoxification bioassay was conducted by applying Antimycin A to the stream at 12 ppb while simultaneously running a detoxification (hereafter, "detox") station at 15 min streamflow travel time downstream (Table 4). Potassium permanganate ($KMnO_4$) was applied at 5 parts per million (ppm). Sentinel fish were held in net bags 15, 30, and 60 min downstream of the detox station. A chlorine meter was used to measure $KMnO_4$ reduction at points 15 and 30 min downstream of the detox station. Reduction occurs as $KMnO_4$ is consumed by instream biological demand and by interaction with Antimycin A. The interaction between the strong oxidizer $KMnO_4$ and Antimycin A effectively stops its effect on aquatic organisms (e.g., detoxifies the piscicide). This detoxification bioassay showed that we could fully neutralize Antimycin A within 30 min of contact with $KMnO_4$.

Five years after the Cherry Creek bioassays were completed, Moore et al. (2008) developed standard procedures for National

Table 4. Remaining $KMnO_4$ (expressed as a percentage of total $KMnO_4$), and fish condition (dead/unstable/upright) of sentinel Rainbow Trout at points downstream of bioassay detoxification station.

Travel time (minutes below $KMnO_4$ station)	Time post $KMnO_4$: Fintrol mixing (h : min)	Percent $KMnO_4$ remaining	Time post initial exposure (h : min)	Fish condition (dead/unstable/upright)
15	2:45	14.6		
	3:45	21.9		
	4:05	25.7		
			5:41	1/0/4
			21:15	3/0/2
30	2:45	8.1		
	3:45	12.4		
	4:05	10.8		
			5:26	0/0/5
			21:00	0/0/5
60			4:56	0/0/5
			20:30	0/0/5

Park Service projects using Antimycin A. They recommend that Antimycin A be applied at concentrations of 5–25 ppb for 7–8 h to conduct an effective treatment. Our bioassay results and treatment dosages comported with these recommendations.

Only a persistence test of rotenone was conducted. Sentinel fish were placed every 30 min out to 4 h of streamflow time, and 50 ppb (0.050 ppm) rotenone was applied to the stream for nearly 8 h. This resulted in complete mortality of fish in less than 4 h at all but the lowest sentinel station (Table 5). The CFT Legumine label allows application up to 4 ppm formula (200 ppb rotenone). In developing standard procedures for rotenone projects, Finlayson et al. (2000) recommended that flowing water bioassays be designed using the expected treatment duration. In the case of Cherry Creek, the results of the persistence bioassay allowed us to reduce the expected treatment time. During treatments, we applied 50 ppb rotenone to the streams at 3-h intervals for a 4 h duration. This conservative spacing and treatment time insured overlap between adjacent stations and prevented areas of potential sublethal concentrations while providing adequate exposure time for complete mortality.

We initially planned to use Fintrol throughout the entire project area, but production issues in 2007 resulted in a lack of availability of reliable Fintrol, so we switched to CFT Legumine. We completed Phases 1 and 2 with Fintrol and used CFT Legumine throughout Phases 3 and 4. In some instances throughout all phases, concentrations of piscicide were reduced in the downstream drip stations to reduce the potential for piscicide concentrations to accumulate or "snowball." We employed this tactic when approaching the downstream end of a phase or on occasions when the piscicide seemed to be carrying farther than bioassays indicated. Generally, daily applications of Fintrol lasted between 6.5 and 7 h (targeted at 8 h) while CFT Legumine applications lasted between 3.5 and 4 h (targeted at 4 h).

We attempted to have sentinel fish deployed throughout a treatment area at least 2 d prior to initiating any piscicide applications. This was precautionary to ensure that sentinel fish mortality was not caused by electrofishing or handling stress. Sentinel fish are useful to determine how far the pi-

Table 5. Results of a 50-ppb rotenone (1 ppm CFT Legumine) bioassay in the East Fork Cherry Creek to determine effective exposure time. Run time of the application station was 7 h 22 min, stream discharge was 0.013 m^3/s. CFT Legumine application was initiated at 0933 hours on July 29, 2007. NA = not applicable.

Sentinel fish station[a]	Time of initial exposure	Time of 100% mortality	Hours of exposure until 100% mortality
30	1003	1050	0:47
60	1033	1255	2:22
90	1103	1255	1:52
120	1133	1400	2:27
150	1203	1455	2:52
180	1233	1615	3:42[b]
210	1303	1615	2:48
240	1333	NA[c]	

[a] Minutes of stream flow time downstream of application station.
[b] Two fish dead, one gravely ill at 1455 hours (2:22 h of exposure).
[c] 100% mortality of sentinel fish was confirmed the following morning at 1145 hours.

scicide traveled each day and how effective each day's treatment was. This daily feedback provided information for us to determine if we should adjust piscicide concentrations higher or lower, or if drip station spacing was appropriate. Over the course of the project we came to prefer flow-through buckets over net bags for holding sentinel fish, as mink, raccoons, and other predators were more likely to tear open the net bags and eat the fish or allow them to escape.

Chemical treatments were conducted annually from 2003 through 2010. Our design called for treating each phase for two consecutive years. We felt that this approach would be more likely to achieve complete eradication of nonnative trout by overcoming errors and issues such as overlooking small streams, springs, and off-channel waters that held fish; localized water chemistry parameters that rendered the piscicide less effective over short reaches; incomplete treatment of interstitial, under bank, and beaver dam habitats; incomplete kill on developing eggs in the gravels; and crew error and inattention caused by fatigue after consecutive long work days over a large area. In hindsight, we believe that at least 2 years of treatment on every reach of stream was absolutely critical to project success, as several of these factors did occur and are discussed later.

Annual treatments were generally confined to a 2–3-week period in August, but circumstances did require us to conduct some treatments at other times. For example, in 2010, two full treatments of Phase 4 were conducted in August and September in an effort to accelerate completion of the project. Also, single-day treatments of certain areas were conducted in September 2007 and 2009, and in October 2008 and 2010 if we either suspected or were certain that unwanted fish remained. Table 6 provides greater detail on annual treatments.

Typically, one phase of the project area was treated each year. Each phase was broken into daily work units that varied in size (e.g., stream length treated) based on remoteness, volume of water, number of drip stations required, number of backpack sprayers needed, and size of field crew. Chemical treatment of a work unit was initiated only after existing fish distributions were confirmed, dye testing was completed, and sentinel fish were deployed. Typically, in headwater streams and tributaries, treatment started at the stream source or several hundred meters above any natural drop structure that, based on appearance and fish distribution surveys, prevented further upstream movement of fish. Treatments were primarily designed to treat a section of the main stem each day, including the tributaries that flowed into that main-stem section. On some occasions, all personnel and equipment were necessary to treat a larger tributary for multiple days. As sections of the main stem or larger tributaries were completed, we used mesh block nets to temporarily prevent upstream movement of fish back into already completed portions. Each treatment day, we ran drip stations about 150 m upstream of all block nets installed the previous day, so we had some overlap between days and did not have untreated water entering the treatment area for the day. We followed this protocol until we had treated the phase downstream to the barrier each year.

Cherry Lake was at the headwaters of Phase 1. It was treated by applying piscicide from a two-person raft and a 53-L plastic tank outfitted with a battery powered electric diastolic pump (Figure 6). The quantity of Fintrol necessary to treat the lake volume was applied to the lake by mixing small quantities of Fintrol with lake water in the 53-L tank and pumping the mixture to various depths throughout the lake through a weighted nozzle manifold.

We applied piscicide to all streams with a steady flow drip station or bucket system dubbed the "Montana Bucket" (Figure 7), by backpack sprayers, and occasionally using a

Table 6. Summary of Cherry Creek piscicide treatments by year, 2003–2010.

Year	Phase[a]	Number of stream kilometers treated[b]	Number of treatment days	Number of worker days[c]	Piscicide quantity used
2003	1[d]	20.3	13	284	18.5 L Fintrol
2004	1	20.3	12	240	24.2 L Fintrol 3.8 L CFT Legumine
2005	2	15.3	8	220	26.5 L Fintrol 3.8 L CFT Legumine 454 g rotenone powder
2006	2	15.3	10	256	22.3 L Fintrol
2007	2	7.2	12	264	34.0 L CFT Legumine
	3	42.4			
2008	3	42.4	8	158	55.3 L CFT Legumine 908 g rotenone powder
2009	3	9.1	2	16	21.6 L CFT Legumine 227 g rotenone powder
2010	3	9.1	10	200	84.8 L CFT Legumine
	4	49.8			

[a] Phase 1 = 20.3 stream kilometers; Phase 2 = 15.3 stream kilometers; Phase 3 = 42.4 stream kilometers; Phase 4 = 22.0 stream kilometers.
[b] Stream length treated includes multiple treatments of a phase or section of a phase within a given year.
[c] Number of worker days includes all preparatory and support activities and treatments.
[d] Phase 1 also includes the 0.3 ha (130,000 m^3) Cherry Lake.

rotenone sand gel mix (rotenone powder label; Finlayson et al. 2010) to create a dough ball that dissolved over time in the water. The Montana Bucket was actually a design used by B. Rosenlund in his Colorado treatments. The system consists of a 13-L plastic bucket, a lid with a polyvinyl chloride screw-cap fitting, an automatically filling pet watering bowl (Figure 8) with a size 53 twist bit hole drilled in the bowl, a 1.2-m section of garden hose, and a common garden hose gate valve. Once at the assigned site on the stream, personnel assembled the bucket, gate valve, hose, and dog bowl. Ideally, the bucket, with lid on, would be firmly leveled (with leveling blocks or on-site materials) on the streambank several centimeters above the stream. The dog bowl was stabilized closely above the stream surface on a stick or rock platform constructed of on-site materials, insuring that the chemical stream from the dog bowl would drain directly into flowing water, not into an eddy or still water area. Once the bucket and dog bowl assembly were well located and stable, the bucket was filled approximately halfway with clean stream water, checked for leaks, dosed with chemical through the fitting in the lid, and completely filled with water. The station attendant then waited until a prescribed time to open the gate valve between the bucket and garden hose to start applying the piscicide to the stream. Drip stations were started each day in an upstream to downstream sequence that coordinated start times to ensure a steady coverage of piscicide at a fish-killing concentration through each day's treatment area.

Each drip station had its own attendant, unless two drip stations were within

Figure 6. Inflatable raft set-up used to apply Fintrol to Cherry Lake.

line of site from each other, where one attendant could observe and tend to both. As required, all attendants used appropriate personal protective equipment when setting up, charging, and tending a drip station. The quantity of chemical (i.e., dose or charge) was specific for each station each day, depending on the stream discharge and the intended application concentration at that site.

Backpack sprayers were used to apply piscicide to off-channel waters, such as disconnected side channels, springs and seeps, or small tributaries that did not have drip buckets, as well as stagnant stream margins that water did not readily mix into. The objective of backpack piscicide application was to treat any water not treated by drip stations and eliminate any peripheral freshwater sources for fish that may reduce or eliminate their exposure to the piscicide. The backpack sprayers were filled in a process similar to the drip buckets. The sprayer was filled half full with a bail bucket, generally 5–10 mL of Fintrol or 10–15 mL of CFT Legumine were used to charge the sprayer, two green dye tablets were placed in the tank, and the tank was filled with water. Care was taken to avoid

Figure 7. The Montana Bucket trickle system and sentinel fish bag on Cherry Lake Creek. The sentinel fish bag is upstream of the piscicide application point to monitor the effectiveness of the station above the one shown here.

Figure 8. Close-up view of the dog waterer trickling piscicide/stream water mixture into the stream during the Cherry Creek project.

incorporating debris in the backpack tank that could plug the spray hose or nozzle. The dye tablets were used to brightly color the solution in the spray tank so that multiple sprayers working an area could identify water that had already been sprayed and gauge how quickly their spray solution was flushed from areas such as stream margins or backwaters. Workers using the backpack sprayers worked in a downstream direction and were generally assigned to cover an area between a specific set of stations along the stream. They did not initiate spraying until at least an hour after the stream treatment with the buckets started. This was to ensure that the piscicide was far enough downstream so that if fish were flushed out of areas sprayed, they would be entering treated water in the stream. Workers were provided with enough piscicide to refill their sprayer several times over the course of the day, if necessary, and to allow them to apply an adequate volume of piscicide to large off-channel pools when encountered.

An extremely critical element of the project was adequate and reliable field communications. The GNF was responsible for providing hand-held radios and training personnel in their use and in proper radio etiquette. Effectively conducting the field operations would have been impossible without adequate field communications. Crew members were frequently spread over several miles of the drainage any given day, and radio communication allowed project leaders to be informed of issues or problems that arose, make informed decisions, alter treatment plans as conditions changed, redirect backpack sprayers, or address any number of situations. Importantly, the radios were critical for crew communications in the potential event of an emergency, such as severe weather, wildfire, bear encounters, or serious injury to a crew member, some of which occurred during the project.

The removal of nonnative fish was assessed with multiple methods, which provided assurance that treatment had been successful. A work unit or phase was typically not considered clean (all fish eradicated) unless no additional fish mortalities were observed during the second or subsequent treatments. In some cases, additional treatments were required (Table 6). After a clean treatment, substantial electrofishing was conducted throughout a phase to further confirm that no fish remained. Finally, environmental DNA (eDNA) samples were collected in 2015 according to methods of Carim et al. (2016) to assess whether nonnative salmonids could be detected. All eDNA samples were negative (Carim et al. 2015), indicating successful eradication of the targeted nonnative salmonids.

Mistakes, learning experiences, and innovations

Relatively few problems arose over 8 years of piscicide treatments with field crews of 16–24 people in the field for 10–14 d annually in a remote field setting. We attribute this to the command structure that was in place, training provided, consistent personnel in supervisory and lead roles for each partner over the course of the project, dedicated field staff, and a little bit of luck. Expected minor problems encountered included improper radio etiquette, disrespectful behavior in camp or towards fellow crew members, and failure to take care of equipment. Occasionally, a crew member failed to complete their assigned duties, failed to follow treatment schedule or procedure, or became distracted in the field. These errors had the potential to be costly in time, materials, and treatment effectiveness but were minimized by the command structure, rotating daily responsibilities (drip station, spraying, sentinel fish, dye testing, etc.), and, when possible, providing days off when individuals or the entire crew requested.

We were not able to achieve complete kill of the YCT in Cherry Lake using Fintrol. Antimycin A was initially applied at 4 ppb throughout the lake on August 4, 2003, and significant fish mortality was observed. Upon completion of the first treatment, we set gill nets to assess the thoroughness of the treatment. Three YCT were captured by gill net 2 d after the initial treatment, so another 4 ppb of Antimycin A was applied. No fish activity was observed, and nothing was captured in gill nets for two more days; however, fish were then observed rising to feed on surface insects. Two additional YCT were captured in gill nets on the morning of August 20, 2003, so the lake was treated for a third time to 4 ppb that day. Rising fish were again observed that evening.

In 2004, we applied 12 ppb of Antimycin A to Cherry Lake on the first day of treatments. Gill nets were again set in the lake, and 16 d later, 16 YCT were found in the gill nets. The gill nets were left in the lake and checked periodically through the fall of 2004. The nets were left in the lake overwinter. On the initial check of the gill nets in July 2005, the decayed carcasses of six fish were found in the nets, but no additional fish were caught the remainder of 2005. Nevertheless, the lake was treated again in August 2005 at a concentration of 8 ppb Antimycin A. No dead fish were observed. Follow-up genetic testing of the restored Cherry Lake population has confirmed that only WCT occupy the lake, indicating that original YCT population was completely eradicated by our multiple treatment and gill netting effort.

We do not definitively know the reason for the failure of the Fintrol to eradicate fish in Cherry Lake, but we surmise that it was due to inadequate mixing of chemical throughout the lake. Accelerated photodegradation of the Antimycin A at the high elevation of this lake is also a possibility; however, it did not appear to be a degradation or water chemistry issue as fish in the inlets and outlet of the lake succumbed to the piscicide as expected. There did not appear to be any significant spring upwellings in the lake basin where fish could access untreated water, as inflow to and outflow from the lake was similar. A scuba inspection of the lake bottom on July 27, 2005 found only one small spring upwelling of relatively insignificant volume.

During the EA process, several members of the public stated that they fished in Cherry Lake every year. To reduce the impact of the treatment on recreational angling in the lake, we agreed to stock the lake with catchable WCT as soon as feasible after removal of the nonnative YCT. Our intent was to initially stock the lake with age-3 and age-4 WCT males retired from MFWP's WCT conservation brood at Washoe Park Hatchery (WPH) so they would be large enough for angling but would not reproduce. We hoped that they would die out as WCT we intended to establish from a wild donor source through fry stocking reached maturity. However, the only WPH WCT available in 2006 were age-2 females, so we stocked 165 of them into Cherry Lake. At that age, visual sex determination is difficult. In July 2009, several size-classes of WCT were observed in Cherry Lake, indicating that reproduction had occurred since 2006. Because we wanted to establish a wild-sourced WCT population in Cherry Lake rather than having the WPH genetics descending from the lake into Phase-1 streams, gill nets were deployed in the lake and a rotenone treatment was conducted in August 2009 in the primary inlet spawning stream. While we do not believe that we completely removed all the WPH WCT from the lake, we did significantly reduce their numbers and removed most, if not all, spawning-sized fish from the lake. After this netting effort, we stocked the first year-class of age-0 wild-sourced WCT into Cherry Lake. The remaining WPH WCT descendants in Cherry Lake would have continued to mature but would have had only one or two spawning seasons prior to the 2009 wild-source male age-0 introductions reaching sexual maturity and beginning to spawn with the remaining WPH WCT.

We failed to locate and treat two tributaries in Phase 1 in 2003. These tributaries were located after treatments were completed in 2003. One of these streams was overlooked because it flowed through tall grass and had a discharge of slightly less than 0.001 m^3/s. It was identified when surrounding vegetation began to dry out and we noticed a narrow strip of green grass on the hillside. The other was a stream of 0.014 m^3/s that simply was not seen in dense forest underbrush during preproject surveys. Nonnative trout occupied both streams, and both were treated with piscicides in 2004. Similarly, in Phase 2, an off-channel spring pool was completely missed by backpack sprayers in 2005 but was found and treated with piscicides in 2006. It was occupied by about 30 Brook Trout. An ephemeral connection between the pool and Cherry Creek provided the Brook Trout access to Cherry Creek. The Brook Trout in the off-channel spring spawned in fall 2005, resulting in dozens of age-0 Brook Trout in the main stem of Cherry Creek that were killed in the 2006 treatment. We retreated the Phase-2 main-stem reach and the lower ends of all tributaries for a third time in 2007, just in case any other fish had survived.

On rare occasions, a drip station tipped over before it was empty, spilling the remaining chemical solution onto the ground or into the stream. In the four times this happened over 8 years of treatments, it was due either to inattentiveness or because of difficult site conditions at the treatment station (e.g., sloping or uneven streambanks) or both. Commonly, as the fluid level in the bucket dropped, the bucket became more unstable and susceptible to tipping, despite being weighted with rocks on the lid. If the bucket tipped into the stream, we refilled it to the level it was when it spilled and charged it with piscicide to achieve one-quarter of the designed treatment concentration. We did this because we wanted to maintain a piscicide concentration that would continue treating that reach of stream but minimize the impact of the spill on nontarget organisms. If piscicide detoxification was occurring downstream, the operators of the detoxification station were notified of the spill and reminded to monitor sentinel fish carefully and be ready to adjust the concentration of potassium permanganate applied, if necessary. Because there were only a few instances of buckets tipping and they generally occurred well upstream of the detoxification site, there was never a need to adjust the level of potassium permanganate in response to a chemical spill. If the bucket tipped onto the ground rather than into the stream, we refilled it to the level it was when it spilled and charged it with piscicide to continue treating the stream at the designed treatment concentration.

In 2003, the first year of eradication efforts, we started treatments by establishing the uppermost drip station approximately 200 m above the end of fish in a given tributary, as defined by electrofishing surveys. Sometimes this was well below the stream source. On two occasions, fish were observed above the uppermost drip station during treatment. In those events, we either instructed backpack sprayers to make sure the area above the drip station was treated or, if streamflow warranted, additional drip stations were deployed upstream. After this happened the second time, we adapted our strategy and made it a rule to start drip station treatment at the stream source, unless there was a clear barrier to fish movement (e.g., waterfall) at a point lower down the drainage and no fish had been found upstream of that barrier during electrofishing surveys.

An unexpected problem occurred in 2008 and 2009 when some adult fish in Phase 4 ascended past the Phase-3 barrier into the previously treated Phase 3, despite modeling showing that the barrier should have been impassable. The fish jump model was based on known swimming speeds for Rainbow Trout, water velocity, depth of water on the apron or splashpad below the barrier, and a range of leaping angles. Phase 3 was treated in its entirety in 2007 and 2008. Again in 2009 several adult fish were discovered above the barrier in the lower part of Phase 3. In early July 2009 fish were observed passing over the Phase-3 barrier. We determined that fish were not leaping over the barrier from the splash pad, but rather jumping from the 25-cm standing wave between the toe of the barrier and the overflowing water, something the modeling effort had not anticipated. An angled screen was constructed and attached to the downstream face of the barrier, and no additional fish made it over the barrier. A third piscicide treatment of the Phase-3 main stem and lower ends of tributaries was conducted in 2009 to remove the fish that had ascended the barrier.

The most significant and disappointing problem occurred in 2010, the highest water year during the project. As previously described, in an average water year, approximately 1 km of main-stem Cherry Creek dewaters upstream of the Phase-3 barrier. Then, starting about 180 m above the barrier, springs begin to recharge streamflow to about 0.45 m^3/s at the barrier. However, during spring runoff in 2010, the Phase-3 barrier area was inundated with flood water, creating overland surface flow between Cherry Creek and the adjacent spring channel. After the initial flood event receded, we sandbagged the margins of Cherry Creek and the spring channel in case water level increased again (Figure 9). Fearing that fish may have been able to swim around the main stem and spring barriers during the flood event, we decided to treat the Phase-3 main stem a fourth time. Additionally, the higher-than-normal late-summer surface flows maintained a weak surface water connection through this typically dewatered reach. Treatment of the main stem of Cherry Creek in 2010 was initiated in early August, with a single drip station applying 50 ppb rotenone at the Phase-2 barrier and another at the lower end of Carpenter Creek, a tributary to Cherry Creek downstream of the Phase-2 barrier. Throughout the day, sentinel fish in main-stem Cherry Creek were observed to determine the effective travel distance of the rotenone. In previous treatments of Phase 3, rotenone applied at the Phase-2 barrier was

Figure 9. Phase-3 barrier after peak discharge receded and sandbagging was completed to prevent further flooding that potentially could allow fish to pass the barrier.

lethal to sentinel fish for no more than about 6.4 km downstream and degraded below a lethal concentration about 1.2–1.6 km above the Phase-3 barrier. The dewatered reach above the Phase-3 barrier also typically prevented rotenone from carrying further, if it even persisted that far. In 2010, sentinel fish several kilometers above and at the Phase-3 barrier showed no signs of rotenone toxicity by late evening, several hours after rotenone should have arrived at those locations. Although a detox station had already been set up and was potentially operable at the mouth of Cherry Creek Canyon (about 1 km below the end of Phase 4 at Cherry Falls or a point that is about 6 km below the Phase-3 barrier and 14.3 km below the rotenone application point at the Phase-2 barrier), it was not activated that evening based on the condition of the sentinel fish located at and above the Phase-3 barrier, and we did not want to release potassium permanganate unnecessarily. The following morning, numerous dead fish were observed several kilometers downstream of the detox station. We determined that most of these fish had not drifted down from Phase 3 or 4 because there were Brown Trout *Salmo trutta* and Mountain Whitefish *Prosopium williamsoni* among the dead fish, species that were not present in the project area.

The decision was made to continue with the treatment that day as scheduled, activate the detoxification station as previously planned, and investigate the extent that rotenone had carried the previous day. All sentinel fish through Phase 4 were already dead, indicating that an unplanned but thorough treatment of Phase 4 had occurred during the night. Additionally, 33 of 37 surplus sentinel fish being held at the detox station were dead, as well as most sentinel fish already deployed at points downstream of the detox site. Through electrofishing, we determined that the rotenone from the previous days treatment (one drip bucket at the Phase-2 barrier, one on Carpenter Creek, and backpack sprayers cumulatively applying 50 ppb rotenone) had remained lethal for approximately 20.9 km downstream to a point 4.8 km upstream of Cherry Creek's confluence with the Madison River and 6.6 km below the site of the detoxification station.

Treatments were completed as scheduled in August 2010, despite the bad publicity generated by the overkill incident. During an investigation to determine why the incident occurred, all calculations and streamflow measurements were double-checked and found to be correct. The rotenone potency was tested by an analytical laboratory and found to be within label specifications. No other obvious application errors, such as an inadvertent spill of rotenone, occurred. Phase 3 does include a large beaver dam complex. The dams had been broken and drained to the extent possible, but several still had modest pools of slow-moving water in them. These pools were also sprayed by backpack sprayers. We can only surmise that a significant amount of rotenone was slowed down in the beaver dam complex and degraded more slowly than anticipated due to reduced agitation in the pools, slightly cooler water temperatures from the higher streamflow, and increasing afternoon cloud cover. Darkness and cooling evening water temperatures would have further slowed rotenone degradation, possibly allowing low dosages of rotenone to flow out of Phase 3 during the night. This same scenario may have occurred in previous years, but the dewatered reach prevented outflow from Phase 3. If the surmised scenario is plausible, it is likely that an extremely low rotenone concentration with up to 12 h of exposure was adequate to cause fish mortality during the night. This incident caused MFWP to develop a formal piscicide policy, including strict detoxification guidelines.

Although projects using rotenone and, in at least one case, toxaphene were conducted in Montana from the 1950s through the 1970s to control nongame or nonnative fish species, these projects were all one-time efforts to reduce the abundance of a target species in a single water body or portion of a drainage. The Cherry Creek project launched a programmatic effort by MFWP and other resource agencies to conserve and restore native Cutthroat Trout throughout their historical range in Montana by eradicating competing or hybridizing species across entire watersheds. The controversy encountered during the planning and initiation of the Cherry Creek Project, as well as the overkill event during the last year of treatment, resulted in the development of a programmatic approach to use of piscicides for native trout conservation and restoration in Montana, as well as a formalized MFWP piscicide use policy developed by members of the Piscicide Technical Committee. Proposed projects are critically assessed to determine feasibility, cost-effectiveness, and biological significance (e.g., hybrids in lakes influencing genetic purity of fish downstream, lack of aboriginal WCT within a larger geographic area, and need to expand small WCT populations). A committee of MFWP fisheries biologists, technicians, and administrators was convened to establish standards for development and certification of personnel to conduct piscicide projects and to ensure compliance with state and federal regulations governing use of piscicides. Using the Cherry Creek EA as a template, this committee developed a boilerplate EA so all projects would use the same format and provide consistent, relevant, and accurate information to the public. The boilerplate EA is updated as new information or issues arise. Information specific to each piscicide project is incorporated into the EA for that project.

Montana law requires that individuals applying restricted use pesticides in Montana be licensed by the Montana Department of Agriculture (DOA). However, due to the relatively few individuals in the state that occasionally used piscicides, training and licensing specific to piscicides was not available. Over the course of the Cherry Creek project, as MFWP's broader piscicide program developed, MFWP worked with the DOA to develop training and certification specific to piscicide use that includes initial certification and then continuing education credits to maintain certification. Other agencies and organizations also use this training process. For example, some Yellowstone National Park biologists and technicians are certified by the DOA and maintain their piscicide applicator certifications through MFWP Piscicide Committee training and adhere to Montana's Piscicide Policy when planning and conducting projects in Yellowstone National Park.

Montana Fish, Wildlife & Park's formal Piscicide Policy directs procedures in six areas: (1) Piscicide Committee membership, (2) applicator responsibilities and training requirements, (3) assistance to projects, (4) treatment procedures and checklist, (5) species of concern and benthic macroinvertebrate sampling protocol, and (6) rotenone detoxification procedures.

Significant elements of the Piscicide Policy include internal review of EA's prior to public release, pre- and posttreatment monitoring of nontarget organisms, piscicide detoxification guidelines, development of various levels of applicators based on experience and frequency of conducting piscicide projects, and assignment of specific duties based on an applicators level of experience. For instance, only the top two levels of applicators are allowed to conduct or supervise detoxification. Additionally, the policy requires appointment of an "independent applicator." This individual is a certified piscicide applicator who is not otherwise involved with the proposed project. This individual is responsible to critically review the EA prior to public release, serve as an on-the-ground observer and devil's advocate for the first day or two of a project, and ensure compliance with established standards for personnel safety and treatment procedures.

Montana's permitting system for piscicide applications also evolved as a result of the Cherry Creek project. Prior to this project, the DEQ did not have an application form for piscicide activities. Their standard process was to have MFWP biologists complete the DEQ application for turbidity exceedance, but to write in information describing the piscicide activity rather than turbidity generating activities. Nevertheless, at one point in 1999, as threats of lawsuits over the Cherry Creek project were made by opponents, a DEQ staff attorney criticized MFWP's lead Cherry Creek biologist for using the DEQ's turbidity exceedance application form to apply for a piscicide permit. Eventually, the DEQ and MFWP worked together to develop an appropriate application form for piscicide applications.

Westslope Cutthroat Trout Translocations

The primary goal of the Cherry Creek project was to establish a large, self-sustaining population of genetically pure WCT by translocating embryos obtained from existing upper Missouri Drainage WCT populations. However, given the scale of the project area, the limited number of wild donor sources in the entire upper Missouri, and research goals, we ultimately decided to use both wild and hatchery sources of WCT, some from outside the upper Missouri, to start the new WCT population in Cherry Creek.

Wild donor populations were selected based on their proximity to Cherry Creek, genetic purity, population abundance, disease-free status, and assumption that they were aboriginal WCT. Potential Cherry Creek donor WCT populations were screened based on their genetic purity (Shepard et al. 2003,

2005) and spatial structuring (Drinan et al. 2011). Pathogen screening was done for all potential donor populations. The standard MFWP disease panel for testing translocation eligibility includes (1) *Aeromonas salmonicida*, which is the bacteria that causes furunculosis; (2) infectious hematopoietic necrosis virus; (3) infectious pancreatic necrosis virus; (4) viral hemorrhagic septicemia virus; (5) *Renicterium salmoninarum*, the bacterium that causes bacterial kidney disease; and (6) *Myxobolus cerebralis*, which is the parasite that causes salmonid whirling disease. Testing fish for pathogens requires sacrificing fish and surrogate trout species were often used, when available. A sample size of 60 fish provides 95% confidence in identifying the presence of pathogens in the population, assuming a 5% infection rate (AFS-FHS 2014), and this level of testing is consistent with MFWP fish health policy. Ultimately, four wild donor streams were selected, none from the Madison River drainage but all from the upper Missouri drainage (Muskrat, Ray, White's, and Bray's Canyon creeks).

Donor streams were electrofished in June, prior to and during spawning, to capture mature, adult WCT, which were held in live cars near their site of capture until they became sexually ripe. Typically, the largest wild WCT in donor streams were 250 mm in length, but average sizes of mature adults were typically 130–150 mm, with some mature males as small as 100 mm. After capture, fish were monitored at least twice weekly, and when ripe, they were spawned on site (Figure 10; Table 7). A nonlethal fin clip was taken from each fish that donated eggs or sperm. Fin clips from each donor adult were sent to the Conservation Genetics Laboratory at the University of Montana and analyzed for genetic purity. This complimented the population level genetic assessment done to identify potential donor streams and insured that all donors were genetically pure. These fin clips were also archived in order to genetically back-assign progeny to parents and donor stock.

Two captive composite populations (broodstocks) were also used to provide eggs or fry for restocking Cherry Creek (Table 7). A small portion of eggs from each individual wild donor stream were kept at the SRH until they hatched. These fry were reared in troughs until approximately mid-September, when they were transferred to the brood pond prior to freeze up. Once the fish in the pond attained sexual maturity, typically 2 years for males and 3 years for females, they were captured by angling or netting and spawned into egg lots, as described above. Lengths of the mature Sun Ranch Pond (SRP) fish were generally 300–400 mm, despite being first-generation descendants from 150 to 250 mm wild parents. No effort was made to determine the natal stream for each adult fish from SRP prior to spawning them because we wished to randomly mix genetics among the different donor stocks held in the SRH. However, genetic samples were taken from individual fish spawned at SRP so the stream of origin from each parent could be back-assigned and progeny from these pairs could be back-assigned to their parents and streams of origin. Fish in the pond became sexually mature in May, about 2 months earlier than adults in the wild donor streams. Because of earlier maturity, SRP embryos were incubated in colder water to slow incubation and better match timing of egg development from wild donor streams that did not arrive at the hatchery until late June or July.

The second composite population that was used for introductions was MFWP WCT conservation broodstock housed at the WPH in Anaconda (Table 7). This broodstock was founded in 1983 and 1984, with wild fish collected from 12 donor streams in the South Fork Flathead River drainage and 2 donor streams in the Montana reach of the Clark Fork River drainage and has periodically been infused with gametes from

Figure 10. Spawning donor Westslope Cutthroat Trout on site at the stream of capture.

the wild founding populations to maintain genetic viability. The WPH WCT were the most genetically diverse source introduced into Cherry Creek. Fish were spawned and incubated on site at the WPH. Similar to SRP eggs, WPH eggs had to be incubated at colder temperatures so eyed eggs would be ready about the same time as wild-sourced eggs. Fin clips were taken from the WPH adults that contributed egg lots to the project.

We followed the same spawning protocol for all fish contributing gametes to Cherry Creek. We stripped each female of eggs and split the eggs between two 1-L insulated bottles. The green eggs in each bot-

Table 7. Number of donor stream crosses and the number of males and females used from each stream.

Donor source	Crosses	Males	Females	Adults
Brays Canyon	7	7	7	14
Muskrat Creek	73	73	37	110
Ray Creek	68	68	35	103
Sun Ranch Pond	29	29	15	44
Washoe Hatchery	60[a]	60	60	120
Whites Creek[b]	28	28	18	46
Total	265	265	172	437

[a] Data sheets indicate 21 parental crosses were combined into three egg lots.

[b] Does not include egg lots used to provide age-0 fry for Cherry Lake introductions in 2009, 2011, and 2014.

tle were fertilized with milt from different males to produce a unique male × female cross, called an "egg lot" (Table 8). Next, we rinsed away remaining milt with freshwater and allowed the fertilized eggs to water-harden in a water-iodophor solution for 30 min. After water hardening was complete, the water-iodophor solution was drained and the eggs were rinsed with freshwater. Wild eggs were packed in coolers for transport to the incubation facility at the SRH, which is located more than 160 km from some of the donor streams.

We held all eggs in vertical tray incubators until the eyed-stage. Eggs from the WPH population were incubated in vertical incubation trays at that hatchery. Incubating eggs were treated with formalin every 3 to 7 d to reduce fungal outbreaks. We controlled water temperatures during early incubation to slow incubation of gametes collected in May and accelerate incubation of gametes collected in late June and early July so that all eyed eggs were released into remote site incubators (RSIs) from July 4 to August 4, except in 2007 when releases occurred from June 19 to July 24. After the eggs had reached the eyed stage, we removed dead eggs, counted the number of survivors, and transported the surviving eggs to the study site in 1-L insulated bottles.

Introduction of genetically pure WCT from selected wild donor populations and brood sources was primarily accomplished by placing eyed eggs in RSIs (Figure 11; B. Shepard, B. B. Shepard and Associates, and colleagues, unpublished manuscript). Tables 9 and 10 detail the number of donor-source eyed eggs and estimated number of donor-source WCT fry released by year, respectively. Tables 11 and 12 show the number of eyed eggs placed in RSIs and the number of fry released, respectively, by phase and year. With the exception of Cherry Lake, Phases 1 and 2 received only eyed eggs (Table 13). For stocking Cherry Lake, we used same-sex WPH broodfish (see discussion above in the Mistakes, Learning Experiences, and Innovations subsection of the Piscicide Treatments section) to jump-start this population for public angling (including by outfitters) and then introduced cohorts of 630, 1,000, and 2,200 age-0 WCT (fry) in 2009, 2011, and 2014, respectively, from one of the four wild-donor streams (i.e., Whites Creek). Because of inlet and outlet stream spawning, trout populations in Cherry Lake are self-sustaining. Phase 3 was restocked with mostly eyed eggs, but 4,000 age-0 SRP fish (fry) were stocked near the bottom of Phase 3 in 2010. Phase 4 was stocked with 4,850 age-0 SRP fry, as well as triploid (ster-

Table 8. Number of egg lots introduced into Cherry Creek by donor stream and year.

Donor source	2006	2007	2008	2009	2010	Total
Brays Canyon					7	7
Muskrat Creek		22	27	24		73
Ray Creek		25	23	20		68
Sun Ranch		13	13		3	29
Washoe Hatchery	4	21[a]	21	12	2	60
White's Creek[b]	3	8	9	8		28
Total	7	89	93	64	12	265

[a] These 21 parental crosses were combined into 3 egg lots for transfer into remote streamside incubators but here are counted as 21 egg lots.

[b] Does not include egg lots used to provide age-0 fry for Cherry Lake introductions in 2009, 2011, and 2014.

Figure 11. Typical remote streamside incubator (RSI) setup. Water enters via pipe from left, upwells through "gravel" (artificial substrate), and then flows out the top of the RSI. In this set-up, the fry exited the RSI voluntarily and were captured in the second bucket to the right so we could calculate the timing of departure and numbers of surviving fry. Without the trap, the fry would drop directly into the stream.

Table 9. Number of donor source eyed-eggs placed in Cherry Creek remote streamside incubators (RSIs), by year.

Donor source	2006	2007	2008	2009	2010	Total
Brays Canyon					1,066	1,066
Muskrat Creek		5,445	3,204	4,004		12,653
Ray Creek		3,467	1,700	1,911		7,078
Sun Ranch		3,075	3,277		398	6,750
Washoe Hatchery	720	1,015[a]	2,645	1,714	154	6,248
White's Creek	725[b]	1,015	974	636		3,350
Total	1,445	14,017	11,800	8,265	1,618	37,145

[a] Confusing data sheets indicate that an additional 108 eggs more than shown here may have been placed in RSIs.

[b] This is the count of eggs made at the Sun Ranch Hatchery prior to transport to the RSIs. Many died during transport so the actual number of eggs placed in the RSIs was significantly less than this number.

Table 10. Estimated[a] number of donor source fry released from Cherry Creek remote streamside incubators (RSIs), by year.

Donor source	2006	2007	2008	2009	2010	Total
Brays Canyon					665	665
Muskrat Creek		4,414	2,354	2,620		9,388
Ray Creek		2,871	1,256	1,336		5,463
Sun Ranch		1,985	2,634		302	4,921
Washoe Hatchery	138	715	2,044	1,296	116	4,309
White's Creek	139	722	781	392		2,034
Total	277	10,707	9,069	5,644	1,083	26,780

[a] Assumes that the survival rate of each donor stock within each RSI was equal, so we applied the overall RSI fry survival rate to the number of eyed eggs from each donor that were placed into each RSI.

Table 11. Number of eyed eggs placed at remote streamside incubator sites, by phase and year. RSI = remote streamside incubator.

Phase	RSI site	2006	2007	2008	2009	2010	Total
1	Cherry Creek	723	7,112				7,835
	Cherry Lake Creek	722	6,905			7,627	
	Pika Creek[a]			5,845			5,845
2	Tributary 1			5,955			5,955
	South Fork Cherry Creek			4,024		4,024	
	Tributary 2					819	819
3	Carpenter Creek 1			4,241	4,241		
	Carpenter Creek 2				799	799	
Total		1,445	14,017	11,800	8,265	1,618	37,145

[a] Informal name given for project purposes to primary Cherry Lake Creek Phase-1 tributary.

ile) WPH WCT and same sex (male) age-3 WPH WCT. The larger triploid and same-sex fish used in Phase 4 were intended to jump-start a recreational fishery on the Flying D Ranch while limiting natural reproduction of these hatchery-origin fish in the system until the eyed eggs and fry released in the upper phases had time to mature and migrate downstream into Phase 4. This project was the first time triploid WCT had been developed and used in Montana. Although the potential supply of eggs from WPH was essentially unlimited, we made a concerted effort not to overwhelm the other sources by stocking WPH eggs in ratios similar to, or less than, the other, more limited wild sources. Hence, when larger fish needed to be stocked for various reasons (e.g., in Cherry Lake and Phase 4), we used same-sex or triploid individuals to prevent reproduction by these WPH-sourced fish as much as practical until wild-source fish were sexually mature.

Sources contributing to the Cherry Creek restocking were (1) four individual wild, aboriginal WCT populations; (2) a mixture of the four wild populations from the WCT broodstock developed at the SRH; and (3) the WCT statewide conservation broodstock from the WPH.

Table 12. Estimated[a] number of fry released from remote streamside incubator (RSI) sites, by phase and year.

Phase	RSI site	2006	2007	2008	2009	2010	Total
1	Cherry Creek	182	5,476				5,658
	Cherry Lake Creek	95	5,231				5,326
	Pika Creek[b]			4,356			4,356
2	Tributary 1			4,713			4,713
	South Fork Cherry Creek				2,811		2,811
	Tributary 2					583	583
3	Carpenter Creek 1				2,833		2,833
	Carpenter Creek 2					500	500
Total		277	10,707	9,069	5,644	1,083	26,780

[a] Assumes that the survival rate of each donor stock within each RSI was equal, so we applied the overall RSI fry survival rate to the number of eyed eggs from each donor that were placed into each RSI.
[b] Informal name given for project purposes to primary Cherry Lake Creek tributary.

Table 13. Summary of phase treatments and Westslope Cutthroat Trout (WCT) introductions. RSI = remote streamside incubator.

Phase	Treatment years	Years of introductions	Number of WCT introduced	Introduction method
1	2003, 2004	2006–2008	15,275	RSI
		2009, 2011, 2014	3,830	Age-0 fry[a]
2	2005, 2006, 2007[b]	2008–2010	8,107	RSI
3	2007, 2008	2009, 2010	3,333	RSI
	2009[b], 2010[b]	2010	4,000	Age-0 fry
4	2010	2011, 2012	4,850	Age-0 fry
		2010, 2012	6,420	Age-1 triploid[c]
		2011	2,000	Age-3 males[c]

[a] Introduced into Cherry Lake.
[b] Mainstem treatment only.
[c] Age-3 Washoe Park Hatchery males and age 1 triploid (sterile) fry were stocked to jump-start a recreational fishery. These fish were isolated from upstream phases by the Phase-3 barrier.

In summary, stocking of the Cherry Creek Project area was as follows:

- 26,780 newly hatched fry into streams in Phases 1–3 via eyed eggs in RSIs from 2006 to 2010.
- 165 age-2 WPH WCT into Cherry Lake in 2006 (most of these were removed by gill netting in 2009).
- 3,830 age-0 Whites Creek fry stocked into Cherry Lake in 2009, 2011, and 2014.
- 8,850 age-0 SRP fry stocked into Phases 3 and 4 in 2010–2012.

- 6,420 age-1 WPH triploids into Phase 4 in 2010 and 2012.
- 2,000 age-3 WPH males into Phase 4 in 2011.

Stocking densities of wild fry were low:

- 0.69 fry/m of stream length via RSI in Phases 1–3 (26,780 age-0 fry; range of 0.5–1.1 fry/m at various monitoring sites).
- 0.14 fry/m in Phases 3 and 4 (8,850 age-0 SRP fry).
- 0.36 fry/m throughout the entire project area (a total of 35,630 age-0 fry was stocked into the 100 stream kilometers where nonnative trout had been eradicated).

Mistakes, learning experiences, and innovations

Locating and confirming WCT populations that met our requirements for introduction to Cherry Creek required significant effort. Genetically pure WCT populations had been previously identified through efforts unrelated to the Cherry Creek project (Shepard et al. 2003). Even so, careful screening and rescreening of potential donor populations was done to confirm their genetic status (no evidence of hybridization, level of genetic diversity, spatial genetic structure relative to the restoration site; Drinan et al. 2011), population abundance, disease status, and timing and locations of spawning. Developing the methods to capture, hold, and spawn WCT from the selected donor populations required forethought, trial and error, and dedicated staff to sustain that effort for a period of several weeks each year. We demonstrated that prespawn, mature, wild-adult WCT could be captured by electrofishing, separated by sex, and held in perforated plastic storage containers in their streams of origin for up to several weeks prior to successfully spawning them. Green-egg to eyed-egg survivals from these spawning pairs was relatively high (average of 79%). Incubating fertilized eggs to the early eyed stage prior to moving them to RSIs resulted in eyed-egg to emergent-fry survivals of greater than 70% (Table 14; Shepard and colleagues, unpublished manuscript). In other words, despite the spawning stress, remoteness of both the source and recipient sites, multiple times they were transported over relatively long distances, and relatively little care given to

Table 14. Donor source contributions of Westslope Cutthroat Trout eggs and age-0 fry introduced into Cherry Creek.

Donor source	Number of contributing adults	Number of viable eggs produced	Estimated[a] number of fry released
Bray's Canyon Creek	14	1,066	665
Muskrat Creek	110	12,653	9,388
Ray Creek	103	7,078	5,463
Sun Ranch brood pond	44	6,750	4,921
Washoe Park Hatchery	120	6,248	4,309
Whites' Creek[b]	46	3,350	2,034
Total	437	37,145	26,780

[a] Assumes that the survival rate of each donor stock within each remote streamside incubator (RSI) was equal, so we applied the overall RSI fry survival rate to the number of eyed eggs from each donor that were placed into each RSI.

[b] Does not include adults used or eggs taken for Cherry Lake stocking.

the RSIs, more than half of the green eggs collected for this project resulted in fry stocked into Cherry Creek. Although not an absolute prescription, we found that locating a few RSIs near the uppermost boundary of potential trout distributions, spacing RSIs about 5–10 km apart, and annually translocating about 5,000–7,000 eyed eggs into RSIs at each site in a downstream progression over 3 to 5 years can adequately seed a vacant habitat the size of Cherry Creek. Translocating embryos or fish from multiple donor sources and high genetic heterozygosity of donor populations likely improves translocation success due to increased genetic diversity (Andrews et al. 2016).

In 2006, the pilot year of translocations, RSIs were located near the headwaters of Cherry Lake Creek and upper Cherry Creek, so as was typical for all back-country RSI sites, the eggs had to be carried to the RSI sites in coolers on pack frames. Pickup of the eyed eggs incubating at SRH was delayed by the Fourth of July holiday weekend; thus, the eggs were too advanced and started hatching during transport to RSIs. The newly hatched sac fry were very fragile, and most died during transport. Consequently, only 95 fry were released from RSIs in Cherry Lake Creek (13% egg-to-fry survival) and 182 fry from RSIs in upper Cherry Creek (25% survival) in 2006. Transport of eyed eggs to RSI sites was not delayed by holidays or weekends in later years.

Monitoring and Research

Monitoring

Complete eradication of nonnative trout was achieved in 100 km of stream and Cherry Lake from 2003 through 2010, followed by successful translocation of native WCT eyed eggs and fry (Figure 11; Table 13). From 2006 through 2010, 37,145 eyed eggs were placed in RSIs in Phases 1, 2, and 3, resulting in 26,780 age-0 fry released into project waters, a 72% egg-to-fry survival rate (Table 14). Remote site incubator stocking density for Phases 1–3, which contain 74 stream km, was 0.36/m; however, as all RSIs were stationed in the approximately 39 km of stream habitat upstream of and in Carpenter Creek, the stocking rate for that area was 0.69/m (Figure 12). Eradication of nonnative trout was confirmed by repeated piscicide treatments that killed no or few fish during the final treatment effort in each phase and by fish monitoring efforts following piscicide treatments. Extensive annual electrofishing sampling throughout the stream network from 2007 to 2016 captured one 200-mm Brook Trout at the foot of the waterfall barrier between Phases 1 and 2 in Cherry Lake Creek during 2009. This Brook Trout was immediately killed. It was the only nonnative fish we found during extensive annual fish sampling after we had completed treatments in any phase. Cherry Lake posttreatment monitoring was accomplished using gill nets and angling, and the inlet and outlet streams to the lake were monitored using electrofishing and direct observations by personnel walking along these streams. After the last piscicide treatment of Cherry Lake in August 2005 killed no fish, we never found nor had reports by anglers of any nonnative trout in Cherry Lake through 2018.

Monitoring the recovery of the new WCT population in Cherry Creek was an important part of assessing the success of the project. Several monitoring sections of varying length (depending on objective) have been established throughout the project area. Three 100-m monitoring sections, one each in Phases 2–4, were first sampled prior to treatment with piscicides in 2001 to provide comparison of pre- and posttreatment population abundance and fish size. These sites have been subsequently sampled every year, unless fish were not present due to treatment. Additional 100-m posttreatment monitoring sites were established in Phases 1 and 2, in part to assess fry survival from RSIs, and several have been sampled regu-

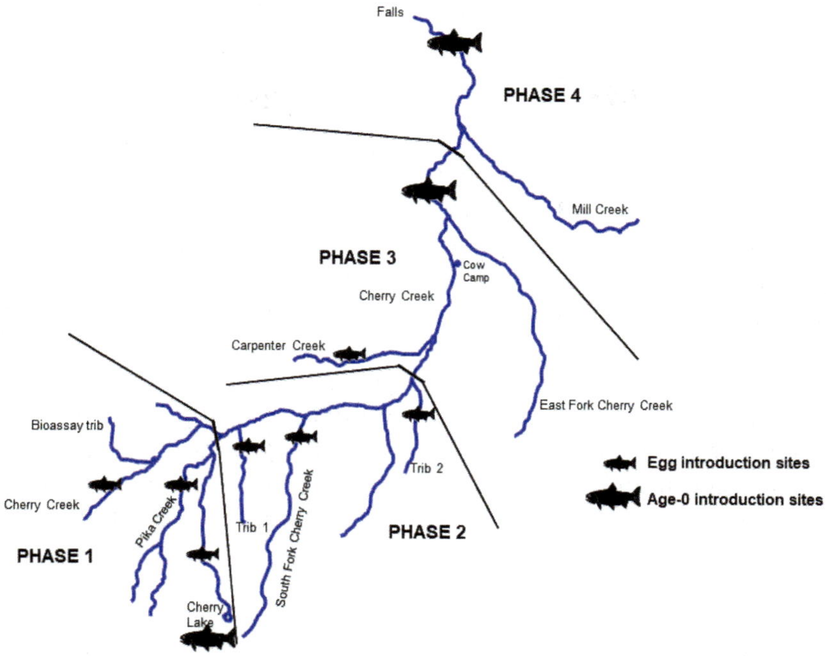

Figure 12. Cherry Creek restoration area showing treatment phases and Westslope Cutthroat Trout introduction sites.

larly since WCT have been introduced. Other preproject fish population estimates were made by Moran (2001) in several sections of Phases 3 and 4 as part of an assessment of the 1994 Arctic Grayling introduction and by the USFS Rocky Mountain Research Station personnel, who took advantage of the piscicide treatments to compare population estimates derived from various sampling methods (e.g., snorkeling and electrofishing) to the actual population of dead fish following treatment in Phases 2 and 3 (R. Thurow, U.S. Forest Service, personal communication). Several of these sites have been resampled.

Pre- and posttreatment fish abundances in monitoring sections indicated that WCT abundance increased rapidly and surpassed nonnative trout abundances 3 to 5 years after they were first translocated as fry (Figure 13). Our data indicates that once the translocated fry matured and successfully spawned, their progeny rapidly filled available habitats. Cherry Creek's base flow wetted width is approximately 5–6 m wide where these monitoring sections are located. Phase 4 was mostly populated by downstream drift of wild and naturally produced WCT originating in the upper three phases. Mean lengths of trout captured in these monitoring sections indicated that when trout abundances were low, the average lengths were highest, but that average lengths of WCT remained higher than average lengths of nonnative trout, even when abundances of WCT were higher than abundances of nonnative trout (Figure 13).

Based on postproject monitoring that expanded population abundances in sample sections to the streams that supported WCT, we estimate that 28,000 WCT occupied the 39 km of stream from Carpenter Creek upstream by 2016 (Figure 14). Expanding these estimates for the upper basin throughout the project area, we estimate that between 50,000 and 100,000 WCT occupied the project area in 2017, almost all of which are descendants of the original fish introduced as eyed eggs or fry. While no gill-net sampling

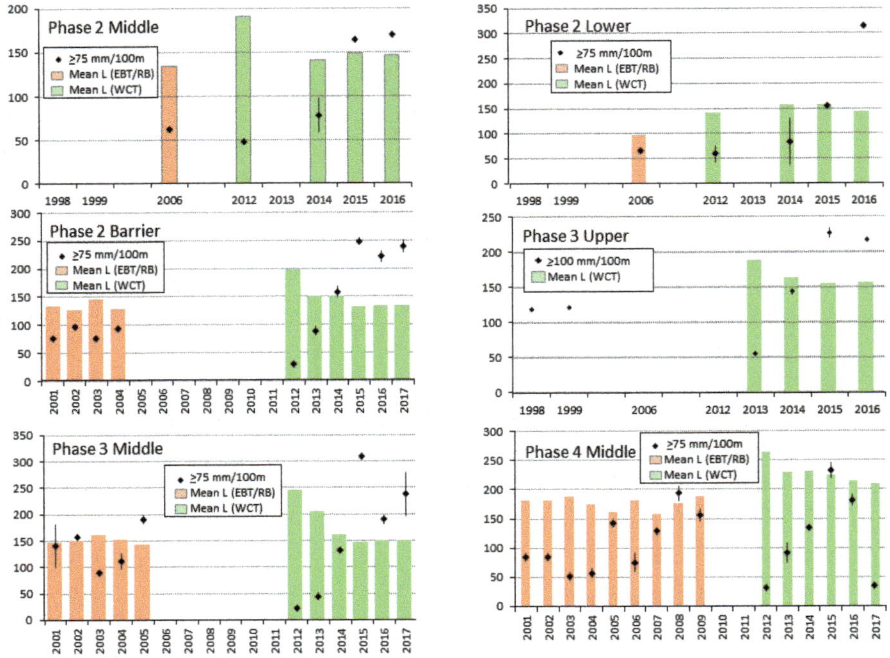

Figure 13. Densities (#/100 m) and mean length of pretreatment nonnative trout (Brook and Rainbow trouts) and posttreatment native trout (Westslope Cutthroat Trout) in six monitoring sections of Cherry Creek.

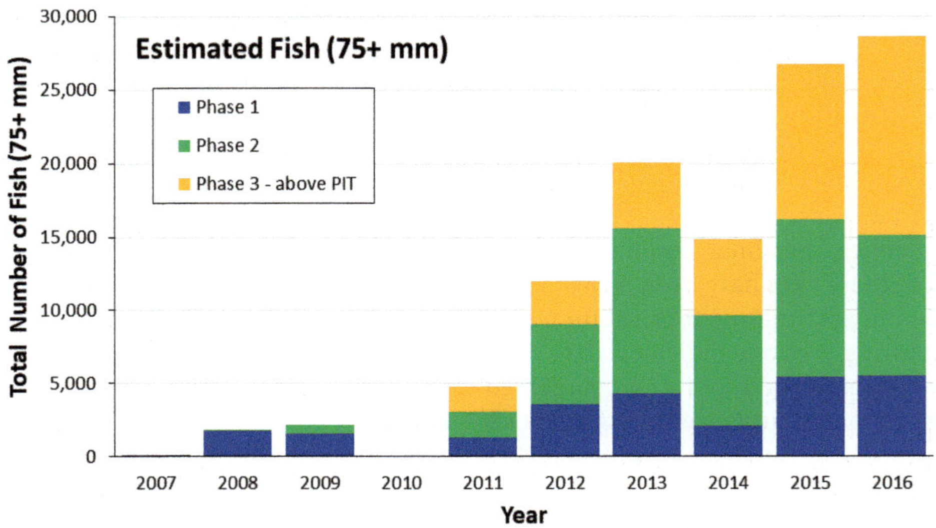

Figure 14. Estimated number of Westslope Cutthroat Trout 75 mm and longer in Cherry Creek basin from 2007 to 2016 by year and location in Phases 1 through 3.

has occurred at Cherry Lake since 2009, angler reports suggest that the lake is currently occupied by a robust population of genetically pure WCT.

Although we expected the new WCT population would thrive, it was unknown how well the stocked fry would survive and how soon they would fill the vacant habitat. We were astonished by how quickly the population expanded. Initially, about 91% of age-1 and 58% of age-2 WCT had moved less than 1,000 m from their RSI release site, mostly in a downstream direction (Andrews et al. 2013), but as habitats near release sites filled and as fish aged, more WCT moved downstream. Throughout the watershed, estimated fish densities increased dramatically about 3 to 5 years following WCT translocations, indicating that natural reproduction was occurring, and survival was high (Figure 14). We expect both numbers and size of WCT to moderate as the population fills the available habitat, and perhaps exceed carrying capacity for a year or two. We are beginning to see WCT in poorer condition, an indication of resource limitation. Based on the data in hand, it appears that pre- and posttreatment fish abundances and average sizes will be similar or that posttreatment populations and sizes will be slightly higher once the initial recovery wave attenuates. This is an important outcome as a common question related to native fish restoration work is how the restored population will compare with the pretreatment, nonnative population.

Apart from Cherry Lake, all live-fish stocking occurred in Phase 4, except the 4,000 age-0 SRP brood progeny stocked into the lower portion of Phase 3 in 2010. Upstream movement of the larger fish stocked into Phase 4 was blocked by the Phase-3 barrier until fall 2011. Even with the mixed-class stocking in Phase 4, it took about 4 years posttreatment for the population density to recover. However, that is not surprising since all the larger fish (same sex and triploids) stocked were not meant to reproduce, but rather to support recreational angling. However, that larger-size component of released WCT is evident in the initial posttreatment monitoring data (Phase 4 Middle; Figure 13). Population recovery in Phase 4 was dependent on the 4,850 age-0 wild sourced Cutthroat Trout stocked in 2011–2012 and downstream movement of wild source fish and their progeny from the upper phases. The 4,850 age-0 fish represent slightly more than one fish per 5 m of stream (0.22 fish/m) in Phase 4. The estimated number of WCT in Phase 4 in 2017 decreased significantly from previous years (Figure 13); however, this estimate was conducted 1 month later in the fall than previous years, so fish had likely already moved out of the sample section seeking pools for overwintering and were unavailable for capture in the monitoring section.

Electrofishing in monitoring sections in Cherry Creek downstream of Cherry Falls and in the Madison River has documented WCT dispersing and establishing downstream of the project area. Montan Fish, Wildlife & Parks has conducted annual mark–recapture electrofishing population estimates in a 6.4-km section of the Madison River immediately adjacent to the Cherry Creek confluence since 1967 to monitor naturalized populations of Rainbow Trout and Brown Trout in the river. Few, if any, Cutthroat Trout were historically captured in that section. An MFWP electrofishing crew began capturing WCT in this river section in 2012 and in March 2016 captured 130 WCT between 180 and 360 mm. Anglers are now pursuing WCT in the river, and many anglers have reported catching WCT, with some proudly sending photos of captured WCT to MFWP. In 2016, anglers reported catching WCT in the river as far as 37 km downstream of Cherry Creek. Although the most severe impacts of whirling disease seen in the Madison River in the early 2000s have abated to some degree (Clancey and Lohrenz 2015), the outcome of the Cherry Creek project is evidence that the strategy implemented by MFWP in the late 1990s to both restore native Cutthroat Trout in tributaries to the Madison River and diversify the fishery in the river was well considered.

Research overview

The Cherry Creek project was first and foremost a conservation effort to expand distribution of wild WCT in the Madison drainage and improve their conservation status, but secondarily, this project provided a unique opportunity to research a variety of restoration-related topics. Combining research efforts with conservation efforts presented some logistical challenges but in some cases also increased efficiencies that improved the project. For instance, RSIs had to be set up, operated, and maintained to translocate the eyed eggs from the selected donor sources to meet the main project objective of establishing a genetically pure WCT population. Simultaneously, to conduct their research on donor fitness, graduate students and their crews needed to operate and monitor RSIs, so they were tasked with conducting this element of the introductions rather than agency employees. We implemented adaptive management by making some changes to WCT translocations during the project based on preliminary research results. Research conducted as part of this restoration project evaluated the translocation, establishment, and expansion of WCT (Shepard and colleagues, unpublished manuscript); the relative fitness, behavior, and plasticity of WCT from different donor stocks (Andrews 2012); and the effects of water temperature on embryo survival and early growth of WCT originating from different stocks that evolved under different thermal regimes (Drinan 2010). We are continuing to collect data to compare the population structure of the pretreatment nonnative salmonid population and the posttreatment native WCT population to inform questions regarding ecological equivalence and recreational opportunity (e.g., size and number of fish; Figure 13). Research was also conducted to assess the effects of piscicide treatments on nontarget organisms such as aquatic invertebrates (Skorupski 2011; C. Kruse, Turner Enterprise, Inc., and D. L. McGuire, McGuire Consulting, unpublished data), amphibians (Billman 2010), and American dippers (also known as water ouzels) *Cinclus mexicanus*, which are a stream-obligate passerine bird (Donnelly 2018). Additionally, research evaluated the effects of environmental variables on piscicide efficacy and persistence (Brown 2010). We encouraged all researchers to publish their results in peer-reviewed journals.

Efficacy of piscicides.—Brown's (2010) doctoral dissertation is entitled "Environmental Conditions Affecting the Efficiency and Efficacy of Piscicides for Use in Nonnative Fish Eradication." He examined the efficacy of piscicides related to fish species and fish size and assessed how sunlight, organic matter, and turbulence influences piscicide effectiveness. He also evaluated the rate of piscicide mixing in a stream related to where it is applied (center of channel or along the edge). He developed models to predict rotenone persistence in various stream channel types. He concluded that increases in sunlight, organic matter, and turbulence all reduced the efficiency of both rotenone and Antimycin A, and his models can be used to better design piscicide treatments (Brown et al. 2011).

Piscicide effects on nontarget organisms.—Skorupski (2011) completed a master's thesis entitled "Effects of CFT Legumine Rotenone on Macroinvertebrates in Four Drainages of Montana and New Mexico." His thesis had two general objectives: (1) to demonstrate the influence CFT Legumine rotenone has on benthic macroinvertebrates for restoration projects in Montana and New Mexico, and (2) to evaluate the immediate response by means of invertebrate drift. His results indicated that effects of piscicide treatments on macroinvertebrates were small and relatively short-term (1 year) for Specimen and Cherry Creek projects in Montana. For the Comanche and Costilla Creek projects in New Mexico, there were more measurable impacts on the aquatic community. He suggested that impacts he observed in three of the four projects may have been related more to potassium permanganate treatments that were used to detoxify rotenone than from the rotenone itself. Nonetheless, aquatic macroinvertebrates in all four projects recovered 1 year after treatment. His study of macroinvertebrate drift found significant increases in macroinvertebrate drift during rotenone application but a delayed drift response compared to previous studies reported in the literature. Rotenone appeared to have the greatest immediate influence on the early life stages of Ephemeroptera and Plecoptera orders. Skorupski (2011) concluded that rotenone impacts on macroinvertebrate communities could be reduced if managers apply the minimal dosages and durations that are needed to eradicate nonindigenous fish species.

Kruse and McGuire (unpublished data) sampled macroinvertebrate communities at several locations in Cherry Creek about 6 years prior to piscicide treatments, the year prior to treatments, immediately before and after treatments in the same year, 2 to 3 years following the final treatment, and most recently in 2017 or 7 years after the last chemical treatment and 20 years after the initial sample. Data are still being analyzed, but as is common with macroinvertebrate data, the results are variable. In general, he found that the number of individuals declined immediately following a treatment and were further suppressed by consecutive years of treatment (average decline was around 40%), but numbers recovered and on average surpassed pretreatment levels within 2 to 3 years after the last treatment. The higher numbers of macroinvertebrates posttreatment were likely due to the lower number of fish in the stream at the time of sampling. Macroinvertebrate recovery may have occurred sooner, but samples were not collected more frequently. Number of taxa also declined, but these results were harder to interpret due to sampling inconsistencies. The number of taxa usually rebounded after the final piscicide treatment but remained lower than pretreatment numbers by about 10% 3 years later.

Billman's (2010) master's thesis is entitled "Investigating Effects of the Piscicide Rotenone on Amphibians in Southwestern Montana through Laboratory Experiments and Field Trials." She tested the effects of up to 2 ppm CFT Legumine rotenone formula (100 ppb rotenone) on various larval and mature stages of native Columbia spotted frogs *Rana luteiventris* and boreal toads *Anaxyrus boreas* in the laboratory (Billman et al. 2011) and in wetlands near Cherry Creek and at High Lake, Yellowstone National Park (Billman et al. 2012). In the laboratory study, she found that tadpole mortality increased with CFT Legumine concentration and exposure period, but effects decreased with age. In the field studies, she found that within 24 h following application of ~50 ppb rotenone at both locations, there was 100% mortality in gill-breathing tadpoles, but non-gill-breathing metamorphs, juveniles, and adults were apparently unaffected. One year following rotenone treatments, tadpoles were found in both sites. Where fish were removed, tadpole abundance was higher than pretreatment abundance after 2 years, which she attributed to reduced predation by fish on macroinvertebrates and tadpoles.

Donnelly (2018) examined the impacts of piscicide treatments on American dippers nesting along Cherry Creek. Dippers feed on a narrow range of aquatic insects (primarily individuals within the orders Ephemeroptera, Plecoptera, and Trichoptera) and small fish. They nest on mid- and near-stream features such as boulders, large woody debris, sheer banks, and bridges. Donnelly examined adult body condition in relation to timing of piscicide application. Body condition of adult dippers was significantly lower the spring immediately following a summer piscicide treatment (i.e., "during treatment") when compared to before and after treatment. This significant effect was attributed to declines in larval aquatic insects in the months following piscicide application. There was no significant difference between pre- and posttreatment body condition after one full year, but the latter was somewhat higher.

A common theme among all the studies that evaluated the effects of piscicide treatments on nontarget organisms was an obvious short-term impact on these nontarget organisms the year or years of the piscicide treatments, followed by a relatively quick recovery in 1 to 2 years. In some cases, it appears the lack of larger fish in the newly establishing translocated WCT population may have allowed amphibians and dippers to rebound relatively quickly and perhaps be in better condition or reach higher abundances due to less fish predation on macroinvertebrates and/or smaller fish providing prey.

Westslope Cutthroat Trout research.— Drinan's (2010) master's thesis was entitled "Thermal Adaptation of Westslope Cutthroat Trout *Oncorhynchus clarki lewisi*," in which he studied thermal adaptation in four wild populations and one hatchery stock of WCT. Mean summer water temperatures in the streams where he collected wild gametes ranged from 6.7°C to 11.2°C. After spawning fish in the wild, embryos were brought into a captive facility and incubated and reared their first summer under three different thermal regimes. Differences in embryonic development, embryonic survival, and juvenile growth were measured. He found that the colder the native stream, the greater the mortality at warm temperatures, but not the converse (Drinan et al. 2012). His findings suggest that in the short term, fisheries managers should consider local adaptations prior to translocating fish between streams (i.e., fish from cold native streams may perform better when moved to cold recipient streams). These results also suggested that because many Montana WCT populations are isolated in cold headwater streams, a fairly rapid increase in water temperature caused by global warming might be detrimental to their long-term persistence. One of the wild populations and the hatchery stock he evaluated were translocated into the Cherry Creek project area.

We integrated Drinans' (2010) findings into our study design for evaluating WCT translocations into RSIs in Cherry Creek. Our research to test the effects of water temperature on fitness of donor stocks from different streams called for splitting the eggs from each donor female used during 2006 and 2007 and placing half of these eggs in Cherry Lake Creek (a cold tributary; Figure 15) and half in upper Cherry Creek (a warm tributary; Figure 15). We spawned each half of these eggs with different males, so we could genetically back-assign progeny. We followed this egg introduction protocol from 2006 to 2010. Tissue samples were taken from each mature adult when we collected gametes. Genetic analyses were used to identify the genetic signature of each donor adult, so when their progeny were captured in the stream they could be back-assigned to each pair using microsatellite genetic analyses.

Because RSIs were such an important tool in our translocation of embryos but had never been used on the scale of a project like this, Shepard and colleagues (unpublished manuscript) evaluated the use of RSIs for translocating WCT in Cherry Creek to provide perspective for others considering this translocation technique. They found that green-egg to eyed-egg survival averaged more than 79% for most egg lots, but survival for a few lots was extremely low, which they related to poor sperm quality for the single male used because survival of

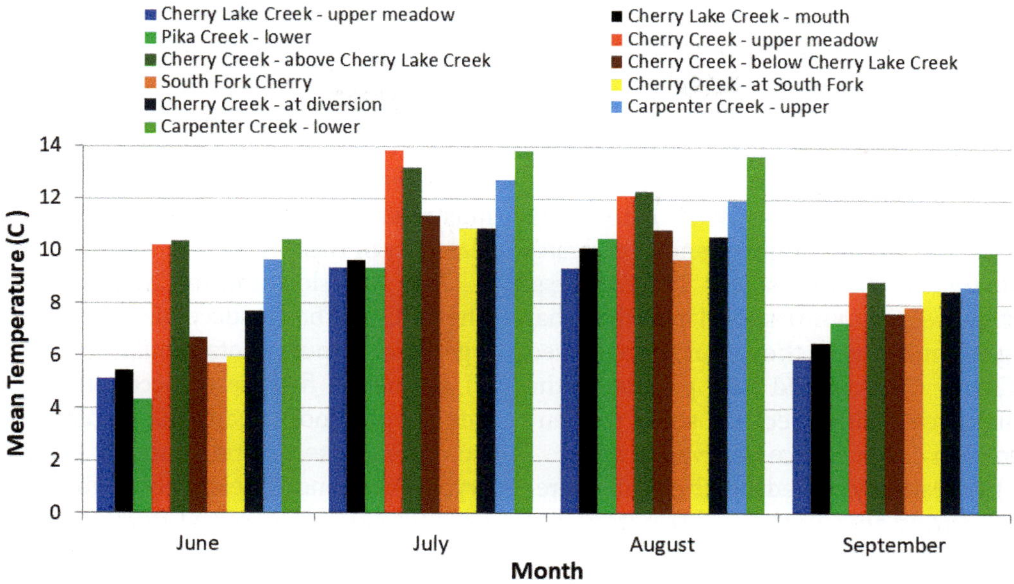

Figure 15. Mean monthly water temperatures at various locations within the Cherry Creek basin from 2008 to 2015. Note the much cooler average water temperatures in the two Cherry Lake Creek sites and in Pika Creek, a tributary to Cherry Lake Creek, and relatively warmer temperatures in upper Cherry Creek above Cherry Lake Creek (upper meadow and above Cherry Lake Creek).

the other half of the same female's eggs was much higher. From 2007 to 2010, eyed-egg to fry survivals averaged greater than 74% and more than 26,000 fry were released from RSIs into Cherry Creek. There were slight but statistically insignificant (Kruskal–Wallis analysis of variance test; $P > 0.10$) differences in eyed-egg to fry survivals among the different donor stocks (average survival ranged from 61% to 73%), but all five donor stocks contributed fry to the new WCT population. We placed RSI sites 5–10 km apart, and the overall fry seeding rates were at least 0.69 fry/m of stream length (range: 0.5–1.1 fry/m). This seeding rate resulted in average densities of about 0.2–1.0 WCT 75 mm and longer per meter of stream length 5 years after eggs were introduced.

Andrews' (2012) doctoral dissertation was entitled "Natural Selection in the Field and the Classroom." In the field portion of her work, she looked for differences in survival, growth, condition, and dispersal following translocation of embryos from five WCT populations to six introduction sites in Cherry Creek. She found that fry emerged from RSIs and dispersed to fill habitat progressively downstream, and to a limited extent upstream, during their first 2 years at large, with wider dispersal occurring the second year (Andrews et al. 2013). Westslope Cutthroat Trout progeny from the captive WPH donors moved shorter distances than progeny originating from wild donor stocks (Andrews et al. 2013). First-generation juveniles that originated from WPH had a higher median survival rate than wild fish. While she found a significant donor source effect for survivals of age-1 and age-2 progeny, she did not find a site effect, indicating that these younger WCT did not perform differently in the warm versus cold stream. However, Drinan's (2010) work suggested that there was an effect of water temperature on incubation success of different stocks, something that was held constant during incubation to the eyed stage for eggs released into Cherry Creek. We may not see an effect of temperature on incubation until we analyze data for the first wild-produced generation of fish that incubate in the cold and warm streams. The higher survivals that Andrews et al. (2016) saw for the captive WPH donor stock may be due to their higher genetic diversity compared to the wild stocks. Nevertheless, Andrews et al. (2016) speculated that that progeny from any of the donor sources used in Cherry Creek would have populated the drainage over time because some progeny from each donor source survived to age 2.

Interestingly, based on the embryo releases during 2007 in the cold tributary that drained Cherry Lake and warmer site in upper Cherry Creek, we found that growth of WCT was much slower, maturity was much later, and longevity was longer for WCT in the colder site than for the other half of the same females' eggs that were released in the warmer water site (Figures 15 and 16). Additionally, successful spawning in this cold tributary was sporadic and only occurred during the warmest years, while in the warmer stream, adult WCT successfully spawned every year (Figure 16). These differences in growth rates, maturation sizes, longevity, and spawning success for the embryos from the same female released into two sites with different thermal regimes illustrates the phenotypic and behavioral plasticity of these fish.

The translocation of WCT into the Cherry Creek basin successfully established a genetically pure metapopulation because (1) nonnative fish were successfully eradicated from 100 stream kilometers and a 130,000-m^3 lake, (2) translocated WCT embryos experienced relatively high survival rates (about 72% survived from eyed egg to fry that dispersed from RSIs) and then successfully reproduced within a few years of their initial release, and (3) all year-classes were represented in WCT populations 5 to 10 years postrelease. Translocations using a mixture of wild sources, which may each individually have relatively low genetic diversity, either together or along with a more genetically diverse captive source, appeared to be a reasonable strategy to conserve overall

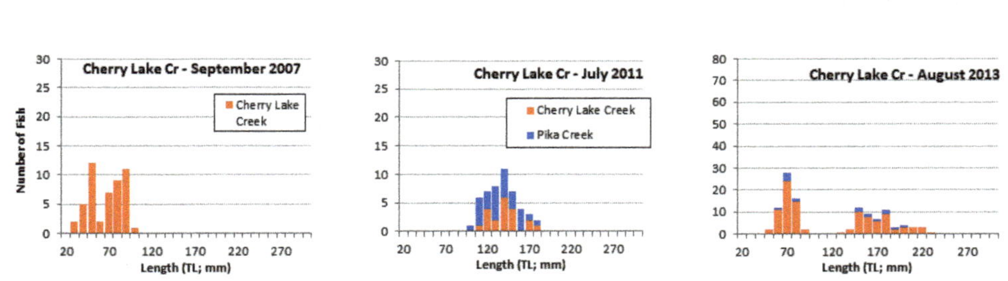

Figure 16. Length–frequency distributions for Westslope Cutthroat Trout captured in upper Cherry Creek (top charts) and Cherry Lake Creek (bottom charts) from 2007 to 2013 showing the growth of the 2007 year-class and evidence of the first successful reproduction in upper Cherry Creek in 2010 and in Cherry Lake Creek in 2013.

genetic diversity of this subspecies in Cherry Creek (Andrews et al. 2016). Using RSIs allowed wild stocks to contribute to translocations but limited the risk of disease and parasites being introduced into the restoration area or brought into captive incubation facilities.

Mistakes, learning experiences, and innovations

Although the scale of the Cherry Creek project was unprecedented and many members of the public and resource agencies were skeptical that it could be done, we demonstrated that removal of a nonnative fish population and reintroduction of native fish can be successful over large spatial scales and complex habitats if implemented methodically across the landscape. The lessons learned and ultimate success of this project led to a more programmatic and mainstream approach to native fish restoration in Montana, as well as across the region, and spawned efforts to attempt additional large, complex restorations.

In reviewing our records of egg transfers to the RSIs, we found some data that are unclear about how many eggs from the WPH were placed in some RSIs, and that some WPH egg lots were combined prior to transfer into RSIs. The number of eggs involved is relatively small and inconsequential to the overall RSI stocking rates and survival of eyed eggs. We have reported the numbers we feel are most reliable and noted them in the pertinent tables. This error is likely due to the fact that several different people were spawning the fish, recording the data, and loading the RSIs over a period of several years, and some may not have clearly understood the detail necessary to fulfil the research objectives. The lesson is that individuals who have accumulated decades of experience and have designed the research need to be ever vigilant that accurate data are being recorded and field operations are properly conducted.

Large-scale native fish restoration efforts provide unique opportunities to conduct research on (1) conservation strategies, such as criteria for selecting donor sources of native species, techniques for eradicating nonnative species and translocating native species, and stocking densities and population recovery expectations; (2) evolutionary ecology, such as local adaptation, inbreeding versus outbreeding depression, and life history plasticity; (3) population demography, such as intrinsic rates of population growth and carrying capacity; and (4) social organization, such as interactions between the public, government agencies and private organizations, management structure to accomplish a large-scale project, and economics of restoration, as a few examples. Involving relevant researchers and statisticians early in the planning stages of native fish restoration efforts ensures that research designs can be incorporated into the project. Integrating research into restoration efforts can provide additional rigor to the planning and implementation of the project and might also provide access to funding sources not otherwise available to complex conservation projects. As described in this and the Research Summary sections, we felt that we were successful at integrating ecological research into the Cherry Creek project. We were less successful in attempting to address some of the social questions related to the project.

Preproject baseline population data, especially abundance, size, and distribution information collected in a statistically rigorous manner, is useful for pre- and postproject comparisons. Surprisingly little information on pre- and postrestoration comparison is available in the literature. Physical habitat, streamflow, and water temperature information are valuable in understanding variations in data over time. We suggest that physical habitat data be collected every 3 to 5 years, but flow and temperature data be collected annually. Fish population data can be collected at whatever frequency is dictated by research or monitoring plans; however, we suggest for trout populations that fish abundance data be collected at least once a generation (every 3 to 5 years) during the first 9 to 15 years after translocations have occurred to document the recovery and expansion of the translocated population.

Collecting and archiving tissue samples from each donor fish used to start a new or supplement an existing native population provides an opportunity to investigate genetic or evolutionary questions in the future. For this project we are using genetic data for back-assignments of progeny to donor pairs or donor populations to assess performance of different donor stocks or pairs (e.g., Andrews et al. 2013, 2016). To successfully assign progeny to parental pairs, a single male's sperm must be used to fertilize each lot of eggs. Depending on research objectives and number of eggs per female, a single females eggs can be divided into multiple lots, each fertilized by a different male. In this "single male" design, there is potential for reduced fertilization rates, embryo survival, or even failure of entire lots of eggs due to poor sperm viability from a single male. However, we believe that this was an acceptable risk given the research objectives on the Cherry Creek Project. Genetic analyses and tagging of individual fish (i.e., using passive integrated transponder tags) can be powerful combination to track the performance of translocated individuals and their progeny. Here, we defined performance as relative survival, growth, and body condition over time among the different donor sources placed in different locations that had different thermal regimes.

Legacy of the Cherry Creek Project

At the implementation of the Cherry Creek project, genetically pure WCT occupied an estimated 836 km (3.0%) of their historical range in the Missouri River drainage in Montana, and nearly all of those populations are in first- or second-order streams, restricted to 8 km or less of habitat, with flows of 0.08 m^3/s or less (Shepard et al. 2003). The Shepard et al. (2003) status review also determined that genetically pure WCT occupied only about 0.7% (13.5 of 1,959.5 km) of their historical range in the Madison River drainage. The Cherry Creek Project increased occupied habitat by 100 km of stream in a fifth-level watershed with as much as 0.57 m^3/s streamflow, thereby increasing occupancy in the Missouri drainage to 3.3% and in the Madison drainage to 5.8% of historic distribution. Successful implementation and completion of the Cherry Creek project launched an ongoing, statewide, programmatic approach to native trout restoration using piscicides in Montana. By 2015, WCT occupied an estimated 5.8% of historical range in Montana's Missouri River drainage, with several populations occupying up to 24 km of habitat with up to 0.34 m^3/s streamflow. The last two known aboriginal Madison drainage WCT populations are being used to establish a new population in Ruby Creek, also in the Madison drainage. Rainbow Trout were chemically eradicated from about 16 km of Ruby Creek from 2012 through 2014, and WCT introductions began in 2015. Ruby Creek has a base flow of 0.28 m^3/s and significant spring influence that maintains favorable water temperatures year-round. Ice rarely forms on Ruby Creek, even in severe winter conditions.

At least partly due to the success of the Cherry Creek project, additional large-scale ambitious piscicide based conservation and restoration projects are now being conducted annually in other regions of Montana where nonnative trout threaten existing WCT populations. Every year, several smaller projects ranging from 2 to 24 km are conducted across western Montana to conserve or restore WCT. In 2017, MFWP completed a 10-year program in the South Fork Flathead River drainage upstream of Hungry Horse Dam to remove hybridized WCT × Rainbow Trout from 21 mountain lakes that were above streams occupied by pure WCT. The smallest of those lakes is 2,960 m^3, the largest is 16 million m^3. Montana Fish, Wildlife & Parks is initiating an effort to remove *Oncorhynchus* hybrid trout with a predominately nonnative Rainbow Trout genetic contribution from the North Fork Blackfoot River above the North Fork Falls to eliminate a source of hybridization to the existing WCT population below the falls and in the main-stem Blackfoot River. It is expected that this project will include treating three mountain lakes and about 73 km of stream flowing up to 1.4 m^3/s.

Yellowstone National Park (YNP) implemented a native trout program in the Madison River drainage within the park in 2013. Yellowstone National Park, with assistance from MFWP, the GNF, TEI, and the USFWS completed a 2-year project to eradicate Rainbow Trout and Brown Trout from approximately 56 km of Grayling Creek, a stream with base discharge at the project barrier of 0.85 m^3/s. They reintroduced native WCT and Arctic Grayling. In 2017, YNP initiated a project to remove nonnative trout from the Gibbon River upstream of Gibbon Falls and introduce WCT and Arctic Grayling. Cherry Creek WCT will be spawned in 2018 to provide eggs for RSI introductions into tributaries of Grebe and Wolf lakes, two of the three lakes in the project area.

Similar to efforts for WCT, MFWP and other partners are conducting piscicide projects to conserve YCT in Montana's Yellowstone River drainage. The largest project completed to date was in Soda Butte Creek, which flows from south-central Montana into the northeastern corner of YNP and is a tributary to the Lamar River where genetically pure YCT reside. A 10-year electrofishing effort to remove nonnative Brook Trout had reduced their numbers in the creek, but their distribution was expanding downstream toward the Lamar River. Montana Fish, Wildlife & Parks, the YNP, the Wyoming Game and Fish Department, the GNF, and the Shoshone National Forest worked together in 2015 and 2016 to salvage YCT from the stream and conduct rotenone treatments in 29 km of stream that successfully eradicated Brook Trout.

Conclusions

The Cherry Creek project is an example of a successful collaboration among state, federal, and private entities to achieve a common goal while allowing each to maintain their legal and statutory authorities, rights, and responsibilities. The success of the project created a pathway for similar efforts to conserve and restore native fish on public and private lands in Montana and elsewhere and emboldened similar efforts in YNP. We believe that the Cherry Creek project was successful for several reasons (see also Kruse et al. 2013):

1. Selection of a restoration site with high-quality and diverse habitat suitable for long-term WCT persistence.
2. Implementation of a systematic approach for nonnative fish eradication over a large scale.
3. An effective collaborative partnership between public resource management agencies and private conservation organizations that created a shared vision and spread financial and logistical obligations.
4. Persistence and mutual support of project partners through social, legal, political, and logistical challenges strengthened the collaborative cohesion and maintained the will to complete the project.
5. Continuity in supervisory personnel for each partner throughout the project, or a transitional period between old and new personnel within an entity, resulted in a smooth, seamless operation.
6. Designation of a point public relations agency and individual that worked cooperatively with the other parties' public relations personnel provided a com-

mon and consistent message to the public and reduced confusion between the cooperating parties.

7. A collaborative process created a trust among the parties and led to open and frank evaluation of issues, problems, and planning.
8. Fair but honest performance reviews winnowed out poorly performing staff, regardless of their affiliation.
9. The communication up the administrative chain in each partner organization carried consistent messaging.
10. The memorandum of understanding and agreements laid the groundwork to build trust among partners and ultimately allowed decisions and actions to be carried out more quickly with less resistance and formality.
11. Use of the best available science and an experimental framework informed project planning, implementation, and monitoring in an adaptive manner that improved the efficiency and outcome.

Beyond serving as a template for successful on-ground efforts to conserve native trout, the Cherry Creek project established procedures and protocols for public involvement, developed and clarified state and federal legal standing to conduct such projects, helped develop public information efforts, and established a model for public/private agency cooperation. In Montana, we are proud of the fact that cooperation between state and federal agencies is the rule rather than the exception, especially when implementing and conducting native fish projects. Private conservation organizations are usually supportive of conservation projects and often comment in favor of projects that are proposed, especially when there is controversy involved. While there may still be public concern and opposition whenever a project is proposed in an area where none have been previously conducted, the public and affected private property owners are becoming more accepting of native fish projects using piscicides as these projects are successfully completed with no negative long-term impacts to nontarget aquatic organisms, wildlife, livestock, water management, public access to waters, and private property issues, and as native fisheries establish and become available for angling.

We estimate the total project cost, including all personnel, planning, implementation, monitoring and research from 1998 to 2018, to be US$2,228,000. We could not document some of the costs between 1998 and 2000, though these costs were minimal due to the relative inactivity of the project, nor could we determine the legal and administrative costs of MFWP and GNF to address the challenges, appeals, and litigation against the project. We estimate project costs as follows:

- TEI
  - o $970,000, which includes
    - § $183,000 to support MFWP during piscicide applications
    - § $90,000 for piscicides and potassium permanganate
    - § $224,000 to support project-related research
    - § $447,000 in general project support, such as proving food, cooks and other support personnel, lodging, equipment purchases, and maintenance.
- National Science Foundation
  - o $800,000, which includes
    - § $100,000 for RSI operation and doctoral candidate support
    - § $700,000 for WCT founder population research, including genetic composition and behavioral research of descendant generations
- The GNF
  - o $165,000
- MFWP
  - o $160,000
- Wildlife Conservation Society
  - o $30,000

- Sun Ranch
  - $30,000
- Montana State University
  - $20,000
- Montana Department of Environmental Quality
  - $15,000
- Various grants
  - $38,000

Acknowledgments

This project would not have been possible without the contributions, persistence and trust of Pat Graham, Jeff Hagener, Ted Turner, Chris Hunter, Larry Peterman, Chris Smith, Becky Dockter, Martha Williams, Bob Lane, Travis Horton, Becky Heath, John Allen, Mary Erickson, Lisa Stoeffler, Wally McClure, Mike Phillips, Russ Miller, Buddy Drake, Wes Orr, Roger Lang, Al Zale, Steven Kalinowski, and the 137 individuals who worked as field crew during the 8 years of piscicide treatments.

References

AFS-FHS (American Fisheries Society-Fish Health Section). 2014. FHS blue book: suggested procedures for the detection and identification of certain finfish and shellfish pathogens, 2014 edition. Accessible at: http://afs-fhs.org/bluebook/bluebook-index.php.

Anderson, J. H., G. R. Pess, R. W. Carmichael, M. J. Ford, T. D. Cooney, C. M. Baldwin, and M. M. McClure. 2014. Planning Pacific salmon and steelhead reintroductions aimed at long-term viability and recovery. North American Journal of Fisheries Management 34:72–93.

Andrews, T. C., B. B. Shepard, A. R. Litt, C. G. Kruse, M. L. Nelson, P. Clancey, A. V. Zale, M. L. Taper, and S. T. Kalinowski. 2016. Performance of juvenile Cutthroat Trout translocated as embryos from five populations into a common habitat. North American Journal of Fisheries Management 36:926–941.

Andrews, T. M. 2012. Natural selection in the field and the classroom. Doctoral dissertation. Montana State University, Bozeman.

Andrews, T. M., B. B. Shepard, A. R. Litt, C. G. Kruse, A. V. Zale, and S. T. Kalinowski. 2013. Juvenile movement among different populations of Cutthroat Trout introduced as embryos to vacant habitat. North American Journal of Fisheries Management 33:795–805.

Behnke, R. J. 1992. Native trout of western North America. American Fisheries Society, Monograph 6, Bethesda, Maryland.

Billman, H. G. 2010. Investigating effects of the piscicide rotenone on amphibians in southwestern Montana through laboratory experiments and field trials. Master's thesis. Idaho State University, Pocatello.

Billman, H. G., S. St-Hilaire, C. G. Kruse, T. S. Peterson, and C. R. Peterson. 2011. Toxicity of the piscicide rotenone to Columbia spotted frog and boreal toad tadpoles. Transactions of the American Fisheries Society 140:919–927.

Billman, H. G., C. G. Kruse, S. St-Hilaire , T. M. Koel, J. L. Arnold, and C. R. Peterson. 2012. Effects of rotenone on Columbia spotted frogs *Rana luteiventris* during field applications in lentic habitats of southwestern Montana. North American Journal of Fisheries Management 32:781–789.

Bramblett, R. G. 1998. Environmental assessment for the Madison River drainage Westslope Cutthroat Trout Conservation and Restoration Program: Cherry Creek Native Fish Introduction Project. Montana Fish, Wildlife and Parks, Helena.

Brown, P. J. 2010. Environmental conditions affecting the efficiency and efficacy of piscicides for use in nonnative fish eradication. Doctoral dissertation. Montana State University, Bozeman.

Brown, P. J., H. Johnson, and A. V. Zale. 2011. Effect of Rainbow Trout size on response to rotenone and Antimycin A. North American Journal of Fisheries Management 31:1146–1152.

Carim, K., M. Young, K. McKelvey, and M. Schwartz. 2015. Environmental DNA sampling for detection of Rainbow Trout in tributaries to Cherry Creek by Turner Enterprises, Inc. final project report. U.S. Forest Service Rocky Mountain Research Station, National Genomics Center for Wildlife and Fish Conservation, Missoula, Montana.

Carim K. J., K. S. McKelvey, M. K. Young, T. M. Wilcox, and M. K. Schwartz. 2016. A protocol for collecting environmental DNA samples from streams. U.S. Forest Service, Rocky Mountain Research Station, General Technical Report RMRS-GTR 355, Fort Collins, Colorado.

Clancey, P. 1997. Madison River/Ennis Reservoir fisheries. 1996 annual report to Montana Power Company, Environmental Division, Butte, Montana.

Clancey, P. 1998a. Madison River/Ennis Reservoir fisheries. 1997 annual report to Montana Power Company, Environmental Division, Butte, Montana.

Clancey, P. 1998b. Cherry Creek Native Fish Introduction Project: decision notice and response to comments. Montana Fish, Wildlife and Parks, Helena.

Clancey, P. and T. Lohrenz. 2015. Madison River Drainage Fisheries and Madison River Drainage Westslope Cutthroat Trout Conservation and Restoration Program. Annual Report to Northwestern Energy Environmental Division, Butte, Montana from Montana Fish, Wildlife and Parks, Ennis.

COSEWIC (Committee on the Status of Endangered Wildlife in Canada). 2006. COSEWIC assessment and update status report on the Westslope Cutthroat Trout *Oncorhynchus clarkii lewisi* (British Columbia population and Alberta population) in Canada. COSEWIC, Ottawa.

Donnelly, R. 2018. Piscicide impact extends beyond targets and toxicity. Restoration Ecology 26:1075–1081.

Drinan, D. P. 2010. Thermal adaptation of Westslope Cutthroat Trout *Onchorhynchus clarkii lewisi*. Master's thesis. Montana State University, Bozeman.

Drinan, D., S. Kalinowski, N. Vu, B. Shepard, C. Muhlfeld, and M. Campbell. 2011. Genetic variation in Westslope Cutthroat Trout *Oncorhynchus clarkii lewisi*: implications for conservation. Conservation Genetics 12:1513–1523.

Drinan, D. P., A. V. Zale, M. A. H. Webb, B. B. Shepard, and S. T. Kalinowski. 2012. Evidence of local adaptation in Westslope Cutthroat Trout. Transactions of the American Fisheries Society 141:872–880.

EPA (U.S. Environmental Protection Agency). 2003. Interim statement and guidance on applications of pesticides to waters of the United States in compliance with FIFRA, July 11, 2003. Federal Register 68:156(13 August 2003):48385–48388.

Finlayson, B., R. Schnick, R. Cailteux, L. DeMong, W. Horton, W. McClay, C. Thompson, and G. Tichacek. 2000. Rotenone use in fisheries management: administrative and technical guidelines manual. American Fisheries Society, Bethesda, Maryland.

Finlayson, B., R. Schnick, D. Skaar, J. Anderson, L, Demong, D. Duffield, W. Horton, and J. Steinkjer. 2010. Planning and standard operating procedures for the use of rotenone in fisheries management—rotenone SOP manual. American Fisheries Society, Bethesda, Maryland.

Gresswell, R. E. 1991. Use of Antimycin A for removal of Brook Trout from a tributary of Yellowstone Lake. North American Journal of Fisheries Management 11:83–90.

Kruse, C. G., P. Clancey, S. Barndt, K. Patten, and B. Shepard. 2013. Setting the stage for conservation success: large-scale watershed renovation and re-introduction of Cutthroat Trout in the Rocky Mountain region of the USA. Pages 26–32 *in* P. S. Soorae, editor. Global re-introduction perspectives: 2013: further case-studies from around the globe. International Union for the Conservation of Nature, Reintroduction Specialist Group, Gland, Switzerland and Environment Agency - Abu Dhabi, Abu Dhabi, UAE.

MFWP (Montana Fish, Wildlife, and Parks). 1999. Memorandum of understanding and conservation agreement for Westslope Cutthroat Trout in Montana. MFWP, Helena.

MFWP (Montana Fish, Wildlife, and Parks). 2007. Memorandum of understanding and conservation agreement for Westslope Cutthroat Trout and Yellowstone Cutthroat Trout in Montana. MFWP, Helena.

Moore, S. E., M. A. Kulp, B. Rosenlund, J. Brooks, and D. Propst. 2008. A field manual for the use of Antimycin A for restoration of native fish populations. National Park Service, Natural Resource Program Center, Natural Resource Report NPS/NRPC/NRR 2008/001, Fort Collins, Colorado.

Moran, S. P. 2001. Survey of habitat and fish communities in segments of Cherry Creek, a proposed site of the re-introduction of Westslope Cutthroat Trout. Master's thesis. Montana State University, Bozeman.

Montana District Court. 2002. Fairhurst v. Montana Department of Environmental Quality, BDV-2000-672 (1st Judicial District March 28, 2002).

Pullin, A. S., and T. M. Knight. 2001. Effectiveness in conservation practice: pointers from medicine and public health. Conservation Biology 15:50–54.

Sheller, F. J., W. F. Fagan, and P. J. Unmack. 2006. Using survival analysis to study translocation success in the Gila Topminnow (*Poeciliopsis occidentalis*). Ecological Applications 16:1771–1784.

Shepard, B. B., B. Sanborn, L. Ulmer, and D. C. Lee. 1997. Status and risk of extinction for Westslope Cutthroat Trout in the upper Missouri River basin, Montana. North American Journal of Fisheries Management 17:1158–1172.

Shepard, B. B., B. E. May, and W. Urie. 2003. Status of Westslope Cutthroat Trout (*Oncorhynchus clarki lewisi*) in the United States: 2002. Montana Fish, Wildlife and Parks for the Westslope Cutthroat Trout Interagency Conservation Team, Helena, Montana.

Shepard, B. B., B. E. May, and W. Urie. 2005. Status and conservation of Westslope Cutthroat Trout within the western United States. North American Journal of Fisheries Management 25:1426–1440.

Skorupski, J. A., Jr. 2011. Effects of CFT Legumine rotenone on macroinvertebrates in four drainages of Montana and New Mexico. Master's thesis. University of North Texas, Denton.

Sloat, M. R., B. Shepard, and P. Clancey. 2000. Survey of tributaries to the Madison River from Hebgen Dam to Ennis, Montana with emphasis on distribution and status of Westslope Cutthroat Trout. Report to Montana Fish, Wildlife and Parks, Helena.

Stuber, R. J., B. D. Rosenlund, and J. R. Bennett. 1988. Greenback Cutthroat Trout recovery program: management overview. Pages 71–74 *in* R. E. Gresswell, editor. Status and management of interior stocks of Cutthroat Trout. Symposium 4, American Fisheries Society, Bethesda, Maryland.

U.S. Court of Appeals for the Ninth Circuit. 2005. Fairhurst v. Hagener, 422 F.3d 1146 (2005).

U.S. District Court for the District of Montana. 2004. Fairhurst v. Hagener, 2004, U.S. District Lexis 30161.

USFS (U.S. Forest Service). 1998. Decision notice and finding of no significant impact. Cherry Creek native fish introduction. Gallatin National Forest, Bozeman Ranger District.

USFS (U.S. Forest Service), Bureau of Land Management, and Association of Fish and Wildlife Agencies. 2006. Policies and guidelines for fish and wildlife management in National Forest and Bureau of Land Management Wilderness (as amended June 2006). U.S. Forest Service, Rocky Mountain Region, Denver, Colorado.

USFWS (U.S. Fish and Wildlife Service). 2003. Reconsidered finding for an amended petition to list the Westslope Cutthroat Trout as threatened throughout its range. Federal Register 68:152(7 August 2003):46989–47009.

USGS (U.S. Geological Survey). 2013. National Hydrography Geodatabase. Available: https://viewer.nationalmap.gov/basic/?basemap=b1&category=nhd&title=NHD%20View. (April 2019).

USGS (U.S. Geological Survey). 2018. Water Boundary Database. Available: https://viewer.nationalmap.gov/basic/?basemap=b1&category=nhd&title=NHD%20View (April 2019).

American Fisheries Society Symposium 91:649–682, 2019

Blackfoot River Restoration: A Thirty-Year Review of a Wild Trout Conservation Endeavor

Ron Pierce*[1]
Montana Fish, Wildlife and Parks
11355 Stella Blue Drive, Lolo, Montana 59847, USA

W. Ladd Knotek and Craig Podner
Montana Fish, Wildlife and Parks
3201 Spurgin Road, Missoula, Montana 59804, USA

Don Peters
Montana Fish, Wildlife and Parks[1]
4955 East Carlton Creek Road, Florence, Montana 59833 USA

Abstract.—The Blackfoot River restoration endeavor is an ongoing collaborative, comprehensive, and successful river conservation story in western North America. This chapter describes the restoration framework and process that shaped the first 30 years of this wild trout conservation story. The program began in the late 1980s when fish population surveys identified widespread habitat degradation and depleted numbers of wild trout throughout lower elevations of the Blackfoot River basin and the precarious status of migratory native trout. These initial findings triggered basinwide protective angling regulations for native trout, followed by fish population surveys in all streams. Fisheries data were then combined with basinwide aquatic habitat assessments to facilitate a collaborative multiscale restoration methodology to improve the ecological integrity of the river and its tributaries. Elements of the restoration framework included (1) basin-scale fish and habitat data collections that helped to identify human-induced limiting factors and promote landowner education/cooperation, (2) pilot restoration projects and prioritizations of tributary restoration work, (3) the site-specific integration of passive restoration (e.g. grazing and revegetation) and active restoration (e.g. fish screens, channel reconstruction) techniques, (4) the application of the reference reach concept within the restoration framework, and (5) the essential role of watershed groups in fundraising, implementation planning, and watershed-scale conservation easement protection, especially on private ranchlands. Finally, this chapter summarizes programmatic elements, specific case studies, and restoration techniques that preceded wild trout population expansion in the tributaries and main-stem Blackfoot River. The purpose of this chapter is to help others understand how we approached and implemented a major watershed restoration program.

* Corresponding author: rpierce12@msn.com
[1] Retired

Introduction

After more than 60 years of stocking hatchery trout in the rivers and streams of western Montana, a 1974 decision by the Montana Fish and Game Commission facilitated an end to stocking practices and ushered in the era of wild trout management (Zackheim 2006). This management philosophy relied on the concept of self-propagating fisheries rather than hatchery supplementation. Decades of stocking also masked a long legacy of mining, dewatering, overgrazing, and other forms of stream degradation throughout western Montana. Once stocking ended, it became increasingly evident that managing for wild trout would not only require reductions in angler harvest, but also restoration of spawning and rearing habitat to recover the natural productivity of damaged trout streams.

Given this context, the Blackfoot River restoration endeavor can be traced to the mid-1980s when public perception of declining wild trout populations in the Blackfoot River prompted Montana Fish, Wildlife and Parks (MFWP) to assess fish populations in the main-stem river and its primary tributaries (Peters and Spoon 1989; Peters 1990). Early investigations confirmed depleted trout populations, overharvest of spawners, precariously low numbers of migratory native trout, and widespread degradation of the tributaries, especially at lower elevations of the watershed (Peters and Spoon 1989; Moore et al. 1991; Pierce et al. 2001, 1997). These initial findings led to basin-scale catch-and release regulations for native trout in 1990, greatly expanded fish population and habitat assessments in tributaries, as well as small-scale, pilot-level restoration projects on private ranchlands (Aitken 1997; Pierce et al. 1997). Increased data collection and early successes on pilot projects led to the incremental development of a stream restoration methodology for the Blackfoot River basin and the expansion of stream restoration from 1990 to the present.

As restoration practices gained social acceptance, the restoration program evolved into an iterative, multiscale native-trout recovery process (Pierce et al. 1997, 2013). Basin-scale fish population information, life-history studies (e.g., movement and habitat use using radiotelemetry), and comprehensive habitat assessments identified anthropogenic limiting factors while directing restoration activities to individual tributary populations. Restoration methods included enhancing instream flows in trout-rearing areas, preventing fish loss in irrigation canals, reconstructing altered streams to naturalize channel form and function, and fencing livestock from riparian areas. The scope and scale of projects gradually expanded outward from the central Blackfoot River valley as human-induced limiting factors were identified and opportunities allowed. Within this framework, monitoring and project evaluation were essential in measuring ecological effectiveness, promoting landowner education, and facilitating adaptive management (Pierce et al. 2013; Penaluna et al. 2016).

This chapter emphasizes the basic restoration framework and phased restoration approaches in the Blackfoot River watershed while providing examples of wild trout responses to the restoration program over a 30-year period. The framework includes (1) watershed-scale fish and habitat data-collection techniques that led to stream prioritizations; (2) strengthened stakeholder relationships through public outreach and fisheries information sharing; (3) how pilot projects, natural channel design methods, and concepts of the reference reach were applied (Harrelson et al. 1994; Rosgen 1996; Pierce et al. 2013); and (4) the essential role of watershed groups. Finally, this chapter summarizes major program elements (landscape connectivity and conservation easement protection, basin-scale and spring creek restoration) using various case studies to help clarify multiscale trends in wild trout response using long-term (>5 years) monitoring data.

Study Area: The Blackfoot River Basin and Its Wild and Native Trout

The Blackfoot River is one of the most scenic, physically diverse, and biologically complex rivers in western Montana. As a headwater basin (6,008 km^2) in the upper Columbia River watershed, it drains the western edge of the Continental Divide via 3,060 km of perennial streams and joins the Clark Fork River near Missoula (Figure 1). Milltown Dam, a run-of-the-river hydroelectric facility, was located at the junction of the Blackfoot and Clark Fork rivers until it was removed in 2008. The Blackfoot River main stem is now free-flowing, 212 km in length, and 1 of 12 renowned blue-ribbon trout rivers in Montana with a publicly (MFWP) held appropriated Murphy instream flow summer water right of 19.8 m^3/s. In 2015, the 1971 priority date of the Murphy water right gained more senior status (i.e., 1904) when the Montana Legislature ratified the Confederated Salish Kootenai Water Compact with Senate Bill 262.

The hydrology of the basin is a snowmelt-dominated regime. As measured at the Bonner gauging station (U.S. Geological

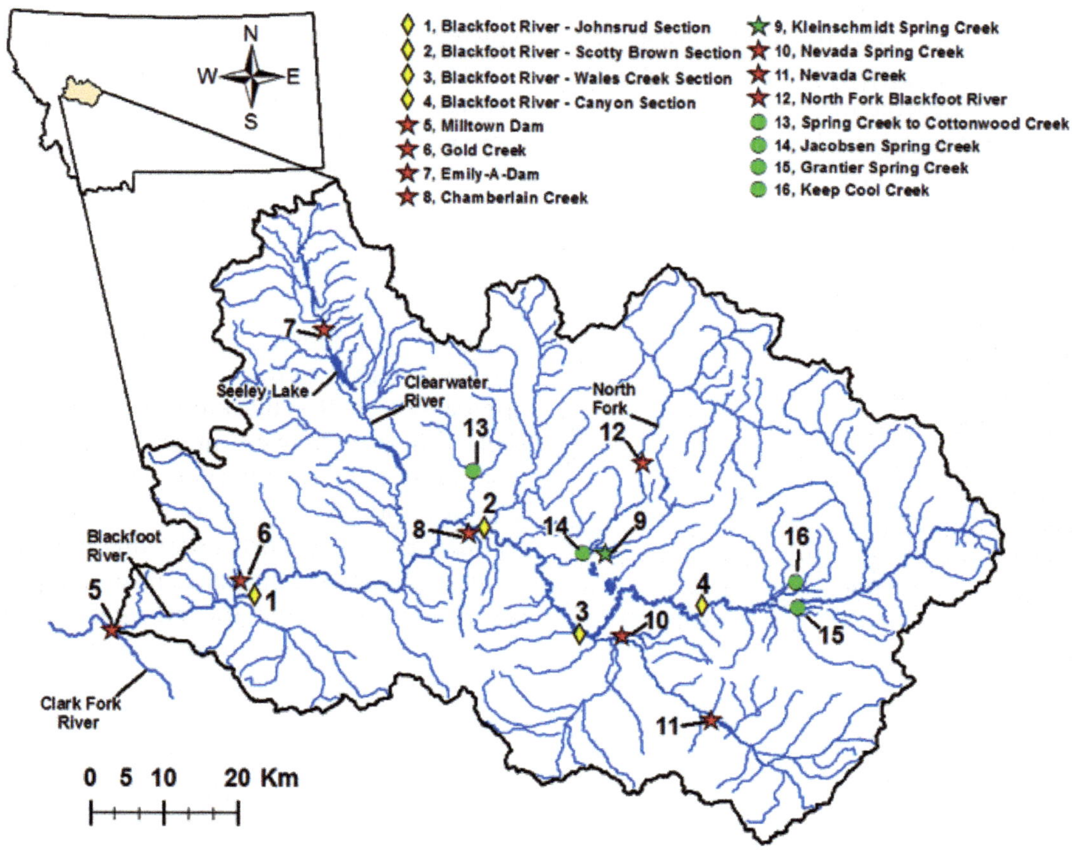

Figure 1. Blackfoot River location map in western Montana including major streams within the basin. Map numbers (1–16) relate to stream names/locations in the legend. The diamonds (1–4) show long-term fish population monitoring sites on the mainstem Blackfoot River. Stars (5–12) refer to tributary restoration case studies described in this report. Green circles (13–16) and the green star (9) show spring creek study sites.

Survey Bonner gage #12340000), river discharge ranges from a high of >140 m^3/s during spring runoff to base flows of 14–20 m^3/s, with a mean annual discharge of about 45 m^3/s. The physical geography of the watershed ranges from high-elevation, glaciated mountains with alpine meadows to timbered forests at the mid-elevations and prairie pothole topography on the valley floor. Glacial landforms, moraine and outwash deposits, and erratic boulders variably cover the floor of the entire Blackfoot and Clearwater River valleys. These features exert a controlling influence on the physical attributes of the Blackfoot River and the lower reaches of most tributaries.

Land ownership in the basin is comprised of public and private holdings: 46% U.S. Forest Service (federal), 36% private, 11% state of Montana, and 7% Bureau of Land Management (federal). In general, public lands and private conservation properties (e.g., The Nature Conservancy holdings) consist of large, forested tracts in mountainous areas, whereas private timber, residential, and agricultural lands are found in the foothills and valley bottoms.

The Blackfoot River watershed supports various cold-, cool- and warmwater fishes. Species richness generally increases with distance downstream. Within this setting, migratory native trout (i.e., Bull Trout *Salvelinus confluentus* and Westslope Cutthroat Trout *Oncorhynchus clarkii lewisi*) have a basinwide distribution and spawn and rear within small- to medium-sized tributaries. As these fish mature, they emigrate to larger, more-productive streams, rivers, and lakes where they grow to maturity before returning to natal tributaries to spawn. Resident trout of these species complete all stages of growth and maturity in small- to medium-sized tributaries and do not migrate to main-stem river or lakes. Predominant introduced wild trout in the main stem and larger tributaries include Rainbow Trout *O. mykiss* and Brown Trout *Salmo trutta*, while Brook Trout *Salvelinus fontinalis* are common in headwater streams.

Wild trout (native and nonnative species) in the watershed depend on tributary habitat during some portion of their life cycle. Therefore, trout species composition and abundance in the Blackfoot River closely reflects the number and quality of nearby tributaries. Biotic relationships among the Blackfoot River and tributary systems vary among river reaches. Some reaches include environments that are naturally (and seasonally) harsh (i.e., intermittent reaches, excessively warm reaches or those prone to winter anchor ice), and/or lack functional tributaries. Within this natural setting, native trout have adapted to a complex, glacially formed riverscape and have developed diverse life-history tactics that include landscape-scale migratory behavior. However, complex life histories and large-scale migrations make native trout especially vulnerable to adverse changes to the aquatic ecosystem. This is particularly true for Bull Trout, a migratory, obligate coldwater char that spawns in discrete upwelling areas and rears in the larger, colder tributaries before moving down stream to larger, more productive rivers and lakes in the Blackfoot River basin (Swanberg 1997; Benson 2009).

Traditional land use in the basin (e.g., mining, timber harvest, and agriculture) have all contributed to habitat degradation and fish-population declines, most of which occurs on the valley floor and foothills of the watershed on agricultural ranch lands. However, a legacy of riparian/aquatic degradation also extends upstream to commercial timber holdings and mining districts, as well as state and federally managed lands. Anthropogenic fisheries impairments were identified on most tributaries (163 of 180) that were surveyed outside of designated wilderness area (Pierce et al. 1997, 2008, 2016). Perturbations varied widely and included culvert crossings and runoff from forest roads, stream dewatering and fish entrainment in irrigation canals, riparian overgrazing and streamside feedlots, channelization of streams, mining disturbance, and streamside residential development (Figure 2).

Restoration Concepts

High-quality stream environments for wild trout include sufficient water quantity and quality, as well as diverse physical channel features that provide food, cover (security) and adequate living space (Orth and White 1999). Stream connectivity provides the mechanism for migratory fish to move among streams or stream reaches and to complete their life cycles. Identification of human-induced limiting factors is essential when attempting to correct impairments and improve wild trout populations. Limiting factors include environmental factors that hold a population below its full potential or natural carrying capacity (Meehan 1991). The concepts of managing for wild trout, focusing on native fish, restoring and connecting habitats, and correcting human-induced limiting factors form the general foundation of the Blackfoot River wild trout restoration program.

Basic restoration planning involves the biogeography of fishes, understanding impacts of habitat impairment, and the role that stakeholders (e.g., private landowners and the angling public) play in restoration outcomes. At a secondary level, restoration methods also consider (1) stream potential (i.e., reference conditions), (2) the relationships of project scale (i.e., stream reach, stream and watershed) to the problem, (3) indirect and/or downstream benefits of restoration actions (e.g., increased flow or improved water quality), and (4) the uncertainty associated with restoration outcomes.

Reducing outcome uncertainty, above all, requires that cooperating parties commit to long-term success and sufficient information on which to base restoration decisions. Project information involves recognizing not only the sources of impairment, but also reasonable assessments of biological poten-

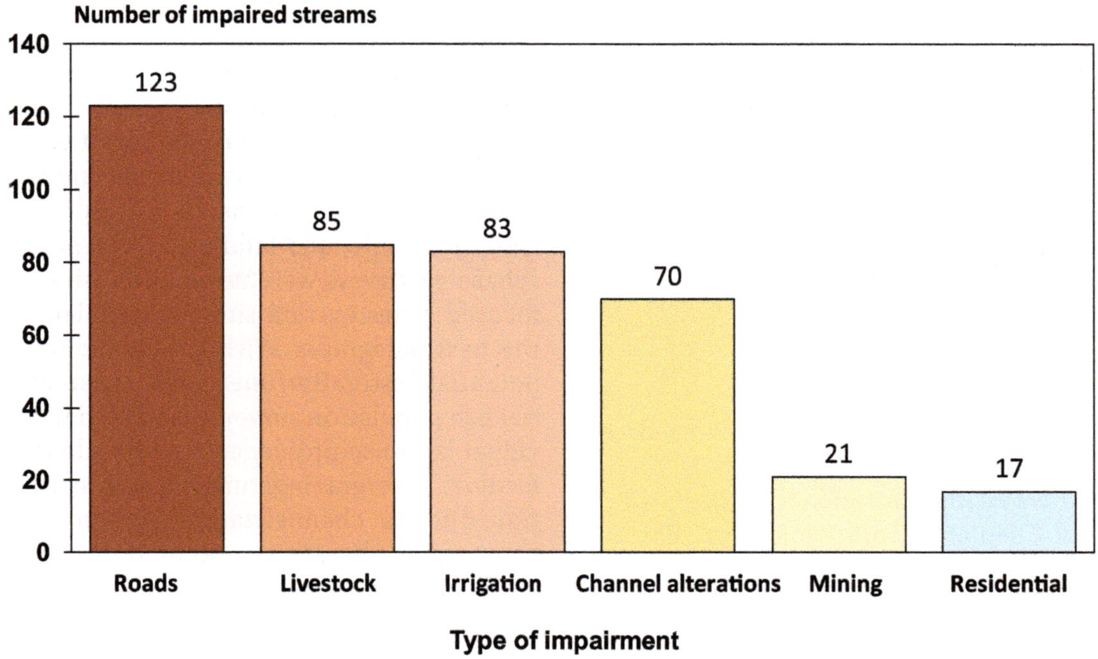

Figure 2. Number of fisheries impaired streams by major land use category.

tial. As described below, obtaining this information usually involves (1) establishing a thorough preproject (fish population and habitat) baseline, incorporating the concept of the reference reach; (2) understanding life history, habitat requirements, anthropogenic impacts, and limiting factors related to target and, in some cases, nontarget species; (3) identifying clear and attainable restoration goals with measurable objectives; (4) developing realistic time frames necessary for project and population recovery; (5) assessing the ability to correct limiting factors; and (6) developing postproject monitoring protocols to ensure projects meet their intended objectives. A willingness to modify subsequent restoration methods based on monitoring results is also important for landowner education, adaptive management, and long-term program success.

Restoration Framework

With these general concepts in mind, the Blackfoot restoration framework included several phased and interrelated elements that began and ended with fish-population data collection (Figure 3). Within this framework, collection of basin-scale fisheries information led to prioritization of tributary projects, which facilitated implementation planning, and ultimately ended with evaluation of restoration outcomes and fish-population response. From the onset, this process engaged stakeholders (e.g., landowners, conservation groups, agencies, and anglers) and informed them with fisheries data. This facilitated a strong educational component that included active and full-time participation of local watershed/conservation groups, including those directly involved with restoration (i.e., the Big Blackfoot Chapter of Trout Unlimited [BBCTU]) and landscape protection (e.g., Blackfoot Challenge and The Nature Conservancy). Within the basic framework, more detailed descriptions of fish population and habitat survey techniques, restoration methodologies, restoration prioritization processes, natural channel design, and reference reach concepts are described below.

Data Collection: Fish Populations, Trout Life Histories, and Stream Habitat Surveys

Fish populations surveys

Longitudinal fish population inventories were completed on all accessible tributaries (1,663 surveys at 772 survey sites on 223 streams) and along the main-stem Blackfoot River from the headwaters to the mouth (Figure 4). Blackfoot River survey sites, established in the late 1980s (prior to restoration work), identified precariously low numbers of native trout and significant recruitment limitations. Tributary fish population sampling began in 1989 with opportunistic surveys that employed consistent, intensive, single-pass electrofishing methods for each survey. When applied consistently, these surveys allowed simple and direct comparisons of fish population metrics over time (i.e., species composition, distribution, relative abundance, and size structure) and within and among small stream sampling locations (Kruse et al. 1998; Pierce et al. 2013). These initial (phase 1) surveys typically began in headwater reference reaches and proceeded downstream in reaches representing the diversity in stream type, land ownership, and land use. Subsequent (phase 2) surveys were more intensive and focused at restoration sites, as outlined in the next paragraph. With an emphasis on potential restoration measures, these initial fish population surveys also noted individual land-use problems (e.g., streamside feedlots, overgrazing, unscreened irrigation ditches, channelization, dewatering, fish passage obstructions, and excessive riparian timber harvest) that potentially impacted fisheries.

Most of the restoration work (75%) occurred on private ranch land. To facilitate

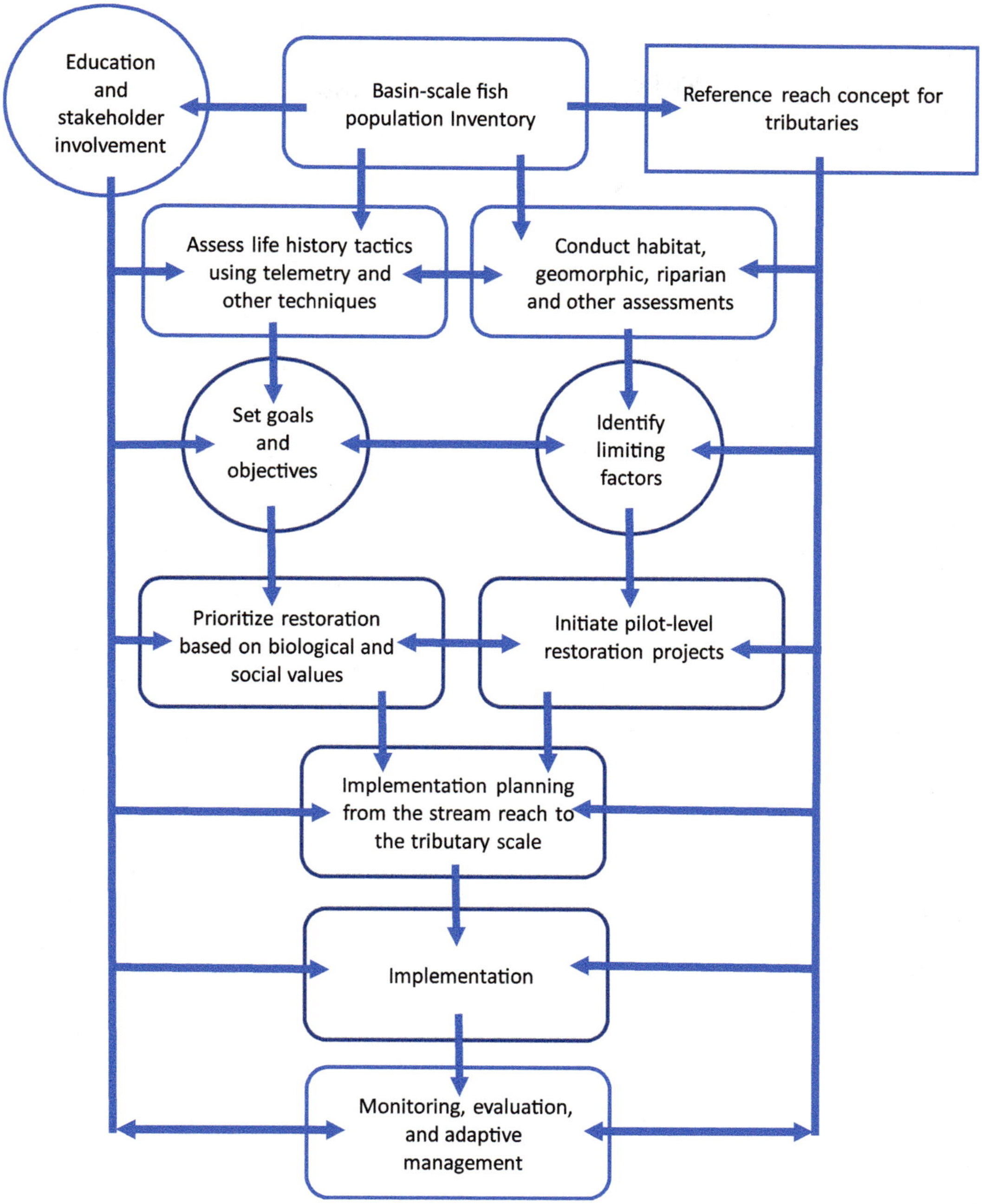

Figure 3. Flow chart summarizing basic framework of the Blackfoot River tributary restoration process. The process engages stakeholders and relies on the reference reach concept throughout all phases of the restoration process.

access to private lands and begin the process of landowner education, field biologists invited private landowners (and their families) to participate in fish sampling (electrofishing) surveys. This fostered initial landowner–agency relationships and basic awareness of stream health issues through private consultation. Other electrofishing surveys were

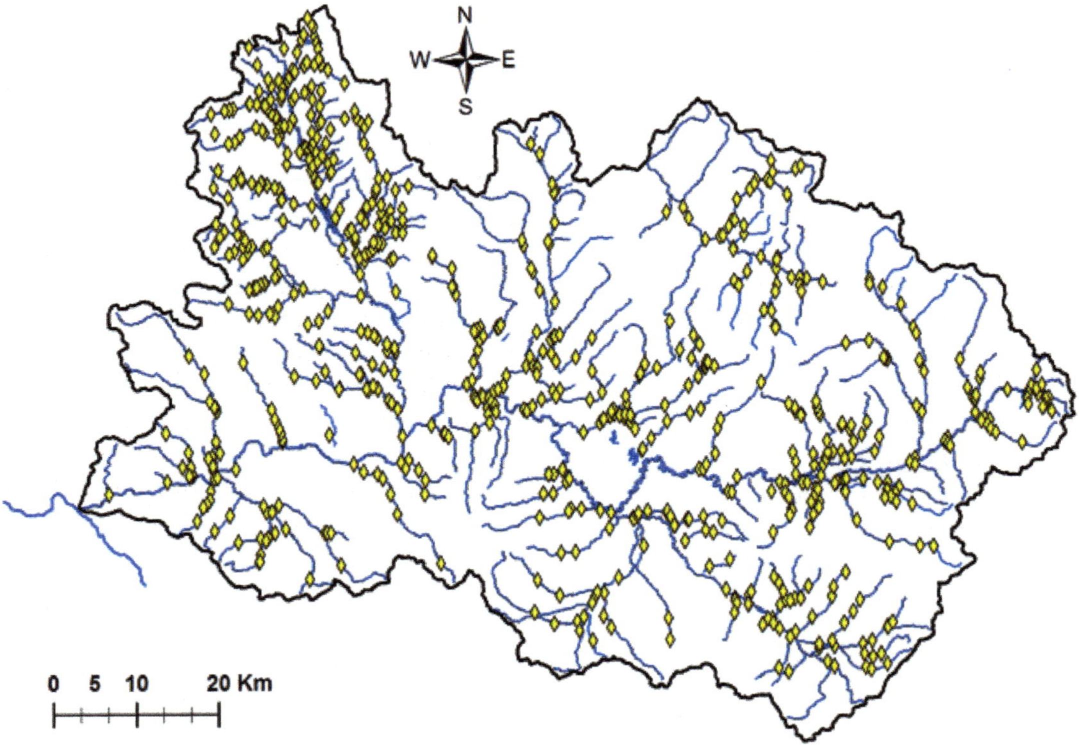

Figure 4. Longitudinal fish populations survey sites (n = 772, yellow diamonds) established on 223 streams between 1988 and 2016 in the Blackfoot River watershed.

used to sample irrigation canals and ditches to assess fish loss, collect genetic samples, and provide fisheries data for research needs. Once streams entered the restoration phase, more quantitative (phase 2) fish-population surveys (e.g., mark-and-recapture or depletion estimates) were established to monitor specific projects (e.g., Pierce et al. 2013).

Fish life-history investigations

In addition to electrofishing surveys, nine radiotelemetry studies were completed between 1996 and 2014 to identify the spawning behavior of adult migratory salmonids (Bull Trout, Westslope Cutthroat Trout, Rainbow Trout, and Mountain Whitefish, *Prosopium williamsoni*) in the Blackfoot River watershed and Clearwater subwatershed (Swanberg 1997; Swanberg and Burns 1997; Schmetterling 2001, 2003; Pierce et al. 2007, 2009, 2012, 2014b; Knotek 2017). Each study investigated the timing of migrations, seasonal habitat use, and spawning locations while also identifying sources of mortality between capture sites and natal tributaries. As with electrofishing surveys, private landowners and local school classes often participated, which included an MFWP Adopt-A-Trout program (Schmetterling and Bernd-Cohen 2002). As technology advanced, genetic techniques allowed identification of nonhybridized, native trout populations and assignment of individual river fish to tributaries of origin using new genetic assignment methods. Using the latter technique, Bull Trout captured in the Blackfoot River or Clearwater chain of lakes were assigned to their natal tributaries with high accuracy (>95%; Pierce and Podner 2016; Knotek 2017). In addition, recent advances in environmental DNA analyses provided reach-specific information on species presence–absence and incidental presence, which can be difficult to reliably detect using standard electrofishing techniques. New techniques and additional life-history information, when merged with standard tributary electrofishing investigations, helped to clarify the status of migratory and stream resident trout populations, as well as identify key factors affecting and limiting them.

Habitat surveys

Physical habitat surveys typically followed fish population surveys. Habitat survey techniques identified habitat features, stream condition (impairments), and restoration opportunities on individual streams, on stream reaches, and across individual land ownerships (Pierce and Peters 1990; Pierce et al. 1990, 1997, 2001, 2002, 2004; Pierce 1991; Koopal 1998). Like electrofishing surveys, habitat surveys typically began at an upstream reference reach and proceeded downstream. These surveys often crossed many land ownerships and covered several kilometers. Depending on expected sample size, every 10th, 4th, or 2nd habitat unit (pool/riffle sequence) was measured for length, wetted width, maximum depth, and bank-full width at riffles; residual pool depth; and pool frequency. Assessment of functional instream wood within the low-flow and bank-full channels along the longitudinal profile was also completed. In addition, riparian vegetation, potential vegetation, and recruitment of instream wood were cataloged. Anthropogenic perturbations in riparian areas (grazing, land clearing, or timber harvest) were noted and/or mapped on 7.5-min quadrangle maps and aerial photographs. Later, high-resolution geographic information system-based aerial photos and multispectral imagery helped evaluate degraded reaches (Fitzgerald 1997; Marler 1998). To characterize survey reaches in greater detail and better identify limiting factors, habitat survey methods adopted more detailed geomorphic survey methods (Rosgen 1994, 1996). These surveys, performed within representative reaches of the habitat surveys, identified reference reach conditions, as well as human-related channel/riparian alterations (e.g., Pierce et al. 2013). Pebble counts (Wolman 1954) and streambed core samples (McNeil and Ahnell 1964) described substrate and spawning area conditions, including anthropogenic sediment (Pierce et al. 2006, 2017; Neudecker et al. 2012; Eby et al. 2015). In addition, stream discharge and continuous water temperature data collection were standard in tributary assessments (Pierce et al. 2009, 2012, 2014a, 2014b). Benthic invertebrate sampling helped clarify certain land use and physical stream habitat relationships (Pierce et al. 2008, 2017). Minimum instream flow assessments, based on the concept of the wetted riffle (Leathe and Nelson 1989), were also completed where water leasing or other conversions to instream flow were pursued.

Pilot Projects and Tributary Prioritization

Pilot projects

With basin-scale data collection fully underway in the early 1990s (Figure 4), reach-scale pilot restoration projects were initiated with landowners willing to demonstrate stream improvement techniques to help educate surrounding landowners. These projects began in the central Blackfoot valley and expanded outward (Figure 5).

Tributary prioritization

To better focus on biological priorities, multicriteria decision trees were periodically developed to prioritize tributaries and guide proposed restoration actions. Matrices integrated fish-population and life-history information for migratory native trout and sport fisheries on the Blackfoot River, stream-health information, and social/financial elements

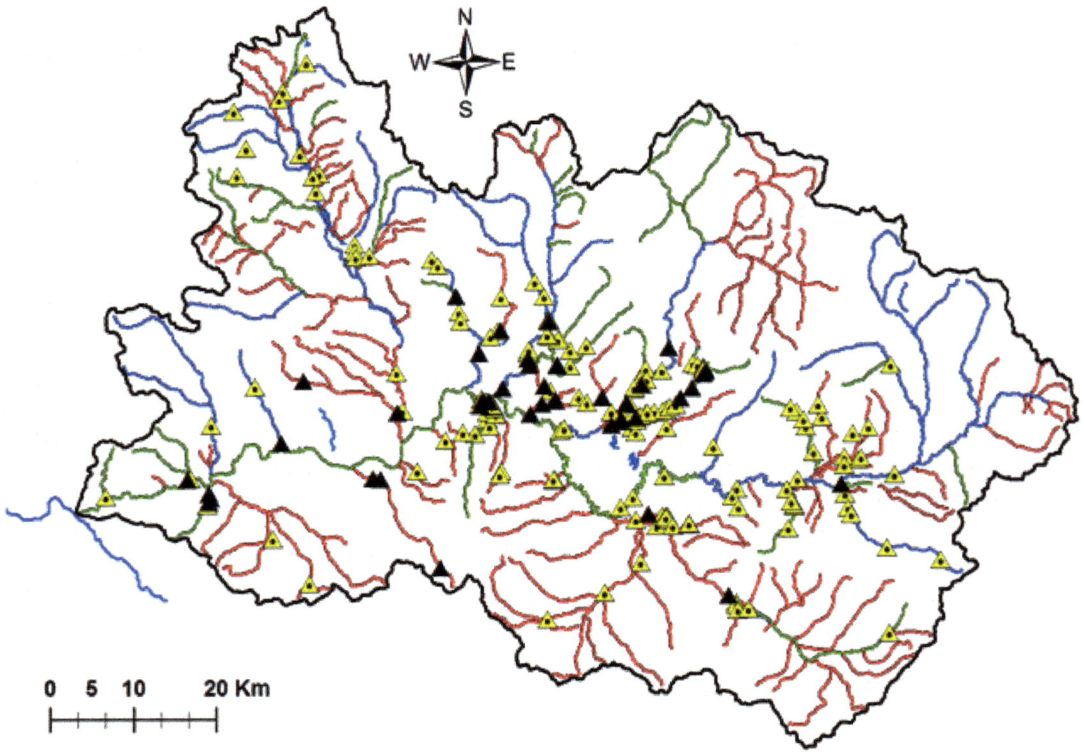

Figure 5. The 2008 Blackfoot River restoration prioritization template for native trout (blue streams = high priority, green streams = moderate priority, and red streams = low priority) along with the location of 199 completed restoration projects (yellow and black triangles) on 75 tributaries. Black triangles represent 42 restoration sites completed between 1990 and 1997; yellow triangles represent 157 projects completed between 1997 and 2017, showing the spatial expansion of the restoration program.

(Table 1). The most recent strategy (Pierce et al. 2007) was developed for 180 tributaries in response to (1) an increasing number of watershed interest groups; (2) state, regional, and federal fisheries management directives; (3) the development of drought, subbasin, and U.S. Environmental Protection Agency water quality improvement plans; (4) Endangered Species Act-designated critical habitat for the recovery of Bull Trout; and (5) recently completed fish population inventories in the Clearwater River subwatershed (Pierce et al. 2008; Knotek 2017). This prioritization provided a native trout-based template for restoration projects, which integrated all fisheries-related improvement programs into a single guiding strategy.

Implementation Planning and the Importance of Watershed Groups

Implementation planning

Stream restoration typically focused on correcting obvious human impacts to fish populations and natural stream function. Within a context of stream priorities, implementation planning typically occurred from the stream reach to the tributary scale. Planning often involved an interdisciplinary team of agency specialists (e.g., biologist, hydrologist, grazing and water rights specialists), nonprofit conservation groups (i.e., 501(c)3: BBCTU and Blackfoot Challenge) and cooperating landowners/managers. Once major projects were selected, fisheries biologists surveyed

Table 1. Restoration prioritization scoring criteria for streams in the Blackfoot River basin. Point values were applied to 182 streams, with scoring weighted toward native trout and biological benefits in identifying the highest priority streams.

	Biological benefits – 150 possible points	Points
1	Bull Trout spawning (yes/no)	20/0
2	Bull Trout rearing (yes/no)	10/0
3	Bull Trout core area (yes/no)	10/0
4	Westslope Cutthroat Trout presence (fluvial/resident/none)	20/10/0
5	Sport fisheries value to the Blackfoot River (multiple species/single species/ none	20/10/0
6	Technically able to address entire stream system (yes/no)	20/0
7	Provides increased stream flow to the Blackfoot River (yes/no)	20/0
8	Improves downstream water quality by reducing sediment (yes/no)	10/0
9	Improves downstream water quality by reducing temperature (yes/no)	10/0
10	Improves downstream water quality by reducing nutrients (yes/no)	10/0
	Social and financial consideration – 50 possible points	
11	Landowner/manager cooperation in the watershed (high/moderate/low)	20/15/10
12	Cost-effectiveness cost/mile (low moderate/high)	20/10/5
13	Demonstration/education value (high/low)	10/5
	Total possible points	200

or updated information on fish populations and habitat conditions to quantify response variables, which usually involved the use of reference (control) reaches. This information supported project design, fundraising, contracting, permitting, landowner agreements, and monitoring. Ultimately, all projects on high-priority streams with this level of planning were funded with the support of regulatory agencies.

The importance of watershed groups

Most of the project administration and fundraising (private donations, foundations, and state and federal grants) were coordinated through BBCTU and agency partners. The nonprofit status of BBCTU and other conservation groups provided a mechanism for generating tax-deductible private funds and cost-effective implementation. In addition to fundraising, BBCTU often obtained local, state, and federal stream permits for private land projects in coordination with agency partners. Project bids (consulting and construction) conformed to state and federal procurement policies that included the development of qualified vendors lists derived through a competitive process. A minimum project cost triggered the qualified vendor list, from which BBCTU solicited bids for consulting and contractor services. Bid contracts were signed between BBCTU and vendors upon bid acceptance. Depending on the specific project, landowners were responsible for certain costs and project maintenance once construction was completed. Written 20-year agreements for project maintenance were signed by landowners and BBCTU on each project. Last, BBCTU supervised contractors during project construction in consultation with agency specialists. This coordinated work often entailed several overlapping techniques that were specifically tailored to the individual problems. These included riparian grazing (36 streams), fish passage improvements (32 streams), channel reconstruction (27 streams), fish screens (18 streams), and instream flow enhancement projects (17 streams). All private land projects sought collective benefits for landowners and aquatic resources and were completed voluntarily. To date, projects have been completed at 199 locations on 75 streams (Figure 5).

Unlike BBCTU, the Blackfoot Challenge focused on broadscale watershed conservation that included organizing educational tours, drought planning, and forest restoration and helped to coordinated conservation easement strategies among cooperating agencies, land trust organizations, and private landowners. The Blackfoot Challenge ultimately helped to increase large-scale land protection (conservation easement and land purchases) in the watershed from 6.5 km^2 in 1975 to 1,623 km^2 in 2016, which included 1,628 km of riparian corridor (Figure 6).

Stream Restoration Techniques

Instream habitat improvement using natural channel design

Restoration techniques were tailored to site-specific problems that required both passive and active methods. Passive methods (e.g., improving riparian grazing and revegetation) relied on natural recovery to address the cause of degradation. Active restoration included new diversions, fish screens, or channel and riparian area modifications with machinery to directly restore habitat features (e.g., natural channel morphology and instream wood) on altered streams that could not otherwise be restored using passive methods alone. Almost all active channel reconstruction projects required pas-

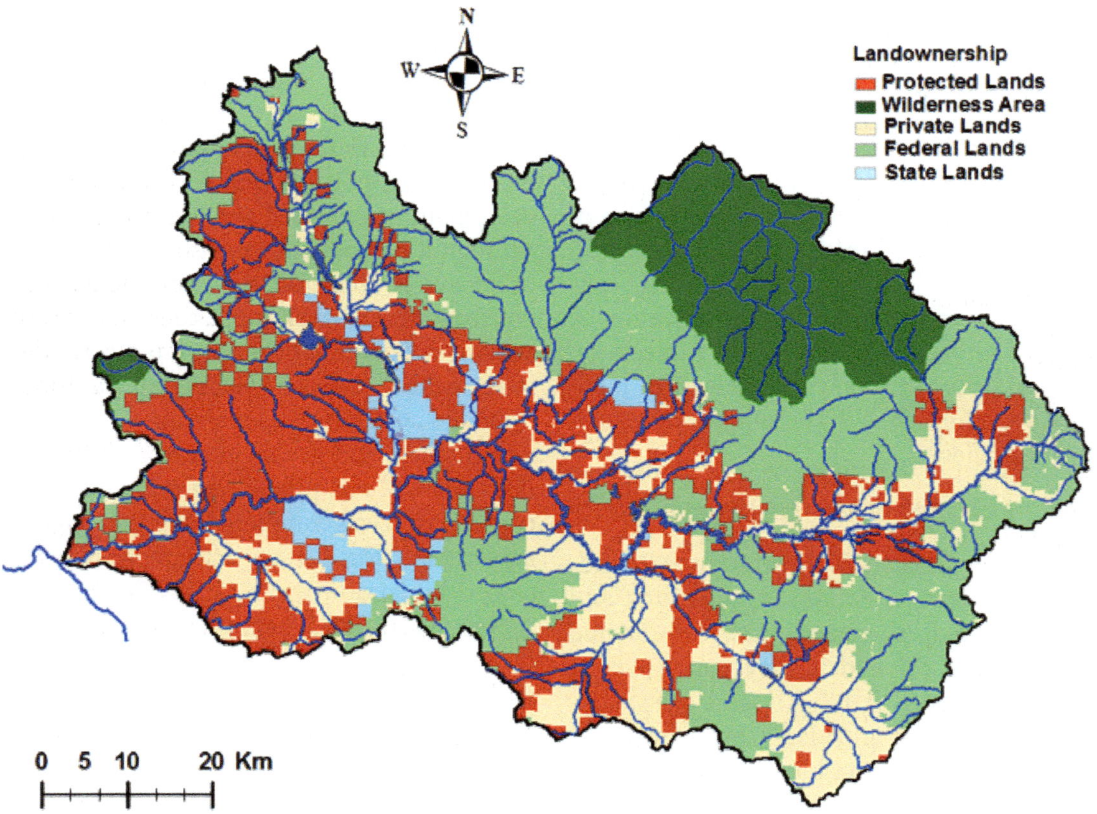

Figure 6. Current land ownership status in the Blackfoot River Watershed. Red areas show either private lands converted to public ownership or private lands with conservation easement protection.

sive methods to ensure vegetative recovery through compatible land use (e.g., managed riparian grazing) once the stream was restored to a more natural condition.

For active channel reconstruction and instream habitat restoration, methods evolved from the use of reference reaches alone to incorporating natural channel design concepts (Rosgen 1994, 1996). Over time, stream classification and related survey techniques became standard in stream assessments, restoration designs, and monitoring programs (Rosgen 1996, 2007, 2011; Pierce et al. 2013). Unlike enhancement techniques, natural channel design involved a geomorphic approach that fits naturally stable streams within the proper valley type. The Rosgen (1994) stream classification system provided the basis of this approach by quantifying channel dimension, pattern, and profile. In addition, riparian health, instream habitat, and fish population surveys, along with measurements of discharge, sediment, and bed and bank stability, permitted the assessment of existing and potential channel (i.e., references) conditions. Geomorphic indicators of the bank-full channel, prediction analyses (reference reaches and dimensionless ratios), and validation (regional curves for western Montana; Lawlor 2002) defined naturally functioning channels and provided the basis for natural channel design.

Final restoration designs sought to emulate naturally complex stream channels capable of conveying flows, transporting sediment, and supporting essential habitat features. Vegetation colonization through intensive seeding, mature shrub, and sod mat transplants, and other bioengineering methods provided immediate fish habitat as well as long-term bank stability. Strategically placed streambank and habitat structures allowed for shrub colonization and further provided long-term channel stability and habitat complexity once vegetation was reestablished. Compatible livestock grazing systems were also an essential component for success (Pierce et al. 2013). Ultimately, restoration projects attempted compatibility with ecologically sound and sustainable land-use practices, conservation of high-quality aquatic habitat, and improvement of native aquatic species.

Programmatic Elements: Landscape Connectivity, Restoration Techniques and Multiscale Trout Response

Four major restoration program elements were selected to illustrate implementation concepts and add relevant long-term case studies to the body of restoration literature: (1) landscape connectivity and Bull Trout recovery, (2) evolution of restoration methods on basin-fed streams that drain from the surrounding mountains, (3) spring creek restoration on the valley floor, and (4) multiscale fisheries response, including trout community response and population trends in the Blackfoot River. Each specific case study included at least 5 years of postrestoration monitoring. All tributary electrofishing metrics were standardized to age-1 and older trout per 30 m, unless otherwise noted. All population estimates of abundance were calculated at the 95% level of confidence (i.e., point estimate ±95% CI). Other analyses of statistical significance varied by project as described (and cited) in the specific case study.

Landscape connectivity and Bull Trout recovery

Functional and resilient Bull Trout populations require habitat that is cold (<13°C), clean (low sediment), complex (with deep pools, overhanging banks, and downed trees), and connected (lacking migration barriers) (USFWS 2010, 2015). Given these species-specific habitat requirements, along with wide-ranging and complex life histories, Bull Trout recovery in the Blackfoot River watershed required a landscape perspective.

In the greater Blackfoot River watershed, major improvements in habitat connectivity included the removal of Milltown Dam at the mouth the Blackfoot River (Figure 7), removal of Emily-A Dam (Figure 8) on the Clearwater River, and fish-friendly improvements at irrigation canals on the north fork of the Blackfoot River (hereafter, North Fork). These measures targeted various native trout life stages (juvenile, subadult, adult) and migratory life histories, including adfluvial (lake dwelling) fish in the

Figure 7. The top photo shows Milltown Dam prior to its removal in 2008. The bottom photo was taken at the same site in 2014.

Figure 8. Before (top) and after (bottom) photos show Emily-A Dam before and after removal, which now allows uninhibited upstream fish passage.

Clearwater subwatershed and fluvial (river dwelling) fish that inhabit stream networks associated with the Blackfoot, Clark Fork, and lower Clearwater rivers.

Like adfluvial populations, fluvial Bull Trout originating in the North Fork are migratory, obligate coldwater char that spawn in discrete groundwater upwelling areas in the larger headwater streams (Swanberg 1997; Pierce et al. 2006). The North Fork system supports the largest population of migratory Bull Trout in the Blackfoot and upper Clark Fork watersheds of western Montana. Adults spawn primarily in undeveloped (e.g., designated wilderness) watersheds and juveniles disperse downstream throughout the river system. Once mature (age 5–7), adult Bull Trout return to spawn in their natal spawning areas.

Beginning in the 1990s, electrofishing, trapping, and telemetry studies identified Bull Trout spawning and rearing areas, spawning behavior, and movements patterns of adult fish (Swanberg 1997; Schmetterling 2003; Benson 2009; Pierce and Podner 2016). In addition, redd counts helped to track population size and trends. Trapping and radiotelemetry identified Milltown Dam as a complete barrier to upstream movement for all fish (including Bull Trout) attempting to ascend the upper Clark Fork and Blackfoot rivers (Schmetterling 2003). In addition, electrofishing identified losses of out-migrant juvenile Bull Trout to five irrigation canals on the main-stem North Fork as a major limiting factor (Pierce et al. 1997). In the Clearwater River system, a large (3 m high), low head dam (Emily-A Dam) constructed in the 1960s completely blocked upstream fish movement upstream of Seeley Lake in the river main stem (Benson 2009). Once identified, habitat connectivity projects were initiated at each site. Fish screens were first placed in all five canals on the North Fork (1990s), Milltown Dam was later removed (2008), and, finally, the Emily-A Dam was removed (2010) and replaced with a step-pool channel that maintained the same crest elevation (Figure 6).

Comprehensive project monitoring confirmed fisheries impacts related to the dams and the importance of restoring connectivity in these main-stem systems. For instance, Bull Trout redd counts increased in the North Fork following protective angling regulation and fish screening in the 1990s. Subsequent removal of Milltown Dam likely also contributed to increasing redd counts after 2008 (Figure 9). Similarly, Bull Trout redds upstream of Emily-A Dam increased from about 19 to 20 redds pre-project to an average of 50 redds after interim (manual capture and release) and permanent fish passage were re-established (Figure 9). Despite positive trends in the strongest remaining migratory populations, small populations of Bull Trout in the lower Blackfoot River watershed have undergone dramatic population declines over the past 30 years (Pierce and Podner 2016). These declines spatially overlap with intensive land use and where warmer stream temperature regimes favor expansion of nonnative competitors (e.g., Brook Trout and Brown Trout; Al-Chokhachy et al. 2016; Pierce al. 2016). According to recent projections (Isaak et al. 2015), regional warming will continue to reduce thermally suitable Bull Trout habitat significantly by 2040, highlighting the importance of high-elevation refugia (e.g., headwater areas of the North Fork; Pierce et al. 2018).

Restoration Methods in Basin-Fed Streams

Unlike broadscale habitat connectivity, small-scale stream restoration projects emphasized natural channel design and reference reach concepts. Within this context, the case studies below highlight fish population response to habitat improvements on basin-fed streams and discuss the evolution of restoration methods to a range of anthropogenic perturbations.

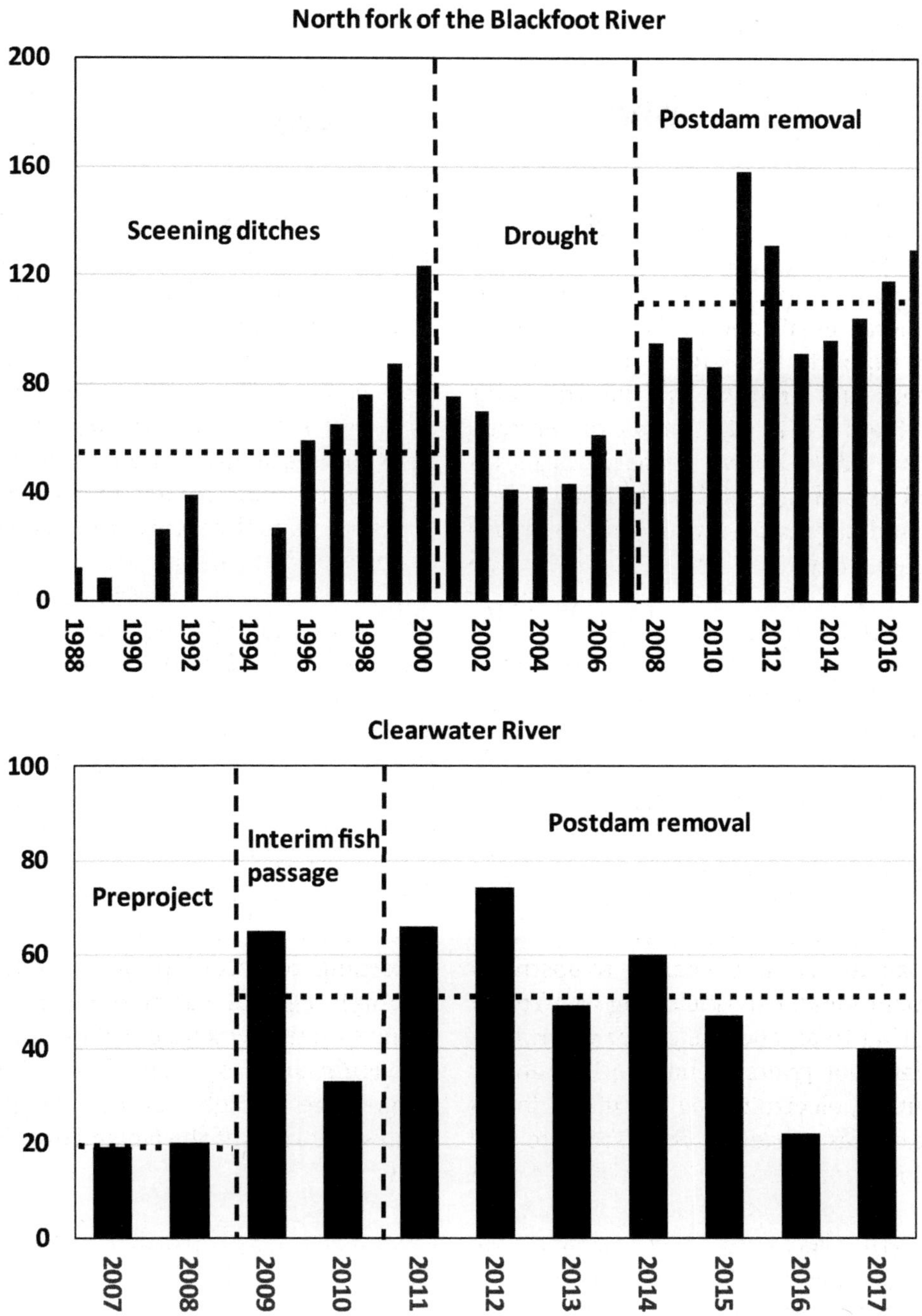

Figure 9. Bull Trout redd counts in the north fork of the Blackfoot River, 1988–2017 (top) and in the Clearwater River upstream of Emily-A Dam, 2007–2017 (bottom). For the north fork, catch-and-release regulations were enacted in 1990, followed by (1) ditch screening installation in the 1990s, (2) 7 years of protracted drought (2001–2007), and (3) the removal of Milltown Dam in 2008. Redd counts upstream of the Emily-A Dam show 2 years of preproject baseline, 2 years of manual interim fish passage, followed by 7 years of full fish passage. The horizontal dotted lines show the yearly mean for redd counts before and after fish passage.

Gold Creek

Lower Gold Creek, encompassed by industrial timberland in the lower Blackfoot River watershed (Figure 1), was the first restoration action to evaluate the performance of habitat structures in a stream with variable channel morphology (Schmetterling and Pierce 1999). Physical habitat structures were placed in two stream types prior to a major flood event: (1) a confined (>2% gradient) channel (i.e., Rosgen B stream type), and (2) an unconfined lower gradient (<2%) channel (i.e., Rosgen C stream type). The project was initiated in 1996 after decades of riparian timber harvest and intentional removal of large wood from the channel led to the dramatic loss of pool habitat and complexity (Pierce 1991; Figure 10). Based on an

Figure 10. Gold Creek: 1996 prerestoration (top) photo shows a B stream type (Rosgen 1994) lacking instream complexity and pools. The postrestoration (bottom) photo shows a wood-and-boulder-formed plunge pool at the same site immediately following installation.

upstream reference reach and a formula for natural pool spacing (Rosgen 1996), a total of 66 pools were created within a 4.8-km section of stream. The project used onsite native material (i.e., large wood and glacial erratic boulders) to construct four types of pool-forming habitat structures (debris collectors, log-formed plunge pools, lateral scour pools, and rock-formed pools; Schmetterling and Pierce 1999).

Once completed, average wetted pool area of the channel increased from 1% (prerestoration) to 14% (postrestoration), which was similar to the reference reach (Schmetterling and Pierce 1999). The project then withstood a ~50-year frequency flood event the following spring. Of the original 66 structures, 85% remained intact and stable. However, structure retention was significantly higher in the B stream type than in the C stream type (94% versus 58%, $P < 0.001$). For the unconfined stream reach, structures emphasizing lateral scour pools had the highest retention rate (75%), whereas rock-formed pools had the lowest retention rates (40%). From these findings, MFWP concluded that improving pool habitat and the ability to withstand major floods was a function of stream type and the type of structure employed. These results changed the way habitat structures were installed in unconfined stream types to those better suited to meandering channels. In the confined reach, long-term (18-years) trout-population monitoring suggested an increase in total trout abundance for age-1 and older trout with a catch per unit effort of 5.6 trout/30 m prerestoration (1996), compared to an average of 12.9 fish/30 m from 1997 to 2015 (MTFWP, unpublished data).

Nevada Creek

Located immediately downstream of a large irrigation storage reservoir, the Nevada Creek demonstration project involved reconstruction of 1.34 km of channel to restore natural channel features to a heavily degraded and unstable section of a laterally extended, meandering (C4) stream type (Rosgen 1994; Figure 11). Prior to restoration, Nevada Creek was relatively incised and overwidened, with eroding banks that contributed an estimated 191 kg/m/year of fine sediment to the channel (Dave Rosgen, Wildland Hydrology, unpublished data). The steam also lacked woody riparian vegetation due to decades of intensive riparian grazing. Channel reconstruction reduced the width-to-depth ratio from 56 to 24, elevated the incised channel, added lateral scour pool habitat structures (instream wood) on the outer stream bends, and placed layered willow cutting to reestablish riparian shrubs along the new floodplain and streambank margins (Figure 11).

The Nevada Creek project also included a three-stage, inner berm channel designed to maintain a low width-to-depth ratio and thereby help mitigate irrigation-related low-flow conditions. In addition to active channel restoration, livestock were fenced from the immediate stream corridor. Like Gold Creek, the Nevada Creek project underwent a significant flood in the spring following project construction with a peak daily flow of 21.6 m^3/s versus the average peak daily flow of 4.02 m^3/s for the 1939–2017 period of record (U.S. Geological Survey provisional data station #12335500). Though the channel adjusted, there was little change in pattern, dimension, or profile, and no lateral pool habitat features failed. Prerestoration (2010) and postrestoration (2016) trout population monitoring indicated a significant increase in abundance (4.6 ± 1.6 to 13.6 ± 2.5 for age-1 and older trout per 30 m) and a more than 200% increase in biomass (1.7 km to 3.6 kg/30 m) (MTFWP, unpublished data).

Chamberlain Creek

Chamberlain Creek was selected for a series of initial, comprehensive restoration projects after fish-population surveys indicated

Figure 11. Nevada Creek restoration photo point showing a lateral scour pool with habitat structures (instream wood) on the outer stream bends that incorporated layered willow cutting to re-establish riparian shrubs along the new floodplain (top photo taken in 2010) and the same streambank after 6 years (2016, bottom photo) of recovery.

that Westslope Cutthroat Trout abundance was 94% lower in downstream disturbed areas relative to an upstream reference reach. Irrigation diversions and dewatering also blocked migratory corridors between Chamberlain Creek and the Blackfoot River (Peters 1990; Pierce 1991; Pierce et al. 1997). Restoration methods included spatially overlapping techniques involving reconstructing a scarified channel (Figure 12) and adding large in-

Figure 12. Chamberlain Creek in 1989 just after it was bulldozed to construct an instream pond (top photo) and in 2008 after stream restoration (bottom photo).

stream wood to a 2-km reach where the loss of riparian conifers from streamside roads reduced wood recruitment. In addition, riparian grazing and irrigation systems were upgraded (consolidation of ditches and installation of a fish ladder) and instream flows were enhanced through water leasing during the low-flow summer irrigation season. Riparian roads were obliterated and culverts were replaced to meet fish standards and natural stream function. All ranch lands in the lower watershed were also enrolled in conservation easement programs that protect open space, ranchlands, and wildlife habitat from residential development (e.g., subdivisions). Last, all private industrial forest land in the middle and upper Chamberlain Creek watershed (77.2 km^2) was transferred to public ownership in 2010 with provisions to remove 8.9 km of riparian roads.

Age-1 and older Westslope Cutthroat Trout increased from a prerestoration of 2.5 trout/30 m to a long-term average of 19.2 trout/30 m after project completion (Pierce et al. 2013). Seven years postrestoration, radiotelemetry research confirmed reconnection for migratory trout when most (7 of 11) Westslope Cutthroat Trout radio-tagged in the Blackfoot River between Gold Creek and the North Fork (65 km) ascended Chamberlain Creek to access spawning areas within and upstream of the restoration reach (Schmetterling 2000, 2001). With the completion of stream restoration, conservation easements, and land exchanges, the Chamberlain Creek project addresses all known major impairments to fisheries and riparian corridors at a scale that encompassed the entire watershed.

Spring Creek Restoration

Spring creeks drain aquifers on the valley floor in areas that spatially overlap with intensive agricultural land use. Because of wet streambanks and low hydraulic energy, spring creeks are prone to disturbance (e.g., intensive grazing) and often require extended recovery periods once disturbed (Pierce et al. 2017). The spring creeks described below were fully reconstructed, managed for full vegetative recovery, and intensively monitored for habitat changes and biotic response.

Kleinschmidt Creek and neighboring spring creeks

Kleinschmidt Creek, a small, alluvial, spring-fed tributary of the North Fork Blackfoot River, was fully reconstructed using natural channel design principles, then closely monitored over a 10-year period (Pierce et al. 2014a, 2015). The project converted a straightened, overwidened stream with a degraded riparian zone (i.e., Rosgen C5 stream type) to a deep, narrow, meandering meadow stream (i.e., Rosgen E4 stream type; Figure 13). This conversion specifically reduced channel wetted surface area by 69%, reduced width-to-depth ratio (150–8.8), increased bank-full mean velocity (0.27–0.67 m/s), increased hydraulic shear stress (7.4–13.7 N/m^2), and increased particle entrainment size (11–21 mm) in riffles. This case study also documented positive trout response trends (increased abundance and higher biomass) associated with placement of instream habitat structure (wood) and riparian recovery (Figure 14). However, the most significant change was water temperature reduction. Summer temperatures in Kleinschmidt Creek declined an average of 3.5°C, with average daily maximum temperatures about 1.5°C colder than those in receiving waters (Figure 15).

Similar to Kleinschmidt Creek, two other nearby spring creeks were fully reconstructed to restore the natural pattern, dimension, and profile of a stream within its valley. Like Kleinschmidt Creek, both spring creeks exhibited similar (3–4°C) reductions in water temperatures following channel renaturalization (Pierce and Podner 2016), which further corroborated the cooling potential of groundwater-dominated streams. In these three examples, water cooling is biologically relevant because these streams enter Endangered Species Act-designated Bull Trout critical habitat in a reach used as thermal (summer) refugia (Swanberg 1997; USFWS 2010). Because these streams are now colder than the receiving waters, they

Figure 13. Kleinschmidt Creek prerestoration (top, 2001) and postrestoration (bottom, 2014) photos. The top photo shows a straightened and overwidened section of channel with a channel-altering rock dam that induced upstream deposition of fine sediment. The bottom depicts the restored E stream type (Rosgen 1994) 13 years after channel reconstruction.

not only enhance habitat quality for Bull Trout, but also add capacity to buffer projected warming trends in the main-stem North Fork (Isaak et al. 2015). Last, the three spring creeks are now protected by conservation easements that prevent future

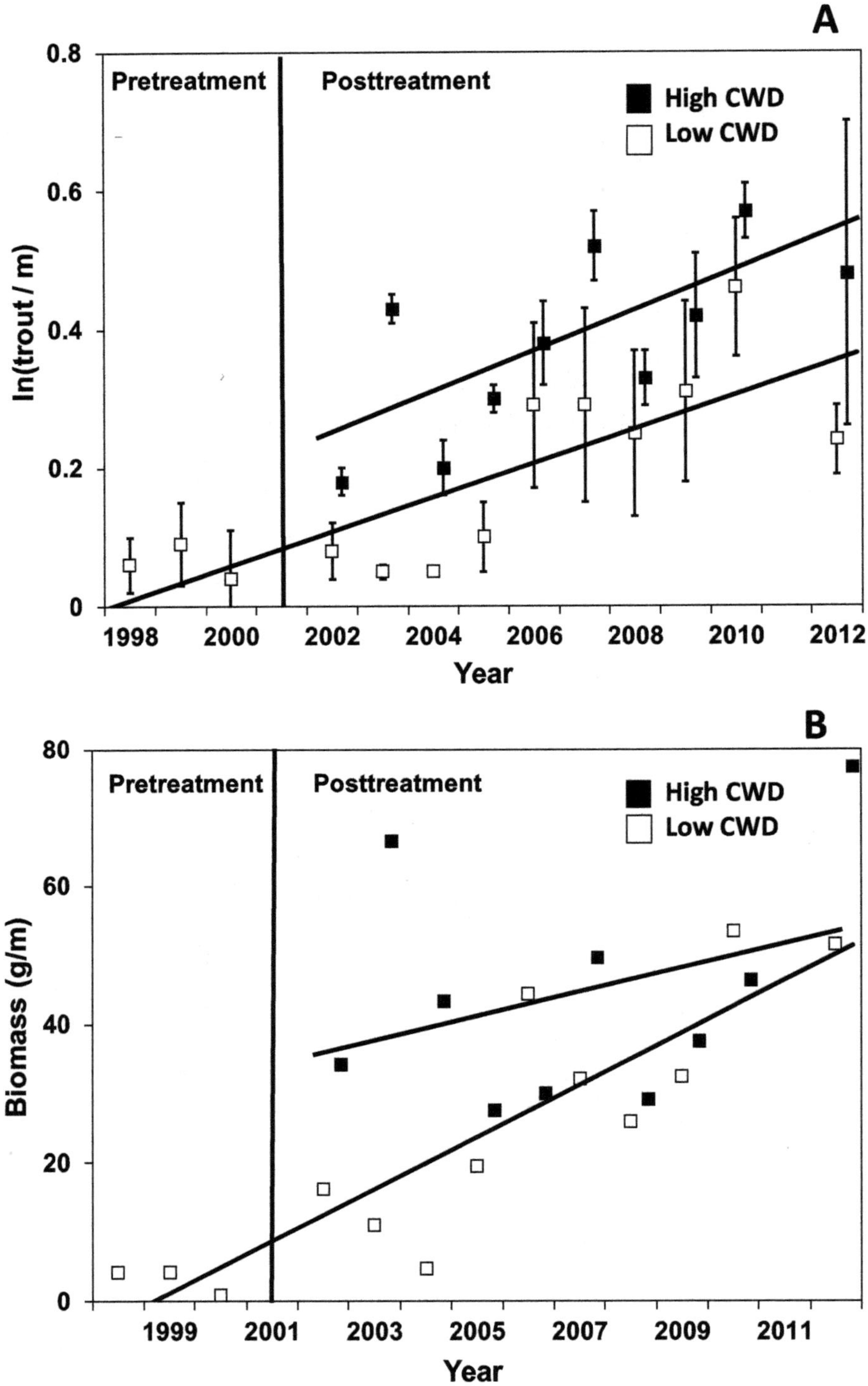

Figure 14. Trout response per linear meter of stream to high and low coarse woody debris (CWD) in Kleinschmidt Creek. From depletion surveys, graph **(A)** estimates age-1+ abundance (±95% confidence interval) and graph **(B)** age 1+ biomass. High and low CWD refers to the amount of coarse woody debris within the channel. The linear lines are best-fit to estimates of abundance and biomass for the two CWD treatment groups (from Pierce et al. 2015).

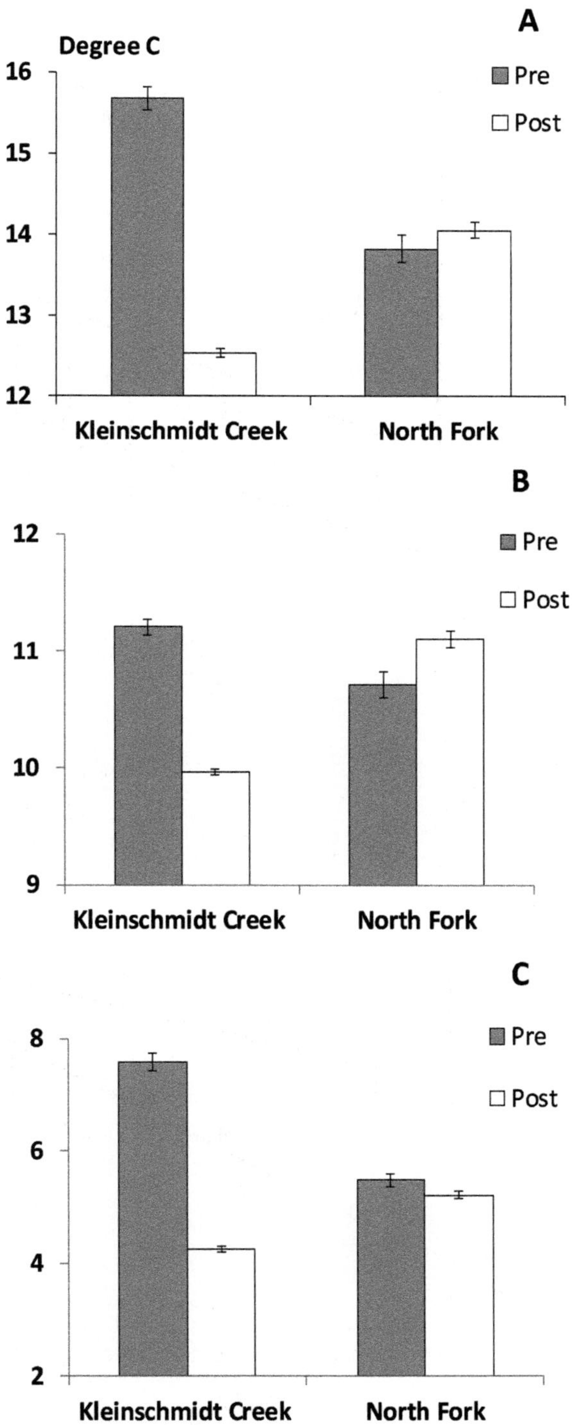

Figure 15. Prerestoration (1998, 1999, and 2001) and postrestoration (2002, 2004, 2010, 2012, and 2013) water temperatures (June through September) for Kleinschmidt Creek (restoration site) and the North Fork Blackfoot River (control site): **(A)** average maximum daily temperatures, **(B)** mean daily temperatures, and **(C)** average daily range of temperatures (from Pierce et al. 2014a).

channel and riparian degradation along Kleinschmidt Creek.

Nevada Spring Creek complex

Nevada Spring Creek flows from an artesian spring, whereas previously discussed spring creeks drain alluvial aquifers. Restoration of the Nevada Spring Creek included complete reconstruction of 7.1 km of stream channel (Pierce et al. 2014b). Like Kleinschmidt Creek, this work converted an incised, overwidened, and overgrazed stream (i.e., a Rosgen C5 stream type) to a deep, narrow, more meandering channel type (i.e., Rosgen E5 stream type; Figure 16). This project reduced average width-to-depth ratio from 22 to 3.2, enhanced instream flows, and reconnected a

Figure 16. Photos of the Nevada Spring Creek project before (1994, top) and after (2012, bottom) reconstruction. Note channel incision, erosion of streambanks, and high channel width prior to restoration.

small, restored Westslope Cutthroat Trout spawning tributary located upstream of the project (Pierce et al. 2013, 2014b). Following restoration of both streams, stream-resident Westslope Cutthroat Trout dispersed downstream into the newly restored habitat (Pierce et al. 2013), re-establishing a migratory Westslope Cutthroat Trout population. Westslope Cutthroat Trout then increased in abundance downstream of the restoration project from zero fish in 2005 to 11.0 ± 2.1 fish/300 m in 2016 (Pierce and Podner 2016). Adult spawners from this migratory population are now returning to their spawning areas (Pierce et al. 2014b). The Nevada Spring Creek property is also protected by conservation easements, with special stipulations that prevent disturbance in the riparian corridor. Last, water from the artesian spring source (up to 0.28 m^3/s) was permanently dedicated to instream flows through a conversion of private water rights to public (MFWP) ownership. Though the project appears promising, recent monitoring indicates that elevated nutrient and sediment issues still need to be addressed (Pierce et al. 2016).

Sediment assessments on spawning riffles in restored spring creeks

Riffle morphology of four degraded spring creeks (Rosgen C4–C5 stream types) was compared to four fully restored and functioning streams with deeper and narrower channel forms (Rosgen E4 stream types) by Pierce et al. (2017) (Figure 17). Specifically, this study examined the correlation of (1) natural channel form with riffle substrates and trout spawning site quality, and (2) riffle/sediment relationships with macroinvertebrate taxa and two biotic indices. Despite no change in channel slope, riffles in restored streams had lower median width-to-depth ratios (11.7 versus 17.4; $P = 0.028$), higher velocities (0.61 versus 0.42 m/s; $P = 0.029$), and a lower-percentage fine sediment <6.3 mm (24% versus 40%; $P = 0.057$). These conditions all indicate higher-quality spawning sites than the degraded (unrestored) group of streams. The study concluded that stream restoration can improve trout spawning substrate by facilitating sediment transport via reduced channel width-to-depth ratios and improved riparian and land management. When all streams were considered, a strong positive correlation between the number of sediment tolerant taxa and the percent fine sediment less than 0.85 mm was identified (rho = 0.901; $P = 0.002$), along with a negative correlation between the percent of sediment less than 6.3 mm and the number of clinger taxa (rho = −0.650; $P = 0.080$). Of the two biotic indices, a negative correlation of the fine sediment biotic index (Relyea et al. 2012) with sediment less than 6.3 mm (rho = −0.622; $P = 0.097$) suggests that it may be a better indicator of spring creek habitat integrity and restoration effectiveness than other fine sediment metrics.

Multiscale Trout Response Trends in Tributaries and the Blackfoot River

Tributary to watershed-scale restoration

In addition to the individual stream case studies, Pierce et al. (2013) compared total trout abundance for a group of 18 stream restoration projects against a group of 23 reference streams to evaluate broad watershed-scale response trends. Most of the streams (11) were actively restored using principles of natural channel design. Compared with prerestoration conditions, actively restored streams revealed common pattern of deeper and narrower channels (width-to-depth ratio reduction of 20.0 verses 9.2), increased sinuosity (1.2 versus 1.5), more pool habitat (21.9% versus 46% pool area) and lower maximum summer water temperatures (20°C versus 16°C) following restoration. In addition, 11 projects that underwent flow enhancement increased minimum summer flows from an average of 0.052 m^3/s prerestoration to an average of 0.144 m^3/s postrestoration.

Figure 17. Example photos for two of eight spring creek study reaches. The top photo displays a wide, shallow degraded (unrestored) stream, whereas the bottom photo shows a deeper, narrow (restored) stream.

Prior to these changes, total trout abundance across all 18 sites was significantly lower than 23 reference sites, with an average of 5.7 trout/30 m for restoration sites compared with 18.6 trout/30 m for reference sites ($P = 0.0001$). By 3 years postrestoration, average abundance for age-1 and older trout in treatment sites had reached 14.1 trout/30 m and were no longer statistically different from reference sites (i.e., 19.5 trout/30 m; $P = 0.012$). Following this initial increase, total trout abundance for all treatment sites remained elevated near the average reference between 4 and 12 years posttreatment.

This rapid initial response was driven mostly by irrigation-based projects (fish screens, fish ladders, and instream flow), whereas streams that were fully reconstructed (e.g., Kleinschmidt Creek) often required extended (>10 year) recovery periods. This study also concluded that consistent monitoring was essential for landowner education and adaptive management necessary to sustain restoration benefits. Most active restoration projects (7 of 13) that included a riparian grazing component required adjustments in riparian grazing practices to protect projects from continued livestock impacts.

Blackfoot River

To evaluate long-term (28 year) trends in trout community composition and abundance in the main-stem Blackfoot River, trout species composition (fish >152 mm) was compared in four river reaches (Figure 1) where population monitoring data predate changes to fishing regulations (1990) and implementation of restoration projects. These data show consistent, long-term, community-level changes that include significant increases in migratory Westslope Cutthroat Trout abundance (Figure 18; Pierce and Podner 2016). These results reflect improved metapopulation function and multiple management strategies (watershed-scale protective regulations, targeted restoration, and habitat connectivity) that collectively improved the Blackfoot River Westslope Cutthroat Trout fishery. These results provide further evidence that significant ecological perturbations can be corrected at multiple spatial scales.

Summary

The Blackfoot River watershed wild trout program represents a comprehensive 30-year case study of progressive river restoration, fisheries management, and landscape protection on both public and private lands. The restoration program began when fisheries data revealed that angling regulations alone could not fully restore wild trout populations and that riparian and aquatic habitat restoration were necessary to restore naturally functioning and biologically productive trout streams. With goals of improving trout recruitment to the Blackfoot River and the status of migratory native trout, restoration practices gained momentum during the 1990s, then expanded outward from the central Blackfoot River valley as updated fisheries, habitat, and water-quality information identified human-induced limiting factors. With an emphasis on key spawning and rearing tributaries and movement corridors, stream prioritization guided allocations of funding and consolidated field work to important habitats used by migratory native trout. Once sites were selected for restoration, reference reach information helped to quantify effects of human impairments relative to functional natural stream conditions. Reference conditions also provided templates for natural channel design, as well as a basis for fundraising, permitting, and postrestoration monitoring and evaluation. Long-term monitoring of habitat changes revealed deeper, narrower, colder stream segments with lower instream sediment levels and improved habitat connectivity. Improved conditions preceded notable examples of fish population recovery, including trout recolonization, re-establishment of migratory life history expression, increases in fish abundance and

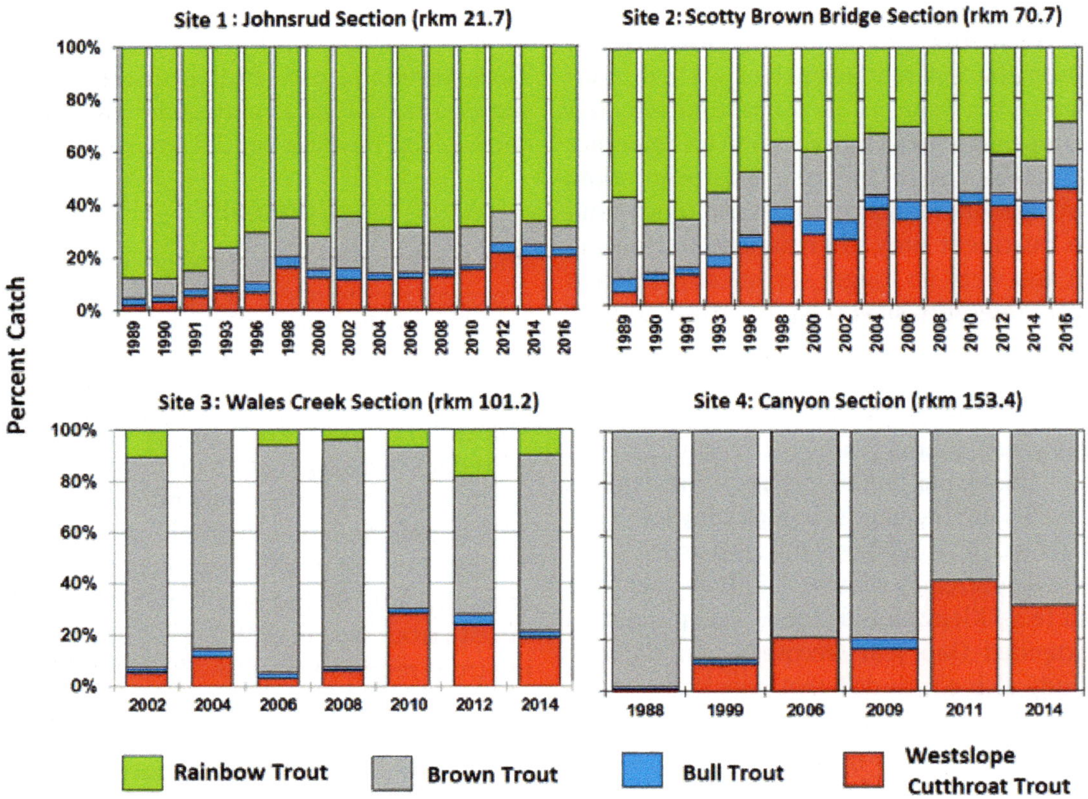

Figure 18. Percent trout species composition (fish >150 mm in total length) for four electrofishing monitoring reaches of Blackfoot River (1988–2016). The graphs show the changing composition of the trout community longitudinally and over time. Mark-and-recapture estimates in the Johnsrud and Scotty Brown Bridge sections show statistically significant increases for Westslope Cutthroat Trout. The acronym "rkm" (river kilometer midpoint) refers to the midpoint of the population survey reach. Monitoring site locations are shown on Figure 1.

biomass, and community-level shifts towards Westslope Cutthroat Trout in the Blackfoot River. The benefits of riparian restoration and conservation easement protection also extend to a myriad of riparian-dependent wildlife species.

Montana Fish, Wildlife and Parks originally estimated that a 50-year restoration process would be required to substantially reverse a century of human damage to the river ecosystem. After 30 years, this chapter was written, in part to promote the continuity of the restoration program. However, restoration alone will not solve many contemporary problems afflicting wild trout. In addition to habitat and fish passage problems described in this chapter, exotic diseases, invasive species, and climate change are now emerging threats to wild trout in the Blackfoot River watershed (e.g., Pierce et al. 2009, 2018; Isaak et al. 2015; Al-Chokhachy et al. 2016). Though targeted restoration can mediate some of these conditions (e.g., whirling disease and anthropogenic warming from land use; Pierce et al. 2009, 2014a; Eby et al. 2015), future wild trout conservation measures will need to consider additional restoration strategies. One strategy, new to the Blackfoot River watershed and specific to native trout, is a developing large-scale (137 km of perennial streams) rotenone project in designated wilderness area upstream of a large waterfall barrier in the headwaters of the North Fork (Pierce et al. 2018). If completed, this project will replace hybrid *Oncorhynchus* (predominantly Rainbow Trout) with genetically pure Westslope Cutthroat Trout and possibly Bull Trout in a large, protected, and pristine conservation area. As the Blackfoot River restoration program transitions from low-elevation habitat restoration projects to this high-elevation replacement project, similar replacement projects should be considered in the high country where historic stocking of nonnative hatchery trout into high elevation lakes has created headwater sources of hybridization.

In closing, it was the early engagement of stakeholders, individually tailored one-on-one solutions to tributary impairments, comprehensive data collections, and a programmatic emphasis on migratory native trout recovery that advanced the scope and scale of restoration program. As the MFWP restoration program matured, both BBCTU and the Blackfoot Challenge have also become more established, effective, and inclusive. Both groups, led by shopkeepers, outfitters, ranchers, and their cooperating agencies facilitated communication, education, and conflict resolution that would not otherwise exist from agency programs alone. This involvement not only helped foster restoration, but also helped secure funds for monitoring while helping to overcome the practice of small-scale projects common to most restoration programs (Bernhardt et al. 2005; Roni 2005; Reeve et al. 2006). After 30 years, this collaboration revealed that river restoration and watershed-scale conservation can both be achieved with long-term vision, dedication, and cooperation.

Acknowledgments

This chapter was the product of a 30-year collaboration among agencies, conservation groups, and private landowners. Several public and private funding initiatives contributed to the restoration work. Major private donors included Orvis, Northwestern Energy's Milltown Mitigation Program, and the Chutney Foundation and private landowners. Public funds were generated from the National Fish and Wildlife Foundation, a special Natural Resources Conservation Service Native Trout Environmental Quality Improvement Program grant, the U.S. Forest Service's Collaborative Forest Landscape Restoration Program, The Montana Legacy Project, U.S. Fish and Wildlife Service Partners for Fish and Wildlife, interagency Native Trout Habitat Conservation Plans, and the Montana Fish, Wildlife, and Parks Future Fisheries Program. Special thanks to Ryen Neudecker, Ryan Kovach, Dan Dauwalter, Jim DeRito, and Brian Hodge for their review of this chapter.

References

Aitken, G. 1997. Restoration of trout waters in the West: Blackfoot River of Montana. Pages 402–424 *in* J. E. Williams, C. A. Wood, and M. P. Dombeck, editors. Watershed restoration: principles and practices. American Fisheries Society, Bethesda, Maryland.

Al-Chokhachy, R., D. Schmetterling, C. Clancy, P. Saffel, R. Kovach, L. Nyce, B. Liermann, W. Fredenberg, and R. Pierce. 2016. Are Brown Trout replacing or displacing Bull Trout populations in a changing climate? Canadian Journal of Fisheries and Aquatic Sciences 73:1395–1404.

Benson, A. 2009. Effects of barriers on migratory Bull Trout and application of a conceptual framework to evaluate tradeoff associates with dam removal in the Clearwater River drainage, Montana. Master's thesis. University of Montana, Missoula.

Bernhardt, E. S., M. A. Palmer, J. D. Allen, G. Alexander, K. Barnas, S. Brooks, J. Carr, S. Clayton, C. Dahm, J. Follstad-Shah, D. Galat, S. Gloss, P. Goodwin, D. Hart, B. Hassett, R. Jenkinson, S. Katz, G. M.

Kondolf, P. S. Lake, R. Lave, J. L. Meyer, T. K. O'Donnell, L. Pagano, B. Powell, and E. Sudduth. 2005. Synthesizing United States river restoration efforts. Science 308:636–637.

Eby, L. A., R. Pierce, M. Sparks, K. Carim, and C. Podner. 2015. Multiscale prediction of whirling disease risk in the Blackfoot River basin, Montana: a useful consideration of restoration prioritization? Transactions of the American Fisheries Society 144:753–766.

Fitzgerald G. 1997. Analysis and inventory of riparian vegetation along Nevada and Monture creeks using ADAR imagery. Master's thesis. University of Montana, Missoula.

Harrelson, C.C., C. L. Rawlins, and J. P. Potyondy. 1994. Stream channel reference sites: an illustrated guide to field technique. U.S. Forest Service, Rocky Mountain Forest and Range Experiment Station, General Technical Report RM-245, Fort Collins, Colorado.

Isaak, D., M. Young, D. Nagel, D. Horan, and M. Groce. 2015. The cold-water climate shield: delineating refugia for preserving salmonid fishes through the 21st century. Global Change Biology 21:2540–2553.

Koopal, M. 1998. Stream habitat analysis of select tributaries to the Blackfoot River drainage, 1998. Report prepared for Montana Fish, Wildlife and Parks, Missoula, Montana.

Knotek, W. L. 2017. Montana statewide fisheries management report—middle Clark Fork and Clearwater River drainages. Montana Fish, Wildlife, and Parks, Missoula.

Kruse, C., G., W. A. Hubert, and F. J. Rahel. 1998. Single pass electrofishing predicts trout abundance in mountain streams with sparse habitat. North American Journal of Fisheries Management 18:949–946.

Lawlor, S. M. 2002. Determination of channel-morphology characteristics, bankfull discharge, and various design-peak discharges in western Montana. U.S. Geological Survey, Scientific Investigations Report 2004-5263, Reston, Virginia.

Leathe, S. A. and F. A. Nelson. 1989. A literature evaluation of Montana's wetted perimeter inflection point method for deriving instream flow recommendations. Montana Department of Fish, Wildlife and Parks, Helena.

McNeil, W. J., and W. H. Ahnell. 1964. Success of Pink Salmon spawning relative to size of spawning bed materials. U.S. Fish and Wildlife Service Special Scientific Report Fisheries 469.

Marler, M. 1998. Riparian vegetation and health on the Blackfoot River between Nevada Creek and the north fork confluence. Montana Fish, Wildlife and Parks, Missoula.

Meehan, W. R. 1991. Influences of forest and rangeland management on salmonid fishes and their habitats. American Fisheries Society, Special Publication 19, Bethesda, Maryland.

Moore, J. N., S. N. Louma, and D. Peters. 1991. Downstream effect of mine effluent on intermontane riparian system. Canadian Journal of Fisheries and Aquatic Sciences 48:222–232.

Neudecker, R. A., T. McMahon, and E. R. Vincent. 2012. Spatial and temporal variation of whirling disease risk in Montana spring creeks. Journal of Aquatic Animal Health 24:201–212.

Orth D. J. and R. J. White. 1999. Stream habitat management. Pages 249–281 *in* C. C. Kohler and W. A. Hubert, editors. Inland fisheries management in North America, 2nd edition. American Fisheries Society, Bethesda, Maryland.

Penaluna, B. E., A. Abadía-Cardoso, J. B. Dunham, F. J. García-Dé León, R. E. Gresswell, A. R. Luna, E. B. Taylor, B. B. Shepard, R. Al-Chokhachy, C. C. Muhlfeld, K. R. Bestgen, K. Rogers, M. A. Escalante, E. R. Keeley, G. M. Temple, J. E. Williams, K. R. Matthews, R. Pierce, R. L. Mayden, R. P. Kovach, J. C. Garza, and K. D. Fausch. 2016. Conservation of native Pacific trout diversity in western North America. Fisheries 41:286–300.

Peters, D. 1990. Inventory of fishery resources in the Blackfoot River and major tributaries to the Blackfoot River. Montana Department of Fish, Wildlife and Parks, Missoula.

Peters, D. and R. Spoon. 1989. Preliminary fisheries inventory of the Big Blackfoot River. Montana Fish, Wildlife and Parks, Missoula.

Pierce, R., 1991. A stream habitat and fisheries analysis for six tributaries to the Blackfoot River. Montana Department of Fish, Wildlife and Parks, Missoula.

Pierce, R., R. Aasheim, and C. Podner. 2007. Fluvial Westslope Cutthroat Trout movements and restoration relationships in the upper Blackfoot basin, Montana. Intermountain Journal of Sciences 13(2):72–85.

Pierce, R., R. Anderson, and C. Podner. 2004. The Big Blackfoot River restoration progress report for 2002 and 2003. Montana Fish, Wildlife and Parks, Missoula.

Pierce, R., M. Davidson, and C. Podner. 2012. Spawning behavior of Mountain Whitefish and co-occurrence of *Myxobolus cerebralis* in the Blackfoot River basin, Montana. Transactions of the American Fisheries Society 141:720–730.

Pierce, R., and D. Peters. 1990. Aquatic investigations in the middle Blackfoot River, Nevada Creek and Nevada Spring Creek corridors, Montana Department of Fish, Wildlife and Parks, Missoula.

Pierce, R., D. Peters, and T. Swanberg. 1997. Blackfoot River restoration progress report. Montana Fish, Wildlife and Parks, Missoula.

Pierce, R. and C. Podner. 2016. Big Blackfoot River fisheries investigations 2012–2015. Montana Fish, Wildlife and Parks, Missoula.

Pierce, R., C. Podner, and K. Carim. 2013. Response of wild trout to stream restoration over two decades in the Blackfoot River basin, Montana. Transactions of the American Fisheries Society 142:68–81.

Pierce, R., C. Podner, and M. Davidson. 2009. Correlation of fluvial Rainbow Trout spawning life history with the severity of infection by *Myxobolus cerebralis* in the Blackfoot River basin, Montana. Transactions of the American Fisheries Society 138:251–263.

Pierce, R., C. Podner, M. Davidson, L. Knotek, and J. Thabes. 2008. Big Blackfoot River fisheries and restoration investigations for 2006 and 2007. Montana Fish, Wildlife and Parks. Missoula.

Pierce, R., C. Podner, and L. Jones. 2015. Long-term increases in trout abundance following channel reconstruction, instream wood placement, and livestock removal from a spring creek in the Blackfoot basin, Montana. Transactions of the American Fisheries Society 144:184–195.

Pierce, R., C. Podner, L. Marczak, and L. Jones. 2014a. Instream habitat restoration and stream temperature reduction in a whirling disease positive spring creek in the Blackfoot River basin, Montana. Transactions of the American Fisheries Society 143:1188–1198.

Pierce, R., C. Podner, and J. McFee. 2001. Blackfoot River fisheries inventory, monitoring and restoration report. Montana Fish, Wildlife and Parks, Missoula.

Pierce, R., C. Podner and J. McFee. 2002. Blackfoot River fisheries inventory, restoration and monitoring report for 2001. Montana Fish, Wildlife and Parks, Missoula.

Pierce, R., C. Podner and P. Saffel. 2018. Aquatic and associated investigations to guide conservation planning for Bull Trout and Westslope Cutthroat Trout in the North Fork Blackfoot River upstream of the North Fork Falls, 2002–2017. Montana Fish, Wildlife and Parks, Missoula.

Pierce, R., C. Podner, T. Selch and S. Begeley. 2016. Nevada Spring Creek aquatic assessment, 1994–2016. Montana Fish, Wildlife and Parks, Missoula.

Pierce, R., C. Podner, and S. Sullivan. 2017. Associations of channel form, spawning riffle quality and macroinvertebrate assemblages in small restored spring creeks of western Montana. Northwest Science 91:283–300.

Pierce, R., C. Podner, T. Wendt, K. Carim, and R. Shields. 2014b. Westslope Cutthroat Trout movements through restored habitat and Coanda diversions in the Nevada Spring Creek complex, Blackfoot basin, Montana.

Transactions of the American Fisheries Society 143:230–239.

Reeve, T., J. Lichatowich, W. Towey, and A. Duncan. 2006. Building science and accountability into community-base restoration: can a new funding approach facilitate effective and accountable restoration. Fisheries 31(1):17–23.

Relyea, C. D., G. W. Minshall, and R. J. Danehy. 2012. Development and validation of an aquatic fine sediment biotic index. Environmental Management 49:242–252.

Roni, P. editor. 2005. Monitoring stream and watershed restoration, American Fisheries Society, Bethesda, Maryland.

Rosgen, D. 1994. A classification of natural rivers. Catena 22:169–199.

Rosgen, D. 1996. Applied river morphology. Wildland Hydrology, Pagosa Springs, Colorado.

Rosgen, D. 2011. Natural channel design: fundamental concepts, assumptions, and methods. Pages 69–93 *in* A. Simon, S. J. Bennett, and J. M. Castro, editors. Stream restoration in dynamic fluvial systems: scientific approaches, analyses, and tools. Geophysical Monograph 194.

Rosgen, D. L. 2007. Rosgen geomorphic channel design. Chapter 11 *in* J. Bernard, J. F. Fripp, and K. R. Robinson, editors. National engineering handbook: Part 654 stream restoration design. U.S. Department of Agriculture, Natural Resources Conservation Service, Washington, D.C.

Schmetterling, D. A. 2000. Redd characteristics of fluvial Westslope Cutthroat Trout in four tributaries of the Blackfoot River, Montana. North American Journal of Fisheries Management 20:776–783.

Schmetterling, D. A. 2001. Seasonal movements of fluvial Westslope Cutthroat Trout in the Blackfoot River drainage, Montana. North American Journal of Fisheries Management 21:507–520.

Schmetterling, D. A. 2003. Reconnecting a fragmented river: movements of Westslope Cutthroat Trout and Bull Trout after transport upstream of Milltown Dam, Montana. North American Journal of Fisheries Management 23:721–731.

Schmetterling, D. A., and T. Bernd-Cohen. 2002. Native species conservation through education: the Adopt-A-Trout program in Montana. Fisheries 27(9):10–15.

Schmetterling, D. A., and R. W. Pierce. 1999. Success of instream habitat structures after a 50-year flood in Gold Creek, Montana. Restoration Ecology 7:369–375.

Swanberg, T. R. 1997. Movements of and habitat use by fluvial Bull Trout in the Blackfoot River. Transactions of the American Fisheries Society 126:735–746.

Swanberg, T. R., and L. Burns. 1997. Movements and habitat use of radio-tagged Bull Trout in the upper Blackfoot River drainage. Montana Fish, Wildlife and Parks, Missoula.

USFWS (U.S. Fish and Wildlife Service). 2010. Endangered and threatened wildlife and plants; revised designation of critical habitat for Bull Trout in the conterminous United States; final rule. Federal Register 75:2269(18 October 2010):63897–64070.

USFWS (U.S. Fish and Wildlife Service). 2015. Recovery unit implementation plan (RUIP) for Bull Trout in the Columbia headwaters recovery unit. USFWS, Portland, Oregon.

Wolman, M. G. 1954. A method of sampling coarse river-bed materials. Transactions, American Geophysical Union 35:951–956.

Zackheim, H. 2006. A history of Montana Fish, Wildlife and Parks Fisheries Division, 1901–2005. Montana Department of Fish, Wildlife and Parks, Helena.

American Fisheries Society Symposium 91:682–693, 2019

It's Not Easy to Save a River

Sky Jones-Lewey*
Nueces River Authority
200 East Nopal, Suite 206, Uvalde, Texas 78801, USA

Abstract.—From off-road vehicle abuse in streambeds to nonnative invasive species in riparian areas, from the threat of wastewater discharge in pristine headwater creeks to proposals for burying low-level radioactive waste and landfilling of fracking byproducts on floodplains, from water marketing plans seeking to export large quantities of groundwater to bulldozing river channels, gravel mining, and diverting stream flows, it's not easy to save a river. It is really all about the people; *Homo sapiens* are an essential species of consideration in any multispecies approach to conservation. The Nueces River Authority (NRA) has served as a guardian of surface-water resources in the Nueces River basin since the agency was created by the Texas Legislature in 1935. The jurisdiction of NRA includes portions of three ecoregions and 22 South Texas counties and encompasses more than 45,300 km2, extending from Rocksprings to the Gulf of Mexico. Over the past eight decades, NRA has strived to develop and implement an adaptive, systems approach that focuses on empowering people with good information and consistent messaging in order to restore and protect the Nueces River.

Introduction

The Nueces River is located in southwest Texas between San Antonio and Mexico. The Nueces River basin extends from the Edwards Plateau ecoregion near Rocksprings to the city of Corpus Christi on the Gulf of Mexico and drains a 45,300-km^2 watershed. In the upper basin, headwater tributaries are clear and spring-fed. These streams contain water-losing reaches, which are major contributors of recharge to the Edwards Aquifer, a water supply for millions of central Texans. An average of 412,476,326 m^3 per year from 1934 to 1996 was recharged from Nueces River basin streams (HDR 1999). The middle basin is known as the "Brush Country," where water holes are precious, oil and gas activities are abundant, and the Nueces River helps to recharge the Carrizo-Wilcox Aquifer. Finally, the river crosses the Gulf Coast Aquifer, again imparting recharge water before delivering freshwater inflows and sediments to Nueces Bay and associated estuary.

The Nueces River Authority (NRA) is a small government agency created by the Texas Legislature in 1935. It has no authority to tax or regulate but does have broad authority to protect, preserve, and develop water resources. In addition to the Nueces River basin, NRA operates within two adjacent coastal basins (i.e., Nueces–Rio Grande and Nueces–San Antonio). Collectively, across all three basins, NRA monitors water quality and engages people to protect the creeks, rivers, bays, and lakes within approximately 77,700 km^2 of south Texas, equaling about 8% of the state. Funding to support operations of NRA is from contracts and contributions, and its work is acutely focused on protecting surface water resources. To cultivate understanding and advocacy for the water resources of the region, NRA uses a multi-

* Corresponding author: slewey@nueces-ra.org

species and watershed approach in schools, with landowners, and with recreational users. To inform NRA's decisions and educate the public, it organizes and assists with research and data collection. Further, NRA recognizes and responds to all threats to aquatic resources to the greatest degree possible.

My 20 years of working with NRA is just a microsecond in the life of a river, but enough time for the river to teach me a thing or two. The Nueces River has been running from its spring-fed headwaters in Edwards and Real counties to the Gulf of Mexico for more than 20,000 years (Richardson and Gold 1995). For most of that time, the impact of humankind on the river was probably negligible. The river functioned without interference, and the natural systems worked during drought and flood and normal flows in between. I have seen how quickly and easily people can alter and disrupt a river, even with the best of intentions. This is a story of how NRA has worked on behalf of the river, with help from many partners and a focus on resource protection and human education. Over the years, I have learned how hard, yet rewarding, it can be to save a river.

No Shortage of Threats

We have seen numerous threats to the river system and its natural function. One by one, NRA has found ways to respond on behalf of the river and preserve the many values it delivers. Some of the primary threats we have seen include

- Off-road vehicles operating in streambeds,
- Commercial gravel mines and recreational bulldozing that alter stream morphology,
- Wastewater discharge and nutrient runoff,
- Large-scale water marketing plans seeking to pump and export groundwater,
- Plans for disposing radioactive and hazardous oilfield wastes within floodplains,
- Inadequate riparian buffers between creeks and agricultural operations,
- Nonnative, invasive species, and
- Profuse litter.

Each of these threats to the system is the result of human activity and a lack of understanding. Sadly, the biggest overall threat to aquatic systems, not just here but everywhere, is people—uninformed, though often well-intentioned people.

Senate Bill 155 to the Rescue

Off-road vehicle recreation in the streambeds of the Nueces basin became an overnight sensation in the late 1990s. Popularity of riverbed off-roading led to four-wheelers coming from near and far to conquer the big rocks, deep gravel, steep cliff sides, and other river "obstacles" in the upper Nueces River basin (Figure 1). One competition involved backing specifically modified trucks into the river until the bed filled with water and "crawling" the truck out of the stream before the water drained out. It was not unusual to observe four-wheeling events with more than a hundred vehicles traversing the riverbeds on a hot summer day. The rivers also provided nice swimming holes, cool water for overheated engines, shade trees for mechanical work, and no laws or rules.

Not only were off-road vehicles destroying the habitat and natural condition of streams in the upper Nueces River basin, but they were also interrupting traditional river recreationists, including families, anglers, and paddlers. The Texas Parks and Wildlife Department (TPWD) underwrote studies of the damage and created a task force to evaluate solutions (Saunders et al. 2004). At the end of the day, it was determined that off-road vehicle operations in Texas rivers were, for the most part, legal, and legislation would be needed to curb it.

The Nueces River Authority, along with a coalition of environmental, wildlife, and

Figure 1. Four-wheeling rally in the Nueces River basin.

agriculture groups, engaged in several years of intensive effort resulting in the passage of Senate Bill 155 during the 78th state legislative session in 2003 (Texas Legislative Council 2003). The new law banned the operation of motorized vehicles in state-owned riverbeds and went into full effect in 2004. Since then, this state law has halted vehicle abuse in Nueces River basin streambeds where it is successfully enforced through a combined effort of TPWD enforcement and landowner vigilance.

Since 2004, more than 550 tickets (TPWD, unpublished data) have been issued in the upper Nueces River basin to offenders of this state law. Still, vehicle damage continues in some remote streambeds where the line between public and private land is obscured.

Dozers, gravel mines, and nutrients

As in most of Texas, large ranches in the upper Nueces River basin are being divided and sold as ranchettes. Landscape fragmentation has led to unprecedented rates of alteration through riverside building and landscaping. Real estate marketers are keen on advertising "live water" and river views with private access.

We have recently seen a proliferation of "recreational" bulldozing in an effort to facilitate building, expand road crossings, or dig deeper swimming holes. Regulations apply to some of these mechanical disturbances but not to others. The rules can be confusing and often seem contradictory to the point that people begin to lose respect for laws and enforcement. Occasionally, in an effort to work around the rules, people lose sight of the reasons disturbance to streambeds or banks are discouraged in the first place.

The cumulative effects of years of "dozer" activities on stream morphology, channel and bank stability, sediment transport, water quality, aquatic habitat, and fish passage have not yet been fully evaluated and warrant further study to guide regulatory decisions.

The Nueces River Authority regularly witnesses the impacts of mechanical disturbances within streambeds. In addition to reporting observations to authorities, we have tried to respond in proactive ways. One such response has been the formation of a Riparian Landowners Network, where riparian experts lead workshops in a neighbor-to-neighbor setting. Workshops focus on riparian function, benefits, and how to maintain a healthy riparian area. Once landowners better understand how the riparian landscape works and the numerous benefits a healthy system delivers, we have seen a decrease in streambed and streambank disturbances.

The development of Eagle Ford shale oil and gas deposits and the proliferation of wind farms in the Nueces River basin are creating an increased demand for caliche and aggregate material. Along the Nueces River, and its gravel-rich tributaries, there has been a significant increase in commercial gravel mining operations. Most aggregate mines are operating just above the gradient boundary that divides public streambeds from private uplands, and therefore they are not regulated. Whether permitted or unpermitted, bed, bank, and floodplain disturbances are damaging to the river system, and their impact is cumulative. When floodwaters cut through a mined area, an unnatural type and amount of sediment can be discharged downstream (Figure 2).

Spring-fed headwater streams, like those in the upper Nueces River basin, are naturally oligotrophic (Elser et al. 1990). Even a small amount of nutrients introduced into these low-nutrient waters can cause damaging algal blooms. Background levels of phosphorus and nitrogen are so low in Nueces River basin headwater streams that even a small spill of livestock mineral supplements (containing phosphorous), a leaking septic tank, or even a highly used animal bedding site near a streambank (Figure 3) can cause observable reactions in algal growth. The regular delivery of treated wastewater effluent, as has recently been proposed for permitting, would ensure unnaturally high rates of primary productivity in headwater streams.

Along the Texas coast, estuaries have long been noted for their productive fisheries (Smith et al. 2013). Important freshwater inflows come from the Nueces River and coastal creeks that drain the Nueces and adjoining coastal basins (HDR and Geraghty & Miller, Inc. 1991). In functional riparian areas and wetlands, pollutants are processed, and sediments are filtered before reaching the waterways that flow directly into the bays and estuaries. Where riparian areas are not functional and have been compromised by development or agriculture, some water bodies are experiencing water quality impairments. The South Texas Land and Water Initiative, another NRA program, grew out of the need to address water quality issues on the ground with riparian landowners. The initiative, currently, is seeking funds for a demonstration project to illustrate farming-out pollutants along Petronila Creek, a tributary to Baffin Bay, by planting specialized riparian plants within and along drainage ways leading to the impaired stream segment. This incentive-based grower program is beginning to be implemented across the United States and bears watching in the years to come.

Landfills and groundwater marketing

A proposal to establish a landfill that would accommodate oil and natural-gas drilling waste on a site within the floodplain of Salt Branch Creek, a tributary of the Frio

Figure 2. A gravel mining operation located on a large alluvial fan within the upper Nueces River basin.

River and Choke Canyon Reservoir in McMullen County, is a recent example of the long-running and ongoing threats to water resources in the Nueces River basin. With a lower population density than much of the state of Texas, the Nueces River basin has long been targeted as a site for controversial projects that pose potential contamination and pollution risks. Over the years, NRA has intervened to successfully oppose numerous applications for waste discharge permits when those projects threatened the surface or groundwater quality of the basin. Three years were spent opposing efforts by the Texas Low-Level Radioactive Waste Disposal Authority to locate the state's first disposal facility in the Nueces River basin. The first proposed site was near Asherton, Texas. The second site was near Tilden, Texas, just upstream from Choke Canyon Reservoir. The Nueces River Authority was instrumental in garnering support that led to the passage of legislation banning the location of this facility within 20 mi (32 km) of a water-supply reservoir (Newberry 1994).

Exportation of large amounts of groundwater from aquifers supplying the Nueces River and its tributaries is another threat

Figure 3. An algal bloom on the upper Nueces River believed to be the result of rainfall runoff from a dense axis deer bedding ground adjacent to the riverbank and following an unusually long dry period.

that has the potential to greatly impact the river. Groundwater export projects could potentially decrease or eliminate spring flows, thus reducing surface flows. This would have adverse effects on downstream surface-water rights and water-supply facilities and impact groundwater supplies needed for current and future domestic, municipal, and agricultural uses. The scientific knowledge of groundwater supplies in the upper Nueces River basin, which is influenced by the highly fractured Edwards and Trinity-Edwards aquifers, is incomplete and at times conflicting. It is unknown how much water could be pumped before impacts would be seen and how those impacts would be manifested. The threat is real, however, and since 2010, several entities have proposed the construction of pipelines to facilitate large-scale exportation of groundwater from the Nueces River basin (EAA 2018). To date, these proposals have been successfully opposed; however, more science is needed to inform legislation and guide protection of surface water resources.

Nonnative invasive species

In 2010, the invasive and nonnative giant reed *Arundo donax* began to rapidly proliferate on four headwater streams in the Nueces River basin. New plants readily propagate from fragments of existing plants, and typical means for dispersion include transportation of road base or other construction material, nutria *Myocastor coypus* (a semiaquatic rodent) cutting stalks, wild hogs gnawing the green shoots, and humans clearing newly fragmented riverfront properties. In a short period of time, giant reed was introduced and expanded to approximately 30% of the floodplain on parts of upper basin streams. The Nueces River Authority responded with a multiyear, multipronged, landowner-directed control effort that was launched with funding from the TPWD, the U.S. Fish and Wildlife Service, private foundations, and concerned individuals.

"Pull.Kill.Plant." became the battle cry of the grassroots effort to control giant reed and restore native riparian vegetation (www.pullkillplant.org). This program is a three-pronged, multiyear effort consisting of hand-pulling new sprouts, herbicide treatment of existing stands using Imazapyr, and planting of native riparian trees in their place. Between 2010 and 2017, education, outreach, and more than 200 cooperating landowners led to treatment of more than 80 ha on 119 km of four headwaters streams. While the giant reed populations are dramatically reduced, they are not eliminated, so ongoing efforts to manage the plant are essential. Our experience in control of giant reed has led us to conclude that prevention of this invasive is preferable, and educating people is the key to success. An important target group for this education effort are road, bridge, and right-of-way construction and maintenance crews who regularly encounter the plant and have the ability to help stop its spread through proper handling. The campaign materials and implementation plan were developed by Nueces River Authority with funding from the TPWD and guidance from the Texas Department of Transportation and Texas A&M Agrilife Extension. Now the prevention effort is being implemented statewide.

In addition to riparian nonnatives, the Nueces River is home to several aquatic nonnative species, with tilapia *Oreochromis* spp. being arguably the most impactful. While no studies have been focused on tilapia in the Nueces River basin, it is locally known that at least one stocking of Blue Tilapia *O. aureus*, a hardy, vegetation-eating, warmwater fish, was conducted in the 1980s behind a low-water dam at a county park on the upper Nueces River in an attempt to help control aquatic vegetation. A 1996 investigation by the Texas Natural Resource Conservation Commission (TNRCC; now Texas Commission on Environmental Quality) documented fine suspended sediments in otherwise pristine clear water and attributed the water-quality issue to spawning tilapia (TNRCC, unpublished data). Tilapia are now abundant throughout the upper Nueces River basin, and suspended sediments are common during the spring spawning season (Figure 4).

As tropical genera, tilapia are intolerant to coldwater temperatures and tend to congregate near spring runs in the Nueces River basin during winter months. It may be possible to use this behavior pattern to capture and remove large quantities using seines, electrofishing, gill nets, cast nets, or other methods. Even artificial heating has been mentioned as a method to stimulate congregational behavior in areas away from springs for a concentrated harvest. Biological controls, including the release of sterilized breeding fish, have also been discussed (Tavares 2009). However, a comprehensive control program has yet to be organized for tilapia in the upper Nueces River basin, though the need for action is imperative.

Among the terrestrial nonnative animals becoming problematic along the Nueces River are axis deer *Axis axis*, aoudad *Ammotragus lervia*, and blackbuck *Antilope cervicapra*. These species have escaped or been released from high-fenced game ranches. They have found refuge within the

Figure 4. Crystal-clear headwater springs and creeks (left) are turned into murky ponds (right) during tilapia spawning season when their nesting behaviors disturb fine sediments and increase overall site turbidity.

wide riparian floodplains where they are often attracted to feeders operated by landowners. Their presence is placing additional stressors on the ecosystem and impacting riparian vegetation growth and regeneration. An adult female axis deer can have two fawns per year, and a single, adult consumes around 680 kg of forage per year. Because the species commonly travels in herds, riparian vegetation is easily decimated. Without riparian vegetation to slow floodwater and trap new sediments, water-storage capacity and riparian function become severely impaired. A comprehensive initiative to address the problem of axis deer, aoudad, and blackbuck impacts on riparian areas of the Nueces has not yet been launched.

More education, more dollars

While it is important to continue to address resource protection through on-the-ground conservation and regulatory enforcement, the only sure way to save a river is through changing people's minds and behaviors. It is really all about the people; *Homo sapiens* are an essential species of consideration in any multispecies approach to conservation. We have found that education is the most powerful tool in our resource protection programs, but also the most difficult to fund. With available funding, most raised locally, NRA works to increase awareness and cultivate knowledge among the public—residents, visitors, and especially youth and landowners—to protect natural function and prevent pollution of streams in the Nueces River basin. Programs include a litter-prevention campaign, classroom presentations using a basin watershed model, and workshops for landowners and land managers about riparian function. Riparian education programs led by NRA have been especially successful and are award-winning (e.g., Environmental Law Institute's National Wetland Award for Outreach and Education 2004; Texas Environmental Excellence Award 2008; Lone Star Land Steward Award 2014; Texas Environmental Excellence Award Finalist 2015) but are still underfunded as an ongoing effort.

More than 13,000 students each year have hands-on learning opportunities with NRA's large, tabletop Nueces River basin model. The 2-m-long topographic model of the Nueces River basin demonstrates watershed function and nonpoint pollution impacts. It also introduces the relationship between river basins and aquifers. It is used to reinforce Texas Essential Knowledge and Skills state curriculum and cultivate understanding of water resources, including watersheds and river basins. Participants are able to identify where they live on the model and see how their activities impact the health of the water resources in the whole basin. It has proven to be an effective tool to show students where their water comes from, where it goes, and what they can do to protect and conserve it.

"Your Remarkable Riparian" is an NRA program that guides landowners in managing land adjacent to creeks and rivers to enhance water supply and improve water quality (Jones-Lewey 2018). Initially funded by private foundations and later supported by state agencies, more than 40 landowner workshops were delivered from 2007 to 2011, and one-on-one riparian consultations were offered throughout the Nueces River basin (Nelle 2011). Because riparian vegetation is the crucial component to riparian health, a plant field guide was needed so that workshop participants could recognize and appreciate these specialized plants for their function, stability rating, and wetland indicator status. The first two editions of the guide were funded by contributions and given away to participants, but demand for them was bigger than the Nueces River basin, with more than 13,000 copies given away. Now in its third edition, *Your Remarkable Riparian: Field Guide to Riparian Plants* found within most of Texas and its companion, *Your Remarkable Riparian: Owner's Manual* are now the leading statewide resources on the subject (Figure 5; NRA 2016). The TPWD's *Troubleshooting Invasive Plants* manual is distributed with NRA's *Remarkable Riparian* book set as well. They are available for sale on the NRA riparian Web site (www.remarkableriparian.org).

Figure 5. *Remarkable Riparian: Field Guide to Riparian Plants* and its companion *Remarkable Riparian: Owner's Manual.*

The *Remarkable Riparian* online learning modules are widely consulted as well. Most recently, NRA developed an online college course currently being piloted at Sul Ross State University in Alpine, Texas. The Nueces River Authority is working to expand the online delivery of riparian modules for continuing education credits at Texas State University in San Marcos, Texas, as well.

River litter prevention took shape as NRA's Up2U campaign in 2005 and continues with the annual production and distribution of branded litter bags and support from local partners. The campaign name and graphic logo underscore a message of personal responsibility for clean rivers and beaches. With a growing number of recreationists now visiting the region, river litter has the potential to become a major pollution issue. More than 75,000 Up2U litter bags are distributed in the region each year, free of charge, providing visitors an active way to become advocates for litter prevention, armed with a handy tool and knowledge that anything left on the ground can wind up in our rivers and oceans.

Conclusions

The resources of the Nueces River basin are clearly better off now as compared to 20 years ago, thanks largely to efforts by NRA. The Nueces River Authority has been actively engaged in saving this river, yet there is so much more to do. The Nueces River Authority must be prepared to address rapidly growing and expanding threats such as excessive groundwater withdrawal that can impact natural flow regimes and dewater creeks and rivers. Nonnative, invasive species are a growing problem in need of creative multipronged solutions. The ongoing fragmentation of waterfront land constantly delivers a new crop of landowners, many of whom are uninformed about the value of riparian ecosystems and who need to be educated. On the Texas coast, farmers may be able to remove costly water-quality impairments before they reach important estuaries, but a program needs to be funded and launched to reward their work.

Despite years of monitoring and reporting through the Texas Commission on Environmental Quality Clean Rivers Program, water quality continues to decline across the basin. Only 4 of 54 classified water bodies remain with no documented water-quality impairments or concerns. So much work needs to be done. Yet, with no dedicated funding sources and few regulatory enforcement opportunities, it is not getting any easier to save this or any other Texas river.

Acknowledgments

We recognize all who have contributed to the success and continuing efforts aimed at saving this river, especially the board of directors of the Nueces River Authority and Con Mims, executive director, for their guidance and leadership in the development of the programs. A special acknowledgment is due the individuals and landowners within the basin who care and continue their stewardship of these river resources. Important funding for this work has been provided by individuals, The Dixon Water Foundation, The Meadows Foundation, the Shield-Ayres Foundation, The Trull Foundation, The Brown Foundation, the Ed Rachal Foundation, the Coastal Bend Bays and Estuaries Program, the Texas Commission on Environmental Quality, and the Texas Parks and Wildlife Department. Individuals who provided substantial guidance and review of this chapter include Gary Garrett, Ph.D. (University of Texas at Austin), Con Mims (Nueces River Authority), Sarah Robertson (Texas Parks and Wildlife Department), and Julie Lewey (Devils River Conservancy).

References

EAA (Edwards Aquifer Authority). 2018. Frequently asked questions: What about

"Western" water? What's that all about? Available: www.edwardsaquifer.net/faqs.html. (January 2018).

Elser, J. J., E. R. Marzolf, and C. R. Goldman. 1990. Phosphorus and nitrogen limitation of phytoplankton growth in the freshwaters of North America: a review and critique of experimental enrichments. Canadian Journal of Fisheries and Aquatic Sciences 47:1468–1477.

HDR (HDR Engineering, Inc.). 1999. Water availability in the Nueces River basin. Final report prepared for the Texas Natural Resource Conservation Commission, Austin.

HDR (HDR Engineering, Inc.) and Geraghty & Miller, Inc. 1991. Regional water supply planning study, Nueces River basin. Final report prepared for the Nueces River Authority, Uvalde, Texas.

Jones-Lewey, S. 2018. Your remarkable riparian: an educational journey toward a field guide to riparian plants. Native Plant Society of Texas News 36:6–7.

Newberry, W. F. 1994. Comparative approaches to siting low-level radioactive waste disposal facilities. U.S. Department of Energy, DOE-LLW-199, Washington, D.C.

Nelle, S. 2011. The remarkable riparian network of the Nueces basin: a model of successful landowner education. Texas Wildlife (March):48–52.

NRA (Nueces River Authority). 2016. Your remarkable riparian: field guide to riparian plants found within most of Texas. NRA, Uvalde, Texas.

Richardson, L. R., and J. R. Gold. Evolution of the *Cyprinella lutrensis* species-complex II: systematics and biogeography of the Edwards Plateau Shiner, *Cyprinella lepida*. Copeia 1995:28–37.

Saunders, K. S., G. W. Linam, T. A. Jurgensen, M. L. Mullins, A. S. Whisenant, and L. Brezina. 2004. An investigation of off-road vehicle impacts on Nueces River fish assemblages in Uvalde and Zavala counties, Texas. Texas Parks and Wildlife Department, Austin.

Smith, E. H., A. Banda, J. W. Tunnell, Jr., and K. Withers. 2013. Ecosystem-based management in the Laguna Madre, western Gulf of Mexico. Pages 131–152 *in* J. W. Day, A. Yáñez-Arancibia, A. Vásquez, W. Arzapalo, M. Baltz, and A. Banda, editors. Gulf of Mexico origin, waters, and biota, volume 4: ecosystem-based management. and Texas A&M University Press, College Station.

Tavares, J. 2009. Invasive fish control and eradication: a preliminary plan for the Kawaiele Bird Sanctuary, Mana Plain Conservation Area, Kaua'I, Hawai'i. Final report to Hawaii Department of Land and Natural Resources, Honolulu.